Reproductive Biology and Phylogeny of Lizards and Tuatara

Reproductive Biology and Phylogeny Series

Series Editor: Barrie G. M. Jamieson

Published:

Vol. 1: Reproductive Biology and Phylogeny of Urodela
(Volume Editor: David M. Sever)

Vol. 2: Reproductive Biology and Phylogeny of Anura
(Volume Editor: Barrie G. M. Jamieson)

Vol. 3: Reproductive Biology and Phylogeny of Chondrichthyes
(Volume Editor: William C. Hamlett)

Vol. 4: Reproductive Biology and Phylogeny of Annelida
(Volume Editor: G. Rouse and F. Pleijel)

Vol. 5: Reproductive Biology and Phylogeny of Gymnophiona
(Caecilians)
(Volume Editor: Jean-Marie Exbrayat)

Vol. 6A: Reproductive Biology and Phylogeny of Birds
(Volume Editor: Barrie G. M. Jamieson)

Vol. 6B: Reproductive Biology and Phylogeny of Birds
(Volume Editor: Barrie G. M. Jamieson)

Vol. 7: Reproductive Biology and Phylogeny of Cetacea
(Volume Editor: D. Miller)

Vol. 8A: Reproductive Biology and Phylogeny of Fishes
(Agnathans and Bony Fishes)
(Volume Editor: Barrie G. M. Jamieson)

Vol. 8B: Reproductive Biology and Phylogeny of Fishes
(Agnathans and Bony Fishes)
(Volume Editor: Barrie G. M. Jamieson)

Vol. 9: Reproductive Biology and Phylogeny of Snakes
(Volume Editor: Robert D. Aldridge and David M. Sever)

Vol. 10: Reproductive Biology and Phylogeny of Lizards and Tuatara
(Volume Editor: Justin L. Rheubert, Dustin S. Siegel and Stanley E. Trauth)

Reproductive Biology and Phylogeny of Lizards and Tuatara

Volume edited by

JUSTIN L. RHEUBERT
Department of Biological Sciences
St. Louis University
St. Louis MO 63139, USA

DUSTIN S. SIEGEL
Department of Biology
Southeast Missouri State University
Cape Girardeau
MO 63701, USA

STANLEY E. TRAUTH
Department of Biological Sciences
Arkansas State University
State University
AR 72467-0599, USA

Volume 10 of Series:
Reproductive Biology and Phylogeny

Series edited by
BARRIE G. M. JAMIESON
School of Biological Sciences
University of Queensland
St. Lucia, Queensland
Australia

THE UNIVERSITY OF QUEENSLAND
AUSTRALIA

CRC Press
Taylor & Francis Group
Boca Raton London New York

CRC Press is an imprint of the
Taylor & Francis Group, an **informa** business
A SCIENCE PUBLISHERS BOOK

CRC Press
Taylor & Francis Group
6000 Broken Sound Parkway NW, Suite 300
Boca Raton, FL 33487-2742

First issued in paperback 2020

ISBN-13: 978-1-4665-7986-6 (hbk)
ISBN-13: 978-0-367-73859-4 (pbk)

Library of Congress Cataloging-in-Publication Data

Reproductive biology and phylogeny of lizards and tuatara / [edited by] Justin L. Rheubert, Dustin S. Siegel, Stanley E. Trauth.
pages cm
Summary: "Reproductive Biology and Phylogeny of Lizards and Tuatara their friendly support and high standards in producing this series. Sincere thanks must be given to the volume editors and to the authors who have freely contributed their chapters in very full schedules. Justin Rheubert, Dustin Siegel and Stan Trauth, the volume editors, are particularly to be thanked for conceiving the present volume and for the diligence and outstanding expertise which they brought to its preparation. The editors and publishers are gratifi ed that the enthusiasm and expertise of these contributors has been refl ected by the reception of the series by our readers. Barrie G. M. Jamieson The School of Biological Sciences University of Queensland"-- Provided by publisher.
Includes bibliographical references and index.
ISBN 978-1-4665-7986-6 (hardback : acid-free paper) 1. Lizards--Sexual behavior. 2. Tuatara--Sexual behavior 3. Evolutionary developmental biology. I. Rheubert, Justin L., editor. II. Siegel, Dustin S., editor. III. Trauth, Stanley E., 1948- editor.

QL666.L2R37 2015
597.95156--dc23 2014026024

Visit the Taylor & Francis Web site at
http://www.taylorandfrancis.com

and the CRC Press Web site at
http://www.crcpress.com

Preface to the Series

This series was founded by the series editor, Barrie Jamieson, in consultation with Science Publishers, in 2001 and is now being distributed by CRC Press (Taylor and Francis).The series bears the title 'Reproductive Biology and Phylogeny' and this title is followed in each volume with the name of the taxonomic group that is the subject of the volume. Each publication has one or more invited volume editors (sometimes the series editor) and a large number of authors of international repute. The level of the taxonomic group which is the subject of each volume varies according, largely, to the amount of information available on the group, the advice of the volume editors, and the interest expressed by the zoological community in the proposed work. The order of publication of taxonomic groups reflects these concerns, and the availability of authors for the various chapters, and does not proceed serially through the animal kingdom. A major aspect of the series is coverage of the phylogeny and classification of the group, as a necessary framework for an understanding of reproductive biology. It is not claimed that a single volume can, in fact, cover the entire gamut of reproductive topics for a given group but it is believed that the series gives an unsurpassed coverage of reproduction and provides a general text rather than being a mere collection of research papers on the subject. Coverage in different volumes varies in terms of topics, though it is clear from the first volume that the standard is uniformly high. The stress varies from group to group; for instance, modes of external fertilization or vocalization, important in one group, might be inapplicable in another.

This is the tenth volume in the series. Previous volumes in the series were devoted to 1. Urodela; 2. Anura; 3. Chondrichthyes: Sharks, Batoids and Chimaeras; 4. Annelida; 5. Gymnophiona (Caecilians); 6A and B. Birds; 7. Cetacea (whales, dolphins and porpoises); 8A and B. Fishes (Agnathans and Bony Fishes); and 9. Snakes.

My thanks are due to the University of Queensland and the School of Biological Sciences for support in many ways since my retirement in 1999 and particularly for excellent library facilities. I sincerely thank my wife, Sheila Jamieson, who has supported me indirectly in so many ways in this work. I am grateful to the publishers, and especially Mr. Raju Primlani, for their friendly support and high standards in producing this series. Sincere thanks must be given to the volume editors and to the authors who have freely contributed their chapters in very full schedules. Justin Rheubert, Dustin Siegel and Stan Trauth, the volume editors, are particularly to be thanked for conceiving the present volume and for the diligence and outstanding expertise which they brought to its preparation. The editors and publishers are gratified that the enthusiasm and expertise of these contributors has been reflected by the reception of the series by our readers.

Barrie G. M. Jamieson
The School of Biological Sciences
University of Queensland

THE UNIVERSITY
OF QUEENSLAND
AUSTRALIA

Date 14th May 2014

Preface to this Volume

> *"Reproduction is so primitive and fundamental a function of vital organisms that the mechanism by which it is assured is highly complex and not yet clearly understood."*
>
> —*Hendry Havelock Ellis*

Reproduction is axiomatic for the existence of life. Through independent evolutionary trajectories, living organisms have evolved spectacular behaviors, physiological mechanisms, and structures to accomplish reproduction. This could not be more true than in the most diverse group of reptiles, the lepidosaurs.

By far, lizards represent the most diverse reptilian group with over 5,000 described species. Considering that the approximately 3,100 species of snakes are no more than a highly specialized group within lizards (collectively termed Squamata), this diversity is even more astounding. Previously, a similar text was dedicated to snakes (Reproductive Biology and Phylogeny of Snakes) and, thus, we felt it necessary to contribute a comparable volume devoted to lizards and the closest extant relative to Squamata, the Rhynchocephalia (comprised of two species of Tuatara).

The great diversity of lepidosaurs (Rhynchocephalia and Squamata) is paralleled by the evolution of remarkable reproductive innovations within this group. For example, the independent evolution of viviparity, the unique structure of extraembryonic membranes surrounding gestating embryos, the many indices of astonishing sexual dimorphism, "virgin birth" (parthenogenesis) within whiptail lizards, etc., attest to the extraordinary evolutionary pathways taken by lizards and tuataras. The attraction to dedicate a large portion of our time over the last couple of years to this book was two-fold: 1) the last major symposium on the reproductive biology of reptiles that included lizards was held in 1981 (Annual Meeting of the Herpetologists' League, in Memphis, Tennessee); and 2) with all of the new tools to investigate the evolution of lizards, we thought it timely to present a up-to-date review of what the world currently knows about lizard reproductive biology in one text so as to facilitate the use of

Reproductive Biology of Reptiles Symposium
Memphis State University
10-12 August 1981

Front row from left to right: Gary Ferguson, Stan Trauth, Gary Packard, Mary Packard, Paul Licht, Françoise Xavier, Robert Aldridge, Richard Vogt, Louis Guillette, John Legler, J. Whitfield Gibbons; middle row from left to right: unknown, David Owens, Robin Andrews, Shuk-mei Ho; back row from left to right: James Jacob, unknown, Richard Jones, Ken Marion, Arthur Dunham, Justin Congdon, Laurie Vitt, R. Wayne Van Devender, William Garstka, Valentine Lance, Henry Fitch.

reproductive biology diversity in future evolutionary studies on lizards. The "Reproductive Biology and Phylogeny" series conceived and edited by Barrie Jamieson was the perfect avenue, in our minds, to accomplish this goal.

The Reproductive Biology of Reptilia symposium of 1981 was a landmark event. We were fortunate that a couple of these authors were able to bridge the generational gap between 1981 and 2015. Many of the presentations from the past symposium were published in Herpetologica (38[1], 1982) and covered diverse topics that align with chapters in the present text, such as gonads and their development (Marion), vitellogenesis (Ho et al.), endocrine cycles (Xavier), reproductive ecology (Vitt and Price), etc… A similar symposium dedicated solely to lizards, featuring many of the authors from the current text, was held in Chattanooga Tennessee at the Joint Meeting of Ichthyologists and Herpetologists (2014). Comparing the modern symposium and text with the past symposium and text highlights our progress on elucidating attributes of the reproductive biology of lizards, but also highlights how much is still unknown.

Justin L. Rheubert
Saint Louis University
Dustin S. Siegel
Southeast Missouri State University
Stanley E. Trauth
Arkansas State University

THE UNIVERSITY OF QUEENSLAND
AUSTRALIA
May 2014

Contents

CHAPTER 1

Lizard Reproduction: A History of Discovery

Laurie J. Vitt

1.1 INTRODUCTION

It is difficult to imagine anything more interesting to biologists and advanced naturalists than understanding origins of the diversity of reproductive patterns among lizards. Among species, some lay eggs, some produce many offspring in a single reproductive episode whereas others produce few offspring, and some make no nest at all whereas others construct nests and attend eggs until hatching. Most species are oviparous and among those, some lay eggs immediately after they are shelled, whereas others retain eggs for extended periods before oviposition. Many species are live-bearing and viviparity has arisen multiple times within squamates. Some reproduce sexually and others asexually. For some the sex of embryos is determined genetically and for others sex is determined by developmental temperature. Differences in reproduction listed above are but a small sampling of the many fascinating aspects of lizard reproduction that have caught the attention of biologists. Moreover, trying to understand how and why suites of reproductive traits vary has resulted in identification of new conceptual issues in evolutionary ecology.

The history of biological investigations into reproduction of lizards began with basic observations on mating behavior and clutch size, largely because these were easy to observe. One would think that Charles Darwin's theory of evolution by natural selection (Darwin 1859, 1871) would have a

Sam Noble Museum, University of Oklahoma, 2401 Chautauqua Avenue, Norman, Oklahoma, 73072 USA.
Present address: 4554 S. Claire Pumpkin Pt., Inverness, Florida 34450 USA.

led to a flurry of activity among biologists seeking to understand how and why animals reproduce, and this may have been true to a certain extent (see below). For the first time, a mechanism-based hypothesis was available to begin to understand variation in reproductive and other traits among organisms. Nevertheless, a long history of often-ridiculous explanations for origins of biological phenomena based on belief rather than empirical data has yet to bite the dustbin of time! However, among the first attempts to understand reproductive related behaviors in an evolutionary framework did not occur until more then 70 years later when G. K. Noble and his colleagues began designing experiments on sexual and brooding behavior of lizards (e.g., Noble and Bradley 1933; Noble and Mason 1933). One hundred and ten years passed before reproduction in lizards was placed in an evolutionary framework (Tinkle 1969; Tinkle *et al.* 1970), even though bird ecologists had been thinking about it since at least the 1940s (e.g., Moreau 1944; Lack 1947, 1948).

It is not intended in this chapter to review all of the works on lizard reproduction. Indeed, several books would be necessary to do so. Rather a brief outline is given of the history of discoveries that have resulted in major insights and controversies into the evolution of reptile reproductive traits in lizards, most of which are applicable not only to other reptiles, but to animals in general. Some of these discoveries in turn have led to entirely new and exciting areas of research. The pattern of interest in lizard reproduction, defined by Internet Google hits (but see below) has varied since the middle of the 19th Century (Fig. 1.1). Several events appear to have had notable impact. First, following Darwin's *Origin of Species* (1859) interest in lizard reproduction appeared to increase. Secondly, following World War I and until the end of World War II, interest in the topic decreased, followed by a steady increase up until 1970. Thirdly, following the paper of Tinkle *et al.* (1970) on reproductive strategies in lizards (which coincided with Henry Fitch's [1970] impressive summary of reproductive studies), interest in lizard reproduction increased at a rate greater than expected based on an exponential model. Given the quality of data, the pattern presented in Fig. 1.1 should be recognized as qualitative because Internet searches produce many duplicates and false positives (papers that are not about lizard reproduction) but nevertheless reveals real trends. Among the many researchers who have made truly exciting discoveries, Donald W. Tinkle is singled out (Fig. 1.2), whose early papers moved the study of lizard reproduction from a basically descriptive science to a science of evolutionary discovery that prevails today.

Tracing the history of reproductive studies in lizards is difficult, largely because a number of non-exclusive approaches exist. The two most revealing approaches may be categorized as topical (e.g., clutch size,

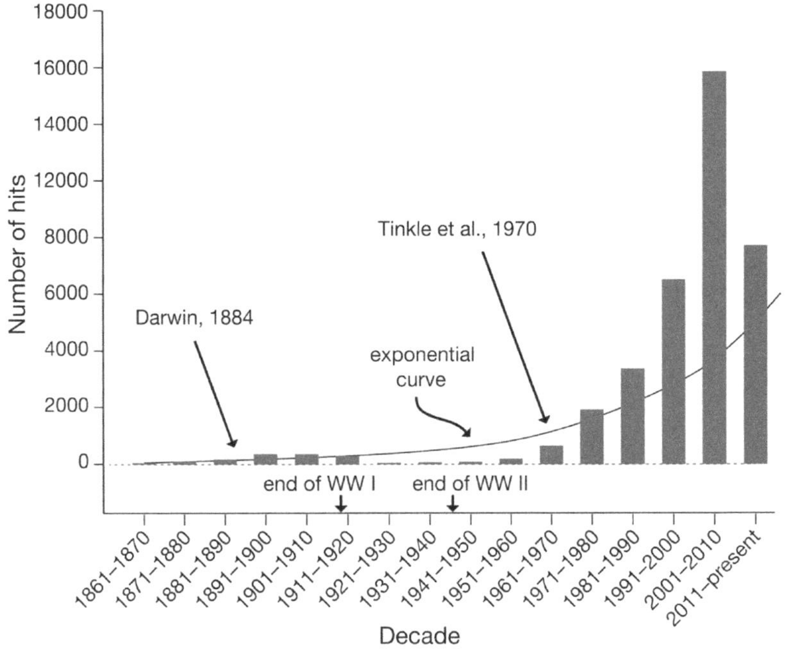

Fig. 1.1 Number of hits on Google Scholar with the search terms "Reproduction" and "Lizards" summarized by decade since 1850. Although this graphic likely indicates the relative interest in lizard reproduction, it should be noted that some hits, especially in the 19th and early 20th century, were not actually papers on lizard reproduction. In more recent decades, especially the last two, many hits stem from papers in which "reproduction" and "lizard" may have been in citations within a paper rather than indicating independent hits. Also, note that in the decade indicated as 2011–present, more than half as many hits occurred in less than one-third of the decade compared with the decade 2001–2010.

parthenogenesis, etc.) and methodological (i.e., descriptive, experimental, comparative, etc.). A summary of topics is presented, followed by comments on different approaches. Six topics are selected to emphasize: 1) seasonality in reproduction and lipid storage, 2) parthenogenesis, 3) evolution of viviparity, 4) placentation, 5) environmental sex determination, and 6) reproductive strategies, reproductive effort, and costs of reproduction. Numerous equally interesting reproduction-related areas of research that have produced new insights include, but are not limited to: 1) hormonal basis for reproduction, 2) direct and indirect behaviors related to reproduction and sex determination, 3) morphology and function of hormone- and pheromone-producing glands, 4) ecological physiology of lizard eggs and, 5) structure and function of eggshells.

Fig. 1.2 Donald W. Tinkle, who used basic lizard reproduction and life history information to address life history theory initiating a question-oriented evolutionary approach to studies of lizard reproduction. Photo by J. D. Congdon, circa 1976.

1.2 TOPICAL SUMMARY

1.2.1 Seasonality in Reproduction and Fat Storage

One of the first studies that included an annual reproductive cycle in lizards was published by Asana in 1931 on the Asian agamid lizard *Calotes versicolor*. This study, like many to follow, demonstrated that reproduction in most but not all lizards is cyclical. We now know, for example, that individuals of most temperate zone species mate in spring and produce eggs or offspring during summer (e.g., Fitch 1954, 1955, 1958; Ballinger 1974; Ballinger *et al.* 1972; Goldberg 1971a; Howard 1974; Howland 1992; Huang 1997a, 1997b; Mayhew 1963, 1965; Ramírez-Bautista *et al.* 1995, 1996; many others). It seems clear that season length and temperature are major limiting factors for lizards in temperate zones. Nevertheless, some high elevation viviparous species ovulate and mate in fall, bask along rock crevices during winter, and produce live young in spring (e.g., *Sceloporus jarrovi*; Goldberg 1971b). Other high elevation species in Mexico and the southwest deserts of the United States also have seasonal patterns that deviate from the typical temperate zone pattern (e.g., *S. scalaris*, Ballinger and Congdon 1981; *S. grammicus*,

Guillette and Casas-Andreu 1980). For most studied lizards, reproduction occurs every year (except during extended droughts). However, some lizards, such as the viviparous African lizard *Cordylus giganteus*, reproduce every other year (Van Wyk 1991).

Annual cycles of tropical species are much more complex. Some species appear to reproduce continually whereas others are seasonal. Not surprisingly, species in tropical rainforests tend to be less cyclical than those in seasonal (wet–dry) tropical habitats, such as the Cerrado and Caatinga in South America (Mesquita and Colli 2010). Most rainforest anoles, gekkonoids, and gymnophthalmids appear to reproduce nearly continually (e.g., Sexton and Turner 1971; Sexton and Brown 1977; Sexton *et al.* 1971; Sherbrooke 1975; Mesquita and Colli 2010). Note that most rainforest species with extended reproductive seasons are species with genetically fixed clutch sizes of either one or two eggs, most of which are capable of producing clutches rapidly (e.g., Fitch 1982; Vitt 1986). Tropical lizards in seasonal environments tend to be seasonal in reproduction (e.g., Magnusson 1987; Ramírez-Bautista *et al.* 2000; Vitt and Goldberg 1983). However, these cannot be considered representative because a wide diversity of reproductive seasonalities exists among species occurring together in some tropical habitats (James and Shine 1988; Vitt 1990, 1992). Unfortunately, very few studies contain full-year cycles for all lizard species at single tropical localities (but see Vitt 1990). Considering that several lizard clades are relatively conservative in respect to reproductive traits (e.g., anoles, gekkonoids, and gymnophthalmids), at least a portion of the variation in seasonality in lizard reproduction is due to evolutionary history (see Colli *et al.* 1997), which has been demonstrated in comparative syntheses (Dunham and Miles 1985; Mesquita and Colli 2010).

Although it may seem intuitively obvious that energy for production of eggs by females and reproductive-related behaviors by males must come at least partially from stored energy, verification did not occur until the mid-1960s when experimental manipulations demonstrated that stored lipids were mobilized for follicular development in the side-blotched lizard, *Uta stansburiana*, and that mobilization of lipids for reproduction was induced by estradiol-17 B (Hahn 1965, 1967; Hahn and Tinkle 1965). Since that time, a number of researchers began weighing fat bodies when conducting necropsies of lizards for reproductive studies.

For typical Spring-breeding Temperate-Zone species, fat is stored during late summer and fall, some of which is used for overwintering (Avery 1970, 1974) with the remainder, supplemented by food intake in early Spring, being used to produce first clutches of the season for females (Derickson 1976). Examples include *Urosaurus ornatus* and *U. graciosus* from the Sonoran Desert (Vitt *et al.* 1978; Van Loben Sels and Vitt 1984) and *Sceloporus undulatus* (referred to as *S. erythrocheilus*) in central western

United States (Ballinger *et al.* 1992). Temperate-Zone lizards that produce a single clutch and have extended maternal care (egg brooding) such as *Plestiodon laticeps* in South Carolina, cycle fat in a similar manner to the species above, but a major portion of lipid reserves are in the tail (Vitt and Cooper 1985), which appears to be the case for many skinks as well as many geckos (e.g., *Coleonyx brevis*, Dial and Fitzpatrick 1981; *Hemidactylus turcicus*, Greene 1969).

For tropical lizards, fat storage is somewhat more complex, largely because most species are active year round. Considering high variation in seasonal patterns of reproduction for tropical lizards at specific localities (e.g., James and Shine 1988; Vitt 1988, 1992), it might appear *a priori*, that fat cycles would be associated with whatever type of environmental seasonality exists (e.g., wet–dry, or warm–cool) and how that seasonality translates into resource availability. However, fat cycles are not necessarily tied to tropical seasonality. For example, although most reproduction occurs during the dry season (July–November) in *Tropidurus hispidus* (formerly *T. torquatus*) and *T. semitaeniatus* in the semiarid and highly seasonal Caatinga of northeastern Brazil, some reproduction occurs throughout the year. Fat bodies cycle inversely with reproduction, being smallest during the primary reproductive season (Vitt and Goldberg 1983). In the same habitat, the teiid lizard *Cnemidophorus ocellifer*, reproduces year round but fat body mass is greatest during late wet season, presumably following peak resources (Vitt 1983). In contrast, female fat body mass is greatest at the beginning of the dry season when implanted embryos are small in the viviparous skink, *Brasiliscincus* (formerly *Mabuya*) *heathi*, at the same locality (Vitt and Blackburn 1983). Throughout the dry season, as embryos undergo rapid growth, fat bodies of female *B. heathi* are depleted, reaching nearly zero by the time parturition occurs late in the dry season. The oviparous lizard, *Polychrus acutirostris*, produces a single large clutch each year at the end of the transition between dry season and wet season, and the fat body cycle is inversely associated with egg production, with fat bodies nearly depleted when oviposition occurs (Vitt and Lacher 1981). The examples given above, suggest that lipid cycles are more closely tied to egg or embryo production than to seasonality of the habitat.

Additional studies on individual species at a variety of tropical or subtropical habitats throughout the world also indicate that lipid cycles are usually associated with reproduction. These include *Chamaeleo jacksoni* and *Chamaeleo hoehnelii* in Africa (Lin 1979a), *Japalura swinhonis* in Taiwan (Lin 1979b), *Liolaemus lutzae* in southeastern Brazil (Rocha 1992), and *Sceloporus variabilis* in Mexico (Benabib 1994). Similarly, reproduction is tied to lipid storage for lizards in montane habitats in subtropical or tropical regions. These include *Eurolophosaurus nanuzae* in southeastern Brazil (Galdino *et al.* 2003) and *Sceloporus grammicus* in Mexico (Guillette and Casas-Andreu 1981).

Detailed energy budgets for lizards also demonstrate the relationship of stored lipids to reproduction (e.g., Congdon *et al.* 1982; Nagy 1983; Dunham *et al.* 1983; Niewiarowski 1994). Stored lipids are used for production of eggs, growth, and maintenance. Reduction of stored lipids from either drought or loss of a lipid-storing appendage (e.g., tails in geckos, skinks, and likely many other lizards) reduces reproductive investment resulting in delayed oviposition or reduced number or size of eggs (e.g., Dial and Fitzpatrick 1981). Clearly, food acquisition influences both reproduction and lipid storage (Ballinger 1977; Ballinger and Congdon 1980; Dunham 1978, 1981; Rose 1982), but it remains to be seen how and when fat is stored across most lizard species.

If a broad generalization can be made regarding lipid storage and reproduction, it is that lipids stored during periods when resources are available are used to support reproduction (either production of eggs in females or reproductive-related behaviors in males) in combination with recently acquired nutrients (Warner *et al.* 2008). Understanding patterns of seasonality in lizard reproduction and fat storage requires relatively long-term data on many species, with particular attention to those in different clades and occurring in different habitat types. Sufficient data do not exist for the kind of analysis that would clearly distinguish among historical (evolutionary conservatism), physiological (energy trade-offs) and ecological (e.g., season length, temperature, nest sites) causes of variation in seasonality of reproduction and fat storage.

1.2.2 Parthenogenesis

Parthenogenesis is cloning by females (see also Neaves Chapter 7, this volume). Although parthenogenesis was well known in invertebrates and fishes, it was unknown in reptiles until Darevsky described it in the Iberian rock lizard, *Darevskia saxicola*, in 1958. This discovery sent herpetologists in Museums in the United States into their collections to look at extensive collections of whiptail (*Aspidoscelis*) lizards, which curators had noted contained mostly females. Examination of collections revealed that, indeed, a number of species (undescribed at the time) were all female. Charles Lowe and his students began detailed studies of these lizards (e.g., Lowe and Wright 1966a, 1966b; Wright and Lowe 1967, 1968; Cole *et al.* 1969), as did several other herpetologists in the western United States (Maslin 1971; Cuellar 1976). Immunocompatability (skin grafting) studies (Cuellar 1976) and rearing of virgin females that began producing offspring when they reached sexual maturity (Cole 1984; Cole and Townsend 1983, 1990) confirmed that parthenogenesis was occurring. Since then more than 50 parthenogenetic squamates have been described (e.g., Cole 1975; Cole *et al.* 1990; Wright 1993; Kearney *et al.* 2006, 2009; Vrijenhoek *et al.* 1989) with a vast majority confirmed to have arisen via hybridization either between two sexual species (diploid parthenogens) or back-crossing between a diploid sexual species and an existing parthenogen producing triploid parthenogens (see Reeder *et al.* 2002). A striking exception is parthenogenic *Lepidophyma flavimaculatum* (Xantusiidae), which appears to not be a hybrid species (Sinclair *et al.* 2010).

A few studies have attempted to examine ecological consequences of parthenogenesis in *Aspidoscelis* (e.g., Congdon *et al.* 1978, 1981; Schall 1978, 1980), producing mixed results. Parthenogenetic species have high genetic diversity within individuals and low genetic diversity among individuals whereas sexual species have relatively low genetic diversity within individuals and high genetic diversity among individuals. Differences in reproductive traits between sexual and asexual species might be expected to compensate for the drastically different potential in terms of population growth. However, such differences remain to be determined. Differences in genomic state of parthenogenetic Australian geckos of reciprocal hybrid origins are reflected in some but not all physiological traits (Roberts *et al.* 2012). Differences in variation of nearly all traits should exist between sexual and parthenogenetic species.

1.2.3 Evolution of Viviparity

Early attempts to identify factors contributing to the evolution (see also Stewart and Blackburn Chapter 13, this volume) of viviparity (defined as production of live young) in squamates did not consider potential phylogenetic effects, largely because appropriate comparative methods were not available (e.g., Harvey and Pagel 1991). The first substantive attempts to understand factors contributing to the evolution of viviparity were based on surveys of extant literature (Tinkle and Gibbons 1977; Pilorge and Barbault 1981; Shine and Berry 1978; Shine and Bull 1979). Because most viviparous species occur at either high elevation or high latitudes, cold climates were considered a major driving force. Since 1977, viviparity has been shown to have originated repeatedly within squamates by using phylogenetic information available at the time (Blackburn 1982, 1985; Shine 1985a). Independent origins exist within numerous smaller clades as well (e.g., *Sceloporus*, Guillette *et al.* 1980).

Identifying origins of viviparity is complicated by the fact that "viviparity" is not the same among different lizard clades. For example, some high latitude or high elevation species retain eggs in the oviduct that effectively hatch within the oviduct thus producing live young (Andrews 1997; Andrews and Rose 1994). Others ovulate large fertilized ova that implant in the oviducts and complete development with no indication that egg shells formed. Yet others ovulate fertilized ova that do not contain sufficient energy for development—for these, various degrees of maternal input occurs (Stewart and Blackburn 1988). Some of these are in largely tropical clades (Blackburn and Vitt 1992). Consequently, a single hypothesis for the evolution of viviparity is unlikely to be applicable to all origins.

Currently, in addition to the "cold climate" hypothesis for the evolution of viviparity (e.g., Tinkle and Gibbons 1977; Shine 1985b), several non-exclusive hypotheses exist (e.g., Andrews *et al.* 1999; Andrews and Mathies 2000; Guillette 1993; Qualls *et al.* 1995, 1997). Some are ecological (e.g., Shine 2002), some physiological (Shine and Guillette 1988; Andrews 2000; Andrews *et al.* 1999), and some morphological, the latter focusing on evolutionary changes in fetal membranes (Blackburn *et al.* 1985; Stewart and Thompson 2000, 2003). The large number of independent origins of viviparity within squamates and among clades that are very old renders squamates excellent models for research aimed at identifying the many potential factors that result in the transition from depositing eggs to production of live young (e.g., Blackburn 1999, 2000, 2006). Primary advantages (but not necessarily causes) of viviparity include: 1) eliminating the unprotected and often prolonged egg stage, which has both ecological (predation) and physiological (e.g., desiccation, variation in developmental rates associated with temperature and moisture) advantages, and 2) providing females the ability to regulate their body temperatures behaviorally and thus physiologically effect developmental rates of their embryos. Costs include risk of predation resulting from lower behavioral performance of females during pregnancy.

1.2.4 Placentation

The first substantive discussion and description of placentation in reptiles was by Weekes in 1935, although simple placentae were known earlier (Weekes 1927, 1929, 1930). At that time, placentae in reptiles were considered to be simple structures providing little or no maternal nutrient transfer to embryos. A paper published by Rebouças-Spieker and Vanzolini (1978) describing parturition in the South American skink *Psychosaura* (formerly *Mabuya*) *macrorhyncha* suggested the possibility that more complex placentae might exist based on observations of dissected females containing embryos. During a one-year field expedition to northeastern Brazil in 1977–78, the present author studied the skink *Brasiliscincus* (formerly *Mabuya*) *heathi*, and, after examining a full year series of embryos implanted in oviducts of females, determined that these skinks had a vascularized placenta that was much more complex than that found in other reptiles. In collaboration with a morphologist and physiologist, the placenta was described in some detail, demonstrating that it was similar in function to the placentae of eutherian mammals (Blackburn *et al.* 1984; Fig. 1.3). It has since been determined that other mainland South American *Brasiliscincus* and *Copeoglossum* contain similar placentae and that the placentae represent a type that had never before been described (Blackburn and Vitt 2002). Most South American *Brasiliscincus, Copeoglossum* and *Psychosaura* have yet to be well studied, but it is likely that specialized placentae occur among all species within the New World mainland clade.

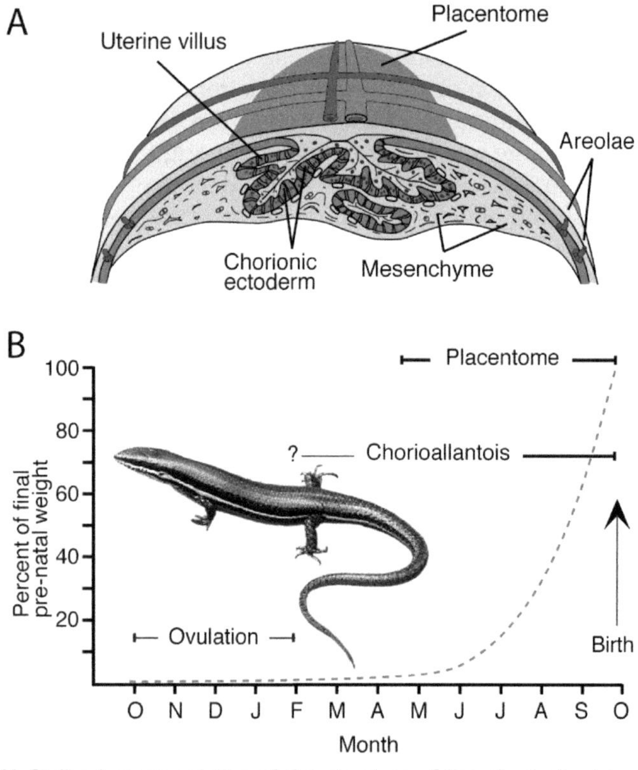

Fig. 1.3 (A) Stylized representation of the structure of the chorioallantoic placenta in *Brasiliscincus* (*Mabuya*) *heathi*. The placenta, which lies above the embryo, consists of hypertrophied uterine (maternal) and chorionic (fetal) tissue forming the placentome, the joint structure for nutrient transfer to the embryo, waste transfer to the female, and gaseous exchange. Transfer and exchange occur in interdigitating chorionic areolae. **(B)** Seasonal growth of implanted embryos (red dashed line). The embryo increases more than 74,000% of its freshly ovulated mass as the result of nutrient uptake from the female. Adapted from Blackburn and Vitt 1992, pp. 150–164 in W. C. Hamlett (ed.). *Reproductive biology of South American vertebrates.* Springer-Verlag: New York, Fig. 11.4 and Vitt and Caldwell 2014, *Herpetology, An Introductory Biology of Amphibians and Reptiles, Fourth Edition.* Academic Press, San Diego, Figs. 5.13 and 5.14.

Color image of this figure appears in the color plate section at the end of the book.

Additional studies on lizard fetal membranes reveal considerable variation in development, morphology, and function of lizard placentae (e.g., Stewart and Blackburn 1988; Blackburn and Vitt 2002). Some species have simple fetal membranes and provide little or no fetal nutrition, whereas others, like *B. heathi*, have eutherian-like placentae that provide as much fetal nutrition as that found in eutherian mammals (Blackburn *et al.* 1984).

1.2.5 Reproductive Strategies, Reproductive Effort, and Costs of Reproduction

During the 1960s two researchers, Fredrick B. Turner and Donald W. Tinkle independently conducted long-term capture-recapture studies aimed at producing life tables using lizards as models. Age-specific fecundity and survivorship data could be used to determine whether populations were increasing or decreasing, among other parameters. Turner's studies (e.g., Turner *et al.* 1969) were directed toward understanding effects of low-level radio-active fallout on animal populations whereas Tinkle's studies were aimed at understanding the evolution of life history traits, and in particular tradeoffs between reproduction and survival (e.g., Tinkle 1967). Both of these studies led to a flurry of activity including several particularly interesting studies that used life tables to examine the influence of food availability on populations of lizards (e.g., Ballinger 1977; Dunham 1978). Long-term capture-recapture studies are time intensive and require a commitment that most present-day researchers are not willing to make, and as a result, long-term capture-recapture studies are becoming rare. Because reproductive output can and usually does vary with age of females, combining survivorship data with reproductive data is necessary to determine causes of differing "reproductive strategies" among species (e.g., Pianka and Parker 1975). A considerable amount of evolutionary life history hypotheses were developed during the 1970s–1980s (Stearns 1976, 1977; Stearns and Koella 1986; Shine 1985; Pianka 1970, 1972, 1976; Hirshfield and Tinkle 1975) and numerous studies were designed to test them.

Because age-specific survivorship schedules exist for only a handful of lizard species, most of which are temperate zone (e.g., *Uta stansburiana*, Tinkle 1967; Turner *et al.* 1973; *Sceloporus virgatus*, Vinegar 1975a; *S. undulatus*, Tinkle 1972; Vinegar 1975b; *S. graciosus*, Tinkle 1973) and relatively closely related, attempts to identify correlates of different life history strategies have been plagued with a lack of complete data. Nevertheless, some interesting analyses using rather basic life-history data have identified clusters of species with similar life histories (e.g., Tinkle *et al.* 1970; Dunham and Miles 1985; Dunham *et al.* 1988; Mesquita and Colli 2010). Many species are early-maturing, invest heavily in reproduction (high effort per reproductive episode), and are short-lived, whereas other species are late maturing, invest less per reproductive episode, and are long-lived. Many investigators (including the present author!) quickly surmised that the former group could be considered *r*-selected and the latter *K*-selected (see Pianka 1970). However, the logic in doing so is circular at best, because *r*- and *K*-selection are defined on the basis of per individual resource availability (unlimited versus limited, respectively), and resource availability data do not exist for most studied populations. Moreover, resource availability is not the only variable affecting the evolution of life history traits (see Wilbur *et al.* 1974). Alternative hypotheses exist (e.g., bet-hedging; Stearns 1977) and phylogenetic conservatism exists with respect to some life history traits (Mesquita and Colli 2010). Nevertheless, it has been possible to test some of the predictions of life history theory by parsing out specific components, such as "reproductive effort" and "costs of reproduction".

The concept of reproductive effort dates back to Fisher (1930) and was then defined in terms of energy allocation (see Williams 1966; Hirshfield and Tinkle 1975). Tinkle (1969) first tied reproductive effort to life histories of lizards, which led to numerous efforts to measure reproductive effort. One of the first studies to attempt to measure reproductive effort in lizards that used ratios of clutch to body calories as a measure of reproductive effort was conducted by Tinkle and Hadley in 1973, followed by a second study in 1975 (Tinkle and Hadley 1975). It was believed at that time that comparisons of reproductive to total production calories (energy) could be used to measure reproductive effort. As it turned out, variation in calories per unit weight of lizard eggs and bodies varied little (Vitt and Congdon 1978), and as a consequence, caloric ratios provided no more information than weight ratios. Moreover, it became clear that species of lizards with squat, robust bodies that were sit-and-wait foragers (e.g., *Phrynosoma*) had much higher ratios of clutch mass to body mass than streamlined, relatively active foraging lizards (e.g., *Aspidoscelis*), suggesting that foraging mode and predator escape modes influenced relative clutch mass to a large degree across species. Future studies would show that differences in foraging mode arose early in the evolutionary history of lizard clades such that differences in relative clutch mass were also historical (Vitt *et al.* 2003). Non-energetic measures of reproductive effort also appeared in the mid 1970s (e.g., Andrews 1979; Andrews and Rand 1974) and later (Rocha 1990; Shine and Schwarzkopf 1992; Shine *et al.* 1996).

One reason for measuring reproductive effort was to determine whether a cost existed to reproduction. Costs of reproduction vary from energetic, behavioral, physiological (inferred from performance traits), to ecological (predation, competition), all of which ultimately feedback on survival. Detailed energy budgets have been used to determine the proportion of total energy diverted to reproduction (e.g., Congdon *et al.* 1982; Nagy 1983;

Dunham *et al.* 1983; i.e., the direct energetic cost of reproduction). A large number of studies have examined non-energetic costs of reproduction experimentally by measuring physiological performance in gravid versus non-gravid females (e.g., Cooper *et al.* 1990; Shine *et al.* 1998; Sinervo *et al.* 1991). Mate guarding increases predation risk in *Plestiodon laticeps* (Cooper and Vitt 2002) and resisting forced copulation entails a cost of reduced reproduction for female *Ctenophorus maculosus* (Olsson 1995). Costs of reproduction vary geographically for the skink *Lampropholis guichenoti* (Qualls and Shine 1997). Producing single-egg clutches does not apparently reduce costs of reproduction in *Anolis* (Cox and Calsbeek 2009; Cox *et al.* 2010).

The most creative and likely most powerful methods to measure costs of reproduction result from technological advances that generally can be described as allometric engineering (Sinervo 1994; Sinervo and Huey 1990). Basically, they involve experimental manipulation of eggs or embryos within females, either by manipulating the amount of yolk in ova within a female's oviduct or removing some eggs or embryos to examine effects on female and/or offspring performance and survival. For example, offspring size can be reduced by removing portions of yolk in oviductal eggs, which not only reduces size of hatchlings, but also reduces the burden that a clutch has on performance of gravid females (e.g., Sinervo 1999; Sinervo *et al.* 1992). In one study conducted in the field, reduction of clutch burden in *Uta stansburiana* by 30% resulted in higher endurance for manipulated females and higher survival to the next season, demonstrating both a performance and survival cost of reproduction (Miles *et al.* 2000). In another study, the number of embryos in females of a viviparous population of the European lacertid *Zootoca vivipara* was surgically reduced to half in an experimental group that, along with sham operated and non-manipulated controls, were maintained in the lab under identical conditions (Bleu *et al.* 2011). Females with reduced litters were relatively heavier than both controls following parturition indicating that reproduction infers a cost to post-parturition females, a cost that could be carried through to the next season. Although treatment did not affect size of individual neonates, individuals from the experimental group grew more rapidly than those in both controls indicating that the cost of reduced litter size is partially offset by more rapid growth in neonates prior to hibernation. Even though survival of released females and neonates to the following season did not differ among groups, the gestational cost of producing more offspring would presumably result in lower fecundity the following season due to reduced growth and thus smaller size. Allometric engineering appears to be the most direct way to measure costs of reproduction, and it will be particularly interesting to see how manipulation of reproductive output impacts reproductive success across several generations in natural populations.

1.2.6 Environmental Sex Determination

Since the discovery of environmental sex determination (ESD) in two turtle species (Pieau 1971, 1975), more than 100 reptile species have been confirmed to have sex of offspring determined by incubation temperature (temperature-dependent sex determination, TSD), including at least 37 lizard squamates (Valenzuela and Lance 2004; Warner and Janzen 2010; Vitt and Caldwell 2014). Establishing threshold temperatures that determine offspring sex during incubation is relatively easy (e.g., Bull 1987; Warner 2011), but because nest temperatures vary considerably under natural conditions in the field, it has been a challenge to identify the set of conditions in natural nests that affect sex ratios. During embryo development, ESD is usually, but not always, associated with a lack of heteromorphic chromosomes (e.g., XY/XX) and ESD usually occurs early during the second trimester of development. As an indication of the interest in ESD in lizards alone, a Google Scholar search with the terms "environmental sex determination, lizard" produced 24,300 hits (note that this search also pulled up papers on other reptiles with ESD)!

One reason that ESD in reptiles has attracted so much attention is because it offers nearly unlimited opportunities to experimentally address questions about the both the evolution of sex (e.g., why not just one sex as in parthenogenetic clones), and, more importantly, under what conditions it is advantageous in terms of individual fitness, to produce uneven sex ratios in offspring. Given that ESD exists in many reptiles, how and why is it maintained, and can maternal or paternal parents control sex of their offspring based on direct or indirect cues about either the operational sex ratio (OSR) in the population or other factors that might provide one sex an advantage over the other in terms of fitness? My brief discussion is of course, an oversimplification of the evolutionary problem, which involves a number of non-mutually exclusive fitness components, each of which is complex in its own right (Fig. 1.4).

Until recently, the adaptive significance of sex determination in lizards (and other reptiles) was poorly understood. In the oviparous Australian agamid, *Amphibolurus muricatus*, TSD results in proportionally more females than males during early spring (Warner and Shine 2005). Those females grow more than either sex produced later in the season and thus enter overwintering at a larger body size. Even though survivorship of the larger females is slightly affected negatively, they produce more eggs than smaller females, thus demonstrating a reproductive advantage to TSD. In addition, nutritional condition of parents can affect sex ratio of offspring. For example, nutrient-deprived females of *Ampibolurus* produce male-biased sex ratios among their offspring, and those offspring are larger than those produced by control females (Warner *et al.* 2007).

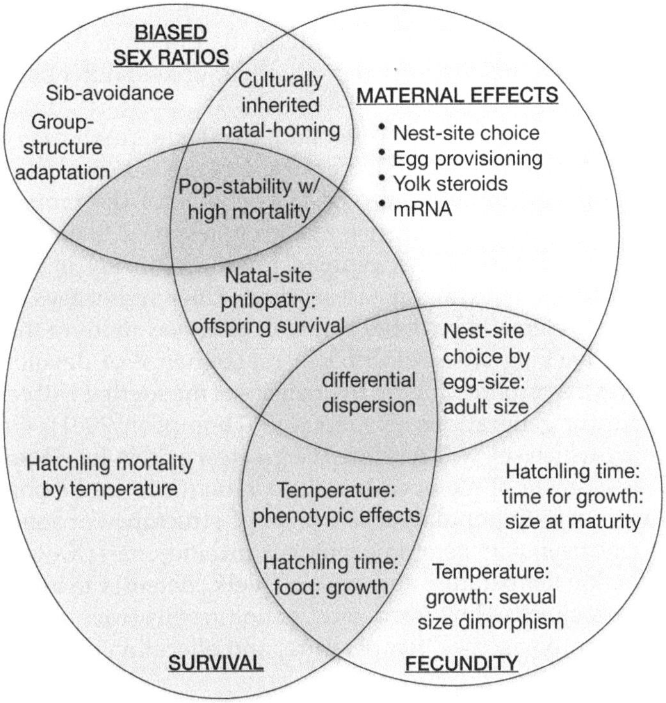

Fig. 1.4 Evolutionary hypotheses to explain TSD in reptiles are complex and non-exclusive. They are best categorized by the finesses component that they address. Redrawn from Valenzuela 2004, pp. 131–147. In N. Valenzuela and V. Lance (eds.), *Temperature-Dependent Sex Determination in Vertebrates*. Smithsonian Books, Washington, DC, Fig. 14.1B.

If there are no fitness differences between males and females based on developmental temperatures, then no advantage exists to ESD over gonadal sex determination, and the expectation would be that sex would be genetically determined. To frame this slightly differently, if male embryos maximize their fitness by developing at a temperature different from the temperature that maximizes fitness for female offspring, then ESD should be at a selective advantage, a scenario that has become known as the Charnov-Bull model (Charnov and Bull 1977).

Demonstrating this empirically has been a challenge, but experiments with *Amphibolurus* add support (Warner and Shine 2008). By measuring lifetime reproductive success in offspring that were incubated as eggs under strictly controlled conditions, Dan Warner and Rick Shine were able to demonstrate that males and females from eggs maintained at temperatures producing a majority of each sex, respectively, generated offspring that had higher lifetime reproductive output than those incubated at temperatures that produced equal numbers of both sexes. These results provide support for the Charnov-Bull model of sex determination, in which ESD will be favored over gonadal sex determination when optimal developmental temperatures (in terms of future reproductive success) differ between the sexes.

Viviparous lizards with ESD offer additional opportunities to examine the adaptive significance of ESD because females can regulate thermal conditions for developing embryos via behavioral thermoregulation. The ability to facultatively adjust sex ratios in response to environmental variables (temperature) is widespread phylogenetically in viviparous Australian skinks, occurring in at least three different lineages (Allsop *et al.* 2006). Gravid females have the ability to affect sex ratios of their litters by selecting basking sites resulting in maintenance of developmental temperatures that produce higher frequencies of the sex that will contribute more to future generations (Robert and Thompson 2001). However, determining which sex will produce the greatest gain in long-term fitness requires assessment of the operational sex ratio (OSR) in the population, which can vary with population size and age structure. For some lizards (*Bassiana duperreyi*) with heteromorphic sex chromosomes (XX versus XY), both incubation temperature and egg size (yolk content) can influence sex ratio within a clutch. Thus sex determination results from an interaction between sex chromosomes, temperature, and allocation of yolk (Radder *et al.* 2009).

Studies on correlates of ESD in lizards have produced fascinating insights, especially during the past ten years. Many more examples exist (e.g., Warner *et al.* 2010; Warner *et al.* 2013), and because of their life history and ecological diversity, lizards will continue to serve as model tetrapod systems to investigate correlates and consequences of ESD. Because measuring lifetime reproductive success is at the heart of understanding the adaptive significance of ESD, short generation times render lizards much better models for these kinds of evolutionary studies than turtles or crocodilians.

1.3 DISCUSSION

During the history of studies on lizard reproduction, descriptive studies have provided data that formed the basis of hypothesis generation. Descriptive studies dominated the literature on lizard reproduction until the early 1970s, when life history theory was being developed and computational advances allowed analyses of large data sets. Early attempts to identify sweeping generalities about lizard reproductive strategies raised more questions than they answered. Discovery of truly fascinating phenomena, such as parthenogenesis and environmental sex determination in lizards led to an increasing use of controlled laboratory and field experiments aimed at teasing out causes of observed phenomena. Comparative methods that take into consideration the evolutionary history of taxa being compared, and thus correcting for a lack of phylogenetic independence, have provided additional insight into the evolution of reproductive traits in the context of life histories.

Although clear time lines for the history of lizard reproductive studies do not exist, the initial descriptive approach to understanding lizard reproduction morphed into correlative studies, synthetic and comparative analyses, and carefully designed experimental studies. As in most areas of biology, the more we learn, the greater the number of interesting questions generated. What is here considered most striking is that seemingly simple and basic natural history observations (e.g., a parthenogenetic lizard, or one that has ESD) open entirely new approaches to some of the most important questions in evolutionary biology (e.g., why have two sexes, or why produce equal numbers of each sex?). At a time when technological and computational advances have made it relatively routine to do previously impossible things (e.g., detailed phylogenies that include divergence time estimates or complex multivariate analyses, respectively), collection of the basic natural history observations that open up new insights is taking a back seat in most research laboratories. Designing experiments based on our limited knowledge of the reproductive biology of lizards, or conducting yet another mega-analysis on existing data fills in many gaps in our overall understanding of reproduction-related phenomena and may be useful. Nevertheless, the many things lizards do that we know nothing about, will not be discovered until basic natural history observations are made in the field. A two-week expedition to collect lizard tissues, for example, is unlikely to tell us anything new about the life history of that species, even though it may fill gaps in an existing phylogeny. Moreover, even short-term field studies are likely to miss potentially game-changing phenomena hidden in the natural histories of individual species.

Considering the nearly blind emphasis in universities on funding and a paucity of funding sources for basic natural history, which is only flashy *after* a discovery is made, and the increasing difficulty involved in acquiring permits to conduct long-term field research, especially internationally, the future of the kinds of reproductive studies that have formed the basis for so much of what we now know seems grim. The single most significant discovery in the history of biology was made by a natural historian (Darwin) as the result of nearly five years in the field when collecting and export permits were not required and none of the technology that we distract ourselves with existed. Natural history may be out of vogue, but basic natural history observations made with knowledge of life-history hypotheses in need of testing remain the starting point for creative science (see Greene 1986, 1994, 1999; Vitt 2013).

1.4 LITERATURE CITED

Allsop, D. J., Warner, D. A., Langkilde, T., Du, W. and Shine, R. 2006. Do operational sex ratios influence sex allocation in viviparous lizards with temperature dependent sex determination? Journal of Evolutionary Biology 19: 1175–1182.

Andrews, R. M. 1979. Reproductive effort of female *Anolis limifrons* (Sauria: Iguanidae). Copeia 1979: 620–626.

Andrews, R. M. 1997. Evolution of viviparity: variation between two sceloporine lizards in the ability to extend egg retention. Journal of Zoology, London 243: 579–595.

Andrews, R. M. 2000. Evolution of viviparity in squamate reptiles (*Sceloporus* spp.): a variant of the cold-climate model. Journal of Zoology, London 250: 243–253.

Andrews, R. M. and Mathies, T. 2000. Natural history of reptilian development: constraints on the evolution of viviparity. BioScience 50: 227–238.

Andrews, R. M., Mathies, T., Qualls, C. P. and Qualls, F. J. 1999. Rates of embryonic development of *Sceloporus* lizards: do cold climates favor the evolution of rapid development? Copeia 1999: 692–700.

Andrews, R. M., and Rand, A. S. 1974. Reproductive effort in anoline lizards. Ecology 55: 1317–1327.

Andrews, R. M. and Rose, B. R. 1994. Evolution of viviparity: constraints on egg retention. Physiological Zoology 67: 1006–1024.

Asana, J. J. 1931. The natural history of *Calotes veriscolor* (Boulenger), the common blood-sucker. Journal of the Bombay Natural History Society 34: 1043–1047.

Avery, R. A. 1970. Utilization of caudal fat by hibernating common lizard, *Lacerta vivipara*. Comparative Biochemistry and Physiology B 37: 119–121.

Avery, R. A. 1974. Storage lipids in the lizard *Lacerta vivipara*: a quantitative sudy. Journal of Zoology (London) 173: 419–425.

Ballinger, R. E. 1974. Reproduction of the Texas horned lizard, *Phrynosoma cornutum*. Herpetologica 30: 321–327.

Ballinger, R. E. 1977. Reproductive strategies: food availability as a source of proximal variation in a lizard. Ecology 58: 628–635.

Ballinger, R. E. and Congdon, J. D. 1980. Food resource limitation of body growth rates in *Sceloporus scalaris* (Sauria: Iguanidae). Copeia 1980: 921–923.

Ballinger, R. E. and Congdon, J. D. 1981. Population ecology and life history of a montane lizard (*Sceloporus scalaris*) in southeastern Arizona. Journal of Natural History 15: 213–222.

Ballinger, R. E., Holy, L., Rowe, J. W., Karst, F., Ogg, C. L. and Stanley-Samuelson, D. W. 1992. Seasonal changes in lipid composition during the reproductive cycle of the red-chinned lizard, *Sceloporus erythrocheilus*. Comparative Biochemistry and Physiology 103B: 527–531.

Ballinger, R. E., Tyler, E. D. and Tinkle, D. W. 1972. Reproductive ecology of a west Texas population of the greater earless lizard, *Cophosaurus texanus*. American Midland Naturalist 88: 419–428.

Benabib, M. 1994. Reproduction and lipid utilization of tropical populations of *Sceloporus variabilis*. Herpetological Monographs 8: 160–180.

Blackburn, D. 1982. Evolutionary origins of viviparity in the Reptilia. I. Sauria. Amphibia-Reptilia 3: 185–205.

Blackburn, D. 1985. Evolutionary origins of viviparity in the Reptilia. II. Serpentes, amphisbaenia, and ichthyosauria. Amphibia-Reptilia 6: 259–291.

Blackburn, D. G. 2000. Reptilian viviparity: past research, future directions, and appropriate models. Comparative Biochemistry and Physiology Part A 127: 391–409.
Blackburn, D. G. 2006. Squamate reptiles as model organisms for the evolution of viviparity. Herpetological Monographs 20: 131–146.
Blackburn, D. G., Evans, H. E. and Vitt, L. J. 1985. The evolution of fetal nutritional adaptations. Fortschritte der Zoologie 30: 437–439.
Blackburn, D. G. and Vitt, L. J. 1992. Reproduction in viviparous South American lizards of the genus *Mabuya*. pp. 150–164. In W. C. Hamlett (ed.), *Reproductive biology of South American vertebrates*. Springer-Verlag, New York.
Blackburn, D. G. and Vitt, L. J. 2002. Specializations of the chorioallantoic placenta in the Brazilian scincid lizard, *Mabuya heathi*: a new placental morphotype for reptiles. Journal of Morphology 254: 121–131.
Blackburn, D. G., Vitt, L. J. and Beuchat, C. A. 1984. Eutherian-like reproductive specializations in a viviparous reptile. Proceedings of the National Academy of Sciences USA 81: 4860–4863.
Bleu, J., Massot, M., Haussy, C. and Meylan, S. 2011. Experimental litter size reduction reveals costs of gestation and delayed effects on offspring in a viviparous lizard. Proceedings of the Royal Society of London B 279: 489–498.
Bull, J. J. 1987. Temperature-sensitive periods of sex determination in a lizard: Similarities with turtles and crocodilians. Journal of Experimental Zoology 241: 143–148.
Charnov, E. L. and Bull, J. J. 1977. When is sex environmentally determined? Nature 266: 828–830.
Cole, C. J. 1975. Evolution of parthenogenetic species of reptiles. pp. 340–355. In R. Reinboth (ed.), *Intersexuality in the Animal Kingdom*. Springer-Verlag, New York.
Cole, C. J. 1984. Unisexual lizards. Scientific American 250: 94–100.
Cole, C. J., Desauer, H. C., Townsend, C. R. and Arnold, M. G. 1990. Unisexual lizards of the genus *Gymnophthalmus* (Reptilia: Teiidae) in the neotropics: genetics, origin, and systematics. American Museum Novitates 2994: 1–29.
Cole, C. J., Lowe, C. H. and Wright, J. W. 1969. Sex chromosomes in teiid whiptail lizards (Genus *Cnemidophorus*). American Museum Novitates 2395: 1–14.
Cole, C. J. and Townsend, C. R. 1983. Sexual behavior in unisexual lizards. Animal Behaviour 31: 724–728.
Cole, C. J. and Townsend, C. R. 1990. Parthenogenetic lizards as vertebrate systems. Journal of Experimental Zoology 4: 174–176.
Colli, G. R., Péres, A. K., Jr. and Zatz, M. G. 1997. Foraging mode and reproductive seasonality in tropical lizards. Journal of Herpetology 31: 490–499.
Congdon, J. D., Dunham, A. E. and Tinkle, D. W. 1982. Energy budgets and life histories of reptiles. pp. 233–271. In C. Gans and F. H. Pough (eds.), *Biology of the Reptilia. Volume 13. Physiology D. Physiological Ecology*. Academic Press, New York.
Congdon, J. D., Vitt, L. J. and Hadley, N. F. 1978. Parental investment: Comparative reproductive energetics in bisexual and unisexual lizard, genus *Cnemidophorus*. American Naturalist 112: 509–521.
Congdon, J. D., Vitt, L. J. and Hadley, N. F. 1981. Reply to Schall. American Naturalist 117: 217–222.
Cooper, W. E., Jr. and Vitt, L. J. 2002. Increased predation risk while mate-guarding as a cost of reproduction for male broad-headed skinks (*Eumeces laticeps*). Acta Ethologica 5: 19–23.
Cooper, W. E., Jr., Vitt, L. J., Hedges, R. and Huey, R. B. 1990. Locomotor impairment and defense in gravid lizards (*Eumeces laticeps*): Behavioral shift in activity may offset costs of reproduction in an active forager. Behavioral Ecology and Sociobiology 27: 153–157.
Cox, R. M. and Calsbeek, R. 2009. Severe costs of reproduction persist in *Anolis* lizards despite the evolution of a single-egg clutch. Evolution 64: 1321–1330.
Cox, R. M., Parker, E. U., Cheney, D. M., Liebl, A. L., Martin, L. B. and Calsbeek, R. 2010. Experimental evidence for physiological costs underlying the trade-off between reproduction and survival. Functional Ecology 24: 1262–1269.

Cuéllar, O. 1976. Intraclonal histocompatibility in a parthenogenetic lizard: Evidence of genetic homogeneity. Science 193: 150–153.
Darevsky, I. S. 1958. Natural parthenogenesis in certain subspecies of rock lizard, *Lacerta saxicola* Eversmann. Doklady Akademii Nauk SSSR. Biological Science 122: 877–879.
Darwin, C. 1859. On the Origin of Species by Means of Natural Selection. John Murray, London.
Darwin, C. 1871. *The Descent of Man, and Selection in Relation to Sex* (2 Volumes). Appleton, New York, pp. 436.
Derickson, W. K. 1976. Ecological and physiological aspects of reproductive strategies in two lizards. Ecology 57: 445–458.
Dial, B. E. and Fitzpatrick, L. C. 1981. The energetic costs of tail autotomy to reproduction in the lizard *Coleonyx brevis* (Sauria: Gekkonidae). Oecologia 51: 310–317.
Dunham, A. E. 1978. Food availability as a proximate factor influencing individual growth rates in the iguanid lizard *Sceloporus merriami*. Ecology 59: 770–778.
Dunham, A. E. 1981. Populations in a fluctuating environment: the comparative population ecology of the iguanid lizards *Sceloporus merriami* and *Urosaurus ornatus*. Miscellaneous Publications of the Museum of Zoology, University of Michigan 158: 1–62.
Dunham, A. E. and Miles, D. B. 1985. Patterns of covariation in life history traits of squamate reptiles: The effects of size and phylogeny reconsidered. American Naturalist 126: 231–257.
Dunham, A. E., Miles, D. B. and Reznick, D. N. 1988. Life history patterns in squamate reptiles. pp. 441–552. In C. Gans and R. B. Huey (eds.), *Biology of the Reptilia. Volume 16. Ecology B. Defense and Life History*. A. R. Liss, New York.
Fisher, R. A. 1930. The Genetical Theory of Natural Selection. Oxford University Press, Oxford, pp. 272.
Fitch, H. S. 1954. Life history and ecology of the five-lined skink, *Eumeces fasciatus*. Museum of Natural History, the University of Kansas, Miscellaneous Publications 8: 1–156.
Fitch, H. S. 1955. Habits and adaptations of the great plains skink (*Eumeces obsoletus*). Ecological Monographs 25: 59–83.
Fitch, H. S. 1958. Natural history of the six-lined racerunner (*Cnemidophorus sexlineatus*). Miscellaneous Publications of the Museum of Natural History of the University of Kansas 11: 11–62.
Fitch, H. S. 1970. Reproductive cycles in lizards and snakes. Miscellaneous Publications of the Museum of Natural History of the University of Kansas 52: 1–247.
Fitch, H. S. 1982. Reproductive cycles in tropical reptiles. Occasional Papers of the Museum of Natural History, the University of Kansas 96: 1–53.
Galdino, C. A. B., Assis, V. B., Kiefer, M. C. and Van Sluys, M. 2003. Reproduction and fat body cycle of *Eurolophosaurus nanuzae* (Sauria; Tropiduridae) from a seasonal montane habitat of southeastern Brazil. Journal of Herpetology 37: 687–694.
Goldberg, S. R. 1971a. Reproduction in the short-horned lizard, *Phrynosoma douglassi* in Arizona. Herpetologica 27: 311–314.
Goldberg, S. R. 1971b. Reproductive cycle of the ovoviviparous iguanid lizard *Sceloporus jarrovi* Cope. Herpetologica 27: 123–131.
Greene, H. W. 1969. Reproduction in a middle American skink, *Leiolopisma cherriei* (Cope). Herpetologica 25: 55–56.
Greene, H. W. 1986. Natural history and evolutionary biology. pp. 99–108. In M. E. Feder and G. V. Lauder (eds.), *Predator–Prey Relationships: Perspectives and Approaches from the Study of Lower Vertebrates*. University of Chicago Press, Chicago.
Greene, H. W. 1994. Systematics and natural history, foundations for understanding and conserving biodiversity. American Zoologist 34: 48–56.
Greene, H. W. 1999. Natural history and behavioural homology. pp. 173–188. In G. R. Bock and G. Cardew (eds.), *Novartis Foundation Symposium 222: Homology*. John Wiley and Sons, New York City.
Guillette, L. J., Jr. 1993. The evolution of viviparity in lizards: Ecological, anatomical, and physiological correlates lead to new hypotheses. BioScience 43: 742–751.

Guillette, L. J., Jr. and Casas-Andreu, G. 1980. Fall reproductive activity in the high altitude Mexican lizard, *Sceloporus grammicus microlepidotus*. Journal of Herpetology 14: 143–147.
Guillette, L. J., Jr. and Casas-Andreu, G. 1981. Seasonal variation in fat body weights of the Mexican high elevation lizard *Sceloporus grammicus microlepidotus*. Journal of Herpetology 15: 366–371.
Guillette, L. J., Jr., Jones, R. E., Fitzgerald, K. T. and Smith, H. M. 1980. Evolution of viviparity in the lizard genus *Sceloporus*. Herpetologica 36: 201–215.
Hahn, W. E. 1965. Lipid mobilization induced by estradiol-17 B in the lizard *Uta stansburiana*. American Zoologist 5: 202.
Hahn, W. E. 1967. Estradiol-induced vitellinogenesis and concomitant fat mobilization in the lizard *Uta stansburiana*. Comparative Biochemistry and Physiology 23: 83–93.
Hahn, W. E. and Tinkle, D. W. 1965. Fat body cycling and experimental evidence for its adaptive significance to ovarian follicle development in the lizard *Uta stansburiana*. Journal of Experimental Zoology 158: 79–86.
Harvey, P. H. and Pagel, M. D. 1991. *The Comparative Method in Evolutionary Biology*. Oxford University Press, New York. pp. 239.
Hirshfield, M. F. and Tinkle, D. W. 1975. Natural selection and the evolution of reproductive effort. Proceedings of the National Academy of Sciences (USA) 72: 2227–2231.
Howard, C. W. 1974. Comparative reproductive ecology of horned lizards (genus *Phrynosoma*) in southwestern United States and northern Mexico. Journal of the Arizona Academy of Sciences 9: 108–116.
Howland, J. M. 1992. Life history of *Cophosaurus texanus* (Sauria: Iguanidae): environmental correlates and interpopulational variation. Copeia 1992: 82–93.
Huang, W.-S. 1997a. Reproductive cycle of the oviparous lizard *Japalura brevipes* (Agamidae: Reptilia) in Taiwan, Republic of China. Journal of Herpetology 31: 22–29.
Huang, W.-S. 1997b. Reproductive cycle of the skink, *Sphenomorphus taiwanensis*, in central Taiwan. Journal of Herpetology 31: 287–290.
James, C. D. and Shine, R. 1988. Life-history strategies of Australian lizards: A comparison between the tropics and the temperate zone. Oecologia 75: 307–316.
Kearney, M., Blacket, M. J., Strasburg, J. L. and Moritz, C. 2006. Waves of parthenogenesis in the desert: Evidence for the parallel loss of sex in a grasshopper and a gecko from Australia. Molecular Ecology 15: 1743–1748.
Kearney, M., Fujita, M. K. and Ridenour, J. 2009. Lost sex in the reptiles: Constraints and correlations. pp. 447–474. In I. Schön, K. Martens, and P. van Dijk (eds.), *Lost Sex, The Evolutionary Biology of Parthenogenesis*. Springer, Dordrecht.
Lack, D. 1947. The significance of clutch-size. Ibis 89: 302–352.
Lin, E. J. I. 1979a. Fatbody and liver cycles in two tropical lizards *Chamaeleo hohneli* and *Chamaeleo jacksoni* (Reptilia, Lacertilia, Chamaeleonidae). Journal of Herpetology 13: 113–117.
Lin, J. 1979b. Ovarian, fat body and liver cycles in the lizard *Japalura swinhonis formosensis* in Taiwan (Lacertilia: Agamidae). Journal of Asian Ecology 1: 29–38.
Lowe, C. H. and Wright, J. W. 1966a. Evolution of parthenogenetic species of *Cnemidophorus* (whiptail lizards) in western North America. Journal of the Arizona Academy of Science 4: 81–87.
Lowe, C. H. and Wright, J. W. 1966b. Chromosomes and karyotypes of cnemidophorine teiid lizards. Mammalian Chromosomes 22: 199–200.
Magnusson, W. E. 1987. Reproductive cycles of teiid lizards in Amazonian savanna. Journal of Herpetology 21: 307–316.
Maslin, T. P. 1971. Parthenogenesis in reptiles. American Zoologist 11: 361–380.
Mayhew, W. 1965. Reproduction in the sand-dwelling lizard *Uma inornata*. Herpetologica 21: 39–55.
Mayhew, W. W. 1963. Reproduction in the granite spiny lizard, *Sceloporus orcutti*. Copeia 1963: 144–152.

Mesquita, D. O. and Colli, G. R. 2010. Life history patterns in tropical South American lizards. pp. 45–71. In O. H. Gallegos, F. R. M. Cruz, and J. F. M. Sánchez (eds.), *Reproducción en Reptiles: Morphologia, Ecología, y Evolución* Universidad Autónoma del Estado de México, México.

Miles, D. B., Sinervo, B. and Frankino, W. A. 2000. Reproductive burden, locomotor performance, and the cost of reproduction in free ranging lizards. Evolution 54: 1386–1395.

Moreau, R. E. 1944. Clutch-size: A comparative study, with special reference to African birds. Ibis 86: 286–347.

Nagy, K. A. 1983. Ecological energetics. pp. 25–54. In R. B. Huey, E. R. Pianka and T. W. Schoener (eds.), *Lizard Ecology: Studies of a Model Organism*. Harvard University Press: Cambridge, MA.

Niewiarowski, P. H. 1994. Understanding geographic life-history variation in lizards. pp. 31–49. In L. J. Vitt and E. R. Pianka (eds.), *Lizard Ecology: Historical and Experimental Perspectives*. Princeton University Press, Princeton, N.J.

Noble, G. K. and Bradley, H. T. 1933. The mating behavior of lizards; its bearing on the theory of sexual selection. Annals of the New York Academy of Sciences 35: 25–100.

Olsson, M. 1995. Forced copulation and costly female resistance behavior in the Lake Eyre dragon, *Ctenophorus maculosus*. Herpetologica 51: 19–24.

Pianka, E. R. 1970. On *r*- and *K*-selection. American Naturalist 104: 592–597.

Pianka, E. R. 1972. *r* and *K* selection or *b* and *d* selection? American Naturalist 106: 581–588.

Pianka, E. R. 1976. Natural selection of optimal reproductive tactics. American Zoologist 16: 775–784.

Pianka, E. R. and Parker, W. S. 1975. Age-specific reproductive tactics. American Naturalist 109: 453–464.

Pieau, C. 1971. Sur la proportion sexuelle chez les embryons des deux Chéloniens (*Testudo graeca* L. et *Emys orbicularis* L.) issus d'oeufs incubés artificiellement. Comptes rendus hebdomadaires des seances de l'Academie des sciences. 272: 3071–3079.

Pieau, C. 1975. Temperature and sex differentiation in embryos of two chelonians, *Emys orbicularis* L. and *Testudo graeca* L. pp. 332–339. In R. Reinboth (ed.), *Intersexuality in the Animal Kingdom*. Springer-Verlag, New York.

Pilorge, T. and Barbault, R. 1981. la viviparite chez les lezards: evolution et adaptation. Acta Oecologia 2: 387–397.

Qualls, C., Shine, R., Donnellan, S. and Hutchison, M. 1995. The evolution of viviparity within the Australian scincid lizard *Lerista bougainvillii*. Journal of Zoology, London 237: 13–26.

Qualls, C. P., Andrews, R. M. and Mathies, T. 1997. The evolution of viviparity and placentation revisited. Journal of Theoretical Biology 185: 129–135.

Qualls, F. J. and Shine, R. 1997. Geographic variation in 'costs of reproduction' in the scincid lizard *Lampropholis guichenoti*. Functional Ecology 11: 757–763.

Radder, R., Pike, D. A., Quinn, A. and Shine, R. 2009. Offspring sex in a lizard depends on egg size. Current Biology 19: 1102–1105.

Ramírez-Bautista, A., Balderas-Valdivia, C. and Vitt, L. 2000. Reproductive ecology of the whiptail lizard *Cnemidophorus lineatissimus* (Squamata: Teiidae) in a tropical dry forest. Copeia 2000: 712–722.

Ramirez-Bautista, A., Guillette, L. J., Jr., Gutierrez-Mayen, G. and Uribe-Peña, Z. 1996. Reproductive biology of the lizard *Eumeces copei* (Lacertilia: Scincidae) from the Eje Neovolcanico, Mexico. Southwestern Naturalist 41: 103–110.

Ramírez-Bautista, A., Uribe-Pena, Z. and Guillette, L. J., Jr. 1995. Reproductive biology of the lizard *Urosaurus bicarinatus bicarinatus* (Reptilia: Phrynosomatidae) from Rio Balsas Basin, Mexico. Herpetologica 51: 24–33.

Rebouças-Spieker, R. and Vanzolini, P. E. 1978. Parturition in *Mabuya macrorhyncha* Hoge, 1946 (Sauria, Scincidae), with a note on the distribution of maternal behavior in lizards. Papéis Avulsos Zoology (São Paulo) 32: 95–99.

Reeder, T. W., Cole, C. J. and Dessauer, H. C. 2002. Phylogenetic relationships of whiptail lizards of the genus *Cnemidophorus* (Squamata: Teiidae): A test of monophyly, reevaluation

of karyotypic evolution, and review of hybrid origins. American Museum Novitates 3365: 1–61.
Robert, K. A. and Thompson, M. B. 2001. Sex determination: Viviparous lizard selects sex of embryos. Nature 412: 698–699.
Roberts, J. A., Vo, H. D., Fujita, M. K., Moritz, C. and Kearney, M. 2012. Physiological implications of genomic state in parthenogenetic lizards of reciprocal hybrid origin. Journal of Evolutionary Biology 25: 252–263.
Rocha, C. F. D. 1990. Reproductive effort in the Brazilian sand lizard *Liolaemus lutzae* (Sauria: Iguanidae). Ciencia e Cultura 42: 1203–1206.
Rocha, C. F. D. 1992. Reproductive and fat body cycles of the tropical sand lizard (*Liolaemus lutzae*) of southeastern Brazil. Journal of Herpetology 26: 17–23.
Rose, B. 1982. Food intake and reproduction in *Anolis acutus*. Copeia 1982: 322–330.
Schall, J. J. 1978. Reproductive strategies in sympatric whiptail lizards (*Cnemidophorus*): two parthenogenetic and three bisexual species. Copeia 1978: 108–116.
Schall, J. J. 1980. Parthenogenetic lizards: r-selected reproductive characteristics? American Naturalist 117: 212–216.
Sexton, O. J. and Brown, K. M. 1977. The reproductive cycle of an iguanid lizard *Anolis sagrei*, from Belize. Journal of Natural History 11: 241–250.
Sexton, O. J., Ortleb, E. P., Hathaway, L. M. and Ballinger, R. E. 1971. Reproductive cycles of three species of anoline lizards from the Isthmus of Panama. Ecology 52: 201–215.
Sexton, O. J. and Turner, O. 1971. The reproductive cycle of a neotropical lizard. Ecology 52: 159–164.
Sherbrooke, W. C. 1975. Reproductive cycle of a tropical teiid lizard, *Neusticurus ecpleopus* Cope, in Peru. Biotropica 7: 194–207.
Shine, R. 1985a. The evolution of viviparity in reptiles: an ecological analysis. pp. 677–680. In C. Gans and F. Billett (eds.), *Biology of the Reptilia*. John Wiley and Sons, Inc.
Shine, R. 1985b. The reproductive biology of Australian reptiles: a search for general patterns. pp. 297–303. In G. Grigg, R. Shine and H. Ehmann (eds.), *Biology of the Australian Frogs and Reptiles*. Surrey Beatty and Sons, New South Wales.
Shine, R. 2002. An empirical test of the 'predictability' hypothesis for the evolution of viviparity in reptiles. Journal of Evolutionary Biology 15: 553–560.
Shine, R. and Berry, J. F. 1978. Climatic correlates of live-bearing in squamate reptiles. Oecologia (Berlin) 33: 261–268.
Shine, R. and Bull, J. J. 1979. The evolution of live-bearing in lizards and snakes. American Naturalist 113: 905–923.
Shine, R., Keogh, S., Doughty, P. and Giragossyan, H. 1998. Costs of reproduction and the evolution of sexual dimorphism in a 'flying lizard' *Draco melanopogon* (Agamidae). Journal of Zoology, London 246: 203–213.
Shine, R. and L. J. Guillette, J. 1988. The evolution of viviparity in reptiles: a physiological model and its ecological consequences. Journal of Theoretical Biology 132: 43–50.
Shine, R. and Schwarzkopf, L. 1992. The evolution of reproductive effort in lizards and snakes. Evolution 46: 62–75.
Shine, R., Schwarzkopf, L. and Caley, M. J. 1996. Energy, risk, and reptilian reproductive effort: a reply to Niewiarowski and Dunham. Evolution 50: 2111–2114.
Sinclair, E. A., Pramuk, J. B., Bezy, R. L., Crandall, K. A. and Sites, J. W., Jr. 2010. DNA evidence for nonhybrid origins of parthenogenesis in natural populations of vertebrates. Evolution 64: 1346–1357.
Sinervo, B. 1994. Experimental tests of reproductive allocation paradigms. pp. 73–90. In L. J. Vitt and E. R. Pianka (eds.), *Lizard Ecology: Historical and Experimental Perspectives*. Princeton University Press, Princeton, N.J.
Sinervo, B. 1999. Mechanistic analysis of natural selection and a refinement of Lack's and William's principles. American Naturalist 154: S26–S42.
Sinervo, B., Doughty, P., R. B. Huey and Zamudio, K. 1992. Allometric engineering: a causal analysis of natural selection on offspring size. Science 258: 1927–1930.

Sinervo, B., Hedges, R. and Adolph, S. C. 1991. Decreased sprint speed as a cost of reproduction in the lizard *Sceloporus occidentalis*: variation among populations. Journal of Experimental Biology 155: 323–366.
Sinervo, B. and Huey, R. B. 1990. Allometric engineering: an experimental test of the causes of interpopulational differences in performance. Science 248: 1106–1109.
Stearns, S. C. 1976. Life-history tactics: a review of the ideas. Quarterly Review of Biology 51: 3–47.
Stearns, S. C. 1977. The evolution of life history traits: a critique of the theory and a review of the data. Annual Review of Ecology and Systematics 8: 145–171.
Stearns, S. C. and Koella, J. C. 1986. The evolution of phenotypic plasticity in life-history traits: predictions of reaction norms for age and size at maturity. Evolution 40: 893–913.
Stewart, J. R. and Blackburn, D. G. 1988. Reptilian placentation: structural diversity and terminology. Copeia 1988: 839–852.
Stewart, J. R. and Thompson, M. B. 2000. Evolution of placentation among squamate reptiles: recent research and future directions. Comparative Biochemistry and Physiology Part A 127: 411–431.
Stewart, J. R. and Thompson, M. B. 2003. Evolutionary transformations of the fetal membranes of viviparous reptiles: a case study of two lineages. Journal of Experimental Zoology 299A: 13–32.
Tinkle, D. W. 1967. The life and demography of the side-blotched lizard, *Uta stansburiana*. Miscellaneous Publications, Museum of Zoology, University of Michigan 132: 1–182.
Tinkle, D. W. 1969. The concept of reproductive effort and its relation to the evolution of life histories of lizards. American Naturalist 103: 501–516.
Tinkle, D. W. 1972. The dynamics of a Utah population of *Sceloporus undulatus*. Herpetologica 28: 351–359.
Tinkle, D. W. 1973. A population analysis of the sagebrush lizard, *Sceloporus graciosus* in southern Utah. Copeia 1973: 284–296.
Tinkle, D. W. 1976. Comparative data on the population ecology of the desert spiny lizard, *Sceloporus magister*. Herpetologica 32: 1–6.
Tinkle, D. W. and Gibbons, J. W. 1977. The distribution and evolution of viviparity in reptiles. Miscellaneous Publications of the Museum of Zoology, University of Michigan 154: 1–55.
Tinkle, D. W. and Hadley, N. F. 1973. Reproductive effort and winter activity in the viviparous montane lizard *Sceloporus jarrovi*. Copeia 1973: 272–277.
Tinkle, D. W. and Hadley, N. F. 1975. Lizard reproductive effort: caloric estimates and comments on its evolution. Ecology 56: 427–434.
Tinkle, D. W., Wilbur, H. M. and Tilley, S. G. 1970. Evolutionary strategies in lizard reproduction. Evolution 24: 55–74.
Turner, F. B., Medica, P. A., Lannom, J. R., Jr. and Hoddenbach, G. A. 1969. A demographic analysis of fenced populations of the whiptail lizard, *Cnemidophorus tigris*, in southern Nevada. Southwestern Naturalist 14: 189–202.
Turner, F. B., Medica, P. A. and Smith, D. D. 1973. Reproduction and survivorship of the lizard, *Uta stansburiana*, and the effects of winter rainfall, density and predation on these processes. Reports of 1973 Progress 3: Process Studies: 117–128.
Valenzuela, N. 2004. Evolution and maintenance of temperature-dependent sex determination. pp. 131–147. In N. Valenzuela and V. Lance (eds.), *Temperature-Dependent Sex Determination in Vertebrates*. Smithsonian Books, Washington, DC.
Valenzuela, N. and Lance, V. (eds.). 2004. *Temperature-Dependent Sex Determination in Vertebrates*. Smithsonian Books, Washington, DC.
Van Loben Sels, R. C. and Vitt, L. J. 1984. Desert lizard reproduction: seasonal and annual variation in *Urosaurus ornatus* (Iguanidae). Canadian Journal of Zoology 62: 1779–1787.
Van Wyk, J. H. 1991. Biennial reproduction in the female viviparous lizard *Cordylus giganteus*. Amphibia-Reptilia 1991: 329–342.
Vinegar, M. B. 1975a. Demography of the striped plateau lizard, *Sceloporus virgatus*. Ecology 56: 172–182.

Vinegar, M. B. 1975b. Life history phenomena in two populations of the lizard *Sceloporus undulatus* in southwestern New Mexico. American Midland Naturalist 93: 388–402.
Vitt, L. J. 1983. Reproduction and sexual dimorphism in the tropical teiid lizard, *Cnemidophorus ocellifer*. Copeia 1983: 359–366.
Vitt, L. J. 1986. Reproductive tactics of sympatric gekkonid lizards with a comment on the evolutionary and ecological consequences of invariant clutch size. Copeia 1986: 773–786.
Vitt, L. J. 1990. The influence of foraging mode and phylogeny on seasonality of tropical lizard reproduction. Papéis Avulsos Zoologia (São Paulo) 37: 107–123.
Vitt, L. J. 1992. Diversity of reproduction strategies among Brazilian lizards and snakes: The significance of lineage and adaptation. pp. 135–149. In W. C. Hamlett (ed.), *Reproductive Biology of South American Vertebrates*. Springer-Verlag, New York.
Vitt, L. J. 2013. Walking the natural-history trail. Herpetologica 69: 105–117.
Vitt, L. J. and Blackburn, D. G. 1983. Reproduction in the lizard *Mabuya heathi* (Scincidae): a commentary on viviparity in New World *Mabuya*. Canadian Journal of Zoology 61: 2798–2806.
Vitt, L. J. and Caldwell, J. P. 2014. *Herpetology, An Introductory Biology of Amphibians and Reptiles, Fourth Edition*. Academic Press, San Diego. pp. 757.
Vitt, L. J. and Congdon, J. D. 1978. Body shape, reproductive effort, and relative clutch mass in lizards: resolution of a paradox. American Naturalist 112: 595–608.
Vitt, L. J. and Cooper, W. E., Jr. 1985. The relationship between reproduction and lipid cycling in *Eumeces laticeps* with comments on brooding ecology. Herpetologica 41: 419–432.
Vitt, L. J. and Goldberg, S. R. 1983. Reproductive ecology of two tropical iguanid lizards: *Tropidurus torquatus* and *Platynotus semitaeniatus*. Copeia 1983: 131–141.
Vitt, L. J. and Lacher, T. E., Jr. 1981. Behavior, habitat, diet, and reproduction of the iguanid lizard *Polychrus acutirostris* in the Caatinga of northeastern Brazil. Herpetologica 37: 53–63.
Vitt, L. J. and Lacher, T. E., Jr. 1981. Behavior, habitat, diet, and reproduction of the iguanid lizard *Polychrus acutirostris* in the caatinga of northeastern Brazil. Herpetologica 37: 53–63.
Vitt, L. J., Pianka, E. R., Cooper, W. E., Jr. and Schwenk, K. 2003. History and the global ecology of squamate reptiles. American Naturalist 162: 44–60.
Vitt, L. J., Van Loben Sels, C. and Ohmart, R. D. 1978. Lizard reproduction: annual variation and environmental correlates in the iguanid lizard *Urosaurus graciosus*. Herpetologica 34: 241–253.
Vrijenhoek, R. C., Dawley, R. M., Cole, C. J. and Bogart, J. P. 1989. A list of the known unisexual vertebrates. pp. 19–23. In R. M. Dawley and J. P. Bogart (eds.), *Evolution and Ecology of Unisexual Vertebrates*. New York State Museum: Albany, NY.
Warner, D. A. 2011. Sex determination in reptiles. pp. 1–38. In D. O. Norris and K. Lopez (eds.), *Hormones and Reproduction of Vertebrates, Volume 3: Reptiles*. Academic Press, San Diego.
Warner, D. A., Bonnet, X., Hobson, K. A. and Shine, R. 2008. Lizards combine stored energy and recently acquired nutrients flexibly to fuel reproduction. Journal of Animal Ecology 77: 1242–1249.
Warner, D. A. and Janzen, F. J. 2010. Diversity of sex-determining mechanisms. pp. 81–83. In D. Westneat and C. W. Fox (eds.), *Evolutionary Behavioral Ecology*. Oxford University Press: Oxford.
Warner, D. A., Lovern, M. and Shine, R. 2007. Maternal nutrition affects reproductive output and sex allocation in a lizard with environmental sex determination. Proceedings of the Royal Society B: Biological Sciences 274: 883–890.
Warner, D. A. and Shine, R. 2005. The adaptive significance of temperature-dependent sex determination: Experimental tests with a short-lived lizard. Evolution 59: 2209–2221.
Warner, D. A. and Shine, R. 2008. The adaptive significance of temperature-dependent sex determination in a reptile. Nature 451: 566–568.
Warner, D. A., Uller, T. and Shine, R. 2013. Transgenerational sex determination: the embryonic environment experienced by a male affects offspring sex ratio. Nature doi: 10.1038/srep02709.

Warner, D. A., Woo, K. L., Van Dyk, D. A., Evans, C. S. and Shine, R. 2010. Egg incubation temperature affects male reproductive success but not display behaviors in lizards. Behavioral Ecology and Sociobiology 64: 803–813.
Weekes, H. C. 1927. Placentation and other phenomena in the scincid lizard *Lygosoma* (*Hinulia*) *quoyi*. Proceedings of the Zoological Society of New South Wales 52: 499–554.
Weekes, H. C. 1929. On placentation in reptiles. I. Proceedings of the Linnean Society of New South Wales: 34–60.
Weekes, H. C. 1930. On placentation in reptiles. II. Proceedings of the Linnean Society of New South Wales 55: 550–576.
Weekes, H. C. 1935. A review of placentation among reptiles with particular regard to the function and evolution of the placenta. Proceedings of the Zoological Society of London Part 3: 625–645.
Wilbur, H. M., Tinkle, D. W. and Collins, J. P. 1974. Environmental certainty, trophic level, and resource availability in life history evolution. American Naturalist 108: 805–817.
Williams, G. C. 1966. Natural selection, the costs of reproduction, and a refinement of Lack's principle. American Naturalist 100: 687–690.
Wright, J. W. 1993. Evolution of the lizards of the genus *Cnemidophorus*. pp. 27–81. In J. W. Wright and L. J. Vitt (eds.), *Biology of whiptail lizards (genus Cnemidophorus)*. Oklahoma Museum of Natural History: Norman, Oklahoma.
Wright, J. W. and Lowe, C. H. 1967. Evolution of the alloploid parthenspecies *Cnemidophorus tesselatus* (Say). Mammalian Chromosomes Newsletter 8: 95–96.
Wright, J. W. and Lowe, C. H. 1968. Weeds, polyploids, parthenogenesis, and the geographical and ecological distributions of all-female species of *Cnemidophorus*. Copeia 1968: 128–138.

CHAPTER 2

The Phylogeny of Lizard Families

*John J. Wiens** and *Shea M. Lambert*

2.1 INTRODUCTION

In this chapter, we will summarize our current knowledge of higher-level lizard phylogeny, focusing on relationships among extant families. Despite recent suggestions that lizard phylogeny is unresolved due to conflicts between molecular and morphological data (e.g., Gauthier *et al.* 2012; Losos *et al.* 2012), we will argue that the phylogeny of extant lizard families is actually becoming increasingly well resolved and well supported. We will first argue that simply pointing out differences between molecular and morphological trees does not help us resolve phylogeny. Instead, the best way to resolve these conflicts is through combined analysis and identification of misleading signals. We will argue that combined analyses of large-scale molecular and morphological datasets clearly support the estimate of higher-level phylogeny from molecular datasets, and that there is strongly misleading signal in the morphological data. We will then review our current knowledge of higher-level squamate phylogeny, emphasizing those aspects that are now agreed upon by large-scale datasets, and those aspects that are still uncertain. We will also present a new time-calibrated phylogeny for squamate reptile families.

Department of Ecology and Evolutionary Biology, University of Arizona, Tucson, Arizona 85721-0088.
* Corresponding author

2.2 MOLECULES VERSUS MORPHOLOGY AND THE WAY FORWARD

Some recent papers have painted a bleak picture of our current state of knowledge of squamate phylogeny, suggesting that the higher-level relationships are largely unresolved. Losos *et al.* (2012) pointed out that a recent analysis of morphological characters alone (Gauthier *et al.* 2012) and a recent analysis of molecular data alone (Wiens *et al.* 2012) gave different results, particularly with regards to the placement of the major clade Iguania. Specifically, morphological analyses place Iguania as the sister group to all other squamates, including recent analyses of large-scale morphological data matrices by Conrad (2008) and Gauthier *et al.* (2012), as well as older studies (e.g., Estes *et al.* 1988). In contrast, recent molecular analyses place Iguania in a clade with Anguimorpha (i.e., Anguidae, Helodermatidae, Lanthanotidae, Shinisauridae, and Xenosauridae) and with snakes, far from the squamate root (e.g., Townsend *et al.* 2004; Vidal and Hedges 2005; Fry *et al.* 2006; Wiens *et al.* 2010, 2012; Mulcahy *et al.* 2012). Losos *et al.* (2012) then concluded that squamate relationships should be considered unresolved because of this difference.

But simply comparing trees from separately analyzed datasets will never make progress in phylogeny. It has been known for years that the trees from molecular and morphological data disagreed (ever since Townsend *et al.* 2004). There has to be a way for conflicts like these to be resolved, and not simply by declaring one dataset victorious or the relationships uncertain. For decades, such conflicts between molecular and morphological data have been resolved through combined analysis of all the relevant character data (for reviews and examples see: Hillis 1987; Kluge 1989; Baker *et al.* 1998; Hillis and Wiens 2000; Nylander *et al.* 2004; Wiens *et al.* 2010).

Two lines of evidence strongly suggest that the higher-level molecular phylogeny is actually correct, at least in the placement of Iguania. First, a combined analysis of large-scale molecular and morphological datasets strongly supports the placement of Iguania with Serpentes and Anguimorpha. Wiens *et al.* (2010) combined the morphological dataset of Conrad (2008; 363 characters) with a dataset of 22 nuclear loci (15,794 characters) for 45 living taxa (with molecular and morphological data) and 19 fossil taxa (morphological data only) and found strong support for the basic relationships suggested in previous molecular studies (e.g., Townsend *et al.* 2004; Fry *et al.* 2006; Vidal and Hedges 2009), including the basal placement of Gekkota and Dibamidae, the placement of amphisbaenians in a clade with lacertids, teiids, and gymnophthalmids, and the placement of iguanians with snakes and anguimorphs. New analyses of the larger molecular dataset of Wiens *et al.* (2012; 44 nuclear loci, 33,717 characters) combined with the larger morphological dataset of Gauthier *et al.* (2012;

610 characters) yield the same conclusion (Reeder, Wiens, and co-authors, unpublished).

Combined analyses are critically important. Otherwise the molecular and morphological data are treated as somehow offering equivalent levels of evidence for every part of the tree. But clearly they do not. For example, in the study of Wiens *et al.* (2010), the morphological dataset contains 363 characters whereas the molecular dataset contains 15,794 characters. Using the logic of Gauthier *et al.* (2012) and Losos *et al.* (2012), the morphological dataset could contain only 1 character and the molecular dataset could contain 1 million, but relationships would be considered unresolved because the two datasets suggest conflicting relationships. Although separate analyses may be useful to identify systematic errors associated with non-independent characters and other issues (e.g., Hillis and Wiens 2000), there is no support for the idea that all 22 nuclear loci for squamates (or 44) are genetically linked and should be treated as a single dataset. In contrast, there is evidence that many of the morphological characters are correlated due to selection associated with similar ecological conditions (e.g., burrowing; Wiens *et al.* 2010).

One might assume that combined-data trees must always be resolved in favor of the dataset with a greater number of characters (e.g., Gauthier *et al.* 2012). But this is demonstrably untrue (Wiens *et al.* 2010). In fact, Bayesian and parsimony analyses of the 22 nuclear loci alone show very strong support for non-monophyly of amphisbaenians (i.e., the amphisbaenian family Rhineuridae as the sister to Lacertidae + other amphisbaenian families; Figs. 3 and 4 of Wiens *et al.* 2010). In contrast, the morphological data strongly support amphisbaenian monophyly (Figs. 1 and 2 of Wiens *et al.* 2010). Remarkably, when the molecular and morphological datasets are combined, monophyly of living amphisbaenians is strongly supported (Figs. 5 and 6 of Wiens *et al.* 2010), despite the much greater number of molecular characters. In fact, despite being dominated by the molecular data, the tree from the combined data is not identical to either the separate molecular or morphological trees (e.g., Hillis 1987).

The second major line of evidence that the molecular tree is correct is that there is misleading signal that has a profound influence on recent morphological datasets. The two large-scale phylogenetic analyses of the morphological data alone (Conrad 2008; Gauthier *et al.* 2012) generate trees that seem to be strongly impacted by morphological convergence in the burrowing taxa (Wiens *et al.* 2010). Specifically both of these analyses unite diverse squamate taxa that show a burrowing lifestyle, body elongation, and limb reduction (i.e., amphisbaenians, dibamids, snakes) into a single clade. The analyses of Conrad (2008) place this burrowing clade inside of Scincidae (his Fig. 55). Gauthier *et al.* (2012) showed this burrowing clade as nested inside Anguimorpha, and also including some anguids (i.e., *Anniella*;

their Figs. 2, 3 and 4) and some burrowing scincids in some analyses (i.e., their Fig. 3). This burrowing clade is strongly rejected by recent molecular and combined-data analyses (Townsend *et al.* 2004; Vidal and Hedges 2009; Wiens *et al.* 2010, 2012; Mulcahy *et al.* 2012; Pyron *et al.* 2013). Furthermore, although not widely appreciated (e.g., Losos *et al.* 2012), the burrowing clade also contradicts traditional taxonomy and hypotheses of phylogeny (e.g., Estes *et al.* 1988). Specifically, these recent morphology-only trees are inconsistent with many traditionally recognized and morphologically based higher taxa, including Anguidae, Anguimorpha, Scincidae, and Scincoidea (e.g., Estes *et al.* 1988). Thus, the morphological data appear to be strongly influenced by this misleading phylogenetic signal.

Interestingly, apart from the basal placement of Iguania and the seemingly erroneous burrowing clade, these large-scale morphological analyses of squamate phylogeny show relatively little resolution among squamate families. For example the Bayesian analyses of Gauthier *et al.* (2012) show squamate relationships as largely unresolved, apart from this burrowing clade (their Fig. 4). The results of the parsimony analyses of both Conrad (2008) and Gauthier *et al.* (2012) are summarized as Adams or majority-rule consensus trees, which suggests that relationships in the fundamental trees are largely unresolved (e.g., a parsimony re-analysis of the dataset of Conrad [2008] using a strict consensus tree shows largely unresolved higher-level relationships; Fig. 1 of Wiens *et al.* 2010).

One might argue that the problem of erroneously clustering the burrowing taxa is completely separate from that of the basal placement of Iguania. However, it is important to note that Iguania does not contain any limb-reduced, elongated, burrowing taxa. Based on the placement of burrowing taxa suggested by the molecular and combined-data analyses, Iguania is the only major clade of lizards that lacks such taxa (e.g., Wiens *et al.* 2006). Thus, it is not entirely surprising that all other clades of lizards would be grouped together to the exclusion of iguanians.

Furthermore, iguanians may be impacted by a similar problem of morphological convergence. Iguanians and scleroglossans (i.e., all other squamates) differ in several traits that are seemingly related to differences in feeding and the development of a more flexible skull (e.g., Vitt *et al.* 2003; Townsend *et al.* 2004). Importantly, the morphological characters that seem to support the basal placement of Iguania (mostly involving changes in the skull; Gauthier *et al.* 2012) may also be non-independent (just as the many characters that place the burrowing taxa together are apparently not independent; Wiens *et al.* 2010).

In theory, it is possible that the molecular data are misleading instead. For example, there is striking convergence in squamate mitochondrial DNA which strongly impacts phylogeny estimation using these data (i.e., placing snake and acrodont iguanians together; Castoe *et al.* 2009).

However, separate analyses of each of the 44 nuclear loci show that none of them actually support the basal placement of Iguania (Wiens *et al.* 2012; Wiens unpubl.). Although there is incongruence among the separately analyzed gene trees, this incongruence is strongly associated with relatively short branch lengths in the combined-data tree (Wiens *et al.* 2012). This incongruence on short branches is the pattern expected given incomplete lineage sorting (e.g., Maddison 1997), not convergence or hidden support for the basal placement of Iguania.

In summary, simply comparing trees from separately analyzed molecular and morphological datasets and declaring the relationships to be uncertain (e.g., Losos *et al.* 2012) is highly problematic. In fact, the morphological data are clearly plagued by strongly misleading signal, yield trees that are mostly unresolved, and offer a small fraction of the character evidence provided by the molecular data (Conrad 2008; Gauthier *et al.* 2012). When the molecular and morphological data are combined (e.g., Wiens *et al.* 2010), the resulting tree is well resolved and supported and is almost identical to the tree from molecular data alone (even though the morphological data are demonstrably capable of overturning relationships from molecular data, despite the smaller number of characters). Most importantly, simply comparing trees from separate datasets offers no way to make progress in understanding phylogeny.

2.3 THE CURRENT STATE OF THE TREE

Here, we summarize our current knowledge of squamate phylogeny (Figs. 2.1 and 2.2). We base this summary primarily on three main sources. First, the most extensive analysis of squamate relationships so far in terms of number of characters (44 nuclear loci [33,717 characters] for 161 extant squamate species, representing nearly all families; Wiens *et al.* 2012). Second, the most extensive analysis of squamate relationships so far in terms of number of taxa (7 nuclear loci, 5 mitochondrial genes for 4161 extant squamate species; Pyron *et al.* 2013). Third, the most extensive combined analysis of molecular and morphological data for squamates so far, including 22 nuclear loci (15,794 characters) and 363 morphological characters for 45 living and 19 fossil taxa (Wiens *et al.* 2010). Note that this review will focus exclusively on the living taxa.

Analyses of 44 nuclear loci place dibamids and gekkotans together as the sister group to all other squamates (Wiens *et al.* 2012). This is also found in combined Bayesian analyses of 22 loci and morphological data (Wiens *et al.* 2010). Analyses of many taxa (Pyron *et al.* 2013) instead suggest that dibamids are sister to all other squamates, with Gekkota the next clade up the tree. This has also been found in earlier analyses using

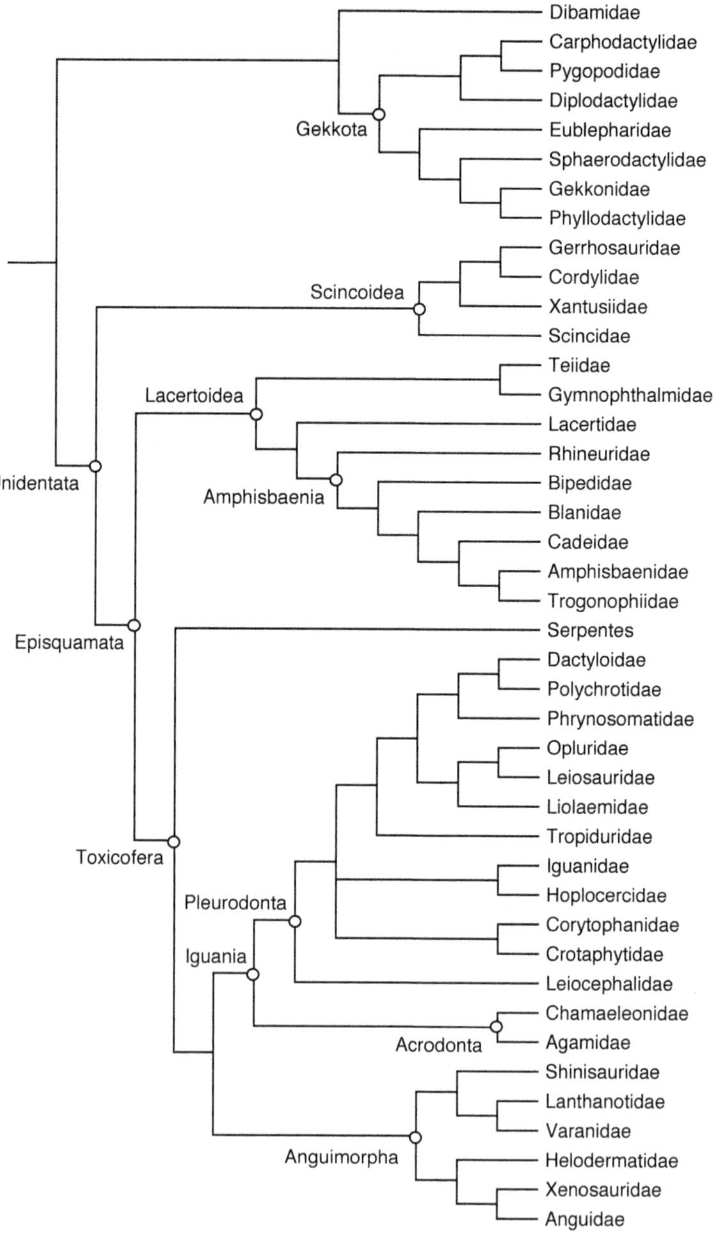

Fig. 2.1 A phylogeny of squamate reptiles, showing relationships among lizard families. This figure is generally based on the maximum likelihood analysis from 44 nuclear loci in Wiens *et al.* (2012). However, relationships within amphisbaenians and gekkotans incorporate additional families not included in that paper, and are based on Pyron *et al.* (2013). The branch lengths are arbitrary. Branch lengths in units of estimated time are provided in Fig. 2.2 and Appendix 2.1.

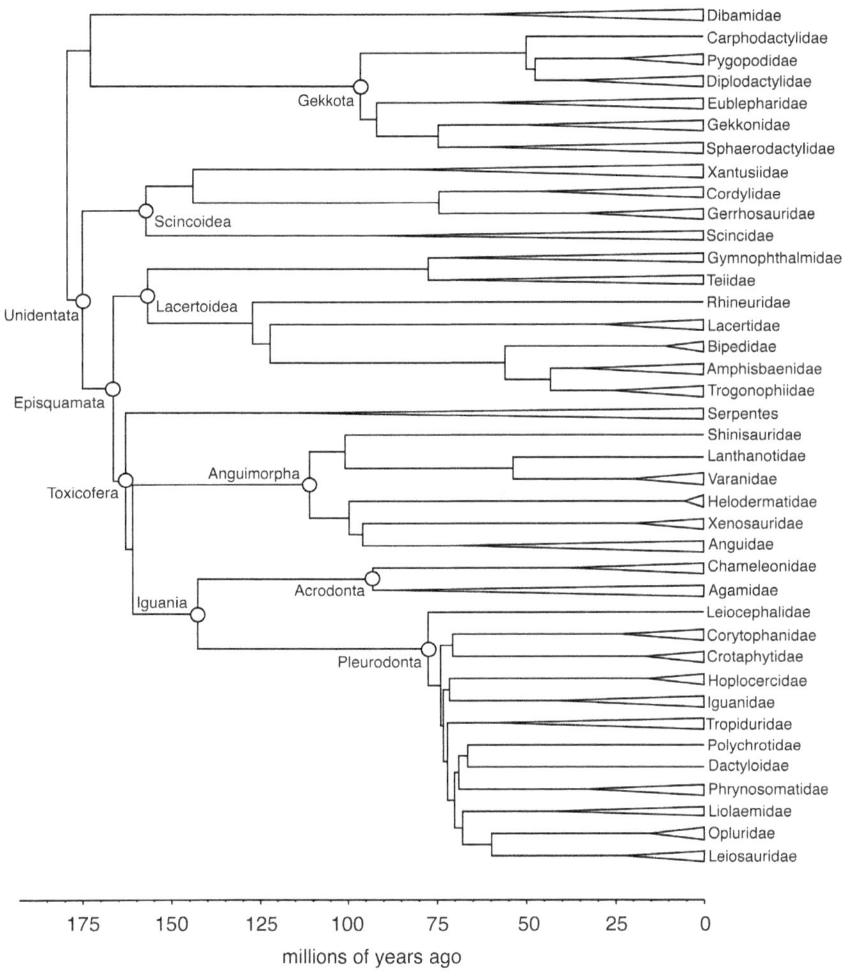

Fig. 2.2 A time-calibrated phylogeny of squamate reptiles, showing relationships among lizard families. Triangles indicate clades represented by multiple species in this analysis. The tree is based on the maximum likelihood analysis from 44 nuclear loci in Wiens *et al.* (2012) and the branch lengths are based on estimated ages from a new analysis using penalized likelihood. The tree is provided in nexus format in Appendix 2.1, including all 161 species (including sampled snakes).

9 nuclear loci (Vidal and Hedges 2009) and 22 nuclear loci (Mulcahy *et al.* 2012), although other analyses have placed Gekkota as sister to all other squamates and Dibamidae as the sister to all other squamates within that clade (e.g., Townsend *et al.* 2004). This issue would benefit from additional study. Nevertheless, the preponderance of current data seem to favor the clade of Dibamidae + Gekkota.

Recent analyses are concordant regarding family-level relationships within Gekkota. Pyron *et al.* (2013) divided Gekkota into two strongly supported clades. One clade includes Diplodactylidae as sister to Carphodactylidae + Pygopodidae. The other clade includes Eublepharidae as sister to Sphaerodactylidae + (Phyllodactylidae + Gekkonidae). Wiens *et al.* (2012) found strong support for these same relationships using 44 nuclear loci, although they did not include any representatives of Phyllodactylidae. These relationships are also consistent with those estimated by Gamble *et al.* (2008a,b, 2011, 2012) using a smaller number of loci, and the taxonomy proposed by Gamble and co-authors is followed here.

Above the level of Gekkota and Dibamidae, the relationships among major squamate clades are relatively well-resolved and consistent among studies (e.g., Townsend *et al.* 2004; Vidal and Hedges 2005; Wiens *et al.* 2010, 2012; Mulcahy *et al.* 2012; Pyron *et al.* 2013). First, squamates are divided into Scincoidea (Scincidae, Xantusiidae, Cordylidae, and Gerrhosauridae), and a clade containing all the remaining squamates (Episquamata).

Relationships within Scincoidea are very strongly supported and concordant among recent studies (e.g., Townsend *et al.* 2004; Vidal and Hedges 2009; Mulcahy *et al.* 2012; Wiens *et al.* 2012; Pyron *et al.* 2013). Scincidae is the sister group to a clade consisting of Xantusiidae (Cordylidae + Gerrhosauridae). Within Scincidae, there is broad agreement that Acontiinae is the sister group to other scincids (e.g., Mulcahy *et al.* 2012; Wiens *et al.* 2012; Pyron *et al.* 2013). There is also support for the monophyly of sampled species of Lygosominae, and for placement of Feylininae inside of Scincinae (e.g., Mulcahy *et al.* 2012; Wiens *et al.* 2012; Pyron *et al.* 2013). Interestingly, an analysis including 683 scincid species (Pyron *et al.* 2013) showed Scincinae to be monophyletic once Feylininae is included. Other recent analyses have suggested that Scincinae is paraphyletic with respect to Lygosominae (e.g., Mulcahy *et al.* 2012; Wiens *et al.* 2012), but with limited taxon sampling and only weak branch support. Analyses including 683 scincids reject the classification recently erected by Hedges and Conn (2012), and show that those authors seem to have erected many non-monophyletic families, subfamilies, and tribes (Pyron *et al.* 2013). Most importantly, there is no phylogenetic justification for subdividing the Scincidae into smaller families in the first place, since the monophyly of Scincidae is strongly supported by molecular data in all recent studies (e.g., Wiens *et al.* 2010, 2012; Mulcahy *et al.* 2012; Pyron *et al.* 2013).

Above the Scincoidea, the Episquamata is consistently divided into two strongly supported clades, the Lacertoidea (or Lacertiformes) and Toxicofera (e.g., Townsend *et al.* 2004; Vidal and Hedges 2009; Mulcahy *et al.* 2012; Wiens *et al.* 2012; Pyron *et al.* 2013). Lacertoidea includes Amphisbaenia, Lacertidae, Teiidae, and Gymnophthalmidae. Toxicofera includes Anguimorpha, Iguania, and Serpentes.

Within Lacertoidea, there is consistently strong support for two clades, one consisting of Lacertidae + Amphisbaenia, the other consisting of Teiidae + Gymnophthalmidae (Townsend *et al.* 2004; Vidal and Hedges 2005; Wiens *et al.* 2010; Mulcahy *et al.* 2012; Wiens *et al.* 2012; Pyron *et al.* 2013). Some molecular analyses suggest that Amphisbaenia is not monophyletic, with rhineurid amphisbaenians as sister to a clade containing Lacertidae and all other amphisbaenians (e.g., likelihood analysis of 44 loci in Wiens *et al.* 2012, but not the Bayesian analysis of these same data, nor the combined molecular-morphological analysis in Wiens *et al.* 2010, nor the likelihood analysis of 25 loci in Mulcahy *et al.* 2012). Other analyses of molecular data alone strongly support monophyly of amphisbaenians (e.g., Pyron *et al.* 2013).

Within Amphisbaenia, there is strong support for Rhineuridae as sister to all other families (Vidal and Hedges 2005; Wiens *et al.* 2012; Pyron *et al.* 2013). Within the strongly supported clade of other amphisbaenians, Bipedidae is consistently placed as sister to the other families (although not all studies have included Blanidae and Cadeidae; i.e., Wiens *et al.* 2010; Mulcahy *et al.* 2012; Wiens *et al.* 2012). Pyron *et al.* (2013) placed Blanidae and Cadeidae as successive outgroups to Amphisbaenidae + Trogonophiidae, whereas Vidal *et al.* (2007) placed Cadeidae as sister to Blanidae but with more limited taxon sampling.

Within Toxicofera there has been uncertainty over the relationships among Anguimorpha, Iguania, and Serpentes. Nevertheless, recent analyses favor Serpentes as the sister to Anguimorpha and Iguania (e.g., Townsend *et al.* 2004 [combined likelihood analysis, Fig. 7b]; Vidal and Hedges 2005; Wiens *et al.* 2010; Mulcahy *et al.* 2012; Wiens *et al.* 2012; Pyron *et al.* 2013). However, it is notable that the support for the Anguimorpha + Iguania clade tends to be relatively weak. In recent years, the major dissenting study has been by Lee (2009), who suggested that snakes are instead sister to Anguimorpha, based on a combined analysis of morphological, nuclear, and mitochondrial data.

Within Anguimorpha, multiple studies are also converging on relatively consistent results. Recent studies suggest that there are two major clades within Anguimorpha (e.g., Townsend *et al.* 2004; Wiens *et al.* 2010, 2012; Pyron *et al.* 2013): one consisting of the predominantly New World families (Helodermatidae, Xenosauridae, Anguidae [including Anniellidae]), the other consisting of predominantly Old World families (Shinisauridae, Lanthanotidae, Varanidae). Within the mostly New World clade, most studies show Helodermatidae as the sister taxon of Xenosauridae + Anguidae (e.g., Townsend *et al.* 2004 [Fig. 7b], Wiens *et al.* 2010, 2012; Mulcahy *et al.* 2012), but Pyron *et al.* (2013) placed Xenosaruidae as sister to Helodermatidae + Anguidae. Within the Old World clade, Shinisauridae is consistently placed as the sister to Lanthanotidae + Varanidae.

Within Iguania, most recent molecular studies (e.g., Townend *et al.* 2011; Wiens *et al.* 2012; Blankers *et al.* 2013; Pyron *et al.* 2013) agree on the split between Acrodonta (Chamaeleonidae, Agamidae) and Pleurodonta (the 12 families that formerly made up Iguanidae, now including Corytophanidae, Crotaphytidae, Dactyloidae, Hoplocercidae, Iguanidae, Leiocephalidae, Leiosauridae, Liolaemidae, Opluridae, Phrynosomatidae, Polychrotidae, and Tropiduridae). Within Acrodonta, there is strong support for monophyly of Agamidae and Chamaeleonidae, and the basal relationships among agamid subfamilies: Uromastycinae as sister to all other subfamilies, followed by Leiolepidinae, and then the clade of Hydrosaurinae, Amphibolurinae, Agaminae, and Draconinae (Townsend *et al.* 2011; Wiens *et al.* 2012; Blankers *et al.* 2013; Pyron *et al.* 2013). All recent analyses have placed Agaminae with Draconinae, but Pyron *et al.* (2013) placed Hydrosaurinae with Amphibolurinae, whereas Townsend *et al.* (2011; using 29 nuclear loci), Blankers *et al.* (2013; the same 29 nuclear loci plus mitochondrial data), and Wiens *et al.* (2012) placed Hydrosaurinae as the sister to Amphibolurinae + (Agaminae + Draconinae).

Relationships among the 12 families of Pleurodonta have become the most consistently challenging aspect of squamate phylogeny. Four recent studies have addressed their relationships in detail (Townsend *et al.* 2011; Wiens *et al.* 2012; Blankers *et al.* 2013; Pyron *et al.* 2013). These studies agree on very few relationships. Specifically, all four agree that Liolaemidae is sister to Opluridae + Leiosauridae. Otherwise, there are many strongly supported disagreements between these trees. For example, Townsend *et al.* (2011) found strong support for Phrynosomatidae as sister to all other pleurodonts. Wiens *et al.* (2012) and Blankers *et al.* (2013) both found strong support for a clade including Phrynosomatidae, Dactyloidae, Polychrotidae, Leiosaridae, Liolaemidae, and Opluridae, but disagreed strongly on the placement of Leiocephalidae (sister to all other pleurodonts [likelihood] or Iguanidae [Bayesian] by Wiens *et al.* [2012] versus sister to Corytophanidae by Blankers *et al.* [2013]). Pyron *et al.* (2013) found strong support for contradictory clades including Crotaphytidae + Phrynosomatidae, Polychrotidae + Hoplocercidae, and Corytophanidae + Dactyloidae. What is particularly disheartening in these results is that strong support values in one analysis seem to have little ability to predict strong support in subsequent analyses.

2.4 TIME SCALE OF THE PHYLOGENY

We also generate a time-calibrated version of the 44-locus likelihood-based concatenated tree from Wiens *et al.* (2012). Given the large number of loci and taxa, use of the Bayesian uncorrelated lognormal relaxed-clock approach (i.e., in BEAST; Drummond *et al.* 2006; Drummond and Rambaut 2007) was not practical for our purposes here (despite many potential advantages), but should be applied in future studies. Instead, we used the penalized likelihood approach (Sanderson 2002). We used the maximum likelihood tree from Wiens *et al.* (2012), rooted at the midpoint, as the input for estimation of divergence times in r8s version 1.8.0 (Sanderson 2003). A total of 14 fossil calibrations were used, all taken from Mulcahy *et al.* (2012). We estimated the optimal smoothing parameter (lowest Chi-square value) using cross-validation, first over a range of $\log_{10}$ values from –1 to 10, using increments of 0.1. We then fine-tuned the estimation using a range of values centered on the optimal value from the first search, using increments of 0.01. The best-fitting value of the smoothing parameter (9.3) was then used in a penalized likelihood analysis utilizing the truncated Newton algorithm and the additive penalty function.

Overall, the estimated clade ages were similar to those from Mulcahy *et al.* (2012), which were generally similar to those in earlier studies (e.g., Wiens *et al.* 2006) but younger than those estimated by Vidal and Hedges (2005). More specifically, the ages estimated here were often intermediate between the ages estimated by r8s (which tend to be older) and BEAST (which tend to be younger) in Mulcahy *et al.* (2012). We illustrate the ages estimated here in Fig. 2.2, and provide the time-calibrated tree for 161 species in nexus format in Appendix 2.1.

Our results suggest that crown-group squamates began to diversify roughly 175 million years ago. The clades Scincoidea, Lacertoidea, Toxicofera, and Iguania each began to diversify around 150 million years ago. Gekkotans and anguimorphs both began to diversify roughly 100 million year ago. The rapid radiation of most major pleurodont clades (ranked as families) occurred around 70 to 80 million years ago.

2.5 AREAS FOR FUTURE RESEARCH

This review highlights three main issues in higher-level lizard phylogeny that remain uncertain. First, the relationships among Gekkota, Dibamidae, and the clade including all other squamates (Unidentata). Second, the relationships among Iguania, Anguimorpha, and Serpentes. Third, the relationships among the 12 families of pleurodont iguanians, which at this point should be considered to be largely unresolved. Resolution of all three problems may be greatly aided by application of both next-generation sequencing techniques (which should increase the number of loci into the thousands; e.g., Faircloth *et al.* 2012) and explicit species-tree methods for dealing with gene incongruence due to incomplete lineage sorting (which may be more accurate than concatenated analysis; e.g., Edwards *et al.* 2007; Heled and Drummond 2010). Such analyses are currently underway.

2.6 ACKNOWLEDGMENTS

J.J.W. thanks his collaborators on the NSF-funded Deep Scaly project (T. Reeder, J. Sites, T. Townsend, D. Mulcahy, B. Noonan), including separate awards to Reeder (EF 0334967), Sites (EF 0334966), and Wiens (EF 0334923). S.M.L. acknowledges the support of an NSF Graduate Research Fellowship.

2.7 LITERATURE CITED

Baker, R. H., Yu, X. and DeSalle, R. 1998. Assessing the relative contribution of molecular and morphological characters in simultaneous analysis trees. Molecular Phylogenetics and Evolution 9: 427–436.

Blankers, T., Townsend, T. M., Pepe, K., Reeder, T. W. and Wiens, J. J. 2013. Contrasting global-scale evolutionary radiations: phylogeny, diversification, and morphological evolution in the major clades of iguanian lizards. Biological Journal of the Linnean Society 108: 127–143.

Castoe, T. A., de Koning, A. P. J., Kim, H. M., Gu, W., Noonan, B. P., Naylor, G., Jiang, Z. J., Parkinson, C. L. and Pollock, D. D. 2009. Evidence for an ancient adaptive episode of convergent molecular evolution. Proceedings of the National Academy of Sciences USA 106: 8986–8991.

Conrad, J. L. 2008. Phylogeny and systematics of Squamata (Reptilia) based on morphology. Bulletin of the American Museum of Natural History 310: 1–182.

Drummond, A. J. and Rambaut, A. 2007. BEAST: Bayesian evolutionary analysis by sampling trees. BMC Evolutionary Biology 7: 214.

Drummond, A. J., Ho, S. Y. W., Phillips, M. J. and Rambaut, A. 2006. Relaxed phylogenetics and dating with confidence. PLoS Biology 4: 699–710.

Edwards, S. V., Liu, L. and Pearl, D. K. 2007. High-resolution species trees without concatenation. Proceedings of the National Academy of Sciences USA 104: 5936–5941.

Estes, R., de Queiroz, K. and Gauthier, J. 1988. Phylogenetic relationships within Squamata. pp. 119–281. In R. Estes and G. Pregill (eds.), *Phylogenetic relationships of the lizard families*. Stanford University Press: Stanford.

Faircloth, B. C., McCormack, J. E., Crawford, N. G., Harvey, M. G., Brumfield, R. T. and Glenn, T. C. 2012. Ultraconserved elements anchor thousands of genetic markers spanning multiple evolutionary timescales. Systematic Biology 61: 717–726.

Fry, B. G., Vidal, N., Norman, J. A., Vonk, F. J., Scheib, H., Ramjan, S. F. R., Kuruppu, S., Fung, K., Hedges, S. B., Richardson, M. K., Hodgson, W. C., Ignjatovic, V., Summerhayes, R. and Kochva, E. 2006. Early evolution of the venom system in lizards and snakes. Nature 439: 584–588.

Gamble, T., Bauer, A. M., Greenbaum, E. and Jackman, T. R. 2008a. Evidence for Gondwanan vicariance in an ancient clade of gecko lizards. Journal of Biogeography 35: 88–104.

Gamble, T., Bauer, A. M., Greenbaum, E. and Jackman, T. R. 2008b. Out of the blue: A novel, trans-Atlantic clade of geckos (Gekkota, Squamata). Zoologica Scripta 37: 355–366.

Gamble, T., Bauer, A. M., Colli, G. R., Greenbaum, E., Jackman, T. R., Vitt, L. J. and Simons, A. M. 2011. Coming to America: multiple origins of New World geckos. Journal of Evolutionary Biology 24: 231–244.

Gamble, T., Greenbaum, E., Jackman, T. R., Russell, A. P. and Bauer, A. M. 2012. Repeated origin and loss of adhesive toepads in geckos. PLoS ONE 7: e39429.

Gauthier, J. A., Kearney, M., Maisano, J. A., Rieppel, O. and Behike, A. D. B. 2012. Assembling the Squamate Tree of Life: perspectives from the phenotype and the fossil record. Bulletin Peabody Museum of Natural History 53: 3–308.

Hedges, S.B. and Conn, C.E. 2012. A new skink fauna from Caribbean islands (Squamata, Mabuyidae, Mabuyinae). Zootaxa 3288: 1–244.

Heled, J. and Drummond, A. J. 2010. Bayesian inference of species trees from multilocus data. Molecular Biology and Evolution 27: 570–580.
Hillis, D. M. 1987. Molecular versus morphological approaches to systematics. Annual Review of Ecology and Systematics 18: 23–42.
Hillis, D. M. and Wiens, J. J. 2000. Molecular versus morphological systematics: Conflicts, artifacts, and misconceptions. pp. 1–19. In J. J. Wiens (ed.), *Phylogenetic analysis of morphological data*. Smithsonian Institution Press: Washington, D.C.
Kluge, A. G. 1989. A concern for evidence and a phylogenetic hypothesis of relationships among *Epicrates* (Boidae, Serpentes). Systematic Zoology 38: 7–25.
Lee, M. S. Y. 2009. Hidden support from unpromising datasets strongly unites snakes and anguimorph lizards. Journal of Evolutionary Biology 22: 1308–1316.
Losos, J. B., Hillis, D. M. and Greene, H. W. 2012. Who speaks with a forked tongue? Science 338: 1428–1429.
Maddison, W. P. 1997. Gene trees in species trees. Systematic Biology 46: 523–536.
Mulcahy, D. G., Noonan, B. P., Moss, T., Townsend, T. M., Reeder, T. W., Sites, J. W., Jr. and Wiens, J. J. 2012. Estimating divergence dates and evaluating dating methods using phylogenomic and mitochondrial data in squamate reptiles. Molecular Phylogenetics and Evolution 65: 974–991.
Nylander, J. A. A., Ronquist, F., Huelsenbeck, J. P. and Nieves-Aldrey, J. L. 2004. Bayesian phylogenetic analysis of combined data. Systematic Biology 53: 47–67.
Pyron, R. A., Burbrink, F. T. and Wiens, J. J. 2013. A phylogeny and revised classification of Squamata, including 4161 species of lizards and snakes. BMC Evolutionary Biology 13: 93.
Sanderson, M. J. 2002. Estimating absolute rates of molecular evolution and divergence times: a penalized likelihood approach. Molecular Biology and Evolution 19: 101–109.
Sanderson, M. J. 2003. r8s: inferring absolute rates of molecular evolution and divergence times in the absence of a molecular clock. Bioinformatics 19: 301–302.
Townsend, T., Larson, A., Louis, E. J. and Macey, J. R. 2004. Molecular phylogenetics of Squamata: the position of snakes, amphisbaenians, and dibamids, and the root of the squamate tree. Systematic Biology 53: 735–757.
Townsend, T., Mulcahy, D. G., Sites, J. W., Jr., Kuczynski, C. A., Wiens, J. J. and Reeder, T. W. 2011. Phylogeny of iguanian lizards inferred from 29 nuclear loci, and a comparison of concatenated and species-tree approaches for an ancient, rapid radiation. Molecular Phylogenetics and Evolution 61: 363–380.
Vidal, N. and Hedges, S. B. 2005. The phylogeny of squamate reptiles (lizards, snakes, and amphisbaenians) inferred from nine nuclear protein-coding genes. Comptes Rendus Biologies 328: 1000–1008.
Vidal, N., Azvolinsky, A., Cruaud, C. and Hedges, S. B. 2007. Origin of tropical American burrowing reptiles by transatlantic rafting. Biology Letters 4: 115–118.
Vidal, N. and Hedges, S. B. 2009. The molecular evolutionary tree of lizards, snakes, and amphisbaenians. Comptes Rendus Biologies 332: 129–139.
Vitt, L. J., Pianka, E. R., Cooper, W. E. and Schwenk, K. 2003. History and the global ecology of squamate reptiles. American Naturalist 162: 44–60.
Wiens, J. J., Brandley, M. C. and Reeder, T. W. 2006. Why does a trait evolve multiple times within a clade? Repeated evolution of snake-like body form in squamate reptiles. Evolution 60: 123–141.
Wiens, J. J., Kuczynski, C. A., Townsend, T., Reeder, T. W., Mulcahy, D. G. and Sites, J. W., Jr. 2010. Combining phylogenomics and fossils in higher level squamate reptile phylogeny: molecular data change the placement of fossil taxa. Systematic Biology 59: 674–688.
Wiens, J. J., Hutter, C. R., Mulcahy, D. G., Noonan, B. P., Townsend, T. M., Sites, J. W., Jr. and Reeder, T. W. 2012. Resolving the phylogeny of lizards and snakes (Squamata) with extensive sampling of genes and species. Biology Letters 8: 1043–1046.

APPENDIX 2.1

Time-calibrated phylogeny of squamate reptiles

((((((((Teius_teyou:34.189276,Aspidoscelis_tigris:34.189276):36.888414,(Tupi
nambis_teguixin:50.754519,Callopistes_maculatus:50.754519):20.32317):6.2
5859,((Pholidobolus_macbrydei:62.162642,Colobosaura_modesta:62.16264
2):10.78068,Alopoglossus_angulatum:72.943322):4.392957):79.276286,(Rhin
eura_floridana:126.957733,((Takydromus_ocellatus:28.89473,Lacerta_virid
is:28.89473):92.994715,((Bipes_canaliculatus:10.811263,Bipes_biporus:10.81
1263):45.091794,((Geocalamus_acutus:34.379576,Amphisbaena_fuliginosa:
34.379576):8.717054,(Trogonophis_wiegmanni:25.65682,Diplometopon_za
rudnyi:25.65682):17.43981):12.806427):65.986389):5.068287):29.654832):9.
764006,(((Leptotyphlops_humilis:107.956482,Typhlops_jamaicensis:107.
956482):7.264883,(Liotyphlops_albirostris:113.226184,(((Trachyboa_boul
engeri:19.613539,Tropidophis_haetianus:19.613539):62.844645,Anilius_sc
ytale:82.458184):10.241816,((((Acrochordus_granulatus:75.447182,(Xeno
dermus_javanicus:66.833535,(Pareas_hamptoni:50.553784,((((Bothrops_
asper:11.072489,(Lachesis_muta:10.378418,Agkistrodon_contortrix:1
0.378418):0.694071):10.494785,Azemiops_feae:21.567274):3.199269,(C
ausus_defillippi:17.726006,Daboia_russelli:17.726006):7.040537):21.79
895,(Homalopsis_buccata:38.578856,((((Atractaspis_irregularis:22.73
6606,Aparallactus_werneri:22.736606):6.836733,(Lamprophis_fuligin
osus:21.785323,Lycophidion_capense:21.785323):7.788016):1.745587,(
Naja_kaouthia:18.48597,((Notechis_scutatus:11.765261,Laticauda_co
lubrina:11.765261):5.933663,Micrurus_fulvius:17.698924):0.787045):1
2.832957):5.033886,((Lampropeltis_getula:17.29987,(Trimorphodon_
biscutatus:13.659574,(Sonora_semiannulata:12.792971,Coluber_co
nstrictor:12.792971):0.866603):3.640295):14.610493,((Imantodes_
cenchoa:22.008486,(Heterodon_platyrhinos:20.289191,Diadophis_punctatus
:20.289191):1.719295):9.1032,((Thamnophis_marcianus:16.537845,Natrix_na
trix:16.537845):7.429826,(Afronatrix_anoscopus:21.373273,(Amphiesma_
stolata:16.300742,Xenochrophis_piscator:16.300742):5.07253):2.5943
99):7.144015):0.798677):4.442449):2.226043):7.986638):3.98829):16.279
751):8.613647):8.793648,(Casarea_dussumieri:78.737483,(Calabaria_
reinhardtii:53.072634,((Eryx_colubrinus:44.015153,(Boa_
constrictor:32.872668,Epicrates_striatus:32.872668):11.142486):2.705897,((Ex
iliboa_placata:32.804083,Ungaliophis_continentalis:32.804083):9.791007,Lic
hanura_trivirgata:42.595089):4.12596):6.351584):25.664849):5.503346):0.4458
59,(Xenopeltis_unicolor:73.838394,(Loxocemus_bicolor:38.714371,(Python_
molurus:20.978856,Aspidites_melanocephalus:20.978856):17.735515):35.124
023):10.848294):0.727074,(Uropeltis_melanogaster:62.639153,Cylindrophis_
rufus:62.639153):22.774609):7.286237):20.526184):1.995181):47.756144,((((Ch

amaeleo_calyptratus:37.947402,Brookesia_brygooi:37.947402):54.993096,(U
romastyx_aegyptus:87.369808,(Leiolepis_belliana:82.973648,(Hydrosaurus_
sp:69.035404,(((Draco_blanfordii:39.351944,(Calotes_
emma:32.817041,Acanthosaura_lepidogaster:32.817041):6.534902):23.747991
,(Phrynocephalus_mystaceus:41.20321,(Agama_agama:34.458137,Trapelus_
agilis:34.458137):6.745073):21.896725):4.82047,(Physignathus_cocinci
nus:27.125928,((Chelosania_brunnea:15.91394,Moloch_horridus:15.91
394):5.317179,(Hypsilurus_boydi:20.325473,(((Rankinia_adelaidensis
:14.069682,Ctenophorus_isolepis:14.069682):2.744743,(Chlamydosaur
us_kingii:13.329024,Pogona_vitticeps:13.329024):3.485401):2.51721,Ph
ysignathus_leseuri:19.331635):0.993838):0.905646):5.89481):40.794476)
:1.114999):13.938244):4.39616):5.570691):49.742262,(Leiocephalus_bara
honensis:77.490835,((((Dipsosaurus_dorsalis:41.218511,(Sauromalus_
obesus:38.259638,Brachylophus_fasciatus:38.259638):2.958873):30.2331
03,(Morunasaurus_annularis:15.721378,Enyalioides_laticeps:15.72137
8):55.730236):1.783021,((Stenocercus_guentheri:58.596898,(Uranoscod
on_superciliosus:33.360027,Tropiduris_plica:33.360027):25.236871):13.48
2029,((((Uma_scoparia:20.64502,Phrynosoma_platyrhinos:20.64502):12.
264075,(Petrosaurus_mearnsi:25.431938,(Uta_stansburiana:23.557532,Sc
eloporus_variabilis:23.557532):1.874406):7.477158):36.00649,(Polychrus_
marmoratus:66.426736,Anolis_carolinensis:66.426736):2.488849):1.215002,((
(Oplurus_cyclurus:15.253908,Chalarodon_madagascariensis:15.253908):44.
581182,((Pristidactylus_torquatus:15.687016,Leiosaurus_catamarcensis:15.6
87016):6.77478,Urostrophus_vautieri:22.461796):37.373293):8.05492,(Phyma
turus_palluma:42.45304,(Liolaemus_bellii:7.105115,Liolaemus_elongatus:7.
105115):35.347925):25.436969):2.240578):1.94834):1.155708):0.719879,((Basilis
cus_basiliscus:23.589111,Corytophanes_cristatus:23.589111):46.973612,(Gam
belia_wislizenii:16.677996,Crotaphytus_collaris:16.677996):53.884727):3.391
792):3.53632):65.191926):18.312384,((Shinisaurus_crocodilurus:100.696159,(
(Varanus_exanthematicus:20.228696,(Varanus_salvator:15.101203,Varanus_
acanthurus:15.101203):5.127493):33.447213,Lanthanotus_borneensis:53.675
908):47.02025):10.16495,((Heloderma_horridum:5.526572,Heloderma_susp
ectum:5.526572):94.073428,((Celestus_enneagrammus:69.06106,(Anniella_
pulchra:63.554116,(Elgaria_multicarinata:42.482205,(Ophisaurus_
ventralis:15.510528,Ophisaurus_apodus:15.510528):26.971677):21.0719
1):5.506945):26.672003,(Xenosaurus_grandis:19.454946,Xenosaurus_pl
atycephalus:19.454946):76.278117):3.866937):11.261109):50.134035):1.98
2364):3.399062):8.529466,((Acontias_meleagris:92.241588,(((Eumeces_
schneideri:24.43028,Scincus_scincus:24.43028):48.435011,(Amphigloss
us_splendidus:51.848426,Feylinia_polylepis:51.848426):21.016865):3.3726
52,((Brachymeles_gracilis:73.102809,((Eugongylus_rufescens:30.112178,S
phenomorphus_solomonis:30.112178):35.512028,(Tiliqua_scincoides:49.5
89052,Trachylepis_quinquetaeniata:49.589052):16.035153):7.478604):1.594

49,(Plestiodon_fasciatus:14.414436,Plestiodon_skiltonianus:14.414436):60
.282864):1.540643):16.003645):64.786337,(((Cordylosaurus_subtesselatus:4
5.787836,Zonosaurus_ornatus:45.787836):28.49618,(Cordylus_mossambic
us:34.115895,Platysaurus_pungweensis:34.115895):40.168121):69.622169,(
Cricosaura_typica:83.248219,(Xantusia_vigilis:34.197378,Lepidophyma_
flavimaculatum:34.197378):49.05084):60.657966):13.1217
4):17.878111):4.264706,(((Saltaurius_cornutus:49.898556,((Delma_
borea:24.52329,Lialis_burtonis:24.52329):22.839482,(Rhacodactylus_au
riculatus:35.873845,Strophurus_cilliaris:35.873845):11.488927):2.535
784):46.416278,((Aeluroscalobates_felinus:60.362449,(Eublepharis_
macularius:56.140762,Coleonyx_variegatus:56.140762):4.221687):31.409316
,((Phelsuma_lineata:49.822486,Gekko_gecko:49.822486):24.676597,(Teratos
cincus_scincus:60.947499,Gonatodes_albogularis:60.947499):13.551584):17.
272682):4.543068):76.340593,(Anelytropsis_papilosus:62.448172,Dibamus_
novaeguineae:62.448172):110.207254):6.515316).

CHAPTER 3

Pheromones and Chemical Communication in Lizards

*José Martín** and *Pilar López*

3.1 INTRODUCTION

Pheromones have been defined, based on entomological studies, as chemicals or semiochemicals produced by one individual that effect a change in the physiology ('primer' pheromone) or behavior ('releaser' pheromone) of conspecifics (Karlson and Lüscher 1959). In insects and many other invertebrates, very often just one or a pair of chemical compounds acts as an exclusive pheromone attracting the opposite sex. In contrast, vertebrates often have multicomponent pheromones with a mixture of many different chemical compounds with distinctly different functions or intended receivers (Müller-Schwarze 2006). However, compounds may be mixed together in specific proportions to determine "odor profiles" of species or individuals (Johnston 2005; Wyatt 2010). The pattern of compounds in the scent of an individual may convey various signals such as sex, age, social status, group, individuality, seasonality, condition, health state, etc. Moreover, in insects, pheromones alone can directly control reproductive behavior, whereas in vertebrates, a combination of different sensory stimuli (visual, tactile, chemical, etc.) is often required to control reproduction. Therefore, in vertebrates, pheromones may be better defined as a group of active compounds in a secretion that supply information to conspecifics that may be relevant for reproductive decisions (for reviews see Mason 1992; Wyatt 2003, 2010; Müller-Schwarze 2006; Apps 2013).

Departamento de Ecología Evolutiva, Museo Nacional de Ciencias Naturales, CSIC, José Gutiérrez Abascal 2, 28006 Madrid, Spain.
* Corresponding author

Reproductive behavior of lizards was traditionally thought to be predominantly based on conspicuous visual signals, whereas the potential role of pheromones in reproduction was not considered in most studies. However, there was considerable evidence of the chemosensory abilities of most lizards, and of the widespread occurrence of multiple types of glands that secrete chemicals with the potential of being pheromones, especially during the reproductive season. Only recently, has it been recognized that in many species of lizards, pheromones (i.e., specific compounds or mixtures of compounds) or "chemical signals" (i.e., undetermined chemical secretions) are very important, and sometimes required, for species and sex recognition, intrasexual relationships between males, social organization, territorial marking, and mate choice of lizards (reviewed in Halpern 1992; Mason 1992; Cooper 1994; Johansson and Jones 2007; Houck 2009; Mason and Parker 2010; Martín and López 2011).

3.2 CHEMOSENSORY ABILITIES OF LIZARDS

Lizards, and most tetrapods, can use their chemosensory senses to detect and discriminate many different scents in their environments coming from prey, conspecifics and/or predators. These abilities are based on the possession of highly developed olfactory and vomeronasal organs (Halpern 1992; Mason 1992; Cooper 1994; Schwenk 1995; Halpern and Martínez-Marcos 2003). The olfactory and vomeronasal systems do not have independent functions, as it was though in the past, but show deep anatomical and functional interrelationships (Halpern and Martínez-Marcos 2003; Ubeda-Banon *et al.* 2011). In many cases, scent stimuli are first received through the nares and processed by the nasal organs, and this triggers tongue-flick mediated vomerolfaction (Halpern 1992; Cooper 1994; Schwenk 1995). The vomeronasal organ sends specific chemical signals to the central nervous system activating accessory olfactory pathways. In particular, chemical compounds with a putative pheromonal function stimulate brain areas involved in sexually dimorphic reproductive behavior.

Associated with chemoreception, tongue-flicking (TF) is a characteristic behavior of lizards and snakes in which the tongue is extruded to sample chemicals from the environment that are delivered into the mouth and transported to the vomeronasal organ (Schwenk 1995). This is an easily observable and quantifiable behavior that has been used as a bioassay for chemosensory discrimination abilities of lizards and snakes (Cooper and Burghardt 1990; Cooper 1994, 1998). Different chemical stimuli impregnated in cotton swabs, tiles or papers are randomly presented close to the snout of experimental subject, and TF rates are measured during a certain time period. Detection of a scent stimulus is inferred by an increase of TF rates in response to the presentation of a given scent above the baseline TF rates observed under the experimental conditions in response to an odorless control (e.g., deionized water). Differential TF rate to different scent stimuli is considered as an indication of discrimination of the different stimuli because these elicit different responses (Cooper and Burghardt 1990; Cooper 1994, 1998). Usually, a higher TF rate indicates a higher "interest" for a given stimulus, which, depending on the context, is often considered a proxy of preference of that scent (e.g., different prey types or potential mates), or an indication that the stimulus is novel and elicits a longer chemosensory investigation (e.g., familiar vs. unfamiliar conspecific recognition). Pungency controls, such as cologne, are often used to assess responses to odorous, readily detectable chemicals, which are not relevant to the discrimination being studied (Cooper 1998; Cooper *et al.* 2003). Differences in latencies to the first TF after presentation of the scent stimulus are also used to indicate detection and discrimination of different stimuli.

In the case of lizards that do not usually tongue-flick during swab tests, such as some iguanids, other similar quantifiable chemosensory behaviors such as labial-licking, chin-rubbing or gular pumping are used (e.g., Wilgers and Horne 2009). To assess the preference, or avoidance, of particular stimuli, such as scent-marks from different individual males, many tests measure changes in behavior (e.g., locomotory activity) or time spent by the experimental lizard in different areas, or refuges, with substrates labeled with different chemical stimuli (e.g., Aragón *et al.* 2001c; Bull *et al.* 2001; Martín and López 2006a).

3.3 CHEMICAL COMPOUNDS AS POTENTIAL LIZARD PHEROMONES

Lizards have several possible sources of chemical compounds that may potentially function as pheromones, such as the skin and secretions by large specialized holocrine glands (e.g., precloacal/preanal or cloacal/urodeal glands) (Mason 1992; Labra *et al.* 2002). Reproductive hormones, such as testosterone, regulate the secretory activity of these glands (Fergusson *et al.* 1985; Mason 1992; Moore and Lindzey 1992), which indicates their role in reproduction. Some studies use gas chromatography coupled with mass spectrometry (GC-MS) for identification and quantification of lipophilic compounds in secretions (Fig. 3.1). Less frequently, the proteinacious fractions of secretions have been studied with different electrophoresis techniques, especially in the past. Studies using both methods have described the mixtures of chemical compounds secreted by lizard glands in a few lizard species from limited taxonomic groups (reviewed in Weldon *et al.* 2008) (Table 3.1).

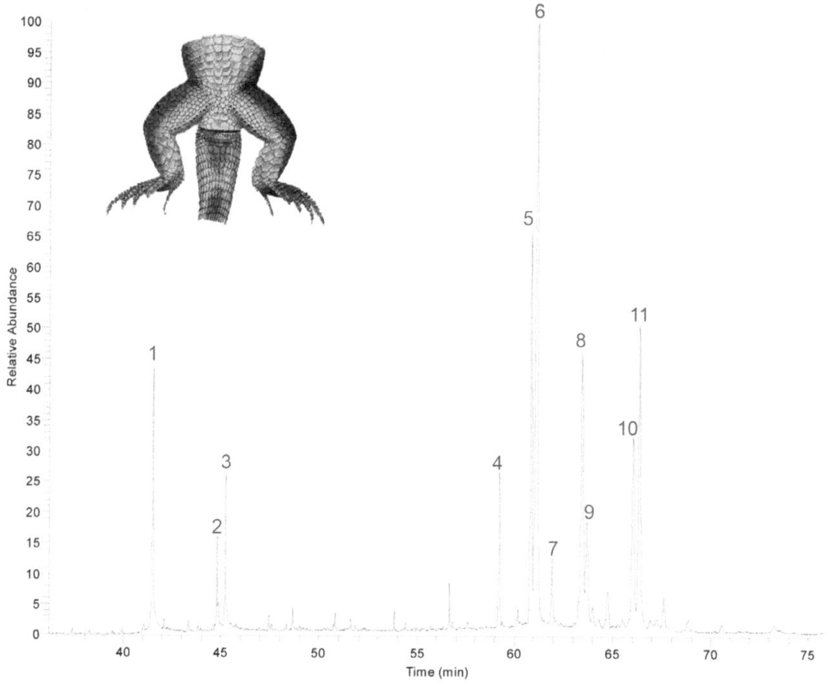

Fig. 3.1 A typical chromatogram of the lipophilic fraction of femoral gland secretions of the ocellated lizard, *Timon lepidus* (Lacertidae). The identification of the peaks of the major compounds are indicated: 1: Hexadecanoic acid; 2: Octadecenoic acid; 3: Octadecanoic acid; 4: γ-Tocopherol; 5: Cholesterol; 6: α-Tocopherol; 7: Cholestan-3-one; 8: Campesterol; 9: Ergostanol; 10: Sitosterol; 11: Ergostanol, methyl derivative. Original.

3.3.1 Skin Lipids

The skin of lizards typically contains mainly fatty acids, hydrocarbons, alcohols, steroids and waxy esters, among others. The characteristic combinations of chemicals and their concentrations vary among species (Roberts and Lillywhite 1980; Weldon and Bagnall 1987; Mason and Gutzke 1990). The main role of these lipids is to protect the skin against water loss (Roberts and Lillywhite 1980). Nevertheless, skin compounds might also be used as pheromones, at least for species, and sex recognition in short-distance interactions in many species. For example, female leopard geckos, *Eublepharis macularius* (Eublepharidae), elicit courtship behavior from males, but when females are shedding the skin, males respond aggressively, as if females were competitor males. These responses may be explained by the presence in the skin of females, but not in the skin of males, of long-

Table 3.1 Lizard and tuatara species for which lipophilic chemical compounds, with the potential of being pheromones, found in gland secretions (mostly femoral or preclocal/preanal glands except those indicated) have been described.

INFRAORDER, Family	Species	Author
IGUANIA:		
Agamidae	*Acanthocercus atricollis*	Martín *et al.* 2013c
	Uromastyx aegyptia	Martín *et al.* 2012
	Uromastyx hardwickii	Chauham 1986
Crotaphytidae	*Crotaphytus bicinctores*	Martín *et al.* 2013b
Iguanidae	*Iguana iguana*	Weldon *et al.* 1990; Alberts *et al.* 1992a,b
	Dipsosaurus dorsalis	Alberts 1990
Liolaemidae	*Liolaemus* spp. (20 species)	Escobar *et al.* 2001
	Liolaemus fabiani	Escobar *et al.* 2003
GEKKOTA:		
Gekkonidae	*Cyrtopodion scabrum*	Khannoon 2012
	Hemidactylus flaviviridis	Chauham 1986; Khannoon 2012
	Hemidactylus turcicus	Khannoon 2012
SCINCOMORPHA:		
Cordylidae	*Cordylus giganteus* (femoral and generational glands)	Louw *et al.* 2007, 2011
Teiidae	*Tupinambis merianae*	Martín *et al.* 2011
Lacertidae	*Acanthodactylus boskianus*	Khannoon *et al.* 2011a, 2013
	Acanthodactylus erythrurus	López and Martín 2005d
	Iberolacerta cyreni (=Lacerta monticola cyreni)	López and Martín 2005c; López *et al.* 2006
	Iberolacerta monticola (=Lacerta monticola monticola)	Martín *et al.* 2007c; López *et al.* 2009a
	Lacerta schreiberi	López and Martín 2006
	Lacerta viridis	Kopena *et al.* 2009
	Podarcis gaigeae	Runemark *et al.* 2011
	Podarcis hispanica (species complex)	Martín and López 2006c; Gabirot *et al.* 2010a, 2012a,b
	Podarcis lilfordi	Martín *et al.* 2013a
	Podarcis muralis	Martín and López 2006c; Martín *et al.* 2008
	Psammodromus algirus	Martín and López 2006d
	Psammodromus hispanicus	López and Martín 2009a
	Timon lepidus (=Lacerta lepida)	Martín and López 2010a
	Zootoca vivipara (=Lacerta vivipara)	Gabirot *et al.* 2008

Table 3.1 contd....

Table 3.1 contd.

INFRAORDER, Family	Species	Author
Scincidae	*Plestiodon laticeps* (=Eumeces laticeps) (urodeal gland)	Cooper and Garstka 1987
	Egernia striolata (feces)	Bull *et al.* 1999a
Amphisbaenidae	*Blanus cinereus*	López and Martín 2005b, 2009b
RHYNCHOCEPHALIA	*Sphenodon punctatus* (cloacal gland)	Flachsbarth *et al.* 2009

chain methyl ketones, which are lost after shedding the skin (Mason and Gutzke 1990). Interestingly, similar methyl ketones are found in the skin of female garter snakes, *Thamnophis sirtalis parietalis,* where they serve as sex attractiveness pheromones (Mason *et al.* 1990).

3.3.2 Compounds in Femoral and Precloacal Gland Secretions

Many lizards have femoral or precloacal/preanal glands, which are probably homologous with each other, differing only in their position in different species (Gabe and Saint Girons 1965). These are holocrine glands that produce an abundant secretion that is slowly secreted through the epidermal pores of femoral, preanal or precloacal glands (Cole 1966; Alberts 1993). Secretion is usually more abundant in males (i.e., it is often absent in females although they have vestigial pores) and during the mating season (Alberts *et al.* 1992b; Martins *et al.* 2006), and production is stimulated by androgenic hormones (e.g., Fergusson *et al.* 1985). Owing to the ventral location of femoral and precloacal pores, secretions are passively deposited on substrates as lizards move, which may serve to scent mark territories (see Section 3.5.1). Moreover, active rubbing of the pores against substrate has been observed.

Both lipophilic and proteinaceous compounds are generally found in femoral (or precloacal) secretions. Proteins may be the major component in secretions. Although they show characteristic and stable species-dependent patterns, minor differences among them among individuals might be used in individual recognition (Alberts 1990, 1991; Alberts and Werner 1993; Alberts *et al.* 1993).

In addition to these roles of proteinaceous compounds, lipophilic compounds may be important for communication in a reproductive context (e.g., Martín and López 2006a). Lipids have the advantage of being more volatile and have a high degree of molecular diversity, which increases the potential information content of a pheromone. In addition, the production of lipids is regulated by the general metabolism, and, thus, secreted lipids could be directly related to, and thereby potentially signal, the characteristics and condition of the signaler. Typical lipophilic compounds in femoral or precloacal gland secretions of lizards are steroids and carboxylic or fatty acids, as major compounds, together with usually minor amounts of alcohols, carboxylic acid esters (=waxy esters), squalene, tocopherol, ketones, aldehydes, furanones, alkanes or amides, and other minor and less frequent compounds (reviewed in Weldon *et al.* 2008).

3.3.2.1 Steroids

Among the lipids found in gland secretions of lizards, steroids are usually the most abundant, with cholesterol being in many cases the main compound (Weldon *et al.* 2008). However, it is likely that cholesterol or other steroids are, at least initially, only useful to form an unreactive apolar "matrix" that holds and protects other lipids in the scent marks (Escobar *et al.* 2003). Nevertheless, the relative amount of cholesterol, for example, may depend on body size in male rock lizards, *Iberolacerta cyreni* (Lacertidae), suggesting a signaling function in male intrasexual relationships (Martín and López 2007; see Section 3.5.2).

Every lizard species seems to have a specific combination of steroids that appear in roughly similar relative proportions in secretions of all individuals, although there is interindividual variability in the exact proportions of each steroid. Cholesterol is the most abundant steroid in many species but not in others. For example, in green lizards, the main steroids are ergostanol and cholestanol in the Schreiber's green lizard, *Lacerta schreiberi* (López and Martín 2006) and cholestanol and cholesterol in the ocellated lizard, *Timon lepidus* (Martín and López 2010a) (Fig. 3.1). Campesterol is the main steroid in *Psammodromus* spp. (Lacertidae) lizards (Martín and López 2006d; López and Martín 2009a). In the green iguana, *Iguana iguana* (Iguanidae), lanosterol is the most abundant steroid, followed by campesterol and cholesterol (Weldon *et al.* 1990; Alberts *et al.* 1992a). In *Liolaemus* spp. (Lioalemidae) lizards, cholesterol and cholestanol are the main steroids (Escobar *et al.* 2001). In the Great Basin collared lizard, *Crotaphytus bicinctores* (Crotaphytidae), in addition to the ubiquitous cholesterol, two triunsaturated steroids, cholesta-2,4,6-triene and cholesta-4,6,8(14)-triene, are the other two main steroids in secretions (Martín *et al.* 2013b). Other steroids, such as cholesta-3,5-diene, stigmasterol, cholestan-3-one and sitosterol are also commonly found in secretions of many lizards in lower proportions (Weldon *et al.* 2008), together with a large variety of derivatives and unidentified (probably unknown) steroids. Many of these steroids are of vegetal (=phytosterols) or microbial origin that have to be obtained from the diet, suggesting a relationship between diet and characteristics of gland secretions. Interestingly, some lizards secrete steroids that are precursors of vitamins, such as cholesta-5,7-dien-3-ol (=dihydrocholesterol; a precursor of vitamin D_3) and ergosterol (provitamin D_2). Thus, diet quality may affect quality of pheromones, which may explain why, in some lacertid lizards, females prefer the scent of males with high proportions of these provitamins (López and Martín 2005a; Martín and López 2006a,b; see Sections 8.5.3 and 8.6).

3.3.2.2 Fatty acids

Fatty or carboxylic acids, both saturated and unsaturated, are abundant in most glandular secretions of lizards. Hexadecanoic (=palmitic) and octadecenoic (=oleic) acids are present in most lizard species. Other fatty acids commonly found, although in lower proportions, are tetradecanoic (=myristic), hexadecenoic (=palmitoleic), octadecanoic (=stearic), 9,12-octadecadíenoic (=linoleic), and eicosanoic (=arachidic) acids, among others (Weldon *et al.* 2008).

The fatty acids are typically found in series, which vary with respect to the number of carbons that form the hydrocarbon chain and vary among species. This has been interpreted as an adaptation to maximize efficacy of substrate scent marks under different microclimatic conditions, with fatty acids of high molecular weight, and, therefore, less volatile, being favored in areas with higher temperatures or greater humidity (Alberts 1992). Fatty acids can also appear in the form of ethyl esters, which confer more stability. For example, in *Iguana iguana*, from warm wet tropics, the chain lengths of fatty acids found in femoral secretions range between C_{14} and C_{26} (Weldon *et al.* 1990; Alberts *et al.* 1992a), while *Iberolacerta cyreni* lizards from cold, dry high mountains, have fatty acids between C_6 and C_{22} (López and Martín 2005c). Also, in the Iberian wall lizard, *Podarcis hispanica* (Lacertidae), populations from relatively dryer habitats with mild temperatures have a higher proportion of fatty acids of low molecular weight (Martín and López 2006c). However, under the warmest and driest conditions, where evaporation rates are higher, *P. hispanica* also have the more stable ethyl ester forms of fatty acids (Gabirot *et al.* 2012a). In contrast, in the Spanish sand lizard, *Psammodromus hispanicus* (Lacertidae), which inhabits grassy substrates where scent marks could be useless, there is a great abundance of fatty acids with a low number of carbons, especially dodecanoic acid, which is highly volatile and, thus, might be more suitable for short-distance communication not requiring a durable signal (López and Martín 2009a).

Within the same species and population, dietary or hormonal differences among individuals might result in different proportions of fatty acids. For example, in *Iberolacerta cyreni*, proportions of oleic acid in femoral secretions of males were positively related to body condition of males, suggesting that the amount of oleic acid secreted may reflect the amount of body fat reserves of a male (Martín and López 2010b). Also, in *Iguana iguana*, the proportion of unsaturated fatty acids increases during the mating season when androgens also increase, which may enhance volatility and detectability of secretions (Alberts *et al.* 1992a,b). Stressful situations, such as increased predation risk levels, may also alter proportion of lipids (fatty acids and steroids) in secretions, probably due to the increase in circulating levels of corticosterone and its effect on lipid metabolism (Aragón *et al.* 2008).

In secretions of the Argentine black and white tegu lizard, *Tupinambis merianae* (Teiidae), there are large (>25%) amounts of 9,12-octadecadienoic acid (= linoleic acid) (Martín *et al.* 2011). This unsaturated fatty acid has been found in secretions of other lizards but always in very small amounts (Weldon *et al.* 2008). Secretion of large amounts of linoleic acid must be costly for lizards because it is one of two essential polyunsaturated fatty acids that many animals must ingest for good health. Given the dietary origin and the important functions of linoleic acid in metabolism, its actual function in femoral secretions of *T. merianae* must be sufficiently important to divert it from metabolism and "secrete" it from the body. It is likely that only males able to get an adequate dietary supply could secrete it. Therefore, the presence of linoleic acid in secretions might signal male quality (see Section 8.6).

3.3.2.3 Alcohols

Glandular secretions of lacertid lizards usually also include some alcohols, but alcohols were absent in secretions of several iguanid species (although relatively few iguanids have been studied). Lacertid lizards usually have alcohols such as hexadecanol or octadecanol in low proportions, but this does not imply that alcohols are unimportant. For example, in rock lizards, *Iberolacerta monticola* (Lacertidae), males with femoral secretions with higher abundances of hexadecanol and octadecanol had higher dominance status, and males respond aggressively to these alcohols (Martín *et al.* 2007c; see Section 3.5.2). In spiny-footed lizards (*Acanthodactylus erythrurus* and *A. boskianus*) (Lacertidae), long-chain alcohols (e.g., hexacosanol and tetracosanol) are the most abundant compounds (hexacosanol is also known as 'ceryl alcohol', and tetracosanol as 'lignoceric alcohol'). These alcohols may form waxy esters that make femoral secretions more cohesive, enhancing durability of pheromonal signals in the dry habitat of these lizards (López and Martín 2005d; Khannoon *et al.* 2011a). These alcohols may also be involved in signaling dominance, as suggested by the avoidance or aggressive behaviors of male *A. boskianus* lizards in response to these compounds (Khannoon *et al.* 2011b).

3.3.2.4 Other lipophilic compounds

Femoral or precloacal secretions of lizards also contain other types of compounds, usually as minor components, but in some cases as major compounds. Even if they are not especially abundant, such compounds are potentially important in communication, either directly or by enhancing the signaling function of other compounds.

Esters of a long chain fatty acid and a long chain fatty alcohol (= waxy esters) are found in secretions of many lizards (Weldon *et al.* 2008). Usually, there are diverse esters of the fatty acids hexadecanoic (=hexadecanoates), octadecenoic (=octadecenoates) and octadecanoic acids (=octadecanoates), linked to alcohols such as tetradecanol, hexadecanol or octadecanol. These are waxy compounds that may confer a greater stability to secretions, allowing scent marks to persist longer in very dry and warm or very humid environments. For example, waxy esters of fatty acids are especially abundant in femoral secretions of *Crotaphytus bicinctores*, in which the high proportion of waxy esters derived from the long chain eicosanoic (=icosanoates) and docosanoic acids (=docosanoates) is noteworthy (Martín *et al.* 2013b). This abundance of more stable waxy esters may protect scent marks from rapid evaporation in the xeric warm conditions in the habitat of this lizard.

Squalene is a hydrocarbon and a triterpene, and is a natural and vital part of the synthesis of all plant and animal sterols, including cholesterol, steroid hormones, and vitamin D. It is a common constituent in secretions of many lizards (Weldon *et al.* 2008), in which it might have a role as an antioxidant. For example, in the common lizard, *Zootoca vivipara* (Lacertidae) (formerly *Lacerta vivipara*), the lipids in femoral secretions would oxidize very quickly under the humid conditions of its environments (e.g., wet meadows, swamps, damp forests, etc.), but squalene might stabilize the other lipid fractions by limiting oxidation (Gabirot *et al.* 2008). Chemosensory discrimination of sex in the fossorial amphisbaenian *Blanus cinereus,* which shows precloacal gland secretions in both sexes, may be based on the much greater proportions of squalene found in secretions of males (López and Martín 2005b). The detection of squalene that "signals" male identity elicits, only in males, aggressive responses similar to those observed in agonistic interactions between males in a reproductive context (López and Martín 2009b).

Tocopherol (=vitamin E) is the main compound in femoral secretions of green lizards (*Lacerta schreiberi, L. viridis* and *Timon lepidus*) (López and Martín 2006; Kopena *et al.* 2009; Martín and López 2010a) (Fig. 3.1), but it is also found in other lizards in much lower amounts (López and Martín 2005d; Martín and López 2006c; Gabirot *et al.* 2008). Tocopherol is a typical antioxidant that may protect other compounds in the secretions from oxidation in wet environments, but it may also have a signaling function in female mate choice (Kopena *et al.* 2011; see Section 3.6).

Ketones can also appear in minor proportions in secretions of some lizards. They might have an important role, as yet untested, in communication in some cases. For example, the presence of a series of C_{17}–C_{25} saturated methyl ketones with mostly odd-numbered carbon chains is noteworthy in preanal gland secretions of male blue-headed tree agamas, *Acanthocercus atricollis* (Agamidae) (Martín *et al.* 2013c). A similar bishomologous series of methyl ketones were found, apparently homplasically, in the femoral gland secretions of the phylogenetically distantly unrelated South African giant girdled lizard, or sungazer, *Cordylus giganteus* (Cordylidae) (Louw *et al.* 2007) and in the skins of females geckos *Eublepharis macularius* (Mason and Gutzke 1990) and female *Thamnophis sirtalis* snakes (Mason *et al.* 1990). In the latter, they have a prominent role in the social and sexual behavior.

Aldehydes, such as tetradecanal or hexadecanal, are also often found in secretions (Weldon *et al.* 2008). These are highly odoriferous compounds that might facilitate detection by conspecifics of femoral secretions after they are deposited. Aldehydes have been found in some lizard species, but not in other phylogenetically related species (e.g., they are abundant in *Psammodromus algirus* but do not appear in *P. hispanicus*; Martín and López 2006d; López and Martín 2009a). This difference suggests the hypotheses that presence of aldehydes in secretions might depend on the environment or social behavior of each species.

Other minor compounds such as amides (e.g., octadecenamide) and furanones (= lactones of fatty acids) have also been found in many lizards (Weldon *et al.* 2008). Furanone derivatives are frequently found in nature as pheromones, flavor compounds or secondary metabolites, but their potential function in lizard secretions is unknown. Also, a large number of homologous long-chain alkanes were identified in the precloacal secretions of 20 *Liolaemus* lizard species (Escobar *et al.* 2001, 2003). However, alkanes have not been found in the secretions of other lizards. In *Liolaemus* the alkanes might have come from the skin surrounding the precloacal pores rather than from the precloacal secretion *per se*. Nevertheless, pentacosane was found in femoral secretion of *Cordylus giganteus* (Louw *et al.* 2007) and octacosane, nonacosane, triacontane and hentriacontane in the Balearic lizard, *Podarcis lilfordi* (Lacertidae) (Martín *et al.* 2013a). Finally, monoglycerides of fatty acids and glycerol monoethers of long chain alcohols were identified in femoral secretions of *Acanthodactylus boskianus* (Khannoon *et al.* 2011a). These compounds have been rarely identified from nature, have not yet been found in other lizards, and their possible signaling function is unknown.

3.3.3 Compounds in Generation Glands

In addition to femoral glands, cordylid lizards have generation glands as a potential source of pheromones (Van Wyk and Mouton 1992). These glands are formed by holocrine secretory cells located in the beta-layer of the epidermis, and may occur in different body locations, such as in the femoral, precloacal, antebrachial (forearm), and dorsal epidermal regions. A chemical analysis of the secretion of generation glands of *Cordylus giganteus*, identified alkenes, carboxylic acids, alcohols, ketones, aldehydes, esters, amides, nitriles and steroids (Louw *et al.* 2011). The most abundant compound was hexadecanoic acid. Interestingly, there are important differences with compounds identified in the femoral gland secretions of this lizard species (Louw *et al.* 2007). Cholesterol, a major component in femoral secretions does not occur in the generation glands, while alkanes do not occur in femoral secretions. These differences were explaining because generation glands are glandular scales, forming part of the lizard's skin.

3.3.4 Compounds in Cloacal Secretions and Feces

Little is known about the functions of the several glands (urodeal, proctodeal, etc.) found in the cloacas of lizards. The initial function of these glands seems to be provision of lubrication to the intestinal tract to facilitate expulsion of excrements or to facilitate mating. However, the glands also are a potential source of pheromones (Trauth *et al.* 1987; Cooper and Trauth 1992). This is a relatively little explored topic, but cloacal secretions might be of great importance in chemosensory communication, especially in the groups of lizards lacking femoral or precloacal glands (e.g., Gonzalo *et al.* 2004), or in species in which females have vestigial pores with little secretion. For example, in the broad-headed skink, *Plestiodon laticeps* (Scincidae), the dorsal cloacal glands may produce a species-identifying pheromone present in both sexes that may be useful to discriminate among conspecific male sexual competitors in *P. laticeps* and closely related skinks (Cooper *et al.* 1986; Cooper and Vitt 1987; Trauth *et al.* 1987). The urodaeal gland of female black-lined plated lizards, *Gerrhosaurus nigrolineatus* (Gerrhosauridae) was hypothesized to be a source of female sex pheromone, while the dorsal and ventral glands may be the source of species-identification or male pheromones (Cooper and Trauth 1992). The precise chemical identity of such pheromones is unknown. However, when different chemical fractions of the whole urodaeal glands of female *P. laticeps* were presented to males, their tongue-flick rates were higher in response to neutral lipids than to other fractions (Cooper and Garstka 1987). Pheromonal activity appears to reside in the neutral lipid fraction, which includes steryl and wax esters and mono-, di- and triacylglycerols, but not in acidic or basic lipids, or in carbohydrate or the protein fractions.

The cloacal gland secretion of the tuatara, *Sphenodon punctatus*, contains a glycoprotein and a complex mixture of triacylglycerols derived from unusual medium chain-length fatty acids as major constituents (Flachsbarth *et al.* 2009). However, it is not clear that these compounds can function as pheromones, because tongue-flicking is not observed in social interactions of tuataras (Gans *et al.* 1984), although it remains possible that olfaction might be used because tuataras respond by biting to swabs impregnated with prey chemicals (Cooper *et al.* 2001).

In several lizard species, chemicals with pheromonal function, probably coming from the cloacal glands, may be secreted onto the surface of the feces or scats as these are deposited by the lizard. Compounds in feces seem to be useful for scent-marking and conspecific recognition. Scent from feces may provide information on familiarity, relatedness, or body size of the producer (Duvall *et al.* 1987; Carpenter and Duvall 1995; López *et al.* 1998; Bull *et al.* 1999a,b, 2001; Aragón *et al.* 2000; Moreira *et al.* 2008; Wilgers and Horne 2009). Compounds from feces with properties of pheromones have not been identified, but they are probably a combination of several lipids as, in the tree skink, *Egernia striolata* (Scincidae), they are contained in scat extracts made with organic solvents (dichloromethane); fractionation of the scats with different solvents (pentane and methanol) led to loss of the unique signals needed for individual recognition (Bull *et al.* 1999a).

Finally, compounds from cloacal glandular secretions that have pheromonal activity may be added to copulatory plugs of males. Male *Iberolacerta monticola*, can distinguish their own copulatory plugs from those of other males and can even assess the dominance status of other males by chemosensory cues from copulatory plugs (Moreira *et al.* 2006). This suggests that copulatory plugs may allow males to "scent-mark" the female body during copulations and that this behavior may influence mating decisions of other males under selective pressures of sperm competition (e.g., a male might recognize and avoid displacing its own plugs while displacing plugs of other males). These hypotheses remain to be tested.

3.4 CHEMOSENSORY RESPONSES TO SPECIFIC CHEMICAL COMPOUNDS

Many studies have shown the ability of lizards to use their chemosensory systems to discriminate the scent of conspecifics from scents of heterospecifics (e.g., Cooper and Vitt 1987; Barbosa *et al.* 2006; Gabirot *et al.* 2010a,b), scents of males from females (e.g., Cooper and Trauth 1992; Cooper and Steele 1997; Cooper *et al.* 1996; Labra and Niemeyer 1999; López and Martín 2001a; Khannoon *et al.* 2010) and reproductive condition of females (e.g., Cooper and Vitt 1984; Cooper and Pérez-Mellado 2002). Lizards also use pheromones to discriminate the scent of familiar from unfamiliar individuals and self-recognition (e.g., Alberts and Werner 1993; Cooper *et al.* 1999; Aragón *et al.* 2001a,b; Carazo *et al.* 2008).

Only a few studies have examined whether lizards can discriminate between the different types of chemical compounds found in these scents. Some studies used the TF rates of lizards to scent stimuli presented on cotton swabs to examine, within a foraging context, discrimination of compounds found in the insect prey of lizards (Cooper and Pérez-Mellado 2001; Cooper *et al.* 2002a,b). *Podarcis lilfordi*, can discriminate between lipids, proteins, and carbohydrates (Cooper *et al.* 2002a) and also among different lipids, such as glycerol, cholesterol, and oleic and hexadecanoic acids (Cooper *et al.* 2002b).

With respect to compounds found in glandular secretions of lizards, another study measured the TF responses of female *Podarcis hispanica* to two lipids (cholesterol and cholesta-5,7-dien-3-ol) (Martín and López 2006e). These steroids are major compounds in femoral secretion of males (Martín and López 2006c). Females discriminate between these two lipids, showing higher TF responses to cholesta-5,7-dien-3-ol and are able to assess differences in its concentration, responding more strongly to higher concentrations. These results, together with the female preference for areas scent marked by males with higher proportions of this steroid in their secretions (López and Martín 2005a; see Section 3.5.3), suggest that cholesta-5,7-dien-3-ol is a "key" pheromonal compound for this lizard.

In some cases, intersexual differences in chemosensory responses suggest that different compounds may carry different messages for males and females. Thus, female *Iberolacerta cyreni* have higher TF responses to cholesta-5,7-dien-3-ol and to ergosterol than to cholesterol, whereas the opposite is found in males (Martín and López 2006a, 2008a). This is probably explained by the preference of females for scent marks of males with higher proportions of cholesta-5,7-dien-3-ol and ergosterol in their secretions, which is related to the "quality" of those males (Martín and López 2006a,b; see Section 3.5.3), whereas in males cholesterol might signal the body size of a potential male opponent (Martín and López 2007; see Section 3.5.2).

Female *Iberolacerta cyreni* can also discriminate among fatty acids found in femoral gland secretions of males, such as oleic acid and hexadecanoic acid, and can assess differences in their concentration (Martín and López 2010b). This discrimination might be important for females because the amount of fatty acids secreted may reflect the amount of body fat reserves of a male. The presence of both saturated (e.g., hexadecanoic) and unsaturated (oleic) fatty acids in the males' secretions might allow the scent signal to function in varying environmental conditions because at ambient temperatures, unsaturated fatty acids may be accessible as liquids, whereas saturated fatty acids may be waxes. Thus, it is likely that females actually responded to the whole mix of fatty acids usually found in males' secretion or that under different temperature conditions, some fatty acids were more effective than others in eliciting chemosensory exploration of females.

Alcohols can also be detected by lizards. Male and female *Podarcis hispanica* lizards can discriminate among alcohols found in secretions of males and vary tongue-flick rates with their concentrations (Gabirot *et al.* 2012c). Male *Iberolacerta monticola* discriminate between different concentrations of hexadecanol, a major compound in glandular secretions of males, from other chemicals (Martín *et al.* 2007c). Moreover, males respond aggressively to hexadecanol, but respond neutrally to other compounds (Martín *et al.* 2007c). These results, together with the relationship observed between femoral secretions with higher proportions of hexadecanol and dominance, suggest that hexadecanol may be a reliable status badge in this lizard. Similarly, male *Acanthodactylus boskianus* show avoidance behavior for substrates marked with cholesterol and long chain alcohol blends (both found in males' secretions), and agonistic behavior towards these stimuli, whereas females do not respond to these chemicals (Khannoon *et al.* 2011b).

The observed chemosensory responses to glandular secretions may be a consequence of response to the combined multiple effects of different compounds. For example, female *Iberolacerta cyreni* discriminate between different concentrations of ergosterol and oleic acid presented alone and exhibit the highest chemosensory exploration to high concentrations of ergosterol, whereas high concentrations of oleic acid elicit tongue-flick (TF) rates of a magnitude similar to those to low concentrations of ergosterol (López and Martín 2012) (Fig. 3.2). Moreover, the highest TF rates are directed to a mixture containing high concentrations of both compounds combined, and there is an upper-shift of the top of the dose-response curve by the combination of the two compounds, suggesting that there are additive or synergic effects of these two compounds (Fig. 3.2).

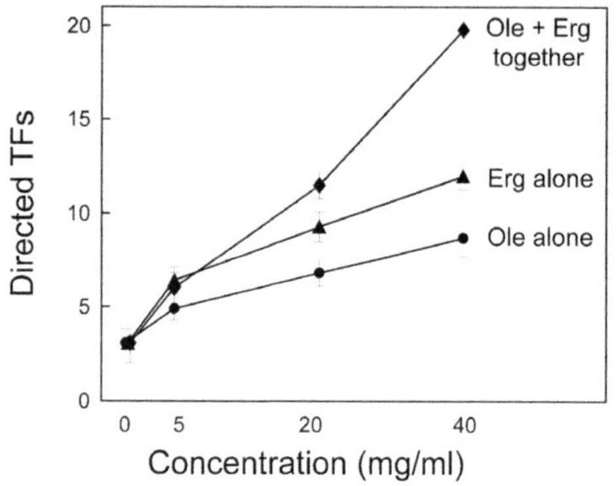

Fig. 3.2 Dose-dependent and additive effects of two compounds from males' femoral gland secretions on chemosensory responses of female Iberian rock lizards (*Iberolacerta cyreni*). Number (mean + SE) of tongue-flicks directed to swabs by female lizards in response to cotton-tipped applicators bearing different concentrations (0, 5, 20, or 40 mg/mL) of oleic acid (Ole) and ergosterol (Erg) (standard compounds) presented alone or together, all dissolved in DCM. From López, P. and Martín, J. 2012. Chemical Senses 37: 47–54. Figure 3.

3.5 REPRODUCTIVE CHEMICAL ECOLOGY IN LIZARDS

Multiple lines of evidences show that lizards have highly developed chemosensory abilities, including strong responses to scent of conspecifics, and that most lizards produce, especially during the mating season, glandular secretions that contain many chemicals that potentially function as signals. Nevertheless, as noted in Section 3.1, the reproductive behavior of lizards was long considered to be mainly based on more conspicuous visual signals. Thus, most research on reproductive behavior of lizards focused on colorful traits or movement displays. Relatively few studies have considered the potential role of chemical signals in reproductive ecology of lizards. These will be summarized in the following sections. We are just starting to understand not only the function of specific chemicals in modulating different behaviors related to reproduction and sexual selection, but also the mechanisms that explain the use and evolutionary persistence of these signals.

3.5.1 Scent-Marking

Glandular secretions, feces, or urine are very often used for scent-marking of substrates by many terrestrial vertebrates, including many lizards. These scent-marks identify territorial boundaries or attract mates (reviewed in Müller-Schwarze 2006; Mason and Parker 2010). Scent-marking a territory can be a simple and effective method to inform conspecifics about the identity and characteristics of the male that defends the marked territory. Many lizards can scent-mark their territories by using femoral or cloacal secretions and/or feces. Semiochemicals in these scent marks are known to convey information on sex, age, body size, dominance status or health condition of the signaler (reviewed in Mason 1992; Mason and Parker 2010; Martín and López 2011).

If the information in the scent-mark is reliable (e.g., Martín and López 2006b; Kopena *et al.* 2011), the signaler will benefit from this advertisement, for example, by repelling rivals or attracting mates (Martín and López 2012). Receivers of the signal may gain benefits by using information about territorial status and dominance obtained from scent-marks into their decisions about aggressive behavior toward the scent-marking male (e.g., Carazo *et al.* 2007, 2008; López and Martín 2011) or about mate choice (e.g., Martín and López 2000, 2006a,b; López *et al.* 2002b; López and Martín 2005a; Olsson *et al.* 2003). Scent-marks may be important in reproductive behavior and sexual selection of many lizards (see below).

3.5.2 Intrasexual Relationships Between Males and Social Organization

When competing for access to mates, males use cues from their rivals to judge relative fighting ability and to evaluate their chances of success in a potential future agonistic contest. In lizards, chemical signals may be a vital component of male-male contests informing males of a rival's quality or intentions. Pheromones may signal a male's dominance status, or characteristics related to fighting ability or dominance such as body size, through rates of production and/or the quality of the glandular secretions (Alberts *et al.* 1992b; López *et al.* 2003b; Martins *et al.* 2006; Martín *et al.* 2007c). In many cases, pheromones also allow lizards to discriminate between familiar and unfamiliar males and may allow true individual recognition (i.e., based on individual identity cues) (Aragón *et al.* 2001a; Carazo *et al.* 2008).

Pheromones affect intrasexual relationships in male lizards in two ways. First, pheromones deposited in substrate scent-marks can provide information in absence of the signaler on the presence of previously known individual rivals or on the fighting potential of unfamiliar individuals (Aragón *et al.* 2000, 2001a; Labra 2006; Carazo *et al.* 2007, 2008). This information may affect behavior and space-use by other males that sample the scent-marks (Alberts *et al.* 1994; Aragón *et al.* 2001c, 2003; Labra 2006). Second, pheromones may be used during actual agonistic encounters, for example, to recognize rival males (Cooper and Vitt 1987; López *et al.*

2002a). This is important because when two males interact, they become familiar and establish their relative dominance relationship, which allows them to decrease the aggressiveness in successive encounters (López and Martín 2001b). In some cases, recognition of familiar lizards or of specific individuals may be predominantly based on pheromonal cues. In *Podarcis hispanica*, resident males respond more aggressively towards unknown or familiar males experimentally impregnated with scents from unfamiliar males than to familiar males or unknown males impregnated with scents of familiar males (López and Martín 2002) (Fig. 3.3).

Chemical rival recognition may be used in other situations. Males of many species of lizards show conspicuous breeding colors, but, in some species, young or competitively inferior males conceal their sexual identity by mimicking a female-like dull coloration that allows them to evade aggression from dominant males and to adopt an alternative satellite-sneaking mating tactic. In two experiments, scent and coloration of satellite males, were manipulated in *Psammodromus algirus* (López *et al.* 2003b) and Augrabies flat lizards, *Platysaurus broadley* (Cordylidae) (Whiting *et al.* 2009). In both species, deceptive coloration was effective in avoiding aggression at long distance. However, at close range, dominant males used chemical signals to identify satellite males, as shown by aggressive responses toward satellites.

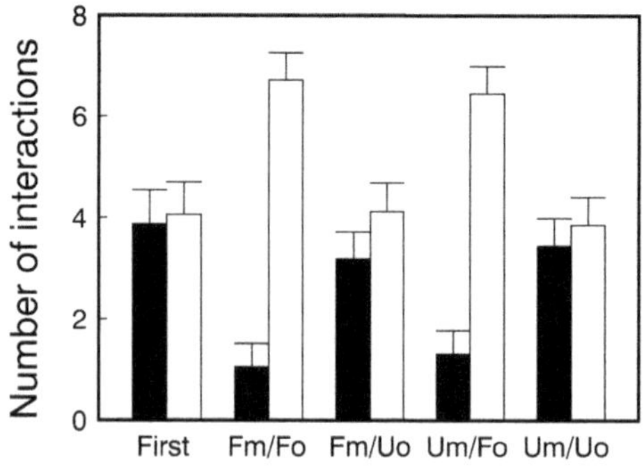

Fig. 3.3 Role of pheromones in intrasexual agonistic behavior of Iberian wall lizards (*Podarcis hispanica*). Number (mean + SE) of aggressive (black bars) and neutral (white bars) interactions in the first contest of a resident male with an intruder male impregnated with his own odor, in posterior contests with the same familiar male bearing his own odor (Fm/Fo) or impregnated with odor of an unfamiliar male (Fm/Uo), and in posterior contests with unfamiliar males impregnated with odor of a familiar male (Um/Fo) or bearing their own odor (Um/Uo). From López, P. and Martín, J. 2002. Behavioral Ecology and Sociobiology 51: 461–465. Figure 1.

In some animals, males may identify territory owners by directly comparing the scent of substrate marks with the scent of any conspecific they encounter nearby, i.e., by scent-matching (Gosling and McKay 1990). This may also occur in lizards. Thus, when an intruding male *Iberolacerta cyreni*, explores substrate scent marks, if he subsequently finds a rival male whose scent experimentally matches that of scent marks (considered presumably to be the territory owner), the intruding male delays time until the first agonistic interaction, reduces the intensity and number of fights, and wins fewer interactions than when encountering other non-matching individual males (López and Martín 2011). Therefore, males may use scent-matching as a mechanism to assess the ownership status of other males, which could contribute to modulation of further intrasexual aggressions.

However, the chemical basis of the assessment of rival dominance status or fighting ability is poorly known. Assessment might be affected by changes in concentrations of some chemicals in scents that are correlated with traits that affect fighting ability. In *Iguana iguana*, femoral glandular productivity, pore size, and the percentage of lipids in the secretions are correlated with plasma testosterone levels in dominant, although not in subordinate, adult males (Alberts *et al.* 1992b). Proportions of cholesterol in femoral secretions of male *Iberolacerta cyreni* increase with body size (López *et al.* 2006). These males discriminate chemically and respond aggressively to cholesterol stimuli presented on swabs (Martín and López 2008a), and, moreover, when cholesterol in the body scent of males is experimentally increased, they win more frequently agonistic interactions (Martín and López 2007), suggesting that high concentration of cholesterol may signal greater fighting ability linked to larger body size. This may be a reliable signal, if higher proportions of cholesterol in secretions indicate higher sex steroid (i.e., testosterone) levels that also determine aggressiveness levels (Alberts *et al.* 1992b; Sheridan 1994).

Similarly, in male *Iberolacerta monticola*, dominant males produce femoral secretions with higher proportions of two alcohols (hexadecanol and octadecanol) (Martín *et al.* 2007c) (Fig. 3.4). Males discriminate different concentrations of hexadecanol from other chemicals found in secretions and respond aggressively towards hexadecanol according to their own dominance status, but respond neutrally to other chemicals. The signal may be reliable because, given that hexadecanol elicits aggressive responses of other males, only truly dominant males with a high fighting potential should chemically signal their status. Also, it might be physiologically costly to produce femoral secretions with high amounts of hexadecanol. Consistent with this view, dominant males are healthier (i.e., have a stronger immune response), which might allow them to afford secreting greater quantities of compounds that signal a high dominance status (Martín *et al.* 2007c).

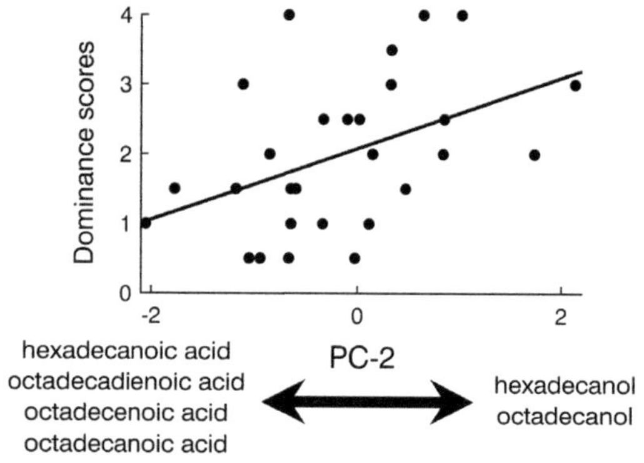

Fig. 3.4 Chemical basis of dominance status signallng in rock lizards (*Iberolacerta monticola*). Relationships between dominance status scores of male lizards and PC scores obtained from a principal components analysis on the relative proportions of chemical compoundss in femoral gland secretions. From Martín, J., Moreira, P. L. and López, P. 2007. Functional Ecology 21: 568–576. Figure 1.

3.5.3 Female Mate Choice

Some field studies suggest that females of some lizard species do not choose males, but base their space-use on the quality of a territory (e.g., thermal characteristics, abundance of food or refuges, etc.) rather than on the quality of the male that defends that territory (e.g., Hews 1993; Calsbeek and Sinervo 2002). Males would only defend these favorable territories from other males to increase their access to females. However, it is still possible that females might be attracted to a territory by male signals that may be used as "public information" to assess the quality of a territory, or through being "lured" by male signals that resemble food. In this context, in lizard species in which males scent-mark territories, pheromones may have an important role in female space-use and mate choice.

Other studies suggest that female lizards of some species might use some chemical compounds in the scent-marks of males to select areas scent marked, and, therefore, occupied by preferred potential mates (Martín and López 2006a, 2012, 2013a; Johansson and Jones 2007). On the other hand, a pre-existing sensory bias for food chemicals might also explain the chemosensory preferences of female lizards for some compounds in the scent-marks of males (Martín and López 2008). However, irrespective of the causes underlying decisions by females to spend more time in a given scent marked area, this decision about use of space will increase the probability of mating with the male that has scent marked the selected area (Martín and López 2012). Females may try to reject mating advances from "undesired" males, but males may obtain many forced matings. Therefore, space-use decisions of female lizards will have direct consequences for their reproductive success, and those space-use strategies that increase the reproductive success of females will be evolutionary selected. As a consequence, space-use decisions of female lizards based on scent marks of males may have the same evolutionary consequences as "direct" mate choice decisions of other animals.

Therefore, male lizards might use scent marks to attract females to their territories, thus increasing the probabilities of mating with these females, whereas females might use scent marks of males to select potential mates or territories of high quality. But this attracting function of the scent-marks of lizards remains little explored. One field study in *Iberolacerta cyreni*, showed that experimentally increasing ergosterol (a compound from femoral secretions of males) on rock substrates inside home ranges of males results, after some days, in increased relative densities of females, but not of males, in those areas. This effectively results in an increase of mating opportunities for resident males (Martín and López 2012) (Fig. 3.5). Also, female *I. cyreni* prefer areas scent-marked by large/old territorial males to those scent-marked by smaller/young satellite-sneaker males (López *et al.* 2003a; Martín and López 2013a) and prefer areas scent-marked by two territorial males to areas of similar size marked by a single territorial male (Martín and López 2013a). The former choice might increase the probability of obtaining multiple copulations with different males, thus favoring sperm competition and cryptic female choice, or may be a way to avoid infertile males (Martín and López 2013a).

To establish whether female mate choice exists, experimenters must select appropriate criteria base on choices used by females to select a mate (or a scent marked territory). For example, female *Podarcis hispanica*, preferentially associate with areas scent-marked by males, but females do not choose territories marked by larger versus smaller males. Taken alone, this might suggest that mate choice by females is absent in this species (Carazo *et al.* 2011). However, other experiments with this species showed that females select scents of males with higher proportions of cholesta-5,7-dien-3-ol (among scents from males of similar size), which are those with a better T-cell-mediated immune response (i.e., with a better health) (López and Martín 2005a). Similarly, female *Iberolacerta cyreni*, select areas scent-marked by males with stronger immune responses, as signaled by high ergosterol proportions in femoral secretions of males (Martín and López 2006a), or with better body condition, as signaled by high proportions of oleic acid in secretions (Martín and López 2010b). In the same way, female *Psammodromus algirus* show higher chemosensory responses to femoral gland secretions of males with low blood parasite loads and stronger immune responses, which is apparently signaled by higher proportions of

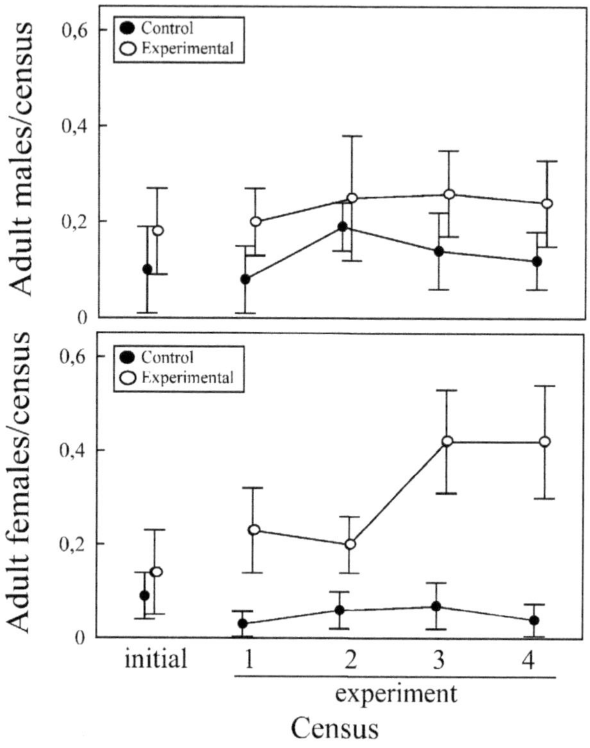

Fig. 3.5 Effects of manipulation of substrate *scent-marks* with pheromones on density of Iberian rock lizards (*Iberolacerta cyreni*). TOP: Numbers (mean + SE) of (a) adult males or (b) adult females observed in each census of the control (black circles) and experimental (open circles) plots before the experiment (initial) and during the four days after rocks were supplemented with ergosterol (experimental) or a control solution. From Martín, J. and López, P. 2012. PLoS One 7: e30108. Figure 1. BELOW: A pair of rock lizards, the territorial male (in front) has approached to a female that was probably attracted to his area by pheromones in scent-marks. Photograph by J. Martín.

two alcohols (octadecanol and eicosanol) and lower proportions of their correspondent carboxylic acids (octadecanoic and eicosanoic acids) (Martín *et al.* 2007b).

Intra- and inter-sexual competition often lead to selection for different secondary sexual traits, which may be reflected in responsiveness to male pheromones by female lizards. Scents of males that signal characteristics that confer competitive advantages to males against other males (such as a larger head or a higher bite force) are often not selected by females. Females may respond more strongly to scents of males that signal traits of a potential mate that are beneficial to females, such as a higher body condition or levels of symmetry (*Iberolacerta cyreni*, López *et al.* 2002b; Dalmatian wall lizard *Podarcis melisellensis* (Lacertidae), Huyghe *et al.* 2012).

Similarly, in sand lizards, *Lacerta agilis* (Lacertidae), females do not seem to mate selectively with larger and/or older males (Olsson and Madsen 1995), but genetic compatibility (based on the major histocompatibility complex, MHC, dissimilarity) is the main characteristic that females select based on the scent of a male (Olsson *et al.* 2003). Similar avoidance of inbreeding based on chemical signals might function in other lizard species (Bull and Cooper 1999; Bull *et al.* 2001). However, the chemical bases of this genetic discrimination remain unknown. Selective mating with non-kin or unrelated pairs may confer genetic benefits because the new combinations of immunocompetence in offspring will defend them more effectively against evolving parasites (Penn and Potts 1999). In contrast, female painted dragons, *Ctenophorus pictus* (Agamidae), do not prefer scents from unrelated males, which might be explained by weak selection against inbreeding in this species (Jansson *et al.* 2005).

3.6 EVOLUTIONARY ORIGIN OF THE CHEMICAL SIGNALING SYSTEM

Although some compounds "preferred" by females in the scent of males, or that affect intrasexual relationships between males, have been identified in a few lizard species, it is not well understood why lizards can be confident in the reliability and honesty of the message in the chemical signal (i.e., that a scent with a specific chemical characteristics always corresponds to a male with some specific characteristics). This is, however, a prerequisite for a signal to persist in evolutionary time.

One possible explanation may reside in the important metabolic organismal functions of some compounds that are, however, secreted by glands to the exterior of the body. Thus, many of the lipids commonly found in glandular secretions of lizards, such as fatty acids and steroids, function as signaling molecules or lipid mediators, show potent biological activity, and are important keys in many metabolic pathways. Some lipophilic compounds in femoral secretions of some lizards appear to be good candidates for conferring honesty to signals. These are α-tocopherol (=vitamin E), cholesta-5,7-dien-3-ol (=dihydrocholesterol; a precursor of vitamin D_3), ergosterol (=pro-vitamin D_2), 9,12-octadecadienoic acid (=linoleic acid) and 5,8,11,14-eicosatetraenoic acid (=arachidonic acid). These are essential components for metabolism that have important physiological functions. In most cases, vertebrates can only obtain them from the diet and their deficiency can cause severe disorders. However, in spite of the importance of these compounds, male lizards divert them from metabolism to allocate them for use in femoral glandular secretions. In such cases, trade-offs must exist between using these essential chemicals in metabolism and using them for scent-marking. Only males that have, or are able to obtain, an adequate supply of vitamins and essential fatty acids could allow diversion of surplus chemicals from metabolism to social signaling. Therefore, the presence of vitamins and essential fatty acids in relative high proportions in secretions might honestly advertise male quality.

For example, cholesta-5,7-dien-3-ol (=pro-vitamin D_3) is often found in the skin, where it transforms into vitamin D_3 after exposure to UV-B irradiation in sunlight. Vitamin D_3 is essential in calcium metabolism and for regulation of the immune system (Fraser 1995). However, very often, the synthesis of vitamin D_3 in the skin is not sufficient to meet physiological requirements, and lizards require dietary intake of vitamin D (Ferguson *et al.* 2005). Under these conditions, vitamin D is an essential nutrient for lizards.

Therefore, when diverting pro-vitamin D from metabolism to femoral secretions, male lizards might need to obtain more vitamin D. After supplementation of dietary vitamin D, male *Iberolacerta cyreni*, increased the proportion of pro-vitamin D_3 in femoral secretions and females preferred areas scent marked by these males with more pro-vitamin D_3 (Martín and López 2006b). This suggests that allocation of this pro-vitamin to secretions is costly and dependent on the foraging ability of a male to obtain enough vitamin D in the diet, or of the food quality within his territory, which may confer honesty to his pheromonal signal and may explain why females select the scent marked territories of these males.

Vitamin E (= α-tocopherol) is the main lipophilic antioxidant involved in membrane defence (Brigelius-Flohe and Traber 1999). This vitamin is of dietary origin and its deficiency has severe pathological consequences, such as infertility, neurological disorders and lung diseases. In the European green lizard, *Lacerta viridis*, males show high proportions of tocopherol in their femoral secretions (Kopena *et al.* 2009). These proportions increased when males were experimentally fed supplementary vitamin E, and females preferred to use areas scent-marked by these males with increased vitamin E secretion levels (Kopena *et al.* 2011) (Fig. 3.6). This suggests that the cost of

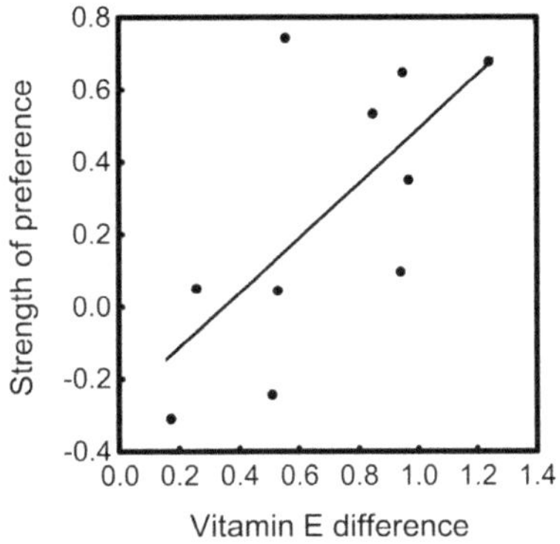

Fig. 3.6 Strength of female preference for male scent in European green lizards (*Lacerta viridis*). Female 'strength of preference' was calculated as the difference in the mean proportion of females that were observed at the areas containing chemical cues from size matched vitamin E supplemented vs. control males. Vitamin E difference is the difference in relative vitamin E content of the femoral secretions of size-matched vitamin E supplemented vs. control males within a male pair. From Kopena, R., Martín, J., López, P. and Herczeg, G. 2011. PLoS One 6: e19410. Figure 2.

allocating vitamin E to secretions, diverting it from its important organismal antioxidant function, may confer reliability to chemical signals of males. Tocopherol may be the chemical signal directly used by females; however, this antioxidant compound might simply increase duration and intensity of information provided by other signaling compounds in secretions.

Similarly, linoleic and arachidonic acids are essential fatty acids that mediate a wide range of physiological responses and maintain homeostasis, and, in spite of their important functions, are also found in secretions of some lizards in relatively high proportions (e.g., Martín *et al.* 2001). This suggests a potential, but untested, signaling function of these essential fatty acids.

Reliability of chemical signals might be also based on other mechanisms. For example, secretion of compounds could be differentially costly for different individual males, because secretion depends on testosterone levels and testosterone may have immunosuppressive effects (Folstad and Karter 1992; Belliure *et al.* 2004; Oppliger *et al.* 2004). Thus, only high quality males could afford the trade-off of producing pheromones while avoiding the detrimental effects of testosterone on their immune system. Female *Podarcis hispanica* showed higher chemosensory responses to scents of males with experimentally supplemented testosterone than to control scents (Martín *et al.* 2007a), probably because production and concentration of glandular secretions increased when testosterone increased. However, testosterone also induced a decrease in proportion of cholesta-5,7-dien-3-ol, probably as a consequence of its immunosuppressive effects (López *et al.* 2009b), and females preferred scent-marks of males that maintained high levels of this compound in secretions independently of the experimental manipulation (Martín *et al.* 2007a).

3.7 FUTURE DIRECTIONS

Although the roles of pheromones in lizard biology have been explored in the studies cited above, there is still much work to do in several fields to fully understand them. First, we need to know the compounds that may be potentially used as pheromones. Many lizard species representing diverse taxonomic lineages and geographical areas have abundant glandular secretions that have never been chemically analyzed, even for species in which glandular secretions are known to be important for chemical communication. Moreover, most available information is from femoral or precloacal glands, whereas almost nothing is known about chemicals from cloacal secretions or feces that may have pheromonal functions. Different compounds might be characteristics of some groups of lizards, but be absent in others. The presence of different compounds might have a phylogenetic effect, but it might also be linked to environmental conditions, or to the way these chemicals are used in communication (e.g., as scent-marks on different substrates, or as chemicals sampled directly from the individuals). This emphasizes the need for more studies analyzing the chemical compounds in glandular secretions of diverse groups of lizards from different habitats. These descriptive studies will allow further comparative studies to clarify the patterns of presence and abundance of compounds observed in each species, and the causes (genetic and environmental) that explain the evolution of gland secretions in lizards.

It is rarely known which specific compounds could act as pheromones. After identification of compounds in glandular secretions, further chemosensory tests should be conducted with the different compounds, alone and combined in different proportions, to understand which compounds are important in communication and how their variability in abundance affects to the chemosensory responses of conspecifics.

The next step would be to identify the "message" of these chemical compounds (i.e., the strategic element of the signal; Guilford and Dawkinds 1991). Behavioral tests are needed that relate differential responses to differences in scents of different individuals for which chemical composition of glandular secretions is known, and to differences in morphological and physiological characteristics. We need to determine which traits are sufficiently important to be signaled in each species and how they are signaled through chemical cues. Also, lizards use other sensory systems in communication, especially vision, and it may be important to understand how information contained in multiple visual and chemical signals is congruent, although sometimes used in different contexts, or different and complementary, as well as the relative importance of one type of signal or other in different species and environments.

Finally, we would need to analyze the mechanisms that allow the origin and evolutionary persistence of a chemical signal, and how the environment may drive the evolution of the chemical signaling system, especially through changes in the design of the signal to maximize its efficacy in different environmental conditions (Guilford and Dawkinds 1991). This is important, on the one hand, because if different populations of the same species use different chemical signals for reproduction, this may later lead to reproductive isolation between populations and speciation processes (e.g., Gabirot *et al.* 2012a). On the other hand, current rapid human-induced changes in the environment, such as global warming or contamination, might affect the efficacy of chemical signals with potential detrimental consequences for reproduction and stability of populations (e.g., Martín and López 2013b).

We encourage researchers to "look" for the roles of chemical signals or pheromones in the social and reproductive behavior of lizards. Not knowing these, we might be missing most of the information contained in sexual displays and would fail to understand many field observations of ecology, behavior, and evolution, of these animals.

3.8 ACKNOWLEDGMENTS

We thank W. E. Cooper and A. Salvador for "inducing" us to study chemical communication in lizards and for their very helpful comments, and to all our friends that have worked with us for helping us to understand pheromones of lizards and other reptiles: L. Amo, P. Aragón, C. Cabido, A. M. Castilla, E. Civantos, M. Cuadrado, M. Gabirot, R. García-Roa, A. Gonzalo, A. Ibáñez, R. Kopena, J. J. Luque-Larena, P. L. Moreira, J. Ortega, K. M. Pilz, V. Pérez-Mellado and N. Polo-Cavia. We also thank "El Ventorrillo" MNCN Field Station for the long term use of their facilities and Nino for taking care of and feeding lizards and researchers. Financial support during writing was provided by the project of Ministerio de Ciencia e Innovación MICIIN-CGL2011-24150/BOS. The editors and reviewers are also thanked for their input.

3.9 LITERATURE CITED

Alberts, A. C. 1990. Chemical properties of femoral gland secretions in the desert iguana *Dipsosaurus dorsalis*. Journal of Chemical Ecology 16: 13–25.

Alberts, A. C. 1991. Phylogenetic and adaptive variation in lizard femoral gland secretion. Copeia 1991: 69–79.

Alberts, A. C. 1992. Constraints on the design of chemical communication systems in terrestrial vertebrates. American Naturalist 139: 62–89.

Alberts, A. C. 1993. Chemical and behavioral studies of femoral gland secretions in iguanid lizards. Brain Behavior and Evolution 41: 255–260.

Alberts, A. C. and Werner, D. I. 1993. Chemical recognition of unfamiliar conspecifics by green iguanas: functional significance of different signal components. Animal Behaviour 46: 197–199.

Alberts, A. C., Sharp, T. R., Werner, D. I. and Weldon, P. J. 1992a. Seasonal variation of lipids in femoral gland secretions of male green iguanas (*Iguana iguana*). Journal of Chemical Ecology 18: 703–712.

Alberts, A. C., Pratt, N. C. and Phillips, J. A. 1992b. Seasonal productivity of lizard femoral glands: Relationship to social dominance and androgen levels. Physiology & Behavior 51: 729–733.

Alberts, A. C., Phillips, J. A. and Werner, D. I. 1993. Sources of intraspecific variability in the protein composition of lizard femoral gland secretions. Copeia 1993: 775–781.

Alberts, A. C., Jackintell, L. A. and Phillips, J. A. 1994. Effects of chemical and visual exposure to adults on growth, hormones, and behavior of juvenile green iguanas. Physiology & Behavior 55: 987–992.

Apps, P. J. 2013. Are mammal olfactory signals hiding right under our noses? Naturwissenschaften 100: 487–506.

Aragón, P., López, P. and Martín, J. 2000. Size-dependent chemosensory responses to familiar and unfamiliar conspecific faecal pellets by the Iberian rock-lizard, *Lacerta monticola*. Ethology 106: 1115–1128.

Aragón, P., López, P. and Martín, J. 2001a. Chemosensory discrimination of familiar and unfamiliar conspecifics by lizards: implications of field spatial relationships between males. Behavioral Ecology and Sociobiology 50: 128–133.

Aragón, P., López, P. and Martín, J. 2001b. Discrimination of femoral gland secretions from familiar and unfamiliar conspecifics by male Iberian rock-lizards, *Lacerta monticola*. Journal of Herpetology 35: 346–350.

Aragón, P., López, P. and Martín, J. 2001c. Effects of conspecific chemical cues on settlement and retreat-site selection of male lizards, *Lacerta monticola*. Journal of Herpetology 35: 681–684.

Aragón, P., López, P. and Martín, J. 2003. Differential avoidance responses to chemical cues from familiar and unfamiliar conspecifics by male Iberian rock-lizards (*Lacerta monticola*). Journal of Herpetology 37: 583–585.

Aragón, P., López, P. and Martín, J. 2008. Increased predation risk modifies lizard scent-mark chemicals. Journal of Experimental Zoology A 309: 427–433.

Barbosa, D., Font, E., Desfilis, E. and Carretero, M. A. 2006. Chemically mediated species recognition in closely related *Podarcis* wall lizards. Journal of Chemical Ecology 32: 1587–1598.

Belliure, J., Smith, L. and Sorci, G. 2004. Effect of testosterone on T cell mediated immunity in two species of Mediterranean Lacertid lizards. Journal of Experiemntal Zoology A 301: 411–418.

Brigelius-Flohe, R. and Traber, M. G. 1999. Vitamin E: function and metabolism. FASEB Journal 13: 1145–1155.

Bull, C. M. and Cooper, S. J. B. 1999. Relatedness and avoidance of inbreeding in the lizard, *Tiliqua rugosa*. Behavioral Ecology and Sociobiology 46: 367–372.

Bull, C. M., Griffin, C. L. and Perkins, M. V. 1999a. Some properties of a pheromone allowing individual recognition from the scats of an Australian lizard, *Egernia striolata*. Acta Ethologica 2: 35–42.

Bull, C. M., Griffin, C. L. and Johnston, G. R. 1999b. Olfactory discrimination in scat-piling lizards. Behavioral Ecology 10: 136–140.

Bull, C. M., Griffin, C. L., Bonnett, M., Gardner, M. G. and Cooper, S. J. 2001. Discrimination between related and unrelated individuals in the Australian lizard *Egernia striolata*. Behavioral Ecology and Sociobiology 50: 173–179.

Calsbeek, R. and Sinervo, B. 2002. Uncoupling direct and indirect components of female choice in the wild. Proceedings of the National Academy of Sciences of the United States of America 99: 14897–14902.

Carazo, P., Font, E. and Desfilis, E. 2007. Chemosensory assessment of rival competitive ability and scent mark function in a lizard (*Podarcis hispanica*). Animal Behaviour 74: 895–902.

Carazo, P., Font, E. and Desfilis, E. 2008. Beyond 'nasty neighbours' and 'dear enemies'? Individual recognition by scent marks in a lizard (*Podarcis hispanica*). Animal Behaviour 76: 1953–1963.

Carazo, P., Font, E. and Desfilis, E. 2011. The role of scent marks in female choice of territories and refuges in a lizard (*Podarcis hispanica*). Journal of Comparative Psychology 125: 362–365.

Carpenter, G. C. and Duvall, D. 1995. Fecal scent marking in the western banded gecko (*Coleonyx variegatus*). Herpetologica 51: 33–38.

Chauhan, N. B. 1986. A preliminary report on the lipid components of pre-anal gland secretion of lizards. *Hemidactylus flaviviridis* and *Uromastix hardwickii*. Journal of Animal Morphology and Physiology 33: 73–76.

Cole, C. J. 1966. Femoral glands in lizards: a review. Herpetologica 22: 199–206.

Cooper, W. E., Jr. 1994. Chemical discrimination by tongue-flicking in lizards: a review with hypotheses on its origin and its ecological and phylogenetic relationships. Journal of Chemical Ecology 20: 439–487.

Cooper, W. E., Jr. 1998. Evaluation of swab and related tests as a bioassay for assessing responses by squamate reptiles to chemical stimuli. Journal of Chemical Ecology 24: 841–866.

Cooper, W. E., Jr. and Burghardt, G. M. 1990. A comparative analysis of scoring methods for chemical discrimination of prey by squamate reptiles. Journal of Chemical Ecology 16: 45–65.

Cooper, W. E., Jr. and Garstka, W. R. 1987. Lingual responses to chemical fractions of urodaeal glandular pheromone of the skink *Eumeces laticeps*. Journal of Experimental Zoology 242: 249–253.

Cooper, W. E., Jr. and Pérez-Mellado, V. 2001. Chemosensory responses to sugar and fat by the omnivorous lizard *Gallotia caesaris* with behavioral evidence suggesting a role for gustation. Physiology & Behavior 73: 509–516.

Cooper, W. E., Jr. and Pérez-Mellado, V. 2002. Pheromonal discrimination of sex, reproductive condition, and species by the lacertid lizard *Podarcis hispanica*. Journal of Experimental Zoology 292: 523–527.

Cooper, W. E., Jr. and Steele, L. J. 1997. Pheromonal discrimination of sex by male and female leopard geckos (*Eublepharis macularius*). Journal of Chemical Ecology 23: 2967–2977.

Cooper, W. E., Jr. and Trauth, A. E. 1992. Discrimination of conspecific male and female cloacal chemical stimuli by males and possession of a probable pheromone gland by females in a cordylid lizard, *Gerrhosaurus nigrolineatus*. Herpetologica 48: 229–236.

Cooper, W. E., Jr. and Vitt, L. J. 1984. Conspecific odor detection by the male broad-headed skink, *Eumeces laticeps*: effects of sex and site of odor source and of male reproductive condition. Journal of Experimental Zoology 230: 199–209.

Cooper, W. E., Jr. and Vitt, L. J. 1987. Intraspecific and interspecific aggression in lizards of the scincid genus *Eumeces*: Chemical detection of conspecific sexual competitors. Herpetologica 43: 7–14.

Cooper, W. E., Jr., Garstka, W. R. and Vitt, L. J. 1986. Female sex pheromone in the lizard *Eumeces laticeps*. Herpetologica 42: 361–366.

Cooper, W. E., Jr., Van Wyk, J. H. and Mouton, P. L. N. 1996. Pheromonal detection and sex discrimination of conspecific substrate deposits by the rock dwelling cordylid lizard *Cordylus cordylus*. Copeia 1996: 839–845.

Cooper, W. E., Jr., Van Wyk, J. H. and Mouton, P. L. N. 1999. Discrimination between self-produced pheromones and those produced by individuals of the same sex in the lizard *Cordylus cordylus*. Journal of Chemical Ecology 25: 197–208.

Cooper, W. E., Jr., Ferguson, G. W. and Habegger, J. J. 2001. Responses to animal and plant chemicals by several iguanian insectivores and the tuatara, *Sphenodon punctatus*. Journal of Herpetology 35: 255–263.

Cooper, W. E., Jr., Pérez-Mellado, V. and Vitt, L. J. 2002a. Responses to major categories of food chemicals by the lizard *Podarcis lilfordi*. Journal of Chemical Ecology 28: 709–720.

Cooper, W. E., Jr., Pérez-Mellado, V. and Vitt, L. J. 2002b. Lingual and biting responses to selected lipids by the lizard *Podarcis lilfordi*. Physiology & Behavior 75: 237–241.

Cooper, W. E., Jr., Pérez-Mellado, V., Vitt, L. J. and Budzynski, B. 2003. Cologne as a pungency control in tests of chemical discrimination: effects of concentration, brand, and simultaneous and sequential presentation. Journal of Ethology 21: 101–106.

Duvall, D., Graves, B. D. and Carpenter, G. C. 1987. Visual and chemical composite signalling effects of *Sceloporus* lizards fecal boli. Copeia 1987: 1028–1031.

Escobar, C. A., Labra, A. and Niemeyer, H. M. 2001. Chemical composition of precloacal secretions of *Liolaemus* lizards. Journal of Chemical Ecology 27: 1677–1690.

Escobar, C. M., Escobar, C. A., Labra, A. and Niemeyer, H. M. 2003. Chemical composition of precloacal secretions of two *Liolaemus fabiani* populations: are they different? Journal of Chemical Ecology 29: 629–638.

Ferguson, G. W., Gehrmann, W. H., Karsten, K. B., Landwer, A. J., Carman, E. N., Chen, T. C. and Holick, M. F. 2005. Ultraviolet exposure and vitamin D synthesis in a sun-dwelling and a shade-dwelling species of *Anolis*: are there adaptations for lower ultraviolet B and dietary vitamin D_3 availability in the shade? Physiological and Biochemical Zoology 78: 193–200.

Fergusson, B., Bradshaw, S. D. and Cannon, J. R. 1985. Hormonal control of femoral gland secretion in the lizard, *Amphibolurus ornatus*. General and Comparative Endocrinology 57: 371–376.

Flachsbarth, B., Fritzsche, M., Weldon, P. J. and Schulz, S. 2009. Composition of the cloacal gland secretion of tuatara, *Sphenodon punctatus*. Chemistry & Biodiversity 6: 1–37.

Folstad, I. and Karter, A. J. 1992. Parasites, bright males and the immunocompetence handicap. American Naturalist 139: 603–622.

Fraser, D. R. 1995. Vitamin D. Lancet 345: 104–107.

Gabe, M. and Saint-Girons, H. 1965. Contribution à la morphologie comparée du cloaque et des glandes épidermoides de la région cloacale chez le lépidosauriens. Memoires du Museum National d'Histoire Naturelle. Nouvelle Serie. Serie A. Zoologie 33: 149–292.

Gabirot, M., López, P., Martín, J., de Fraipont, M., Heulin, B., Sinervo, B. and Clobert, J. 2008. Chemical composition of femoral secretions of oviparous and viviparous types of male Common lizards *Lacerta vivipara*. Biochemical Systematics and Ecology 36: 539–544.

Gabirot, M., Castilla, A. M., López, P. and Martín, J. 2010a. Differences in chemical signals may explain species recognition between an island lizard, *Podarcis atrata*, and related mainland lizards, *P. hispanica*. Biochemical Systematics and Ecology 38: 521–528.

Gabirot, M., Castilla, A. M., López, P. and Martín, J. 2010b. Chemosensory species recognition may reduce the frequency of hybridization between native and introduced lizards. Canadian Journal of Zoology 88: 73–80.

Gabirot, M., López, P. and Martín, J. 2012a. Differences in chemical sexual signals may promote reproductive isolation and cryptic speciation between Iberian wall lizard populations. International Journal of Evolutionary Biology Article ID 698520.

Gabirot, M., López, P. and Martín, J. 2012b. Interpopulational variation in chemosensory responses to selected steroids from femoral secretions of male lizards, *Podarcis hispanica*, mirrors population differences in chemical signals. Chemoecology 22: 65–73.
Gabirot, M., López, P. and Martín, J. 2012c. Chemosensory responses to alcohols found in femoral gland secretions of male Iberian wall lizards *Podarcis hispanica*. Herpetological Journal 22: 139–145.
Gans, C., Gillingham, J. C. and Clark, D. L. 1984. Courtship, mating, and male combat in Tuatara, *Sphenodon punctatus*. Journal of Herpetology 18: 194–197.
Gonzalo, A., Cabido, C., Martín, J. and López, P. 2004. Detection and discrimination of conspecific scents by the anguid slow-worm *Anguis fragilis*. Journal of Chemical Ecology 30: 1565–1573.
Gosling, L. M. and McKay, H. V. 1990. Competitor assessment by scent matching: an experimental test. Behavioral Ecology and Sociobiology 26: 415–420.
Guilford, T. and Dawkins, M. S. 1991. Receiver psychology and the evolution of animal signals. Animal Behaviour 42: 1–14.
Halpern, M. 1992. Nasal chemical senses in reptiles: structure and function. pp. 423–523. In C. Gans and D. Crews (eds.), *Biology of the Reptilia, vol. 18*. University of Chicago Press, Chicago, Illinois.
Halpern, M. and Martínez-Marcos, A. 2003. Structure and function of the vomeronasal system: an update. Progress in Neurobiology 70: 245–318.
Hews, D. K. 1993. Food resources affect female distribution and male mating opportunities in the iguanian lizard *Uta palmeri*. Animal Behaviour 46: 279–291.
Houck, L. D. 2009. Pheromone communication in Amphibians and Reptiles. Annual Review of Physiology 71: 161–176.
Huyghe, K., Vanhooydonck, B., Herrel, A., Tadic, Z. and Van Damme, R. 2012. Female lizards ignore the sweet scent of success: Male characteristics implicated in female mate preference. Zoology 115: 217–222.
Jansson, N., Uller, T. and Olsson, M. 2005. Female dragons, *Ctenophorus pictus*, do not prefer scent from unrelated males. Australian Journal of Zoology 53: 279–282.
Johansson, B. G. and Jones, T. M. 2007. The role of chemical communication in mate choice. Biological Reviews 82: 265–289.
Johnston, R. E. 2005. Communication by mosaic signals: individual recognition and underlying neural mechanisms. pp. 269–282. In R. T. Mason, M. P. LeMaster and D. Müller-Schwarze (eds.), *Chemical Signals in Vertebrates, vol. 10*. Springer, New York.
Karlson, P. and Lüscher, M. 1959. Pheromones: a new term for a class of biologically active substances. Nature 183: 55–56.
Khannoon, E. R. 2012. Secretions of pre-anal glands of house-dwelling geckos (Family: Gekkonidae) contain monoglycerides and 1,3-alkanediol. A comparative chemical ecology study. Biochemical Systematics and Ecology 44: 341–346.
Khannoon, E. R., Breithaupt, T., El-Gendy, A. and Hardege, J. D. 2010. Sexual differences in behavioural response to femoral gland pheromones of *Acanthodactylus boskianus*. Herpetological Journal 20: 225–229.
Khannoon, E. R., Flachsbarth, B., El-Gendy, A., Mazik, J., Hardege, J. D. and Schulz, S. 2011a. New compounds, sexual differences, and age-related variations in the femoral gland secretions of the lacertid lizard *Acanthodactylus boskianus*. Biochemical Systematics and Ecology 39: 95–101.
Khannoon, E. R., El-Gendy, A. and Hardege, J. D. 2011b. Scent marking pheromones in lizards: cholesterol and long chain alcohols elicit avoidance and aggression in male *Acanthodactylus boskianus* (Squamata: Lacertidae). Chemoecology 21: 143–149.
Khannoon, E. R., Lunt, D. H., Schulz, S. and Hardege, J. D. 2013. Divergence of scent pheromones in allopatric populations of *Acanthodactylus boskianus* (Squamata: Lacertidae). Zoological Science 30: 380–385.

Kopena, R., López, P. and Martín, J. 2009. Lipophilic compounds from the femoral gland secretions of male Hungarian green lizards, *Lacerta viridis*. Zeitschrift für Naturforschung C 64: 434–440.
Kopena, R., Martín, J., López, P. and Herczeg, G. 2011. Vitamin E supplementation increases the attractiveness of males' scent for female European green lizards. PLoS One 6: e19410.
Labra, A. 2006. Chemoreception and the assessment of fighting abilities in the lizard *Liolaemus monticola*. Ethology 112: 993–999.
Labra, A. and Niemeyer, H. M. 1999. Intraspecific chemical recognition in the lizard *Liolaemus tenuis*. Journal of Chemical Ecology 25: 1799–1811.
Labra, A., Escobar, C. A., Aguilar, P. M. and Niemeyer, H. M. 2002. Sources of pheromones in the lizard *Liolaemus tenuis*. Revista Chilena de Historia Natural 75: 141–147.
López, P. and Martín, J. 2001a. Pheromonal recognition of females takes precedence over the chromatic cue in male Iberian wall lizards, *Podarcis hispanica*. Ethology 107: 901–912.
López, P. and Martín, J. 2001b. Fighting rules and rival recognition reduce costs of aggression in male lizards, *Podarcis hispanica*. Behavioral Ecology and Sociobiology 49: 111–116.
López, P. and Martín, J. 2002. Chemical rival recognition decreases aggression levels in male Iberian wall lizards, *Podarcis hispanica*. Behavioral Ecology and Sociobiology 51: 461–465.
López, P. and Martín, J. 2005a. Female Iberian wall lizards prefer male scents that signal a better cell-mediated immune response. Biology Letters 1: 404–406.
López, P. and Martín, J. 2005b. Intersexual differences in chemical composition of precloacal gland secretions of the amphisbaenian, *Blanus cinereus*. Journal of Chemical Ecology 31: 2913–2921.
López, P. and Martín, J. 2005c. Chemical compounds from femoral gland secretions of male Iberian rock lizards, *Lacerta monticola cyreni*. Zeitschrift für Naturforschung C 60: 632–636.
López, P. and Martín, J. 2005d. Age related differences in lipophilic compounds found in femoral gland secretions of male spiny-footed lizards, *Acanthodactylus erythrurus*. Zeitschrift für Naturforschung C 60: 915–920.
López, P. and Martín, J. 2006. Lipids in the femoral gland secretions of male Schreiber's green lizards, *Lacerta schreiberi*. Zeitschrift für Naturforschung C 61: 763–768.
López, P. and Martín, J. 2009a. Lipids in femoral gland secretions of male lizards, *Psammodromus hispanicus*. Biochemical Systematics and Ecology 37: 304–307.
López, P. and Martín, J. 2009b. Potential chemosignals associated with male identity in the amphisbaenian *Blanus cinereus*. Chemical Senses 34: 479–486.
López, P. and Martín, J. 2011. Male iberian rock lizards may reduce the costs of fighting by scent-matching of the resource holders. Behavioral Ecology and Sociobiology 65: 1891–1898.
López, P. and Martín, J. 2012. Chemosensory exploration of male scent by female rock lizards result from multiple chemical signals of males. Chemical Senses 37: 47–54.
López, P., Aragón, P. and Martín, J. 1998. Iberian rock lizards (*Lacerta monticola cyreni*) assess conspecific information using composite signals from faecal pellets. Ethology 104: 809–820.
López, P., Martín, J. and Cuadrado, M. 2002a. Pheromone mediated intrasexual aggression in male lizards, *Podarcis hispanicus*. Aggressive Behavior 28: 154–163.
López, P., Muñoz, A. and Martín, J. 2002b. Symmetry, male dominance and female mate preferences in the Iberian rock lizard, *Lacerta monticola*. Behavioral Ecology and Sociobiology 52: 342–347.
López, P., Aragón, P. and Martín, J. 2003a. Responses of female lizards, *Lacerta monticola*, to males' chemical cues reflect their mating preference for older males. Behavioral Ecology and Sociobiology 55: 73–79.
López, P., Martín, J. and Cuadrado, M. 2003b. Chemosensory cues allow male lizards *Psammodromus algirus* to override visual concealment of sexual identity by satellite males. Behavioral Ecology and Sociobiology 54: 218–224.
López, P., Amo, L. and Martín, J. 2006. Reliable signaling by chemical cues of male traits and health state in male lizards, *Lacerta monticola*. Journal of Chemical Ecology 32: 473–488.

López, P., Moreira, P. L. and Martín, J. 2009a. Chemical polymorphism and chemosensory recognition between *Iberolacerta monticola* lizard color morphs. Chemical Senses 34: 723–731.
López, P., Gabirot, M. and Martín, J. 2009b. Immune activation affects chemical sexual ornaments of male Iberian wall lizards. Naturwissenschaften 96: 65–69.
Louw, S., Burger, B. V., Le Roux, M. and Van Wyk, J. H. 2007. Lizard epidermal gland secretions I: chemical characterization of the femoral gland secretion of the sungazer, *Cordylus giganteus*. Journal of Chemical Ecology 33: 1806–1818.
Louw, S., Burger, B. V., Le Roux, M. and Van Wyk, J. H. 2011. Lizard epidermal gland secretions II. Chemical characterization of the generation gland secretion of the sungazer, *Cordylus giganteus*. Journal of Natural Products 74: 1364–1369.
Martín, J. and López, P. 2000. Chemoreception, symmetry and mate choice in lizards. Proceedings of the Royal Society of London. Series B 267: 1265–1269.
Martín, J. and López, P. 2006a. Links between male quality, male chemical signals, and female mate choice in Iberian rock lizards. Functional Ecology 20: 1087–1096.
Martín, J. and López, P. 2006b. Vitamin D supplementation increases the attractiveness of males' scent for female Iberian rock lizards. Proceedings of the Royal Society of London. Series B 273: 2619–2624.
Martín, J. and López, P. 2006c. Interpopulational differences in chemical composition and chemosensory recognition of femoral gland secretions of male lizards *Podarcis hispanica*: implications for sexual isolation in a species complex. Chemoecology 16: 31–38.
Martín, J. and López, P. 2006d. Age-related variation in lipophilic chemical compounds from femoral gland secretions of male lizards *Psammodromus algirus*. Biochemical Systematics and Ecology 34: 691–697.
Martín, J. and López, P. 2006e. Chemosensory responses by female iberian wall lizards, *Podarcis hispanica* to selected lipids found in femoral gland secretions of males. Journal of Herpetology 40: 556–561.
Martín, J. and López, P. 2007. Scent may signal fighting ability in male Iberian rock lizards. Biology Letters 3: 125–127.
Martín, J. and López, P. 2008a. Intersexual differences in chemosensory responses to selected lipids reveal different messages conveyed by femoral secretions of male Iberian rock lizards. Amphibia-Reptilia 29: 572–578.
Martín, J. and López, P. 2008b. Female sensory bias may allow honest chemical signaling by male Iberian rock lizards. Behavioral Ecology and Sociobiology 62: 1927–1934.
Martín, J. and López, P. 2010a. Multimodal sexual signals in male ocellated lizards *Lacerta lepida*: vitamin E in scent and green coloration may signal male quality in different sensory channels. Naturwissenschaften 97: 545–553.
Martín, J. and López, P. 2010b. Condition-dependent pheromone signalling by male rock lizards: more oily scents are more attractive. Chemical Senses 35: 253–262.
Martín, J. and López, P. 2011. Pheromones and reproduction in Reptiles. pp. 141–167. In D. O. Norris and K. H. Lopez (eds.), *Hormones and Reproduction in Vertebrates, vol. 3. Reptiles*. Academic Press, San Diego, California.
Martín, J. and López, P. 2012. Supplementation of male pheromone on rock substrates attracts female rock lizards to the territories of males: a field experiment. PLoS One 7: e30108.
Martín, J. and López, P. 2013a. Responses of female rock lizards to multiple scent marks of males: effects of male age, male density and scent over-marking. Behavioural Processes 94: 109–114.
Martín, J. and López, P. 2013b. Effects of global warming on sensory ecology of rock lizards: increased temperatures alter the efficacy of sexual chemical signals. Functional Ecology 27: 1332–1340.
Martín, J., López, P., Gabirot, M. and Pilz, K. M. 2007a. Effects of testosterone supplementation on chemical signals of male Iberian wall lizards: consequences for female mate choice. Behavioral Ecology and Sociobiology 61: 1275–1285.

Martín, J., Civantos, E., Amo, L. and López, P. 2007b. Chemical ornaments of male lizards *Psammodromus algirus* may reveal their parasite load and health state to females. Behavioral Ecology and Sociobiology 62: 173–179.

Martín, J., Moreira, P. L. and López, P. 2007c. Status-signalling chemical badges in male Iberian rock lizards. Functional Ecology 21: 568–576.

Martín, J., Amo, L. and López, P. 2008. Parasites and health affect multiple sexual signals in male common wall lizards, *Podarcis muralis*. Naturwissenschaften 95: 293–300.

Martín, J., Chamut, S., Manes, M. E. and López, P. 2011. Chemical constituents of the femoral gland secretions of male tegu lizards (*Tupinambis merianae*) (fam. Teiidae). Zeitschrift für Naturforschung C 66: 434–440.

Martín, J., Castilla, A. M., Lopez, P., Al Jaidah, M. and Mohtar, R. 2012. Lipophilic compounds in femoral gland secretions of spiny-tailed lizard, dhub, *Uromastyx aegyptia microlepis* (Reptilia, Agamidae) from the Qatar desert. Qatar Foundation Annual Research Forum EEP53.

Martín, J., López, P., Garrido, M., Pérez-Cembranos, A. and Pérez-Mellado, V. 2013a. Inter-island variation in femoral secretions of the Balearic lizard, *Podarcis lilfordi* (Lacertidae). Biochemical Systematics and Ecology 50: 121–128.

Martín, J., Ortega, J. and López, P. 2013b. Lipophilic compounds in femoral secretions of male collared lizards, *Crotaphytus bicinctores* (Iguania, Crotaphytidae). Biochemical Systematics and Ecology 47: 5–10.

Martín, J., Ortega, J. and López, P. 2013c. Chemical compounds from the preanal gland secretions of the male tree agama (*Acanthocercus atricollis*) (fam. Agamidae). Zeitschrift für Naturforschung C 68: 253–258.

Martins, E. P., Ord, T. J., Slaven, J., Wright, J. L. and Housworth, E. A. 2006. Individual, sexual, seasonal, and temporal variation in the amount of sagebrush lizard scent marks. Journal of Chemical Ecology 32: 881–893.

Mason, R. T. 1992. Reptilian pheromones. pp. 114–228. In C. Gans and D. Crews (eds.), *Biology of the Reptilia, vol. 18*. University of Chicago Press, Chicago, Illinois.

Mason, R. T. and Gutzke, W. H. N. 1990. Sex recognition in the leopard gecko, *Eublepharis macularius* (Sauria: Gekkonidae). Possible mediation by skin-derived semiochemicals. Journal of Chemical Ecology 16: 27–36.

Mason, R. T. and Parker, M. R. 2010. Social behavior and pheromonal communication in reptiles. Journal of Comparative Physiology A 196: 729–749.

Mason, R. T., Jones, T., Fales, H., Pannell, L. and Crews, D. 1990. Characterization, synthesis, and behavioral responses to sex attractiveness pheromones of red-sided garter snakes *(Thamnophis sirtalis parietalís)*. Journal of Chemical Ecology 16: 2353–2369.

Moore, M. C. and Lindzey, J. 1992. The physiological basis of sexual behavior in male reptiles. pp. 70–113. In C. Gans and D. Crews (eds.), *Biology of the Reptilia, vol. 18*. University of Chicago Press, Chicago, Illinois.

Moreira, P. L., López, P. and Martín, J. 2006. Femoral secretions and copulatory plugs convey chemical information about male identity and dominance status in Iberian rock lizards (*Lacerta monticola*). Behavioral Ecology and Sociobiology 60: 166–174.

Moreira, P. L., López, P. and Martín, J. 2008. Discrimination of conspecific fecal chemicals and spatial decisions in juvenile Iberian rock lizards (*Lacerta monticola*). Acta Ethologica 11: 26–33.

Müller-Schwarze, D. 2006. *Chemical Ecology of Vertebrates*. Cambridge University Press, Cambridge, pp. 563.

Olsson, M. and Madsen, T. 1995. Female choice on male quantitative traits in lizards—why is it so rare? Behavioral Ecology and Sociobiology 36: 179–184.

Olsson, M., Madsen, T., Nordby, J., Wapstra, E., Ujvari, B. and Wittsell, H. 2003. Major histocompatibility complex and mate choice in sand lizards. Proceedings of the Royal Society of London. Series B (Supplement) 270: S254–S256.

Oppliger, A., Giorgi, M. S., Conelli, A., Nembrini, M. and John-Alder, H. B. 2004. Effect of testosterone on immunocompetence, parasite load, and metabolism in the common wall lizard (*Podarcis muralis*). Canadian Journal of Zoology 82: 1713–1719.

Penn, D. J. and Potts, W. K. 1998. Chemical signals and parasite mediated sexual selection. Trends in Ecology and Evolution 13: 391–396.

Roberts, J. B. and Lillywhite, H. B. 1980. Lipid barrier to water exchange in reptile epidermis. Science 207: 1077–1079.

Runemark, A., Gabirot, M. and Svensson, E. I. 2011. Population divergence in chemical signals and the potential for premating isolation between islet- and mainland populations of the Skyros wall lizard (*Podarcis gaigeae*). Journal of Evolutionary Biology 24: 795–809.

Schwenk, K. 1995. Of tongues and noses: chemoreception in lizards and snakes. Trends in Ecology and Evolution 10: 7–12.

Sheridan, M. A. 1994. Regulation of lipid-metabolism in poikilothermic vertebrates. Comparative Biochemistry and Physiology B 107: 495–508.

Trauth, S. E., Cooper, W. E., Jr., Vitt, L. J. and Perrill, S. A. 1987. Cloacal anatomy of the broad-headed skink, *Eumeces laticeps*, with a description of a female pheromone gland. Herpetologica 43: 458–466.

Ubeda-Banon, I., Pro-Sistiaga, P., Mohedano-Moriano, A., Saiz-Sanchez, D., de la Rosa-Prieto, C., Gutierrez-Castellanos, N., Lanuza, E., Martinez-Garcia, F. and Martinez-Marcos, A. 2011. Cladistic analysis of olfactory and vomeronasal systems. Frontiers In Neuroanatomy 5: 3. doi 10.3389/fnana.2011.00003.

Van Wyk, J. H. and Mouton, P. F. N. 1992. Glandular epidermal structures of cordylid lizards. Amphibia-Reptilia 13: 1–12.

Weldon, P. J. and Bangall, D. 1987. A survey of polar and nonpolar skin lipids from lizards by thin-layer chromatography. Comparative Biochemestry and Physiology B 87: 345–349.

Weldon, P. J., Dunn, B. S., McDaniel, C. A. and Werner, D. I. 1990. Lipids in the femoral gland secretions of the green iguana (*Iguana iguana*). Comparative Biochemestry and Physiology B 95: 541–543.

Weldon, P. J., Flachsbarth, B. and Schulz, S. 2008. Natural products from the integument of nonavian reptiles. Natural Products Report 25: 738–756.

Wilgers, D. J. and Horne, E. A. 2009. Discrimination of chemical stimuli in conspecific fecal pellets by a visually adept iguanid lizard, *Crotaphytus collaris*. Journal of Ethology 27: 157–163.

Whiting, M. J., Webb, J. K. and Keogh, J. S. 2009. Flat lizard female mimics use sexual deception in visual but not chemical signals. Proceedings of the Royal Society of London. Series B 276: 1585–1591.

Wyatt, T. D. 2003. *Pheromones and Animal Behaviour*. Cambridge University Press, Cambridge, pp. 391.

Wyatt, T. D. 2010. Pheromones and signature mixtures: defining species-wide signals and variable cues for identity in both invertebrates and vertebrates. Journal of Comparative Physiology A 196: 685–700.

CHAPTER 4

Sexual Selection and Sexual Dimorphism

*Robert M. Cox** and *Ariel F. Kahrl*

4.1 INTRODUCTION

Sexual selection arises from variance in mating success that occurs due to intrasexual (male-male or female-female competition) and intersexual mechanisms (mate choice), as well as from postcopulatory processes (sperm competition, cryptic female choice) that influence fertilization success. Lizards figure prominently in historical and current research in each of these areas, and much of this work focuses on the role of sexual selection in generating the dramatic sex differences in size, shape, coloration, behavior, physiology, and life history that characterize this group. Sexual dimorphism has also been studied with respect to selection arising from variance in fecundity, and though some definitions of sexual selection include fecundity selection (Cornwallis and Uller 2010), we retain the traditional Darwinian distinction between these terms (Darwin 1871; Andersson 1994; Fairbairn *et al.* 2007). In this chapter, we briefly review the major patterns of sexual selection in lizards and tuatara, their roles in shaping sexual dimorphism, and the evolutionary dynamics that ensue. Our goal is not to thoroughly review the vast literature on these topics, but to provide a brief overview of the state of the field while highlighting several exiting new directions in which studies of lizards are advancing evolutionary theory. In particular, we focus on lizards as models for the study of intralocus sexual conflict, alternative reproductive tactics, and speciation.

Department of Biology, University of Virginia, PO Box 400328, Charlottesville, VA 22904 (USA).
* Corresponding author

4.2 SEXUAL SELECTION IN LIZARDS

As an empirical backdrop for this review, we compiled a dataset of sexual selection gradients from wild lizard populations, including only those reports in which standard procedures for selection-gradient analyses were followed (Lande and Arnold 1983; Arnold and Wade 1984). Despite the popularity of lizards for field studies of behavioral ecology and demography, surprisingly few quantitative estimates of phenotypic sexual selection are available for wild populations (e.g., data from only 11 species in Table 4.1). In part, this reflects the historical difficulty of measuring reproductive success prior to the availability of molecular markers for paternity (Wapstra and Olsson Chapter 14, this volume), though Trivers' (1976) study of mating success in the Jamaican anole (*Anolis garmani*) still stands as one of the classic field studies of sexual selection in any taxon. In this and other species, estimating fitness from behavioral observations of mating success reveals a common pattern of strong sexual selection favoring large male body size (Arnold and Wade 1984; Hews 1990; Wikelski and Trillmich 1997; Olsson *et al.* 2002). This same pattern of selection for large male body size occurs in more recent studies using genetic paternity analysis to quantify reproductive success, which should also include any potential effects of postcopulatory selection in mitigating or enhancing precopulatory selection (Abell 1997; John-Alder *et al.* 2009; Olsson *et al.* 2011; Noble *et al.* 2013). In fact, we only found one negative selection gradient reported for sexual selection on body size of male lizards (Fig. 4.1; Table 4.1), and the median strength of sexual selection on body size in male lizards (0.52) is noticeably higher than the median strength of sexual selection on body size across all other organisms (0.13) (Kingsolver and Pfennig 2004, 2007). Moreover, several additional studies in lizards and tuatara provided evidence for sexual selection favoring large size, but could not be included in our quantitative summary because they did not calculate standardized selection gradients (Hofmann and Henle 2006; Salvador *et al.* 2008; Miller *et al.* 2009).

Sexual selection also tends to strongly favor large head and jaw morphology, and large territories or home ranges (Fig. 4.1). The median strength of sexual selection across all phenotypic traits in lizards (0.34) is nearly double the median strength of sexual selection reported across diverse traits in other organisms (0.18) (Kingsolver *et al.* 2001). Though these patterns may be driven in part by publication bias and spuriously high estimates of selection resulting from relatively low samples sizes in some cases, strong sexual selection is nonetheless observed in studies involving hundreds or even thousands of individual male lizards (Olsson *et al.* 2002; Fitze and Le Galliard 2008; Olsson *et al.* 2011). Hence, available data indicate that sexual selection is a powerful evolutionary force in male lizards. Competition for mates may also favor the largest females, as observed in

Table 4.1. Standardized sexual selection gradients (β or *i*) on male phenotypes in wild lizard populations. Fitness measures are classified as behavioral (B) or genetic (G), the latter indicating that reproductive success was assessed by genetic paternity analysis.

Species	Fitness Measure	Phenotypic Trait	β or i	n	Reference
Amblyrhynchus cristatus	copulatory success (B)	snout-vent length	0.77	147	Wikelski and Trillmich (1997)
Amblyrhynchus cristatus	copulatory success (B)	snout-vent length	0.42	343	Wikelski and Trillmich (1997)
Anolis garmani	copulations (B)	snout-vent length	0.63	523	Trivers (1976)
Eulamprus quoyii	offspring sired (G)	snout-vent length	1.30	49	Noble *et al.* (2013)
Eulamprus quoyii	offspring sired (G)	home range	0.44	49	Noble *et al.* (2013)
Eulamprus quoyii	offspring sired (G)	days active	0.46	49	Noble *et al.* (2013)
Eulamprus quoyii	offspring sired (G)	time spent moving	0.13	49	Noble *et al.* (2013)
Lacerta agilis	offspring sired (G)	snout-vent length	0.51	2251	Olsson *et al.* (2011)
Lacerta vivipara	number of mates (B)	snout-vent length	0.12 to 0.32	225	Fitze *et al.* (2008)
Niveoscincus microlepidotus	predicted offspring (B)	snout-vent length	0.38	381	Olsson *et al.* (2002)
Niveoscincus microlepidotus	predicted offspring (B)	interlimb length	−0.19	381	Olsson *et al.* (2002)
Niveoscincus microlepidotus	predicted offspring (B)	head length	−0.19	381	Olsson *et al.* (2002)
Podarcis gaigeae	mating probability (B)	body size (PC1)	−0.17	78	Runemark and Svensson (2012)
Sauromalus obesus	females in territory (B)	home-range area	−0.10 to 0.34	33	Kwiatkowski and Sullivan (2002)
Sauromalus obesus	females in territory (B)	territory quality	0.44 to 0.81	33	Kwiatkowski and Sullivan (2002)
Sauromalus obesus	females in territory (B)	snout-vent length	0.24 to 0.57	33	Kwiatkowski and Sullivan (2002)
Sauromalus obesus	females in territory (B)	tail length	0.07 to 0.27	33	Kwiatkowski and Sullivan (2002)
Sauromalus obesus	females in territory (B)	jaw length	0.14 to 0.75	33	Kwiatkowski and Sullivan (2002)
Sauromalus obesus	females in territory (B)	head width	0.13 to 0.46	33	Kwiatkowski and Sullivan (2002)
Sauromalus obesus	females in territory (B)	head depth	−0.17 to 0.28	33	Kwiatkowski and Sullivan (2002)
Sauromalus obesus	females in territory (B)	body mass	0.07 to 0.55	33	Kwiatkowski and Sullivan (2002)
Sauromalus obesus	females in territory (B)	color saturation	0.32 to 0.41	33	Kwiatkowski and Sullivan (2002)
Sauromalus obesus	females in territory (B)	color brightness	0.17 to 0.58	33	Kwiatkowski and Sullivan (2002)

Sauromalus obesus	females in territory (B)	PC1	0.08 to 0.74	33	Kwiatkowski and Sullivan (2002)
Sauromalus obesus	females in territory (B)	PC2	0.02 to 0.19	33	Kwiatkowski and Sullivan (2002)
Sauromalus obesus	females in territory (B)	PC3	0.03 to 0.98	33	Kwiatkowski and Sullivan (2002)
Sceloporus undulatus	offspring sired (G)	snout-vent length	0.29	36	John-Alder *et al.* (2009)
Sceloporus undulatus	offspring sired (G)	endurance	0.24	36	John-Alder *et al.* (2009)
Sceloporus undulatus	offspring sired (G)	home-range area	0.56	36	John-Alder *et al.* (2009)
Sceloporus undulatus	offspring sired (G)	plasma testosterone	0.39	36	John-Alder *et al.* (2009)
Sceloporus undulatus	offspring sired (G)	plasma corticosterone	0.32	36	John-Alder *et al.* (2009)
Sceloporus virgatus	proximity to female (B)	body size (PC1)	0.08 to 0.73	55	Abell (1997)
Sceloporus virgatus	proximity to female (B)	blue patch size (PC2)	0.03 to 0.22	55	Abell (1997)
Sceloporus virgatus	proximity to female (B)	Color intensity (PC3)	−0.15 to −0.22	55	Abell (1997)
Uta palmeri	copulations (B)	snout-vent length	0.55	33	Hews (1990)
Uta palmeri	copulations (B)	body mass	0.75	33	Hews (1990)
Uta palmeri	copulations (B)	head width	0.29	33	Hews (1990)
Uta palmeri	copulations (B)	jaw length	0.33	33	Hews (1990)
Uta palmeri	copulations (B)	head depth	0.61	33	Hews (1990)
Uta palmeri	copulations (B)	territory quality	0.74	33	Hews (1990)
Uta palmeri	copulations (B)	PC1	−0.08	33	Hews (1990)
Uta palmeri	copulations (B)	PC2	1.24	33	Hews (1990)
Uta palmeri	copulations (B)	PC3	1.08	33	Hews (1990)

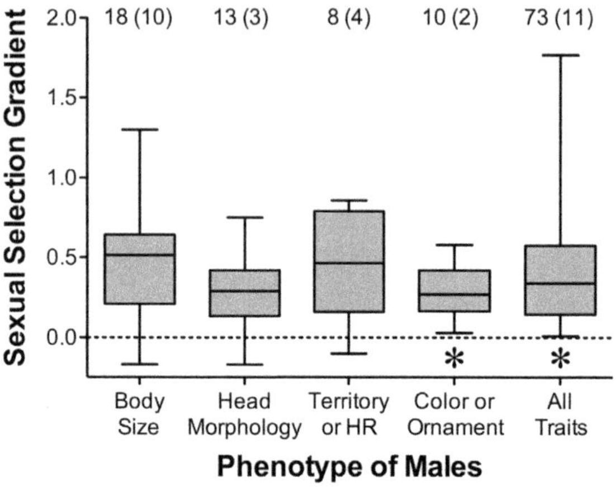

Fig. 4.1 Summary of standardized selection gradients for sexual selection acting on body size, head morphology (jaw length, head width or depth), territory or home range (HR) size or quality, coloration or ornament size, and all phenotypic traits in male lizards. Selection gradients in the first three columns (body size, head morphology, territory or HR) can take positive (favoring larger phenotypes) or negative values (favoring smaller phenotypes). Asterisks indicate that selection gradients in the final two columns (color or ornament, all traits) are expressed as absolute values to estimate the average strength of selection irrespective of directionality, which is uninformative for some traits (e.g., components of color such as brightness or saturation). Box-and-whisker plots show the median, 25–75 percentiles, and range of selection estimates. Numbers atop each column indicate number of selection estimates (number of species).

several species (Trivers 1976; Fitze and Le Galliard 2008). However, it is not clear whether this actually translates into variance in reproductive success, which is more likely determined by fecundity. Fecundity selection for large body size is probably ubiquitous across species with variable clutch or litter sizes, and perhaps even in species with fixed one- or two-egg clutches (Cox and Calsbeek 2011), though most published data simply report slopes from clutch or litter size regressed on female body size without standardizing these values to conduct formal selection-gradient analyses (but see Olsson *et al.* 2002).

4.2.1 Intrasexual Selection

The selection gradients presented above do not empirically distinguish between intra- and intersexual selection, but it is likely that most reflect a substantial contribution of intrasexual selection in the form of male-male competition, which has long been a focus of research on lizards (Stamps 1983). In both territorial and non-territorial species, male-male competition commonly takes the form of stereotyped behavioral displays (push-ups, head-bobs, full-shows, gaping, dewlap extensions) that can escalate into wrestling and biting. To assess the generality of this type of intrasexual selection and identify its phenotypic targets, we reviewed the literature for evidence of the role of various male traits as predictors of dominance in social interactions, success in acquiring territories, or increased access to mates (Fig. 4.2A). In general, male body size is a strong predictor of success in these aspects of intrasexual competition (Fig. 4.2A), as are bite force (Huyghe *et al.* 2005; Lappin and Husak 2005; Husak *et al.* 2006) and other performance traits, such as endurance or sprint speed (Garland *et al.* 1990; Robson and Miles 2000; Perry *et al.* 2004). These morphological and performance traits may interact to increase male territory size, which predicts access to females in many species (Robson and Miles 2000; John-Alder *et al.* 2009). Not surprisingly, body size, head and jaw morphology, and territory size are the same traits we found to be under strong sexual selection in lizards (Fig. 4.1; Table 4.1). Body size and related traits may also mediate dominance in female-female competition (While *et al.* 2009), though scant empirical data are available to assess whether this actually generates sexual selection through variance in reproductive success, or whether it primarily results in natural selection through variance in resource acquisition and subsequent survival or fecundity.

Traits such as body size, jaw morphology, bite force, and locomotor performance have intuitive links to dominance and mate acquisition, but the considerable costs associated with overt combat should favor the evolution of signals that convey dominance or fighting ability to preempt unnecessary physical altercations (López and Martín 2001). Consistent with this idea, we found broad support for the roles of coloration, social displays, and chemosensory cues in mediating the outcome of male-male competition (Fig. 4.2A). Head or throat color signals dominance and predicts success in male-male contests in color-polymorphic species (Thompson and Moore 1991; Whiting *et al.* 2006; Healey *et al.* 2007), and continuous variation in coloration, dewlap size, or display behaviors can serve a similar function (Huyghe *et al.* 2005; Lailvaux and Irschick 2007; Martín and López 2009). In species that rely heavily on sensory modalities other than vision, dominance may be signaled by chemical cues (Martín and López 2007; Martin *et al.* 2007; see also Martin and Lopez Chapter 3, this volume), or even by vocalizations, as in barking geckos, *Ptenoptus garrulus* (Hibbitts *et al.* 2007).

4.2.2 Intersexual Selection

Although male-male competition has long been viewed as an important form of sexual selection in lizards, the historical view of female choice is that it is rare (or rarely documented) in lizards (Olsson and Madsen 1995; Tokarz

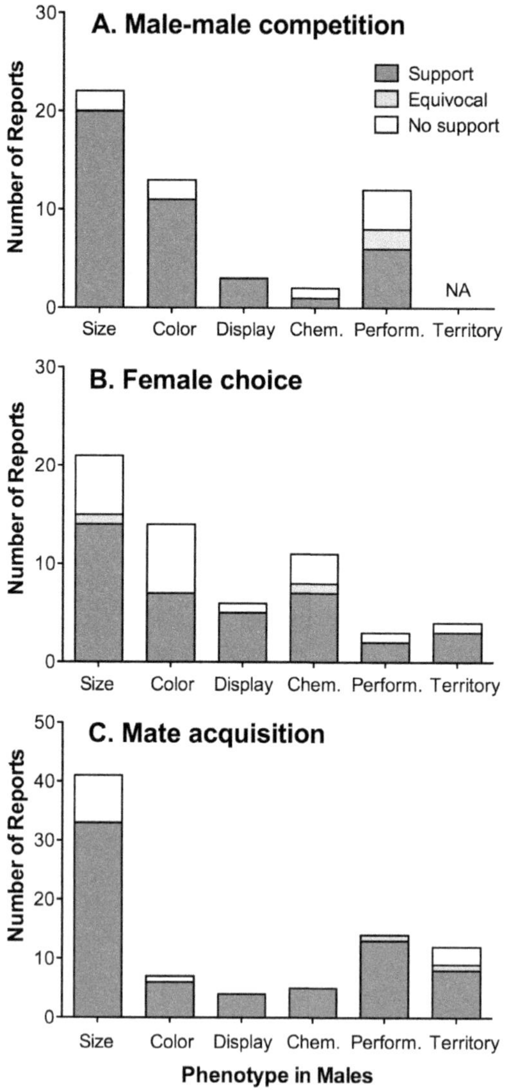

Fig. 4.2 Summary of reported evidence for sexually selected traits in male lizards. Studies are categorized according to whether they provide support for six classes of phenotypic trait as **(A)** predictors of dominance in male-male competition, **(B)** targets of female choice, and **(C)** correlates of mate acquisition or reproductive success, which could reflect joint combinations of intra- and intersexual selection. Chem. = chemical cues or major histocompatibility complex (MHC) haplotypes. Perform. = performance (bite force, locomotor performance, and so forth).

1995). This may reflect a low potential for direct (non-genetic) benefits due to the absence of paternal investment, an inherent difficulty in assessing indirect (genetic) benefits due to unreliable phenotypic indicators of genetic quality (Olsson and Madsen 1995), or a tendency for female choice to be context-dependent (Fitze *et al.* 2010). However, our review of the current evidence for female preference suggests that it may be more prevalent than previous evidence suggested (Fig. 4.2B). Females preferentially visit or copulate with larger males in many species (Cooper and Vitt 1993; Censky 1997; Wikelski *et al.* 2001; López *et al.* 2002), though females exhibit little preference for body size in others (Andrews 1985; Baird *et al.* 1997; Olsson 2001; Stuart-Smith *et al.* 2007). Female preferences for colorful display traits, such as dewlaps and badges, have proven more difficult to detect (Tokarz 1995; Smith and Zucker 1997; Lebas and Marshall 2001; Olsson 2001; Tokarz 2002; Tokarz *et al.* 2005), but do occur in some species (Baird *et al.* 1997; Bajer *et al.* 2010; Runemark and Svensson 2012). Studies of tree lizards (*Urosaurus ornatus*) illustrate the caveat that univariate analyses may often fail to uncover complex interactions among size, coloration, and display behaviors that females use as the basis for mate choice (Hamilton and Sullivan 2005). Studies of mate choice by males are less common, but reveal preferences for novel females (Tokarz 1992; Orrell and Jenssen 2002), large or fecund females (Olsson 1993; Whiting and Bateman 1999; Wymann and Whiting 2003; John-Alder *et al.* 2009), and coloration indicating quality or receptivity (Watkins 1997; LeBas and Marshall 2000; Weiss 2002; Baird 2004; Weiss *et al.* 2009).

A major source of ongoing debate in evolutionary biology is whether adaptive mate choice is driven by direct benefits (non-genetic factors such as territory quality), indirect benefits (good genes), or genetic compatibility (avoidance of inbreeding or hybridization). Lizards provide support for each scenario. When the spatial distribution of territory quality is altered in populations of side-blotched lizards (*Uta stansburiana*), females relocate to experimentally improved territories that provide direct benefits by enhancing offspring growth and survival but they also retain preferences for individual males that provide indirect genetic benefits to progeny (Calsbeek and Sinervo 2002). Female Iberian wall lizards (*Podarcis hispanicus*) and rock lizards (*Iberolacerta monticola*) select mates based on the chemical composition of their femoral pore secretions (Martín and López 2000), and these scent profiles may signal good genes in the form of better cell-mediated immune response and improved health (López and Martín 2005; López *et al.* 2006; Martin and Lopez 2006; Martin and Lopez Chapter 3, this volume). By contrast, female sand lizards (*Lacerta agilis*) from inbred populations appear to select males with dissimilar major histocompatibility complex (MHC) genotypes that improve genetic compatibility by reducing inbreeding (Olsson *et al.* 2003), though this may also reflect good-genes benefits (Olsson *et al.* 2005). Cunningham's skinks (*Egernia cunninghami*) in highly fragmented habitats may avoid inbreeding depression via mating preferences for unrelated individuals (Stow and Sunnucks 2004), but female painted and ornate dragons (*Ctenophorus pictus* and *C. ornatus*) exhibit no clear preference for genetic similarity or dissimilarity (Jansson *et al.* 2005), even in inbred populations (Lebas 2001). Female side-blotched lizards (*Uta stansburiana*) prefer males of the same genetic color morph during the production of their first clutch, though condition-dependent reversals to preferences for dissimilar males can occur for later clutches (Bleay and Sinervo 2007). Dynamic patterns of mate choice with respect to genetic similarity also occur in the common lizard, *Lacerta vivipara* (Richard *et al.* 2009). Thus, as in other organisms, the precise benefits that females obtain from mate choice appear to vary among lizard species.

4.2.3 Postcopulatory Selection

In the past two decades, studies of sexual selection in lizards have increasingly focused on postcopulatory selection, which occurs when male ejaculates compete for fertilization (sperm competition) or when females bias paternity (cryptic female choice) after copulation. The opportunity for postcopulatory selection is likely high in most lizards due to the scarcity of true monogamy (Bull 2000) and the apparent ubiquity of both sperm storage (Sever and Hamlett 2002) and multiple paternity (Uller and Olsson 2008). Though specialized sperm storage structures have not been described in tuatara (Sever and Hamlett 2002), viable progeny are produced up to ten months after mating (Holt and Lloyd 2010) and multiple paternity occurs at low levels (8–19%) in wild and captive-bred populations, despite the prevalence of social monogamy (Moore *et al.* 2008; Moore *et al.* 2009). Across lizard taxa, the time over which viable sperm can be stored ranges from several weeks to over a year (Birkhead and Moller 1993; Holt and Lloyd 2010), and the incidence of multiple paternity ranges from 2% to 87% of clutches or litters, frequently exceeding 50% (Uller and Olsson 2008; Wapstra and Olsson Chapter 14, this volume).

Potential adaptations driven by postcopulatory selection include copulatory plugs, mate guarding, increased testis size, and altered sperm morphology. Though many lacertids produce sperm plugs, those of *Iberolacerta monticola* have no clear "defensive" role in preventing fertilization by subsequent males (Moreira and Birkhead 2003, 2004), but may instead act as "offensive" mechanisms that displace prior ejaculates (Moreira *et al.* 2007). The sperm masses that occur in *Anolis* (Conner and Crews 1980; A. F. Kahrl, pers. obs.), *Psammophilus* (Srinivas *et al.* 1995), and other genera may primarily safeguard against sperm leakage, rather than preventing remating by females, but this has yet to be tested (Olsson and Madsen 1998). Mate guarding has been described in many lizards, often as a tactic that varies with the age, size or reproductive strategy of the male, or with the size, fecundity, or reproductive history of the female (Cuadrado 1998). Testis size fluctuates dramatically in seasonally breeding lizards (Mendez de la Cruz Chapter 10, this volume) but is characteristically larger (relative to body size) for species at high latitudes, presumably reflecting a greater intensity of sperm competition during shorter reproductive seasons (Olsson and Madsen 1998; Uller *et al.* 2010). Relatively little is known about the functional morphology of lizard sperm (Uller *et al.* 2010; Gribbins and Rheubert Chapter 11, this volume), though our recent work indicates significant variation in morphology of *Anolis* sperm occurs among individuals, among species, and in response to dietary manipulation. Fertilization success in *Anolis sagrei* improves with male body condition and females exhibit a strong first-male fertilization bias in sequential matings (Duryea *et al.* 2013), but the mechanisms that underlie these patterns are unknown.

Postcopulatory selection can also lead to complex evolutionary dynamics, often through interactions with precopulatory selection. In polymorphic male side-blotched lizards (*Uta stansburiana*) and painted dragons (*Ctenophorus pictus*), morphs specialized for sperm competition can persist as evolutionarily stable alternatives to strategies such as mate guarding and territorial-defense polygyny (Sinervo and Lively 1996; Zamudio and Sinervo 2000; Olsson *et al.* 2007). Multiple paternity in *U. stansburiana* also facilitates another form of postcopulatory selection in which females are more likely to produce sons with sperm from large males, but daughters with sperm from small males (Calsbeek and Sinervo 2004). Similar biases in progeny sex allocation as a function of paternal size and body condition occur in both captive and free-living populations of *Anolis sagrei* (Calsbeek and Bonneaud 2008; Cox and Calsbeek 2010a; Cox *et al.* 2011), in which long-term sperm storage results in high levels of multiple paternity (80%) across successive, single-egg clutches (Calsbeek *et al.* 2007). It is unknown whether this "cryptic sex-ratio bias" arises from sperm competition, cryptic female choice, or a combination of these processes, but it appears to provide indirect genetic benefits by increasing progeny viability in both *Uta* and *Anolis* (Calsbeek *et al.* 2007; Cox and Calsbeek 2010a).

4.3 SEXUAL DIMORPHISM IN LIZARDS

Lizards have historically served as important models for the evolution of sexual dimorphism (Schoener 1967; Trivers 1976; Stamps 1983), providing an ideal group in which to integrate broad phylogenetic patterns and population-level selection analyses with mechanistic studies of the underlying physiology and genetics of sexual differentiation (Hews and Quinn 2003; Cox 2005; Cox *et al.* 2007; Cox *et al.* 2009). Studies of sexual size dimorphism (SSD) dominate this literature, though studies of sex differences in shape (Cooper and Vitt 1989; Braña 1996; Sanger *et al.* 2013), coloration (Wiens 1999; Wiens *et al.* 1999; Cox *et al.* 2005b), bite force and weaponry (Herrel *et al.* 2007), behavior (Johnson and Wade 2010), neural anatomy and physiology (Oakes 1992), stress physiology (Grassman and Hess 1992; Carsia and John-Alder 2003; Cartledge and Jones 2007; Carsia *et al.* 2008), demography (Perry and Garland 2002), parasitism (Klukowski and Nelson 2001; Cox and John-Alder 2007b; Václav *et al.* 2007), energetics (Nagy 1983), and exercise physiology (Garland and Else 1987; Cullum 1998; Lailvaux *et al.* 2003) are also common. In light of this substantial body of research, we narrow our present focus to sex differences in size and coloration because their evolutionary dynamics have been studied in greatest detail, and because their links to sexual selection are best understood.

4.3.1 Ultimate Causes of Sexual Dimorphism

Sexual size dimorphism (SSD) is highly variable across lizards, reaching extremes in which adult males average 50% longer (snout-vent length) than females in some dactyloids (*Anolis*), tropidurids (*Tropidurus*), iguanids (*Amblyrhynchus*), and varanids (*Varanus*), whereas females exceed males by up to 20% in some polychrotids (*Polychrus*), skinks (*Mabuya*), and pygopodids (*Aprasia*) (Cox *et al.* 2007). Variation in the direction and magnitude of SSD is typical within most families and also within some genera (*Anolis, Lacerta, Mabuya, Sceloporus*) and even within some geographically widespread species (Zamudio 1998; Corl *et al.* 2010a; Cox and Calsbeek 2010b), implying a high degree of evolutionary lability (Fitch 1976, 1978, 1981; Cox *et al.* 2003; Cox *et al.* 2007). Broadly, Gekkota and Scinciformata are characterized by a tendency for female-biased SSD and monomorphism, whereas Teiiodea, Anguimorpha, and Iguania are increasingly characterized by male-biased SSD (Fig. 4.3). This stands in stark contrast to the prevalence of female-biased SSD in snakes (Serpentes), which are sister to either Anguimorpha or Iguania.

From the standpoint of ultimate causation, SSD presents a complex problem because sexual selection is only one of several factors, along with fecundity selection and viability selection, expected to influence the evolution of body size (Anderson and Vitt 1990; Cox *et al.* 2003). Nonetheless, comparative analyses indicate that male aggression, territoriality, and female density are evolutionarily correlated with male-biased SSD (Carothers 1984; Stamps *et al.* 1997; Cox *et al.* 2003). However, these presumed correlates of sexual selection explain only a small proportion of the interspecific variance in SSD across all lizards (Cox *et al.* 2003), and evolutionary changes in SSD can occur independent of the evolution of male aggression within

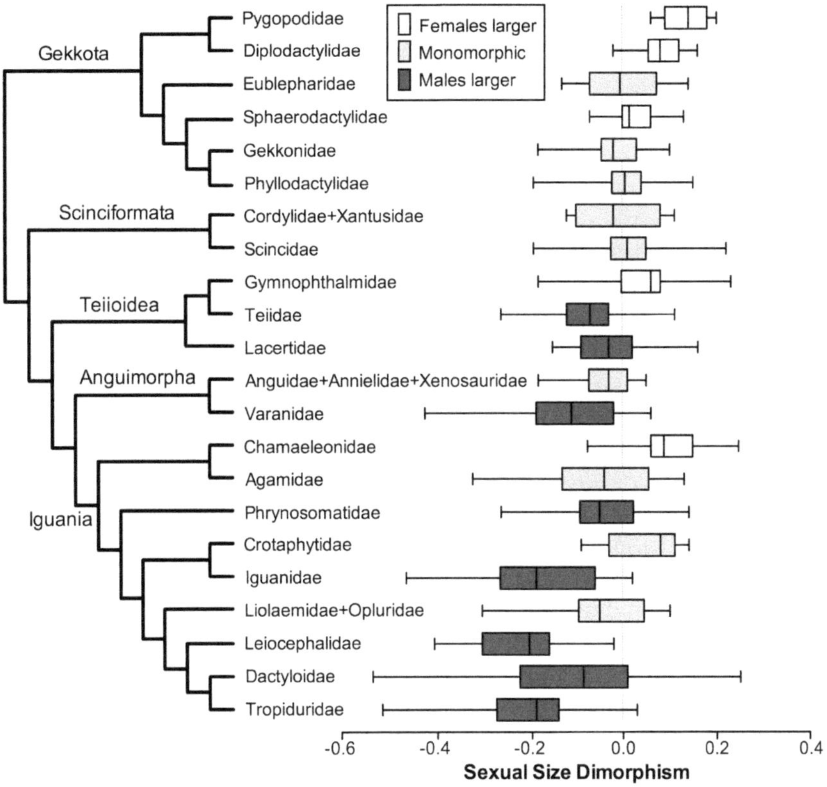

Fig. 4.3 Patterns of sexual size dimorphism (SSD) across major lizard lineages, redrawn from data in Cox *et al.* 2003; Cox *et al.* 2007. SSD is calculated as (size of larger sex/size of smaller sex)-1, using mean snout-vent length of adults and arbitrarily expressing this value as negative when males are the larger sex. Box-and-whisker plots show the median, 25–75 percentiles, and range of SSD within each lineage using individual species or populations as observations. Shading indicates primary trends within each lineage based on *t*-tests for significant deviations from zero (monomorphism). Phylogeny based on Vitt and Caldwell (2013).

some lineages (Stamps 1999; Kratochvil and Frynta 2002). The fairly weak explanatory power of sexual selection with respect to SSD probably stems in part from the use of imprecise proxies for intrasexual selection in comparative analyses, as well as an inability to test for either the effects of intersexual selection or advantages due to small male size. However, given the apparent ubiquity of sexual selection for large body size in males (Figs. 4.1 and 4.2), the key to explaining interspecific variation in SSD may instead lie in the counterbalancing effects of fecundity selection and natural selection.

Fecundity selection can influence the evolution of SSD by favoring large females when clutch or litter size increases with female size (Braña 1996), as is observed in most lizards with variable clutch or litter sizes (Fitch 1970). In support of this "fecundity advantage hypothesis", evolutionary increases in clutch size and fecundity slope (the slope of clutch size regressed on female size) are correlated with evolutionary shifts toward female-biased SSD, as are the evolution of viviparity (Blackburn and Stewart Chapter 13, this volume) and reduced reproductive frequency (Braña 1996; Cox *et al.* 2003; Cox *et al.* 2007). Thus, comparative studies indicate that the evolution of SSD in lizards broadly reflects a balance between fecundity selection favoring large females and sexual selection favoring large males, and selection analyses in wild populations suggest similar interactions with respect to body shape (Olsson *et al.* 2002).

Natural selection for viability can counteract SSD arising from sexual or fecundity selection, as observed when the largest male marine iguanas (*Amblyrhynchus cristatus*), which are favored by sexual selection in a lekking mating system, experience the greatest mortality during El Niño events (Wikelski and Trillmich 1997). It can also create or reinforce SSD, as observed in the highly dimorphic brown anole (*Anolis sagrei*). Viability selection is strongly directional in adult males, favoring the largest individuals, but it is primarily quadratic (stabilizing) in adult females, favoring females near the lower extremes of the size distribution for males (Cox and Calsbeek 2010b). Indeed, *Anolis* provided the inspiration for pioneering "ecological hypotheses" for SSD, which view dimorphism as the product of natural selection for reduced intraspecific competition over food and other resources (Rand 1967; Schoener 1967). Subsequent studies revealed that patterns of SSD across one- and two-species communities in the Lesser Antilles were remarkably similar to those predicted by optimal-foraging theory, and inconsistent with predictions based on sexual selection as the primary driver of SSD (Schoener 1969a,b, 1977). More recent work has shown that the convergent evolution of "ecomorphs" on major islands of the West Indies is associated with the convergent evolution of SSD. Whereas trunk-ground and trunk-crown ecomorphs consistently evolve extreme male-biased SSD, crown-giant, grass-bush, trunk, and twig ecomorphs tend to evolve lower SSD (Butler *et al.* 2000; Butler and Losos 2002; Butler *et al.* 2007).

Similar correlations between SSD and ecomorphological variation are observed in dwarf chameleons (*Bradypodion*), which differ from anoles in that females are typically the larger sex (Stuart-Fox and Moussalli 2007). Although resource partitioning is the primary force behind character displacement in the *Anolis* radiation, the extent to which ecomorph convergence in SSD reflects intraspecific resource partitioning, as opposed to convergence in mating systems and sexual selection, is largely unknown. Surprisingly, intersexual competition coefficients for trophic resources in the bronze anole (*Anolis aeneus*) are actually predicted to be higher in the presence of extreme SSD than in its absence (Stamps *et al.* 1997). This underscores the inherent difficulty in establishing whether differences in resource utilization are the ultimate drivers of SSD, or simply downstream consequences of sexual divergence resulting from other selective pressures (Schoener *et al.* 1982; Vitt and Cooper 1985; Shine 1989).

The ultimate causes of sexual dichromatism would appear to be more straightforward, as elaborate dewlaps, bright ventral patches, and nuptial coloration all function primarily in social communication. However, the extent to which male signals convey information about individual variation to mediate ongoing male-male competition or female choice (Thompson and Moore 1991; Olsson 1994; Baird *et al.* 1997; Vanhooydonck *et al.* 2005; Martín and López 2009; Bajer *et al.* 2010), as opposed to simply facilitating sex or species recognition (Losos 1985; Cooper and Burns 1987; Cooper Jr. and Vitt 1988; Robertson and Rosenblum 2010), is often unclear (Losos and Chu 1998; Ord *et al.* 2001; Ord and Martins 2006). Predation is also expected to counterbalance the evolution of bright coloration, and comparative analyses of agamids suggest that the evolution of sexual dichromatism has been influenced by a combination of sexual selection and predator-mediated natural selection (Stuart-Fox and Ord 2004). Phylogenetic analyses also reveal a recurrent tendency for repeated evolutionary transitions between sexual monomorphism in which both sexes lack colorful displays, sexual dichromatism in which only males exhibit colorful displays, and sexual monomorphism in which both sexes express colorful displays, as observed for the dewlaps of anoles (Harrison and Poe 2012) and the blue ventral patches of *Sceloporus* (Wiens 1999). Whether the presence of colorful displays in both sexes reflects a common function (e.g., species recognition or social communication), separate roles in sexual selection within each sex (e.g., reciprocal mate choice), or evolutionary constraint arising from shared genetic and physiological regulation, is an open question. In females, orange or red nuptial coloration is generally thought to signal receptivity (Watkins 1997; LeBas and Marshall 2000; Baird 2004), but it may also signal female quality and thus evolve via sexual selection (Weiss 2002; Weiss 2006; Weiss *et al.* 2009). Other forms of female-specific coloration, such as the sex-limited dorsal-pattern polymorphisms observed in many anole species (Paemelaere *et al.* 2011a), have no clear sex-specific function (Calsbeek *et al.* 2010a; Cox and Calsbeek 2011; Paemelaere *et al.* 2011b). However, morph differences in immune function and predator susceptibility may partially explain the maintenance of the polymorphism itself, if not its sex-limited expression (Schoener and Schoener 1976; Calsbeek *et al.* 2008; Calsbeek and Cox 2012).

4.3.2 Proximate Mechanisms for Sexual Dimorphism

A major conceptual focus of modern biology is to understand the intermediate links between genotype and phenotype, and sexual dimorphism presents an informative context in which to address this problem, as two separate phenotypes must be produced from essentially the same underlying genome. Hence, recent research on lizards and other taxa has attempted to link the evolution of sexual dimorphism to the underlying developmental, physiological, and genetic mechanisms that facilitate sexual differentiation. Sex differences in body size are typically the result of differences in the rate or duration of growth (Stamps 1993; Watkins 1996; Cox and John-Alder 2007a), raising the question of how growth is differentially regulated in males and females. One possibility is that sex steroids such as testosterone, which is secreted in age- and sex-specific fashion by the gonads, act as modifiers of the expression of shared growth-regulatory genes and pathways. In *Sceloporus*, testosterone has a classic stimulatory effect on growth in species with male-biased SSD, but identical manipulations reveal that testosterone has an inhibitory effect on growth in congeners with female-biased SSD (Abell 1998; Cox and John-Alder 2005; Cox *et al.* 2005a; John-Alder *et al.* 2007). Data from other squamate lineages support the generality of this mechanism, with testosterone stimulating growth in brown anoles (*Anolis sagrei*) and inhibiting growth in garter snakes (*Thamnophis sirtalis*) (Crews *et al.* 1985; Lerner and Mason 2001; Cox *et al.* 2009). Thus, the evolutionary coupling and decoupling of testosterone from various genes and pathways involved in growth regulation may facilitate considerable evolutionary lability in SSD itself.

Due to their systemic circulation, sex steroids can exert pleiotropic effects on multiple tissues, making them excellent candidates for organism-wide integration of sexually dimorphic phenotypes. Androgens often regulate the expression of orange facial or dorsal coloration in male lizards (Cooper *et al.* 1987; Rand 1992; Salvador *et al.* 1996; Cox *et al.* 2005b), whereas androgens, estrogens, and progesterone induce orange nuptial coloration in females of other species (Cooper and Ferguson 1972; Medica *et al.* 1973; Cooper and Clarke 1982). The blue ventral patches of *Sceloporus* lizards offer an intriguing case-study due to the multiple evolutionary gains and losses of coloration in one or both sexes (Wiens 1999). In sexually dimorphic species, such as the eastern fence lizard (*Sceloporus undulatus*), castration feminizes the coloration of males and testosterone induces bright blue and black coloration in castrated males and juvenile females (Cox *et al.* 2005b). Both sexes also respond to testosterone in species where blue coloration is present in each sex, but the loss of coloration in both sexes has apparently been achieved through the wholesale decoupling of color expression from testosterone (Abell 1998; Hews and Quinn 2003; Quinn and Hews 2003;

Cox *et al.* 2008). Testosterone also stimulates development of polymorphic dewlap coloration in juvenile tree lizards, *Urosaurus ornatus* (Hews *et al.* 1994; Hews and Moore 1995). Though our recent work suggests that testosterone also stimulates male-typical enlargement and darkening of the dewlap in both male and female brown anoles (*Anolis sagrei*), endocrine regulation has yet to be incorporated into a phylogenetic framework to investigate the mechanisms underlying repeated evolutionary shifts in sexual dimorphism and monomorphism of the *Anolis* dewlap (Harrison and Poe 2012). The two major patterns that emerge from studies of the endocrine regulation of sexual dimorphism in lizards are that hormones can be evolutionarily coupled to and uncoupled from various tissue-specific responses, thus facilitating evolutionary lability in sexual dimorphism, and that females and males are often similarly responsive to hormone manipulation, suggesting that sexual dimorphism reflects sex differences in circulating hormone levels more so than sex differences in tissue sensitivity.

Studies of evolutionary developmental biology and quantitative genetics represent promising new avenues for research on sexual dimorphism in lizards. For example, across *Anolis* species, two distinct clades have evolved sexual dimorphism in cranial morphology in which males possess dramatically elongated heads, but the specific developmental patterns that underlie this convergence in sexual shape dimorphism are unique to each clade (Sanger *et al.* 2013). The availability of an annotated *Anolis* genome and the expansion of related developmental and genetic resources are allowing phenomena such as this to be recast in a more mechanistic framework addressing the evolution of sex-specific patterns of gene expression and developmental regulation (Schneider 2008; Alfioldi *et al.* 2011; Kusumi *et al.* 2011). The quantitative genetics of sexual dimorphism have been investigated using parent-offspring regression in *Anolis sagrei* (Calsbeek and Bonneaud 2008), which revealed that traits such as body size and limb length are heritable within each sex, but weakly or even negatively genetically correlated between sexes. More powerful paternal half-sib breeding designs have largely confirmed this pattern and are now being used to derive between-sex genetic (co)variance matrices for suites of dimorphic traits and integrate this genetic perspective with developmental biology, sex-specific gene expression, and endocrine regulation.

4.4 LIZARDS AS MODELS FOR EVOLUTIONARY DYNAMICS

Sexual selection is of broad biological significance because it can generate conflict within the genome of a species, deplete or maintain genetic variance in fitness, provide alternative pathways for "evolutionarily successful" males and females, and create both phenotypic and biological diversity by driving rapid evolution. Lizards provide excellent models for each of these phenomena because they are amenable to a variety of evolutionary approaches, ranging from phylogenetic comparisons and population-level studies of current selection down to phenotypic manipulations that address sub-organismal mechanisms and genetic or genomic studies that identify the molecular basis of evolution. Here, we present case studies of three areas in which lizards have made particularly important contributions to our understanding of evolutionary dynamics driven by sexual selection.

4.4.1 Intralocus Sexual Conflict

One key consequence of sexual selection is its potential to generate intralocus sexual conflict (Rice and Chippindale 2001; Bonduriansky and Chenoweth 2009). This conflict arises because males and females share an autosomal genome, which can constrain them from evolving toward the separate optima defined by sex differences in natural or sexual selection. This genomic conflict is gradually resolved through the evolution of mechanisms that facilitate the sex-specific expression of loci subject to sexual conflict (e.g., hormonal regulation, see above), which facilitates the evolution of sexual dimorphism (Cox and Calsbeek 2009; Svensson *et al.* 2009). The few available studies directly comparing phenotypic selection between male and female lizards reveal that sex differences in current selection are still observed for sexually dimorphic traits such as body size and shape (Trivers 1976; Wikelski and Trillmich 1997; Olsson *et al.* 2002; Cox and Calsbeek 2010b), which implies that the evolution of sexual dimorphism has yet to fully resolve sexual conflict, since males and females are inferred to be in the process of moving toward their respective phenotypic optima (Cox and Calsbeek 2009). However, traits such as body size in *Anolis sagrei* and immune function in *Uta stansburiana* exhibit weak or even negative between-sex genetic correlations, suggesting that intralocus sexual conflict has already been greatly reduced at a genomic level (Calsbeek and Bonneaud 2008; Svensson *et al.* 2009).

One potential mechanism for the resolution of sexual conflict that has thus far been described in only two lizards is the tendency for females to produce more sons with sperm from large or high-condition mates, but more daughters with sperm from small or low-condition mates (Calsbeek and Sinervo 2004; Calsbeek and Bonneaud 2008; Cox and Calsbeek 2010a; Cox *et al.* 2011). Though it was initially hypothesized that this mechanism might resolve sexual conflict by preferentially allowing "good male genes" to pass from large males to their sons and "good female genes" to pass from small males to their daughters, actual analyses of progeny fitness are more consistent with the interpretation that this sex-ratio bias is beneficial to sons (indirect genetic benefits), but neutral to daughters (Cox and Calsbeek 2010a). Lizards may prove ideal for addressing many other unanswered questions regarding intralocus sexual conflict, particularly with respect to sexually antagonistic selection at the level of the genome, since most studies of wild populations have only inferred genomic conflict indirectly from phenotypic selection.

4.4.2 Alternative Reproductive Tactics

Color polymorphisms have been described in many lizards and often correspond to discrete differences in reproductive strategies, termed alternative reproductive tactics (ARTs) (Calsbeek and Sinervo 2008). Lizards are particularly well-suited to the study of ARTs because the bright colors used in social displays of many species create distinctive phenotypes that are easily recognizable by researchers (Calsbeek *et al.* 2010b), thus overcoming the problems associated with detecting more "cryptic" variation in reproductive tactics (Noble *et al.* 2013). In both males and females, these color polymorphisms and their associated phenotypes are often genetically determined (Sinervo and Lively 1996; Vercken *et al.* 2007; Calsbeek *et al.* 2010a), which potentially sets the stage for complex evolutionary dynamics driven by sexual selection.

One such dynamic may have implications for the classic evolutionary puzzle of how genetic variance for fitness is maintained in the face of strong selection, which should deplete this variance. ARTs in lizards provide several examples in which males specialized for female mimicry, sperm competition, or non-territorial strategies persist as alternatives to territory defense and mate guarding (Sinervo and Lively 1996; Moore *et al.* 1998; Zamudio and Sinervo 2000; Olsson *et al.* 2007; Whiting *et al.* 2009). For example, orange-throated *Uta stansburiana* males are "ultra-dominant" and maintain large territories that encompass multiple females, blue-throated "mate guarders" maximize their reproductive success by defending smaller territories but improving their paternity within clutches, and yellow-throated "sneakers" achieve reproductive success through a high incidence of posthumous fertilization and shared paternity (Zamudio and Sinervo 2000). In addition to demonstrating that there is often more than one way to be an evolutionarily successful male, this system illustrates how genetic variance in the traits that distinguish each morph (e.g., behavior, throat color, immune function, endocrinology) is maintained by the evolutionary equivalence of the ARTs. Morph differences in these and other traits that determine male reproductive success have been documented in a variety of color-polymorphic lizards (Thompson and Moore 1991; Healey *et al.* 2007; Huyghe *et al.* 2007; Sacchi *et al.* 2007; Huyghe *et al.* 2009; Calsbeek *et al.* 2010b).

The evolutionary stability of ARTs is often the result of negative frequency-dependent selection, which occurs if particular tactics are most successful when rare in the population. In the case of side-blotched lizards, this frequency dependence arises from "rock-paper-scissors" dynamics in which each male tactic is successful when rare and competing primarily against a common tactic that it can exploit, but unsuccessful when common and itself exploited by a rare tactic (Sinervo and Lively 1996; Sinervo *et al.* 2000; Sinervo 2001; Sinervo *et al.* 2007). The result is often periodic oscillation in morph frequencies, which perpetually stirs the pot of underlying genetic variation in a population. Alternative reproductive strategies and resultant patterns of frequency- or density-dependent selection are not restricted to males, but also occur in color-polymorphic females of some species (Forsman and Shine 1995; Sinervo *et al.* 2000; Vercken *et al.* 2007; Vercken *et al.* 2010). However, ARTs in females are usually thought to represent different strategies for reproductive investment (e.g., offspring quality versus quantity), rather than alternatives favored by sexual selection *per se* (though the latter possibility has not been rigorously investigated in lizards). Other polymorphisms in females, such as the dorsal pattern morphs observed in *Anolis sagrei*, have no clear link to ARTs (Cox and Calsbeek 2011). When ARTs occur in both sexes, morph-assortative mating can ensue (Stapley and Keogh 2005), possibly to prevent the breakup of coadapted gene complexes, and this could theoretically result in divergence toward reproductive isolation and speciation (Alonzo and Sinervo 2001; Sinervo and Svensson 2002; Gray and McKinnon 2006).

4.4.3 Speciation and Species Recognition

Perhaps the most important evolutionary consequence of sexual selection is its potential to drive rapid reproductive divergence and speciation, thus generating both phenotypic and biological diversity. From a macroevolutionary perspective, comparative analyses reveal that proxies for sexual selection, such as SSD and sexual dichromatism, are strong predictors of species richness across agamid genera (Stuart-Fox and Owens 2003). Within the genus *Anolis*, the adaptive radiation of convergent ecomorphs in the West Indies is driven by ecological character displacement due to natural selection, but sex differences also comprise a substantial portion of the total phenotypic variation (Butler *et al.* 2007), and are thought to be partially due to convergent patterns of sexual selection (Butler and Losos 2002). At the species and population level, rapid divergence in sexual dichromatism among lineages in the *Sceloporus jarovii* complex (in which males from different populations range in dorsal color from brilliant blues, greens, and reds to subtler blacks, browns, and yellows) is also consistent with diversification by sexual selection (Wiens *et al.* 1999). Mitochondrial DNA haplotypes sampled across geographic variation in dewlap color of several Hispaniolan anoles (*A. distichus, A. brevirostris*) reveal that "subspecies" traditionally classified by dewlap coloration actually represent deeply divergent clades, implicating mate recognition via dewlap coloration as a potential factor in reproductive isolation and diversification (Ng and Glor 2011; Glor and Laport 2012). Comparative analyses of populations in the *Uta stansburiana* complex reveal that evolutionary losses of color morphs, which alter the dynamics of sexual selection via alternative reproductive tactics, are repeatedly associated with the emergence of distinct morphological species or subspecies, as well as increased rates of evolution in body size and SSD (Corl *et al.* 2010a; Corl *et al.* 2010b).

Evolutionary snapshots of "speciation in action" also reveal a key role of sexual selection and sexual dimorphism in reproductive isolation, even in situations where divergence itself is driven primarily by ecological processes. In the gypsum dunes of White Sands, New Mexico, little striped whiptails (*Aspidoscelis inornata*), eastern fence lizards (*Sceloporus undulatus*), and lesser earless lizards (*Holbrookia maculata*) each exhibit convergent evolution of blanched dorsal coloration driven by natural selection for crypsis on the stark white dunes, which differ from the darker soil habitats of ancestral populations (Rosenblum 2006; Rosenblum *et al.* 2010; Rosenblum and Harmon 2011). Across the sand/soil ecotone, each species has also diverged in social coloration used in intra- and intersexual signaling, with *Aspidoscelis* and *Sceloporus* males displaying more conspicuous blue gular and ventral coloration and *Holbrookia* females displaying duller and less conspicuous orange gular coloration on the white dunes, relative to the dark soil (Robertson and Rosenblum 2009). Whether this divergence is a correlated byproduct of natural selection for crypsis or an adaptive mechanism for maintaining reproductive isolation and avoiding gene flow across the ecotone is not fully resolved. However, the latter scenario is supported by the observation that both *Holbrookia* and *Sceloporus* males from White Sands preferentially court females with White Sands coloration (Rosenblum 2008; Hardwick *et al.* 2013). The more conspicuously colored *Sceloporus* males from White Sands also elicit stronger aggressive responses from other males, whereas males from dark soil habitats are mistakenly courted by White Sands males due to the similarity of their ventral coloration to that of White Sands females (Robertson and Rosenblum 2010). Striking divergence in coloration of *Anolis marmoratus* males is also observed across a sharp mesic/xeric ecotone on Guadeloupe, but whereas this phenotypic divergence may be maintained by divergent sexual and/or natural selection, gene flow is still occurring and there is no evidence for sexual selection driving reproductive isolation across the ecotone (Muñoz *et al.* 2013). One generality that emerges from these examples is that most studies of lizards have focused on ecological speciation as the primary agent of divergence, with sexual selection playing a secondary role in reproductive isolation. Given the apparent strength and ubiquity of sexual selection in lizards (Figs. 4.1 and 4.2), future work may find it informative to directly explore its potential role as an engine for speciation (Stuart-Fox and Owens 2003; Camargo *et al.* 2010; Corl *et al.* 2010b; Runemark and Svensson 2012).

4.5 ACKNOWLEDGMENTS

We thank Malcolm Augat, Michael Hague, Amanda Hanninen, Aaron Reedy, Brian Sanderson, Gabriela Toledo, and Corlett Wood for helpful suggestions on earlier drafts of this chapter. Laurie Vitt provided expert advice on the illustration of phylogenetic patterns in sexual size dimorphism in Fig. 4.3. Many of the ideas in this chapter arose from collaborations with Ryan Calsbeek, Henry John-Alder, and Joel McGlothlin.

4.6 LITERATURE CITED

Abell, A. J. 1997. Estimating paternity with spatial behaviour and DNA fingerprinting in the striped plateau lizard, *Sceloporus virgatus* (Phrynosomatidae). Behavioral Ecology and Sociobiology 41: 217–226.

Abell, A. J. 1998. The effect of exogenous testosterone on growth and secondary sexual character development in juveniles of *Sceloporus virgatus*. Herpetologica 54: 533–543.

Alfioldi, J., Palma, F. D., Grabherr, M., Williams, C., Kong, L., Mauceli, E., Russell, P., Lowe, C. B., Glor, R. E., Jaffe, J. D., Ray, D. A., Boissinot, S., Shedlock, A. M., Botka, C., Castoe, T. A., Colbourne, J. K., Fujita, M. K., Moreno, R. G., ten Hallers, B. F., Haussler, D., Heger, A., Heiman, D., Janes, D. E., Johnson, J., de Jong, P. J., Koriabine, M. Y., Lara, M., Novick, P. A., Organ, C. L., Peach, S. E., Poe, S., Pollock, D. D., de Queiroz, K., Sanger, T., Searle, S., Smith, J. D., Smith, Z., Swofford, R., Turner-Maier, J., Wade, J., Young, S., Zadissa, A., Edwards, S. V., Glenn, T. C., Schneider, C. J., Losos, J. B., Lander, E. S., Breen, M., Ponting, C. and Lindblad-Toh, K. 2011. The genome of the green anole lizard and a comparative analysis with birds and mammals. Nature 477: 587.

Alonzo, S. H. and Sinervo, B. 2001. Mate choice games, context-dependent good genes, and genetic cycles in the side-blotched lizard, *Uta stansburiana*. Behavioral Ecology and Sociobiology 49: 176–186.

Anderson, R. A. and Vitt, L. J. 1990. Sexual selection versus alternative causes of sexual dimorphism in teiid lizards. Oecologia 84: 145–157.

Andersson, M. 1994. *Sexual Selection*. Princeton University Press, New Jersey, pp. 624.

Andrews, R. M. 1985. Mate choice by females of the lizard, *Anolis carolinensis*. Journal of Herpetology 19: 284–289.

Arnold, S. J. and Wade, M. J. 1984. On the measurement of natural and sexual selection: applications. Evolution 38: 720–734.

Baird, T. A. 2004. Reproductive coloration in female collared lizards, *Crotaphytus collaris*, stimulates courtship by males. Herpetologica 60: 337–348.

Baird, T. A., Fox, S. F. and McCoy, J. K. 1997. Population differences in the roles of size and coloration in intra- and intersexual selection in the collared lizard, *Crotaphytus collaris*: influence of habitat and social organization. Behavioral Ecology 8: 506–517.

Bajer, K., Molnár, O., Török, J. and Herczeg, G. 2010. Female European green lizards (*Lacerta viridis*) prefer males with high ultraviolet throat reflectance. Behavioral Ecology and Sociobiology 64: 2007–2014.

Birkhead, T. R. and Moller, A. P. 1993. Sexual selection and the temporal separation of reproductive events: sperm storage data from reptiles, birds and mammals. Biological Journal of the Linnean Society 50: 295–311.

Bleay, C. and Sinervo, B. 2007. Discrete genetic variation in mate choice and a condition-dependent preference function in the side-blotched lizard: implications for the formation and maintenance of coadapted gene complexes. Behavioral Ecology 18: 304–310.

Bonduriansky, R. and Chenoweth, S. F. 2009. Intralocus sexual conflict. Trends in Ecology and Evolution 24: 280–288.

Braña, F. 1996. Sexual dimorphism in lacertid lizards: male head increase vs. female abdomen increase. Oikos 75: 511–523.

Bull, C. M. 2000. Monogamy in lizards. Behavioural Processes 51: 7–20.

Butler, M. A. and Losos, J. B. 2002. Multivariate sexual dimorphism, sexual selection, and adaptation in Greater Antillean *Anolis* lizards. Ecological Monographs 72: 541–559.

Butler, M. A., Sawyer, S. A. and Losos, J. B. 2007. Sexual dimorphism and adaptive radiation in *Anolis* lizards. Nature 447: 202–205.

Butler, M. A., Schoener, T. W. and Losos, J. B. 2000. The relationship between sexual size dimorphism and habitat use in Greater Antillean *Anolis* lizards. Evolution 54: 259–272.

Calsbeek, R. and Bonneaud, C. 2008. Postcopulatory fertilization bias as a form of cryptic sexual selection. Evolution 62: 1137–1148.

Calsbeek, R. and Cox, R. 2012. An experimental test of the role of predators in the maintenance of a genetically based polymorphism. Journal of Evolutionary Biology 25: 2091–2101.

Calsbeek, R. and Sinervo, B. 2002. Uncoupling direct and indirect components of female choice in the wild. Proceedings of the National Academy of Science USA 99: 14897–14802.

Calsbeek, R. and Sinervo, B. 2004. Within-clutch variation in offspring sex determined by differences in sire body size: cryptic mate choice in the wild. Journal of Evolutionary Biology 17: 464–470.

Calsbeek, R. and Sinervo, B. 2008. Alternative reproductive tactics in reptiles. pp. 332–342. In R. F. Oliviera, M. Taborsky and H. J. Brockman (eds.), *Alternative reproductive tactics: an integrative approach*. Cambridge University Press: Cambridge.

Calsbeek, R., Bonneaud, C., Prabhu, S., Manoukis, N. and Smith, T. B. 2007. Multiple paternity and sperm storage lead to increased genetic diversity in the Cuban anole, *Anolis sagrei*. Evolutionary Ecology Research 9: 495–503.

Calsbeek, R., Bonneaud, C. and Smith, T. B. 2008. Differential fitness effects of immunocompetence and neighborhood density in alternative female lizard morphs. Journal of Animal Ecology 77: 103–109.

Calsbeek, R., Bonvini, L. A. and Cox, R. M. 2010a. Geographic variation, frequency-dependent selection, and the maintenance of a female-limited polymorphism. Evolution 64: 116–125.

Calsbeek, B., Hasselquist, D. and Clobert, J. 2010b. Multivariate phenotypes and the potential for alternative phenotypic optima in wall lizard (*Podarcis muralis*) ventral color morphs. Journal of Evolutionary Biology 23: 1138–1147.

Camargo, A., Sinervo, B. and Sites, J. W. 2010. Lizards as model organisms for linking phylogeographic and speciation studies. Molecular Ecology 19: 3250–3270.

Carothers, J. H. 1984. Sexual selection and sexual dimorphism in some herbivorous lizards. The American Naturalist 124: 244–254.

Carsia, R. V. and John-Alder, H. 2003. Seasonal alterations in adrenocortical cell function associated with stress-responsiveness and sex in the eastern fence lizard (*Sceloporus undulatus*). Hormones and Behavior 43: 408–420.

Carsia, R. V., McIlroy, P. J., Cox, R. M., Barrett, M. and John-Alder, H. B. 2008. Gonadal modulation of *in vitro* steroidogenic properties of dispersed adrenocortical cells from *Sceloporus* lizards. General and Comparative Endocrinology 158: 202–210.

Cartledge, V. A. and Jones, S. M. 2007. Does adrenal responsiveness vary with sex and reproductive status in *Egernia whitii*, a viviparous skink? General and Comparative Endocrinology 150: 132–139.
Censky, E. J. 1997. Female mate choice in the non-territorial lizard *Ameiva plei* (Teiidae). Behavioral Ecology and Sociobiology 40: 221–225.
Conner, J. and Crews, D. 1980. Sperm transfer and storage in the lizard, *Anolis carolinensis*. Journal of Morphology 163: 331–348.
Cooper Jr., W. E. and Vitt, L. J. 1988. Orange head coloration of the male broad-headed skink (*Eumeces laticeps*), a sexually selected social cue. Copeia 1988: 1–6.
Cooper, Jr. W. E., and Burns, N. 1987. Social significance of ventrolateral coloration in the fence lizard, *Sceloporus undulatus*. Animal Behaviour 35: 526–532.
Cooper, Jr. W. E. and Clarke, R. F. 1982. Steroidal induction of female reproductive coloration in the keeled earless lizard, *Holbrookia propinqua*. Herpetologica 38: 425–429.
Cooper, Jr. W. E. and Ferguson, G. W. 1972. Relative effectiveness of progesterone and testosterone as inductors of orange spotting in female collared lizards. Herpetologica 28: 64–65.
Cooper, Jr. W. E. and Vitt, L. J. 1989. Sexual dimorphism of head and body size in an iguanid lizard: paradoxical results. The American Naturalist 133: 729–733.
Cooper, Jr. W. E.and Vitt, L. J. 1993. Female mate choice of large male broad-headed skinks. Animal Behaviour 45: 683–693.
Cooper, Jr. W. E., Mendonca, M. T. and Vitt, L. J. 1987. Induction of orange head coloration and activation of courtship and aggression by testosterone in the male broad-headed skink (*Eumeces laticeps*). Journal of Herpetology 21: 96–101.
Corl, A., Davis, A. R., Kuchta, S. R., Comendant, T. and Sinervo, B. 2010a. Alternative mating strategies and the evolution of sexual size dimorphism in the side-blotched lizard, *Uta stansburiana*: a population-level comparative analysis. Evolution 64: 79–96.
Corl, A., Davis, A. R., Kuchta, S. R. and Sinervo, B. 2010b. Selective loss of polymorphic mating types is associated with rapid phenotypic evolution during morphic speciation. Proceedings of the National Academy of Sciences USA 107: 4254–4259.
Cornwallis, C. K. and Uller, T. 2010. Towards an evolutionary ecology of sexual traits. Trends in Ecology and Evolution 25: 145–152.
Cox, R. M. 2005. Integrating proximate and ultimate causes of sexual size dimorphism in lizards. Ph.D. thesis, Rutgers University, New Brunswick, New Jersey.
Cox, R. M. and Calsbeek, R. 2009. Sexually antagonistic selection, sexual dimorphism, and the resolution of intralocus sexual conflict. The American Naturalist 173: 176–187.
Cox, R. M. and Calsbeek, R. 2010a. Cryptic sex-ratio bias provides indirect genetic benefits despite sexual conflict. Science 328: 92–94.
Cox, R. M. and Calsbeek, R. 2010b. Sex-specific selection and intraspecific variation in sexual size dimorphism. Evolution 64: 798–809.
Cox, R. M. and Calsbeek, R. 2011. An experimental test for alternative reproductive tactics underlying a female-limited polymorphism. Journal of Evolutionary Biology 24: 343–353.
Cox, R. M. and John-Alder, H. B. 2005. Testosterone has opposite effects on male growth in lizards (*Sceloporus* spp.) with opposite patterns of sexual size dimorphism. Journal of Experimental Biology 208: 4679–4687.
Cox, R. M. and John-Alder, H. B. 2007a. Growing apart together: the development of contrasting sexual size dimorphisms in sympatric *Sceloporus* lizards. Herpetologica 63: 245–257.
Cox, R. M. and John-Alder, H. B. 2007b. Increased mite parasitism as a cost of testosterone in male striped plateau lizards, *Sceloporus virgatus*. Functional Ecology 21: 327–334.
Cox, R. M., Skelly, S. L. and John-Alder, H. B. 2003. A comparative test of adaptive hypotheses for sexual size dimorphism in lizards. Evolution 57: 1653–1669.
Cox, R. M., Skelly, S. L. and John-Alder, H. B. 2005a. Testosterone inhibits growth in juvenile male eastern fence lizards (*Sceloporus undulatus*): Implications for energy allocation and sexual size dimorphism. Physiological and Biochemical Zoology 78: 531–545.

Cox, R. M., Skelly, S. L., Leo, A. and John-Alder, H. B. 2005b. Testosterone regulates sexually dimorphic coloration in the eastern fence lizard, *Sceloporus undulatus*. Copeia 2005: 597–608.

Cox, R. M., Butler, M. A. and John-Alder, H. B. 2007. The evolution of sexual size dimorphism in reptiles. pp. 38–49. In D. J. Fairbairn, W. U. Blanckenhorn and T. Szekely (eds.), *Sex, size and gender roles: evolutionary studies of sexual size dimorphism*. Oxford University Press: London.

Cox, R. M., Zilberman, V. and John-Alder, H. B. 2008. Testosterone stimulates the expression of a social color signal in Yarrow's spiny lizard, *Sceloporus jarrovii*. Journal of Experimental Zoology 309A: 505–514.

Cox, R. M., Stenquist, D. S. and Calsbeek, R. 2009. Testosterone, growth, and the evolution of sexual size dimorphism. Journal of Evolutionary Biology 22: 1586–1598.

Cox, R. M., Duryea, M. C., Najarro, M. and Calsbeek, R. 2011. Paternal condition drives progeny sex-ratio bias in a lizard that lacks parental care. Evolution 65: 220–230.

Crews, D., Diamond, M. A., Whittier, J. and Mason, R. 1985. Small male body size in snakes depends on testes. American Journal of Physiology 18: R62–R66.

Cuadrado, M. 1998. The influence of female size on the extent and intensity of mate guarding by males in Chamaeleo chamaeleon. Journal of Zoology 246: 351–358.

Cullum, A. J. 1998. Sexual dimorphism in physiological performance of whiptail lizards (Genus *Cnemidophorus*). Physiological Zoology 71: 541–552.

Darwin, C. 1871. *The Descent of Man, and Selection in Relation to Sex*. J. Murray, London, pp. 903.

Duryea, M. C., Kern, A., Cox, R. and Calsbeek, R. 2013. A novel application of approximate bayesian computation for detecting male reproductive advantages due to mating order. Behavioral Ecology and Sociobiology 67: 1867–1875.

Fairbairn, D. J., Blanckenhorn, W. U. and Szekely, T. 2007. *Sex, Size and Gender Roles: Evolutionary Studies of Sexual Size Dimorphism*. Oxford University Press, London, pp. 266.

Fitch, H. S. 1970. Reproductive cycles of lizards and snakes. University of Kansas Museum of Natural History, Miscellaneous Publications 52: 1–247.

Fitch, H. S. 1976. Sexual size differences in the mainland anoles. University of Kansas Museum of Natural History, Occasional Papers 50: 1–21.

Fitch, H. S. 1978. Sexual size differences in the genus *Sceloporus*. The University of Kansas Science Bulletin 51: 441–461.

Fitch, H. S. 1981. Sexual size differences in reptiles. University of Kansas Museum of Natural History, Miscellaneous Publications 70: 1–72.

Fitze, P. S., Cote, J. and Clobert, J. 2010. Mating order-dependent female mate choice in the polygynandrous common lizard Lacerta vivipara. Oecologia 162: 331–341.

Fitze, P. S. and Le Galliard, J. -F. 2008. Operational sex ratio, sexual conflict and the intensity of sexual selection. Ecology Letters 11: 432–439.

Forsman, A. and Shine, R. 1995. The adaptive significance of colour pattern polymorphism in the Australian scincid lizard *Lampropholis delicata*. Biological Journal of the Linnean Society 55: 273–291.

Garland, T. and Else, P. L. 1987. Seasonal, sexual, and individual variation in endurance and activity metabolism in lizards. American Journal of Physiology—Regulatory, Integrative and Comparative Physiology 252: R439–R449.

Garland, T., Hankins, E. and Huey, R. 1990. Locomotor capacity and social dominance in male lizards. Functional Ecology 4: 243–250.

Glor, R. E. and Laport, R. G. 2012. Are subspecies of *Anolis* lizards that differ in dewlap color and pattern also genetically distinct? A mitochondrial analysis. Molecular Phylogenetics and Evolution 64: 255–260.

Grassman, M. and Hess, D. L. 1992. Sex differences in adrenal function in the lizard *Cnemidophorus sexlineatus*: I. Seasonal variation in the field. Journal of Experimental Zoology 264: 177–182.

Gray, S. M. and McKinnon, J. S. 2006. Linking color polymorphism maintenance and speciation. Trends in Ecology and Evolution 22: 71–79.

Hamilton, P. S. and Sullivan, B. K. 2005. Female mate attraction in ornate tree lizards, *Urosaurus ornatus*: a multivariate analysis. Animal Behaviour 69: 219–224.
Hardwick, K. M., Robertson, J. M. and Rosenblum, E. B. 2013. Asymmetrical mate preference in recently adapted White Sands and black lava populations of *Sceloporus undulatus*. Current Zoology 59: 20–30.
Harrison, A. and Poe, S. 2012. Evolution of an ornament, the dewlap, in females of the lizard genus *Anolis*. Biological Journal of the Linnean Society 106: 191–201.
Healey, M., Uller, T. and Olsson, M. 2007. Seeing red: morph-specific contest success and survival rates in a colour-polymorphic agamid lizard. Animal Behaviour 74: 337–341.
Herrel, A., McBrayer, L. D. and Larson, P. M. 2007. Functional basis for sexual differences in bite force in the lizard *Anolis carolinensis*. Biological Journal of the Linnean Society 91: 111–119.
Hews, D. K. 1990. Examining hypotheses generated by field measures of sexual selection on male lizards, *Uta palmeri*. Evolution 44: 1956–1966.
Hews, D. K. and Moore, M. C. 1995. Influence of androgens on differentiation of secondary sex characters in tree lizards, *Urosaurus ornatus*. General and Comparative Endocrinology 97: 86–102.
Hews, D. K. and Quinn, V. S. 2003. Endocrinology of species differences in sexually dimorphic signals and aggression: using the organization and activation model in a phylogenetic framework. pp. 253–277. In S. F. Fox, T. A. Baird and J. C. McCowy (eds.), *Lizard social behavior*. Johns Hopkins University Press: Baltimore.
Hews, D. K., Knapp, R. and Moore, M. C. 1994. Early exposure to androgens affects adult expression of alternative male types in tree lizards. Hormones and Behavior 28: 96–115.
Hibbitts, T., Whiting, M. and Stuart-Fox, D. 2007. Shouting the odds: vocalization signals status in a lizard. Behavioral Ecology and Sociobiology 61: 1169–1176.
Hofmann, S. and Henle, K. 2006. Male reproductive success and intrasexual selection in the common lizard determined by DNA-microsatellites. Journal of Herpetology 40: 1–6.
Holt, W.V. and Lloyd, R.E. 2010. Sperm storage in the vertebrate female reproductive tract: How does it work so well? Theriogenology 73: 713–722.
Husak, J. F., Fox, S. F., Lovern, M. B. and Bussche, R. A. V. D. 2006. Faster lizards sire more offspring: sexual selection on whole-animal performance. Evolution 60: 2122–2130.
Huyghe, K., Vanhooydonck, B., Scheers, H., Molina-Borja, M. and Van Damme, R. 2005. Morphology, performance and fighting capacity in male lizards, *Gallotia galloti*. Functional Ecology 19: 800–807.
Huyghe, K., Vanhooydonck, B., Herrel, A., Tadić, Z. and Van Damme, R. 2007. Morphology, performance, behavior and ecology of three color morphs in males of the lizard *Podarcis melisellensis*. Integrative and Comparative Biology 47: 211–220.
Huyghe, K., Husak, J. F., Herrel, A., Tadić, Z., Moore, I. T., Van Damme, R. and Vanhooydonck, B. 2009. Relationships between hormones, physiological performance and immunocompetence in a color-polymorphic lizard species, *Podarcis melisellensis*. Hormones and Behavior 55: 488–494.
Jansson, N., Uller, T. and Olsson, M. 2005. Female dragons, *Ctenophorus pictus*, do not prefer scent from unrelated males. Australian Journal of Zoology 53: 279–282.
John-Alder, H. B., Cox, R. M. and Taylor, E. N. 2007. Proximate developmental mediators of sexual size dimorphism: case studies from squamate reptiles. Integrative and Comparative Biology 47: 258–271.
John-Alder, H. B., Cox, R. M., Haenel, G. J. and Smith, L. C. 2009. Hormones, performance and fitness: natural history and endocrine experiments on a lizard (*Sceloporus undulatus*). Integrative and Comparative Biology 49: 1–15.
Johnson, M. A. and Wade, J. 2010. Behavioural display systems across nine *Anolis* lizard species: sexual dimorphisms in structure and function. Proceedings of the Royal Society B: Biological Sciences 277: 1711–1719.
Kingsolver, J. G. and Pfennig, D. W. 2004. Individual-level selection as a cause of Cope's rule of phyletic size increase. Evolution 58: 1608–1612.

Kingsolver, J. G. and Pfennig, D. W. 2007. Patterns and power of phenotypic selection in nature. BioScience 57: 561–572.
Kingsolver, J. G., Hoekstra, H. E., Hoekstra, J. M., Berrigan, D., Vignieri, S. N., Hill, C. E., Hoang, A., Gibert, P. and Beerli, P. 2001. The strenght of phenotypic selection in natural populations. The American Naturalist 157: 245–261.
Klukowski, M. and Nelson, C. E. 2001. Ectoparasite loads in free-ranging northern fence lizards, *Sceloporus undulatus hyacinthinus*: effects of testosterone and sex. Behavioral Ecology and Sociobiology 49: 289–295.
Kratochvil, L. and Frynta, D. 2002. Body size, male combat, and the evolution of sexual dimorphism in eublepharid geckos (Squamata: Eublepharidae). Biological Journal of the Linnean Society 76: 303–314.
Kusumi, K., Kulathinal, R., Abzhanov, A., Boissinot, S., Crawford, N., Faircloth, B., Glenn, T., Janes, D., Losos, J. and Menke, D. 2011. Developing a community-based genetic nomenclature for anole lizards. BMC Genomics 12: 554.
Lailvaux, S. P. and Irschick, D. J. 2007. The evolution of performance-based male fighting ability in Caribbean *Anolis* lizards. The American Naturalist 170: 573–586.
Lailvaux, S. P., Alexander, G. J. and Whiting, M. J. 2003. Sex-based differences and similarities in locomotor performance, thermal preferences, and escape behaviour in the lizard *Platysaurus intermedius wilhelmi*. Physiological and Biochemical Zoology 76: 511–521.
Lande, R. and Arnold, S. J. 1983. The measurement of selection on correlated characters. Evolution 37: 1210–1226.
Lappin, A. K. and Husak, J. F. 2005. Weapon performance, not size, determines mating success and potential reproductive output in the collared lizard (*Crotaphytus collaris*). The American Naturalist 166: 426–436.
Lebas, N. R. 2001. Microsatellite determination of male reproductive success in a natural population of the territorial ornate dragon lizard, Ctenophorus ornatus. Molecular Ecology 10: 193–203.
LeBas, N. R. and Marshall, N. J. 2000. The role of colour in signalling and male choice in the agamid lizard *Ctenophorus ornatus*. Proceedings of the Royal Society of London. Series B: Biological Sciences 267: 445–452.
Lebas, N. R. and Marshall, N. J. 2001. No evidence of female choice for a condition-dependent trait in the agamid lizard, *Ctenophorus ornatus*. Behaviour 138: 965–980.
Lerner, D. T. and Mason, R. T. 2001. The influence of sex steroids on the sexual size dimorphism in the red-spotted garter snake, *Thamnophis sirtalis concinnus*. General and Comparative Endocrinology 124: 218–225.
López, P. and Martín, J. 2001. Fighting rules and rival recognition reduce costs of aggression in male lizards, *Podarcis hispanica*. Behavioral Ecology and Sociobiology 49: 111–116.
López, P. and Martín, J. 2005. Female Iberian wall lizards prefer male scents that signal a better cell-mediated immune response. Biology Letters 1: 404–406.
López, P., Muñoz, A. and Martín, J. 2002. Symmetry, male dominance and female mate preferences in the Iberian rock lizard, *Lacerta monticola*. Behavioral Ecology and Sociobiology 52: 342–347.
López, P., Amo, L. and Martín, J. 2006. Reliable signaling by chemical cues of male traits and health state in male lizards, *Lacerta monticola*. Journal of Chemical Ecology 32: 473–488.
Losos, J. B. 1985. An experimental demonstration of the species-recognition role of *Anolis* dewlap color. Copeia 1985: 905–910.
Losos, J. B. and Chu, L. -r. 1998. Examination of factors potentially affecting dewlap size in Caribbean anoles. Copeia 1998: 430–438.
Martín, J. and López, P. 2000. Chemoreception, symmetry and mate choice in lizards. Proceedings of the Royal Society of London. Series B: Biological Sciences 267: 1265–1269.
Martín, J. and López, P. 2006. Links between male quality, male chemical signals, and female mate choice in Iberian rock lizards. Functional Ecology 20: 1087–1096.
Martín, J. and López, P. 2007. Scent may signal fighting ability in male Iberian rock lizards. Biology Letters 3: 125–127.

Martín, J. and López, P. 2009. Multiple color signals may reveal multiple messages in male Schreiber's green lizards, *Lacerta schreiberi*. Behavioral Ecology and Sociobiology 63: 1743–1755.

Martín, J., Moreira, P. L. and López, P. 2007. Status-signalling chemical badges in male Iberian rock lizards. Functional Ecology 21: 568–576.

Medica, P. A., Turner, F. B. and Smith, D. D. 1973. Hormonal induction of color change in female leopard lizards, *Crotaphytus wislizenii*. Copeia 1973: 658–661.

Miller, H. C., Moore, J. A., Nelson, N. J. and Daugherty, C. H. 2009. Influence of major histocompatibility complex genotype on mating success in a free-ranging reptile population. Proceedings of the Royal Society B: Biological Sciences 276: 1695–1704.

Moore, J. A., Nelson, N., Keall, S. and Daugherty, C. 2008. Implications of social dominance and multiple paternity for the genetic diversity of a captive-bred reptile population (tuatara). Conservation Genetics 9: 1243–1251.

Moore, J. A., Daugherty, C. H., Godfrey, S. S. and Nelson, N. J. 2009. Seasonal monogamy and multiple paternity in a wild population of a territorial reptile (tuatara). Biological Journal of the Linnean Society 98: 161–170.

Moore, M. C., Hews, D. K. and Knapp, R. 1998. Hormonal control and evolution of alternative male phenotypes: generalizations of models for sexual differentiation. American Zoologist 38: 133–151.

Moreira, P. L. and Birkhead, T. R. 2003. Copulatory plugs in the Iberian rock lizard do not prevent insemination by rival males. Functional Ecology 17: 796–802.

Moreira, P. L. and Birkhead, T. R. 2004. Copulatory plug displacement and prolonged copulation in the Iberian rock lizard (*Lacerta monticola*). Behavioral Ecology and Sociobiology 56: 290–297.

Moreira, P. L., Nunes, V. L., Martín, J. and Paulo, O. S. 2007. Copulatory plugs do not assure high first male fertilisation success: sperm displacement in a lizard. Behavioral Ecology and Sociobiology 62: 281–288.

Muñoz, M. M., Crawford, N. G., McGreevy, T. J., Messana, N. J., Tarvin, R. D., Revell, L. J., Zandvliet, R. M., Hopwood, J. M., Mock, E., Schneider, A. L. and Schneider, C. J. 2013. Divergence in coloration and ecological speciation in the *Anolis marmoratus* species complex. Molecular Ecology 22: 2668–2682.

Nagy, K. A. 1983. Ecological energetics. pp. 24–54. In R. B. Huey, E. R. Pianka and T. W. Schoener (eds.), *Lizard ecology: studies of a model organism*. Harvard University Press: Cambridge.

Ng, J. and Glor, R. E. 2011. Genetic differentiation among populations of a Hispaniolan trunk anole that exhibit geographical variation in dewlap colour. Molecular Ecology 20: 4302–4317.

Noble, D. W., Wechmann, K., Keogh, J. S. and Whiting, M. J. 2013. Behavioral and morphological traits interact to promote the evolution of alternative reproductive tactics in a lizard. The American Naturalist 182: 726–742.

Oakes, E. J. 1992. Lekking and the evolution of sexual dimorphism in birds: comparative approaches. The American Naturalist 140: 655–684.

Olsson, M. 1993. Male preference for large females and assortative mating for body size in the sand lizard (*Lacerta agilis*). Behavioral Ecology and Sociobiology 32: 337–341.

Olsson, M. 1994. Nuptial coloration in the sand lizard, Lacerta agilis: an intra-sexually selected cue to lighting ability. Animal Behaviour 48: 607–613.

Olsson, M. 2001. No female mate choice in Mallee dragon lizards, *Ctenophorus fordi*. Evolutionary Ecology 15: 129–141.

Olsson, M. and Madsen, T. 1995. Female choice on male quantitative traits in lizards—why is it so rare? Behavioral Ecology and Sociobiology 36: 179–184.

Olsson, M. and Madsen, T. 1998. Sexual selection and sperm competition in reptiles. pp. 503–578. In T. R. Birkhead and A. P. Moller (eds.), *Sperm competition and sexual selection*. Academic Press: Cambridge.

Olsson, M., Shine, R., Wapstra, E., Ujvari, B. and Madsen, T. 2002. Sexual dimorphism in lizard body shape: the roles of sexual selection and fecundity selection. Evolution 56: 1538–1542.

Olsson, M., Madsen, T., Nordby, J., Wapstra, E., Ujvari, B. and Wittsell, H. 2003. Major histocompatibility complex and mate choice in sand lizards. Proceedings of the Royal Society of London. Series B: Biological Sciences 270: S254–S256.
Olsson, M., Madsen, T., Wapstra, E., Silverin, B., Ujvari, B. and Wittzell, H. 2005. MHC, health, color, and reproductive success in sand lizards. Behavioral Ecology and Sociobiology 58: 289–294.
Olsson, M., Healey, M. O., Wapstra, E., Schwartz, T., Lebas, N. and Uller, T. 2007. Mating system variation and morph fluctuations in a polymorphic lizard. Molecular Ecology 16: 5307–5315.
Olsson, M., Wapstra, E., Schwartz, T., Madsen, T., Ujvari, B. and Uller, T. 2011. In hot pursuit: fluctuating mating system and sexual selection in sand lizards. Evolution 65: 574–583.
Ord, T. J. and Martins, E. P. 2006. Tracing the origins of signal diversity in anole lizards: phylogenetic approaches to inferring the evolution of complex behaviour. Animal Behaviour 71: 1411–1429.
Ord, T. J., Blumstein, D. T. and Evans, C. S. 2001. Intrasexual selection predicts the evolution of signal complexity in lizards. Proceedings of the Royal Society of London. Series B: Biological Sciences 268: 737–744.
Orrell, K. S. and Jenssen, T. A. 2002. Male mate choice by the lizard *Anolis carolinensis*: a preference for novel females. Animal Behaviour 63: 1091–1102.
Paemelaere, E. A. D., Guyer, C. and Dobson, F. S. 2011a. A phylogenetic framework for the evolution of female polymorphism in anoles. Biological Journal of the Linnean Society 104: 303–317.
Paemelaere, E. A. D., Guyer, C. and Dobson, F. S. 2011b. Survival of alternative dorsal-pattern morphs in females of the *Anole norops humilis*. Herpetologica 67: 420–427.
Perry, G. and Garland, T., Jr. 2002. Lizard home ranges revisited: effects of sex, body size, diet, habitat, and phylogeny. Ecology 83: 1870–1885.
Perry, G., Levering, K., Girard, I. and Garland, T., Jr. 2004. Locomotor performance and social dominance in male *Anolis cristatellus*. Animal Behavior 67: 37–47.
Quinn, V. S. and Hews, D. K. 2003. Positive relationship between abdominal coloration and dermal melanin density in phrynosomatid lizards. Copeia 2003: 858–864.
Rand, A. S. 1967. Ecology and social organization in *Anolis lineatopus*. Proceedings of the United States National Museum 122: 1–79.
Rand, M. S. 1992. Hormonal control of polymorphic and sexually dimorphic coloration in the lizard *Sceloporus undulatus erythrocheilus*. General and Comparative Endocrinology 88: 461–468.
Rice, W. R. and Chippindale, A. K. 2001. Intersexual ontogenetic conflict. Journal of Evolutionary Biology 14: 865–693.
Richard, M., Losdat, S., Lecomte, J., de Fraipont, M. and Clobert, J. 2009. Optimal level of inbreeding in the common lizard. Proceedings of the Royal Society B: Biological Sciences 276: 2779–2786.
Robertson, J. M. and Rosenblum, E. B. 2009. Rapid divergence of social signal coloration across the White Sands ecotone for three lizard species under strong natural selection. Biological Journal of the Linnean Society 98: 243–255.
Robertson, J. M. and Rosenblum, E. B. 2010. Male territoriality and 'sex confusion' in recently adapted lizards at White Sands. Journal of Evolutionary Biology 23: 1928–1936.
Robson, M. and Miles, D. 2000. Locomotor performance and dominance in male tree lizards, *Urosaurus ornatus*. Functional Ecology 14: 338–344.
Rosenblum, E. B. 2006. Convergent evolution and divergent selection: lizards at the White Sands ecotone. American Naturalist 167: 1–15.
Rosenblum, E. B. 2008. Preference for local mates in a recently diverged population of the lesser earless lizard (*Holbrookia maculata*) at White Sands. Journal of Herpetology 42: 572–583.
Rosenblum, E. B. and Harmon, L. J. 2011. "Same same but different": replicated ecological speciation at White Sands. Evolution 65: 946–960.

Rosenblum, E. B., Römpler, H., Schöneberg, T. and Hoekstra, H. E. 2010. Molecular and functional basis of phenotypic convergence in white lizards at White Sands. Proceedings of the National Academy of Sciences 107: 2113–2117.

Runemark, A. and Svensson, E. I. 2012. Sexual selection as a promoter of population divergence in male phenotypic characters: a study on mainland and islet lizard populations. Biological Journal of the Linnean Society 106: 374–389.

Sacchi, R., Rubolini, D., Gentilli, A., Pupin, F., Razzetti, E., Scali, S., Galeotti, P. and Fasola, M. 2007. Morph-specific immunity in male *Podarcis muralis*. Amphibia-Reptilia 28: 408–412.

Salvador, A., Díaz, J. A., Veiga, J. P., Bloor, P. and Brown, R. P. 2008. Correlates of reproductive success in male lizards of the alpine species *Iberolacerta cyreni*. Behavioral Ecology 19: 169–176.

Salvador, A., Veiga, J. P., Martin, J., Lopez, P., Abelenda, M. and Puerta, M. 1996. The cost of producing a sexual signal: testosterone increases the susceptibility of male lizards to ectoparasitic infestation. Behavioral Ecology 7: 145–150.

Sanger, T. J., Sherratt, E., McGlothlin, J. W., Brodie, E. D., Losos, J. B. and Abzhanov, A. 2013. Convergent evolution of sexual dimorphism in skull shape using distinct developmental strategies. Evolution 67: 2180–2193.

Schneider, C. J. 2008. Exploiting genomic resources in studies of speciation and adaptive radiation of lizards in the genus *Anolis*. Integrative and Comparative Biology 48: 520–526.

Schoener, T. W. 1967. Ecological significance of sexual dimorphism in size in the lizard *Anolis conspersus*. Science 155: 474–477.

Schoener, T. W. 1969a. Models of optimal size for solitary predators. American Naturalist 103: 277–313.

Schoener, T. W. 1969b. Size patterns in West Indian *Anolis* lizards: I. Size and species diversity. Systematic Zoology 18: 386–401.

Schoener, T. W. 1977. Competition and the niche. Biology of the Reptilia 7: 35–136.

Schoener, T. W. and Schoener, A. 1976. The ecological context of female pattern polymorphism in the lizard *Anolis sagrei*. Evolution 30: 650–658.

Schoener, T. W., Slade, J. B. and Stinson, G. H. 1982. Diet and sexual dimorphism in the very catholic lizard genus *Leiocephalus* of the Bahamas. Oecologia 53: 160–169.

Sever, D. M. and Hamlett, W. C. 2002. Female sperm storage in reptiles. Journal of Experimental Zoology 292: 187–199.

Shine, R. 1989. Ecological causes for the evolution of sexual dimorphism: a review of the evidence. Quarterly Review of Biology 64: 419–461.

Sinervo, B. 2001. Runaway social games, genetic cycles driven by alternative male and female strategies, and the origin of morphs. Genetica 112-113: 417–434.

Sinervo, B. and Lively, C. M. 1996. The rock-paper-scissors game and the evolution of alternative male reproductive strategies. Nature 380: 240–243.

Sinervo, B. and Svensson, E. 2002. Correlational selection and the genomic architecture. Heredity 89: 329–338.

Sinervo, B., Svensson, E. and Comendant, T. 2000. Density cycles and an offspring quantity and quality game driven by natural selection. Nature 406: 985–988.

Sinervo, B., Heulin, B., Surget-Groba, Y., Clobert, J., Miles, D. B., Corl, A., Chaine, A. and Davis, A. 2007. Models of density-dependent genic selection and a new rock-paper-scissors social system. American Naturalist 170: 663–680.

Smith, J. M. and Zucker, N. 1997. Do female tree lizards, *Urosaurus ornatus*, exhibit mate choice? Journal of Herpetology 31: 179–186.

Srinivas, S. R., Hegde, S. N., Sarkar, H. B. D. and Shivanandappa, T. 1995. Sperm storage in the oviduct of the tropical rock lizard, *Psammophilus dorsalis*. Journal of Morphology 224: 293–301.

Stamps, J. A. 1983. Sexual selection, sexual dimorphism and territoriality. pp. 169–204. In E. R. Pianka and T. W. Schoener (eds.), *Lizard ecology: studies of a model organism*. Harvard University Press: Cambridge.

Stamps, J. A. 1993. Sexual size dimorphism in species with asymptotic growth after maturity. Biological Journal of the Linnean Society 50: 123–145.
Stamps, J. A. 1999. Relationships between female density and sexual size dimorphism in samples of *Anolis sagrei*. Copeia 1999: 760–765.
Stamps, J. A., Losos, J. B. and Andrews, R. M. 1997. A comparative study of population density and sexual size dimorphism in lizards. American Naturalist 149: 64–90.
Stapley, J. and Keogh, J. S. 2005. Behavioral syndromes influence mating systems: floater pairs of a lizard have heavier offspring. Behavioral Ecology 16: 514–520.
Stow, A. J. and Sunnucks, P. 2004. Inbreeding avoidance in Cunningham's skinks (*Egernia cunninghami*) in natural and fragmented habitat. Molecular Ecology 13: 443–447.
Stuart-Fox, D. and Moussalli, A. 2007. Sex-specific ecomorphological variation and the evolution of sexual dimorphism in dwarf chameleons (*Bradypodion* spp.). Journal of Evolutionary Biology 20: 1073–1081.
Stuart-Fox, D. M. and Ord, T. J. 2004. Sexual selection, natural selection and the evolution of dimorphic coloration and ornamentation in agamid lizards. Proceedings of the Royal Society of London. Series B: Biological Sciences 271: 2249–2255.
Stuart-Fox, D. and Owens, I. P. F. 2003. Species richness in agamid lizards: chance, body size, sexual selection or ecology? Journal of Evolutionary Biology 16: 659–669.
Stuart-Smith, J., Swain, R. and Wapstra, E. 2007. The role of body size in competition and mate choice in an agamid with female-biased size dimorphism. Behaviour 144: 1087–1102.
Svensson, E. I., McAdam, A. G. and Sinervo, B. 2009. Intralocus sexual conflict over immune defense, gender load, and sex-specific signaling in a natural lizard population. Evolution 63: 3124–3135.
Thompson, C. W. and Moore, M. C. 1991. Throat colour reliably signals status in male tree lizards, *Urosaurus ornatus*. Animal Behaviour 42: 745–753.
Tokarz, R. R. 1992. Male mating preference for unfamiliar females in the lizard, *Anolis sagrei*. Animal Behaviour 44: 843–849.
Tokarz, R. R. 1995. Mate choice in lizards: a review. Herpetological Monographs 9: 17–40.
Tokarz, R. R. 2002. An experimental test of the importance of the dewlap in male mating success in the lizard *Anolis sagrei*. Herpetologica 58: 87–94.
Tokarz, R. R., Paterson, A. V. and McMann, S. 2005. Importance of dewlap display in male mating success in free-ranging brown anoles (*Anolis sagrei*). Journal of Herpetology 39: 174–177.
Trivers, R. L. 1976. Sexual selection and resource-accruing abilities in *Anolis garmani*. Evolution 30: 253–269.
Uller, T. and Olsson, M. 2008. Multiple paternity in reptiles: patterns and processes. Molecular Ecology 17: 2566–2580.
Uller, T., Stuart-Fox, D. and Olsson, M. 2010. Evolution of primary sexual characters in reptiles. pp. 426–452. In J. L. Leonard and A. Cordoba-Aguilar (eds.), *The evolution of primary sexual characters in animals*. Oxford University Press: Oxford.
Václav, R., Prokop, P. and Fekiač, V. 2007. Expression of breeding coloration in European green lizards (*Lacerta viridis*): variation with morphology and tick infestation. Canadian Journal of Zoology 85: 1199–1206.
Vanhooydonck, B., Herrel, A. Y., Van Damme, R. and Irschick, D. J. 2005. Does dewlap size predict male bite performance in Jamaican *Anolis* lizards? Functional Ecology 19: 38–42.
Vercken, E., Massot, M., Sinervo, B. and Clobert, J. 2007. Colour variation and alternative reproductive strategies in females of the commong lizard *Lacerta vivipara*. Journal of Evolutionary Biology 20: 221–232.
Vercken, E., Clobert, J. and Sinervo, B. 2010. Frequency-dependent reproductive success in female common lizards: a real-life hawk-dove-bully game? Oecologia 162: 49–58.
Vitt, L. J. and Caldwell, J. P. 2013. *Herpetology: an Introductory Biology of Amphibians and Reptiles*. Academic Press, London, pp. 776.
Vitt, L. J. and Cooper, W. E., Jr. 1985. The evolution of sexual dimorphism in the skink *Eumeces laticeps*: an example of sexual selection. Canadian Journal of Zoology 63: 995–1002.

Watkins, G. G. 1996. Proximate causes of sexual size dimorphism in the iguanian lizard *Microlophus occipitalis*. Ecology 77: 1473–1482.
Watkins, G. G. 1997. Inter-sexual signalling and the functions of female coloration in the tropidurid lizard *Microlophus occipitalis*. Animal Behaviour 53: 843–852.
Weiss, S. L. 2002. Reproductive signals of female lizards: pattern of trait expression and male response. Ethology 108: 793–813.
Weiss, S. L. 2006. Female-specific color is a signal of quality in the striped plateau lizard (*Sceloporus virgatus*). Behavioral Ecology 17: 726–732.
Weiss, S. L., Kennedy, E. A. and Bernhard, J. A. 2009. Female-specific ornamentation predicts offspring quality in the striped plateau lizard, *Sceloporus virgatus*. Behavioral Ecology 20: 1063–1071.
While, G. M., Sinn, D. L. and Wapstra, E. 2009. Female aggression predicts mode of paternity acquisition in a social lizard. Proceedings of the Royal Society B: Biological Sciences 276: 2021–2029.
Whiting, M. J. and Bateman, P. W. 1999. Male preference for large females in the lizard *Platysaurus broadleyi*. Journal of Herpetology 33: 309–312.
Whiting, M. J., Stuart-Fox, D. M., O'Connor, D., Firth, D., Bennett, N. C. and Blomberg, S. P. 2006. Ultraviolet signals ultra-aggression in a lizard. Animal Behaviour 72: 353–363.
Whiting, M. J., Webb, J. K. and Keogh, J. S. 2009. Flat lizard female mimics use sexual deception in visual but not chemical signals. Proceedings of the Royal Society B, Biological Sciences 276: 1585–1591.
Wiens, J. J. 1999. Phylogenetic evidence for multiple losses of a sexually selected character in phrynosomatid lizards. Proceedings of the Royal Society of London B Biological Sciences 266: 1529–1536.
Wiens, J. J., Reeder, T. W. and Montes de Oca, A. N. 1999. Molecular phylogenetics and evolution of sexual dichromatism among populations of the Yarrow's spiny lizards (*Sceloporus jarrovii*). Evolution 53: 1884–1897.
Wikelski, M. and Trillmich, F. 1997. Body size and sexual dimorphism in marine iguanas fluctuates as a result of opposing natural and sexual selection: an island comparison. Evolution 51: 922–936.
Wikelski, M., Carbone, C., Bednekoff, P. A., Choudhury, S. and Tebbich, S. 2001. Why is female choice not unanimous? Insights from costly mate sampling in marine iguanas. Ethology 107: 623–638.
Wymann, M. and Whiting, M. 2003. Male mate preference for large size overrides species recognition in allopatric flat lizards (*Platysaurus broadleyi*). Acta Ethologica 6: 19–22.
Zamudio, K. R. 1998. The evolution of female-biased sexual size dimorphism: a population-level comparative study in horned lizards (*Phrynosoma*). Evolution 52: 1821–1833.
Zamudio, K. R. and Sinervo, B. 2000. Polygyny, mate-guarding, and posthumous fertilization as alternative male mating strategies. Proceedings of the National Academy of Sciences 97: 14427–14432.

CHAPTER 5

Cues for Reproduction in Squamate Reptiles

James U. Van Dyke

5.1 INTRODUCTION

To maximize fitness, animals should initiate reproduction based on information from suites of cues that communicate three variables critical to reproductive success: 1) environmental conduciveness for successful reproduction, and survival of offspring and (usually) parents; 2) physiological capability of parents to reproduce; and 3) likelihood of successful mating. Squamates vary widely in reproductive mode (egg-laying, or oviparity, vs. live birth, or viviparity), reproductive frequency (including reproducing only once, i.e., semelparity), and output (Tinkle *et al.* 1970; Dunham *et al.* 1988), all of which may alter the phenology of gametogenesis and embryonic development relative to season, physiological state (i.e., body condition), courtship, and mating. These phenomenological differences necessitate divergent reproductive decision-making approaches that may be informed by different suites of cues. In addition, specific components of reproduction, including gametogenesis and mating behavior, may not be stimulated by the same environmental or physiological cues.

The purpose of this review is to discuss the current state of knowledge of the mechanisms squamates use as cues for the decision to reproduce. Here, the decision to reproduce is defined as analogous to a life-history allocation decision (e.g., Dunham *et al.* 1989), rather than as a result of conscious thought processes. The endocrine connections of the

School of Biological Sciences, Heydon-Laurence Bldg A08, University of Sydney, New South Wales, 2006, Australia.

hypothalamic-pituitary-gonadal axis are briefly reviewed because they are critical to communicating information from reproductive cues to the brain, gonads, and accessory reproductive organs. Associated and dissociated reproduction are also briefly introduced because both strategies might have ramifications for the evolution of reproductive cues in squamates. The bulk of the review then focuses on cues that provide squamates with information regarding season and environmental conditions, resource availability, and the likelihood of mating. In cases where phenomena have been suggested to act as cues for reproduction, but few physiological explanations have been examined in squamates, relevant literature from mammals and birds is discussed in order to suggest possible avenues for future research. Throughout, a heuristic model of hypothetical signaling mechanisms that communicate detected cues to the hypothalamic-pituitary-gonadal axis (Figs. 5.1–5.4) is constructed to clearly identify hypothetical mechanisms linking reproductive cue detection to the hypothalamic-gonadal-pituitary axis. The review concludes with a discussion of how recent advances in

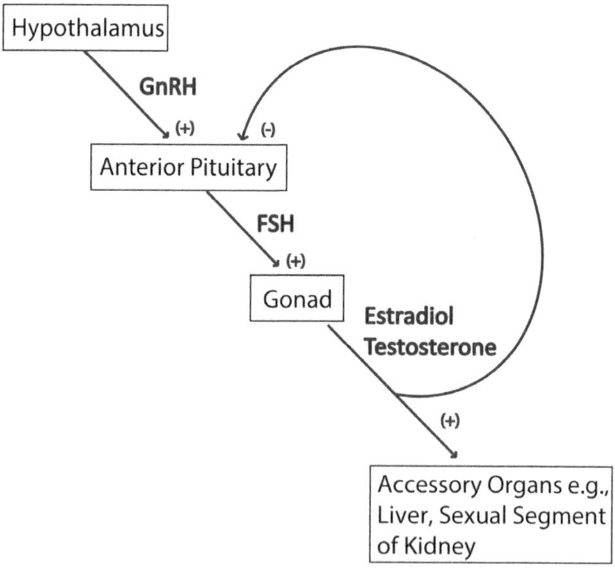

Fig. 5.1 Generalized structure of the vertebrate hypothalamic-pituitary-gonadal axis (HPGA). The HPGA forms the backbone of a heuristic model of the signaling mechanisms that communicate reproductive cues to the brain and result in reproductive decisions. Organs are represented by black boxes, while hormones are blue. Addition signs (+) indicate that a hormone stimulates upregulation of the receiving organ, while subtraction symbols (–) indicate that it stimulates downregulation of the receiving organ. It remains unclear how the HPGA behaves differently in associated or dissociated breeders.

Color image of this figure appears in the color plate section at the end of the book.

next-generation sequencing, candidate-gene approaches, and proteomics might be used to further elucidate the molecular mechanisms underlying reproductive cues. Although this volume focuses on lizards, snakes are included in this review because they are highly specialized lizards (Vidal and Hedges 2005; Pyron *et al.* 2013), and because they have been a major model system for investigations of the effects of resource availability and mating on reproduction, and the roles of the pineal gland and melatonin, in squamates. Where possible, relevant information from tuatara, *Sphenodon punctatus,* is also reviewed.

5.2 HYPOTHALAMIC-PITUITARY-GONADAL AXIS

As in other vertebrates, the hypothalamic-pituitary-gonadal axis (HPGA) communicates reproductive cues from the brain to the gonads and accessory reproductive organs in squamate reptiles (Fig. 5.1; Licht 1979; Bona-Gallo *et al.* 1980; Krohmer and Lutterschmidt 2011). Reproduction is initiated by production of gonadotropin-releasing hormones (GnRHs) by the hypothalamus (King and Millar 1980). Secretion of GnRHs is likely pulsatile (Licht and Porter 1987), but the only evidence for pulsatile release in reptiles is from turtles (Tsai and Licht 1993). Gonadotropin-releasing hormones stimulate the production of gonadotropins in the anterior pituitary, which are carried in blood plasma to the gonads (Eyeson 1971; Licht 1979). In response to stimulation by gonadotropins, gonads recruit gametes and synthesize the steroid sex hormones estradiol (in females) and testosterone (in males), which activate accessory reproductive organs (Hahn 1967; Courty and Dufaure 1979; Aldridge 1982; Ho *et al.* 1982).

Estradiol in particular acts to initiate production of yolk (vitellogenesis) in the liver of female squamates (Hahn 1967; Ho *et al.* 1982; Wallace 1985; Ho 1987), while in males of many species testosterone stimulates the hypertrophy of the sexual segment of the kidney and mating behavior, and likely plays a role in spermatogenesis (Prasad and Sanyal 1969; Weil and Aldridge 1981; Weil 1985; Aldridge *et al.* 1990; Aldridge *et al.* 2009; Aldridge *et al.* 2011). Both estradiol and testosterone may also inhibit gonadotropin secretion by the anterior pituitary (reviewed by Licht 1979), producing a negative-feedback loop once reproduction is initiated (Fig. 5.1).

Although the general structure of the HPGA described here is largely conserved across vertebrates, the endocrine components of the squamate HPGA are unique. Most notably, luteinizing hormone (LH), which is an important gonadotropin in many vertebrates, including turtles and crocodilians, does not exist in the squamate HPGA (Licht 1974; Licht *et al.* 1974; Licht and Crews 1975). Instead, follicle-stimulating hormone (FSH) appears to be the only functional gonadotropin in squamates (Licht and

Crews 1975; Licht 1979). Furthermore, the peptide structures of GnRH and FSH (and their receptors) differ across taxa (Licht 1983; Licht *et al.* 1984; Powell *et al.* 1986; Licht and Porter 1987; Borrelli *et al.* 2001). Thus, commercially-available hormone assays, which are designed to target mammal or chicken hormones, are unable to reliably measure titers of GnRH or FSH in squamates. In addition, injections of heterologous GnRH, LH, or FSH into squamates, although seemingly effective (e.g., Jones *et al.* 1973; Sinervo and Licht 1991), may not be reliable indicators of hormone function (Licht 1983). Prolactin may also play a role in the squamate HPGA, but studies examining the effects of prolactin on reproduction have produced contradictory results (reviewed by Mazzi and Vellano 1987). The HPGA of the tuatara, *Sphenodon punctatus* has not been thoroughly examined, but males and females appear to utilize testosterone and estradiol similarly to squamates and other vertebrates (Cree *et al.* 1991; Cree *et al.* 1992).

Difficulties in accurately measuring GnRH and gonadotropin secretion and function have constrained studies of reproductive cues in squamate reptiles. Perhaps as a result, studies of GnRH, gonadotropins, other potential hormonal regulating systems (e.g., inhibins; Licht and Porter 1987), and reproductive cues in squamates have declined in frequency since the early 1990s. In contrast to GnRH and FSH, steroid hormone structures are largely conserved across taxa, and assays targeting mammalian or chicken estradiol, testosterone, and progesterone have been widely used in studies of squamate reproduction (e.g., Naulleau *et al.* 1987; Naulleau and Fleury 1990; Saint Girons *et al.* 1993; Swain and Jones 1994; Schuett *et al.* 1997; Edwards and Jones 2001; Martinez-Torres *et al.* 2003; Zaidan *et al.* 2003; Almeida-Santos *et al.* 2004; Taylor *et al.* 2004; Yamanouye *et al.* 2004; Lind *et al.* 2010; Van Dyke *et al.* 2012).

Elucidating the endocrine components of the HPGA is of paramount importance to understanding the cues for reproduction in squamates. Without this knowledge, we can only determine that a particular event is a cue for reproduction and how it is detected, but we cannot effectively determine how detection of the event is transduced into a reproductive response. We also cannot determine how multiple cues (e.g., temperature, resource availability) are simultaneously transduced in the brain and/or HPGA.

5.3 ASSOCIATED VS. DISSOCIATED REPRODUCTION

Squamate reptiles exhibit remarkable diversity in the timing of gonadal recrudescence, gametogenesis, and mating. In associated reproduction, gonadal recrudescence and gametogenesis occur simultaneously with mating (modified from Crews 1984; Crews 1999; Aldridge *et al.* 2009). In dissociated reproduction, gametogenesis occurs months prior to mating behavior, and gametes are stored until use (modified from Crews 1984; Crews 1999; Aldridge *et al.* 2009). For the purposes of this review, this distinction is critical because mating and gametogenesis are likely stimulated by the same cues and HPGA hormonal cascade in associated breeders, but may be stimulated by different cues in dissociated breeders. Krohmer and Lutterschmidt (2011) suggest that dissociated reproduction occurs in male snakes when the HPGA stimulates a peak of plasma testosterone concentrations during gametogenesis, and metabolic clearance slowly declines until the mating season. In contrast, gametogenesis and mating may occur at such distant points in time (e.g., late summer vs. the following spring, respectively) that they could even be stimulated by different HPGA hormonal cascades, but this hypothesis has never been tested due to the difficulties of measuring cycles in GnRH and gonadotropin concentrations, and corresponding receptor densities, in reptiles. However, other studies have documented two peaks in plasma testosterone concentration in male snakes with dissociated reproduction, one during spermatogenesis and one during the mating season (Aldridge 1979b; Weil and Aldridge 1981; Aldridge *et al.* 1990). Interestingly, the peak during the mating season is often higher than during spermatogenesis. Weil (1985) suggests that, during spermatogenesis, testicular production of testosterone might not be reflected by plasma concentrations due to capture and metabolic clearance within testicular tissues prior to transport to the bloodstream.

Although many studies consider associated and dissociated reproduction only in males (e.g., Krohmer and Lutterschmidt 2011), I follow Crews (1984) in applying the terms to both males and females. Under this definition, one sex within a species may exhibit associated reproduction, while the other exhibits dissociated reproduction. For example, most male North American colubrids undergo spermatogenesis in the summer, but mate during the following spring (reviewed in Aldridge *et al.* 2009), which is an example of dissociated reproduction. In contrast, all female North American colubrids examined thus far initiate vitellogenesis in the spring, and also mate during the same spring (reviewed in Aldridge *et al.* 2009), which exemplifies associated reproduction. Reproduction in the sympatric North American pit vipers differs greatly from the colubrid condition. Like colubrids, male North American pit vipers all undergo spermatogenesis in the summer, but males of different species mate in the late summer/early autumn immediately after spermatogenesis (associated reproduction), store sperm to mate in the spring long after spermatogenesis (dissociated reproduction), or both (Aldridge and Duvall 2002). In contrast, all female North American pit vipers examined thus far initiate vitellogenesis in the late summer/early autumn, but do not complete vitellogenesis and ovulate until spring (Aldridge and Duvall 2002). Thus, because vitellogenesis may last from summer to the following spring and mating occurs during either or both seasons, all female North American pit vipers exhibit associated reproduction, regardless of when mating occurs. However, in females that mate in the spring, it is likely that mating behavior and gametogenesis are stimulated by different cues, even though they technically exhibit associated reproduction. While both sexes of most lizards are associated breeders (Aldridge *et al.* 2011), some temperate lizards, including Australian spotted skinks, *Niveoscincus ocellatus* (Jones *et al.* 1997) and tussock skinks, *Pseudemoia entrecasteauxii* (Murphy *et al.* 2006), exhibit a reproductive cycle similar to that of North American pit vipers. The tuatara, *Sphenodon punctatus*, exhibits a unique reproductive cycle: males initiate spermatogenesis just prior to or concurrent with autumn mating, while females initiate vitellogenesis up to three years prior to a given reproductive event (Cree *et al.* 1992; Brown *et al.* 1994). Thus, males appear to be associated breeders, while females are dissociated breeders, but exhibit a reproductive schedule vastly different from dissociated breeding in squamates.

Associated and dissociated reproduction are more thoroughly discussed elsewhere in the primary literature (Crews 1984; Crews 1999; Aldridge and Duvall 2002; Aldridge *et al.* 2009). They are mentioned here to illustrate that cues for reproduction may differ within and among even closely related or sympatric squamate taxa, and may also differ between gametogenesis and mating behavior. In the pit viper example above, the cues for spermatogenesis and vitellogenesis may also stimulate mating behavior in populations that mate in autumn, but gametogenesis and mating behavior may be stimulated by different cues in populations that mate in spring. Similarly, if males of a species initiate spermatogenesis in autumn but females initiate vitellogenesis in spring, then the cues for gonadal recrudescence and gametogenesis, or the physiological processes that detect and transduce those cues, may differ between sexes.

5.4 SEASONAL CUES: PHOTOPERIOD, TEMPERATURE, AND MOISTURE

Reproduction must be coordinated with suitable environmental conditions to maximize developmental success, parental survival, and offspring survival. In squamates, developmental success is directly related to temperature and, especially in oviparous (egg-laying) species, moisture (Packard *et al.* 1982; Gutzke and Packard 1987; Andrews *et al.* 2000; Rock *et al.* 2000; Ji and Du 2001). In temperate and tropical systems, temperature and/or moisture vary seasonally throughout the year. Therefore, squamates can only maximize reproductive success if reproduction is initiated based on cues that provide information on time of year (i.e., season). Seasonal changes in temperature and rainfall provide direct cues regarding environmental conditions favorable to reproduction, while photoperiod provides an indirect metric of season, and should be correlated, at least partially, with seasonal changes in temperature and moisture. Accordingly, squamates appear to use all three factors as seasonal cues for reproduction (e.g., zeitgebers; Tinkle and Irwin 1965; Licht and Porter 1987; Brown and Shine 2006). Within species, inter-individual variation in the sensitivity of season detecting mechanisms may also play a role in determining how early reproduction occurs (Wapstra *et al.* 1999), which may be critical to maximizing offspring fitness (Olsson and Shine 1997).

5.4.1 Phenology of Seasonal Cues: Temperate vs. Tropical Taxa

In temperate squamates, seasonal changes in environmental temperature are the primary seasonal cue for initiating reproduction, while seasonal changes in photoperiod appear to be a secondary cue (Saint Girons 1982). Most temperate squamates (but not all; Moore *et al.* 1984) appear to require a period of exposure to cold environmental temperatures during winter, followed by warm temperatures in the spring or summer to initiate gonadal recrudescence, reproductive behavior, or both (Marion 1970; Licht 1972; Botte *et al.* 1978; Marion 1982; Crews 1983; Lutterschmidt 2012). Spermatogenesis in particular appears to be stimulated by warm temperatures in North American colubrids and rattlesnakes, while photoperiod has little effect (Aldridge 1975; Aldridge 1979b; Weil and Aldridge 1979). In captivity, many temperate squamate species only breed after winter exposure to cold temperatures followed by spring warming (Osborne 1982). Delays in seasonal spring warming during spring may delay the initiation of reproduction (e.g., Castilla *et al.* 1992; Smith *et al.* 1995). The seasonal cues for reproduction remain unknown in *Sphenodon*, but males initiate spermatogenesis in late summer/autumn. Females initiate vitellogenesis in the spring following oviposition, yet vitellogenesis may last three years until ovulation occurs, and mating occurs in autumn (Brown *et al.* 1991; Cree *et al.* 1992).

Temperate species that do not require exposure to cold temperatures still exhibit greater reproductive responses to seasonal changes in temperature than to changes in photoperiod. These species exhibit "refractory periods" after reproduction during which changes in neither photoperiod nor temperature stimulate gonadal recrudescence. After the refractory period ends, only increases in temperature, and not day length, can stimulate reproduction (Tinkle and Irwin 1965; Licht *et al.* 1969; Cuellar and Cuellar 1977; Lofts 1978; Cuellar 1984). The animals used in most of these studies (excluding male *Naja naja* in Lofts 1978) were female lizards from species

(*Aspidoscelis uniparens, Lacerta* sp., and *Uta stansburiana*) that produce multiple clutches each year. Multi-clutching species do not experience cold temperatures between clutches in a single season, and refractory periods may be necessary to allow females to "recover" in some physiological condition, possibly stored nutrient reserves, prior to subsequent reproductive bouts. Thus, the evolution of post-reproductive refractory periods may be partly associated with the evolution of repeated reproduction in a single season.

Tropical squamates exhibit considerable diversity in the seasonality of reproduction, but relatively few species reproduce continuously throughout the year (reviewed by Brown and Shine 2006). Although tropical squamates do not experience substantial cool periods during winter, slight decreases in temperature, in concert with reduced day length, are sufficient to stimulate reproduction in some species (e.g., tropical anoles; Gorman and Licht 1974). Tropical tropidurine lizards, while potentially capable of reproducing throughout the year, exhibit a peak of reproduction during the wet season (Vitt and Goldberg 1983). In the tropical spiny lizard, *Sceloporus utiformis*, testicular recrudescence is associated with increasing temperature and precipitation, while ovarian recrudescence is associated only with increasing photoperiod (Ramirez-Bautista and Gutierrez-Mayen 2003). In tropical Australian Skinks (*Ctenotus* sp.), reproduction is highly correlated with rainfall (James 1991). Tropical Australian water pythons, *Liasis fuscus*, nest late in the dry season thus maximizing incubation temperatures during development, and coil around their eggs to prevent desiccation, while the sympatric keelback, *Tropidonophis mairii*, nests immediately after the end of the wet season, possibly to reduce egg desiccation (Brown and Shine 2006). In contrast, tropical populations of several Australian skinks (*Carlia pectoralis, Cryptoblepharus virgatus, Heteronotia binoei*, and *Lampropholis delicata*) do not appear to entrain reproduction with seasonal changes in temperature, photoperiod, or rainfall, and may rely on different environmental factors, such as solar light spectrum (Clerke and Alford 1993). Likewise, monitor lizards (*Varanus* sp.) in the Australian wet tropics exhibit considerable diversity in timing of reproduction relative to wet and dry seasons (James *et al*. 1992). In the absence of natural cues, common boas, *Boa constrictor*, can be induced to reproduce in captivity via reductions in day length, temperature, and feeding for several weeks (De Vosjoli *et al*. 2005). Notably, few of these studies directly investigate the cues that stimulate gonadal recrudescence or reproductive behavior in tropical squamates, but only report when reproductive events occur relative to seasonal environmental conditions. Together, these studies illustrate the potential diversity of seasonal environmental cues that temperate and tropical squamates can use to initiate reproduction. Further study is necessary to determine how environmental cue-detection mechanisms co-evolve with different reproductive cycles in both tropical and temperate ecosystems.

5.4.2 Physiology of Seasonal Cue Detection and Transduction

Reproductive cues derived from photoperiod are transduced by the parietal-pineal complex in squamates and in *Sphenodon* (Fig. 5.2; Firth *et al.* 1989; Underwood 1989; Tosini *et al.* 2001). In many lizards the parietal eye, which is dense in photoreceptors (reviewed by Tosini 1997), appears to be the

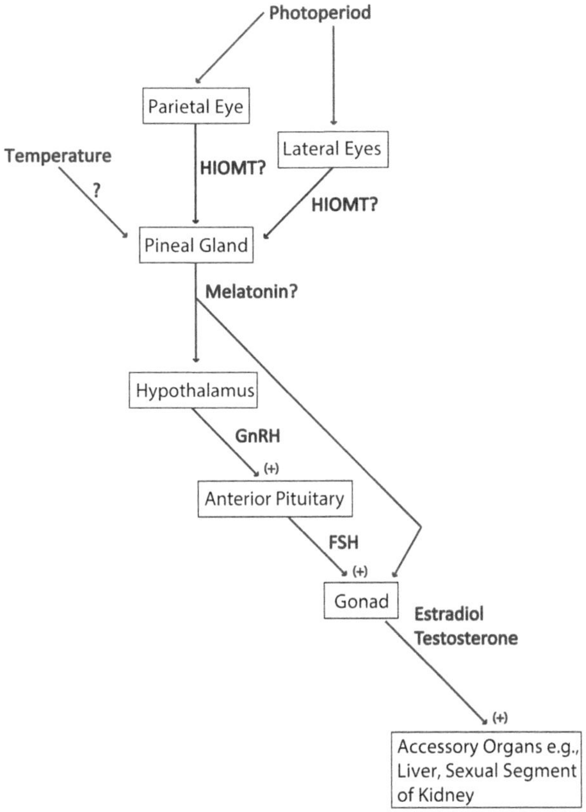

Fig. 5.2 Hypothetical mechanisms squamates might use to detect environmental conduciveness for reproduction are added to the generalized HPGA. Question marks indicate that detection or signaling mechanisms are unknown (temperature), are hypothetical (HIOMT), or are not fully understood (melatonin). Temperature is likely to be more important than photoperiod in stimulating reproduction, but the mechanisms for detecting and communicating temperature to the pineal gland are unknown. HIOMT represents hydroxyindole-O-methyltransferase, but serotonin or norepinephrine could also act as signaling factors between both the parietal and lateral eyes, and the pineal gland. The pineal also directly innervates the pretectal and tegmental areas of the brain, which could provide a mechanism for neuronal signaling to the HPGA. The estradiol/testosterone negative feedback loop to the anterior pituitary is removed for clarity. Organs are represented by black boxes, hormones are blue, and cues are green.

Color image of this figure appears in the color plate section at the end of the book.

primary detector of photoperiod. For example, circadian melatonin cycles are eliminated after parietalectomy in shingleback skinks, *Tiliqua rugosa* (Firth and Kennaway 1980). In contrast, snakes and geckos have secondarily lost the parietal eye (Quay 1979). Perception of day length may primarily occur in the lateral eyes in these taxa, but it is notable that even blinding does not prevent circadian entrainment in geckos (Underwood and Groos 1982); similar studies have not been performed on snakes.

Regardless of whether photoperiod is detected by the lateral eyes, parietal eye, or both (Tosini *et al.* 2001), perception of day length is communicated to the body via secretion of melatonin by the pineal organ (Underwood 1989; Mendonça *et al.* 1995; Mendonça *et al.* 1996b). The pineal organ also directly innervates the pretectal and tegmental areas of the brain (Tosini 1997), which may allow direct neuronal communication between the pineal gland and the HPGA. How light cues from the parietal eye and lateral eyes are communicated to the pineal gland is not well understood, but could occur via secretion of serotonin, norepinephrine, or hydroxyindole-O-methyltransferase (HIOMT; Tosini 1997).

The pineal organ is also responsible for transducing seasonal information from body temperature. It is not known whether the pineal organ itself detects body temperature, or only transduces thermal signals from other detecting mechanisms (Underwood 1989). Regardless, the pineal organ appears to be responsible for simultaneously transducing seasonal information based on both photoperiod and body temperature via secretion of melatonin. Daily melatonin secretions in squamates cycle between maxima synchronized by cool temperatures during scotophase, and minima synchronized by warm temperatures during photophase (Firth *et al.* 1979; Underwood and Calaban 1987; Firth *et al.* 1989; Tilden and Hutchinson 1993). Together, seasonal increases in day length and temperature are accompanied by overall diel decreases in plasma melatonin concentration, and vice-versa (Firth *et al.* 1979; Mendonça *et al.* 1995). Interestingly, photic suppression of melatonin secretion during long photoperiods is less pronounced in *Anolis* lizards than in other vertebrates (Moore and Menaker 2011). It remains unclear whether circadian cycles in circulating melatonin in *Sphenodon* are entrained primarily by temperature or photoperiod, but the presence of a well-developed parietal eye and thermal effects on melatonin secretion suggest both mechanisms are possible (Dendy 1911; Firth *et al.* 1989). After secretion, circulating melatonin is bound by receptors throughout the brain (Wiechmann and Wirsig-Wiechmann 1994), and possibly in the gonads (Mayer *et al.* 1997). In mammals, high-affinity melatonin receptors in the hypothalamus may play a role in activating the HPGA (Reppert *et al.* 1994), but the existence of melatonin receptors in the squamate hypothalamus has not been tested.

High serum melatonin concentrations generally suppress reproduction in male squamates. Pinealectomy, which eliminates most melatonin secretion, induces testicular recrudescence in male green anoles, *Anolis carolinensis*, in autumn, but not in summer (Underwood 1981). Injected melatonin does not inhibit testicular recrudescence in pinealectomized males, but does so in intact males (Underwood 1981). Similarly, summer injections of melatonin in male oriental garden lizards, *Calotes versicolor*, inhibit testicular recrudescence (Haldar and Thapliyal 1981), while pinealectomy induces testicular recrudescence during both summer and winter (Thapliyal and Haldar 1979). Fall pinealectomy inhibits spring reproductive behavior in male *Thamnophis sirtalis* (Mendonça *et al.* 1996b), while spring pinealectomy does not (Mendonça *et al.* 1996a). This suggests that pineal transduction of thermal and photoperiod cues for reproduction occurs during or prior to hibernation in male *Thamnophis sirtalis* (Mendonça *et al.* 1996a). Seasonal differences in the effect of pinealectomy and melatonin injection may indicate physiological mechanisms that explain refractory periods, or why exposure to cold temperatures during winter is critical to successful reproduction in many temperate squamates.

Studies of melatonin effects on reproduction in female squamates have been less frequent. Pinealectomy induces ovarian recrudescence in female *Anolis carolinensis*, while subsequent melatonin injections inhibit it outside of the reproductive season; however, the effects of pinealectomy decrease during the reproductive season (Levey 1973). Parietalectomy accelerates gonadal recrudescence in female *Sceloporus* lizards, but not males, indicating that the role of the parietal eye (and presumably, the pineal gland and melatonin) in regulating reproduction might differ between sexes (Stebbins and Cohen 1973).

The mechanisms squamates use to detect seasonal changes in moisture or rainfall remain unknown. Rainfall and soil moisture levels may not act as cues for the initiation of gametogenesis, but may act as cues for specific reproductive behaviors, especially oviposition (Stamps 1976; Brown and Shine 2006). In captive reptiles, availability of moist nesting environments is critical to successful oviposition, and, in dry conditions, females may retain eggs for so long that they develop dystocia (e.g., egg binding; DeNardo 2006). Nest moisture is especially important for successful development in many oviparous squamates because many taxa produce leathery eggshells with relatively high water permeability (Packard *et al.* 1982). In contrast, female pythons, many of which actively incubate their eggs via encirclement with their bodies, can limit egg desiccation even in dry nest environments (Lourdais *et al.* 2007).

Reproduction might also be correlated with rainfall if increased rainfall is associated with increased food abundance, which could maximize resources available for reproduction, or could maximize offspring growth and survival rates (but there is limited evidence for this pattern in squamates; James and Shine 1985; Brown and Shine 2006; Shine and Brown 2008). Therefore, food availability, rather than rainfall *per se*, might be the primary cue for reproduction (Wikelski *et al.* 2000). If food availability is limiting to offspring survival after reproduction, females might over-allocate nutrients to yolk in order to provide offspring with energy reserves to sustain them until they can successfully forage (e.g., residual yolk; Troyer 1983; Van Dyke *et al.* 2011).

Most studies of reproductive cues and phenology in response to environmental conditions have been conducted on iteroparous squamates with no embryonic diapause. Environmental cues for reproduction might vary among squamate taxa depending on whether they are semelparous (reproduce only once in their lifetime) or iteroparous (reproduce multiple times throughout life), produce single or multiple litters/clutches per year, or exhibit embryonic diapause. Indeed, the selection pressure to tightly coordinate reproduction with suitable environmental conditions should be strongest in semelparous and single-litter/clutch species because the risks of total reproductive failure are higher than in iteroparous and multi-clutching species.

Species that exhibit embryonic diapause (e.g., Chameleons; Andrews 2004) might be able to reproduce in spite of poor environmental conditions. In chameleons, embryonic diapause may also allow delayed hatching until seasonal environmental conditions maximize offspring survival (Andrews and Donoghue 2004). In addition, some viviparous lizards that inhabit cold climates, including the Australian southern snow skink, *Niveoscincus microlepidotus*, and the New Zealand common gecko, *Hoplodactylus maculatus*, carry fully-developed offspring over winter and give birth in the following spring (Olsson and Shine 1999; Girling *et al.* 2002; Rock 2006). In these taxa, delaying birth via embryonic diapause has been suggested to maximize potential offspring foraging and growth during the summer prior to the subsequent winter (Olsson and Shine 1998). Thus, embryonic diapause might allow the dissociation of environmental constraints on offspring survival from parents' decisions to reproduce. Taken together, the diversity of reproductive strategies exhibited among squamates could promote the utilization of different environmental cues for reproduction, and phylogenetic analyses are needed to understand how divergent mechanisms of cue detection and transduction have evolved as a response.

5.5 RESOURCE AVAILABILITY CUES: INCOME AND CAPITAL

Reproduction requires a significant investment of energy and nutrients. Gametes, especially eggs, are rich in energy and nutrients and expensive to produce (Olsson *et al.* 1997; Van Dyke and Beaupre 2011), and females may incur additional metabolic costs of gravidity or pregnancy (Birchard *et al.* 1984; DeMarco and Guillette 1992; Angilletta and Sears 2000; Robert and Thompson 2000; Schultz *et al.* 2008). In addition, reproductive behaviors, including mate-searching, courtship, mating, and nest brooding, may be energetically expensive (Harlow and Grigg 1984; Olsson *et al.* 1997; Friesen *et al.* 2009). As a result, the availability of resources, including both food and stored reserves, is an important constraint on reproduction, and detection of resource availability may function as a cue for the decision to reproduce.

5.5.1 Phenology of Resource Availability Cues

Individuals of many squamate species do not reproduce annually (Bull and Shine 1979), and multi-year gaps between reproductive events have been suggested to be the result of adults requiring multiple years to accumulate the resources necessary for reproduction (Hahn and Tinkle 1966; Aldridge 1979a; Slip and Shine 1988; Gloyd and Conant 1990; Brown 1991; Van Wyk 1991; Naulleau *et al.* 1999; Olsson and Shine 1999; Diller and Wallace 2002; Ernst and Ernst 2003; Ibargüengoytía 2004). Even within a given active season, low resource availability can delay the initiation of reproduction until sufficient resources are accumulated (Vitt *et al.* 1978; Bauwens and Verheyen 1985; Abell 1999), which may reduce offspring fitness (Olsson and Shine 1997). The capital-income dichotomy predicts that animals base the decision to reproduce on either the magnitude of stored reserves (capital) or the rate of nutrient assimilation from diet (income; Drent and Daan 1980). In squamates, this prediction has been frequently tested in studies that compare body condition indices between reproductive and non-reproductive individuals in tropical spiny lizards, *Sceloporus mucronatus* (de la Cruz *et al.* 1988), viviparous lizards, *Zootaca vivipara* (Bleu *et al.* 2013), smooth snakes, *Coronella austriaca* (Reading 2004), garter snakes, *Thamnophis sirtalis* (Gregory 2006), and Aspic vipers, *Vipera aspis* (Naulleau and Bonnet 1996; Aubret *et al.* 2002). Other studies have correlated body condition indices with total reproductive output in collared lizards, *Crotaphytus collaris* (Telemeco and Baird 2011), and water skinks, *Eulamprus tympanum* (Doughty and Shine 1998). In these comparisons, body condition is usually (but not always; Doughty and Shine 1997) defined either as the ratio of body mass to body length, or via a regression of body mass and body length. Regardless, reproductive individuals often exhibit higher body conditions than do non-reproductive individuals, which is frequently interpreted as evidence for body-condition thresholds for reproduction (e.g., capital breeding; Bonnet *et al.* 1998).

Unfortunately, most studies of the effects of body condition on reproduction in squamates are correlative, and do not directly distinguish between contributions made by fat mass and muscle mass to total body condition. Whereas fat mass is often described as the primary cause of fluctuations in body condition (e.g., Bonnet *et al.* 1998; Aubret *et al.* 2002), changes in muscle mass may also drive fluctuations in body condition (Lourdais *et al.* 2004). In addition, other studies report that reproductive allocation decisions, particularly clutch or litter size, are at least partially dependent upon food availability during reproduction (income) rather than capital, in marine iguanas, *Amblyrhynchus cristatus* (Rubenstein and Wikelski 2003), Asian northern grass lizards, *Takydromus septentrionalis* (Du 2006), swamp snakes, *Seminatrix pygaea* (Winne *et al.* 2006), and Aesculapian snakes, *Zamensis longissimus* (Naulleau and Bonnet 1995), as well as in so-called capital breeders like *Vipera aspis* (Madsen and Shine 1999; Bonnet *et al.* 2001). In *Sphenodon*, gravidity rates do not differ across populations in correlation with population differences in body condition (Tyrrell *et al.* 2000), so magnitude of capital may not be a cue for reproduction. Notably, most studies of capital and income effects on reproduction in squamates have focused on females, and male reproductive decisions may be made using different cues (Aubret *et al.* 2002).

5.5.2 Physiology of Resource Availability Detection and Transduction

The possibility that reproductive decisions depend on magnitude of capital or rate of income assimilation raises the implicit hypothesis that squamates have physiological mechanisms for detecting the magnitude of their stored reserves (capital) of lipid, protein, total energy, and/or specific limiting nutrients, or the rate of assimilation (income) of at least one of those resources. To date, few experiments have tested this hypothesis, but ghrelin (Unniappan 2010) and leptin (Niewiarowski *et al.* 2000), or similar hormones, might play roles in communicating income and capital resource abundance to the brain. As a regulator of serum glucose, insulin might also play a role in communicating nutritional status to the brain, but insulin has not been investigated in reptiles aside from structural comparisons with other taxa (Conlon and Hicks 1990). In addition, it is not clear how the HPGA functions when opportunities for reproduction are skipped. Females of some species (e.g., *Vipera aspis*) appear to skip opportunities for reproduction by not initiating vitellogenesis at all (Aubret *et al.* 2002), which suggests that hormonal cascades in the HPGA either do not occur or are arrested prior to reaching the ovary. Females of other species (e.g., *Nerodia sipedon* and *Tantilla coronata*) appear to initiate vitellogenesis, presumably via activation of the HPGA, but do not ovulate and instead undergo follicular atresia (Aldridge 1982; Aldridge and Semlitsch 1992). It is not clear whether atresia occurs because available resources become limiting after vitellogenesis is initiated. Other factors, such as courtship or mating, could play a role in ensuring that vitellogenesis is completed (discussed in the next section).

In mammals, ghrelin is a multifunctional peptide hormone that is primarily secreted by the stomach in response to stomach emptying and filling (Murakami *et al.* 2002). Ghrelin regulates appetite across vertebrate taxa, but is also linked to gonadotropin production in fish (Kaiya *et al.* 2008). Ghrelin has been identified in the slider turtle, *Trachemys scripta*, in which it shares most of its tertiary structure with other vertebrates (Kaiya *et al.* 2004). Ghrelin has not been identified in squamates and its functions in reptiles remain untested. Regardless, the role of ghrelin in regulating appetite, as well as its secretion in response to changes in stomach filling, suggests that it could communicate the rate of income (i.e., food) acquisition in squamate reptiles. If squamate ghrelin also stimulates gonadotropin production, as in fish, then ghrelin might be an important mechanistic link between the rate of food acquisition and the decision to reproduce. Stomach fullness and ghrelin secretion may therefore be a physiological mechanism by which food availability acts as a cue for reproduction in some squamate taxa. The apparent conservation of the tertiary structure of ghrelin across vertebrate taxa also suggests that the role that ghrelin plays in reproduction could be easily tested using standard immunoassays.

In mammals, leptin is a peptide hormone secreted by adipose tissue in proportion to adipose tissue mass (Havel 2000). In the Italian wall lizard, *Podarcis sicula*, serum leptin concentrations are highest in spring immediately prior to vitellogenesis, when adipose tissue mass is greatest, and fall as adipose tissue mass decreases during vitellogenesis (Paolucci *et al.* 2001). Serum leptin concentrations are also usually correlated with adipose tissue mass in the fence lizard, *Sceloporus undulatus,* but are notably lowest in autumn, when adipose tissue mass is greatest (Spanovich *et al.* 2006).

In mammals, serum leptin concentrations are directly correlated with the release of GnRH, gonadotropin, and estradiol (Caprio *et al.* 2001), and maternal allocations to offspring increase with increasing serum leptin (French *et al.* 2009). In addition, birds can be induced to lay additional clutches as a result of artificially increased serum leptin concentrations (Lõhmus and Björklund 2009). Thus, leptin may be a mechanism for communicating adipose tissue mass (i.e., fat/capital storage) to the brain, and may allow fat mass to serve as a cue for reproduction in apparent capital breeders, but this hypothesis has not been tested in squamates. Importantly, the temporal disconnect between adipose tissue mass and serum leptin concentration that occurs in autumn in *Scleropus undulatus* (and in some mammals; Kronfeld-Schor *et al.* 2000), if conserved across taxa, would compromise studies of direct relationships between leptin and reproduction in species that initiate reproduction in autumn. In bats, leptin production declines in autumn, regardless of fat mass, to maximize pre-hibernatory fattening (Kronfeld-Schor *et al.* 2000), and could follow a similar trend in temperate squamates. Thus, in temperate squamates that initiate reproduction in the autumn, any "signal" for reproduction transduced from fat mass could be received by the brain prior to the autumn down-regulation of leptin production. Alternatively, reception or interpretation of the leptin "signal" in the brain could be modified by transduction of photoperiod or temperature. Indeed, leptin sensitivity, presumably modulated by receptor density in target tissue, is regulated by photoperiod in both hamsters and voles (Klingenspor *et al.* 2000; Krol *et al.* 2006). If similar mechanisms occur in squamate reptiles, then leptin might not promote reproduction on its own. Instead, reproduction might be promoted by a decrease in serum leptin simultaneous with increased melatonin production. Possible interactions between cues for time-of-year and resource availability illustrate that reproduction is promoted by multifaceted regulatory systems (Fig. 5.3), possibly tuned by natural selection to initiate reproduction only when multiple cues indicate that reproductive success is likely to be maximized.

As noted in the prior section, even when reproduction is initiated, female squamates appear to adjust clutch and litter sizes depending on resource availability. Clutch/litter size may be determined at two points during the reproductive cycle: first, females recruit a given number of ovarian follicles for primary vitellogenesis at the start of a reproductive cycle; secondly, females selectively allocate yolk to, and ovulate, a fraction of the "committed" follicles, while the rest undergo atresia (Aldridge 1982; Aldridge and Semlitsch 1992). Selective atresia of follicles, in particular, appears to be widespread in squamate reptiles (Shine 1977; Jones *et al.* 1978; Trauth 1978; Etches and Petitte 1990; Mendez-De La Cruz *et al.* 1993).

Whether clutch/litter size is ultimately determined at the initiation of vitellogenesis or by selective ovulation/atresia remains unknown, but exogenous FSH has been shown to increase clutch size in several species (Sinervo and Licht 1991; Jones and Swain 2000). The correlations between body condition (or fat body mass) and clutch/size, along with the observation that artificial increases in FSH stimulate increases in clutch size, suggest that hormones that signal resource abundance to the brain, if they exist, act in a dose-dependent manner. This could provide a mechanism for females to adjust clutch/litter size to resource availability. Selection could be expected to act strongly on this ability, because overcommitting resources to reproduction could lead to starvation prior to parturition, while undercommitting could prevent females from realizing their actual reproductive value (e.g., Fisher 1930; Lack 1954).

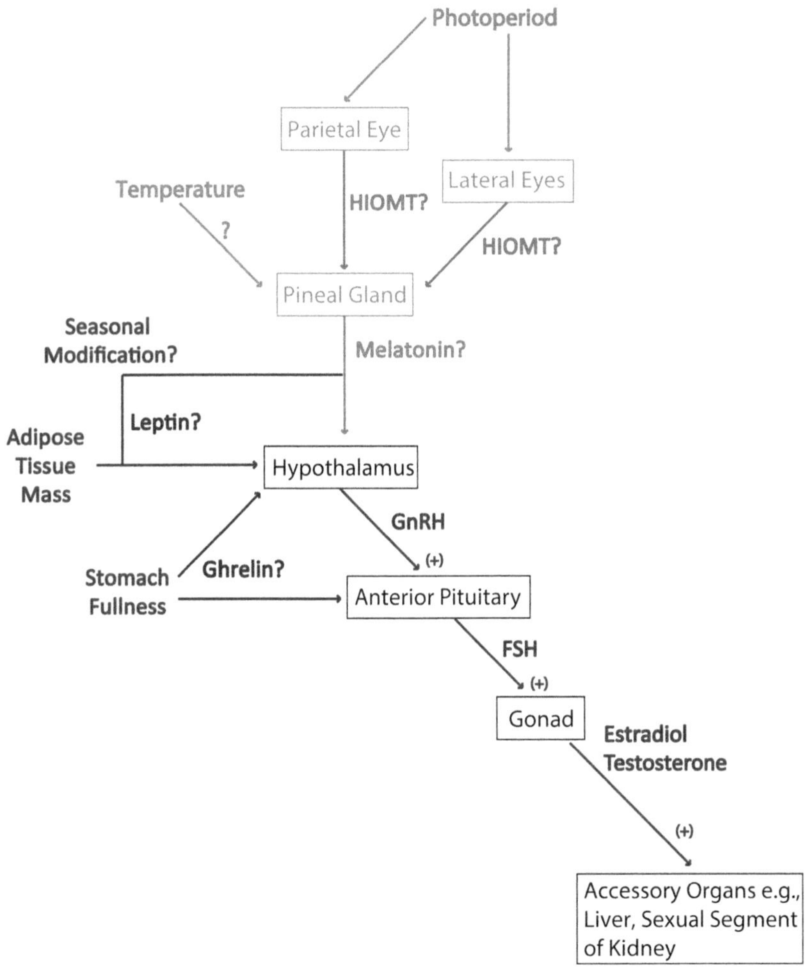

Fig. 5.3 Mechanisms squamates might use to detect resource availability are added to the generalized HPGA. Question marks indicate that leptin and ghrelin signaling mechanisms are hypothetical indicators of resource availability to the HPGA, and that melatonin may seasonally modify the action of leptin. Organs are represented by black boxes, hormones are blue, and cues are green. Mechanisms of detecting environmental conduciveness, introduced in Fig. 2, are obscured to enhance clarity yet emphasize that multiple detection systems might interact to inform the decision to reproduce. The pathway linking melatonin directly to the gonad has been removed for clarity.

Color image of this figure appears in the color plate section at the end of the book.

5.6 SOCIAL CUES: MATING AND COURTSHIP

Cues that communicate likelihood of successful mating are likely to be important in the decision to reproduce in species that experience high fitness and/or energy costs of reproduction. Cues communicating likelihood of successful mating may be especially important in females, who may incur both high energetic costs of gametogenesis and gestation (Robert and Thompson 2000; Schultz *et al.* 2008; Van Dyke and Beaupre 2011), and high fitness and energetic costs of transport while gravid or pregnant (Shine 1980; Webb and Lannoo 2004). As a result, females may be under strong selective pressure to avoid potential costs of reproduction unless mating and fertilization are likely.

5.6.1 Phenology of Mating and Courtship Cues

Induced ovulation, in which mating physically or chemically stimulates ovulation in the female (Taymor 1978), is the most likely cue for communicating the likelihood of successful mating. Induced ovulation is widespread among eutherian mammals (Lariviere and Ferguson 2003), but among reptiles has only been reported in sea turtles (Manire *et al.* 2008). Although induced ovulation has not been reported in squamates, females of some species, especially *Thamnophis sirtalis*, appear to initiate ovarian recrudescence and/or vitellogenesis only after mating (Bona-Gallo and Licht 1983; Whittier and Crews 1986b; Mendonça and Crews 1990; Mathies *et al.* 2004). If the disparity between mating-induced ovulation and ovarian recrudescence represents a taxonomic difference between mammals and squamates, it could be because vitellogenesis is more energetically expensive than pregnancy or gravidity in squamates (Van Dyke and Beaupre 2011), while pregnancy and lactation are more energetically expensive than vitellogenesis in mammals (Gittleman and Thompson 1988). Thus, squamates may be more likely to avoid initiating vitellogenesis unless mating has occurred, whereas mammals are more likely to avoid ovulation unless mating has occurred. Long-term oviducal sperm storage is also common in squamates (Sever and Hamlett 2002), and allows females to store sperm from matings that occur prior to vitellogenesis.

In addition to mating, courtship and/or the presence of males alone may be sufficient to induce reproduction in females of some squamate species, possibly because it confers a reasonable expectation of successful mating. Captive female blood pythons, *Python curtus*, are more likely to initiate vitellogenesis when housed with males (DeNardo and Autumn 2001) and male courtship plays a role in environmentally-induced ovarian recrudescence during vitellogenesis in green anoles, *Anolis carolinensis* (Crews 1975). Similarly, male courtship is linked to ovulation in sexual whiptail lizards (*Aspidoscelis* sp.; Crews *et al.* 1986). Interestingly, parthenogenetic whiptails like *Aspidoscelis uniparens* retain this response, and ovulation can be stimulated by courtship and pseudocopulation by male-like females (Moore *et al.* 1985; Crews *et al.* 1986).

Despite the preceding examples, mating and/or courtship may not be important cues for reproduction in all female squamates. Female viviparous lizards, *Zootoca vivipara*, ovulate eggs regardless of whether they have mated or experienced courtship, and may produce unfertilized eggs as a result (Bleu *et al.* 2011). Likewise, female leopard geckos, *Eublepharis macularius*, initiate vitellogenesis regardless of exposure to males (LaDage and Ferkin 2008). Female tuataras, *Sphenodon punctatus*, initiate vitellogenesis years prior to mating (Brown *et al.* 1991; Cree *et al.* 1992), but it is not clear whether mating affects the decision to ovulate. Furthermore, observed relationships between mating and/or courtship and vitellogenesis or ovulation in squamates are, with some exceptions (e.g., Mendonça and Crews 1990), largely correlative. In female *Thamnophis sirtalis*, both vitellogenesis and pheromone production are stimulated by elevated serum estradiol concentrations (Parker and Mason 2012). Therefore, vitellogenesis may not be stimulated by courtship; instead, vitellogenesis and attractiveness (and subsequent courtship) may be simultaneous consequences of females initiating reproduction as a result of courtship-independent cues.

5.6.2 Physiology of Mating and Courtship Cue Detection and Transduction

Nearly all studies of the physiology underlying the effects of mating and courtship on ovarian recrudescence and vitellogenesis in squamates have focused on *Thamnophis sirtalis*. Female *T. sirtalis* exhibit a pronounced surge in serum estradiol and prostaglandin (PG-F2α) in response to mating (Whittier *et al.* 1987; Mendonça and Crews 2001), which is probably the result of a neuroendocrine cascade initiated by physical stimulation of stretch receptors in the cloaca during intromission (Whittier and Crews 1986b). Spinal transection and injection of both lidocaine and tetracaine near the cloaca inhibit the mating-induced estradiol surge, but only tetracaine inhibits post-mating ovarian recrudescence in *T. sirtalis* (Mendonça and Crews 1990). Similar phenomena have been observed in cats, which induce ovulation after physical stimulation of the vagina during mating (Greulich 1934), as a result of direct neuronal communication with the brain (Rose 1978).

Mendonça and Crews (1990, 2001) suggested that tactile stimulation of the skin during both mating and courtship could also contribute to the ovarian response to mating in *Thamnophis sirtalis*, but it is not clear whether estradiol or PG-F2α mediate this response. Many male squamates bite females during intromission (Stamps 1975; Carpenter 1977; Gillingham 1979; Pandav *et al.* 2007), and male boas and pythons are well-known for using their pelvic spurs to "stimulate" females during courtship (Gillingham and Chambers 1982). These tactile behaviors could provide alternative means of physical neuronal stimulation in addition to intromission. A relationship between tactile stimulation of the skin and ovarian recrudescence could also provide a mechanism for explaining why male presence and/or courtship alone can stimulate vitellogenesis and/or ovulation in other squamate taxa.

Most research on mating-induced ovarian recrudescence and ovulation in squamates has focused on physical stimulation during intromission, but other mechanisms have been investigated in mammals and could exist in squamates. Ovulation-inducing factors (OIF), which stimulate LH surges in females after mating, have been identified in the seminal fluid of camels (Adams *et al.* 2005), and similar mechanisms may also exist in rabbits, mice, pigs, and horses (Bogle *et al.* 2011; Silva *et al.* 2011). In pigs, seminal prostaglandins may play a role in ovulation induction (Ratto *et al.* 2011). In squamates, unmated female *Thamnophis sirtalis*, whose cloacae were smeared with male seminal fluid, exhibited reduced attractiveness to other males (Shine *et al.* 2000). Female attractiveness in snakes is primarily mediated by pheromone production, which is stimulated by estradiol (Parker and Mason 2012), so these data suggest that male seminal fluid might alter estradiol production in females that have mated. If that is the case, then the possibility remains that male seminal fluid contains factors that could also interact with female HPGA to initiate or regulate vitellogenesis, possibly in an interaction with physical stimuli during intromission and/or courtship (Fig. 5.4). Prostaglandins are abundant in snake seminal fluid (Whittier and Crews 1986a), and may be a useful candidate mechanism for investigating the possibility of chemical induction of vitellogenesis in female squamates (Friesen 2012). Finally, females of many species store sperm in oviducal crypts after mating (Sever and Hamlett 2002). These crypts could produce a neuronal or hormonal signal that notifies the brain and/or HPGA whether healthy sperm are present at the appropriate time for reproduction to be initiated.

5.7 FUTURE DIRECTIONS

Squamates are excellent model organisms for studying the function and evolution of reproductive cue-detecting mechanisms, but studies have been limited by difficulty in accurately measuring the secretion and function of peptide hormones, especially GnRH and gonadotropins. As a result, we still

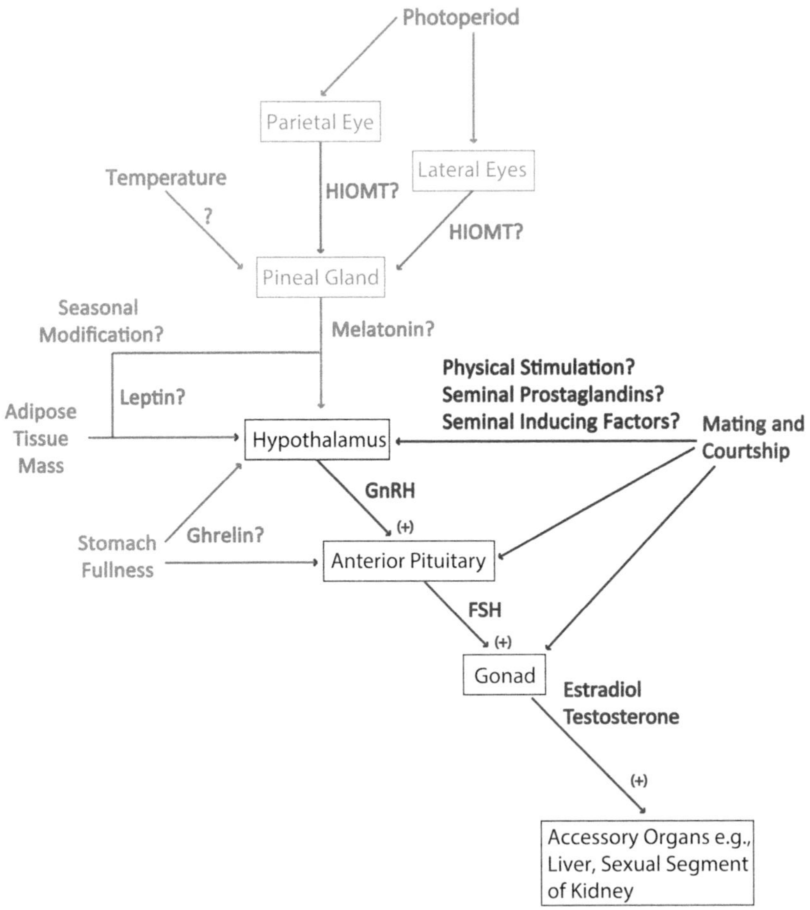

Fig. 5.4 Mechanisms female squamates might use to determine whether fertilization is likely are added to the generalized HPGA. Question marks indicate that all signaling mechanisms that might communicate mating and/or courtship to the HPGA are hypothetical. It is also unclear whether any of these hypothetical mechanisms act directly on the hypothalamus, anterior pituitary, or gonads. Organs are represented by black boxes, hormones and neural signaling are blue, and cues are green. Previously introduced cue detection mechanisms are obscured to enhance clarity yet emphasize that multiple detection systems might interact to inform the decision to reproduce.

Color image of this figure appears in the color plate section at the end of the book.

lack a comprehensive understanding of the phenomena squamates use as cues for reproduction, and how those cues are detected and communicated to the HPGA. Indeed, the mechanisms summarized in this review are largely untested in reptiles, and it is possible that alternative mechanisms remain to be discovered. The advent of low-cost next-generation transcriptome sequencing (e.g., Mardis 2008; Ekblom and Galindo 2011; Brandley *et al.* 2012), along with the growing availability of genomes for non-model organisms, including squamates (Castoe *et al.* 2013; Vonk *et al.* 2013), is poised to significantly advance our ability to identify both hypothetical and novel hormonal pathways, as well as other molecular mechanisms involved in cue detection and transduction. In particular, a transcriptome sequencing approach examining total gene expression in the organs involved in both the HPGA and reproductive cue detection could determine 1) what hormones are involved in communication of reproductive cues to and within the HPGA; 2) where those hormones are produced; 3) what organs produce receptors for those hormones and might be important in transduction of reproductive cues; and 4) whether dissociated breeders initiate gametogenesis and mating using separate cues, or even separate HPGA hormonal cascades.

Both nucleotide and peptide sequences can be identified from transcriptome sequencing, thereby allowing researchers to design functional experiments using standard molecular methods, including both gene expression (e.g., quantitative PCR or *in situ* hybridization) and immunological approaches (e.g., western blot, immunohistochemistry) if suitable antibodies are readily available. Furthermore, the combination of transcriptome sequencing and proteomics allows simultaneous examination of gene expression and protein translation (e.g., Wong *et al.* 2012; Margres *et al.* 2014), which would be a powerful method for examining hormonal cascades throughout cue detection, stimulation of the HPGA, and upregulation of reproductive physiology. Similar approaches are already being used to study multiple aspects of the biology of non-model organisms, including the evolution of extreme digestive physiology, venom, and viviparity in squamates (Murphy and Thompson 2011; Brandley *et al.* 2012; Margres *et al.* 2014; Castoe *et al.* 2013), thus they represent powerful tools for an area of research that has been historically constrained by the limitations of traditional methods.

Elucidating how reproductive cue-detection mechanisms function in concert with diverse reproductive strategies and environmental conditions is a critical component to understanding how species respond to environmental change. Indeed, mismatches between reproductive cues and optimal reproductive conditions, caused by global climate change, have already been reported in migratory birds (Visser *et al.* 2004). As environmental temperatures have warmed, vegetation and insect production has advanced earlier in the year, while reproduction, which is cued by photoperiod in some birds, has not (Visser *et al.* 1998). The potential for similar mismatches in squamate reptiles is unclear because they potentially utilize multiple simultaneous cues for reproduction, including both photoperiod and temperature. Nevertheless, some squamate species may be highly vulnerable to global climate change (Huey *et al.* 2010; Sinervo *et al.* 2010), and assessing the role of mismatches between reproductive cue detection and changes in environmental conditions is currently not possible in most taxa. As a result, integrated investigations of reproductive cue detection mechanisms, utilizing genomic, molecular, and ecological methods, may hold great promise for improving the conservation of these species. More broadly, these investigations are needed to advance understanding of how animals make reproductive decisions (e.g., when to reproduce, reproductive frequency, clutch size), which are critical components of life-history evolution.

5.8 ACKNOWLEDGMENTS

I thank the editors for granting the opportunity to contribute to this volume. The ideas presented in this chapter were influenced by discussions and debates with Steven J. Beaupre, David L. Kreider, Craig M. Lind, M. Rockwell Parker, and Christopher R. Friesen. I am grateful to Matthew C. Brandley, Oliver W. Griffith, and Camilla M. Whittington for sharing their expertise on gene sequencing and gene expression techniques, and to Christopher R. Murphy, Laura A. Lindsay, and Jacqueline F. Herbert for training in immunostaining and candidate gene analysis. I thank Robert D. Aldridge, Michael B. Thompson, M. C. Brandley, O. W. Griffith, C. M. Whittington, and C. R. Friesen for critical comments on early drafts of the manuscript. Support was provided by the National Science Foundation (IRFP #1064803).

5.9 LITERATURE CITED

Abell, A. J. 1999. Variation in clutch size and offspring size relative to environmental conditions in the lizard *Sceloporus virgatus*. Journal of Herpetology 33: 173–180.

Adams, G. P., Ratto, M. H., Huanca, W. and Singh, J. 2005. Ovulation-inducing factor in the seminal plasma of alpacas and llamas. Biology of Reproduction 73: 452–457.

Aldridge, R. D. 1975. Environmental control of spermatogenesis in the rattlesnake *Crotalus viridis*. Copeia 1975: 493–496.

Aldridge, R. D. 1979a. Female reproductive cycles of the snakes Arizona elegans and *Crotalus viridis*. Herpetologica 35: 256–261.

Aldridge, R. D. 1979b. Spermatogenesis in sympatric *Crotalus viridis* and Arizona elegans in New Mexico. Journal of Herpetology 13: 187–192.

Aldridge, R. D. 1982. The ovarian cycle of the watersnake *Nerodia sipedon*, and effects of hypophysectomy and gonadotropin administration. Herpetologica 38: 71–79.

Aldridge, R. D. and Duvall, D. 2002. Evolution of the mating system in the pitvipers of North America. Herpetological Monographs 16: 1–25.

Aldridge, R. D., Goldberg, S. R., Wisniewski, S. S., Bufalino, A. P. and Dillman, C. B. 2009. The reproductive cycle and estrus in the colubrid snakes of temperate North America. Contemporary Herpetology 2009: 1–31.

Aldridge, R. D., Greenhaw, J. J. and Plummer, M. V. 1990. The male reproductive cycle of the rough green snake (*Opheodrys aestivus*). Amphiba-Reptilia 11: 165–172.
Aldridge, R. D., Jellen, B. C., Siegel, D. S. and Wisniewski, S. S. 2011. The sexual segment of the kidney. pp. 477–509. In R. D. Aldridge and D. M. Sever (eds.), *Reproductive Biology and Phylogeny of Snakes*. Science Publishers, Inc., Enfield, New Hampshire.
Aldridge, R. D. and Semlitsch, R. D. 1992. Female reproductive biology of the southeastern crowned snake (*Tantilla coronata*). Amphibia-Reptilia 13: 209–218.
Almeida-Santos, S. M., Abdalla, F. M. F., Silveira, R., Yamanouye, N., Breno, M. C. and Salomao, M. G. 2004. Reproductive cycle of the neotropical *Crotalus durissus terrificus*: I. Seasonal levels and interplay between steroid hormones and vasotocinase. General and Comparative Endocrinology 139: 143–150.
Andrews, R. M. 2004. Patterns of embryonic development. pp. 75–102. In D. C. Deeming (ed.), *Reptilian Incubation: Environment, Evolution, and Behaviour*. Nottingham University Press, Nottingham, United Kingdom.
Andrews, R. M. and Donoghue, S. 2004. Effects of temperature and moisture on embryonic diapause of the veiled chameleon (*Chamaeleo calyptratus*). Journal of Experimental Zoology Part A: Comparative Experimental Biology 301: 629–635.
Andrews, R. M., Mathies, T. and Warner, D. A. 2000. Effect of incubation temperature on morphology, growth, and survival of juvenile *Sceloporus undulatus*. Herpetological Monographs 14: 420–431.
Angilletta, M. J., Jr. and Sears, M. W. 2000. The metabolic cost of reproduction in an oviparous lizard. Functional Ecology 14: 39–45.
Aubret, F., Bonnet, X., Shine, R. and Lourdais, O. 2002. Fat is sexy for females but not males: the influence of body reserves on reproduction in snakes (*Vipera aspis*). Hormones and Behavior 42: 135–147.
Bauwens, D. and Verheyen, R. F. 1985. The timing of reproduction in the lizard *Lacerta vivipara*: differences between individual females. Journal of Herpetology 19: 353–364.
Birchard, G. F., Black, C. P., Schuett, G. and Black, V. 1984. Influence of pregnancy on oxygen consumption, heart rate, and hematology in the garter snake: implications for the "cost of reproduction" in live bearing reptiles. Comparative Biochemistry and Physiology 77A: 519–523.
Bleu, J., Le Galliard, J. -F., Fitze, P. S., Meylan, S., Clobert, J. and Massot, M. 2013. Reproductive allocation strategies: a long-term study on proximate factors and temporal adjustments in a viviparous lizard. Oecologia (Berlin) 171: 141–151.
Bleu, J., Le Galliard, J. -F., Meylan, S., Massot, M. and Fitze, P. S. 2011. Mating does not influence reproductive investment, in a viviparous lizard. Journal of Experimental Zoology Part A: Ecological Genetics and Physiology 313: 458–464.
Bogle, O. A., Ratto, M. H. and Adams, G. P. 2011. Evidence for the conservation of biological activity of ovulation-inducing factor in seminal plasma. Reproduction 142: 277–283.
Bona-Gallo, A. and Licht, P. 1983. Effects of temperature on sexual receptivity and ovarian recrudescence in the garter snake, *Thamnophis sirtalis parietalis*. Herpetologica 39: 173–182.
Bona-Gallo, A., Licht, P., MacKenzie, D. S. and Lofts, B. 1980. Annual cycles in levels of pituitary and plasma gondotropin, gonadal steroids, and thyroid activity in the Chinese Cobra (*Naja naja*). General and Comparative Endocrinology 42: 477–493.
Bonnet, X., Bradshaw, D. and Shine, R. 1998. Capital versus income breeding: an ectothermic perspective. Oikos 83: 333–342.
Bonnet, X., Naulleau, G., Shine, R. and Lourdais, O. 2001. Short-term versus long-term effects of food intake on reproductive output in a viviparous snake, *Vipera aspis*. Oikos 92: 297–308.
Borrelli, L., De Stasio, R. and Filosa, S. 2001. Molecular cloning, sequence and expression of follicle-stimulating hormone receptor in the lizard *Podaricis sicula*. Gene (Amsterdam) 275: 149–156.
Botte, V., Angelini, F. and Picariello, O. 1978. Autumn photothermal regimes and spring reproduction in the female lizard, *Lacerta sicula*. Herpetologica 34: 298–302.

Brandley, M. C., Young, R. L., Warren, D. L., Thompson, M. B. and Wagner, G. P. 2012. Uterine gene expression in the live-bearing lizard, *Chalcides ocellatus*, reveals convergence of squamate reptile and mammalian pregnancy mechanisms. Genome Biology and Evolution 4: 394–411.

Brown, G. P. and Shine, R. 2006. Why do most tropical animals reproduce seasonally? Testing hypotheses on an Australian snake. Ecology 87: 133–143.

Brown, M. A., Cree, A., Chambers, G. K., Newton, J. D. and Cockrem, J. F. 1991. Variation in plasma constituents during the natural vitellogenic cycle of tuatara, *Sphenodon punctatus*. Comparative Biochemistry and Physiology B Comparative Biochemistry 100: 705–710.

Brown, M. A., Cree, A., Daugherty, C. H., Dawkins, B. P. and Chambers, G. K. 1994. Plasma concentrations of vitellogenin and sex steroids in female tuatara (*Sphenodon punctatus* punctatus) from northern New Zealand. General and Comparative Endocrinology 95: 201–212.

Brown, W. S. 1991. Female reproductive ecology in a northern population of the timber rattlesnake, *Crotalus horridus*. Herpetologica 47: 101–115.

Bull, J. J. and Shine, R. 1979. Iteroparous animals that skip opportunities for reproduction. The American Naturalist 114: 296–303.

Caprio, M., Fabbrini, E., Isidori, A. M., Aversa, A. and Fabbri, A. 2001. Leptin in reproduction. Trends in Endocrinology and Metabolism 12: 65–72.

Carpenter, C. C. 1977. Communication and displays of snakes. American Zoologist 17: 217–223.

Castilla, A. M., Barbadillo, L. J. and Bauwens, D. 1992. Annual variation in reproductive traits in the lizard *Acanthodactylus erythrurus*. Canadian Journal of Zoology 70: 395–402.

Castoe, T. A., de Koning, A. P. J., Hall, K. T., Card, D. C., Schield, D. R., Fujita, M. K., Ruggiero, R. P., Degner, J. F., Daza, J. M., Gu, W., Reyes-Velasco, J., Shaney, K. J., Castoe, J. M., Fox, S. E., Poole, A. W., Polanco, D., Dobry, J., Vandewege, M. W., Li, Q., Schott, R. K., Kapusta, A., Minx, P., Feschotte, C., Uetz, P., Ray, D. A., Hoffmann, F. G., Bogden, R., Smith, E. N., Chang, B. S. W., Vonk, F. J., Casewell, N. R., Henkel, C. V., Richardson, M. K., Mackessy, S. P., Bronikowsi, A. M., Yandell, M., Warren, W. C., Secor, S. M. and Pollock, D. D. 2013. The Burmese python genome reveals the molecular basis for extreme adaptation in snakes. Proceedings of the National Academy of Sciences 110: 20645–20650.

Clerke, R. B. and Alford, R. A. 1993. Reproductive biology of four species of tropical Australian lizards and comments on the factors regulating lizard reproductive cycles. Journal of Herpetology 27: 400–406.

Conlon, J. M. and Hicks, J. W. 1990. Isolation and structural characterization of insulin, glucagon and somatostatin from the turtle, *Pseudemys scripta*. Peptides (New York) 22: 461–466.

Courty, Y. and Dufaure, J. P. 1979. Levels of testosterone in the plasma and testis of the viviparous lizard (*Lacerta vivipara* jacquin) during the annual cycle. General and Comparative Endocrinology 39: 336–342.

Cree, A., Cockrem, J. F. and Guillette, L. J., jr. 1992. Reproductive cycles of male and female tuatara (*Sphenodon punctatus*) on Stephens Island, New Zealand. Journal of Zoology (London) 226: 199–217.

Cree, A., Guillette, L. J., jr., Brown, M. A., Chambers, G. K., Cockrem, J. F. and Newton, J. D. 1991. Estradiol-induced vitellogenesis in the tuatara, *Sphenodon punctatus*. Physiological Zoology 64: 1234–1251.

Crews, D. 1975. Effects of different components of male courtship behaviour on environmentally induced ovarian recrudescence and mating preferences in the lizard, *Anolis carolinensis*. Animal Behaviour 23: 349–356.

Crews, D. 1983. Alternative reproductive tactics in reptiles. Bioscience 33: 562–566.

Crews, D. 1984. Gamete Production, Sex Hormone Secretion, and Mating Behavior Uncoupled. Hormones and Behavior 18: 22–28.

Crews, D. 1999. Reptilian reproduction, overview. pp. 254–259. In E. Knobil and J. D. Neill (eds.), *Encyclopedia of Reproduction*. Academic Press, New York, New York, USA.

Crews, D., Grassman, M. and Lindzey, J. 1986. Behavioral facilitation of reproduction in sexual and unisexual whiptail lizards. Proceedings of the National Academy of Sciences of the United States of America 83: 9547–9550.

Cuellar, H. S. and Cuellar, O. 1977. Refractoriness in female lizard reproduction: a probable circannual clock. Science 197: 495–497.

Cuellar, O. 1984. Reproduction in a parthenogenetic lizard: with a discussion of optimal clutch size and a critique of the clutch weight/body weight ratio. American Midland Naturalist 111: 242–258.

de la Cruz, F. R. M., Guillette, L. J., jr., Santa Cruz, V. M. and Casas-Andreu, G. 1988. Reproductive and fat body cycles of the viviparous lizard, *Sceloporus mucronatus* (Sauria: Iguanidae). Journal of Herpetology 22: 1–12.

De Vosjoli, P., Klingenberg, R. and Ronne, J. 2005. The Boa Constrictor Manual (Advanced Vivarium Systems). Bowtie Press, Irvine, California, pp. 104.

DeMarco, V. and Guillette, L. J., jr. 1992. Physiological cost of pregnancy in a viviparous lizard (*Sceloporus jarrovi*). Journal of Experimental Zoology 262: 383–390.

DeNardo, D. 2006. Reproductive biology. pp. 376–390. In D. R. Mader (ed.), *Reptile Medicine and Surgery*. Saunders Elsevier, St Louis, Missouri.

DeNardo, D. F. and Autumn, K. 2001. Effect of male presence on reproductive activity in captive female blood pythons, *Python curtus*. Copeia 2001: 1138–1141.

Dendy, A. 1911. On the Structure, Development and Morphological Interpretation of the Pineal Organs and Adjacent Parts of the Brain in the Tuatara (*Sphenodon punctatus*). Philosophical Transactions of the Royal Society of London. Series B, Containing Papers of a Biological Character 201: 227–331.

Diller, L. V. and Wallace, R. L. 2002. Growth, reproduction, and survival in a population of *Crotalus viridis oreganus* in north central Idaho. Herpetological Monographs 16: 26–45.

Doughty, P. and Shine, R. 1997. Detecting life history trade-offs: measuring energy stores in "capital" breeders reveals costs of reproduction. Oecologia (Berlin) 110: 508–513.

Doughty, P. and Shine, R. 1998. Reproductive energy allocation and long-term energy stores in a viviparous lizard (*Eulamprus tympanum*). Ecology 79: 1073–1083.

Drent, R. H. and Daan, S. 1980. The prudent parent: energetic adjustments in avian breeding. Ardea 68: 225–252.

Du, W. -g. 2006. Phenotypic plasticity in reproductive traits induced by food availability in a lacertid lizard, *Takydromus septentrionalis*. Oikos 112: 363–369.

Dunham, A. E., Grant, B. W. and Overall, K. L. 1989. Interfaces between biophysical and physiological ecology, and the population ecology of terrestrial vertebrate ectotherms. Physiological Zoology 62: 335–355.

Dunham, A. E., Miles, D. B. and Reznick, D. N. 1988. Life history patterns in squamate reptiles. pp. 441–552. In C. Gans and R. B. Huey (eds.), *Biology of the Reptilia*. Academic Press, Liss, New York.

Edwards, A. and Jones, S. M. 2001. Changes in plasma progesterone, estrogen, and testosterone concentrations throughout the reproductive cycle in female viviparous blue-tongued skinks, *Tiliqua nigrolutea* (Scincidae), in Tasmania. General and Comparative Endocrinology 122: 260–269.

Ekblom, R. and Galindo, J. 2011. Applications of next generation sequencing in molecular ecology of non-model organisms. Heredity 107: 1–15.

Ernst, C. H. and Ernst, E. M. 2003. Snakes of the United States and Canada. Smithsonian Institution, Washington, D.C., pp. 680.

Etches, R. J. and Petitte, J. N. 1990. Reptilian and avian follicular hierarchies: models for the study of ovarian development. Journal of Experimental Zoology Supplement 4: 112–122.

Eyeson, K. N. 1971. Pituitary control of ovarian activity in the lizard, *Agama agama*. Journal of Zoology (London) 165: 367–372.

Firth, B. T. and Kennaway, D. J. 1980. Plasma melatonin levels in the scincid lizard *Trachydosaurus rugosus*: the effects of parietal eye and lateral eye impairment. Journal of Experimental Biology 85: 311–321.

Firth, B. T., Kennaway, D. J. and Rozenbilds, M. A. M. 1979. Plasma melatonin in the scincid lizard, *Trachydosaurus rugosus*: diel rhythm, seasonality, and the effect of constant light and constant darkness. General and Comparative Endocrinology 37: 493–500.

Firth, B. T., Thompson, M. B., Kennaway, D. J. and Belan, I. 1989. Thermal sensitivity of reptilian melatonin rhythms: "cold" tuatara vs. "warm" skink. American Journal of Physiology—Regulatory Integrative and Comparative Physiology 256: R1160–R1163.

Fisher, R. A. 1930. *The Genetical Theory of Natural Selection*. Clarendon Press, Oxford, pp. 298.

French, S. S., Greives, T. J., Zysling, D. A., Chester, E. M. and Demas, G. E. 2009. Leptin increases maternal investment. Proceedings of the Royal Society Biological Sciences Series B 276: 4003–4011.

Friesen, C. R. 2012. Patterns and mechanisms: postcopulatory sexual selection and sexual conflict in a novel mating system. Ph.D. thesis, Oregon State University, Corvalis, Oregon.

Friesen, C. R., Powers, D. R. and Mason, R. T. 2009. Cost of male courtship: using whole group metabolic rate to assess cost of courtship. Society for Integrative and Comparative Biology Annual Meeting. Boston, Massachusetts.

Gillingham, J. C. 1979. Reproductive behavior of the rat snakes of eastern North America, genus *Elaphe*. Copeia 1979: 319–331.

Gillingham, J. C. and Chambers, J. A. 1982. Courtship and pelvic spur use in the Burmese python, Python *Molurus bivittatus*. Copeia 1982: 193–196.

Girling, J. E., Jones, S. M. and Swain, R. 2002. Delayed ovulation and parturition in a viviparous alpine lizard (*Niveoscincus microlepidotus*): morphological data and plasma steroid concentrations. Reproduction, Fertility, and Development 14: 43–53.

Gittleman, J. L. and Thompson, S. D. 1988. Energy allocation in mammalian reproduction. American Zoologist 28: 863–875.

Gloyd, H. K. and Conant, R. 1990. Snakes of the *Agkistrodon* complex. Society for the Study of Amphibians and Reptiles, St. Louis, Missouri, pp. 614.

Gorman, G. C. and Licht, P. 1974. Seasonality in ovarian cycles among tropical anolis lizards. Ecology 55: 360–369.

Gregory, P. T. 2006. Influence of income and capital on reproduction in a viviparous snake: direct and indirect effects. Journal of Zoology (London) 270: 414–419.

Greulich, W. W. 1934. Artificially induced ovulation in the cat (*Felis domestica*). Anatomical Record 58: 217–224.

Gutzke, W. H. N. and Packard, G. C. 1987. Influence of the hydric and thermal environments on eggs and hatchlings of Bull Snakes *Pituophis melanoleucus*. Physiological Zoology 60: 9–17.

Hahn, W. E. 1967. Estradiol-induced vitellogenesis and concomitant fat mobilization in the lizard *Uta stansburiana*. Comparative Biochemistry and Physiology 23: 83–93.

Hahn, W. E. and Tinkle, D. W. 1966. Fat body cycling and experimental evidence for its adaptive significance to ovarian follicle development in the lizard, *Uta stansburiana*. Journal of Experimental Zoology 158: 79–86.

Haldar, C. and Thapliyal, J. P. 1981. Effect of melatonin on the testes and the renal sex segment in the garden lizard, *Calotes versicolor*. Canadian Journal of Zoology 59: 70–74.

Harlow, P. and Grigg, G. 1984. Shivering thermogenesis in a brooding diamond python, *Python spilotes spilotes*. Copeia 1984: 959–965.

Havel, P. J. 2000. Role of adipose tissue in body-weight regulation: mechanisms regulating leptin production and energy balance. Proceedings of the Nutrition Society 59: 359–371.

Ho, S. -m., Kleis, S., McPherson, R., Heisermann, G. J. and Callard, I. P. 1982. Regulation of vitellogenesis in reptiles. Herpetologica 38: 40–50.

Ho, S. 1987. Endocrinology of vitellogenesis. pp. 355–384. In N. Greenberg, J. Wingfield, D. Norris and R. Jones (eds.), *Hormones and Reproduction in Fishes, Amphibians, and Reptiles*. Plenum Press, New York, New York.

Huey, R. B., Losos, J. B. and Moritz, C. 2010. Are lizards toast? Science 328: 832–833.

Ibargüengoytía, N. R. 2004. Prolonged cycles as a common reproductive pattern in viviparous lizards from Patagonia, Argentina: reproductive cycle of *Phymaturus patagonicus*. Journal of Herpetology 38: 73–79.

James, C. and Shine, R. 1985. The seasonal timing of reproduction: a tropical-temperate comparison in Australian lizards. Oecologia (Berlin) 67: 464–474.

James, C. D. 1991. Annual variation in reproductive cycles of scincid lizards (*Ctenotus*) in central Australia. Copeia 1991: 744–760.

James, C. D., Losos, J. B. and King, D. R. 1992. Reproductive biology and diets of goannas (Reptilia: Varanidae) from Australia. Journal of Herpetology 26: 128–136.

Ji, X. and Du, W. -g. 2001. The effects of thermal and hydric environments on hatching success, embryonic use of energy and hatchling traits in a colubrid snake, *Elaphe carinata*. Comparative Biochemistry and Physiology Part A Molecular & Integrative Physiology 129: 461–471.

Jones, R. E., Fitzgerald, K. T. and Duvall, D. 1978. Quantitative analysis of the ovarian cycle of the lizard *Lepidodactylus lugubris*. General and Comparative Endocrinology 35: 70–76.

Jones, R. E., Roth, J. J., Gerrard, A. M. and Kiely, R. G. 1973. Endocrine control of clutch size in reptiles I. Effects of FSH on ovarian follicular size-gradation in *Leiolopisma laterale* and *Anolis carolinensis*. General and Comparative Endocrinology 20: 190–198.

Jones, S. M. and Swain, R. 2000. Effects of exogenous FSH on follicular recruitment in a viviparous lizard *Niveoscincus metallicus* (Scincidae). Comparative Biochemistry and Physiology Part A Molecular & Integrative Physiology 127: 487–493.

Jones, S. M., Wapstra, E. and Swain, R. 1997. Asynchronous male and female gonadal and plasma steroid concentrations in a viviparous lizard, *Niveoscincus ocellatus* (Scincidae), from Tasmania. General and Comparative Endocrinology 108: 271–281.

Kaiya, H., Miyazato, M., Kangawa, K., Peter, R. E. and Unniappan, S. 2008. Ghrelin: a multifunctional hormone in non-mammalian vertebrates. Comparative Biochemistry and Physiology Part A Molecular & Integrative Physiology 149: 109–128.

Kaiya, H., Sakata, I., Kojima, M., Hosoda, H., Sakai, T. and Kangawa, K. 2004. Structural determination and histochemical localization of ghrelin in the red-eared slider turtle, *Trachemys scripta elegans*. General and Comparative Endocrinology 138: 50–57.

King, J. A. and Millar, R. P. 1980. Comparative aspects of luteinizing hormone-releasing hormone structure and function in vertebrate phylogeny. Endocrinology 106: 707–717.

Klingenspor, M., Niggeman, H. and Heldmaier, G. 2000. Modulation of leptin sensitivity by short photoperiod acclimation in the Djungarian hamster, *Phodopus sungorus*. Journal of Comparative Physiology B Biochemical Systemic and Environmental Physiology 170: 37–43.

Krohmer, R. W. and Lutterschmidt, D. I. 2011. Environmental and neuroendocrine control of reproduction in snakes. pp. 289–346. In R. D. Aldridge and D. M. Sever (eds.), *Reproductive Biology and Phylogeny of Snakes*. Science Publishers, Enfield, New Hampshire and CRC Press, Fl, NL, UK.

Krol, E., Duncan, J. S., Redman, P., Morgan, P. J., Mercer, J. G. and Speakman, J. R. 2006. Photoperiod regulates leptin sensitivity in field voles, *Microtus agrestis*. Journal of Comparative Physiology B Biochemical Systemic and Environmental Physiology 176: 153–163.

Kronfeld-Schor, N., Richardson, C., Silvia, B. A., Kunz, T. H. and Widmaier, E. P. 2000. Dissociation of leptin secretion and adiposity during prehibernatory fattening in little brown bats. American Journal of Physiology—Regulatory Integrative and Comparative Physiology 279: R1277–R1281.

Lack, D. 1954. *The Natural Regulation of Animal Numbers*. London, UK, Oxford University Press.

LaDage, L. D. and Ferkin, M. H. 2008. Do conspecific cues affect follicular development in the female leopard gecko (*Eublepharis macularius*)? Behaviour 145: 1027–1039.

Lariviere, S. and Ferguson, S. H. 2003. Evolution of induced ovulation in North American carnivores. Journal of Mammalogy 84: 937–947.

Levey, I. L. 1973. Effects of pinealectomy and melatonin injections at different seasons on ovarian activity in the lizard *Anolis carolinensis*. Journal of Experimental Zoology 185: 169–174.

Licht, P. 1972. Environmental physiology of reptilian breeding cycles: role of temperature. General and Comparative Endocrinology Supplement 3: 477–488.
Licht, P. 1974. Luteinizing hormone (LH) in the reptilian pituitary gland. General and Comparative Endocrinology 22: 463–469.
Licht, P. 1979. Reproductive endocrinology of reptiles and amphibians: gonadotropins. Annual Review of Physiology 41: 337–351.
Licht, P. 1983. Evolutionary divergence in the structure and function of pituitary gonadotropins of tetrapod vertebrates. American Zoologist 23: 673–683.
Licht, P. and Crews, D. P. 1975. Stimulation of ovarian and oviducal growth and ovulation in female lizards by reptilian (turtle) gonadotropins. General and Comparative Endocrinology 25: 467–471.
Licht, P., Hoyer, H. E. and Van Oordt, P. G. W. J. 1969. Influence of photoperiod and temperature on testicular recrudescence and body growth in lizards, *Lacerta sicula* and *Lacerta muralis*. Journal of Zoology (London) 157: 467–501.
Licht, P., Millar, R., King, J. A., McCreery, B. R., Mendonça, M. T., Bona-Gallo, A. and Lofts, B. 1984. Effects of chicken and mammalian gonadotropin-releasing hormones (GnRH) on *in vivo* pituitary gonadotropin release in amphibians and reptiles. General and Comparative Endocrinology 54: 89–96.
Licht, P., Papkoff, H., Goldman, B. D., Follett, B. K. and Scanes, C. G. 1974. Immunological relatedness among reptilian, avian, and mammalian pituitary luteinizing horomones. General and Comparative Endocrinology 24: 168–176.
Licht, P. and Porter, D. A. 1987. Role of gonadotropin-releasing hormone in regulation of gonadotropin secretion from amphibian and reptilian pituitaries. pp. 61–85. In D. O. Norris and R. E. Jones (eds.), *Hormones and Reproduction in Fishes, Amphibians, and Reptiles*. Plenum Press, New York, New York.
Lind, C. M., Husak, J. F., Eikenaar, C., Moore, I. T. and Taylor, E. N. 2010. The relationship between plasma steroid hormone concentrations and the reproductive cycle in the Northern Pacific Rattlesnake, *Crotalus oreganus*. General and Comparative Endocrinology 166: 590–599.
Lofts, B. 1978. Reptilian reproductive cycles and environmental regulators. pp. 37–43. In I. Assenmacher and D. S. Farner (eds.), *Environmental Endocrinology*. Springer-Verlag Press, New York, New York.
Lõhmus, M. and Björklund, M. 2009. Leptin affects life history decisions in a Passerine bird: a field experiment. PLos One 4: e4602.
Lourdais, O., Brischoux, F., DeNardo, D. and Shine, R. 2004. Protein catabolism in pregnant snakes (*Epicrates cenchria maurus* Boidae) compromises musculature and performance after reproduction. Journal of Comparative Physiology B Biochemical Systemic and Environmental Physiology 174: 383–391.
Lourdais, O., Hoffman, T. C. M. and DeNardo, D. F. 2007. Maternal brooding in the children's python (*Antaresia childreni*) promotes egg water balance. Journal of Comparative Physiology B Biochemical Systemic and Environmental Physiology 177: 569–577.
Lutterschmidt, D. I. 2012. Chronobiology of reproduction in garter snakes: neuroendocrine mechanisms and geographic variation. General and Comparative Endocrinology 176: 448–455.
Madsen, T. and Shine, R. 1999. The adjustment of reproductive threshold to prey abundance in a capital breeder. Journal of Animal Ecology 68: 571–580.
Manire, C. A., Byrd, L., Therrien, C. L. and Martin, K. 2008. Mating-induced ovulation in loggerhead sea turtles, *Caretta caretta*. Zoo Biology 27: 213–225.
Mardis, E. R. 2008. Next-generation DNA sequencing methods. Annual Review of Genomics and Human Genetics 9: 387–402.
Margres, M. J., McGivern, J. J., Wray, K. P., Seavy, M., Calvin, K. and Rokyta, D. R. 2014. Linking the transcriptome and proteome to characterize the venom of the eastern diamondback rattlesnake (*Crotalus adamanteus*). Journal of Proteomics 96: 145–158.

Marion, K. R. 1970. Temperature as a reproductive cue for the female fence lizard *Sceloporus undulatus*. Copeia 1970: 562–564.
Marion, K. R. 1982. Reproductive cues for gonadal development in temperate reptiles: temperature and photoperiod effects on the testicular cycle of the lizard *Sceloporus undulatus*. Herpetologica 38: 26–39.
Martinez-Torres, M., Hernandez-Caballero, M. E., Alvarez-Rodriguez, C., Luis-Diaz, J. A. and Ortiz-Lopez, G. 2003. Luteal development and progesterone levels during pregnancy of the viviparous temperate lizard *Barisia imbricata imbricata* (Reptilia: Anguidae). General and Comparative Endocrinology 132: 55–65.
Mathies, T., Franklin, E. A. and Miller, L. A. 2004. Proximate cues for ovarian recrudescence and ovulation in the brown treesnake (*Boiga irregularis*) under laboratory conditions. Herpetological Review 35: 46–49.
Mayer, I., Bornestaf, C. and Borg, B. 1997. Melatonin in non-mammalian vertebrates: physiological role in reproduction? Comparative Biochemistry and Physiology Part A Physiology 18: 515–531.
Mazzi, V. and Vellano, C. 1987. Prolactin and reproduction. pp. 87–115. In D. O. Norris and R. E. Jones (eds.), *Hormones and Reproduction in Fishes, Amphibians, and Reptiles*. Plenum Press, New York, New York.
Mendez-De La Cruz, F. R., Guillette, L. J., jr. and Villagran-Santa Cruz, V. 1993. Differential atresia of ovarian follicles and its effect on the clutch size of two populations of the viviparous lizard *Sceloporus mucronatus*. Functional Ecology 7: 535–540.
Mendonça, M. T. and Crews, D. 1990. Mating-Induced Ovarian Recrudescence in the Red-Sided Garter Snake. Journal of Comparative Physiology A Sensory Neural and Behavioral Physiology 166: 629–632.
Mendonça, M. T. and Crews, D. 2001. Control of attractivity and receptivity in female red-sided garter snakes. Hormones and Behavior 40: 43–50.
Mendonça, M. T., Tousignant, A. J. and Crews, D. 1995. Seasonal changes and annual variability in daily plasma melatonin in the red-sided garter snake (*Thamnophis sirtalis pareitalis*). General and Comparative Endocrinology 100: 226–237.
Mendonça, M. T., Tousignant, A. J. and Crews, D. 1996a. Courting and noncourting male red-sided garter snakes, *Thamnophis sirtalis parietalis*: plasma melatonin levels and the effects of pinealectomy. Hormones and Behavior 30: 176–185.
Mendonça, M. T., Tousignant, A. J. and Crews, D. 1996b. Pinealectomy, melatonin, and courtship behavior in male red-sided garter snakes (*Thamnophis sirtalis parietalis*). Journal of Experimental Zoology 274: 63–74.
Moore, A. F. and Menaker, M. 2011. The effect of light on melatonin secretion in the cultured pineal glands of *Anolis* lizards. Comparative Biochemistry and Physiology Part A Molecular & Integrative Physiology 160: 301–308.
Moore, M. C., Whittier, J. M., Billy, A. J. and Crews, D. 1985. Male-like behaviour in an all-female lizard: relationship to the ovarian cycle. Animal Behaviour 33: 284–289.
Moore, M. C., Whittier, J. M. and Crews, D. 1984. Environmental control of seasonal reproduction in a parthenogenetic lizard *Cnemidophorus uniparens*. Physiological Zoology 57: 544–549.
Murakami, N., Hayashida, T., Kuroiwa, T., Nakahara, K., Ida, T., Mondal, M. S., Nakazato, M., Kojima, M. and Kangawa, K. 2002. Role for central ghrelin in food intake and secretion profile of stomach ghrelin in rats. Journal of Endocrinology 174: 283–288.
Murphy, B. F. and Thompson, M. B. 2011. A review of the evolution of viviparity in squamate reptiles: the past, present and future role of molecular biology and genomics. Journal of Comparative Physiology B Biochemical Systemic and Environmental Physiology 181: 575–594.
Murphy, K., Hudson, S. and Shea, G. 2006. Reproductive seasonality of three cold-temperate viviparous skinks from southeastern Australia. Journal of Herpetology 40: 454–464.

Naulleau, G. and Bonnet, X. 1995. Reproductive ecology, body fat reserves and foraging mode in females of two contrasted snake species: *Vipera aspis* (terrestrial, viviparous) and *Elaphe longissima* (semi-arboreal, oviparous). Amphibia-Reptilia 16: 37–46.

Naulleau, G. and Bonnet, X. 1996. Body condition threshold for breeding in a viviparous snake. Oecologia (Berlin) 107: 301–306.

Naulleau, G., Bonnet, X., Vacher-Vallas, M., Shine, R. and Lourdais, O. 1999. Does less-than-annual production of offspring by female vipers (*Vipera aspis*) mean less-than-annual mating? Journal of Herpetology 33: 688–691.

Naulleau, G. and Fleury, F. 1990. Changes in Plasma Progesterone in Female *Vipera aspis* L. (Reptilia, Viperidae) During the Sexual Cycle in Pregnant and Nonpregnant Females. General and Comparative Endocrinology 78: 433–443.

Naulleau, G., Fleury, F. and Boissin, J. 1987. Annual Cycles in Plasma Testosterone and Thyroxine in the Male Aspic *Viper vipera* aspis L. (Reptilia, Viperidae), in Relation to the Sexual Cycle and Hibernation. General and Comparative Endocrinology 65: 254–263.

Niewiarowski, P. H., Balk, M. L. and Londraville, R. L. 2000. Phenotypic effects of leptin in an ectotherm: a new tool to study the evolution of life histories and endothermy? Journal of Experimental Biology 203: 295–300.

Olsson, M., Madsen, T. and Shine, R. 1997. Is sperm really so cheap? Costs of reproduction in male adders, *Vipera berus*. Proceedings of the Royal Society of London B Biological Sciences 264: 455–459.

Olsson, M. and Shine, R. 1997. The seasonal timing of oviposition in sand lizards (*Lacerta agilis*): why early clutches are better. Journal of Evolutionary Biology 10: 369–381.

Olsson, M. and Shine, R. 1998. Timing of parturition as a maternal care tactic in an alpine lizard species. Evolution 52: 1861–1864.

Olsson, M. and Shine, R. 1999. Plasticity in frequency of reproduction in an alpine lizard, *Niveoscincus microlepidotus*. Copeia 1999: 794–796.

Osborne, S. T. 1982. The captive breeding of colubrid snakes: Part II. Annual cycles and breeding techniques. San Diego Herpetological Society Newsletter 4.

Packard, M. J., Packard, G. C. and Boardman, T. J. 1982. Structure of eggshells and water relations of reptilian eggs. Herpetologica 38: 136–155.

Pandav, B. N., Shanbhag, B. A. and Saidapur, S. K. 2007. Ethogram of courtship and mating behaviour of garden lizard, *Calotes versicolor*. Current Science (Bangalore) 93: 1164–1167.

Paolucci, M., Rocco, M. and Varricchio, E. 2001. Leptin presence in plasma, liver and fat bodies in the lizard *Podarcis sicula* fluctuations throughout the reproductive cycle. Life Sciences 69: 2399–2408.

Parker, M. R. and Mason, R. T. 2012. How to make a sexy snake: estrogen activation of female sex pheromone in male red-sided garter snakes. Journal of Experimental Biology 215: 723–730.

Powell, R. C., Ciarcia, G., Lance, V., Millar, R. P. and King, J. A. 1986. Identification of diverse molecular forms of GnRH in reptile brain. Peptides (New York) 7: 1101–1108.

Prasad, M. R. N. and Sanyal, M. K. 1969. Effect of sex hormones on the sexual segment of kidney and other accessory reproductive organs of the Indian house lizard *Hemidactylus flaviviridis* Ruppell. General and Comparative Endocrinology 12: 110–118.

Pyron, R. A., Burbrink, F. T. and Wiens, J. J. 2013. A phylogeny and revised classification of Squamata, including 4161 species of lizards and snakes. BMC Evolutionary Biology 13: 93.

Quay, W. B. 1979. The parietal eye-pineal complex. pp. 245–406. In C. Gans, R. G. Northcutt and P. Ulinski (eds.), *Biology of the Reptilia*. Academic Press, London, United Kingdom.

Ramirez-Bautista, A. and Gutierrez-Mayen, G. 2003. Reproductive ecology of *Sceloporus utiformis* (Sauria: Phrynosomatidae) from a tropical dry forest of Mexico. Journal of Herpetology 37: 1–10.

Ratto, M. H., Delbaere, L. T. J., Leduc, Y. A., Pierson, R. A. and Adams, G. P. 2011. Biochemical isolation and purification of ovulation-inducing factor (OIF) in seminal plasma of llamas. Reproductive Biology and Endocrinology 9: 24.

Reading, C. J. 2004. The influence of body condition and prey availability on female breeding success in the smooth snake (*Coronella austriaca* Laurenti). Journal of Zoology (London) 264: 61–67.

Reppert, S. M., Weaver, D. R. and Ebisawa, T. 1994. Cloning and characterization of a mammalian melatonin receptor that mediates reproductive and circadian responses. Neuron 13: 1177–1185.

Robert, K. A. and Thompson, M. B. 2000. Energy consumption by embryos of a viviparous lizard, *Eulamprus tympanum*, during development. Comparative Biochemistry and Physiology Part A Molecular & Integrative Physiology 127: 481–486.

Rock, J. 2006. Delayed parturition: constraint or coping mechanism in a viviparous gekkonid? Journal of Zoology (London) 268: 355–360.

Rock, J., Andrews, R. M. and Cree, A. 2000. Effects of reproductive condition, season, and site on selected temperatures of a viviparous gecko. Physiological and Biochemical Zoology 73: 344–355.

Rose, J. D. 1978. Distribution and properties of diencephalic neuronal responses to genital stimulation in the female cat. Experimental Neurology 61: 231–244.

Rubenstein, D. R. and Wikelski, M. 2003. Seasonal changes in food quality: a proximate cue for reproductive timing in marine iguanas. Ecology 84: 3013–3023.

Saint Girons, H. 1982. Reproductive cycles of male snakes and their relationships with climate and female reproductive cycles. Herpetologica 38: 5–16.

Saint Girons, H., Bradshaw, S. D. and Bradshaw, F. J. 1993. Sexual Activity and Plasma Levels of Sex Steroids in the Aspic Viper *Vipera aspis* L. (Reptilia, Viperidae). General and Comparative Endocrinology 91: 287–297.

Schuett, G. W., Harlow, H. J., Rose, J. D., VanKirk, E. A. and Murdoch, W. J. 1997. Annual cycle of plasma testosterone in male copperheads, *Agkistrodon contortrix* (Serpentes, Viperidae): Relationship to timing of spermatogenesis, mating, and agonistic behavior. General and Comparative Endocrinology 105: 417–424.

Schultz, T. J., Webb, J. K. and Christian, K. A. 2008. The physiological cost of pregnancy in a tropical viviparous snake. Copeia 2003: 637–642.

Sever, D. M. and Hamlett, W. C. 2002. Female sperm storage in reptiles. Journal of Experimental Zoology 292: 187–199.

Shine, R. 1977. Reproduction in Australian elapid snakes II. Female reproductive cycles. Australian Journal of Zoology 25: 655–666.

Shine, R. 1980. "Costs" of reproduction in reptiles. Oecologia (Berlin) 46: 92–100.

Shine, R. and Brown, G. P. 2008. Adapting to the unpredictable: reproductive biology of vertebrates in the Australian wet–dry tropics. Philosophical Transactions of the Royal Society of London B Biological Sciences 363: 363–373.

Shine, R., Olsson, M. M. and Mason, R. T. 2000. Chastity belts in gartersnakes: the functional significance of mating plugs. Biological Journal of the Linnean Society 70: 377–390.

Silva, M., Nino, A., Guerra, M., Letelier, C., Valderrama, X. P., Adams, G. P. and Ratto, M. H. 2011. Is ovulation-inducing factor (OIF) present in the seminal plasma of rabbits? Animal Reproduction Science 127: 213–221.

Sinervo, B. and Licht, P. 1991. Hormonal and Physiological Control of Clutch Size, Egg Size, and Egg Shape in Side-Blotched Lizards (*Uta stansburiana*)—Constraints on the Evolution of Lizard Life Histories. Journal of Experimental Zoology 257: 252–264.

Sinervo, B., Mendez-De La Cruz, F. R., Miles, D. B., Heulin, B., Bastiaans, E., Villagran-Santa Cruz, V., Lara-Resendiz, R., Martinez-Mendez, N., Calderon-Espinosa, M. L., Meza-Lazaro, R. N., Gadsden, H., Avila, L. J., Morando, M., De la Riva, I. J., Sepulveda, P. V., Duarte Rocha, C. F., Ibarguengoytia, N., Puntriano, C. A., Massot, M., Lepetz, V., Oksanen, T. A., Chapple, D. G., Bauer, A. M., Branch, W. R., Clobert, J. and Sites, J. W., jr. 2010. Erosion of lizard diversity by climate change and altered thermal niches. Science 328: 894–899.

Slip, D. J. and Shine, R. 1988. The reproductive biology and mating system of diamond pythons, *Morelia spilota* (Serpentes: Boidae). Herpetologica 44: 396–404.

Smith, G. R., Ballinger, R. E. and Rose, B. R. 1995. Reproduction in *Sceloporus virgatus* from the Chiricahua Mountains of southeastern Arizona with emphasis on annual variation. Herpetologica 51: 342–349.
Spanovich, S., Niewiarowski, P. H. and Londraville, R. L. 2006. Seasonal effects on circulating leptin in the lizard Sceloporus undulatus from two populations. Comparative Biochemistry and Physiology Part B Biochemistry & Molecular Biology 143: 507–513.
Stamps, J. A. 1975. Courtship patterns, estrus periods and reproductive condition in a lizard, *Anolis aeneus*. Physiology & Behavior 14: 531–535.
Stamps, J. A. 1976. Egg retention, rainfall and egg laying in a tropical lizard *Anolis aeneus*. Copeia 1976: 759–794.
Stebbins, R. C. and Cohen, N. W. 1973. The effect of parietalectomy on the thyroid and gonads in free-living western fence lizards, *Sceloporus occidentalis*. Copeia 1973: 662–668.
Swain, R. and Jones, S. M. 1994. Annual Cycle of Plasma Testosterone and Other Reproductive Parameters in the Tasmanian Skink, *Niveoscincus metallicus*. Herpetologica 50: 502–509.
Taylor, E. N., Denardo, D. F. and Jennings, D. H. 2004. Seasonal steroid hormone levels and their relation to reproduction in the western diamond-backed rattlesnake, *Crotalus atrox* (Serpentes: Viperidae). General and Comparative Endocrinology 136: 328–337.
Taymor, M. L. 1978. The induction of ovulation. pp. 373–381. In D. B. Crighton, G. R. Foxcroft, N. B. Haynes and G. E. Lamming (eds.), *Control of Ovulation*. Butterworths, London, United Kingdom.
Telemeco, R. S. and Baird, T. A. 2011. Capital energy drives production of multiple clutches whereas income energy fuels growth in female collared lizards *Crotaphytus collaris*. Oikos 120: 915–921.
Thapliyal, J. P. and Haldar, C. 1979. Effect of pinealectomy on the photoperiodic gonadal response of the Indian garden lizard, *Calotes versicolor*. General and Comparative Endocrinology 39: 79–86.
Tilden, A. R. and Hutchinson, V. H. 1993. Influence of photoperiod and temperature on serum melatonin in the diamondback water snake, *Nerodia rhombifera*. General and Comparative Endocrinology 92: 347–354.
Tinkle, D. W. and Irwin, L. N. 1965. Lizard reproduction: refractory period and response to warmth in *Uta stansburiana* females. Science 148: 1613–1614.
Tinkle, D. W., Wilbur, H. M. and Tilley, S. G. 1970. Evolutionary strategies in lizard reproduction. Evolution 24: 55–74.
Tosini, G. 1997. The pineal complex of reptiles: physiological and behavioral roles. Ethology Ecology & Evolution 9: 313–333.
Tosini, G., Bertolucci, C. and Foa, A. 2001. The circadian system of reptiles: a multioscillatory and multiphotoreceptive system. Physiology & Behavior 72: 461–471.
Trauth, S. E. 1978. Ovarian cycle of *Crotaphytus collaris* (Reptilia, Lacertilia, Iguanidae) from Arkansas with emphasis on corpora albicantia, follicular atresia, and reproductive potential. Journal of Herpetology 12: 461–470.
Troyer, K. 1983. Posthatching yolk energy in a lizard: utilization pattern and interclutch variation. Oecologia 58: 340–344.
Tsai, P. -s. and Licht, P. 1993. GnRH-induced desensitization of *in vitro* luteinizing hormone secretion in the turtle, *Trachemys scripta*. General and Comparative Endocrinology 89: 238–247.
Tyrrell, C. L., Cree, A. and Towns, D. R. 2000. Variation in reproduction and condition of northern tuatara (*Sphenodon punctatus punctatus*) in the presence and absence of kiore. New Zealand Department of Conservation, Wellington, New Zealand, pp. 42.
Underwood, H. 1981. Effects of pinealectomy and melatonin on the photoperiodic gonadal response of the male lizard *Anolis carolinensis*. Journal of Experimental Zoology 217: 417–422.
Underwood, H. 1989. The pineal and melatonin: regulators of circadian function in lower vertebrates. Experientia (Basel) 45: 914–922.

Underwood, H. and Calaban, M. 1987. Pineal melatonin rhythms in the lizard *Anolis carolinensis*: I. Response to light and temperature cycles. Journal of Biological Rhythms 2: 179–193.
Underwood, H. and Groos, G. 1982. Vertebrate circadian rhythms: retinal and extraretinal photoreception. Experientia (Basel) 38: 1013–1021.
Unniappan, S. 2010. Ghrelin: an emerging player in the regulation of reproduction in non-mammalian vertebrates. General and Comparative Endocrinology 167: 340–343.
Van Dyke, J. U. and Beaupre, S. J. 2011. Bioenergetic components of reproductive effort in viviparous snakes: costs of vitellogenesis exceed costs of pregnancy. Comparative Biochemistry and Physiology Part A Molecular & Integrative Physiology 160: 504–515.
Van Dyke, J. U., Beaupre, S. J. and Kreider, D. L. 2012. Snakes allocate amino acids acquired during vitellogenesis to offspring: are capital and income breeding consequences of variable foraging success? Biological Journal of the Linnean Society 106: 390–404.
Van Dyke, J. U., Plummer, M. V. and Beaupre, S. J. 2011. Residual yolk energetics and postnatal shell growth in Smooth Softshell Turtles, *Apalone mutica*. Comparative Biochemistry and Physiology Part A Molecular & Integrative Physiology 158: 37–46.
Van Wyk, J. H. 1991. Biennial reproduction in the female viviparous lizard *Cordylus giganteus*. Amphiba-Reptilia 12: 329–342.
Vidal, N. and Hedges, S. B. 2005. The phylogeny of squamate reptiles (lizards, snakes, and amphisbaenians) inferred from nine nuclear protein-coding genes. Comptes Rendus Biologies 328: 1000–1008.
Visser, M. E., Both, C. and Lambrechts, M. M. 2004. Global climate change leads to mistimed avian reproduction. Advances in Ecological Research 35: 89–110.
Visser, M. E., van Noordwijk, A. J., Tinbergen, J. M. and Lessells, C. M. 1998. Warmer springs lead to mistimed reproduction in great tits (*Parus major*). Proceedings of the Royal Society of London B Biological Sciences 265: 1867–1870.
Vitt, L. J. and Goldberg, S. R. 1983. Reproductive ecology of two tropical iguanid lizards: *Tropidurus torquatus* and Platynotus semitaeniatus. Copeia 1983: 131–141.
Vitt, L. J., Van Loben Sels, R. C. and Ohmart, R. D. 1978. Lizard reproduction: annual variation and environmental correlats in the iguanid lizard *Urosaurus graciosus*. Herpetologica 34: 241–253.
Vonk, F. J., Casewell, N. R., Henkel, C. V., Heimberg, A. M., Jansen, H. J., McCleary, R. J. R., Kerkkamp, H. M. E., Vos, R. A., Guerreiro, I., Calvete, J. J., Wüster, W., Woods, A. E., Logan, J. M., Harrison, R. A., Castoe, T. A., de Koning, A. P. J., Pollock, D. D., Yandell, M., Calderon, D., Renjifo, C., Currier, R. B., Salgado, D., Pla, D., Sanz, L., Hyder, A. S., Ribeiro, J. M. C., Arntzen, J. W., van den Thillart, G. E. E. J. M., Boetzer, M., Pirovano, W., Dirks, R. P., Spaink, H. P., Duboule, D., McGlinn, E., Kini, R. M. and Richardson, M. K. 2013. The king cobra genome reveals dynamic gene evolution and adaptation in the snake venom system. Proceedings of the National Academy of Sciences 110: 20651–20656.
Wallace, R. A. 1985. Vitellogenesis and oocyte growth in nonmammalian vertebrates. pp. 127–177. In L. W. Browder (ed.), *Developmental Biology, a Comprehensive Synthesis*. Plenum Press, New York, New York.
Wapstra, E., Swain, R., Jones, S. M. and O'Reilly, J. 1999. Geographic and annual variation in reproductive cycles in the Tasmanian spotted snow skink, *Niveoscincus ocellatus* (Squamata: Scincidae). Australian Journal of Zoology 47: 539–550.
Webb, J. K. and Lannoo, M. J. 2004. Pregnancy decreases swimming performance of female northern death adders (*Acanthophis praelongus*). Copeia 2004: 357–363.
Weil, M. R. 1985. Comparison of plasma and testicular testosterone levels during the active season in the common garter snake, *Thamnophis sirtalis* (L.). Comparative Biochemistry and Physiology 81A: 585–587.
Weil, M. R. and Aldridge, R. D. 1979. The effect of temperature on the male reproductive system of the common water snake (*Nerodia sipedon*). Journal of Experimental Zoology 210: 327–332.
Weil, M. R. and Aldridge, R. D. 1981. Seasonal androgenesis in the male water snake, *Nerodia sipedon*. General and Comparative Endocrinology 44: 44–53.

Whittier, J. M. and Crews, D. 1986a. Effects of prostaglandin F2α on sexual behavior and ovarian function in female garter snakes (*Thamnophis sirtalis parietalis*). Endocrinology 119: 787–792.

Whittier, J. M. and Crews, D. 1986b. Ovarian development in red-sided garter snakes, *Thamnophis sirtalis parietalis*: relationship to mating. General and Comparative Endocrinology 61: 5–12.

Whittier, J. M., Mason, R. T. and Crews, D. 1987. Plasma Steroid Hormone Levels of Female Red-Sided Garter Snakes, *Thamnophis sirtalis parietalis*: Relationship to Mating and Gestation. General and Comparative Endocrinology 67: 33–43.

Wiechmann, A. F. and Wirsig-Wiechmann, C. R. 1994. Melatonin receptor distribution in the brain and retina of a lizard, *Anolis carolinensis*. Brain Behavior and Evolution 43: 26–33.

Wikelski, M., Hau, M. and Wingfield, J. C. 2000. Seasonality of reproduction in a neotropical rain forest bird. Ecology 81: 2458–2472.

Winne, C. T., Willson, J. D. and Gibbons, J. W. 2006. Income breeding allows an aquatic snake *Seminatrix pygaea* to reproduce normally following prolonged drought-induced aestivation. Journal of Animal Ecology 75: 1352–1360.

Wong, E. S. W., Morgenstern, D., Mofiz, E., Gombert, S., Morris, K. M., Temple-Smith, P., Renfree, M. B., Whittington, C. M., King, G. F., Warren, W. C., Papenfuss, A. T. and Belov, K. 2012. Proteomics and Deep Sequencing Comparison of Seasonally Active Venom Glands in the Platypus Reveals Novel Venom Peptides and Distinct Expression Profiles. Molecular & Cellular Proteomics 11: 1354–1364.

Yamanouye, N., Silveira, P., Abdalla, F. M. F., Almeida-Santos, S. M., Breno, M. C. and Salomao, M. G. 2004. Reproductive cycle of the Neotropical *Crotalus durissus terrificus*: II. Establishment and maintenance of the uterine muscular twisting, a strategy for long-term sperm storage. General and Comparative Endocrinology 139: 151–157.

Zaidan, F., III., Kreider, D. L. and Beaupre, S. J. 2003. Testosterone cycles and reproductive energetics: Implications for northern range limits of the Cottonmouth (*Agkistrodon piscivorus leucostoma*). Copeia: 231–240.

CHAPTER 6

Female Reproductive Anatomy: Cloaca, Oviduct and Sperm Storage

Dustin S. Siegel,[1,*] *Aurélien Miralles,*[2] *Justin L. Rheubert*[3] and *David M. Sever*[4]

6.1 OVERVIEW

The following chapter is a review of the female reproductive anatomy of lizards. We limit our discussions to the anatomy of the cloacae, oviducts, and sperm storage receptacles in female lizards, as other chapters focus attention toward topics such as ovarian development/ovarian cycle (Ramirez-Pinilla et al. Chapter 8, this volume) and placental morphology/eggs shelling (Stewart and Blackburn Chapter 15, this volume).

[1] Department of Biology, Southeast Missouri State University, Cape Girardeau, MO 63701, USA.
[2] CNRS-UMR5175 CEFE, Centre d'Ecologie Functionnelle et Evolutive, 1919 route de mende, 34293 Montpellier cedex 5.
[3] College of Sciences, The University of Findlay, Findlay, Ohio 45840, USA.
[4] Department of Biological Sciences, Southeastern Louisiana University, Hammond, LA 70402, USA.
* Corresponding author

6.2 THE CLOACA

6.2.1 Overview

Few investigators have surveyed the morphology of the lizard cloaca by means of histological examination. Prominent previous studies were accomplished by Gabe and Saint-Girons (1965), Whiting (1969), Hardy and Cole (1981), Trauth *et al.* (1987), Sánchez-Martínez *et al.* (2007), Gharzi *et al.* (2013), and Siegel *et al.* (2013). As Siegel *et al.* (2011a) indicated in a review on the cloacal anatomy of snakes, gross examination of snake cloacae that pre-dated histological studies (e.g., Gadow 1887) confused many of the distinct cloacal regions and their orientation. Thus, we focus the majority of our review on historical literature that utilized histological examination. Although not technically lizards, details of the cloacal anatomy of *Sphenodon* will also be reviewed (i.e., Gabe and Saint-Girons 1964, 1965), as in depth details on the cloacal anatomy of the closest extant relative to squamates may be pertinent to the understanding of cloaca evolution in lizards (e.g., out-group comparison; see also Cree Chapter 16, this volume).

Considering many historical studies provide inadequate details on the morphology of the cloaca of lizards, we examined cloacae from three individuals of Yarrow's Spiny Lizard (*Sceloporus jarrovi*; collected from Cochise and Pima counties, Arizona, April, 2013; permit #M587954). Specimens were euthanased via a lethal injection of MS-222 and urogenital tracts were removed, fixed in buffered formalin, dehydrated in ascending concentrations of ethanol, cleared in toluene, embedded in paraffin, sectioned sagittally or transversely, affixed to albumenized slides, stained with hematoxylin and eosin for general histological examination, and examined with light microscopy. This novel collection was used to construct detailed micrographs to aid in the visualization of the female lizard cloaca, and confirm anatomical accounts from previous assessments.

6.2.2 *Sceloporus jarrovi* Cloaca

The vent of female *Sceloporus jarrovi* opens dorsally through the ventral body wall and into the proctodaeum (Fig. 6.1a). Sagittaly, the proctodaeum proceeds dorsal and slightly caudal before its dorsal extreme, at which point the proctodaeum abruptly bends cranial to a horizontal plane and then slightly ventral (Fig. 6.1a-d). Transversely, the lumen of the proctodaeum is dorso-laterally flattened (Fig. 6.2A,B). The lumen of the proctodaeum is lined by a conspicuous stratified squamous epithelium (~6 cell layers thick) with a keratinized superficial cell layer. A large ventral gland complex empties into the lumen of the proctodaeum cranial to the bend of the proctodaeum

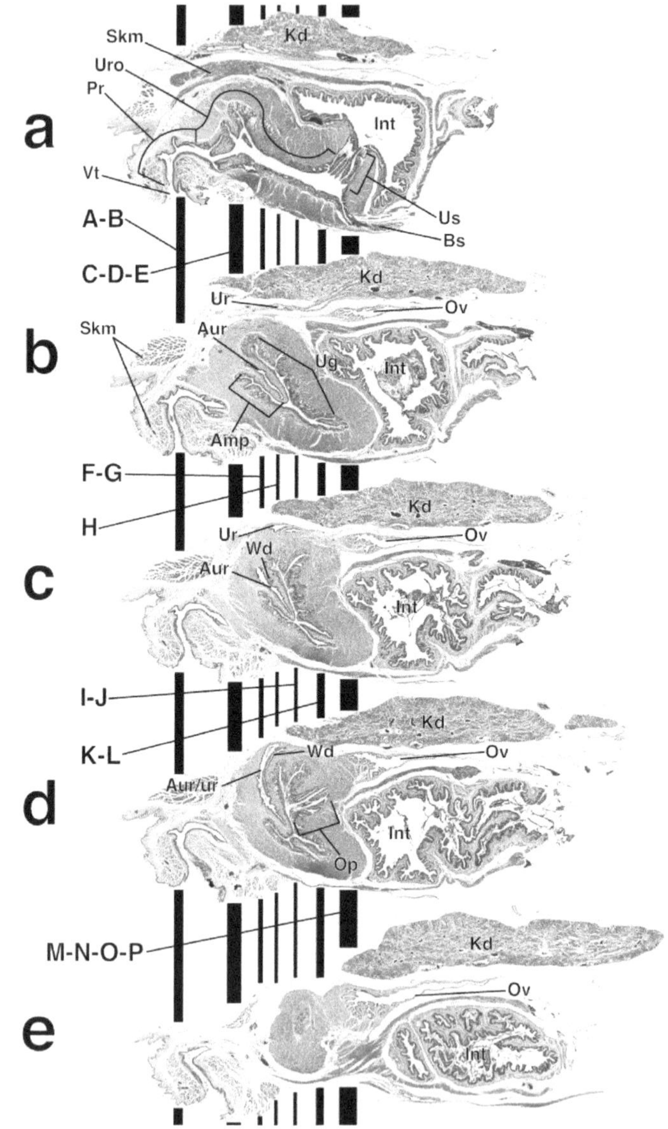

Fig. 6.1 Representative sagittal sections of the cloaca of female *Sceloporus jarrovi*. **a-e.** Sections from the mid-sagittal plane of the cloaca **(a)** to the left periphery **(e)** of the cloaca (hematoxylin and eosin; scale bar [top right horizontal line] = 1,000 μm). Sections are orientated so that their left extreme represents the caudal extreme of the cloacal apparatus and the right extreme represents the cranial extreme of the cloacal apparatus. See text for description of figure. Amp, ampullary papilla; Aur, ampulla ureter; Aur/ur, ampulla ureter/ureter communication; Bs, bladder stalk; Int, intestine; Kd, kidney; Op, oviducal papilla; Ov, oviduct; Pr, proctodaeum; Skm, skeletal muscle; Ug, urodaeal gland; Uro, urodaeum; Us, urodaeal sphincter; Vt, vent; Wd, Wolffian duct.

Color image of this figure appears in the color plate section at the end of the book.

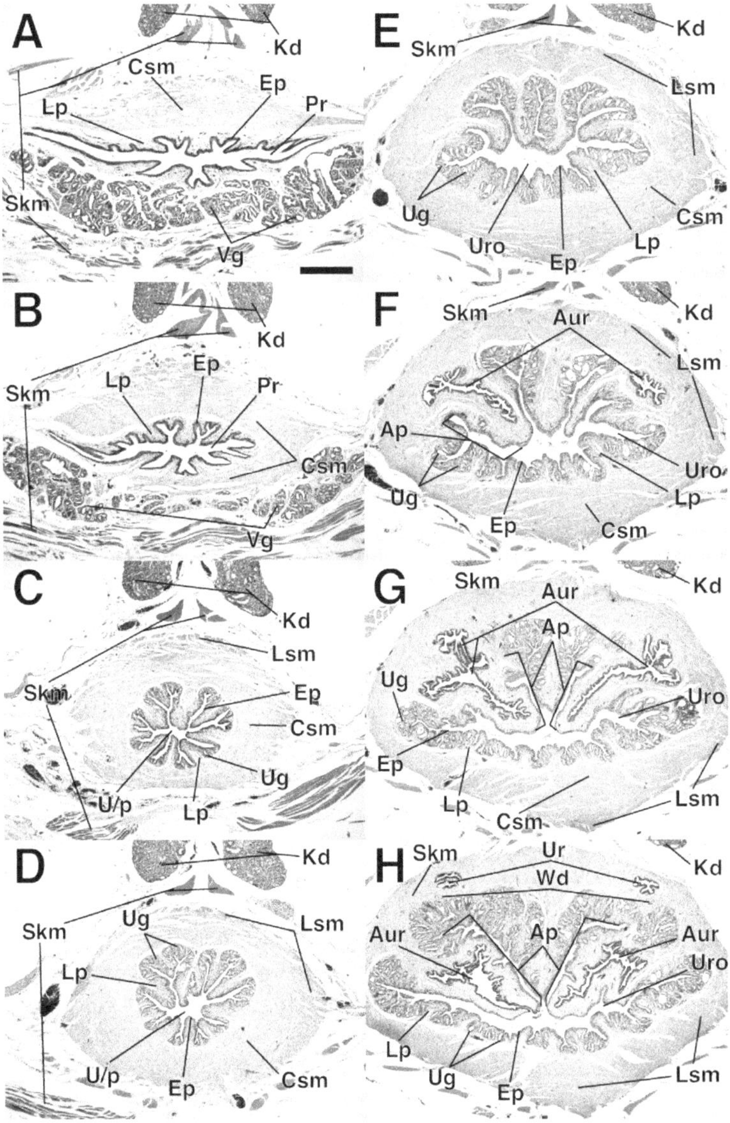

Fig. 6.2 Representative transverse sections from the cloaca of female *Sceloporus jarrovi*. Sections are organized from most caudal **(A)** to most cranial **(H)**. Individual micrograph lettering represents the capital lettering on Fig. 9.1, as demonstrative of the approximate location that each representative transverse section originated (hematoxylin and eosin; scale bar = 500 μm). See text for description of figure. Ap, ampullary papilla; Aur, ampulla ureter; Csm, circular smooth muscle; Ep, epithelium; Kd, kidney; Lp, lamina propria; Lsm, longitudinal smooth muscle; Pr, proctodaeum; Skm, skeletal muscle; Ug, urodaeal gland; U/p, urodaeal/proctodaeal transition; Ur, ureter; Uro, urodaeum; Vg, ventral gland; Wd, Wolffian duct.

Color image of this figure appears in the color plate section at the end of the book.

to a horizontal plane (Fig. 6.1a; 6.2A,B). This gland complex rests in the submucosa space of the proctodaeum between skeletal muscle and the lamina propria and is complex tubular in morphology. The epithelium lining the gland tubules is simple columnar with the cytoplasm of tubular ends of the glandular cells containing numerous eosinophilic granules (Fig. 6.2A,B). The histology of the lamina propria of the proctodaeum is consistent with that of a loose areolar connective tissue, and the dorsal projection of the proctodaeum is encompassed exteriorly by skeletal muscle (Fig. 6.1c). As the proctodaeum bends cranially, skeletal muscle dorsal to the proctodaeum terminates and is replaced by smooth muscle fibers circulating around the cloaca (Fig. 6.2A-D).

As the proctodaeum persists cranially, the dorso-ventrally flattened appearance decreases, and internal folding of the mucosa increases and results in rugae extension into the lumen (Fig. 6.2C,D). The epithelium lining the lumen remains stratified (~3 cell layers thick), however, the superficial cell lining transitions to cuboidal with eosinophilic epithelial cell cytoplasm. Notably, the mucosa is filled with simple and branched tubular glands emptying onto the surface of the epithelium (Fig. 6.1b; Fig. 6.2C,D). These glands are lined by a simple columnar epithelium with eosinophilic granules scattered in individual epithelial cell cytoplasm. At this plane of section, the cloaca is now entirely encompassed by a thick muscularis of mainly circular muscle fibers, but also contains longitudinal fibers dorsally and laterally (Fig. 6.2C,D). We consider this region a transitional region between the proctodaeum and urodaeum; however, the majority of the charateristics of this region are more similar to that of the urodaeum.

More cranially, the lumen of the cloaca expands dramatically (Fig. 6.1a,b; Fig. 6.2E), and this expansion delineates the urodaeum proper. At this juncture of the cloaca, urodaeal mucosal glands are numerous and the epithelium lining the lumen of the urodaeum is stratified (~3 cell layers thick) with a columnar superficial cell layer. Individual epithelial cell cytoplasm stains eosinophilic, as observed in the transition between the proctodaeum and urodaeum. A thick layer of circular smooth muscle encompasses the entire urodaeum, and a thinner layer of longitudinal smooth muscle covers the circular layer externally (Fig. 6.2E). Just cranial to the transition to the urodaeum proper, two ampullary papilla carry the ampullae ureters to the lumen of the urodaeum from a dorsolateral position (Fig. 6.1b-d; Fig. 6.2F-H). The ampullary papillae project cranially into the cloaca and, thus, the ureters and Wolffian ducts travel caudal to their point of communication with the urodaeum (Fig. 6.2H) and then travel cranially through the ampullary papilla (Fig. 6.1b-d). Before traveling through the ampullary papillae, the ureters and Wolffian ducts unite (Fig. 6.1d; Fig. 6.2H), forming the ampullae ureters. The lumina of the ampullae ureters and ureters are both lined by a simple and highly basophilic columnar epithelium, whereas, the lumina of the Wolffian ducts are lined by a simple cuboidal epithelium that stains eosinophilic.

Immediately cranial to the communication of the ampullae ureters with the urodaeum, the distal tips of the oviducal papillae are observed, through which the oviducts communicate with the urodaeum (Fig. 6.1d; Fig. 6.3I,H). The distal extremities of the oviducal papillae are free from the wall of the urodaeum, but more cranially they are embedded in the urodaeal wall through a dorso-lateral invasion into the urodaeum (Fig. 6.1d,e; Fig. 6.3J-L). Cranial to the invasion of the oviducal papillae, the urodaeum is not bifurcated, and the lumen decreases in size dramatically (Fig. 6.3L). At this plane of section, the cloaca still possesses numerous urodaeal glands in the mucosa lining the dorsal wall of the cloaca, but not the ventral (Fig. 6.3L). More cranially, the lumen of the urodaeum decreases in size and the bladder stalk, with a simple columnar and eosinophilic epithelium, can be observed branching from the ventral wall of the urodaeum (Fig. 6.1a; Fig. 6.3M,N).

Cranial to the branching of the bladder stalk, the cloaca transitions to a sphincter lined solely by circular muscle, with a very thin mucosa and no urodaeal glands (Fig. 6.1a; 6.3N-P). The epithelium transitions from a stereotypical urodaeal epithelium to an epithelial lining typical of the intestine; i.e., simple columnar with eosinophilic enterocytes intermixed with highly basophilic goblet cells. The terminal portion of the intestine envelops the urodaeal sphincter and, thus, the urodaeal sphincter opens into the intestine through the most caudal extremity of the intestine (Fig. 6.3P).

6.2.3 Historical Overview of the Cloacae

Proctodaeum (region a). In all previous studies on squamates the proctodaeum was defined as the cloacal region that communicates the outside environment to the urodaeum. From Gabe and Saint-Girons (1965), the opening of the cloaca through the ventral body wall leads to the proctodaeum. Sagitally, the lumen of the proctodaeum is "S" shaped in the Common Wall Lizard (*Podarcis muralis*), Tuatara (*Sphenodon punctatus*), and Western Whiptail (*Aspidoscelis tigris*), due to folding of the caudal and cranial walls of the proctodaeum, a similar finding to that observed in *Sceloporus jarrovi* (see Section 6.2.2). In other taxa previously examined, the proctodaeum does not possess this "S" shaped folding along its ventral to dorsal aspect. The epithelium lining the proctodaeal lumen is stratified squamous, and keratinized at its caudal extremity in all lizards previously examined, concordant with the histological analysis of *Sceloporus jarrovi* (see Section 6.2.2, this volume). More cranially, keratinization of the superficial epithelial cells decreases, and the cytoplasm of the squamous cells stains intensely for mucopolysaccharides, indicative of a mucous membrane.

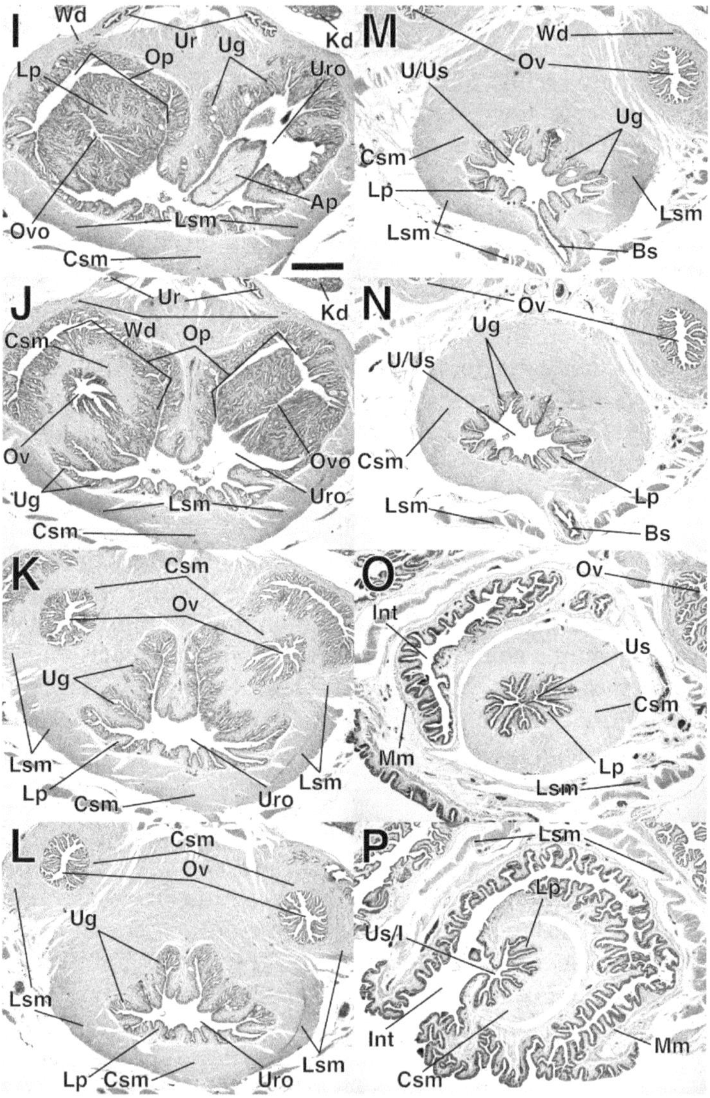

Fig. 6.3 Representative transverse sections from the cloaca of female *Sceloporus jarrovi*. Sections are organized from most caudal **(I)** to most cranial **(P)**. Individual micrograph lettering represents the capital lettering on Fig. 9.1, as demonstrative of the approximate location that each representative transverse section originated (hematoxylin and eosin; scale bar = 500 µm). See text for description of figure. Ap, ampullary papilla; Bs, bladder stalk; Csm, circular smooth muscle; Int, intestine; Lsm, longitudinal smooth muscle; Lp, lamina propria; Mm, muscularis mucosae; Op, oviducal papilla; Ov, oviduct; Ovo, opening of oviduct; Ug, urodaeal gland; Ur, ureter; Uro, urodaeum; U/Us, urodaeal/urodael sphincter transition; Us/I, urodaeal sphincter opening to intestine; Wd, Wolffian duct.

Color image of this figure appears in the color plate section at the end of the book.

In *Sceloporus jarrovi*, we termed this region the transitional zone between the proctodaeum and urodaeum (see Section 6.2.2, this volume). In the African Chameleon (*Chamaeleo africanus*), Checkerboard Worm Lizard (*Trogonophis wiegmanni*), Flat-tail Horned Lizard (*Phrynosoma mcalli*), Mediterranean Worm Lizard (*Blanus cinereus*), Morocco Cylindrical Skink (*Chalcides mionecton*), *Podarcis muralis*, Sandfish Skink (*Scincus scincus*), *Sphenodon punctatus*, Spotted Anole (*Anolis stratulus*), and Western Forest Feylinia (*Feylinia currori*), the proctodaeum remains stratified along its entire length, while in *Aspidoscelis tigris*, Burton's Snake-lizard (*Lialis burtonis*), California Legless Lizard (*Anniella pulchra*), Common Wall Gecko (*Tarentola mauritanica*), Desert Agama (*Trapelus mutabilis*), Desert Monitor (*Varanus griseus*), Fraser's Delma (*Delma fraseri*), Green Crested Lizard (*Bronchocela cristatella*), Koelliker's Glass Lizard (*Ophisaurus koellikeri*), Madagascar Day Gecko (*Phelsuma madagascariensis*), Pacific Gecko (*Dactylocnemis pacificus*), Prickly Gecko (*Heteronotia binoei*), Southern Alligator Lizard (*Elgaria multicarinata*), Varied Dtella (*Gehyra variegata*), and Western Banded Gecko (*Coleonyx variegatus*), the proctodaeum transitions to a bistratified mucous membrane. In the Carpet Chameleon (*Furcifer lateralis*), the proctodaeal epithelium is stratified, but not covered superficially by keratinized epithelial cells, like in other taxa, more caudally along the length of the proctodaeum. In the Coachella Valley Fringe-toed Lizard (*Uma inornata*), Collared Lizard (*Crotaphytus collaris*), Granite Night Lizard (*Xantusia henshawi*), and Sagebrush Lizard (*Sceloporus graciosus*) the proctodaeum is keratinized along its entire length, but is relatively short compared to the previous taxa examined. Although examined by Gabe and Saint-Girons (1965), the epithelium lining the proctodaeum of the Desert Iguana (*Dipsosaurus dorsalis*) was not discussed.

Sánchez-Martínez *et al.* (2007) described the proctodaeum in lizards (see their Table 4, page 296) as lined by a bistratified to stratified epithelium with superficial squamous cells more caudally and superficial cuboidal to columnar cells more cranially (see their Table 4). Hardy and Cole (1981; multiple species of whiptail lizards [*Aspidoscelis*]), Trauth *et al.* (1987; Broad-headed Skink [*Plestiodon laticeps*]), and Gharzi *et al.* (2013; Bosk's Fringe-fingered Lizard [*Acanthodactylus boskianus*]) provided similar histological descriptions of the proctodaeum from multiple different taxa. The caudal aspect of the proctodaeum was lined by a stratified squamous epithelium that transitions to a more stratified cuboidal epithelium cranially. These authors did not mention increased keratinzation of the proctodaeum more caudally, as indicated by Gabe and Saint-Girons (1965) and observed in our examination of *Sceloporus jarrovi* (see Section 6.2.2, this volume). As observed in our novel examination, the connective tissue (Sánchez-Martínez *et al.* 2007; Gharzi *et al.* 2013) and muscularis (Gharzi *et al.* 2013) deep to the epithelium was reduced in thickness in comparison to other regions of the cloaca.

Based on the overall similarity between the proctodaeum of lizards reviewed above, and the proctodaeum of snakes recently reviewed by Siegel *et al.* (2011a), we hypothesize that the proctodaeum is homologous across all species of Lepidosauria. To eliminate potential conflicting terminology (although historical terminology for this cloacal region has been quite consistent since Gadow 1887), Siegel *et al.* (2011a) termed this region of the snake cloaca "region a". We feel that a similar designation is appropriate for the proctodaeum of lizards.

Urodaeum (region b). Like the proctodaeum, little controversy exists over the delineation of the urodaeum in lizards and snakes. In both, the urodaeum is the main body of the cloaca that accepts urinary waste, fecal waste, and reproductive material before evacuation through the proctodaeum and to the outside environment. Thus, the urodaeum acts as the main functional body of the cloaca in all Lepidosauria. Siegel *et al.* (2011a) coined this region of the cloaca "region b" to eliminate any potential confusion in terminology.

The main chamber of the urodaeum has been reported by Gabe and Saint-Girons (1965) as stratified cuboidal or columnar in *Anolis stratulus, Blanus cinereus, Coleonyx variegatus, Chalcides mionecton, Chamaeleo africanus, Crotaphytus collaris, Dactylocnemis pacificus, Delma fraseri, Dipsosaurus dorsalis, Elgaria multicarinatus, Feylinia currori, Furcifer lateralis, Gehyra variegata, Lialis burtonis, Ophisaurus koellikeri, Phrynosoma mcalli, Podarcis muralis, Sceloporus graciosus, Scincus scincus, Sphenodon punctatus, Trogonophis wiegmanni, Uma inornata*, and *Xantusia henshawi*; stratified squamous and often degrading in *Aspidoscelis tigris, Heteronotia binoei*, and *Phelsuma madagascariensis*; or bistratified cuboidal or columnar in *Anniella pulchra, Bronchocela cristatella, Tarentola mauritanica, Trapelus mutabilis*, and *Varanus griseus*. Other studies provided similar results; i.e., stratification of the urodaeal lining with either cuboidal or columnar apical cells (Hardy and Cole 1981; Trauth *et al.* 1987; Saint-Girons and Eneich 1992; Sánchez-Martínez *et al.* 2007). See Tables 2 and 3 in Sánchez-Martínez *et al.* (2007; page 294) for variation in urodaeum stratification for a variety of lizard taxa.

Cranially, the stratified epithelium is replaced with a simple columnar epithelium in only *Anolis stratulus* (Gabe and Saint-Girons 1965), as observed in some taxa of Colubroides (Giacomini 1893; Gabe and Saint-Girons 1965; Sánchez-Martínez *et al.* 2007; Siegel *et al.* 2011a; Siegel *et al.* 2012). In Colubroides, this region of the urodaeum was termed the urodaeal pouch, acknowledging the transition of epithelial composition and noticeable widening of the cranial urodaeal chamber (Siegel *et al.* 2011a; Siegel *et al.* 2012).

The ipsolateral ureters and Wolffian ducts combine, forming the ampullae ureters, and empty into urodaeum caudal to the opening of the oviducts into the cloaca. However, besides our novel description of these ducts from the cloaca of *Sceloporus jarrovi*, we know of no study that discusses the interaction of those ducts in females, besides a small note by Gabe and Saint-Girons (1965) where they describe the condition we observed in *S. jarovii* as consistent in all lizards. Thus, we assume similar associations exist in all lizards. Furthermore, the communication of these ducts was observed in all taxa of snakes (Siegel *et al.* 2011a,b) except snakes that possess an association of these ducts through common ipsolateral ampullae uriniferous papillae (most members of Colubroides; Siegel *et al.* 2011a,b), but that morphology has never been observed in lizards. In this book, Rheubert *et al.* (Chapter 10, this volume) note a similar finding in male lizards of *Aspidoscelis* (i.e., ampullae urogenital papillae; union of the ipsolateral ureters and vasa deferentia through common ampullae urogenital papillae), but females of this taxon have yet to be examined.

In *Sphenodon punctatus*, the ureters empty into the cloaca individually via the paired oviducal papillae, just caudal to the opening of the oviducts through the same papillae (Gabe and Saint-Girons 1965; see also Cree Chapter 17, this volume). This morphology has only been observed in *Sphenodon*. In the Eastern Spotted Whiptail (*Aspidoscelis gularis*; Brooks 1906), *Elgaria multicarinata* (Blackburn 1998), and Texas Alligator Lizard (*Gerrhonotus infernalis*; Brooks 1906) the ureters appear to empty into the urodaeum via openings on a common oviducal papilla. From the extensive review by Gabe and Saint-Girons (1965), the ureters empty into the urodaeum through independent ampullary papillae in *Anniella pulchra* (from a dorso-lateral urodaeal invasion), *Anolis stratulus* (from a dorso-lateral urodaeal invasion), *Chamaeleo africanus* (from a dorsal urodaeal invasion), *Coleonyx variegatus* (from a dorsal urodaeal invasion), *Dactylocnemis pacificus* (from a dorsal urodaeal invasion), *Delma fraseri* (from a dorso-lateral urodaeal invasion), *Elgaria multicarinatus* (from a dorsal urodaeal invasion), *Furcifer lateralis* (from a dorsal urodaeal invasion), *Gehyra variegata* (from a dorsal urodaeal invasion), *Heteronotia binoei* (from a dorsal urodaeal invasion), *Ophisaurus koellikeri* (from a dorso-lateral urodaeal invasion), *Phelsuma madagascariensis* (from a dorsal urodaeal invasion), *Sceloporus graciosus* (from a lateral urodaeal invasion), *Tarentola mauritanica* (from a dorsal urodaeal invasion), and *Varanus griseus* (from a dorso-lateral urodaeal invasion). The only exceptions to this rule were the absence of ampullary papillae in *Xantusia henshawi*, the presence of a single ampullary papilla that acts as passage of the ureters into the urodaeum in *Aspidoscelis tigris*, *Bronchocela cristatella*, *Lialis burtonis*, and *Trapelus mutabilis* (all from a dorsal urodaeal invasion), and a poor description of the presence/morphology of ampullary papillae from *Blanus cinereus*, *Chalcides mionecton*, *Crotaphytus collaris*, *Dipsosaurus dorsalis*, *Feylinia currori*, *Phrynosoma mcalli*, *Podarcis muralis*, *Sceloporus graciosus*, *Scincus scincus*, *Trogonophis wiegmanni*, and *Uma inornata*. Other studies reported similar findings; i.e., ureter communication with the urodaeaum through paired ampullary papillae from a dorsal and lateral position in *Aspidoscelis exsanguis* (Hardy and Cole 1981) and *Plestiodon laticeps* (Trauth *et al.* 1987). Gharzi *et al.* (2013) noted the union of the ureters before emptying into the urodaeum via a common opening through a single median ampullary papilla in *Acanthodactylus boskianus.*

From Gabe and Saint-Girons (1965), the ureters are lined by a bistratified cuboidal or columnar epithelium in *Sphenodon*; a simple cuboidal or columnar epithelium in *Anniella pulchra, Anolis stratulus, Aspidoscelis tigris, Blanus cinereus, Bronchocela cristatella, Coleonyx variegatus, Chalcides mionecton, Chamaeleo africanus, Crotaphytus collaris, Dactylocnemis pacificus, Delma fraseri, Dipsosaurus dorsalis, Elgaria multicarinatus, Feylinia currori, Furcifer lateralis, Gehyra variegata, Heteronotia binoei, Lialis burtonis, Phelsuma madagascariensis, Phrynosoma mcalli, Sceloporus graciosus, Scincus scincus, Tarentola mauritanica, Trapelus mutabilis, Trogonophis wiegmanni, Uma inornata,* and *Xantousia henshawi*; a pseudostratified cuboidal or columnar epithelium in *Elgaria multicarinatus* and *Ophisaurus koellikeri*; or a stratified cuboidal or columnar epithelium in *Podarcis muralis.* No description of the ureter epithelium was provided for *Varanus griseus.* Sánchez-Martínez *et al.* (2007) organized epithelial histology descriptions from multiple taxa of lizards in their Table 5 (page 296).

Cranial to the invasion of the ureters through the ampullary papilla/ae, the oviducts empty into the urodaeum before cranial termination of the cloaca. The cloacal/oviducal junction will be discussed in detail in Section 6.2.4, this volume. However, the cranial termination of the cloaca necessitates discussion, as this termination is variable in different species; i.e., terminates with or without bifurcation of the urodaeum. From Gabe and Saint-Girons (1965), the cranial extremity of the urodaeum is not bifurcated in *Anniella pulchra, Chalcides mionecton, Crotaphytus collaris, Delma fraseri, Elgaria multicarinatus, Feylinia currori, Ophisaurus koellikeri, Phrynosoma mcalli, Sceloporus graciosus, Scincus scincus, Sphenodon punctatus, Uma inornata,* and *Xantusia henshawi*; and bifurcated in *Anolis stratulus, Aspidoscelis tigris, Blanus cinereus, Bronchocela cristatella, Chamaeleo africanus, Coleonyx variegatus, Dactylocnemis pacificus, Dipsosaurus dorsalis, Furcifer lateralis, Gehyra variegata, Heteronotia binoei, Lialis burtonis, Phelsuma madagascariensis, Podarcis muralis, Tarentola mauritanica, Trapelus mutabilis, Trogonophis wiegmanni,* and *Varanus griseus.*

To highlight intrafamiliar variation, many taxa of Scincidae were reported as having a non-bifurcated urodaeum (e.g., *Chalcides mionecton*; Gabe and Saint-Girons 1965), while others have a bifurcated urodaeum (e.g., *Plestiodon laticeps*; Trauth *et al.* 1987). Like Gabe and Saint-Girons (1965), we also found that the urodaeum of *Sceloporus* was not bifurcated; however, Sánchez-Martinez *et al.* (2007) described the urodaeum as bifurcated in all taxa of *Sceloporus* that they examined (Cozumel Spiny Lizard [*S. cozumelae*],

Gadow's Spiny Lizard [*S. gadoviae*], and Horrible Spiny Lizard [*S. horridus*]), also highlighting intrageneric variation. Furthermore, Sánchez-Martinez *et al.* (2007) found a bifurcated cloaca in every lizard examined besides the Desert Night Lizard [*Xantusia vigilis*], a similar finding in this genus through examination of a different species by Gabe and Saint-Girons (1965). Gharzi *et al.* (2013) also reported a non-bifurcated urodaeum in *Acanthodactylus boskianus*. Concurrent with Gabe and Saint-Girons (1965) findings, the urodaeum was also described as bifurcated in the genera *Aspidoscelis* (Hardy and Cole 1981) and *Coleonyx* (Sánchez-Martinez *et al.* 2007).

Coprodaeum. Unlike in snakes (for review see Siegel *et al.* 2011a), there does not seem to be as much controversy as to what constitutes a "coprodaeum" in lizards. In snakes, some studies indicated that the intestine was the coprodaeum, others indicated that a sphincter that adjoins the caudal extremity of the intestine to the urodaeal/proctodael junction through a ventral communication was the coprodaeum, while still others used the term coprodaeum to apparently describe both of these cloacal regions. Because of this confusion, Siegel *et al.* (2011a), following the suggestion of Gabe and Saint-Girons (1965), termed the sphincter the "urodaeal sphincter" and the intestine the "posterior intestine". "Urodaeal" was used in the terminology because the sphincter carries the same epithelial lining as the urodaeum along the majority of its length, and transitions to an epithelial lining more consistent with the intestine at only its most cranial aspect (see Table 6 [page 297] in Sánchez-Martínez *et al.* 2007 for a review of urodaeal sphincter epithelial linings). Thus, Siegel *et al.* (2011a) hypothesized that the sphincter developed from the same anlage as the urodaeum. A similar morphology was found for *Sceloporus jarrovi* (see Section 6.2.2, this volume).

In every previous examination of lizard cloacae that we could find, we observed no controversy concerning the coprodaeal region, as all studies delineated the coprodaeum as the sphincter that adjoins the intestine to the other regions of the cloaca, except for one. Seshadri (1959) described a coprodaeum branching from the ventral side of the proctodaeum in a monitor lizard (*Varanus* sp.) that communicates with the caudal extremity of the intestine via a sphincter. Seshadri (1959) did not utilize histological examination for the *V.* sp. cloacal descriptions, and we feel that he was describing the ventral branching of the proctodaeal/urodaeal junction and subsequent transition into the urodaeal sphincter.

To maintain consistency with the recent review of female cloacal anatomy in snakes (Siegel *et al.* 2011a), we adopt the terminology proposed (i.e., urodaeal sphincter and posterior intestine) as the terminology for the region that typically includes the coprodaeum in lizards. To eliminate the inconsistencies from terminology, Siegel *et al.* (2011a) also termed the urodaeal sphincter and posterior intestine "region c" and "region d", respectively.

Urodaeal sphincter (region c) and posterior intestine (region d). In many lizards the urodaeal sphincter branches midventrally from the caudal region of the urodaeum, immediately ventral to where ureters open through common or paired ampullary papilla(ae). We term this morphology of the urodaeal sphincter the "branched" morphology. The branched morphology results in the urodaeum proper continuing cranially in a dorsal position to the urodaeal sphincter. Thus, in a transverse section from the cranial urodaeum, both the urodaeum proper and urodaeal sphincter can be observed. This is the morphology for the majority of snakes examined (for review see figures from Siegel *et al.* 2011a,b,c; Siegel *et al.* 2012). In other lizards, the urodaeal sphincter is fairly continuous with the most cranial region of the urodaeum, as observed in *Sceloporus jarrovi* (see Section 6.2.2, this volume). Thus, the urodaeum proper and urodaeal sphincter will never be seen simultaneously in transverse section. We term this morphology of the urodaeal sphincter the "continuous" morphology, as the urodaeal sphincter is continuous with the urodaeum proper without discrete branching.

It is difficult to determine lizards that posses the branched vs. continuous morphology in the text of Gabe and Saint-Girons (1965). From their figures, it is clear that *Aspidoscelis tigris*, *Coleonyx variegatus*, *Elgaria multicarinatus*, *Feylinia currori*, *Furcifer lateralis*, *Podarcis muralis*, and probably *Chamaeleo africanus* (described as similar to *F. lateralis*), *Dactylocnemis pacificus* (described as similar to *C. variegatus*), *Gehyra variegata* (described as similar to *C. variegatus*), and *Heteronotia binoei* (described as similar to *C. variegatus*), possess the branched morphology due to enlarged cranial extensions (often bifurcated) of the urodaeum. Similar findings were observed in *Acanthodactylus boskianus* (Gharzi *et al.* 2013), Chihuahuan Spotted Whiptail (*Aspidoscelis exsanguis*; Hardy and Cole 1981), *Plestiodon laticeps* (Trauth *et al.* 1987), and *Varanus* sp. (Seshadri 1959). Based on the figures from Gabe and Saint-Girons (1965), *Anolis stratulus*, *Sceloporus graciosus*, *Sphenodon punctatus*, *Trogonophis wiegmanni*, and probably *Crotaphytus collaris* (described as similar to *S. graciosus*), *Phrynosoma mcalli* (described as similar to *S. graciosus*), and *Uma inornata* (described as similar to *S. graciosus*), possess the continuous urodaeal sphincter morphology. Detailed descriptions and figures from Gabe and Saint-Girons (1965) are lacking for *Anniella pulchra*, *Blanus cinereus*, *Bronchocela cristatella*, *Chalcides mionecton*, *Delma fraseri*, *Dipsosaurus dorsalis*, *Lialis burtonis*, *Ophisaurus koellikeri*, *Phelsuma madagascarensis*, *Scincus scincus*, *Tarentola mauritanica*, *Trapelus mutabilis*, *Xantusia henshawi*, and *Varanus griseus*; although, description of the placement of the genital tubercle in *D. fraseri* and *X. henshawi* (anterior extremity of the urodaeum and behind the urodaeal sphincter) is most indicative of the continuous morphological condition. Furthermore, the possibility of a continuous urodaeal sphincter is eliminated with a cranially bifurcated urodaeum and, thus, we assume that *B. cinereus*, *B. cristatella*, *D. dorsalis*, *L. burtonis*, *T. mutabilis*, and *V. griseus* possess a branched urodaeal sphincter morphology based on this restriction.

The bladder stalk (Mulaik 1946) communicates the urinary bladder with the urodaeum, and branches from the most caudal portion of the ventral wall of the urodaeal sphincter (Gabe and Saint-Girons 1965; Trauth *et al.* 1987; Sánchez-Martínez *et al.* 2007; Gharzi *et al.* 2013; Siegel *et al.* 2013). Some lizards do not possess urinary bladders, but still possess a bladder stalk (Mulaik 1946; Beuchat 1986). Beuchat (1986) reviewed the phylogenetic distribution of urinary bladders, and tabulated the results in his Table 1 (page 513). Gabe and Saint-Girons (1965) described the epithelium of the urinary bladder as very similar to that of the urodaeum/caudal urodaeal sphincter in *Anolis stratulus*, *Aspidoscelis tigris*, and *Sphenodon punctatus*; pseudostratified with ciliated and mucous cells in *Coleonyx variegatus*, *Feylinia currori*, *Gehyra variegata*, *Heteronotia binoei*, and *Phelsuma madagascarensis*; or bistratified or simple with ciliated and mucous cells in *Anniella pulchra*, *Blanus cinereus*, *Chalcides mionecton*, *Chamaeleo africanus*, *Elgaria multicarinatus*, *Furcifer lateralis*, *Ophisaurus koellikeri*, *Podarcis muralis*, *Scincus scincus*, and *Trogonophis wiegmanni*. Sánchez-Martínez *et al.* (2007) also provided histological description of the urinary bladder epithelium for a variety of taxa, and organized these descriptions into their Table 6 (page 297). A urinary bladder or stalk was described as absent by Gabe and Saint-Girons (1965) in *Dactylocnemus pacificus*, *Delma fraseri*, and *Varanus griseus*, a reduced urinary bladder (probably a stalk) was described in *Aspidoscelis tigris*, *Crotaphytus collaris*, *Dipsosaurus dorsalis*, *Lialis burtonis*, *Phrynosoma mcalli*, *Sceloporus graciosus*, *Uma inornata*, and *Xantusia henshawi*, and no mention of the urinary bladder or stalk was reported for *Agama inermis*, *Calotes cristatellus*, and *Tarentola mauritanica*. Sánchez-Martínez *et al.* (2007) reported that every taxon of lizard that they examined possessed a urinary bladder.

While moving cranially through the urodaeum, the posterior intestine can first be observed cranial to the termination of the urodaeum (continuous condition) or often sandwiched between the urodaeum and urodaeal sphincter (branched condition). In either condition, the urodaeal sphincter communicates with the posterior intestine cranial to the terminal wall (i.e., *Anolis stratulus*; Gabe and Saint-Girons 1965) or continuously through terminal wall (i.e., *Feylinia currori* and *Sceloporus graciousus*; Gabe and Saint-Girons 1965) of the posterior intestine. The former condition was observed in transverse micrographs from *Anolis stratulus* (Gabe and Saint-Girons 1965) and *Heloderma suspectum* (Siegel *et al.* 2013) and results in a caudal caecum that is often observed in transverse section. The later condition was observed in transverse micrographs from *Feylinia currori* (Gabe and Saint-Girons 1965), *Sceloporus graciosus* (Gabe and Saint-Girons 1965), and

Sceloporus jarrovi (see Section 6.2.2) and results in no caecum, or a folding of the intestine symmetrically around the cranial extremity of the urodaeal sphincter.

In only two taxa did Gabe and Saint-Girons (1965) discuss the histology of the urodaeal sphincter/posterior intestine junction. The intestine is lined by a stratified columnar epithelium in *Sphenodon punctatus* that is composed entirely of mucocytes. A muscularis mucosa as well as an external muscularis with inner circular and outer longitudinal layers are both present. These features are in line with those observed in *Sceloporus jarrovi*; however, two types of epithelial cells were found in the simple epithelium lining the posterior intestine in *S. jarovii*: mucocytes and eosinophilic cells with apical microvilli (see Section 6.2.2, this volume). A muscuarlis mucosae is apparently absent in *Coleonyx variegatus*, but like in *S. jarovii*, two epithelial cell types were observed. The epithelium of the posterior intestine in *C. variegatus* was also described as simple to pseudostratified.

6.2.4 Cloacal/Oviducal Junction

Oviducal papillae invade the cranial region of the urodaeum through bifurcated diverticuli of the urodaeum or into a common chamber of the urodaeum (see Section 6.2.3, this volume). The oviducal papillae can be either large and invade deep into the urodaeum ('salient genital tubercle'; Raynaud and Pieau 1985) or hardly make contact with the urodaeum ('non-salient genital tubercle'; Raynaud and Pieau 1985).

Histological evidence that unambiguously demonstrates a common oviducal opening into the urodaeal chamber has been reported only in *Acanthodactylus boskianus* (Gharzi *et al.* 2013), *Elgaria multicarninata* (Blackburn 1998), and *Heloderma suspectum* (Siegel *et al.* 2013), though Gabe and Saint-Girons do not provide this description for their one female specimen of *Elgaria multicarinata*. Grossly, two studies note adjoining of the oviducts in *Aspidoscelis gularis* (Brooks 1906), *Gerrhonotus infernalis* (Brooks 1906), and *Lophura* (no species was provided, and we assume that *Lophura* is a synonym for *Hydrosaurus* [sailfin lizards]; Gadow 1887) before entrance into the urodaeum, but no evidence is provided that the lumina of the two oviducts communicate before emptying into the urodaeum. In opposition to Brooks (1906), Gabe and Saint-Girons (1965) and Hardy and Cole (1981) describe the oviducts emptying individually into cranial urodaeal arms of *Aspidoscelis*. In all other reports from lizards, the oviducts empty into the urodaeum individually.

In general, the histology of the oviducal papillae mucosal membranes and external layers of muscularis matches that of the non-glandular uteri and, thus, we refer the readers to Section 6.3.4, this volume, for histological description of this most caudal region of the oviducts. Text and figures from Gabe and Saint-Girons (1965) demonstrate a dorsal, cranial, and slightly lateral to the midline invasion of the urodaeum by the oviducal papillae in *Anniella pulchra, Aspidoscelis tigris, Chalcides mionecton, Crotaphytus collaris, Delma fraseri, Feylinia currori, Sceloporus graciosus, Scincus scincus, Sphenodon punctatus*, and *Xantusia henshawi*. The oviducal papillae invade the non-bifurcated urodaeal chamber from a ventral and lateral position in *Elgaria multicarinatus, Ophisaurus koellikeri*, and *Uma inornata*. In *Phrynosoma mcalli* and *Varanus griseus*, the oviducal papillae invade the urodaeal chamber midlaterally. The oviducal papilla invade the cranial bifurcations of the urodaeaum (urodaeal arms; Siegel *et al.* 2011a) from a midcranial and dorsal position in *Coleonyx variegatus, Dactylocnemus pacificus, Gehyra variegata, Heteronotia binoei, Phelsuma madagascarensis*, and *Tarentola mauritanica*, and extend deep into the urodaeal chamber caudal to urodaeal bifurcation. This is similar to the condition found in *Blanus cinereus* and *Trogonophis wiegmanni* but with ventral lateral invasions of the oviducal papillae into the urodaeal arms. In *Anolis stratulus, Bronchocela cristatella, Chamaeleo africanus, Furcifer lateralis, Lialis burtonis*, and *Trapelus mutabilis*, the oviducal papillae invade urodaeal arms from a dorsal and slightly lateral position and do not extend caudal to the plane of urodaeal bifurcation. A similar condition was found in *Plestiodon laticeps* by Trauth *et al.* (1987). The oviducal papillae are described as making direct contact with the urodaeal arms of *Dipsosaurus dorsalis*, but no further details were provided by Gabe and Saint-Girons (1965). *Podarcis muralis* was described as possessing no oviducal papillae; i.e., a continuous transition from the urodaeal arms to the non-glandular uteri, a similar characteristic that has been observed in some advanced snakes (Siegel *et al.* 2011a, 2012).

Sánchez-Martínez *et al.* (2007) described multiple different morphologies of the oviducal/cloacal junction for their cladistic analysis. Formation of a prominent oviducal papilla ("intrusion of the vaginal tube into the urodaeum"), central or lateral communication of the oviducts with the urodaeum ("place of the vaginal intrusion into the anterior urodaeal chamber"), dorsal or ventral communication of the oviducts with the urodaeum ("position of the vaginal tube on the horizontal axis of the urodaeum"), and communication of the oviducts parallel to the midsagittal plane of the urodaeum or oblique ("place of vaginal tube regarding the middle line of the body"). These morphological variations can be determined for each taxon by matching their Table 1 (page 291) with the "definition of characters" in their text. In general, most taxa examined had prominent oviducal papillae, a lateral communication of the oviducts with the urodaeum, and oviducts communicating with the urodaeum from a dorsal position. Whether the oviducts invaded the urodaeum parallel to the mid-sagittal plane of the urodaeum or from an oblique position was quite variable. However, we are not sure if we fully understand this last character as, to us, it would seem that taxa that possessed oviducts that communicated with the urodaeum from a lateral position would inherently possess oviducts that communicated with the urodaeum from an oblique angle.

6.2.5 Cloacal Glands

Cloacal glands are diverse in squamates. Functionally, cloacal glands have been implicated in reproduction (Regamey 1935; Trauth *et al.* 1987) because of the correlation of their secretory activity with the reproductive cycle (mainly urodaeal glands) and communication (precloacal glands and dorsal/ventral glands; Cole 1966a); i.e., lizards are capable of discriminating cloacal "odors" from conspecifics (Cooper and Vitt 1984a,b; Cooper and Vitt 1986; Cooper and Trauth 1992; see also Martin and Lopez Chapter 3, this volume). Other lesser known hypotheses include for tactile stimuli (mainly precloacal glands; Cole 1966a), functionless (precloacal glands; Cole 1966a), as an additive to seminal fluid (mainly dorsal/ventral glands; Disselhorst 1904; Goin and Goin 1962), lubrification (mainly dorsal and ventral glands; Lereboullet 1851; Disselhorst 1904), or sperm storage (mainly postcloacal "sacs"; Wellborn 1933).

In snakes, cloacal glands are restricted to glandular masses outside of the cloacal wall (for review see Gabe and Saint-Girons 1965; Whiting 1969; Young *et al.* 1999; Siegel *et al.* 2011a) and empty their contents into the lumen of the cloaca (primarily the proctodaeum) via ducts that traverse the cloacal wall. Glands described in snakes are found dorsal to the proctodaeum (median cloacal gland and dorsal glands), caudal to the proctodaeum (anal glands), and ventral to the proctodaeum (ventral glands), although the presence of the ventral glands is highly questionable (Siegel *et al.* 2011a). Extracloacal gland masses are typically large and complex, except for the anal glands, which have a unique histology (refer to Young *et al.* 1999; Siegel *et al.* 2011a). Whiting (1969) tabulated the occurence of multiple gland types in squamates in her Appendix (table begins on her page 76).

Other lepidosaurs besides snakes have similar complex extracloacal gland masses, but also possess glandular invaginations of the proctodaeal and urodaeal epithelial linings into the lamina propria (i.e., mucosal glands), forming simpler glandular components of the cloaca. From Gabe and Saint-Girons (1965), proctodaeal mucosal glands have been identified only in *Coleonyx variegatus* and *Sphenodon punctatus*. Our report of this finding from Gabe and Saint-Girons (1965) was only from the appearance of these glands in two freehand drawings, but Sánchez-Martínez *et al.* (2007) reported similar glands in the proctodaeum of multiple taxa (see their Table

4; page 296). Gabe and Saint-Girons (1965) reported urodaeal mucosal glands in *Anniella pulchra, Aspidoscelis tigris, C. variegatus, Dactylocnemis pacificus, Delma fraseri, Dipsosaurus dorsalis, Elgaria multicarinatus, Feylinia currori, Gehyra variegata, Heteronotia binoei, Ophisaurus koellikeri, Phelsuma madagascarensis, Phrynosoma mcalli, Sceloporus graciosus, Tarentola mauritanica, Trogonophis wiegmanni, Uma inornata, Xantusia henshawi*, and *Varanus griseus*. Similar mucosal glands of the urodaeaum were also observed in *A. exsanguis* (Hardy and Cole 1981), *Plestiodon laticeps* (Trauth *et al.* 1987), *Lepidodactylus lugubris* (Saint-Girons and Ineich 1992), and the majority of lizards examined by Sánchez-Martínez *et al.* (2007; see their Table 1; page 291). Whiting (1969) confirmed Gabe and Saint-Girons' (1965) discoveries and also added alligator lizards (*Gerrhonotus* sp.) and the Tokay Gecko (*Gekko gecko*) to their list. Gabe and Saint-Girons (1965) provided few details on the histology of mucosal glands but they consistently referred to mucosal glands as "mucosal crypts". In our examination of these crypts in *Sceloporus jarrovi*, we found that the cytoplasm of glandular cells contained eosinophilic granules, indicating a possible protein based secretion. Gabe and Saint-Girons (1965) reported histochemical properties of the urodaeal mucosal glands of *Aspidoscelis tigris, Elgaria multicarinatus*, and *Feylinia currori*, which they reported contained granules that stained positive with the periodic acid-Schiff's procedure (for neutral carbohydrate moieties) and negative (*F. currori*) or positive (*A. tigris* and *E. multicarinatus*) with alcian blue (for acidic mucoid substances). In *Varanus griseus*, Gabe and Saint-Girons (1965) described the epithelia of the urodaeal mucosal glands as mucous-secreting linings, identical to that of the urodaeum proper. In *Lepidodactylus lugubris*, Saint-Girons and Eneich (1992) described urodaeal mucosal glands as simple tubular with little mucous. During the mating season, mucosal glands of the urodaeum hypetrophy in *Plestiodon laticeps* and possess secretions that are acidic or basic in composition (Trauth *et al.* 1987).

There are multiple types of extracloacal glands in lizards but all empty into the proctodaeum or at the lips of the cloacal opening. Precloacal glands (possibly homologous to femoral glands; for review see Whiting 1969) empty into the anterior lip of the cloaca and have been described in all amphisbaenids (Whiting 1969), *Coleonyx variegatus* (Greenberg 1943), *Crotaphytus collaris* (Cole 1966b), the European Green Lizard (*Lacerta viridis*; Disselhorst 1904), and possibly in the Common Chameleon (*Chamaeleo chamaeleon*; Disselhorst 1904). Dorsal and ventral gland masses are as named, found dorsal and ventral to the proctodaeum, empty into the proctodaeum from a dorsal or ventral position, and have been described in numerous lizard taxa (discussed in detail in subsequent text). It has been hypothesized that dorsal and ventral glands stem from the same anlage, as indicated by their similarity in morphology and secretory product (for review see Whiting 1969). Postcloacal "sacs" are found in the ventor of the tail immediately caudal to the cloacal opening; however, these glands are hypothesized to be functionless in female lizards, as indicated by their highly reduced structure/activity by Kluge (1967). Considering dorsal/ventral glands are the only prominent extracloacal glands in female lizards that definitively arise from cloacal tissues, only these glands will be discussed further; however, we urge readers to see Whiting (1969) for a review of the other gland types.

Whiting (1969) hypothesized that dorsal and ventral extracloacal glands were derived from the same gland complex, and the main difference between the two morphological conditions was the extent that these glands encompassed the proctodaeum, as noted previously by Disselhorst (1904), Regamey (1935), and Gabe and Saint-Girons (1965). Described by Gabe and Saint-Girons (1965), extracloacal (dorsal/ventral) glands were found emptying into the proctodaeum from a ventral and lateral position in *Coleonyx variegatus* (four distinct masses), *Gehyra variegata* (number of masses could not be determined but were reported as similar to *C. variegatus*), *Heteronotia binoei* (number of masses could not be determined but reported as similar to *C. variegatus*), *Sphenodon punctatus* (two distinct masses), and *Tarentola mauritanica* (four distinct masses; one set more dorsal). In *Dactylocnemis pacificus* and *Phelsuma madagascarensis*, two gland masses were found dorsal and lateral to the proctodaeum, and also a single gland mass in a median position dorsal to the cloaca. Gland masses were described as surrounding the entire proctodaeum in *Bronchocela cristatella*, *Chalcides mionecton*, *Chamaeleo africanus*, *Furcifer lateralis*, *Scincus scincus*, *Trapelus mutabilis*, and *Xantusia henshawi*, but the number of extracloacal gland masses was not reported. *Blanus cinereus*, *Delma fraseri*, *Lialis burtonis*, *Sceloporus graciosus*, and *Uma inornata* have two gland masses dorsal and lateral to the proctodaeum and one median gland mass ventral to the proctodaeum. Only one ventral gland mass was described in *Aspidoscelis tigris*, *Crotaphytus collaris*, and *Phrynosoma mcalli* (with small scattered dorsal groups), and three ventral gland masses were described in *Dipsosaurus dorsalis*. Two gland masses are positioned lateral to the proctodaeum in *Feylinia currori*. Apparently, *Podarcis muralis* possesses a single median ventral and dorsal gland, but this was difficult to determine from the text of Gabe and Saint-Girons (1965). In *Ophisaurus koellikeri*, six gland masses are positioned ventral to the proctodaeum, while two are positioned dorsal and lateral to the proctodaeum. A similar finding was reported for *Anniella pulchra* but with eight gland masses ventral to the proctodaeum. *Elgaria multicarinata* possesses two masses dorsal and lateral to the proctodaeum and four masses ventral to the proctodaeum. *Trogonophis wiegmanni* possesses two gland masses positioned side-by-side and dorsal to the proctodaeum. *Anolis stratulus* does not possess extracloacal glands, a finding also presented for *Lepidodactylus lugubris* (Saint-Girons and Ineich 1992), and the cloaca of *Varanus griseus* was not sampled properly for determination of extracloacal gland content; however, Auffenberg (1994) noted the presence of dorsal and ventral glands in the Bengal Monitor (*V. bengalensis*).

Whiting (1969) noted a continuous ring of dorsal and ventral gland masses in the Brown Basilisk (*Basiliscus vittatus*), Eastern Garden Lizard (*Calotes versicolor*), and *Gekko gecko*. Dorsal and ventral glands were also found in *Bipes canaliculatus*, *Coleonyx variegatus*, and Yellow Monitor (*Varanus flavescens*) but these glands did not form a continuous ring around the cloaca. Apparently, *Uma inornata* and *V. flavescens* possess only a ventrolateral gland. We can only assume that a typo exists in Whiting (1969) and that the first varanid with dorsal and ventral glands was actually a male Common Water Monitor (*V. salvator*), but in general, it is difficult to assess whether Whiting (1969) was discussing glands in females or males in her text. Gland masses found only lateral to the cloaca were observed in *Gerrhonotus* sp., *Plestiodon laticeps*, and house geckos (*Hemidactylus* sp.), which she has recorded as the Mediterranean Gecko (*H. turcicus*) in her Table II (page 33). Ventral glands were described only for the Beaded Lizard (*Heloderma horridum*) and Black Tegu (*Tupinambis teguixin*) but it is unclear if these actually represent glands homologous to typical dorsal/ventral glands, as they were found at the border of the cloacal lip. The presence of only dorsal glands was restricted to the amphisbaenians, such as the Angled Worm Lizard (*Agamodon anguliceps*), *Bipes canaliculatus*, Crooked Worm Lizard (*Amphisbaena camura*), Gonave Worm Lizard (*A. gonavensis*), and Zarudnyi's Worm Lizard (*Diplometopon zarudnyi*). Obviously, intraspecific variation has been either observed or occurred via misinterpretation when comparing Whiting's (1969) work with Gabe and Saint-Girons' (1965), as extracloacal glands in some taxa were described differently; e.g., *Coleonyx variegatus* possesses both dorsal and ventral glands in Whiting's (1969) study, but only ventral glands in Gabe and Saint-Girons' (1965) study.

Paired extracloacal dorsal glands emptying into the proctodaeum were also observed in *Acanthodactylus boskianus* (Gharzi *et al.* 2013), *Aspidoscelis exsanguis* (Hardy and Cole 1981), and *Plestiodon laticeps* (Trauth *et al.* 1987). Interestingly, in *Aspidoscelis*, extracloacal glands were described as both ventral (*A. tigris*; Gabe and Saint-Girons 1965) and dorsal (*A. exsanguis*; Hardy and Cole 1981), highlighting potential intrageneric variation in glandular content. Sanchez-Martinez *et al.* (2007) also reviewed extracloacal gland absence or presence and categorized these data in their Table 5 (page 296); however, they provide little detail on the position of the glands to the cloaca. Only one study that we know of, on *A. boskianus*, found extracloacal glands emptying into the urodaeum from a dorsal position, lateral to the communication of the ureters with the urodaeum (Gharzi *et al.* 2013).

In general, Gabe and Saint-Girons (1965) described the gland tubules of the extracloacal glands as tubular, or tubuloacinar, and either branched or simple with epithelia that are simple or pseudostratified. Stratified epithelial linings were observed only in *Chamaeleo chamaeleon* by Disselhorst (1904). From Gabe and Saint-Girons (1965), the epithelia lining the gland tubules always reacted positively with eosin. They also reported that the secretion product and cytoplasmic granules of the glandular epithelia of extracloacal glands always reacted positively with periodic acid-Schiff's (except in *Ophisaurus koellikeri* and *Podarcis muralis*), with a more varied reaction with alcian blue. In some taxa, mucous and serous glands are present in the same gland complex (for review see Gabe and Saint-Girons 1965). It is of importance to note that Gabe and Saint-Girons (1965) describe the extracloacal glands of *Sphenodon punctatus* as "sebaceous glands" and determined that this gland type was only found in *S. punctatus* within Lepidosauria. However, in *Acanthodactylus boskianus* (Gharzi *et al.* 2013) and *Plestiodon laticeps* (Trauth *et al.* 1987), dorsal extracloacal glands stained intensely with eosin and were more secretory during reproductive activity (Trauth *et al.* 1987), while Sánchez-Martínez *et al.* (2007) reported eosinophilic and basophilic reactions (taxa designations were not provided). Urodaeal extracloacal gland cytoplasm also stained intensely eosinophilic in *A. boskianus* (Gharzi *et al.* 2013).

6.2.6 Conclusions for Cloacae

In general, the cloacae of all squamates are fairly similar; i.e., a proctodaeum communicating the internals of the squamate gonads, gut, and urinary tracts with the outside environment, and a urodaeum to which the ducts that carry these internals empty (for review see Siegel *et al.* 2011a).

Out of the 42 current lizard families recognized, 29 have been examined in terms of some aspect of cloacal anatomy, just from the literature reviewed in this chapter (Table 6.1). Thus, in terms of comparative biology, the squamate cloaca seems to be a potentially useful tool for testing relationships between different squamate lineages. Unfortunately, it is obvious that the variation observed within squamate lineages is too prominent for any robust analysis of squamate phylogeny. However, the rampant amount of possible convergence in cloacal characteristics highlights potential avenues for comparative studies that aim to elucidate the selective pressures that result in high amounts of homoplasy. Of note is the lack of mucosal glands in any known species of snake (Gabe and Saint-Girons 1965; Whiting 1969; Siegel *et al.* 2011a) and the lack of urodaeal glands in many lineages of lizards (see Sánchez-Martínez *et al.* 2007).

Table 6.1 Literature on cloacal anatomy and cloacal glands discussed in this review. We refer readers to the appendix of Whiting (1969; page 76) for a more thorough review of the literature.

Family	Species	Reference	Cloacae	Glands
Sphenodontidae	*Sphenodon punctatus*	Gabe and Saint-Girons 1964, 1965	X	X
Agamidae	*Trapelus mutabilis*	Gabe and Saint-Girons 1965	X	X
	Bronchocela cristatella	Gabe and Saint-Girons 1965	X	X
	Hydrosaurus sp.	Gadow 1887	X	
Anguidae	*Anguis fragilis*	Rathke 1839; Leydig 1872; Disselhorst 1897, 1904; Unterhössel 1902	X	X
	Barisia imbricata	Sánchez-Martínez *et al.* 2007	X	X
	Elgaria multicarinata	Gabe and Saint-Girons 1965; Blackburn 1998	X	X
	Gerrhonotus infernalis	Brooks 1906	X	
	Gerrhonotus sp.	Whiting 1969		X
	Ophisaurus koellikeri	Gabe and Saint-Girons 1965	X	X
	Ophisaurus sp.	Blackburn 1998	X	
Amphisbaenidae	*Amphisbaena camura*	Whiting 1969		X
	Amphisbaena fugilinosa	Disselhorst 1904	X	X
	Amphisbaena gonavensis	Whiting 1969		X
Anniellidae	*Anniella pulchra*	Gabe and Saint-Girons 1965	X	X
Bipedidae	*Bipes canaliculatus*	Whiting 1969		X
Blanidae	*Blanus cinereus*	Bons and Saint-Girons 1963; Gabe and Saint-Girons 1965	X	X
Chamaeleonidae	*Chamaeleo africanus*	Gabe and Saint-Girons 1965	X	X
	Chamaeleo chamaeleon	Disselhorst 1904; Whiting 1969	X	X
	Chamaeleo sp.	Blackburn 1998	X	
	Furcifer lateralis	Saint-Girons 1962; Gabe and Saint-Girons 1965	X	X
Corytophanidae	*Corytophanes* sp.	Blackburn 1998	X	

Table 6.1 contd....

Table 6.1 contd.

Family	Species	Reference	Cloacae	Glands
Crotaphytidae	*Crotaphytus collaris*	Brooks 1906; Gabe and Saint-Girons 1965	X	X
	Crotaphytus collaris	Blackburn 1998	X	
	Gambelia wislizenii	Sánchez-Martínez *et al.* 2007	X	X
Dactyloidae	*Anolis* sp.	Blackburn 1998	X	
	Anolis stratulus	Gabe and Saint-Girons 1965	X	X
Diplodactylidae	*Dactylocnemis pacificus*	Gabe and Saint-Girons 1965	X	X
Eublepharidae	*Coleonyx* sp.	Blackburn 1998	X	
	Coleonyx variegatus	Noble and Bradley 1933; Greenberg 1943; Gabe and Saint-Girons 1965; Whiting 1969; Sánchez-Martinez *et al.* 2007	X	X
Gekkonidae	*Gekko gecko*	Whiting 1969		X
	Gehyra variegata	Gabe and Saint-Girons 1965	X	X
	Hemidactylus sp.	Blackburn 1998	X	X
	Hemidactylus turcicus	Whiting 1969; This study	X	X
	Heteronotia binoei	Gabe and Saint-Girons 1965	X	X
	Lepidodactylus lugubris	Saint-Girons and Ineich 1992	X	X
	Phelsuma madagascariensis	Gabe and Saint-Girons 1965	X	X
	Phelsuma sp.	Blackburn 1998	X	
Helodermatidae	*Heloderma horridum*	Whiting 1969		X
	Heloderma suspectum	Siegel *et al.* 2013	X	
Iguanidae	*Ctenosaura* sp.	Blackburn 1998	X	
	Dipsosaurus dorsalis	Gabe and Saint-Girons 1965; Blackburn 1998	X	X
	Iguana sp.	Blackburn 1998	X	
	Sauromalus klauberi	Blackburn 1998	X	

Table 6.1 contd....

Table 6.1 contd.

Family	Species	Reference	Cloacae	Glands
Lacertidae	*Acanthodactylus boskianus*	Gharzi *et al.* 2013	X	X
	Lacerta agilis	Lereboullet 1851; Leydig 1872; Disselhorst 1904; Krause 1922; Mtthey 1929; Regamey 1933, 1935; Mathews and Marshall 1956	X	X
	Lacerta viridis	Disselhorst 1904; Herlant 1933	X	X
	Podarcis muarlis	Disselhorst 1904; Gabe and Saint-Girons 1965	X	X
	Zootoca vivipara	Disselhorst 1904	X	X
Liolaemidae	*Liolaemus albiceps*	Sánchez-Martínez *et al.* 2007	X	X
	Liolaemus quilmes	Sánchez-Martínez *et al.* 2007	X	X
Phrynosomatidae	*Callisaurus draconoides*	Blackburn 1998	X	
	Cophosaurus texanus	Brooks 1906; Blackburn 1998	X	
	Holbrookia sp.	Blackburn 1998	X	
	Phrynosoma mcalli	Gabe and Saint-Girons 1965	X	X
	Phrynosoma sp.	Brooks 1906; Blackburn 1998	X	
	Sceloporus cozumelae	Sánchez-Martínez *et al.* 2007	X	X
	Sceloporus gadoviae	Sánchez-Martínez *et al.* 2007	X	X
	Sceloporus horridus	Sánchez-Martínez *et al.* 2007	X	X
	Sceloporus jarrovi	This study	X	X
	Sceloporus graciosus	Gabe and Saint-Girons 1965	X	X
	Sceloporus olivaceus	Brooks 1906	X	
	Sceloporus sp.	Blackburn 1998	X	
	Uma inornata	Gabe and Saint-Girons 1965; Whiting 1969	X	X
	Uma sp.	Blackburn 1998	X	
	Urosaurus sp.	Blackburn 1998	X	
	Uta sp.	Blackburn 1998	X	

Table 6.1 contd....

Table 6.1 contd.

Family	Species	Reference	Cloacae	Glands
Phyllodactylidae	*Tarantola mauritanica*	Braun 1878; Gabe and Saint-Girons 1965	X	X
	Thecadactylus rapicauda	Sánchez-Martínez *et al.* 2007	X	X
Polychrotidae	*Polychrus marmoratus*	Sánchez-Martínez *et al.* 2007	X	X
Pygopodidae	*Delma fraseri*	Gabe and Saint-Girons 1965	X	X
	Lialis burtonis	Gabe and Saint-Girons 1965	X	X
Scincidae	*Chalcides mionecton*	Gabe and Saint-Girons 1965	X	X
	Chalcides sp.	Blackburn 1998	X	
	Emoia sp.	Blackburn 1998	X	
	Feylinia currori	Gabe and Saint-Girons 1965	X	X
	Mabuya sp.	Blackburn 1998; Sánchez-Martínez *et al.* 2007	X	X
	Plestiodon copei	Sánchez-Martínez *et al.* 2007	X	X
	Plestiodon laticeps	Trauth *et al.* 1987	X	X
	Plestiodon sp.	Blackburn 1998	X	
	Scincus scincus	Gabe and Saint-Girons 1965	X	X
Sphaerodactylidae	*Gonatodes albogularis*	Sánchez-Martínez *et al.* 2007	X	X
Teiidae	*Aspidoscelis exsanguis*	Hardy and Cole 1981	X	X
	Aspidoscelis gularis	Brooks 1906	X	
	Aspidoscelis lemniscatus	Sánchez-Martínez *et al.* 2007	X	
	Aspidoscelis tigris	Gabe and Saint-Girons 1965	X	X
	Tupinambis teguixin	Whiting 1969		X
Trogonophiidae	*Agamodon anguliceps*	Whiting 1969		X
	Diplometopon zarudnyi	Whiting 1969		X
	Trogonophis wiegmanni	Bons and Saint-Girons 1963; Gabe and Saint-Girons 1965	X	

Table 6.1 contd....

Table 6.1 contd.

Family	Species	Reference	Cloacae	Glands
Tropiduridae	*Tropidurus etheridgei*	Sánchez-Martínez *et al.* 2007	X	
Varanidae	*Varanus griseus*	Gabe and Saint-Girons 1965	X	X
	Varanus sp.	Seshadri 1959	X	
	Varanus bengalensis	Auffenberg 1994		X
Xantusiidae	*Xantusia henshawi*	Gabe and Saint-Girons 1965	X	X
	Xantusia vigilis	Sánchez-Martínez *et al.* 2007	X	X
Xenosauridae	*Senosaurus newmanorum*	Sánchez-Martínez *et al.* 2007	X	X

6.3 THE OVIDUCTS

The most recent in depth review of lizard oviducts was completed by Blackburn (1998), and since this review, few studies have added details that alter his stereotypical conclusions on the oviduct of squamates. Girling (2002) offered another recent review that incorporates much of Blackburn's (1998) findings for comparison across all reptiles. In an effort not to "reinvent the wheel," much of our review of lizard oviducal anatomy is synthesized from these works. However, we employ terminology provided by Siegel *et al.* (2011a) for oviducal regions in snakes in order to maintain consistency amongst lepidosaurs, as the general oviduct structure in snakes is similar to that of lizards (Blackburn 1998; Girling 2002). Blackburn (1998) tabulated studies on the histology of the oviducts in his Table 2 (page 571), and more recent studies include those on the Cool-temperate Water-skink (*Eulamprus tympanum*; Hosie *et al.* 2003; Adams *et al.* 2007a), Cuban Brown Anole (*Anolis sagrei*; Sever and Hamlett 2002), Eastern Bearded Dragon (*Pogona barbata*; Amey and Whittier 2000), Ground Skink (*Scincella lateralis*; Sever and Hopkins 2004), *Hemidactylus turcicus* (Eckstut *et al.* 2009a,b), House Gecko (*Hemidactylus mabouia*; Nogueira *et al.* 2011), Ocellated Skink (*Chalcides ocellatus*; Corso *et al.* 2000), Pale-flecked Garden Sun Skink (*Lampropholis guichenoti*; Adams *et al.* 2004), Pelagic Gecko (*Nactus pelagicus*; Eckstut *et al.* 2009c), South-eastern Slider (*Lerista bougainvillii*; Adams *et al.* 2007b), Southern Forest Cool-skink (*Niveoscincus conventryi*; Ramírez-Pinilla *et al.* 2012), Tussock Cool-skink (*Pseudemoia entrecasteauxii*; Biazik *et al.* 2013), Viviparous Lizard (*Zootoca vivipara*; Heulin *et al.* 2005), Yellow-bellied Three-toed Skink (*Saiphos equalis*; Stewart *et al.* 2010), and multiple taxa tabulated in Table 1 of Sánchez-Martínez *et al.* (2007; page 291).

We examined oviducts from three individuals of *Hemidactylus turcicus* and multiple representatives of *Sceloporus* to facilitate review of the lizard oviduct. Specimens of *H. turcicus* were provided from the private collection of J. L. Rheubert. Specimens of *Sceloporus* (*S. jarovii*, Southern Prairie Lizard [*S. consobrinus*], and Striped Plateau Lizard [*S. virgatus*]) were captured in Cochise and Pima counties, Arizona, April, 2013 (permit #M587954) or Jackson County, Illinois, May (permit #NH13.5672). In *H. turcicus*, urogenital tracts were removed, fixed in buffered formalin, dehydrated in ascending concentrations of ethanol, cleared in toluene, embedded in paraffin, sectioned sagittally or transversely, affixed to albumenized slides, stained with hematoxylin and eosin for general histological examination, and examined with light microscopy. Specimens of *Sceloporus* were used for gross examination.

6.3.1 Gross Morphology

Brooks (1906) provides one of the most general descriptions of the gross structure of the lizard oviducts in his review of multiple lizard taxa urogenital systems: a thin ostium cranially that adjoins a convoluted region (infundibulum), which smooths caudally (glandular uterus) just before entering the cloaca as a short, heavily muscularized straight region (non-glandular uterus). Of course, these regions are much more difficult to observe in practice when dissecting lizards from multiple different life history stages (compare Fig. 6.4A-D to Fig. 6.4E,F). As juveniles, the oviducts appear as homogenously straight organs that are highly translucent (Fig. 6.4A). As non-vitellogenic adults, the regions become more observable as the infundibulum and glandular uterus become coiled, and the oviduct appears more opaque (Fig. 6.4B). As vitellogenic adults, the entire oviduct is highly coiled and opaque, causing gross delimination of the different oviducal regions difficult (Fig. 6.4C,E). As gravid adults, the regions are easily observed as "functional regions", as observed by Blackburn (1998); i.e., the oviducal region cranial to eggs in the glandular uterus is the infundibulum, whereas the oviducal region caudal to eggs in the glandular uterus is the non-glandular uterus (Fig. 6.4F).

6.3.2 *Hemidactylus turcicus* Oviduct Histology

The oviducts of *Hemidactylus turcicus* communicate with cranial arms of the urodaeum through oviducal papillae (Fig. 6.5). Inside the urodaeal arm chambers, the oviducts possess a thick muscularis primarily composed of circular smooth muscle. As the oviducts exit the urodaeal arms cranially, the circular muscle decreases in depth and is covered externally by a

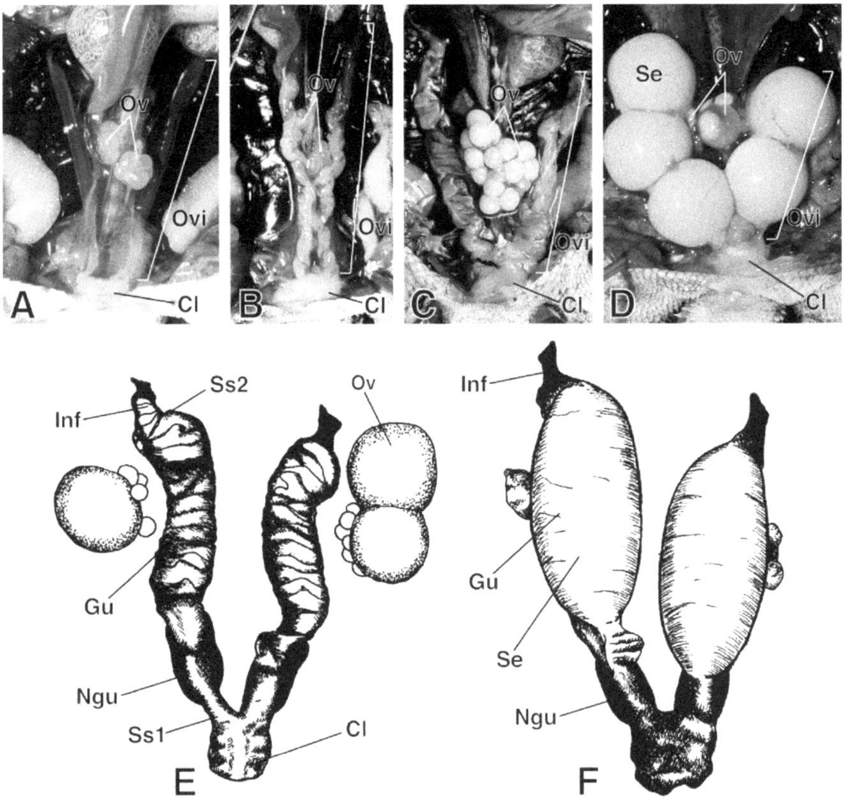

Fig. 6.4 Gross morphology of the oviduct from immaturity (**A**; *Sceloporus virgatus*) to adulthood **(B-F)**. Adult micrographs represent non-reproductive (**B**; *S. virgatus*), vitellogenic (**C** [*S. consobrinus*], **E** [*Scincella lateralis*]), and post-ovulation (**D** [*S. jarrovi*], **F** [*S. lateralis*]) oviducal morphology. Note that Ss2 (cranial sperm storage location) is for diagrammatic purposes only, as *S. lateralis* only stores sperm in the Ss1 (caudal sperm storage location) region. Cl, cloaca; Gu, glandular uterus; Inf, infundibulum; Ngu, non-glandular uterus; Ov, ovary; Ovi, oviduct; Se, eggs shelling in the uterus; Ss1, site of caudal sperm storage location; Ss2, site of cranial sperm storage location.

Color image of this figure appears in the color plate section at the end of the book.

thin layer of longitudinal smooth muscle and visceral pleuroperitoneum (Fig. 6.6A). The junction between the oviducts and the cloaca marks the caudal extremity of the non-glandular uteri (Figs. 6.5 and 6.6A). The non-glandular uteri are lined by a simple columnar epithelium with regular crypts; i.e., invaginations of the epithelium toward the basal lamina without invading the lamina propria (Fig. 6.6A). The cytoplasm of epithelial cells stains eosinophilic, and nuclei are found in a basal position in each epithelial cell (Fig. 6.6A). The majority of epithelial cells appear ciliated. The mucosa is highly folded and the lamina propria is thin (Fig. 6.6A).

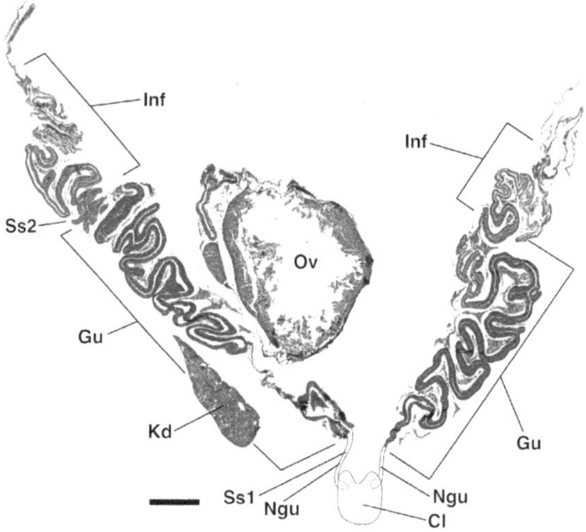

Fig. 6.5 Frontal sections from the oviduct of female *Hemidactylus turcicus* demonstrating the histological regions of the squamate oviduct (hematoxylin and eosin; scale bar = 1350 µm). Note that Ss1 (caudal sperm storage location) is for diagrammatic purposed only, as *H. turcicus* only stores sperm in the Ss2 (cranial sperm storage location) region. Cl, cloaca; Gu, glandular uterus; Inf, infundibulum; Kd, kidney; Ngu, non-glandular uterus; Ov, ovary; Ss1, site of caudal sperm storage location; Ss2, site of cranial sperm storage location.

Color image of this figure appears in the color plate section at the end of the book.

More cranially, the non-glandular uteri transition into the glandular uteri (Fig. 6.5). The glandular uteri possess a relatively narrow muscularis with circular and longitudinal muscle fibers encased by the visceral pleuroperitoneum (Fig. 6.6B). The epithelium lining the lumen is simple columnar and stains eosinophilic (Fig. 6.6B). Nuclei are found in a basal position in all epithelial cells and the majority of epithelial cells appear ciliated (Fig. 6.6B). No crypts of the epithelium were observed in the glandular uteri like those found in the non-glandular uteri (Fig. 6.6B). The lamina propria is thick compared to that of the non-glandular uteri due to the presence of numerous simple tubular uterine glands (Fig. 6.6B). The glands are often coiled in appearance and completely fill the lamina propria (Fig. 6.6B). The epithelial cells of uterine glands are filled with secretory granules that stain intensely with eosin (Fig. 6.6B).

The glandular uteri transition into the infundibula more cranially (Fig. 6.5). Unlike the non-glandular uteri and glandular uteri, the infundibula vary along their length. More caudally, two layers of smooth muscle (circular and longitudinal) are observed lining each infundibulum with a thin visceral pleuroperitoneum covering the muscularis (Fig. 6.6C). The lamina propria is thick, but not as thick as observed in the glandular uteri,

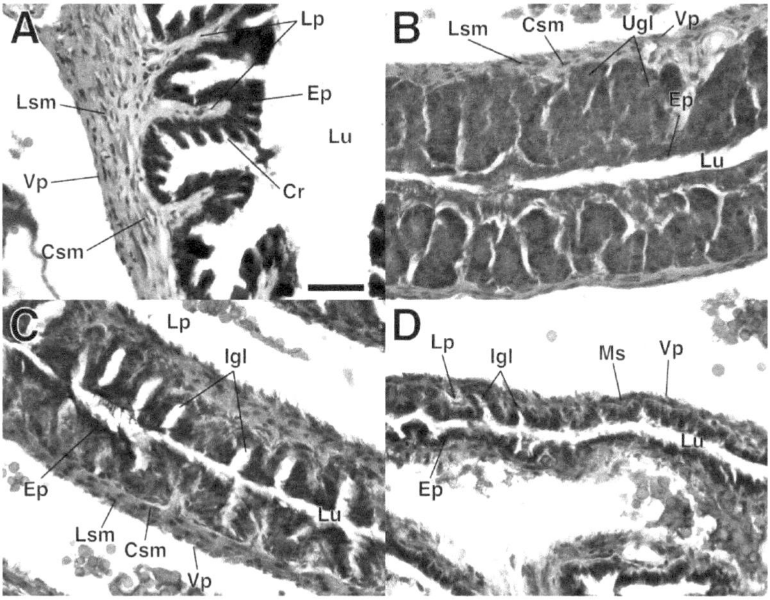

Fig. 6.6 High magnification histological sections from the oviduct (caudal to cranial) of *Hemidactylus turcicus* demonstrating four histologically unique regions (hematoxylin and eosin; scale bar = 50 µm). **(A)** The non-glandular uterus demonstrating a simple columnar epithelium with regular crypts lining the lumen, and a rather thick muscularis compared to more cranial regions of the oviduct. **(B)** The glandular uterus demonstrating numerous uterine glands in the lamina propria, a simple epithelium lining the lumen, and a thin muscularis compared to that of the non-glandular uterus. **(C)** The posterior infundibulum demonstrating infundibular glands, a simple epithelium lining the lumen, and a muscularis of similar thickness to that of the glandular uterus. **(D)** The anterior infundibulum demonstrating a decrease in infundibular glands, a simple epithelium lining the lumen, and a thin muscularis compared to more caudla regions of the oviduct. Cr, crypt; Csm, circular smooth muscle; Ep, epithelium; Igl, infundibular gland; Lp, lamina propria; Lsm, longitudinal smooth muscle; Lu, lumen; Ms, muscularis; Ugl, uterine gland; Vp, visceral pleuroperitoneum.

Color image of this figure appears in the color plate section at the end of the book.

due to the presence of numerous simple tubular glands (Fig. 6.6C). These infundibular glands are not coiled and are only simple invaginations of the epithelium into the lamina propria (Fig. 6.6C). The epithelial cells of the glandular invaginations and those lining the lumen of the infundibula are mostly cuboidal with eosinophilic cytoplasm (Fig. 6.6C). Most nuclei are found near the basal lamina. More cranially, infundibular glands are absent and, thus, the lamina propria is much thinner (Fig. 6.6D). Furthermore, the muscularis is reduced to a point that determining multiple layers becomes difficult (Fig. 6.6D).

6.3.3 Historical Overview of the Oviducts

Non-glandular uterus. In all taxa examined previously (for review see Blackburn 1998; Girling 2002), the oviducts communicate with the urodaeum through paired oviducal papillae, besides a few, in which the oviducts unite after entering the urodaeum (see Section 6.2.4, this volume). In *Anniella*, the left oviduct is highly reduced, but still present (Coe and Kunkel 1905). This most caudal region of the oviduct was traditionally termed the vagina in squamates (for review see Blackburn 1998; Siegel *et al.* 2013) but, as Blackburn (1998) summarized, this terminology was quite unfortunate; i.e., the most caudal portion of the oviduct is probably not functionally or developmentally equivalent to the vagina of other amniotes. Thus, Siegel *et al.* (2011a) renamed this portion of the oviduct in snakes the "non-glandular uterus," a term that we have adopted for this portion of the oviduct in this review, as this portion of the oviduct is most likely homologous across squamate taxa. However, we are fully aware that the consistent usage of vagina for this region of the oviduct dating back to at least Giersberg (1922) could result in the absence of "non-glandular uterus" appearing in future studies besides our own. Furthermore, this term may not be as appropriate for lizards, as multiple lizard taxa contain "crypts" in the non-glandular uterus that function in sperm storage (see Section 6.4, this volume).

Little variation is present in the non-glandular uterus among taxa. As observed in *Hemidactylus turcicus*, the non-glandular uterus in lizards is demarcated by its thick muscularis at the point of junction with the cloaca and highly folded mucosa (Cueller 1966; Botte 1973; Adams and Cooper 1988; Uribe *et al.* 1988; Shanthakumari *et al.* 1990; Saint-Girons and Ineich 1992; Shanthakumari *et al.* 1992; Girling *et al.* 1997, 1998; Backburn 1998; Girling 2002; Sánchez-Martínez *et al.* 2007; Eckstut *et al.* 2009b). Ciliated crypts where sperm are often stored (see Section 6.4), like those that were described for *H. turcicus*, are common in lizards as invaginations of the epithelial lining (Saint-Girons 1975; Conner and Crews 1980; Shanthakumari *et al.* 1990; Adams and Cooper 1988; Amey and Whittier 2000; Girling 2002; Sever and Hamlett 2002; Sever and Hopkins 2004; Eckstut *et al.* 2009a). Like the apices of the crypt epithelia, the majority of the epithelial lining of the non-glandular uterus is ciliated, with occasional secretory cells (Boyd 1942; Fox 1977; Adams and Cooper 1988; Uribe *et al.* 1988; Saint-Girons and Ineich 1992; Palmer *et al.* 1993; Girling *et al.* 1997, 1998; Blackburn 1998; Girling 2002; Adams *et al.* 2004; Sánchez-Martínez *et al.* 2007; Eckstut *et al.* 2009b; Nogueira *et al.* 2011). Secretory cells of the non-glandular uterus epithelium produce mainly mucous and/or neutral carbohydrates as indicated by their strong reaction with alcian blue and/or periodic acid-Schiff's (Picariello 1989; Shanthakumari *et al.* 1992; Palmer *et al.* 1993; Girling *et al.* 1997, 1998;

Sever and Hopkins 2004; Nogueira *et al.* 2011). The non-glandular uterus exhibits little variation throughout the reproductive and non-reproductive seasons and any variation observed is typically described as increased epithelium height during the reproductive season (Picariello 1989; Girling *et al.* 1997); although, some authors have also noted a thickening of the muscularis (Uribe *et al.* 1988) and increased ciliation (Girling *et al.* 1997).

It has been an assumption of Siegel that the non-glandular uterus acts as a sphincter (owing to the rather highly developed circular smooth muscle of the muscularis externa in this oviducal region), and tightening during egg development in the glandular uterus aids in the correct positioning of eggs in the glandular uterus. This idea is not novel, and was based on a statement from Girling (2002): "The vagina [non-glandular uterus] is a thick, muscular region that acts as a sphincter during gravidity or gestation." In snakes, the non-glandular uterus is notably variable in size, resulting in eggs in one glandular uterus positioned cranially (typically the right) in comparison to eggs in the other, which are positioned more caudally (typically the left). Thus, variability in length of the non-glandular uterus could be a direct result of evolution to a more stream-lined bauplan; i.e., eggs are positioned cranial to caudal in juxtapositioned oviducts instead of side-by-side. Of course, this hypothesis still needs tested.

Glandular uterus. Like the non-glandular uterus, we chose to use the term "glandular uterus" for the middle region of the oviduct where eggs receive their shell (in oviparous species) after Siegel *et al.* (2011a). But traditionally, this region has merely been termed the uterus in squamates (for review see Blackburn 1998; Girling 2002; Siegel *et al.* 2011a). In all taxa examined, this region of the oviduct is demarcated by dense aggregations of uterine glands, more so in oviparous species than viviparous species, in which latter uterine glands responsible for eggshelling are reduced in density (for review see Blackburn 1998; Sánchez-Martínez *et al.* 2007).

As observed in the non-glandular uterus, the glandular uterus is encompassed by a muscularis composed of an inner circular layer and outer longitudinal layer (Guillette and Jones 1985; Adams and Cooper 1988; Girling *et al.* 1988, 1997; Blackburn 1998; Blackburn *et al.* 1998; Girling *et al.* 1998; Corso *et al.* 2000; Heulin *et al.* 2005). The epithelial lining of the mucosa, which is highly obscured by mucosal glands, is comprised of a simple/pseudostratified cuboidal to columnar epithelium with secretory (microvillus) cells and ciliated cells (Botte 1973; Guillette and Jones 1985; Adams and Cooper 1988; Uribe *et al.* 1988; Saint-Girons and Ineich 1992; Palmer *et al.* 1993; Girling *et al.* 1997, 1998; Guillette *et al.* 1989; Blackburn *et al.* 1998; Corso *et al.* 2000; Hosie *et al.* 2003; Adams *et al.* 2004; Heulin *et al.* 2005; Adams *et al.* 2007a,b; Nogueira *et al.* 2011; Ramírez-Pinilla *et al.* 2012). In the majority of taxa examined, this epithelial lining was described as simple, but was reported stratified in *Hemidactylus turcicus* by Girling *et al.* (1998; a finding contrary to the histological examination for this chapter; see Section 6.3.2, this volume), and the Leaf-tailed Gecko (*Saltuarius wyberba*; Gerling *et al.* 1998). Epithelia lining the lumen of the uterus is mostly secretory/microvillus (Hosie *et al.* 2003; Adams *et al.* 2004; Sánchez-Martínez *et al.* 2007; Adams *et al.* 2007b; Biazik *et al.* 2012; Ramírez-Pinilla 2012), and numerous authors described changes in epithelial surface composition in multiple lizard taxa based on parity mode and seasonal reproductive cycles (e.g., Corso *et al.* 2000; Hosie *et al.* 2003; Adams *et al.* 2004, 2007a,b; Biazik *et al.* 2012; Ramírez-Pinilla *et al.* 2012). Secretory cells have been shown to react positively with both alcian blue and periodic acid-Schiff's staining procedures (Botte 1973; Shanthakumari *et al.* 1992; Palmer *et al.* 1993; Girling *et al.* 1997, 1998; Heulin *et al.* 2005; Stewart *et al.* 2010; Nogueira *et al.* 2011). The lamina propria of the glandular uterus is highly vascularized in all lizards (regardless of parity mode), and vascularization increases during reproductive activity (Guillette and Jones 1985; Masson and Guillette 1987; Picariello *et al.* 1989).

The glandular uterus exhibits the most variation between taxa because of variation in gland abundance/development, which is due mainly to parity mode (for review see Blackburn 1998; Girling 2002). Since Giacomini (1893), a trend has been noted that oviparous taxa possess denser or more developed glands in the glandular uterus of squamates than viviparous taxa. This has been confirmed in lizards by multiple independent authors (e.g., Guillette and Jones 1985; Girling *et al.* 1997, 1998; Heulin *et al.* 2005) on studies comparing closely related taxa (or different forms of the same species) with differing modes of parity.

Morphological variation also exists in gland morphology between lizard taxa. Glands were described as simple or branched tubular in *Chalcides chalcides* (Blackburn *et al.* 1998), Great Plains Skink (*Plestiodon obsoletus*; Guillette *et al.* 1989), *Hemidactylus mabouia* (Nogueira *et al.* 2011), *Hemidactylus turcicus* (this study; see Section 6.3.2, this volume), *Sphenomorphus fragilis* (no common name known; Guillette 1992), and *Tarentola mauritanica* (Picariello *et al.* 1989). Others described glands as alveolar or saccular in *Chalcides ocellatus* (Corso *et al.* 2000), *Crotaphytus collaris* (Guillette *et al.* 1989), *Hoplodactylus maculatus* (Girling *et al.* 1997), and *Saiphos equalis* (Stewart *et al.* 2010). Still, other studies described glands as tubulo-alveolar in Southern Bunchgrass Lizard (*Sceloporus aeneus*; Guillette and Jones 1985), Trans Volcanic Bunchgrass Lizard (*Sceloporus bicanthalis*; Guillette and Jones 1985), and Western Spiny-tailed Iguana (*Ctenosaura pectinata*; Uribe *et al.* 1988), and one study described glands as branched acinar and tubular in Florida Scrub Lizard (*Sceloporus woodi*; Palmer *et al.* 1993). Obviously, morphological designation of glands does not correlate with parity mode, as multiple oviparous and viviparous taxa are found in each grouping above.

Epithelial linings of uterine gland invaginations have been described as pseudostratified (Picariello *et al.* 1989) or simple pyramidal/polyhedrical/cuboidal (Uribe *et al.* 1988; Blackburn *et al.* 1998; Corso *et al.* 2000). During reproductive activity, the epithelium lining the lumen of the glandular uterus increases in height (Picariello *et al.* 1989), with a concurrent increase in height of epithelial cells lining the mucosal glands (Picariello *et al.* 1989). In oviparous species, gland hypertrophy is so great that the lamina propria is obscured (Picariello *et al.* 1989). Staining characteristics of the secretions produced by uterine glands vary between taxa. Multiple studies indicated a negative reaction to both periodic acid-Schiff's and alcian blue in agamids (Shanthakumari *et al.* 1992), gekkonids (Girling *et al.* 1997; Girling *et al.* 1998) and scincids (Stewart *et al.* 2010), whereas a positive reaction to both of these stains was observed in a lacertid (Heulin *et al.* 2005) and to only periodic acid-Schiff's in a phrynosomatid (Palmer *et al.* 1993). Other positive staining characteristics include dihydroxy-6, 6'disulfide-dinapthyl for proteins rich in cysteine and cystine in *Saiphos equalis* (Stewart *et al.* 2010) and *Zootoca vivipara* (Heulin *et al.* 2005); ninhydrin-Schiff's for aliphatic amines in *Chalcides ocellatus* (Corso *et al.* 2000); procollagen-like compounds in *Plestiodon obsoletus* but not in *Crotaphytus collaris* (Guillette *et al.* 1989); keratin in the Italian Wall Lizard (*Podarcis siculus*; Botte 1973); and mercuric bromophenol blue for general proteins in *Calotes versicolor* (Shanthakumari *et al.* 1992). In conclusion, glands of the glandular uterus are well known for the production of proteins that are involved in shell membrane formation (for review see Blackburn 1998).

Infundibulum. The infundibulum is the most cranial portion of the oviduct and is delineated by a very thin mucosa and muscularis (for review see Blackburn 1992; Girling 2002). As reviewed by Blackburn (1998), multiple layers of the muscularis (both circular and longitudinal) are difficult to differentiate as the infundibulum approaches its most cranial extremity. In many squamates, lizards included (see Section 6.2.3, this volume), the infundibulum can be divided into two discrete regions: the caudal region where sperm storage occurs in some taxa (for review see Sever and Hamlett 2002; Eckstut *et al.* 2009a; Section 6.4, this volume) and the more cranial region that terminates at the ostium (for review see Siegel *et al.* 2011a). The main difference in these infundibular regions is the presence (caudal region) or absence (cranial region) of distinct infundibular glands (see Section 6.2.3, this volume) or a simple thickening of the more posterior infundibular wall. Some authors have divided the infundibulum into the infundibulum proper, tuba (equivalent to the caudal extremity of the infundibulum described here) and the isthmus (see Picariello *et al.* 1989); however, these designations are difficult to delineate, and following Blackburn (1998), we use the terminology of the infundibulum to encompass all of these regions with posterior and anterior designation to delineate minor differences from caudal to cranial. Aspects of the infundibulum have been discussed in the context of sperm storage for some taxa. Review of this material is found in Section 6.4, this volume.

For the most part, the entire infundibulum is folded and lined by a primarily ciliated epithelium with the occasional non-ciliated/microvillus cell (Botte 1973; Guillette and Jones 1985; Uribe *et al.* 1988; Guillette *et al.* 1989; Guillette 1992; Saint-Girons and Ineich 1992; Palmer *et al.* 1993; Girling *et al.* 1997, 1998; Blackburn 1998; Adams *et al.* 2004; Nogueira *et al.* 2011). However, in other taxa, e.g., *Heteronotia binoei* (Whittier *et al.* 1994) and Keeled Earless Lizard (*Holbrookia propinqua*; Adams and Cooper 1988), only ciliated cells were identified. Girling *et al.* (1998) noted regional differences in the occurence of ciliated versus non-ciliated/microvillus cells in gekkonids. Epithelia linings varied between simple cuboidal and columnar within taxa, were described as higher in the more posterior regions of the infundibulum, and increased in height during reproductive activity (e.g., Guillette *et al.* 1989). Saint-Girons and Ineich (1992) described the epithelial lining of the anterior infundibulum in *Lepidodactylus lugubris* as "undifferentiated".

Non-ciliated/microvillus cells of the luminal epithelium stain positive with the periodic acid-Schiff's and alcian blue procedures in *Calotes versicolor* (Shanthakumari *et al.* 1992), *Hoplodactylus duvaucelii* (Girling *et al.* 1998; posterior region only), *H. maculatus* (Girling *et al.* 1998; posterior region only), *Hemidactylus mabouia* (Nogueira *et al.* 2011; posterior region only), *H. turcicus* (Girling *et al.* 1998; posterior region only), *Saltuarius wyberba* (Girling *et al.* 1998; posterior region only), and *Sceloporus woodi* (Palmer *et al.* 1993). Only alcian blue positive non-ciliated/microvillus cells were observed in the anterior region of the infundibulum of *H. mabouia* (Nogueira *et al.* 2011) and only periodic acid-Schiffs reactivity was observed in the anterior region of *H. duvaucelii, H. maculatus,* and *Saltuarius wyberba* (Girling *et al.* 1998).

As described for *Hemidactylus turcicus* in Section 6.3.2, this volume, the major difference between the anterior and posterior regions of the infundibulum is the presence of glandular-like invaginations in some taxa. The invaginations were referred to as "crypts" or "sacs" in Keeled Earless Lizard (*Holbrookia propinqua*; Adams and Cooper 1988), *Lepidodactylus lugubris* (Saint-Girons and Ineich 1992), *Podarcis siculus* (Botte 1973), *Sceloporus aeneus* (Guillette and Jones 1985), and *Scelporus bicanthalis* (Guillette and Jones 1985). The lining of these crypts was reported as ciliated and cuboidal (Guillette and Jones 1985; Adams and Cooper 1988) and the epithelia react positively to the periodic acid-Schiff's procedure in at least *Podarcis siculus* (Botte 1973). In other taxa, these invaginations were reported as more complex; i.e., branched tubular glands in *Hemidactylus mabouia* (Nogueira *et al.* 2011), alveolar glands in *Hoplodactylus maculatus* (Girling *et al.* 1997), and "glands" in *H. duvaucelii, Hemidactylus turcicus, Saltuarius wyberba*, and *Sphenomorphus fragilis* (Guillette 1992). The epithelia lining these glands were described as ciliated and non-ciliated (Girling *et al.* 1998; Nogueira *et al.* 2011) and reacted negatively with the periodic acid-Schiffs and alcian blue staining procedures (Girling *et al.* 1997; Girling *et al.* 1998). Glands/crypts often contained sperm (e.g., Whittier *et al.* 1994; Girling *et al.* 1998; Nogueira *et al.* 2011) and will be discussed in more detail in Section 6.4, this volume.

6.3.4 Conclusions for Oviducts

The oviducts of all lizards are very similar with three easily discernable regions. From caudal to cranial: Non-glandular uterus (often termed vagina), glandular uterus (often termed uterus), and infundibulum. The infundibulum is typically divided into distinct posterior (often termed tube) and anterior regions. The oviducal wall contains three distinct layers. From external to internal: muscularis (with outer longitudinal and inner circular smooth muscle fibers; often hard to distinguish in the infundibulum), lamina propria, and epithelium. The entire oviduct is encompassed by the visceral pleuroperitoneum that is continuous with the parietal pleuroperitoneum of the body cavity through dorsal mesentery. All three regions of the oviduct have crypts or mucosal glands depending on the taxa, and only the glandular uterus has glandular modifications in every taxon examined. Little meaningful variation exists in the overall structure of the oviduct in terms of phylogeny (at least that is obvious with the few taxa examined to date), but uterine gland morphology and activity appears to vary between taxa with different parity modes. This finding is not novel, and makes sense considering the multiple independent origins of viviparity in squamates (for review see Blackburn 1985, 1999).

6.4 SPERM STORAGE AND TRANSPORT

6.4.1 Overview

Literature on female sperm storage in reptiles, including lizards, was reviewed relatively recently by Sever and Hamlett (2002) and Eckstut *et al.* (2009a). Female sperm storage tubules (Ssts) in lizards were first reported in three species of chameleons (*Chameleo*) by Saint Girons (1962) and in the Green Anole (*Anolis carolinensis*) by Fox (1963). Most research on anatomy of sperm storage in female lizards has been limited to light microscopy, and at least 27 species representing eight families have been studied in that manner (Table 6.2). As indicated by Table 6.2, the location of Ssts is quite variable.

Table 6.2 Literature on sperm storage in female lizards and the location of sperm storage tubules (SSTs).

Family	**Species**	**Location of SSTs**	**Reference**
Agamidae	*Psammophilus dorsalis*	Anterior non-glandular uterus	Srinivas *et al.* 1995
Agamidae	*Calotes versicolor*	Transition between non-glandular and glandular uterus	Kumari *et al.* 1990
Chameleonidae	*Chameleo* sp.	Non-glandular uterus	St. Girons 1962
Crotaphytidae	*Crotaphytus collaris*	Anterior and middle non-glandular uterus	Cuellar 1966
Gekkonidae	*Coleonyx variegatus*	Infundibulum	Cuellar 1966
	Phyllodactylus homolepidurus	Infundibulum	Cuellar 1966
	Hemidactylus frenatus	Infundibulum and glandular uterus junction	Murphy-Walker and Haley 1996
	Hemidactylus turcicus	Infundibulum and glandular uterus junction	Eckstut *et al.* 2009a,b
	Nactus multicarinatus	Posterior infundibulum	Eckstut *et al.* 2009c
Lacertidae	*Acanthodactylus scutellatus*	Infundibulum	Bou-Resli *et al.* 1981
Phrynosomatidae	*Urosaurus microsctatus*	Anterior and middle non-glandular uterus	Cuellar 1966
	Uta stansburiana	Anterior and middle non-glandular uterus	Cuellar 1966
	Uta squamata	Anterior and middle non-glandular uterus	Cuellar 1966
	Uta palmeri	Anterior and middle non-glandular uterus	Cuellar 1966
	Callisaurus dragonoides	Infundibulum and non-glandular uterus	Cuellar 1966
	Holbrookia propinqua	Anterior and middle non-glandular uterus	Adams and Cooper 1988
	Holbrookia elegans	Infundibulum and non-glandular uterus	Cuellar 1966
	Phrynosoma cornutum	Anterior and middle non-glandular uterus	Cuellar 1966
	Sceloporous undulatus	Anterior and middle non-glandular uterus	Cuellar 1966
	Sator grandaevus	Anterior and middle non-glandular uterus	Cuellar 1966

Table 6.2 contd....

Table 6.2 contd.

Family	Species	Location of SSTs	Reference
Polychrotidae	*Anolis carolinensis*	Transition between non-glandular and glandular uterus	Conner and Crews 1980
	Anolis sagrei	Transition between non-glandular and glandular uterus	Sever and Hamlett 2002
	Anolis pulchellus	Posterior infundibulum (ampulla)	Ortiz and Morales 1974
Scincidae	*Hemiergis peronii*	Oviduct, region not specified	Smyth and Smith 1968
	Eumeces egregius	Posterior non-glandular uterus	Schaefer and Roeding 1973
	Mabuya scincoides	Non-glandular uterus	Sarkar and Shivanandappa 1989
	Scincella lateralis	Posterior non-glandular uterus	Sever and Hopkins 2004

Ultrastructural studies are limited to scanning electron microscopy (SEM) of the Sst area in *A. carolinensis* (Conner and Crews 1980) and several species in the Gekkonidae (Girling *et al.* 1997, 1998), SEM and transmission electron microscopy (TEM) of the Ssts of *Hemidactylus turcicus* (Eckstut *et al.* 2009b) and the Nidua Fringe-fingered Lizard (*Acanthodactylus scutellatus*; Bou-Resli *et al.* 1981), TEM of *Anolis sagrei* (Sever and Hamlett 2002), and SEM and TEM of *Scincella lateralis* (Sever and Hopkins 2004). Ssts were described earlier in snakes (Fox 1956), but the only ultrastructural studies on snakes are TEM on the Common Gartersnake (*Thamnophis sirtalis*; Hoffman and Wimsatt 1972), SEM on the Ringneck Snake (*Diadophis punctatus*; Perkins and Palmer 1996), and TEM and SEM on the Black Swamp Snake (*Seminatrix pygaea*; Sever and Ryan 1999) and Cottonmouth (*Agkistrodon piscivorus*; Siegel and Sever 2008).

Eckstut *et al.* (2009a) conducted phylogenetic analyses of sperm storage characters in female squamates, including Sst location, cytology of the Sst linings, secretory activity in the Ssts, and length of sperm storage. They used three competing squamate phylogenies, one morphological (Lee 2005) and two molecular, including Vidal and Hedges (2005) and a new hypothesis using the nuclear c-mos gene from 611 taxa. Thus, Eckstut *et al.* (2009a) also tested the hypothesis that sperm storage may offer reciprocal illumination in choosing among hypotheses of squamate phylogeny. They found, however, minimal value for phylogenetic inference from sperm storage characters. Several of these characters are relatively conserved across the sampled squamates (including presence of sperm storage and sperm storage tubules (Ssts), embedding of sperm, and sperm storage with eggs *in utero*). Alternatively, location of Ssts, secretions within the Ssts, and length of sperm storage are highly variable traits. Eckstut *et al.* (2009a) suggest that these differences are correlated with ecological differences in the reproductive tactics of each species. Because of limited data available for many of the taxa, however, statements regarding the ecological utility of the traits are highly speculative. The relationship between Sst secretion location and sperm storage length, however, indicates correlations between long term sperm storage (four months or greater) and secretory cells throughout, moderate sperm storage lengths (two and a half months) and distal secretory cells, and short term sperm storage (two weeks) and proximal secretory cells.

The location of Ssts was the most highly variable character analyzed, with six recorded character states (Eckstut *et al.* 2009a). The morphological phylogeny suggests that the basal squamate clade retained Ssts in the non-glandular uterus (as indicated in Section 6.3.3, this volume; often termed the vagina), with one evolution of Ssts located both in the non-glandular uterus and infundibulum, and no occurrences of solely infundibular Ssts in this clade (Fig. 6.7A). The analysis of character state evolution lacks a substantial amount of resolution in both molecular phylogenies as a result of the lack of data on several taxa (Fig. 6.7B,C). However, the most parsimonious resolution of the nuclear phylogeny (Vidal and Hedges 2005) depicted the non-glandular uterus Sst clade (which is consistent with the morphological non-glandular uterus Sst clade) as derived from the infundibular Ssts.

We will briefly review studies dealing with the ultrastructure of the Ssts of *Anolis sagrei* (Sever and Hamlett 2002), *Scincella lateralis* (Sever and Hopkins 2004), and *Hemidactylus turcicus* (Eckstut *et al.* 2009a,b), as these species show striking variation in location and cytology of Ssts.

6.4.2 *Anolis sagrei*

Fox (1963) was the first to report sperm in Ssts of the non-glandular uterus in a lizard, *Anolis carolinensis*, which is closely related to *A. sagrei*. Fox reported that the oviduct consists of three parts. The proximal infundibulum is thin-walled and lined by ciliated and non-ciliated columnar epithelium that is apparently aglandular. The glandular uterus contains coiled tubular shell glands. The non-glandular uterus (their vagina) consists of a thin, anterior tube and a thick, posterior pouch. From the bottom of longitudinal grooves in the anterior two-thirds of the non-glandular uterus (tube; Fox 1963) found small tubules, Ssts, extending deep in the thick tunica propria and passing more or less parallel to the folds. In many of these tubules, depending on the time elapsed since the last insemination, are bundles of sperm. The ciliated cells associated with the rest of the lining do not extend into the distal areas of the Ssts where sperm are stored.

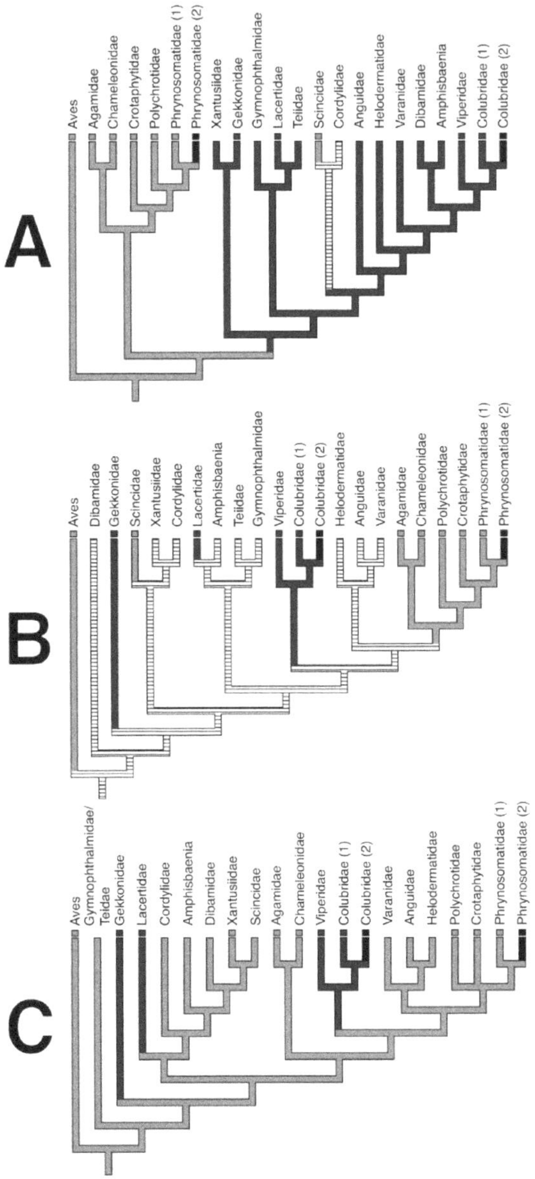

Fig. 6.7 Location of SSTs in squamate reptiles mapped on **(A)** a morphological phylogeny from Lee (2005), **(B)** a molecular phylogeny from Vidal and Hedges (2005), and **(C)** a molecular phylogeny from Eckstut *et al.* (2009b). Green indicates Ssts in the non-glandular uterus, blue indicates infundibular Ssts, and black indicates both non-glandular uterus and infundibular Ssts. Cross hatching represents equivocal ancestral state reconstructions. Modified from Eckstut *et al.* (2009a, Amphibia-Reptilia, 30: 45–56, Fig. 4).

Color image of this figure appears in the color plate section at the end of the book.

Conner and Crews (1980) refined the description of the non-glandular uterus of *Anolis carolinensis* provided by Fox (1963). They noted that the anterior portion of the non-glandular uterus of Fox (1963) consists of two portions, the transition between the non-glandular and glandular uterus and an area just posterior to the transition, which consists of a thin tube flanked on two sides by openings that extend anteriorly and end blindly in the lamina propria. The entire anterior half of the non-glandular uterus is highly folded with ciliated and non-ciliated cells alternating in the epithelium. The Ssts are located in the non-glandular and glandular uterus transition.

Whether one refers to the Sst area in *Anolis* as the anterior non-glandular uterus tube (Fox 1963) or the non-glandular/glandular uterus transition area (Conner and Crews 1980) could be significant as further comparative work emanates. Sever and Hamlett (2002) reported that the Ssts of *A. sagrei* are clearly limited to the region immediately adjacent to the glandular uterus where eggs develop prior to oviposition (Fig. 6.8A). This is equivalent to the region where Ssts are found in birds, and Bakst (1987) and others have adopted the term "uterovaginal junction [non-glandular/glandular uterus transition]" to describe the location of such glands. It is unclear whether Ssts are limited to this junctional area in some other lizards in which non-glandular uterus Ssts are known (Cuellar 1966; Adams and Cooper 1988; Kumari *et al.* 1990). Thus, to be as precise as possible, Sever and Hamlett (2002) referred to the Sst region in *A. sagrei* as the "uterovaginal junctional area [junction between the non-glandular and glandular uterus]"; use of the term "transition," is inaccurate because this region is unique by virtue of the Ssts and does not represent a gradation between the glandular uterus and the remainder of the non-glandular uterus.

Many species of *Anolis* are known to lay single eggs every 10–14 days during a two or three month breeding period, and the ovaries and oviducts alternate in this process (Conner and Crews 1980). This situation apparently exists in *A. sagrei* Sever and Hamlett (2002) collected in the Florida Keys, and all the females that were dissected from their April samples contained one well-developed egg in one uterus, as well as a more recently ovulated egg in the other uterus (Fig. 6.8A). The egg is shelled by the time it reaches the non-glandular uterus, so sperm apparently migrate up the oviduct to the infundibulum to fertilize eggs; sperm have been found in the infundibulum 6–24 hrs after mating (Conner and Crews 1980). Sperm stored in the Ssts may be used to fertilize successive eggs over several months. The tubules thus provide protection for sperm during passage of eggs through the non-glandular uterus (Conner and Crews 1980). Recently, Uller *et al.* (2013) showed that sperm remains viable in the female reproductive tract across ovarian cycles in the Dragon Lizard, *Ctenophorus fordi* (Agamidae), but stored sperm were generally less likely to fertilize eggs than recently inseminated sperm.

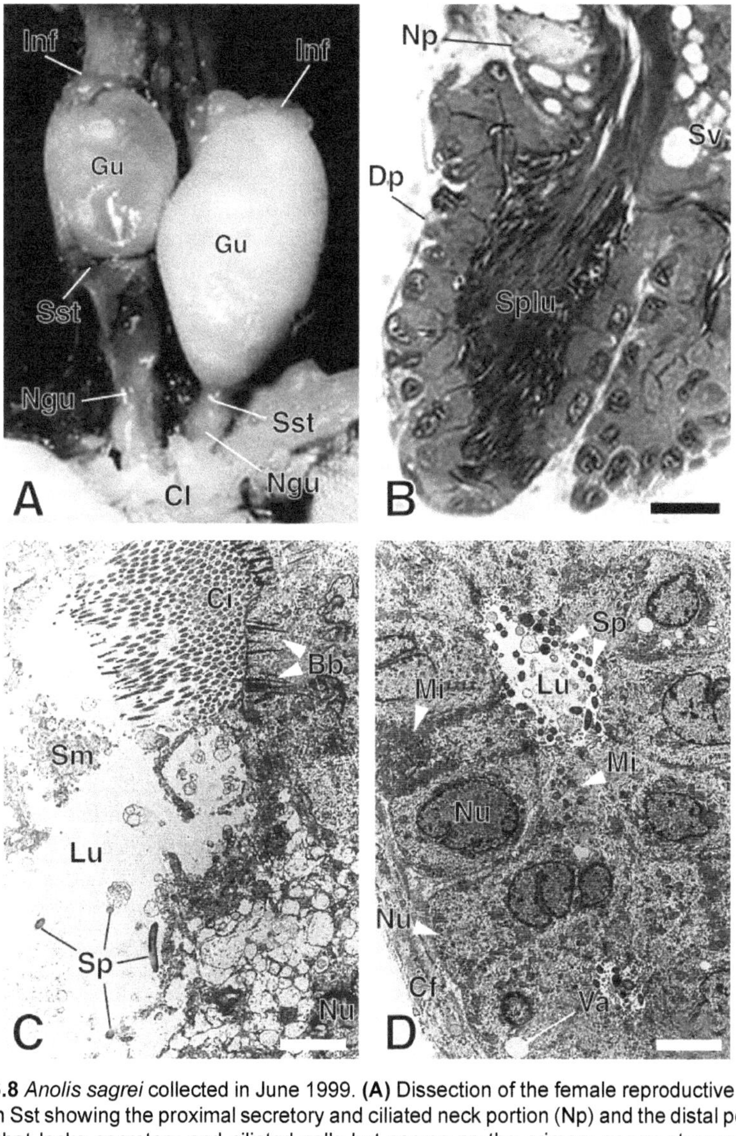

Fig. 6.8 *Anolis sagrei* collected in June 1999. **(A)** Dissection of the female reproductive tract. **(B)** An Sst showing the proximal secretory and ciliated neck portion (Np) and the distal portion (Dp) that lacks secretory and ciliated cells but serves as the primary sperm storage area (scale bar = 150 μm). Note crowded cluster of sperm aligned together along their long axes. **(C)** Sperm in the neck portion of an Sst (scale bar = 5 μm). **(D)** Sperm in the distal portion of an Sst (scale bar = 5 μm). Bb, basal bodies; Cf, collagen fibers; Ci, cilia; Gu, glandular uterus; Inf, infundibulum; Lu, lumen; Mi, mitochondria; Ngu, non-glandular uterus; Nu, nuclei; Sm, secretory material; Sp, sperm; Splu, sperm in the lumen; Sst, sperm storage tubules; Sv, secretory vacuoles; Va, vacuole; Vp, visceral pleuroperitoneum. Modified from Sever and Hamlett (2002, Journal of Experimental Zoology, 292: 187–199, Fig. 6A,C and Fig. 7C,D).

As reported by Fox (1963) and Conner and Crews (1980), the lining of the oviduct in the non-glandular/glandular uterus junction area consists of ciliated cells and nonciliated secretory cells, a typical situation in vertebrate oviducts. This lining extends into the neck portion of the Ssts (Fig. 6.8B,C). Some variation occurs among females in extent of secretory activity in the Ssts and adjacent oviducal lining, and this variation seems related to the amount of sperm present. Secretory vacuoles are larger, and perhaps contain a different product, in specimens in which the Ssts contain more sperm. Parts of a few sperm were found embedded in secretory cells of the neck region, but the significance of embedded sperm (phagocytosis? nutrition?) is unknown.

The distal portion of the Ssts, however, contain cells that lack secretory vacuoles and cilia (Fig. 6.8B,D). The distal ends contain the most sperm and are presumably the primary storage area (Fox 1963). Sperm may be clustered in bundles or align themselves more randomly, perhaps depending on the numerical density of the sperm.

No contractile elements are associated with the basal lamina of the Ssts. Collagen fibers occur superficial to the Ssts (Fig. 6.8D) and the muscularis is thin. The role, if any, of the tissues surrounding the Ssts in sperm transport into and out of the Ssts remains unresolved (Conner and Crews 1980; Bou-Resli *et al.* 1981).

The epithelium of the distal portions lacks synthetic organelles and, indeed, appears almost syncytial (Fig. 6.8D). The intercellular canaliculi are marked by thickened, felt-like areas (Fig. 6.8D), which, on closer examination, reveal a lack of continuity of the plasma membranes between desmosomal areas. Scattered oval, dense mitochondria, and a few clear vacuoles are the only cytoplasmic structures noted in most cells (Fig. 6.8D).

6.4.3 *Scincella lateralis*

In *Scincella lateralis*, sperm storage also occurs in the non-glandular uterus, but unlike *Anolis sagrei*, the location is the posterior portion of the non-glandular uterus, and specialized Ssts do not occur. Stored sperm occur in the lumen of the posterior non-glandular uterus, in between rugae and in the crypts in the folds of the rugae (Fig. 6.9A). The orientation of sperm in the lumen and in between rugae is disorderly, but sperm in the crypts usually appear to have their nuclei aligned and orientated into the crypts (Fig. 6.9B-D). The lining of the non-glandular uterus consists of ciliated simple columnar epithelial cells interspersed with occasional secretory cells. The bases of the crypts, however, consist strictly of secretory cells (Fig. 6.9D). Sperm in the lumen of the non-glandular uterus are often seen mixed in secretory product, although this is not apparent in the crypts. Whether mixed with secretory product or not, these sperm appear normal in cytology.

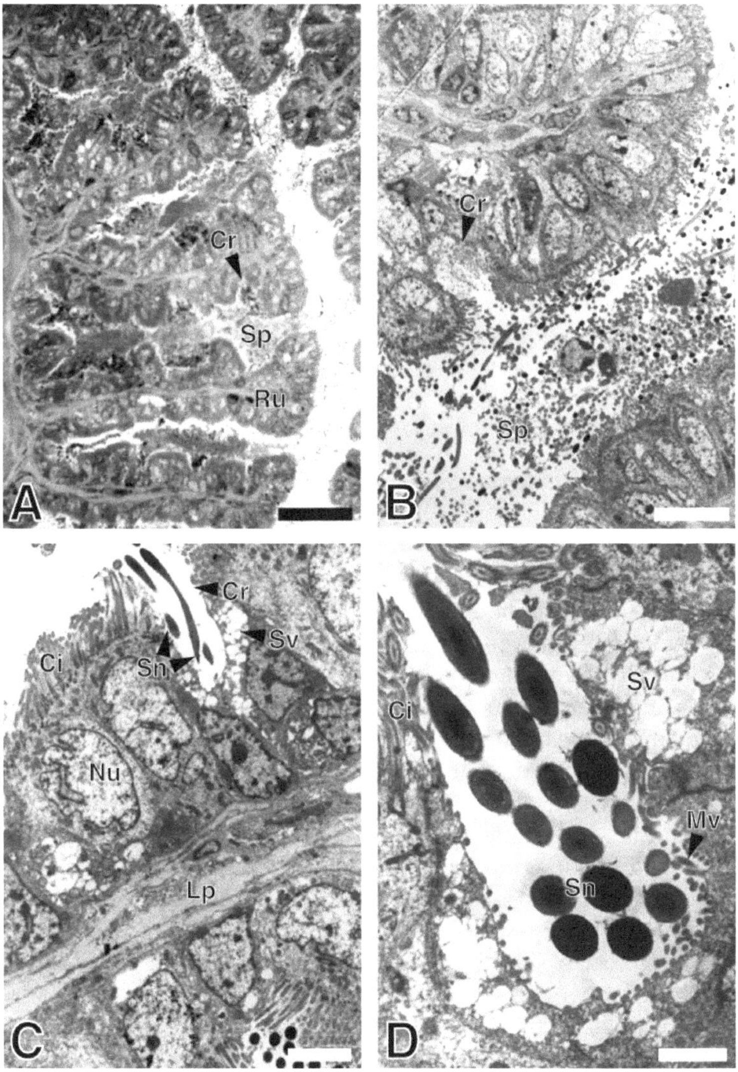

Fig. 6.9 Transmission electron-micrographs through the posterior region of the non-glandular uterus from a 36.4 mm SVL *Scincella laterale* collected in April with enlarged ovarian follicles, no eggs in the glandular uterus, and sperm in the posterior non-glandular uterus. **(A)** Sperm associated with ciliated and secretory cells and secretory material in the lumen (scale bar = 40 µm). **(B)** Portions of sperm cells associated with luminal secretory material (scale bar = 10 µm). **(C)** Apical cytoplasm of a secretory cell (scale bar = 3 µm). **(D)** Apical cytoplasm of a ciliated cell (scale bar = 1 µm). Ci, cilia; Cr, crypt; Ru, rugae; Mv, microvilli; Nu, epithelial cell nucleus; Lp, lamina propria; Sn, sperm nuclei; Sp, sperm; Sv, secretory vacuoles. Modified from Sever and Hopkins (2004, Journal of Experimental Zoology, 301A: 599–611, Fig. 3A-D).

The secretion is a mucoid substance that gives positive tests for periodic acid-Schiff's and alcian blue 8GX at pH 2.5, but not bromophenol blue (proteins). The release of the product is apparently a merocrine process, and some membrane fusion between vacuoles is often evident prior to release (Fig. 6.9D). Synthetic organelles (endoplasmic reticulum, Golgi bodies) are not prominent in the cytoplasm. Intercellular canaliculi are very narrow and have tight junctions extending inward from the luminal borders. The ciliated cells are characterized by elongate, densely staining mitochondria in the apical cytoplasm. Both ciliated and secretory cells have irregular, euchromatic nuclei with prominent nucleoli. Occasionally, sperm are found embedded in the cytoplasm of secretory cells and associated with clear vacuoles, or embedded in secretory vacuoles. The embedded portions of sperm cells generally appear normal in cytology.

6.4.4 *Hemidactylus turcicus*

In *Hemidactylus turcicus,* sperm were located in the posterior infundibulum of the oviduct in Ssts. Sperm were embedded deep in the glands and orientated parallel to one another (Fig. 6.10A). No sperm was found in more posterior regions of the oviduct, probably because no specimens were collected immediately after mating. Sperm stained basophilic in hematoxylin, alcian blue positive, and slightly bromophenol blue positive. With transmission electron microscopy, sperm were found in the lumen of Ssts and were not embedded in the epithelium (Fig. 6.10B). Scanning electron microscopy provided an overview of the spacing of numerous gland orifices and the parallel arrangement of sperm (see Eckstut *et al.* 2009b).

Large aggregations of sperm were found deep within the sperm storage tubules between the months of May and August and were found both with the presence of enlarged ovarian follicles and with eggs *in utero,* which were found to occur simultaneously in some specimens. Few sperm were found in September and November, and such sperm were likely residual.

Eckstut *et al.* (2009b) reviewed seasonal changes in histochemistry and ultrastructure of the oviduct in *Hemidactylus turcicus.* During the reproductive season, active oviducts have large amounts of secretory vacuoles and secretory products. The infundibulum has two abundant types of secretory products during the breeding season: glycoproteins, which are characterized by a core of electron-dense proteins in a vacuole containing polysaccharides; and mucoidal substances, which have less electron dense material that is diffused uniformly across the vacuole. Secretions located in the glandular uterus and non-glandular uterus are primarily mucoidal

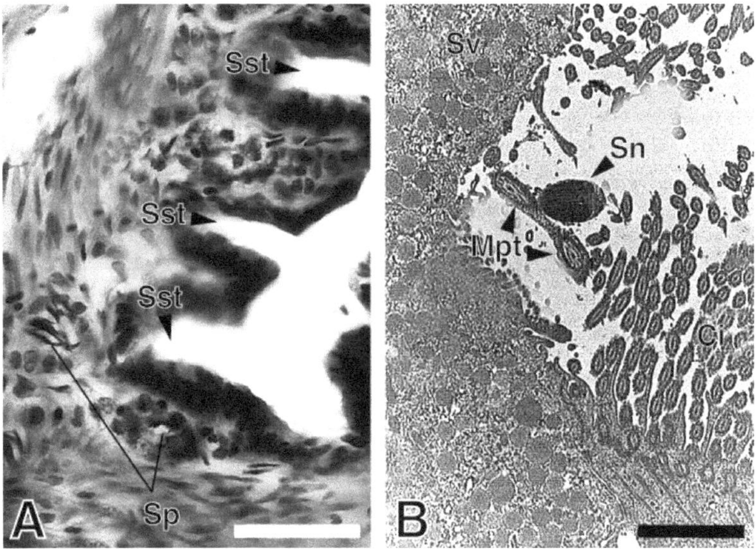

Fig. 6.10 Histology and ultrastructure of stored sperm in the oviduct of *Hemidactylus turcicus*. **(A)** Sperm in oviductal Ssts of a post-ovulatory female from May stained with hematoxylin-eosin for general cytology (scale bar = 50 µm). **(B)** Transmission electron micrograph of sperm in the lumen of a sperm storage tubule from a post-ovulatory female from June (scale bar = 2 µm). Ci, cilia; Mpt, middle piece of tail; Sn, sperm nucleus; Sp, sperm; Ssts, sperm storage tubules; Sv, secretory vacuoles. Modified from Eckstut *et al.* (2009a, Amphibia-Reptilia, 30: 45–56, Fig. 2A,D).

substances. The epithelial lining of the glandular uterus is especially active in the production of secretory material with Golgi complexes and rough endoplasmic reticulum associated with condensing vacuoles.

Throughout the active oviduct, mitochondria are abundant in the epithelium, and nuclei of epithelial cells are irregularly shaped and contain electron-dense nucleoli. Additionally, nuclei of secretory cells are largely heterochromatic, with large amounts of electron-dense material interspersed, whereas nuclei of ciliated cells are more euchromatic and thus possess more uniformity in electron density throughout the nucleus.

Inactive oviducts are characterized by fewer secretory vacuoles throughout the oviduct. Although epithelial nuclei are still irregularly shaped, their nucleoli are less prominent. Furthermore, mitochondria are less prominent and scarce, and intercellular canaliculi are wider. Golgi complexes occur in high abundance in order to produce the secretions needed during the reproductively active season. Some secretory mucoidal substances are found throughout the oviduct but the secretory vacuoles are more electron lucent and substantially smaller and less abundant in reproductively inactive females.

6.4.5 Conclusions for Sperm Storage/Transport

In summary, female (as well as male) sperm storage is an obligatory part of the reproductive cycle of lizards that have been studied for this trait. A considerable amount of variation has been found in the ultrastructure of female sperm storage in the few lizards that have been examined and the data from light microscopy on other lizards hint that the full panoply of variation has yet to unfold. With information available on fewer than one-hundredth of all known lizard species, it is certainly premature to construct any robust hypotheses concerning phylogeny of female sperm storage in this group. As with other symplesiomorphic morphological characters, discerning specializations that are phyletically relevant from those that are homoplastic is a challenge. Phylogeny is likely to be obfuscated by convergence ascribable to similar functional adaptations and perhaps design constraints imposed by the basic structure of the vertebrate oviduct and sperm.

6.5 ACKNOWLEDGMENTS

We thank Matt Ritch and Lauren Law for preparing lizard tissues for histological examination. We also acknowledge our home institutions for continued support of our research.

6.6 LITERATURE CITED

Adams, C. S. and Cooper, W. E. 1988. Oviductal morphology and sperm storage in the keeled earless lizards, *Holbrookia propinqua*. Herpetologica 44: 190–197.

Adams, S. M., Hosie, M. J., Murphy, C. R. and Thompson, M. B. 2004. Changes in oviductal morphology of the skink, *Lampropholis guichenoti*, associated with egg production. Journal of Morphology 262: 536–544.

Adams, S. M., Lui, S., Jones, S. M., Thompson, M. B. and Murphy, C. R. 2007a. Uterine epithelial changes during placentation in the viviparous skink *Eulamprus tympanum*. Journal of Morphology 268: 385–400.

Adams, S. M., Biazik, J., Stewart, R. L., Murphy, C. R. and Thompson, M. B. 2007b. Fundamentals of viviparity: Comparison of seasonal changes in the uterine epithelium of oviparous and viviparous *Lerista bougainvillii* (Squamata: Scincidae). Journal of Morphology 268: 624–635.

Amey, A. P. and Whittier, J. M. 2000. The annual reproductive cycle and sperm storage in the bearded dragon, *Pogona barbata*. Australian Journal of Zoology 48: 411–419.

Auffenberg, W. 1994. The Bengal Monitor. University Press of Florida: Florida.

Bakst, M. R. 1987. Anatomical basis of sperm-storage in the avian oviduct. Scanning Microscopy 1: 1257–1266.

Beuchat, C. A. 1986. Phylogenetic distribution of the urinary bladder in lizards. Copeia 1986: 512–517.

Biazik, J. M., Parker, S. L., Murphy, C. R. and Thompson, M. B. 2012. Uterine epithelial morphology and progesterone receptors in a mifepristone-treated viviparous lizard

Psuedemoia entrecasteauxii (Squamata: Scincidae) during gestation. Journal of Experimental Zoology (Molecular, Development, and Ecology) 318: 148–158.
Blackburn, D. G. 1985. Evolutionary origins of viviparity in the Reptilia. II. Serpentes, Amphisbaenia, and Icthyosauria. Amphibia-Reptilia 6: 259–291.
Blackburn, D. G. 1998. Structure, function, and evolution of the oviducts of squamate reptiles, with special reference to viviparity and placentation. Journal of Experimental Zoology 282: 560–617.
Blackburn, D. G. 1999. Viviparity and oviparity: evolution and reproductive strategies. pp. 994–1003. In T. E. Knobil and J. D. Neill (eds.), *Encyclopaedia of Reproduction*. Academic Press: New York.
Blackburn, D. G., Francisco, S. K. -S. and Callard, I. P. 1998. Histology of abortive egg sites in the uterus of a viviparous, placentotrophic lizard, the skink *Chalcides chalcides*. Journal of Morphology 235: 97–108.
Bon, J. and Saint-Girons, H. 1963. Ecologie et cycle sexuel des amhiseniens du Maroc. Bulletin de la Société des Sciences Naturelles (et Physiques) du Maroc 43: 117–170.
Botte, V. 1973. Morphology and histochemistry of the oviduct in the lizard, *Lacerta sicula*. The annual cycle. Bolletino di Zoologia 40: 305–314.
Bou-Resli, M. N., Bishaw, L. F. and Al-Zaid, N. S. 1981. Observations on the fine structure of the sperm storage crypts in the lizard *Acanthodactylus scutellatus hardyi*. Archives of Biology (Bruxelles) 92: 287–298.
Braun, M. 1878. Das urogenitalsystem der einheimischen reptilien entwicklungsgeschichtliche und anatomisch bearbeitet. Arbeiten aus dens Zoologischen–Zootomischen Institut in Wurzburg 4: 113–230.
Brooks, B. 1906. The anatomy and internal urogenital organs of certain North American lizards. Transactions of the Texas Academy of Science 8: 23–38.
Cole, C. 1966a. Femoral glands in lizards: A review. Herpetologica 22: 199–206.
Cole, C. 1966b. Femoral glands of the lizard, *Crotaphytus collaris*. Journal of Morphology 118: 119–136.
Conner, J. and Crews, D. 1980. Sperm transfer and storage in the lizard, *Anolis carolinensis*. Journal of Morphology 130: 331–348.
Cooper, W. E. and Trauth, S. E. 1992. Discrimination of conspecific male and female cloacal chemical stimuli by males and possession of a probably pheromone gland by females in a cordylid lizard, *Gerrhosaurus nigrolineatus*. Herpetologica 48: 229–236.
Cooper, W. E. and Vitt, L. J. 1984a. Conspecific odor detection by the male broad-headed skink, *Eumeces laticeps*: Effects of sex and site of odor source and of male reproductive condition. Journal of Experimental Zoology 230: 199–209.
Cooper, W. E. and Vitt, L. J. 1984b. Detection on conspecific odors by the female broad-headed skink, *Eumeces laticeps*. Journal of Experimental Zoology 229: 49–54.
Cooper, W. E. and Vitt, L. J. 1986. Interspecific odour discriminations among syntopic congeners in scincid lizards (genus *Eumeces*). Behavior 97: 1–9.
Corso, G., Delitala, G. M. and Carcupino, M. 2000. Uterine morphology during the annual cycle in *Chalcides ocellatus tiligugu* (Gmelin) (Squamata: Scincidae). Journal of Morphology 243: 153–165.
Cueller, O. 1966. Oviductal anatomy and sperm storage structures in lizards. Journal of Morphology 130: 129–136.
Disselhorst, R. 1897. *Die accessorischen Geschlectsdrüüsen der Wirbeltiere. Eine Vergleichend-anatomische Untersuchung*. Thesis, Tübingen.
Disselhorst, R. 1904. Ausführapparat und anhangsdrüsen der männlichen geschlechtsorgane. pp. 1–386. In A. Oppel (ed.), *Lehrbuch der Vergleichenden Mikroskopischen Anatomie der Wirbeltiere, Vol. 12*. Fischer: Jenna.
Eckstut, M. E., Sever, D. M., White, M. E. and Crother, B. I. 2009a. Phylogenetic analysis of sperm storage in female squamates. pp. 185–215. In L. T. Dahnof (ed.), *Animal Reproduction: New Research Developments*. Nova Science Publishers, Inc.: Hauppauge.

Eckstut, M. E., Lemons, E. and Sever, D. M. 2009b. Annual dynamics of sperm production and storage in the Mediterranean gecko, *Hemidactylus turcicus*, in the southeastern United States. Amphibia-Reptilia 30: 45–56.
Eckstut, M. E., Hamilton, A. M., Austin, C. C. and Sever, D. M. 2009c. Asynchronous oviductal seasonal variation in the unisexual-bisexual *Nactus multicarinatus* complex from Vanuatu Archipelago (Reptilia: Squamata: Gekkonidae). pp. 295–307. In L. T. Dahnof (ed.), *Animal Reproduction: New Research Developments*. Nova Science Publishers, Inc.: Hauppauge.
Fox, W. 1956. Seminal receptacles of snakes. Anatomical Record 124: 519–539.
Fox, W. 1963. Special tubules for sperm storage in female lizards. Nature 198: 500–501.
Fox, H. 1977. The urogenital system of reptiles. pp. 1–157. In C. Gans (ed.), *Biology of the Reptilia, Vol. 6*. Academic Press: New York.
Gabe, M. and Saint-Girons, H. 1964. *Histologie de Sphenodon punctatus*. Centre National de la Recherche Scientifique, Paris, pp. 148.
Gabe, M. and Saint-Girons, H. 1965. Contribution à la morphologie comparée du cloaque et des glandes épidermoïdes de la région cloacale chez les lépidosauriens. Mémoires du Muséum National d'Histoire Naturelle XXXIII: 149–332.
Gadow, H. 1887. Remarks on the cloaca and on the copulatory organs of the Amniota. Philosophical Transactions of the Royal Society of London B 178: 5–37.
Gharzi, A., Yari, A. and Rastergar pouyani, N. 2013. Cloacal anatomy and histology of female Bosc's fringe-toed lizard, *Acanthdactylus boskianus*. Asian Journal of Experimetnal Biology 4: 297–301.
Giacomini, E. 1893. Sull'ovidutto del Sauropsidi. Monitore Zoologico Italiano 4: 202–265.
Giersberg, H. 1922. Untersuchungen über physiologie und histologie des eileiters der reptilien und vögel; nebst einem beitrag zue fasergenese. Zeitchrift für Wissenshchaftliche Zoologie 70: 1–97.
Girling, J. E., Cree, A. and Guillette, L. J. 1997. Oviductal structure in a viviparous New Zealand gecko, *Hoplodactylus maculates*. Journal of Morphology 234: 51–68.
Girling, J. E., Crree, A. and Guillette, L. J. 1998. Oviducts structure in four species of gekkonid lizard differing in parity mode and eggshell structure. Reproducxtion, Fertility, and Development 10: 139–154.
Girling, J. E. 2002. The reptilian oviduct: a review of structure and function and direction for future research. Journal of Experimental Zoology 293: 141–170.
Goin, C. J. and Goin, O. B. 1962. Introduction to Herpetology. W. H. Freeman and Co., San Fransisco.
Greenberg, B. 1943. Social behavoir of the western banded gecko, *Coleonyx variegatus* Baird. Physiological Zoology 16: 110–122.
Guillette, L. J. 1992. Morphology of the reproductive tract in a lizard exhibiting incipient viviparity (*Sphenomorphus fragilis*) and its implications for the evolution of the reptilian placenta. Journal of Morphology 212: 163–173.
Guillette, L. J. and Jones, R. E. 1985. Ovarian, oviductal, and placental morphology of the reproductively bimodal lizard, *Sceloporus aeneus*. Journal of Morphology 184: 85–98.
Guillette, L. J., Fox, S. L. and Palmer, B. D. 1989. Oviductal morphology and egg shelling in the oviparous lizards *Crotaphytus collaris* and *Eumeces obsoletus*. Journal of Morphology 201: 145–159.
Hardy, L. M. and Cole, C. J. 1981. Parthenogentic reproduction in lizards: Histological evidence. Journal of Morphology 170: 215–237.
Herlant, M. 1933. Recherches histologiques et expérimentales sur les variations cycliques du testicule et des caracteres sexuels secondaires chez les reptiles. Archives de Biologie 44: 347–468.
Heulin, B., Stewart, J. R., Surget-Groba, Y., Bellaud, P., Jouan, F., Lancien, G. and Deunft, J. 2005. Development of the uterine shell glands during preovulator and early gestation periods in oviparous and viviparous *Lacerta vivipara*. Journal of Morphology 266: 80–93.

Hoffman, L. H. and Wimsatt, W. A. 1972. Histochemical and electron microscopic observations on the sperm receptacles in the garter snake oviduct. American Journal of Anatomy 134: 71–96.
Hosie, M. J., Adams, S. M., Thompson, M. B. and Murphy, C. R. 2003. Viviparous lizard, *Eulamprus tympanum*, shows changes in the uterine surface epithelium during early pregnancy that are similar to plasma membrane transformation of mammals. Journal of Morphology 258: 346–357.
Kumari, T. R. S., Sarkar, H. B. D. and Shivanandappa, T. 1990. Histology and histochemistry of the oviductal sperm storage pockets of the agamid lizard *Calotes versicolor*. Journal of Morphology 203: 97–106.
Kluge, A. 1967. Higher taxonomic categories of gekkonid lizards and their evolution. Bulletin of the American Museum of Natural History 135: 1–60.
Krause, R. 1922. *Mikroskopische Anatomie der Wirbeltiere in Einzeldarstellungen*. Walter de Gruyter and Co.: Berlin.
Leydig, F. 1872. *Die in Deutschland Lebenden Arten der Saurier*. Tübingen, pp. 312.
Lee, M. S. Y. 2005. Squamate phylogeny, taxon sampling, and data congruence. Organisms, Diversity, & Evolution 5: 25–45.
Lereboullet, A. 1851. Rechereches aur l'anatomie des organs génitaux des animaux vertébrés. L'Académie Impériale de Curieux de la Nature 23: 1–228.
Masson, G. R. and Guillette, L. J. 1987. Changes in oviducal vascularity during the reproductive cycle of three oviparous lizards (*Eumeces obsoletus*, *Sceloporus undulatus* and *Crotaphytus collaris*). Journal of Reproduction and Fertility 80: 361–371.
Matthews, L. H. and Marshall, F. H. A. 1956. Cyclical changes in the reproductive organs of lower vertebrates. Reptiles. pp. 189–204. In F. H. A. Marshall (ed.), *Physiology of Reproduction*. Green and Co.: London.
Matthey, R. 1929. Caractères sexuels secondaires du lezard male (*Lacerta agilis*). Bulletin de la Societe Vaudoise des Sciences Naturelles 57: 71–81.
Mulaik, G. 1946. A comparative study of the urogenital systems of an oviparous and two ovoviviparous species of the lizard genus *Sceloporus*. Bulletin of the University of Utah Biology Series 37: 3–24.
Murphy-Walker, S. and Haley, S. R. 1996. Functional sperm storage duration in female *Hemidactylus frenatus* (family Gekkonidae). Herpetologica 52: 365–373.
Noble, G. K. and Bradley, H. T. 1933. The mating behavior of lizards, its bearing on the theory of sexual selection. Annals of the New York Academy of Sciences 35: 25–100.
Nogueira, K. O. P. C., Rodrigues, S. S., Araújo, A. and Neves, C. A. 2011. Oviductal structure and ultrastructure of the oviparous gecko, *Hemidactylus mabouia* (Moreau De Jonnès 1818). The Anatomical Record 294: 883–892.
Ortiz, E. and Morales, M. 1974. Development and function of the female reproductive tract of the tropical lizards, *Anolis pulchellus*. Physiological Zoology 41: 207–217.
Palmer, B. D., DeMarco, V. G. and Guillette, L. J. 1993. Oviductal morphology and eggshell formation in the lizard, *Sceloporus woodi*. Journal of Morphology 217: 205–217.
Perkins, J. M. and Palmer, B. D. 1996. Histology and functional morphology of the oviduct of an oviparous snake, *Diadophis punctatus*. Journal of Morphology 227: 67–79.
Ramírez-Pinilla, M. P., Parker, S. L., Murphy, C. R. and Thompson, M. B. 2012. Uterine and chorioallantoic angiogenesis and changes in the uterine epithelium during gestation in the viviparous lizard, *Niveoscincus conventryi* (Squamata: Scincidae). Journal of Morphology 273: 8–23.
Rathke, H. 1839. *Enwickelungsgeschichte der Natter* (Coluber natrix). Konigsberg.
Regamey, J. 1933. Les differences sexuelles du cloaque chez le lezard *Lacerta agilis* Linne. Buletin de la Societe Vaudoise des Sciences Naturelles 58: 185–186.
Regamey, J. 1935. Les caracteres sexuels du Lezard (*Lacerta agilis* L.). Revue Suisse de Zoologie 42: 87–168.

Raynaud, A. and Pieau, C. 1985. Embryonic development of the genital system. pp. 149–300. In C. Gans and F. Billett (eds.), *Biology of the Reptilia. Vol. 15B*. John Wiley & Sons, Inc: New York.
Saint-Girons, H. 1962. Presence de receptacles seminaux chez les cameleons. Beaufortia 9: 165–172.
Saint-Girons, H. and Ineich, I. 1992. Histology of the reproductive tract of hybrids between gonochoristic males and parthenogenetic females of *Lepidodactylus lugubris* in French Polysesia (Reptilia, Gekkonidae). Journal of Morphology 212: 55–64.
Sánchez-Martínez, P. M., Ramírez-Pimilla, M. P. and Miranda-Esquivel, D. R. 2007. Comparative histology of the vaginal-cloacal region in Squamata and its phylogenetic implications. Acta Zoologica 88: 289–307.
Sarker, H. B. D. and Shivanandappa, T. 1989. Reproductive cycle of reptiles. pp. 225–272. In S. K. Saidapur (ed.), *Reproductive Cycles of Indian Vertebrates*. Allied Publishers: Bombay.
Schaefer, G. C. and Roeding, C. E. 1973. Evidence for vaginal sperm storage in the mole skink, *Eumeces egregious*. Copeia 1973: 346–347.
Seshadri, C. 1959. Functional morphology of the cloaca of *Varanus monitor* (Linnaeus) in relation to water economy. Proceedings of the National Institute of Science B 25: 101–106.
Sever, D. M. and Hamlett, W. C. 2002. Female sperm storage in reptiles. Journal of Experimental Zoology 292: 187–199.
Sever, D. M. and Hopkins, W. A. 2004. Oviductal sperm storage in the ground skink *Scincella laterale* Holbrook (Reptilia: Scincidae). Journal of Experimental Zoology 301A: 599–611.
Sever, D. M. and Ryan, T. J. 1999. Ultrastructure of the reproductive system of the black swamp snake (*Seminatrix pygaea*). I. Evidence for oviducal sperm storage. Journal of Morphology 241: 1–18.
Shanthakumari, T. R., Sarkar, H. B. D. and Shivanadappa, T. 1990. Histology and histochemistry of the oviductal sperm storage pockets of the agamid lizard, *Calotes versicolor*. Journal of Morphology 203: 97–106.
Shanthakumari, T. R., Sarkar, H. B. D. and Shivanandappa, T. 1992. Histological, histochemical and biochemical changes in the annual oviduct cycle of the agamid, *Calotes versicolor*. Jounral of Morphology 211: 295–306.
Siegel, D. S. and Sever, D. M. 2008. Sperm aggregations in female *Agkistrodon piscivorus* (Reptilia: Squamata): A histological and ultrastructural investigation. Journal of Morphology 269: 189–206.
Siegel, D. S., Miralles, A., Chabarria, R. E. and Aldridge, R. D. 2011a. Female reproductive anatomy: Cloaca, oviducts, and sperm storage. pp. 347–409. In R. D. Aldridge and D. M. Sever (eds.), *Reproductive Biology and Phylogeny of Snakes*. CRC Press: Florida.
Siegel, D. S., Trauth, S. E., Sever, D. M. and Aldridge, R. D. 2011b. The phylogenetic distribution of the ampulla ureter and ampulla urogenital/urinigerous papilla in the Serpentes. Journal of Zoological Systematics and Evolutionary Research 49: 160–168.
Siegel, D. S., Miralles, A., Trauth, S. E. and Aldridge, R. D. 2012. The phylogenetic distribution and morphological variation of the "pouch" in female snakes. Acta Zoologica 93: 400–408.
Siegel, D. S., Miralles, A., Rheubert, J. L. and Aldridge, R. D. 2013. *Heloderma suspectum* (Gila Monster). Female reproductive anatomy. Herpetological Review 44: 142–143.
Smyth, M. and Smith, M. J. 1968. Obligatory sperm storage in the skink *Hemiergis peronii*. Science 161: 575–576.
Srinivas, S. R., Shinvanandappa, T., Hedge, S. N. and Sarkar, H. B. D. 1995. Sperm storage in the oviduct of the tropical rock lizard, *Psammophilus dorsalis*. Journal of Morphology 224: 293–301.
Stewart, J. R., Mathieson, A. N., Ecay, T. W., Herbert, J. F., Parker, S. L. and Thompson, M. B. 2010. Uterine eggshell structure and histochemistry in a lizard with prolonged uterine egg retention (Lacertilia, Scincidae, *Saiphos*). Journal of Morphology 271: 1342–1351.
Trauth, S. E., Cooper, W. E., Vitt, L. J. and Perrill, S. A. 1987. Cloacal anatomy of the broad-headed skink, *Eumeces laticeps*, with a description of a female pheromonal gland. Herpetologica 43: 458–466.

Uller, T., Schwartz, T., Koglin, T. and Olsson, M. 2013. Sperm storage and sperm competition across ovarian cycles in the dragon lizard, *Ctenophorus fordi*. Journal of Experimental Zoology A: Ecology, Genetics, and Physiology 319: 404–408.

Unterhössel, P. 1902. Die eideschsen und schlangen. pp. 541–581. In A. Fleischmann (ed.), *Morhpologische Studien über Koake und Phallus der Amnioten*. Morphologisches Jahrbuch 30: 539–675.

Uribe, M. C. A., Velasko, S. R., Guillette, L. J. and Estrada, E. F. 1988. Oviduct histology of the lizard, *Ctenosaura pectinata*. Copeia 1988: 1035–1042.

Vidal, N. and Hedges, S. B. 2005. The phylogeny of squamate reptiles (lizards, snakes, and amphisbaenians) inferred from nine nuclear protein-coding genes. C. R. Biologies 328: 1000–1008.

Whiting, A. M. 1969. *Squamate Cloacal Glands: Morphology, Histology and Histochemistry*. Dissertation from The Pennsylvania State University. Zoology.

Young, B. A., Marsit, C. and Meltzer, K. 1999. Comparative morphology of the cloacal scent gland in snakes (Serpentes: Reptilia). The Anatomical Record 256: 127–138.

CHAPTER 7

Discovery of Parthenogenesis in Lizards

William Neaves

7.1 INTRODUCTION

The propagation of life on earth fascinated people long before they understood the processes of reproduction. Personal experience along with observation of domestic species indicated that gender played a critical role. In human beings and other mammals, females gestated developing fetuses and eventually gave birth to immature offspring, but this did not happen in the absence of intact males. Similarly in domestic fowl, reliable production of eggs capable of embryonic development into viable hatchlings required a fertile male. More than two millennia ago, Aristotle emphasized the essential role of both sexes, male and female, in the reproduction of higher animals. He wrote:

> "Speaking generally, however, we may say that in the case of all those animals which have the power of locomotion, whether they are adapted to be swimmers, or fliers, or walkers, male and female are found; and this applies not only to the blooded animals but to some of the bloodless ones as well."
>
> (Peck 1942, pp. 5–7)

Only in recent centuries has microscopy revealed the contribution of males to be a myriad of individual motile cells, known as spermatozoa. In mammals, other vertebrates, and even non-vertebrate animals, the female

Stowers Institute for Medical Research, 1000 East 50th Street, Kansas City, Missouri 64110-2262, USA.

counterpart—the ovum—fuses with a spermatozoon to initiate embryonic development. As recently as the early decades of the twentieth century, this type of propagation—sexual reproduction—was considered universal among vertebrates. Up until the middle of the 20th century, the only proven exception in nature was a clonally reproducing species of fish, and even in that situation, eggs of the single-sex species required activation by spermatozoa from a sympatric bisexual species (Hubbs and Hubbs 1932).

Although the discovery of an all-female fish species did not occur until the 20th century, Aristotle acknowledged the possibility long ago:

> "If there is any class of animal which is female and has no separate male, it is possible that this generates offspring from itself. This has not so far been reliably observed, it is true, but some instances in the class of fishes give cause to suspect that it may be the case. Thus, of the fish known as *erythrinus* not a single male specimen has so far been observed, whereas female ones have been, full of fetations (roe)."
>
> (Peck 1942, p. 205)

Aristotle referred to the Common Pandora, named *Pagellus erythrinus* by Linnaeus in 1758, and females apparently do predominate. Among 1,850 adults recently trawled from the Aegean Sea, the female-to-male sex ratio was twelve-to-one (Gülner *et al.* 2011). Although females greatly outnumber males, the pandora is not an all-female species. Until the 20th century, no unisexual vertebrate species was known to exist. If zoologists had found such a species, and if there were truly no role for males in its propagation, the mode of reproduction would have been termed "parthenogenesis"—Greek for "virgin birth".

Mendelian genetics and Darwinian evolution have conspired to make sexual reproduction seem virtually mandatory in most animals. Crossing-over within chromosomes during meiotic prophase and independent assortment of chromosomes into haploid gametes during meiotic division enable fertilization to produce diploid zygotes with recombined alleles that fuel natural selection and ensure survival of the fittest. While some plants and a few non-vertebrate animals can persist while foregoing sex, many biologists have assumed that "highly evolved" species owe their status to sexual reproduction.

7.2 THE DISCOVERY

In this intellectual context, a report appeared in the Bulletin of the Society of Zoology of France during 1936 that described the vain search by L. A. Lantz for males of a Caucasian rock lizard, *Lacerta saxicola armeniaca* (now known as *Darevskia armeniaca*; Arribas 1999, p. 18) in the vicinity of Lake Sevan (called Lake Gok-Tsha in the publication) between the Black Sea and the Caspian Sea in the Caucasus Mountains of Armenia (Lantz and Cyren 1936). Lantz and his co-author, O. Cyren, wrote:

> "*L. s. armeniaca* is a well-characterized subspecies with little variation. What is strange is that we do not know the male. One of us (L.A.L.) searched in vain in the region near Lake Gok-Tsha where *armeniaca* abounds. In addition to forty specimens which were captured, at least an equal number was observed without finding among them a male, although the young were many. A review of the material in the British Museum has revealed that of four males identified by Boulenger, two were females, and the other two belonged to a subspecies other than *armeniaca*. It is to be observed that among *L. s. bithynica*, the male also seems to be very rare, because there are only two among approximately eighteen known specimens. We are confronted with a very strange biological anomaly which is even more remarkable when one considers that among lizards, the males are usually more numerous than females."
>
> (Lantz and Cyren 1936, p. 167)

Sergius Chernov, Curator of the Department of Herpetology at the Zoological Institute of the USSR Academy of Sciences in Leningrad conducted field studies of the *Lacerta saxicola* complex in Armenia during 1937–1939 and was aware of the 1936 publication of Lantz and Cyren (Chernov 1939, p. 113). In Chernov's synopsis of the herpetofauna of the Caucasus published in 1939, he examined 180 *L. s. armeniaca* and found only seven males (Chernov 1939, p. 115). World War Two intervened before Chernov could encourage his young Ph.D. student, Ilya Darevsky (also spelled Darevskii), to undertake field work with the *L. saxicola* complex in Armenia and address the "very strange biological anomaly" two decades after publication of its discovery.

Darevsky completed his undergraduate degree in biology at Moscow State University in 1953 and travelled to Armenia to begin nine years of research in the Caucasus Mountains around Lake Sevan and in the Zoological Institute of the Armenian Soviet Socialist Republic Academy of Science in Yerevan (Editorial Board 2009). Based on studying *Lacerta saxicola* subspecies in their natural habitats during 1954 and 1955, Darevsky concluded that some were in fact all-female taxa. His first paper published on *Lacerta* in the Caucasus designated, in addition to *L. s. armeniaca*, two newly described subspecies, *L. s. dahli* and *L. s. rostombekovi*, as all-female (Darevskii 1957).

Continuing his studies in 1956 and 1957, Darevsky captured many hatchlings as well as adults of *L. s. armeniaca*, *L. s. dahlia*, and *L. s. rostombekowi* without finding any males (Darevskii 1958). In contrast, he found slightly more males than females among both adults and hatchlings of the other three local subspecies, *L. s. defilippii*, *L. s. terentjevi*, and *L. s. portschinskii*. During the breeding season, microscopic examination revealed sperm in the cloaca and reproductive ducts of females of the latter three subspecies, but none were detected in the former (Darevskii 1958, p. 877 and Table 3). Darevsky believed the presence of only females among hatchlings of these three subspecies strongly indicated that the bias against males reflected a developmental phenomenon rather than difficulty in detecting them due to behavioral differences or habitat preferences. He tested this possibility directly.

Darevsky focused on the Semenovsky Pass in the Caucasus Mountains of Northern Armenia where none of the three bisexual *saxicola* subspecies occurred. In September 1956, he captured 30 *L. s. armeniaca* (all females) at this location. Half were sexually immature. He placed them in a cage filled with moss and lowered it into a deep, damp fissure in the rock to approximate a hibernaculum used by such lizards during winter. Darevsky retrieved the lizards in April 1957 and placed them in a secure, open-air enclosure. In early summer, 16 surviving lizards began laying eggs, and by the end of August, 56 hatchlings had emerged. All were female. Included among the hatchlings were the offspring of 6 individuals that were sexually immature when captured the previous September. Darevsky concluded, "Thus, the existence of natural parthenogenesis in *L. s. armeniaca* was proven" (Darevskii 1958, p. 879).

On May 5, 1958, Darevsky submitted his paper reporting proof of natural parthenogenesis in the *Lacerta saxicola armeniaca* captured during September 1956. Darevsky must have contemplated this reproductive possibility well in advance of planning the hibernaculum experiment, and he most likely discussed it with his mentor, Sergius Chernov. However, no record has emerged regarding the inception and development of the hypothesis that Darevsky so persuasively validated in his 1958 paper.

Neither Lantz and Cyren (1936) nor Chernov (1939) mentioned parthenogenesis as a possible explanation for all-female subspecies of *L. saxicola*. Regardless of how the concept evolved in the context of *Lacerta saxicola*, Darevsky's 1958 paper left no doubt about the reality of parthenogenic reproduction in lizards, and as awareness of his research spread, it catalyzed decades of research during which many dozens of saurian taxa around the globe were found to be all-female.

Independently and coincidentally, an American herpetologist doing field work in the Big Bend of Texas during 1955 began thinking about parthenogenesis in lizards. In his autobiography published 45 years later, Sherman Minton wrote:

"Big Bend also has a rich lizard fauna. Hardest to identify were the whiptail lizards of the genus *Cnemidophorus*. I counted five species in the region, and for one of them, the checkered whiptail, I had only females. Remembering Jefferson's salamanders in Indiana, I thought there might be a similar situation with these desert lizards. When I suggested this in a manuscript I submitted on Big Bend reptiles, the editors quite reasonably crossed it out, for I didn't have many specimens, and the odd sex distribution could have been pure chance. However, time and a lot of work by other people proved I was essentially correct."

(Minton 2001, p. 58)

The Checkered Whiptail mentioned by Minton was *Cnemidophorus tesselatus* [the genus *Cnemidophorus* is now known as *Aspidoscelis* (Reeder *et al.* 2002)], and his finding only females of this species reminded him of an all-female form of Jefferson's salamander in the vicinity of Ann Arbor, Michigan (Clanton 1934). Minton described two common types of mole salamanders collected in Indiana (Minton 1954). He considered the larger, lighter form to be *Ambystoma jeffersonianum* and the smaller, darker form to be *A. laterale*; males and females occurred in both forms. He also identified apparent hybrids between the two distinctive species, and he believed these *jeffersonianum-laterale* hybrids could be the Indiana counterpart of the all-female form of *jeffersonianum* so well documented in Michigan by Wesley Clanton 30 years earlier.

Minton suspected unusual reproduction in the all-female hybrids of the *A. jeffersonianum* complex, and ten years later, Herbert Macgregor and Thomas Uzzell showed that they reproduced by a process identified as sperm-dependent gynogenesis (Macgregor and Uzzell 1964). Although not true parthenogenesis, the biological consequences are similar: all-female taxa propagate clonally without retaining male DNA in the germline. Minton's knowledge of all-female *Ambystoma* stimulated his interpretation of what he observed in Big Bend. He became the first in the Western Hemisphere to note the absence of males in a lizard species, and he proposed a hybrid origin of *Cnemidophorus tesselatus* in a manuscript on Big Bend reptiles submitted to Herpetologica early in 1956, but the editors elected not to publish it (Minton 2001, pp. 58–59). On June 20, 1956, Hobart M. Smith sent a sympathetic letter regarding the manuscript, but he expressed reservations about Minton's hypothesis. Smith wrote:

"The idea of *tessellatus* being a hybrid is certainly novel to me. It would explain certain facts, yet I am not by any means willing to regard the possibility as being too strong. The best information on habits I've found is in Strecker's paper on the Panhandle published around 1905. He records seeing females lay eggs, but he didn't

> describe any mating or courtship activity, and unfortunately failed to record the sexes of the specimens whose measurements he published. It is an odd situation and certainly worth further investigation and thought."
>
> (Minton 2001, p. 59)

When Minton eventually published his paper on Big Bend reptiles late in 1958 (it was actually printed early in 1959), his experience with the Herpetologica submission and his high regard for the comments by Hobart Smith restrained his speculation about *C. tesselatus*. He only wrote, "All specimens are females. I have never seen a male of this species from any part of the range." (Minton 1958, p. 44).

Nevertheless, Minton's observation stimulated a young professor at Texas Tech University, Donald Tinkle, to examine the specimens deposited in the school's collections and to encourage his student, James Peebles, to collect more *C. tesselatus* from the Texas Panhandle. Late in 1959, Tinkle published a brief paper documenting the results of gender determination in this species and noted, "The most remarkable aspect of the biology of these lizards is the apparent absence of males in the populations" (Tinkle 1959, p. 196). Tinkle added a few thoughts, but he was as circumspect as Minton had been. Tinkle wrote:

> "No explanation is advanced for the absence of males because it is felt that continued collecting of large series from all seasons will reveal the presence of males. In a series of 65 specimens, however, one should expect to find some males, particularly because these specimens are from several different months, so that sampling errors would not seem to account for the absence of males."
>
> (Tinkle 1959, p. 197)

Meanwhile, thanks to the translation of Darevsky's work with *Lacerta saxicola armeniaca* into English, the idea that some *Cnemidophorus* species may completely lack males and may reproduce by true parthenogenesis suddenly became not just acceptable, but extremely seductive. A young herpetologist, Richard Zweifel, Assistant Curator of Amphibians and Reptiles at the American Museum of Natural History, spread the news of Darevsky's discovery and his demonstration of parthenogenetic reproduction in all-female *Lacerta* (Darevskii 1958) to colleagues across the U.S. (Zweifel 2013).

During 1959, Zweifel wrote letters to collaborators, co-authors, and friends in the field, including Paul Maslin in Colorado and Charles Lowe in Arizona, who, together with their students, would play leading roles in documenting and explaining parthenogenesis in *Cnemidophorus*. By then, Minton had moved to Karachi to teach Pakistani students in a joint program with Indiana University School of Medicine. Zweifel sent a letter to Minton in Pakistan dated August 20, 1959 that included:

> "Along the lines of your inability to find any males among your specimens of *Cnemidophorus tessellatus*, some Russians have just reported a subspecies of *Lacerta* that consists wholly of females reproducing parthenogenetically. If we can accept the work and translation, it looks as if they have eliminated all reasonable alternatives to parthenogenesis. I suspect this will trigger a burst of investigation of *Cnemidophorus* for the same phenomenon. Curiously, the report is that the parthenogenetic *Lacerta* hybridizes with populations with normal reproduction."
>
> (Minton 2001, p. 80)

Zweifel's prediction of a "burst of investigation" of parthenogenesis in *Cnemidophorus* proved to be a self-fulfilling prophesy, aided substantially by his alerting herpetologists in the U.S. to Darevsky's work. Among those who responded promptly was Paul Maslin, Curator of Herpetology at the University of Colorado Museum. On October 2, 1961, Maslin submitted a paper published in Science on January 19, 1962 reporting evidence indicating that six species of *Cnemidophorus* were "all-female or virtually all-female" (Maslin 1962). The six taxa listed by Maslin as all-female were: *C. deppei cozumelus* [reclassified later as *C. cozumelus* (McCoy and Maslin 1962)]; *C. costatus exsanguis* [reclassified later as C. exsanguis (Duellman and Zweifel 1962, pp. 184–186)]; "western populations" of a subspecies of *C. inornatus* [described as a new species, *C. uniparens* (Wright and Lowe 1965, pp. 167–168)]; *C. perplexus* [commonly known as *C. neomexicanus* after its karyotype was published (Lowe and Wright 1966)]; *C. velox*; and *C. tesselatus*.

Maslin examined 223 museum specimens of *C. tesselatus* and found only one male (Maslin 1962, p. 212). In late summer of 1959, he collected 60 hatchling *tesselatus* in Pueblo County, Colorado, and all were females (Maslin 1962, p. 213). Although Darevsky's hibernaculum experiment showed true parthenogenetic reproduction in *Lacerta saxicola armeniaca*, Maslin remained cautious in the absence of such direct evidence in *Cnemidophorus* and acknowledged that "parthenogenesis in vertebrates is extremely rare" (Maslin 1962, p. 213). He wrote:

> "Two possible explanations of this phenomenon can be made. First, it may be assumed that males do exist, that collecting was biased, and that the apparent lack of males has an ecological explanation. Or, second, it may be assumed that, during the breeding season at least, males do not exist, that females reproduce by some parthenogenetic mechanism, and the explanation of the phenomenon is genetic."
>
> (Maslin 1962, p. 212)

At the time Maslin wrote his paper late in 1961, only Darvesky had proved true parthenogenetic reproduction in a vertebrate, and gynogenesis would not be documented in the *Ambystoma jeffersonianum* complex for another two years (Macgregor and Uzzell 1964). So when Maslin's paper appeared in January 1962, a clonal mode of reproduction in an all-female vertebrate species other than *Lacerta saxicola armeniaca* was firmly known to occur only in an Amazon molly fish, *Mollienesia formosa* (Hubbs and Hubbs 1932), but it was suspected in other poeciliid fish species lacking males (Miller and Schultz 1959).

In February 1962, William Duellman and Richard Zweifel published a synopsis of the *Cnemidophorus sexlineatus* group in which they recognized 17 species and 31 forms—the largest species group in the genus (Duellman and Zweifel 1962). They acknowledged Darevsky's proof of parthenogenesis in *Lacerta* and work by Minton, Tinkle, and Maslin supporting the all-female status of *C. tesselatus* (Duellman and Zweifel 1962, pp. 168–171). However, they had not seen Maslin's Science paper when their synopsis went to press, and even if they had, they might still have concluded:

> "Because of the uncertainty of our meager knowledge regarding the possibility of parthenogenesis of some populations, we have not stressed this point in determining species relationships. If it is established that parthenogenesis truly occurs in *Cnemidophorus*, this genus notorious for its taxonomic difficulties will be subjected to additional complications."
>
> (Duellman and Zweifel 1962, p. 171)

Maslin's paper (Maslin 1962) notwithstanding, publications reflecting the "burst of investigation" of all-female *Cnemidophorus* predicted by Zweifel in 1959 (Minton 2001, p. 80) paused before picking up speed. Nothing was published about all-female species of *Cnemidophorus* in 1963. The following year, Charles Lowe and his graduate student, John Wright, described two new all-female species, *C. flagellicaudus* and *C. sonorae* (Lowe and Wright 1964).

It remained relatively quiet in the Western Hemisphere until July 1965 when Maslin's graduate student, Lewis Pennock, published results of his study of mitotic figures from bone marrow of four all-female species and one bisexual species of *Cnemidophorus* (Pennock 1965). He found triploidy in three of the all-female taxa—*C. exsanguis*, *C. velox*, and *C. tesselatus*—and diploidy both in the bisexual species *C. tigris* and in an all-female species from New Mexico designated as *C. perplexus* but commonly known later as *C. neomexicanus* (Lowe and Wright 1966). Pennock titled his paper "Triploidy in parthenogenetic species of the Teiid lizard, genus *Cnemidophorus*." He explained:

> "There are no reports of the hatchling *Cnemidophorus* being produced by an individual hatched and reared in complete isolation. While such conclusive evidence is lacking, the term parthenogenesis is here used without qualification because all available evidence supports the viewpoint that the all-female populations of *Cnemidophorus* are parthenogenetic rather than gynogenetic as in the case of triploid *Ambystoma*."
>
> (Pennock 1965, p. 540)

The diploid species that Pennock identified as *C. perplexus* has a convoluted systematic history (Wright and Lowe 1967a). It was one of the all-female species reported by Maslin (1962), and although its name has been persistently controversial, Maslin's identification of it as a unisexual species has never been challenged. Lowe and Wright (1966) published this lizard's karyotype and established the practice of referring to this taxon as *C. neomexicanus*, a species described early in the previous decade (Lowe and Zweifel 1952).

A female was selected as the type specimen for *C. neomexicanus* and deposited in the Museum of Vertebrate Zoology at Berkeley (Lowe and Zweifel 1952, p. 230), but data from one animal listed as a male appeared in Table 1 (Lowe and Zweifel 1952, p. 235). In retrospect, it seems likely that this male was a hybrid between C. neomexicanus and a male *C. inornatus*, which was sympatric at the collection site in Socorro County, New Mexico (Lowe and Zweifel 1952, p. 231). Two hybrid males between *C. inornatus* and *C. perplexus* = *neomexicanus* were subsequently found in New Mexico (Taylor and Medica 1966). The following year, Wright and Lowe reported collecting three male and three female hybrids between *C. inornatus* and *C. neomexicanus* at various sites in New Mexico (Wright and Lowe 1967a). The hybrids resembled the maternal species closely enough to explain the male included in Table 1 of Lowe and Zweifel (1952).

Data from 20 individual *C. neomexicanus* contributed to Table 3 (Lowe and Zweifel 1952, p. 243), but gender was not indicated, and nothing in that paper suggested that it might be an all-female species (Lowe and Zweifel 1952). Fifty-one years later, Richard Zweifel mused, "I wonder what might have followed if Chuck Lowe and I had slit the bellies of our specimens of *neomexicanus* rather than just injecting preservative and found only eggs and no testes" (Zweifel 2013).

Pennock reported that three all-female species were triploid—*C. velox*, *C. exsanguis*, and *C. tesselatus* (Pennock 1965). Each species merits further comment in the context of his paper. In 1955, Lowe resurrected an old name, *C. velox*, for a new species found on the Colorado Plateau of northeastern Arizona, southeastern Utah, southwestern Colorado, and northwestern New Mexico (Lowe 1955). At least 81 individual specimens contributed morphometric data to the paper, but gender was not mentioned, even when describing the type specimen deposited in the Museum of Comparative Zoology at Harvard (Lowe 1955, p. 4). In their synopsis of the *C. sexlineatus* group published in 1962, Duellman and Zweifel concluded, "*Cnemidophorus velox* is one of the species in which males appear to be lacking" (Duellman and Zweifel 1962, p. 206). This statement can be attributed to the authors' awareness of Maslin's work, even though they were not yet able to cite his paper (Maslin 1962). They wrote: "T. Paul Maslin, University of Colorado, is engaged in a study of this problem with reference to *C. tessellatus* and other forms suspected of lacking males" (Duellman and Zweifel 1962, p. 171).

Maslin also established *C. exsanguis* as an all-female taxon, although he referred to it as a subspecies of *C. costatus* (Maslin 1962). Lowe described *exsanguis* as a new subspecies of *C. sacki* during the previous decade (Lowe 1956). The holotype, deposited in the Department of Zoology at the University of California, Los Angeles (Lowe 1956, p. 138), was a female, but even though 24 individuals contributed data to Table 1 and "large numbers" were collected, the ratio of females to males was not mentioned, and there was no suggestion that *exsanguis* might be all-female (Lowe 1956). In 1962, *exsanguis* was elevated to full species status (Duellman and Zweifel 1962, pp. 184–186), and the authors added, "...all adult specimens examined by us are females. We suspect that *exsanguis* is one of those species that lack males" (Duellman and Zweifel 1962, p. 186).

With regard to *C. tesselatus*, Maslin firmly established it as an all-female species (Maslin 1962), but he and others around that time spelled it "tessellatus". Zweifel (1965, p. 38) reviewed the issue and concluded, "There seems to be no justification for emending the original spelling." By the time Pennock's paper appeared in 1965, the spelling had reverted to Say's original (James 1823, pp. 50–51) and so remained.

After Pennock (1965) reported triploidy in some all-female species of *Cnemidophorus* and explained why parthenogenetic reproduction was highly likely, subsequent publications during the decade of the 1960s generally accepted his assumption and focused on understanding the origin of each unisexual species. *C. tesselatus* will serve as an example.

In the same year that Pennock's triploidy paper appeared, Zweifel (1965) described six color pattern classes of *C. tesselatus*, denoted as A, B, C, D, E, and F (Fig. 7.1). Two years later, Wright and Lowe (1967b) reported that C, D, E, and F were diploid (Fig. 7.2) while A and B were triploid (Fig. 7.3). Color pattern class F was soon reclassified as *C. dixoni* (Scudday 1973), and the triploids were later designated as *C. neotesselatus* (Walker *et al.* 1997) to distinguish them from diploid *C. tesselatus*.

Wright and Lowe (1967b) published the karyotypes of diploid and triploid *C. tesselatus* together with karyotypes of three bisexual species—*C. septemvitattus*, *C. tigris*, and *C. sexlineatus*. The karyotypes were consistent

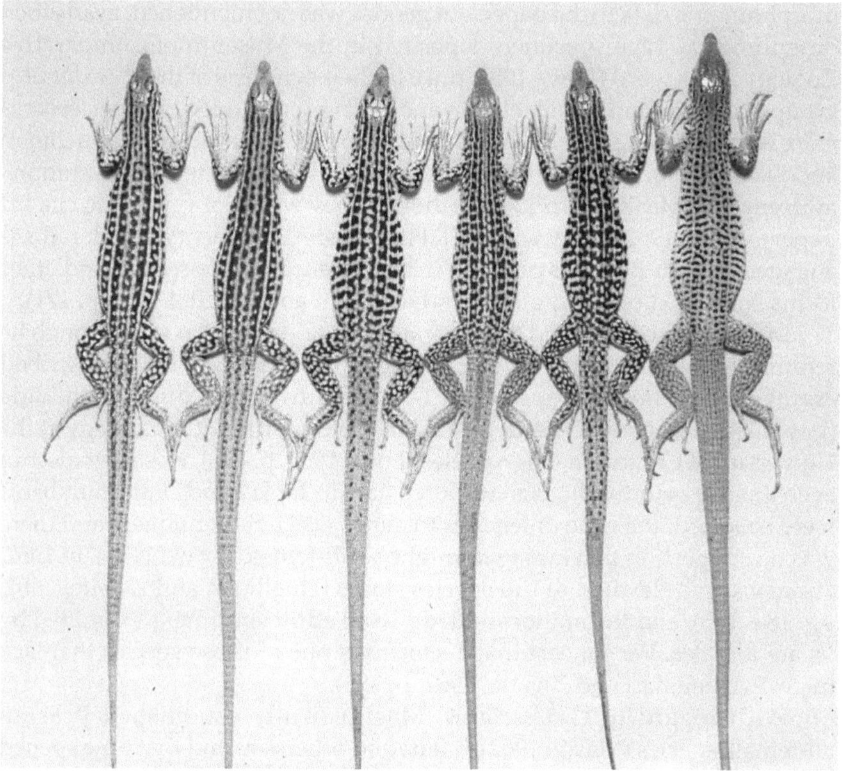

Fig. 7.1 *Cnemidophorus tesselatus* color pattern classes **A, B, C, D, E**, & **F** (Zweifel 1965). Left to Right: Color pattern class **A** captured 7/4/1968 in Pueblo County, Colorado; **B** captured 7/2/1968 in Otero County, Colorado; **C** captured 7/10/1968 in San Miguel County, New Mexico; **D** captured 7/3/1968 in Otero County, Colorado; **E** captured 8/22/1968 in Debaca County, New Mexico; **F** captured 8/1/1968 in Hidalgo County, New Mexico. Figs. 1–3 photographed in sunlight using Ektachrome B sheet film 10/1/1968 in Boston, Massachusetts after chilling lizards in refrigerator one hour at 4° Fahrenheit.

Color image of this figure appears in the color plate section at the end of the book.

with the ancestry proposed by Wright and Lowe (1967b), which suggested that diploid *C. tesselatus* originated as a hybrid of *C. septemvitattus* and *C. tigris* (Fig. 7.2) and that triploid *C. tesselatus* arose when *C. sexlineatus* and diploid *C. tesselatus* hybridized (Fig. 7.3). The proposed ancestry of diploid and triploid *C. tesselatus* was supported by adenosine deaminase alleles that enabled discrimination among bisexual species with similar karyotypes (Neaves 1969).

While most researchers in the U.S. embraced Pennock's assumption about parthenogenesis in all-female *Cnemidophorus*, some investigators continued to seek evidence that would directly support this interpretation.

Fig. 7.2 Bisexual parental ancestral species of diploid *Cnemidophorus tesselatus* as proposed (Wright and Lowe 1967b). Left to Right: *C. septemvitattus*, parental ancestral species, captured 8/8/1968 in Brewster County, Texas; *C. tesselatus*, color pattern class E, captured 8/22/1968 in Debaca County, New Mexico; *C. tigris*, parental ancestral species, captured 8/4/1968 in Luna County, New Mexico.

Color image of this figure appears in the color plate section at the end of the book.

In 1966, Maslin reported four years of effort to hatch eggs laid in the laboratory by adults of four all-female species captured in the field [*C. exsanguis*; *C. perplexus*—actually *C. neomexicanus*; *C. tesselatus*—actually *C. tesselatus* and *C. neotesselatus*; and *C. velox*—actually *C. uniparens* and *C. velox* (Maslin 1966)]. He incubated 469 eggs laid by 144 individuals, and 111 of the eggs hatched, representing all four species; all the hatchlings were females (Maslin 1966, p. 375). Maslin also reported that only female hatchlings of these four species plus *C. cozumelus* could be found in the field (Maslin 1966, p. 376).

Fig. 7.3 Parental ancestral species of triiploid *Cnemidophorus tesselatus* as proposed (Wright and Lowe 1967b). Left to Right: *C. sexlineatus*, parental ancestral species, captured 7/1/1968 in Las Animas County, Colorado; *C. tesselatus*, color pattern class A, captured 7/4/1968 in Pueblo County, Colorado; *C. tesselatus*, color pattern class E, parental ancestral species, captured 8/22/1968 in Debaca County, New Mexico.

Color image of this figure appears in the color plate section at the end of the book.

In 1968, Orlando Cuellar, a graduate student with Maslin, reported finding sperm from copulation with males in the reproductive tracts of six of nine *C. tigris* females collected during the annual breeding season; under similar circumstances, he could find no evidence of sperm in 36 adult females representing five all-female species: *C. exsanguis*, *C. neomexicanus*, *C. tesselatus*, *C. uniparens*, and *C. velox* (Cuellar 1968). Cuellar concluded:

> "The fact that six of the nine bisexual females had spermatozoa in their reproductive tracts while 36 unisexual females lacked similar evidence of courtship, reasonably excluded gynogenesis, a cryptic male behavior, and differential male and female activity periods as interpretations for the all-female condition, and supports true parthenogenesis as the most probable mechanism."
>
> (Cuellar 1968, p. 148)

Definitive evidence of true parthenogenetic reproduction in *Cnemidophorus*—comparable to what Darevsky accomplished in *Lacerta* with his hibernaculum experiment in 1956–1957—did not appear until 1971 when Maslin updated the results of his effort to hatch eggs in the laboratory (Maslin 1971). Maslin began his report by stating:

> "To date the contention that the all-female species of *Cnemidophorus* are parthenogenetic has been based on indirect evidence alone. While this evidence is overwhelming, it is not conclusive."
>
> (Maslin 1971, p. 156)

After reviewing the "overwhelming" but "not conclusive" evidence, Maslin wrote:

> "The ultimate proof, however, could be provided if hatchlings of these forms, raised to maturity in the absence of males, laid viable eggs. This has finally been accomplished with *C. neomexicanus*, *C. tesselatus*, and *C. uniparens*."
>
> (Maslin 1971, p. 157)

Over a nine-year period spanning most of the decade of the 1960s, captive individuals of various all-female species laid more than 1,200 eggs in Maslin's laboratory (Maslin 1971, p. 157). A few hatched after incubation and produced females that matured in the lab. Nine of these (one *C. neomexicanus*, one *C. tesselatus*, and seven *C. uniparens*) reached reproductive maturity and laid 29 eggs in the laboratory. One of two *C. tesselatus* eggs, two of four *C. neomexicanus* eggs, and 12 of 20 *C. uniparens* eggs showed early, intermediate, or later stages of embryonic development upon examination, usually after terminating incubation due to spoilage. The remaining three eggs from laboratory-hatched and reared *C. uniparens* "… developed to term and hatched naturally" (Maslin 1971, p. 158).

Darevsky tested parthenogenic reproduction in *L. s. armeniaca* far more efficiently by avoiding a laboratory environment, devising a natural hibernaculum for the 30 lizards captured in September 1956, and keeping the survivors in open-air enclosures the following summer. Sixteen lizards began laying eggs in mid-July 1957, and "…between the middle and the end of August altogether 56 lizards were hatched, all of which turned out to be females" (Darevskii 1958, p. 879).

7.3 CONCLUSION

The record of discovery of parthenogenesis in lizards in the Eastern Hemisphere began with the report of apparent absence of males in *L. s. armeniaca* (Lantz and Cyren 1936) and concluded twenty-two years later with published documentation of parthenogenesis in *L. s. armeniaca* (Darevskii 1958). In the Western Hemisphere, the record of discovery began with Minton's report of apparent absence of males in *C. tesselatus* (Minton 1958) and ended thirteen years later with Maslin's paper announcing three *C. uniparens* hatchlings whose mothers were themselves hatched and reared in the laboratory (Maslin 1971).

The publication of Darevsky's work in 1958, and Zweifel's dissemination of it in 1959, ensured that the observation published by Minton in 1958 would stimulate similar work in *Cnemidophorus*.

Darvesky's discovery laid the foundation for a half-century of progress in understanding the origins of parthenogenetic lizards, the relationships of all-female taxa to bisexual species, the mechanisms of clonal propagation, the genetic implications of this reproductive mode, and the evolution of parthenogenetic species (see Neaves and Baumann 2011). Ilya Darevsky remained at the forefront of this new field of vertebrate biology until his death in 2009 at age 85.

A few herpetologists have a species or subspecies of amphibian or reptile named to honor them. Many species were named in Darevsky's honour. Very rarely, an individual's name is given to a reptilian genus. It is most fitting that the entire genus of Caucasian rock lizards now bears his name, and the taxon in which he proved the existence of true vertebrate parthenogenesis is known as *Darevskia armeniaca*.

7.4 ACKNOWLEDGMENTS

Chuck Lowe, Paul Maslin, and Don Tinkle oriented the author to the biology of whiptail lizards at various times during 1968. Priscilla Neaves collaborated in the capture of lizards during the summer of 1968. Jay Cole, John Wright, and Dick Zweifel recently shared their recollections of the events reviewed in this chapter. Boris Rubenstein ensured interpretation of the untranslated papers of Chernov and Darevsky. Peter and Diana Baumann commented helpfully on the manuscript.

7.5 LITERATURE CITED

Arribas, O. J. 1999. Phylogeny and relationships of the mountain lizards of Europe and Near East (*Archaeolacerta* Mertens, 1921, *sensu lato*) and their relationships among the Eurasian lacertid radiation. Russian Journal of Herpetology 6: 1–22.

Chernov, S. A. 1939. Herpetological fauna of Armenia and Nachichevan ASSR. Zoology Sbornik Academy Nauk Armenian SSR 7: 77–194.

Clanton, W. 1934. An unusual situation in the salamander *Ambystoma jeffersonianum* (Green). Occasional Papers of the Museum of Zoology, University of Michigan 290: 1–14.

Cuellar, O. 1968. Additional evidence for true parthenogenesis in lizards of the genus *Cnemidophorus*. Herpetologica 24: 146–150.

Darevskii, I. S. 1957. Systematic and ecological aspects of the dispersal of the lizard *Lacerta saxicola* Eversmann in Armenia. Zoology Sbornik Academy Nauk Armenian SSR 10: 27–57.

Darevskii, I. S. 1958. Natural parthenogenesis in certain subspecies of rocky lizard, *Lacerta axicola* Eversmann. Doklady Akademy Nauk S.S.S.R., Biological Sciences 122: 877–879. (Russian to English translation of Doklady Akademy Nauk S.S.S.R., Biological Sciences 122: 730–732).

Duellman, W. E. and Zweifel, R.G. 1962. A synopsis of the lizards in the *sexlineatus* group (genus *Cnemidophorus*). Bulletin of the American Museum of Natural History 123: 155–210.

Editorial Board 2009. Ilya Sergeevich Darevsky (1924–2009). Russian Journal of Herpetology 16: 241–242.

Hubbs, C. L. and Hubbs, L. C. 1932. Apparent parthenogenesis in nature, in a form of fish of hybrid origin. Science 76: 628–630.

James, E. 1823. *Account of an Expedition from Pittsburg to the Rocky Mountains.* H.C. Carey and I. Lea, Philadelphia, Vol. II, pp. 442.

Lantz, L. A. and Cyren, O. 1936. Contribution à la connaissance de *Lacerta saxicola* Eversmann. Bulletin Societe Zoologie France 61: 159–181.

Lowe, C. H. 1955. A new species of whiptailed lizard (Genus *Cnemidophorus*) from the Colorado Plateau of Arizona, New Mexico, Colorado, and Utah. Breviora 47: 1–7.

Lowe, C. H. 1956. A new species and a new subspecies of whiptailed lizards (Genus *Cnemidophorus*) of the inland Southwest. Bulletin of the Chicago Academy of Science 10: 137–150.

Lowe, C. H. and Zweifel, R. G. 1952. A new species of whiptailed lizard (Genus *Cnemidophorus*) from New Mexico. Bulletin of the Chicago Academy of Science 9: 229–247.

Lowe, C. H. and Wright, J. W. 1964. Species of the *Cnemidophorus exsanguis* subgroup of whiptail lizards. Journal of the Arizona Academy of Science 3: 78–80.

Lowe, C. H. and Wright, J. W. 1966. Evolution of parthenogenetic species of *Cnemidophorus* (whiptail lizards) in western North America. Journal of the Arizona Academy of Science 4: 81–87.

Macgregor, H. C. and Uzzell, T. M., Jr. 1964. Gynogenesis in salamanders related to *Ambystoma jeffersonianum*. Science 143: 1043–1045.

Maslin, T. P. 1962. All-female species of the lizard genus *Cnemidophorus*, Teiidae. Science 135: 212–213.

Maslin, T. P. 1966. The sex of hatchlings of five apparently unisexual species of whiptail lizards (*Cnemidophorus*, Teiidae). American Midland Naturalist 76: 369–378.

Maslin, T. P. 1971. Conclusive evidence of parthenogenesis in three species of *Cnemidophorus* (Teiidae). Copeia 1971: 156–158.

McCoy, C. J. and Maslin, T. P. 1962. A review of the lizard *Cnemidophorus cozumelus* and the recognition of a new race, *Cnemidophorus cozumelus rodecki*. Copeia 1962: 620–627.

Metin, G., İlkyaz, A. T., Soykan, O. and Kinacigİl, H. T. 2011. Biological characteristics of the common pandora, *Pagellus erythrinus* (Linnaeus 1758), in the central Aegean Sea. Turkish Journal of Zoology 35: 307–315.

Miller, R. R. and Schultz, R. J. 1959. All-female strains of the teleost fishes of the genus *Poeciliopsis*. Science 130: 1656–1657.
Minton, S. A. 1954. Salamanders of the *Ambystoma jeffersonianum* complex in Indiana. Herpetologica 10: 173–179.
Minton, S. A. 1958. Observations on amphibians and reptiles of the Big Bend Region of Texas. Southwestern Naturalist 3: 28–54.
Minton, S. A. 2001. *Life, Love, and Reptiles*. Krieger Publishing, Malabar, Florida, pp. 217.
Neaves, W. B. 1969. Adenosine deaminase phenotypes among sexual and parthenogenetic lizards in the genus *Cnemidophorus* (Teiidae). Journal of Experimental Zoology 71: 175–183.
Neaves, W. B. and Baumann, P. 2011. Unisexual reproduction among vertebrates. Trends in Genetics 27: 81–88.
Peck, A. L. 1942. *Aristotle—Generation of Animals, with an English Translation by A. L. Peck.* Harvard University Press, Cambridge, Massachusetts, pp. 608.
Pennock, L. A. 1965. Triploidy in parthenogenetic species of the Teiid lizard, Genus *Cnemidophorus*. Science 149: 539–540.
Reeder, T. W., Cole, C. J. and Dessauer, H. 2002. Phylogenetic relationships of whiptail lizards of the genus *Cnemidophorus* (Squamata: Teiidae): A test of monophyly, reevaluation of karyotypic evolution, and review of hybrid origins. American Museum Novitates 3365: 1–61.
Scudday, J. F. 1973. A new species of lizard of the *Cnemidophorus tesselatus* group from Texas. Journal of Herpetology 7: 363–371.
Taylor, H. L. and Medica, P. A. 1966. Natural hybridization of the bisexual teiid lizard *Cnemidophorus inornatus* and the unisexual *Cnemidophorus perplexus* in southern New Mexico. University of Colorado Studies, Series in Biology 22: 1–9.
Tinkle, D. W. 1959. Observations on the lizards *Cnemidophorus tigris*, *Cnemidophorus tessellatus*, and *Crotaphytus wislizeni*. Southwestern Naturalist 4: 195–200.
Walker, J. M., Cordes, J. E. and Taylor, H. L. 1997. Parthenogenetic *Cnemidophorus tesselatus* complex (Sauria: Teiidae): a neotype for diploid *C. tesselatus* (Say 1823), redescription of the taxon, and description of a new triploid species. Herpetologica 53: 233–259.
Wright, J. W. and Lowe, C. H. 1965. The rediscovery of *Cnemidophorus arizonae* Van Denburgh. Journal of the Arizona Academy of Science 3: 164–168.
Wright, J. W. and Lowe, C. H. 1967a. Hybridization in nature between parthenogenetic and bisexual species of whiptail lizards (genus *Cnemidophorus*). American Museum Novitates 2286: 1–36.
Wright, J. W. and Lowe, C. H. 1967b. Evolution of the alloploid parthenospecies Cnemidophorus tesselatus (Say)". Mammalian Chromosome Newsletter 8: 95–96.
Zweifel, R. G. 1965. Variation in and distribution of the unisexual lizard, *Cnemidophorus tesselatus*. American Museum Novitates 2235: 1–49.
Zweifel, R. G. 2013. Personal communication by e-mail and telephone March 10–15, 2013.

CHAPTER 8

Oogenesis and the Ovarian Cycle

Martha Patricia Ramírez-Pinilla,[1,*] *Gloria R. de Pérez*[2] and *Camilo Alvarado-Ramírez*[2]

8.1 OVERVIEW

Oogonesis is the process by which the ovarian germ cells mature into eggs to be fertilized. The process occurs through several phases: 1) specification and formation of the primary germ cells (PGCs) in the extraembryonic area and migration of these cells to the ovary; 2) sex differentiation where germ cells are transformed into primary and then secondary oogonia; 3) onset of meiosis where oogonia are now committed to be oocytes; 4) folliculogenesis where diplotene oocytes are recruited to constitute ovarian follicles; 5) intrafollicular growth where the oocyte is under meiotic arrest; 6) vitellogenesis and follicular maturation; 7) ovulation where the oocyte is expelled from its follicle. The end result of oogenesis is PGCs that have attained the competency of forming viable embryos after fertilization.

In lizard embryos, PGCs are formed in the posterior part of the extra-embryonic area. They reach the ovary after interstitial migration by amoeboid movements in Gekkota, Iguania, and Lacertidae (Hubert 1970, 1971a), or by vascular migration in Scincidae, Chamaeleonidae, Anguidae, and Agamidae (Hubert 1985). As in other reptiles, germ cell differentiation occurs in the middle third of embryonic development (Pieau 1996). At first, a brief period of gonad bipotentiality occurs, regardless of the sex determining mechanism. The undifferentiated gonads keep both a cortex and medulla with PGCs. Differentiation starts when the cortex thickens and the PGCs in

[1] Escuela de Biología, Universidad Industrial de Santander, Bucaramanga, Colombia.
[2] Departamento de Biología, Universidad Nacional de Colombia, Bogotá, Colombia.
* Corresponding author

the cortical zone become more abundant, whereas medullary PGCs become scarce and the medulla transforms into the loose mesenchymal tissue, within which the ovarian cavity appears. In the developing ovary, the ovarian cortex is almost completely filled with oogonia and naked oocytes, and some somatic cells. A thin connective tissue layer develops between the cortex and the medulla. Later in ovary development, somatic cells surround the germ cells (at diplotene) in the cortex layer forming primordial follicles; however, they are not present in the ovary until sometime after hatching (Raynaud and Pieau 1985; Austin 1988; Doddamani 1994; Neaves *et al.* 2006).

8.2 THE OVARY

The ovary of lizards has been studied in a relatively wide range of species; however, its morphology and physiology are well known only from a few model species such as the European Common lizard (*Zootoca vivipara*; Lacertidae; e.g., Hubert 1971b,c, 1973, 1976; Xavier 1982), the Italian Wall lizard (*Podarcis siculus*; Lacertidae; e.g., Taddei 1972; Andreuccetti *et al.* 1978; Filosa *et al.* 1979; Andreuccetti 1992; Motta *et al.* 1995, 1996, 2001; Raucci and Di Fiore 2011), the Common Garden Lizard (*Calotes versicolor*; Agamidae; e.g., Varma 1970; Gouder *et al.* 1979; Shanbhag and Prasad 1993), and the Green Anole (*Anolis carolinensis*; Dactyloidae; e.g., Neaves 1971; Jones *et al.* 1975, 1976, 1984; Laughran *et al.* 1981).

Ovarian morphology (see also Siegel *et al.* Chapter 6, this volume) is highly conserved despite the high diversity of lizard species, body morphologies and sizes, clutch sizes and fecundities. Lizard females possess a pair of ovaries attached to the dorsal body wall through a thin mesovarium. The ovaries are hollow sacs consisted of the cortex covered by a thin wall made by the surface ovarian epithelium and the tunica albuginea (Guraya 1989). The epithelial lining is simple squamous and the tunica albuginea is a dense regular connective tissue (Klosterman 1983; Guraya 1989). The oocytes are produced within germinal beds (GBs). They are small regions located on the dorsal side of the ovary, near the ovarian hilum at the insertion of the mesovarium (Figs. 8.1A,B), and consist of dividing oogonia, undifferentiated somatic cells, oocytes partially surrounded by a few somatic cells, and primordial follicles (Jones *et al.* 1982; Klosterman 1983). Growing follicles are recruited from the GBs and located in the ovarian cortex where follicles at different stages of development (follicles in a broad spectrum of sizes), corpora lutea, and atretic follicles are observed in the reproductively active ovary. The ovarian stroma is composed of highly vascularized connective tissue surrounding the ovarian follicles.

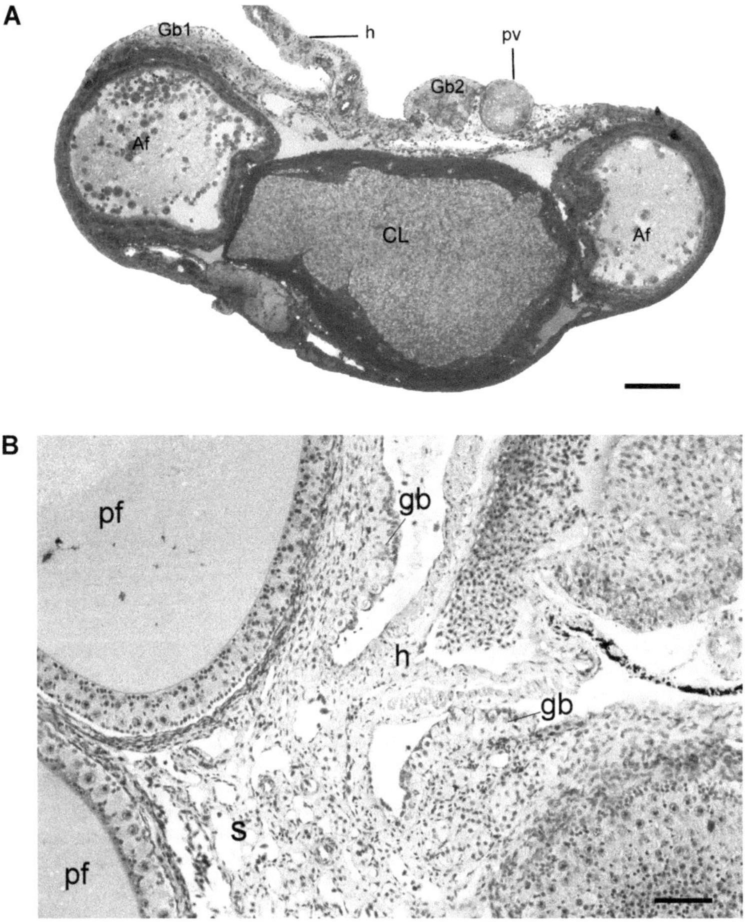

Fig. 8.1 (A) General view of the *Mabuya* sp. ovary during pregnancy. h, ovarian hilum; pv, an early previtellogenic follicle; Gb1 and Gb2, two germinal beds; CL, an active corpus luteum and two atretic follicles (Af). Scale bar = 80 μm. **(B)** General view of the *Liolaemus scapularis* ovary. Two germinal beds (gb) near to the ovarian hilum (h) and two previtellogenic follicles (pf) are observed. s, ovarian stroma. Scale bar = 50 μm.

Color image of this figure appears in the color plate section at the end of the book.

8.2.1 Germinal Beds

The GBs are composed of germ cells (primary and secondary oogonia and primary oocytes), as well as somatic cells. Oogonia are diploid cells that divide mitotically. Many primary oogonia persist in the adult ovary and consequently there is a continual renewal of germ cells in the lizard ovary. Oogonia are large globoid cells with lightly stained cytoplasm and round nuclei with prominent nucleoli (Fig. 8.2A). The oocytes initially lack an envelope of somatic cells (i.e., they are naked oocytes) and, have clear cytoplasm and round nuclei arrested in prophase I.

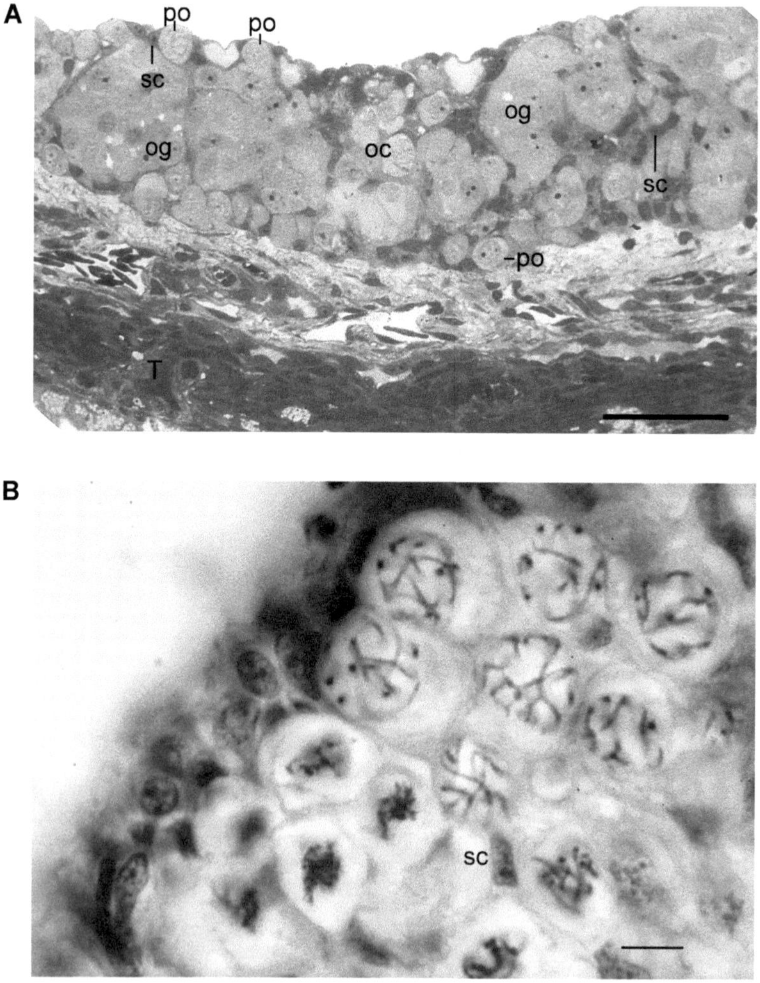

Fig. 8.2 (A) Germinal bed of a pregnant female *Mabuya* sp. exhibiting the prevalence of clusters of secondary oogonia (og) that are discernible by their larger size and pale cytoplasm. oc, primary oocytes; po, primary oogonia; sc, somatic cells (darker cytoplasm); T, theca of an adjacent corpus luteum. Scale bar = 50 µm. **(B)** Germinal bed of a reproductively active female of *Liolaemus scapularis* with prevalence of oocytes that are discernible by their nuclei in different stages of prophase I (zygotene and pachytene). Somatic cells (sc) have a darker cytoplasm and small nuclei. Scale bar = 8 µm.

Color image of this figure appears in the color plate section at the end of the book.

Hubert (1970) described the formation of PGCs and their migration during the embryonic development in *Zootoca vivipara*. Primary oocytes start the meiotic process and are present in GBs from prophase I up to diplotene (Fig. 8.2B) stages. In pachytene cells synaptonemal complexes are observed (Klosterman 1983). Oocytes are progressively surrounded by somatic cells with darker cytoplasm (granulosa cells or follicular epithelium) and this association constitutes the first stage of follicular development; i.e., the primordial follicles (Fig. 8.3A,B), which eventually leave the GB and are enveloped by a thin layer of connective tissue (the follicular theca; Varma 1970; Klosterman 1983; Vieira *et al.* 2010).

Oogonial proliferation and follicular recruitment from the pool of pre-follicular oocytes in the GBs is under control of gonadotropins (GTH) regulated by the hypothalamic gonadotropin releasing hormone (GnRH), both progesterone and estradiol, and favorable environmental conditions; in contrast, oogonial proliferation is inhibited by corticosterone (Tokarz 1978; Nijagal and Yajuervedi 1999; Motta *et al.* 2004). Mammalian follicle stimulating hormone (FSH) stimulates oogonial proliferation, oogenesis and folliculogenesis, resulting in increased numbers of pre-follicular oocytes and determines the recrudescence of ovarian activity in species with seasonal reproductive activity (Jones *et al.* 1975; Motta *et al.* 1995). Gonadotropins stimulate primary oogonia proliferation giving rise to clusters of secondary oogonia, whereas a diffusible recruiting factor from granulosa cells of large previtellogenic follicles stimulates the differentiation of oogonia into oocytes (enter meiosis) and thus, oocyte recruitment (Sica *et al.* 2001). Hundreds of germ cells start meiosis but very few are transformed into follicular oocytes and even fewer are ovulated. Therefore, oocytes undergo a dramatic selection by atresia and most of them die within the GBs (Andreuccetti *et al.* 1990; Motta *et al.* 1995).

The number of GBs per ovary varies both intra-specifically and inter-specifically and ranges from one to six among lizard species (Jones *et al.* 1982; Radder *et al.* 2008). Some studies have suggested that the number of GBs per ovary may be linked to clutch size and breeding frequency. Jones *et al.* (1982) found a pattern that relates the number of GBs with clutch size. Only one GB per ovary is found in most of the species that ovulate a single egg from each ovary each ovulation event (monoallochronic ovulation; i.e., ovulating a single egg from each ovary alternatively as in species of the family Dactyloidae) and species with low fixed clutch sizes and monoautochronic ovulation (ovulating a single egg from both ovaries simultaneously; e.g., Gekkota, Jones and Summers 1984; Gymnophtalmidae, Vitt 1982); whereas, two or more well spatially separated GBs in each ovary are found in high fecundity lizards (polyauthochronic ovulation with multiple and large clutches; Radder *et al.* 2008).

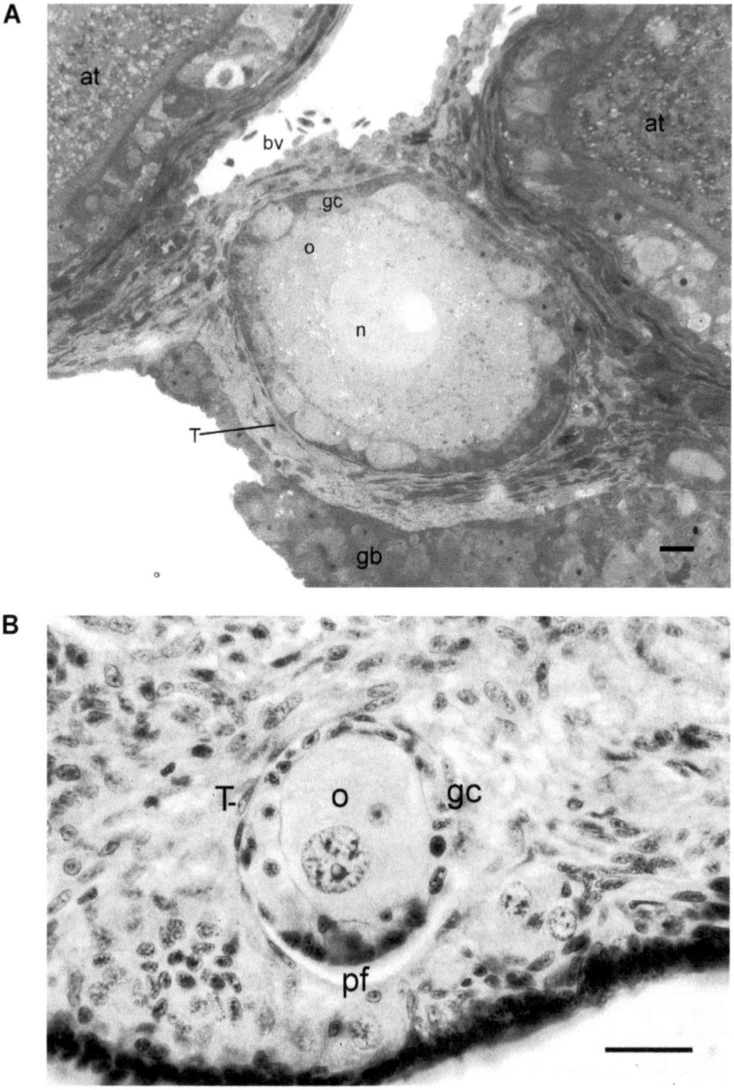

Fig. 8.3 (A) Young previtellogenic follicule of *Mabuya* sp. The theca (T) is thin. The follicular epithelium is monolayered and composed of cuboidal epithelial cells (gc), some of which are larger and possess a clear cytoplasm. bv, blood vessel; gb, germinal bed; n, nucleus; o, oocyte; at, adjacent atretic follicles. Scale bar = 20 µm. **(B)** Primary follicule within the germinal bed of *Liolaemus scapularis*. The theca (T) is very thin. The follicular epithelium is monolayered and composed of cuboidal epithelial cells (gc), some of which are larger and possess a clear cytoplasm. o, diplotene oocyte. Scale bar = 30 µm.

Color image of this figure appears in the color plate section at the end of the book.

The relationship between GBs per ovary and clutch size/breeding frequency is not always observed. Two or more GBs per ovary have been found in some species with only one egg per ovulation (e.g., Blue-throated Rainbow-skink; *Carlia rhomboidalis*; Wilhoft 1963) or low clutch sizes (4–6 GBs and 1–2 ovulated eggs per ovary in Cope's Skink; *Plestiodon copei*; Jones *et al.* 1982); whereas, other species with large clutch sizes have only one GB per ovary (e.g., Three-lined Skink, *Bassiana duperreyi*, Radder and Shine 2007a; Bibron's Agama, *Agama impalearis*, Radder *et al.* 2008). Several and spatially separated GBs were correlated with slender bodies and arboreality (Jones *et al.* 1982; Radder *et al.* 2008). Additional GBs may allow growing follicles to develop separately at multiple locations instead of in groups (Radder *et al.* 2008). This hypothesis has garnered support because in the elongated bodies of snakes, GBs are distributed as scattered irregular patches over the outer border of the ovarian stroma (Lance and Lofts 1978).

Intraspecific variation in the number of GBs has been reported only in *Mabuya* sp. This taxon displays one or two GBs per ovary (Gómez and Ramírez-Pinilla 2004). In the Bosc's Fringe-toed lizard, *Acanthodactylus boskianus*, one elongated GB was observed comprising two partially fused GBs (Gharzi *et al.* 2012). Consequently, it has been suggested that one GB per ovary in species with large clutches may have been the result of merging of at least two GBs into one (e.g., in the Southern Rock Agama, *Agama atra*, Radder and Shine 2007a; *Acanthodactylus boskianus*, Gharzi *et al.* 2012).

8.2.2 Folliculogenesis

Folliculogenesis is the process of formation and maturation of the ovarian follicles. The ovarian follicle is the functional unit within the ovary. An ovarian follicular hierarchy is a characteristic feature of reptilian ovaries (Etches and Pettite 1990). This hierarchy includes follicles at all stages of maturation. Growing follicles are grouped in size classes and the number of follicles growing simultaneously in each size class is related to the instantaneous fecundity of each species.

Previtellogenic follicles. Primordial follicles move from the region of the GBs to the ovarian cortical stroma, where folliculogenesis continues. In the youngest follicles that leave the GBs, the follicular epithelium is single-layered and cuboidal and the theca is a thin layer formed of fibroblasts surrounded by bundles of collagen fibers (Fig. 8.2B). The zona pellucida (ZP) is slightly distinguishable from the cytoplasm of the oocyte, which is filled with a medium-dense homogeneous substance. An aggregation of RNA, lipid, and protein inclusions is found in the ooplasm as an irregular and lightly stained region adjacent to the nucleus. The aggregate, together with lipids and membranous organelles such as Golgi complexes, endoplasmic reticulum cisternae, and mitochondria, form an evident structure termed the Balbiani's vitelline body (yolk nucleus), which is dispersed in the cytoplasm with the growth of the oocyte (Guraya 1968, 1989).

Oocyte growth during previtellogenesis is remarkable (more than 1,000 fold in volume) and occurs by accumulation of a variety of organelles and macromolecules in the cytoplasm (e.g., enzymes, histones, tubulin, regulatory factors, mitochondria, ribosomes, and dormant mRNA) that are utilized during early embryogenesis. The oocyte nucleus also increases in volume due to lampbrush chromosome activity, and is now called the germinal vesicle (Guraya 1989).

The follicular epithelium of previtellogenic follicles is a dynamic structure that changes in morphology and function along with the growth and maturation of the follicle. During this process the number of follicle cells increases and large round vesicular cells appear; consequently, the epithelium acquires a multilayered and polymorphic organization. Therefore, in contrast to most vertebrates, the follicular epithelium of the ovary of lizards is multilayered and composed of three distinct cell types: 1) small, 2) intermediate, and 3) pyriform cells (Figs. 8.4A,B). The follicular epithelium is formed by 4 (e.g., Northern Alligator Lizard, *Elgaria coerulea*, Klosterman 1987; Central American Mabuya, *Mabuya unimarginata*, Hernández-Franyutti *et al.* 2005; Sudan Mabuya, *Trachylepis brevicollis*, Al-Dokhi 1998) to 6 cell layers (e.g., *Calotes versicolor*, Varma 1970; *Podarcis siculus*, Filosa *et al.* 1979; *Anolis carolinensis*, Laugrhan *et al.* 1981; Western Spiny-tailed Iguana, *Ctenosaura pectinata*, Uribe *et al.* 1996), and remains relatively constant in cell depth throughout previtellogenesis. The thickening of the follicular epithelium during this phase is related to the increased number of small cells and to a further enlargement of the differentiated cells (intermediate and pyriform cells).

Small cells have irregularly shaped and highly heterochromatic nuclei. They are located both near the basal lamina and at the oocyte surface. Only basal small cells are mitotically active and, as a result, all the other cell types differentiate from small cells (Filosa *et al.* 1979; Klosterman 1987).

As named, intermediate cells are intermediate in size between small and pyriform cell types. Their morphology suggests that they are in the process of transforming from small cells to pyriform cells. Intermediate cells are spherical with lightly granular cytoplasm and round nuclei. They are often observed near to the small cells at both basal and apical sides of the follicular epithelium (Neaves 1971; Andreuccetti *et al.* 1978).

Pyriform cells are a unique feature of squamate ovarian follicles. The pyriform cells are conspicuous by having a bulbous shape that contacts the basal side of the epithelium and a slender apical projection toward the oocyte, extending across the entire epithelial lining (Fig. 8.4B). In some lizards, however, the larger cells are spherical more than flame-shaped but

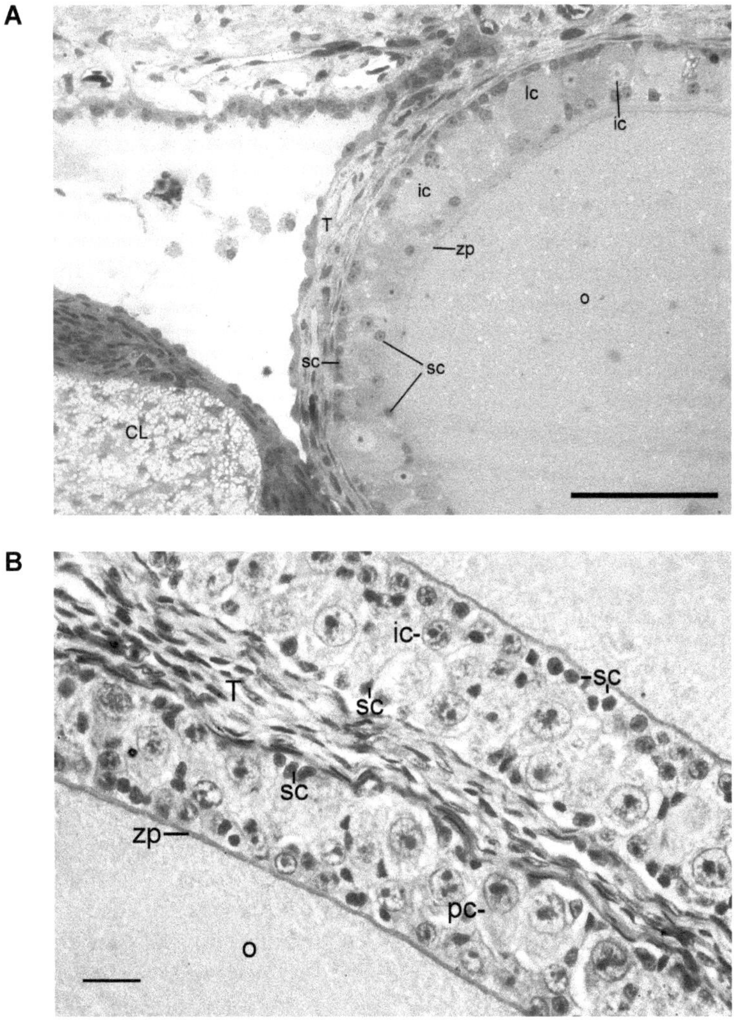

Fig. 8.4 (A) Partial view of a corpus luteum (CL) and a previtellogenic follicle (right) of *Mabuya* sp. The previtellogenic follicle is enveloped by the theca (T). Three types of characteristic follicular epithelial cells are observed at this stage: small (sc), intermediate (ic) and large (lc). In the apical region of the follicular cells the zona pellucida (zp) is conspicuous. O, oocyte; CL, corpus luteum in early luteolysis. Scale bar = 50 μm. **(B)** Partial view of the follicular wall of two adjacent previtellogenic follicles of *Liolaemus scapularis*. Three types of characteristic follicular epithelial cells are observed at this stage: small (sc), intermediate (ic) and pyriform (pc). In the apical region of the follicular cells the zona pellucida (zp) is conspicuous. O, oocyte; T, thecae. Scale bar = 12 μm.

Color image of this figure appears in the color plate section at the end of the book.

share all other features and are homologous to the pyriform cells of other squamates (Fig. 8.4A; e.g., in scincids, *Mabuya* sp., Gómez and Ramírez-Pinilla 2004; Vieira *et al.* 2010; *M. unimarginata*, Hernández-Franyutti *et al.* 2005; Ocellated Skink, *Chalcides ocellatus*, Corso *et al.* 1978; Corso and Pala 1987; *Trachylepis brevicollis*, Al-Dokhi 1998; in cordylids, Transvaal Girdled Lizard, *Cordylus vitiffer*, and Common Flat Lizard, *Platysaurus intermedius*, Andreuccetti 1992). They have a relatively pale cytoplasm and a large round nucleus containing clumps of heterochromatin and a dense and prominent nucleolus, which in older cells forms fibrillar micronucleoli that are extruded from the nucleolus and adhere to the nuclear envelope.

The cytoplasm of pyriform cells has most of the mitochondria located adjacent to the nucleus and a few near to the apical membrane of the cell where both granular endoplasmic reticulum and well developed Golgi complexes with lipid inclusions are grouped (Neaves 1971; Klosterman 1987; Hubert 1985).

Although not apparent in thick sections, the apical projection of the pyriform cells in lizard previtellogenic follicles passes through the zona pellucida and comes into contact with the oocyte through an intercellular bridge of 0.5 to 5 μm width. The intercellular bridges were observed by several investigators (Neaves 1971; Hubert 1971b; Bou-Resli 1974; Filosa and Taddei 1976; Andreuccetti *et al.* 1978; Filosa *et al.* 1979; Klosterman 1987; Ibrahim and Wilson 1989; Maurizii and Taddei 2012) as a continuity of a narrow strand of cytoplasm between pyriform cells and the oocyte to form a cytoplasmic bridge between the two cells (Fig. 8.5). Some studies have described the pyriform cell fine structure with particular reference to its relationship with the oocyte (Neaves 1971; Bou-Resli 1974; Klosterman 1987; Maurizii and Taddei 2012) demonstrating continuity between the oolemma and the apical membrane of the pyriform cell. The heterologous fusion of somatic follicle cells and the oocyte leads to the formation of an integrated system (Andreuccetti 1992; Maurizii *et al.* 2004). Pyriform cells do not appear to be present in the ovaries of any other class of terrestrial vertebrate; however, intercellular bridges between follicle cells and an oocyte have been described in the previtellogenic and vitellogenic follicles of elasmobranchs (Hamlett *et al.* 1999; Andreuccetti *et al.* 1999; Davenport *et al.* 2011).

The cytoplasm in the bridge contains ribonucleoprotein particles and glycogen granules, a dense fibro-granular material that may be exchanged between the two cells, and bundles of intermediate filaments (cytokeratins and vimentin) coursing along the axis of the bridge.

These filaments plus the presence of actin and tubulin inside the cytoplasmic bridges form a network that crosses the ZP and appears to be continuous with a cortical ring of cytokeratin in the oocyte underlying the

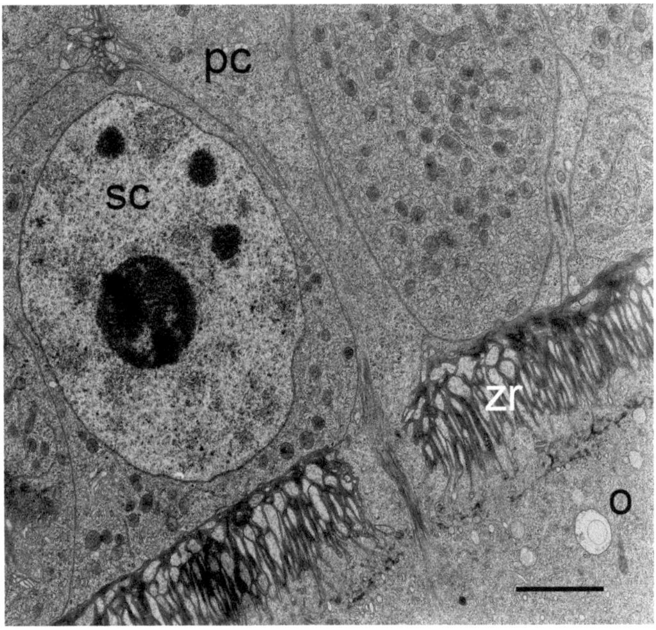

Fig. 8.5 Intercellular bridges in previtellogenic follicles of *Podarcis siculus*. Sections of ovarian follicles of about 1 mm in diameter. Transmission electron micrograph, the bundles of filaments crossing the intercellular bridges are made by cytokeratins. Pyriform cell (pc), oocyte (o), nucleus of a small apical cell (sc), zona radiata (zr). Scale bar = 2 μm. Image provided by Prof. Carlo Taddei.

oocyte microvilli (Fig. 8.6A,B; Neaves 1971; Maurizii and Taddei 1996, 2012; Maurizii *et al.* 1997, 2000, 2004).

Different from other intercellular bridges that are formed by incomplete cytokinesis, the intercellular bridges in the lizard ovarian follicles are formed by cell contact and membrane fusion between pyriform cells of the follicular epithelium and the oocyte (Varma 1970; Neaves 1971). The connections with the oocyte may already be formed in the intermediate cells when they are transforming into pyriform cells (Andreuccetti *et al.* 1978).

It has been suggested that the intercellular bridges in previtellogenic follicles may be involved in the maintenance of oocytes in latency and in their preparation for the intense metabolic activity during yolk accumulation (Neaves 1971). In fact, pyriform cells are considered as nurse cells that contribute to oocyte growth by producing and sending macromolecules, mitochondria, ribosomes, vesicles, DNA and RNA via intercellular bridges (Taddei 1972; Andreuccetti *et al.* 1978; Motta *et al.* 1995). The DNA coming from the pyriform cells is neutralized in the oocyte via fragmentation by endonucleases; thereby it is recycled (De Caro *et al.* 1998).

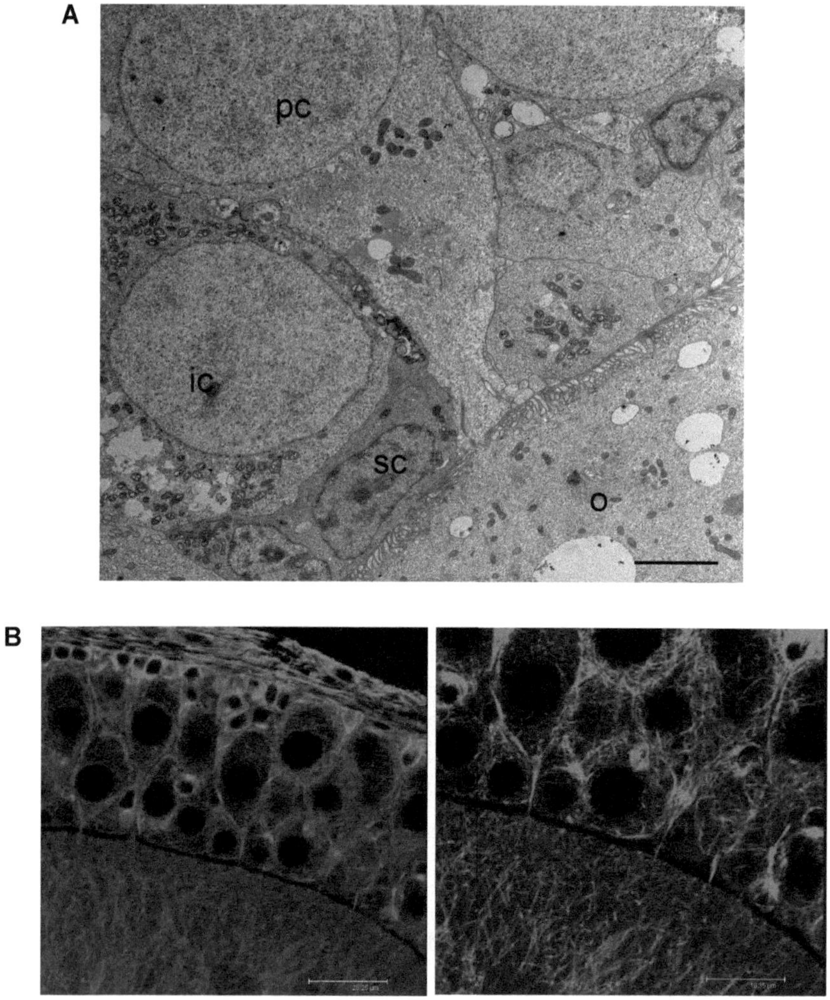

Fig. 8.6 (A) Transmission electron micrograph of a pyriform cell with the apex pointed to the oocyte surface in a follicle of about 350 μm in diameter of *Podarcis siculus*. Observe the pyriform cell (pc), an intermediate (ic) and a small apical cell (sc), and the ooplasm of the oocyte (o). Scale bar = 2 μm. Image provided by Prof. Carlo Taddei. **(B)** Confocal images of the follicular epithelium and oocyte of *Podarcis siculus* immunostained for α-Tubulin with DM1A monoclonal antibody. The image at the right is an enlargement of the left image showing the filaments crossing the intercellular bridges to the oocyte cytoplasm. Scale bar left image = 28.3 μm; right image = 18.4 μm. Images provided by Prof. Carlo Taddei.

Color image of this figure appears in the color plate section at the end of the book.

Pyriform cells develop a distinctive protein related to germ cells; i.e., the synthesis of Vasa protein, a specific germ cell marker. Transport of Vasa from the oocyte to the early differentiating pyriform cells could be the basis for the transformation of pyriform cells into germ-line nurse cells (Maurizii *et al.* 2009). Therefore, pyriform cells dump their contents in the growing oocyte, and thereby function similarly to *Drosophila* nurse cells; however, they are somatic in origin.

Changes in the organization of the follicular epithelium are accompanied by changes in the follicle cell apical surface. The distribution of lectin-binding sites over the follicular epithelium changes with pyriform cell differentiation. The progressive appearance of glycoproteins with terminal α-N-acetylgalactosamine residues on the cell surface seems to be involved in the differentiation and maintenance of the pyriform cells and in the fusion between the oocyte and the follicle cell. In contrast, unlike the follicular epithelium, the ZP did not show changes in carbohydrate composition during oocyte development (Andreuccetti *et al.* 2001). The ZP is clearly observed as a lightly stained non-cellular layer formed at the interface between the oolemma and the apical membranes of the granulosa cells, and as a result, it separates the follicular epithelium from the oocyte. It is a network of fine, filamentous material that is synthesized and secreted interiorly by the oocyte and exteriorly by granulosa cells (Andreuccetti and Carrera 1987; Andreuccetti *et al.* 2001).

The ZP consists of a polysaccharide–protein complex characterized by the presence of GalNAc, GluNAc, Man, and Gal (Andreuccetti *et al.* 2001), is positive to the periodic acid-Schiff (PAS) reaction (a hyaluronidase-labile metachromasia), and fixes alcian blue (Uribe *et al.* 1996; Maurizii *et al.* 2004). In mammals, the ZP has been related to sperm-egg recognition and barriers to interspecies fertilization; nevertheless, there are no data on the recognition and binding between spermatozoa and the egg envelopes in lizards. As follicular development advances, the ZP is divided into two distinct layers: an outer homogeneous layer adjacent to the follicular epithelium and an inner striated layer (zona radiata) against the oolemma (Laughran *et al.* 1981; Uribe *et al.* 1996). This layer looks striated because it is penetrated by pleomorphic macrovilli from the granulosa cells and microvilli from the oocyte.

The ooplasm of previtellogenic follicles acquires several small clear vacuoles that progressively grow and fuse and are observed as larger vacuoles in the central ooplasm (van Wyk 1984; Hubert 1985; Uribe *et al.* 1995; Gómez and Ramírez-Pinilla 2004). Guraya (1989) indicated that this ooplasmic vacuolization occurs during the post-lampbrush stage of previtellogenesis and that during the first stage of vitellogenesis the first primordial yolk spheres will be developed in these vacuoles.

The basal lamina of the previtellogenic follicles separates the epithelium from the underlying stroma. Surrounding the ovarian follicle are the thecal layers, the theca interna (Fig. 8.7A) and theca externa, composed of connective tissue. Small blood and lymphatic capillaries separate both thecae, and fibroblasts-secretory cells are adjacent to blood vessels (Figs. 8.4A,B).

Previtellogenic follicles represent an important source of estradiol and their pyriform cells are the location of its synthesis (Endo and Park 2005; Hammouche *et al.* 2007).

Estradiol exerts an autocrine and paracrine action in the previtellogenic follicles; in fact, estrogens act intrafollicularly by stimulating the aromatase activity synergistically with GTH (Hammouche *et al.* 2007). The expression of the 17β-estradiol and P450 aromatase in the granulosa pyriform cells of the North African Spiny-tailed Lizard (*Uromastyx acanthinura*) and their absence in the cells of theca interna indicate a functional differentiation in the 17β-estradiol biosynthesis and support the bicellular theory described in mammals; i.e., androgens are produced by cells of the theca interna under luteinizing hormone (LH) stimulation and are then transported to the granulosa, where they are aromatized into estrogens under the influence of FSH (Hammouche *et al.* 2009).

Limatola *et al.* (2002) detected the presence of an immunoreactive (ir)-inhibin-like protein in the cytoplasm of the pyriform cells of previtellogenic follices and in the cortical cytoplasm of late previtellogenic oocytes of *Podarcis siculus*, which suggest transfer of inhibin from the follicle pyriform cells to the oocyte through intercellular bridges. The presence of an inhibin-like protein in the ovary of this lizard would be related to the maintenance of different responsiveness of the follicles to gonadotropins and therefore, to the control of follicular hierarchy.

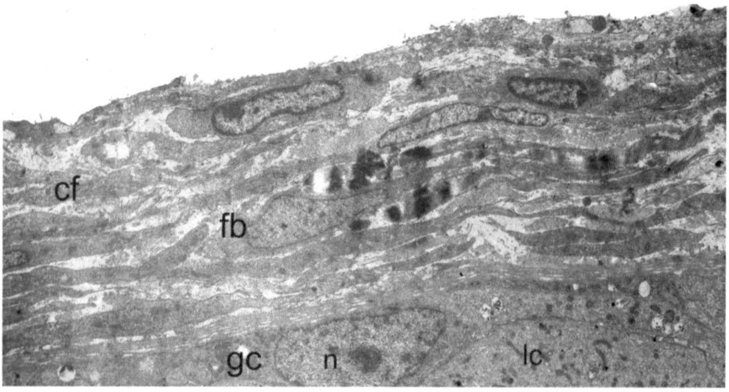

Fig. 8.7 Low magnification transmission electron micrograph of the theca of a previtellogenic follicle of *Mabuya* sp. Cf, collagen fibers; fb, fibroblast; gc, granulosa small cell; n, nucleus of the small cell; lc, large granulosa cell. Magnification = 6000x.

In *Uromastyx acanthinura*, β-endorphin is strongly distributed in the granulosa cells and oocyte cytoplasm of the previtellogenic follicles in sexually quiescent lizards when steroidogenesis is interrupted, whereas it is absent in the vitellogenic and previtellogenic follicles (Hammouche *et al.* 2009). β-endorphin is released in response to stress; it inhibits ovarian recrudescence and germinal bed activity, and prevents vitellogenic follicular development in the Keeled Indian Mabuya (*Eutropis carinata*; Ganesh and Yajurvedi 2002). Therefore this opioid peptide also acts in the modulation of ovarian steroidogenesis of lizards.

At the end of the previtellogenic stage, during transition to vitellogenic follicles, the granulosa layer changes considerably. It diminishes in width as the pyriform cells become reduced in size. The follicular epithelium gradually undergoes changes as a result of apoptosis of the intermediate and pyriform cells (Motta *et al.* 1995, 1996). In those cells several modifications can be observed: the shrinkage of the nucleus and of the cytoplasm, extensive vacuolization of the cytoplasm; blebbing of the nuclear envelope, chromatin margination, condensation, and fragmentation, vesicle dilation, and formation of apoptotic bodies (Motta *et al.* 1996). During initial apoptosis the mitochondria of the pyriform cells are translocated unaltered to the oocyte (Tammaro *et al.* 2007). Therefore, the pyriform cells and the intercellular bridges that they constitute disappear immediately before the onset of the yolk deposition in the oocyte cytoplasm. At this time, a great thickening of the ZP coincides with the separation of intercellular bridges. The pyriform cells lose their apical processes, changing their shape to round cells that resemble the intermediate cells. Shortly after, the form and size of the cells that constitute the follicular epithelium becomes homogeneous. The follicular epithelium turns monolayered and only small cells persist as a unique component of the epithelium until ovulation (Filosa 1973). In late previtellogenic follicles the cellular theca interna is separated from a fibrous theca externa by a net of small blood vessels and capillaries.

Homogeneous, lightly stained vacuoles occupy the peripheral ooplasm and may open into the ZP. On the oocyte surface, microvilli are numerous. Many coated pits are situated along the base of the oocyte microvilli, and coated vesicles, ribosome granules and other granular clusters are present in the peripheral ooplasm (Laughran *et al.* 1981; Vieira *et al.* 2010). The ooplasm contains fibrillar granules mainly located at the periphery and the center exhibits a finely granular cytoplasm. A distinct nucleus is localized near the center of the oocyte.

Multioocytic follicles, also known as polyovular follicles, have been found in some lizards in which each ovarian follicle hosts two or more oocytes (e.g., *Chalcides ocellatus*, Ibrahim and Wilson 1989; *Mabuya* sp., Gómez and Ramírez-Pinilla 2004; Central American Mabuya, *M. unimarginata*, Hernández-Franyutti *et al.* 2005). Although this phenomenon occurs naturally at a low rate, it can be induced by exposure of the developing ovary to exogenous estrogens (Iguchi *et al.* 1990; Guillette *et al.* 1994).

Vitellogenic follicles. In most lizards the eggs are telolecithal, and therefore follicular development is typically associated with extensive vitellogenesis. The oocyte requires the storage of large quantities of yolk in the ooplasm, causing an increase more than 20-fold in egg volume during development (Ho *et al.* 1982). Oocyte growth by vitellogenesis occurs during meiotic arrest by cytoplasmic accumulation of lipoproteic nutrients synthesized in the liver, secreted in the blood stream, and finally incorporated into oocytes (Limatola and Filosa 1989; Romano and Limatola 2000). These nutrients are represented by vitellogenin (VTG) and very low-density lipoproteins (VLDL). In the Puerto Rican Anole (*Anolis pulchellus*) lipovitellins are a combination of distinct peptides related to various lipovitellin domains of multiple plasma VTGs (Morales *et al.* 1996, 2002). Vitellogenin is a lipoglycophosphoprotein that is conserved phylogenetically (Chen *et al.* 1997).

Estradiol is essential for the production of the hepatic vitellogenin and for oviduct development. Estradiol 17β plays a key role in the liver synthesis of VTG (Tokarz 1977; Morales *et al.* 1996). However, in the Bearded Dragon (*Pogona barbata*) very low plasma concentrations of estradiol were measured during vitellogenesis (Amey and Whittier 2000), suggesting that a different estrogen is more important in this species. Alternative estrogens are also synthetized in some lizard species (e.g., in the Blotched Blue-Tongued Lizard; *Tiliqua nigrolutea*; Edwards *et al.* 2002).

Gonadal steroids exert their physiological effects by binding to specific tissue receptors. Two major forms of the estrogen receptor, ERα and ERβ, with differential ligand activation and different tissue distributions have been found in vertebrates. Verderame and Limatola (2010) cloned and characterized estrogen receptors ERα and ERβ at molecular level in the lizard *Podarcis siculus* and found their differential expression in the liver during the reproductive cycle. Both ERα and ERβ were expressed in the liver of vitellogenic females, whereas only ERβ is expressed in the liver of previtellogenic females. Therefore, during vitellogenesis the hepatic expression of ERα could be added to ERβ by higher levels of estradiol (E2). Progesterone receptors (PRs) have also been identified in tissues of the lizard *Podarcis siculus* (Paolucci and Di Cristo 2002). Progesteone receptors are detectable when VTG is not synthesized, suggesting a mechanism through which progesterone inhibits vitellogenesis.

Once incorporated into the oocytes, VTG and VLDL are cleaved into a characteristic set of polypeptides forming yolk platelets (De Stasio *et al.* 1999). Vitellogenesis requires the mobilization of important maternal reserves and is then dependent upon a threshold level of body condition.

In many temperate and subtropical lizards, an inverse relationship exists between gonadal size and the amount of fat reserves (Derickson 1976). Hadley and Christie (1974) found that the lipids of the egg yolk are similar to the lipids in the abdominal fat bodies in the Yarrow's Spiny Lizard (*Sceloporus jarrovii*). Activation of the hypothalamic-pituitary-gonadal axis to initiate vitellogenesis might involve leptin (a peptide produced mainly by adipocytes) as observed in other vertebrates. Plasma leptin levels in *Podarcis siculus* increased before the onset of vitellogenesis when the mass of abdominal fat bodies decreased. Therefore, fat body depletion during the period of active synthesis of vitellogenin seems to be due to their utilization in support of reproduction which in turn causes a decrease in the circulating leptin levels (Paolucci *et al.* 2001). In the tropics, several species of lizards that inhabit areas with uniform climates and have aseasonal reproductive cycles do not appear to store abdominal fat bodies or their storage varies little and is not clearly related to reproductive activity (e.g., Giant Ameiva, *Ameiva ameiva*, Rainbow lizard, *Cnemidophorus lemniscatus*, Striped Kentropyx, *Kentropyx striata*, Magnusson 1987; Yellow-headed Gecko, *Gonatodes albogularis*, Serrano-Cardozo *et al.* 2007), suggesting that resources for reproduction are sufficient for most of the individuals, because the storage of lipids is usually a direct function of food availability (Derickson 1976).

The flow of exogenous yolk precursors toward the oocyte takes the route from blood stream through intercellular spaces between follicular cells and the zona pellucida (Neaves 1972). Oocyte uptake of the yolk exogenous precursor occurs by receptor-mediated endocytosis (micropinocytosis) and takes place through the formation of multiple coated pits and vesicles. Romano and Limatola (2000) demonstrated the occurrence of a VTG binding protein in the oocyte plasma membrane of *Podarcis siculus*. These membrane receptors necessary for micropinocytosis are available only in vitellogenic oocytes and are not present in previtellogenic follicles, which suggest an endovarian control of vitellogenesis (Limatola and Filosa 1989; Romano and Limatola 2000). The absorptive surface is greatly increased by development of microvilli from the surfaces of both the oocyte and follicle cells. Associated with these microvilli are many micropinocytotic vesicles. Therefore, throughout vitellogenesis the granulosa is composed of a single layer of small cuboidal to squamous cells that maintain junctions with the oocyte surface through pleomorphic macrovilli that traverse the zona pellucida and are attached to the oocyte by adhering zonules (Laughran *et al.* 1981). At this juncture, the intercellular bridges are absent. Granulosa cells accumulate lipid-like droplets and whorls of filaments in their cytoplasm and have large amounts of RER and nuclei with dispersed heterochromatin and dense nucleoli.

Beneath the basal lamina of the follicular epithelium lie the thecae, which correspond to layers of connective tissue containing mostly fibroblasts and collagen fibers. The difference between the fibrous theca externa and the cellular theca interna becomes obvious in large follicles. Thecal cells become long and attenuate with follicular growth. The thecal layers seem to have several functions. Fibroblasts of the theca externa produce collagen and also can differentiate into myofibroblasts that probably aid in ovulation. Fibroblasts within the theca interna have abundant SER (Laughran *et al.* 1981) suggesting that they are stereidogenic and secrete androgens under GTH stimulation as in mammals (Jones 2011).

During early vitellogenesis (the yolk vesicle stage of vitellogenesis of Guraya 1989), the first primordial yolk spheres are developed in vacuoles and consist of proteid yolk bodies that contain proteins, lipoproteins, triglycerides, and some carbohydrates. Larger vesicles at the base of the oocyte microvilli contain electron dense material that is presumably yolk precursors. Those vesicles taken up by the oocyte by micropinocytosis accumulate peripherally in yolk globules (Fig. 8.8A).

At this time, plasma levels of estradiol, progesterone, and testosterone are low, whereas high levels of these hormones occur at late vitellogenesis and ovulatory stages (e.g., Carnevali *et al.* 1991; Cree *et al.* 1991, 1992; Rhen *et al.* 2000; Edwards and Jones 2001; Al-Amri *et al.* 2012). In the ooplasm, the vitellogenin is enzymatically cleaved into phosvitins and lipovitellin forming yolk platelets. It has been demonstrated in the lizard *Podarcis siculus* that the intraoocytic processing of yolk proteins is mediated by an aspartic endopeptidase, cathepsin D (de Stasio *et al.* 1999). Besides the yolk platelets, yolk DNA is present in vitellogenic oocytes and comes from the apoptotic bodies resulting from granulosa cell regression at the end of previtellogenesis (Motta *et al.* 2001). The majority of the yolk platelets form a dense and narrow concentric ooplasm layer near the oocyte periphery, but the region rich in yolk platelets is separated from the oolemma and the ZP by a peripherical platelet-free zone where in the larger vitellogenic oocytes, the cytokeratin cytoskeleton constitutes a thick cortical layer (Maurizii *et al.* 1997). The observed variation in yolk platelet shape and size among lizards suggests that yolk platelets might be formed by different sequestration methods and that this diversity may also correspond to vesicles having different contents (Uribe *et al.* 1996). Lipid-like bodies and yolk granules almost completely fill the deeper ooplasm and the nucleus of the oocyte (the germinal vesicle) migrates from the center to the surface.

Oocytes resume the first meiotic division and differentiate into mature eggs through final maturation; therefore, recommencement of meiosis marks the onset of oocyte maturation and seems to be stimulated by progesterone. Diakinesis and metaphase I occur rapidly and immediately

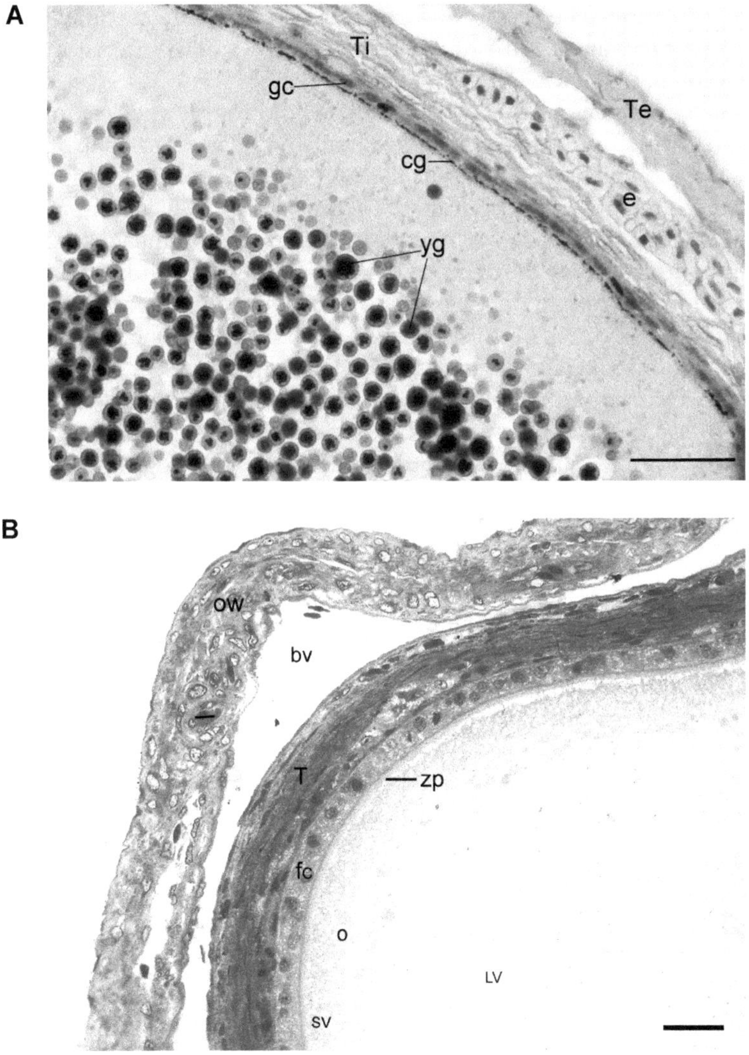

Fig. 8.8 (A) Vitellogenic follicle of the oviparous gymnophtalmid *Ptychoglossus bicolor*. The granulosa is monolayered with squamous epithelial cells (gc). Dark cortical granules (cg) are observed near oolemma. Abundant yolk granules (yg) are observed in the ooplasm. e, erythrocytes; Ti, theca interna; Te, theca externa. Scale bar = 50 µm. **(B)** Early vitellogenic follicle of *Mabuya* sp. The granulosa cells (fc) constitute a cubic monolayered epithelium surrounded by a thick fibrous theca (T). A thin zona radiata is present. The clear vesicles (SV) are smaller near the zona radiata and larger at the center (LV) of the ooplasm (o). No yolk platelets are observed. bv, blood vessel; ow, ovarian wall; zp, zona pellucida. Scale bar = 80 µm.

Color image of this figure appears in the color plate section at the end of the book.

prior to ovulation. The second meiotic division probably starts when the oocytes are still ovarian, either immediately before or at the precise moment of ovulation (Cuellar 1971).

Viviparous species of lizards include lecithotrophic species, in which embryonic nutrition depends on egg yolk reserves, and matrotrophic species, in which the nutrition of the embryos depends principally on extravitelline maternal contribution (Blackburn 1992). New World species of the skinks of the genus *Mabuya* are one of the few lizard lineages that have evolved matrotrophy to the degree where the placenta provides most of the nutrients for embryonic development (placentotrophy) (see also Stewart and Blackburn Chapter 14, this volume). The size of the oocyte in *Mabuya* is very small (microlecithal egg, diameter 1–2 mm) because it contains very little yolk. Ovarian morphology and folliculogenesis are known for two species (*Mabuya* sp., Gómez and Ramírez-Pinilla 2004; Vieira *et al.* 2010; *M. unimarginata*, Hernández-Franyutti *et al.* 2005). The histological appearance and ultrastructure of the ovary during previtellogenic stages is very similar to that described for lecitotrophic species; however, in preovulatory follicles whereas the migration of the nucleus toward the animal pole is preserved, only one stage of vitellogenesis is observed. Primordial yolk vacuoles, electron-lucent vesicles, and small cortical granules are deposited in the ooplasm instead of fatty yolk platelets (Fig. 8.8B).

Consequently, the evolution of placentotrophy in the family Scincidae is related to the reduction in the contribution of yolk lipids to the embryo in placentotrophic species as a result of the reciprocal reduction in egg size (Speake and Thompson 2000). Studying the nutrient content of recently ovulated microlecithal eggs in *Mabuya* sp. established that the eggs contain the entire set of nutrients that are present in lecithotrophic eggs of other lizards, although in small quantities and different proportions. Particularly, yolk lipids are not only reduced in absolute quantities affecting egg size but also in relative content of the different lipid classes; i.e., reduced proportions of energetically rich lipids (triacylglycerols) and increased proportions of energetically less-rich lipids (phospholipids and cholesteryl esters) (Ramírez-Pinilla *et al.* 2011).

Like other matrotrophic vertebrates, the small size attained by the ovulated oocytes of *Mabuya* is related to suppression and modification of the vitellogenic process. Brawand *et al.* (2008) found that the vitellogenin-encoding genes (*VIT*) have lost much or all of the function in marsupials and eutherian mammals, suggesting that *VIT* genes can be used as indicators for the switch from predominately yolk-based embryo nourishment to the rise of placentotrophy. It would be interesting to assess whether the *VIT* genes show similar patterns of inactivation in the placentotrophic scincid lizards.

In lizards ovarian sex steroid production is directly related to positive and/or negative feedback of the hypothalamal–pituitary axis. Production depends mainly on GTHs and is modulated in the ovary through local agents acting as autocrine and/or paracrine mechanisms (Raucci and Di Fiore 2011). In vertebrates, the main regulators during vitellogenesis and oocyte maturation are the pituitary gonadotropins (LH and FSH) and sex steroids (Jones 2011).

However, squamate reptiles seem to have a single gonadotropin that combines FSH and LH activity. Borrelli *et al.* (2001) described the full nucleotide sequence encoding a gonadotropin receptor that shares a high level of similarity with FSH receptor of other vertebrates in *Podarcis siculus*. By using immunocytochemical methods of specific antisera against human FSH (hFSH) and LH (hLH), Hammouche *et al.* (2007) identified only FSH-like containing cells in the pars distalis from the pituitary gland of adult female desert lizards (*Uromastyx acanthinura*); however, two populations of gonadotropic cells expressing FSH-like or both hormones (FSH and LH) were revealed suggesting that as in other vertebrates, gonadotropic cells can be unequivocally classified as bihormonal because they contain both glycoprotein hormones, FSH and LH.

In *Calotes versicolor* a gonadotropin-releasing hormone (GnRH-1) and bradykinin, and their respective receptors, have been identified in the ovaries that can have autocrine or paracrine actions (Singh *et al.* 2007). Although GnRH was originally identified as a hypothalamic neurohormone, several forms of GnRHs are now known to be expressed in a wide range of tissues, including the ovary (Jones 2011). In the Common Leopard Gecko (*Eublepharis macularius*) two GnRH forms (GnRH-I and GnRH-II) and three receptor types (GnRHR1, GnRHR2, and GnRHR3) were identified in the ovary (Ikemoto and Park 2007). Similarly, bradykinin that stimulates GnRH release in the hypothalamus is also produced in the granulosa cells of previtellogenic or early vitellogenic follicles, which suggest having a role in regulation of follicular development (Singh *et al.* 2007, 2008).

Maternal sex hormones have been detected in the egg yolks of some lizards (Lovern and Wade 2003a,b; Rhen *et al.* 2006; Radder and Shine 2007b). These hormones might originate from the ovary, blood plasma, or the embryo itself. During vitellogenesis steroid hormones can be deposited into the yolk (from thecal and granulosa cells) or originate from the mother's plasma after the egg has been ovulated (Moore and Johnston 2008). In *Anolis carolinensis* it has been found that maternal steroids can cross to the oviductal eggs through egg-shells before oviposition (Cohen and Wade 2010) and that changes in deposition of yolk steroids in response to diet may be minimal (Lovern and Adams 2008). Therefore, maternal plasma may have a greater influence on yolk steroid levels in oviductal eggs than on yolking follicles.

Ovulation. Ovulation is the release of the mature oocyte from the follicle. At the time of ovulation eggs are arrested in metaphase II (Cuellar 1971).

Ovulation involves enzymatic degradation and disruption of the follicular wall. The stigma is the point of follicular rupture during ovulation. It is visible externally as a pale region and corresponds to the area between the follicle and the ovarian surface that begins to thin and weaken.

Immediately before rupture of the stigma, the short microvilli of the granulosa lose their connections with the oocyte surface and do not contain pits or vesicles (Laughran *et al.* 1981). The collagen fibers dissociate and the theca accumulates extracellular fluids between the tunica albuginea and the theca externa. Extracellular fluid also accumulates between the granulosa cells and their basal lamina, with a reduction in follicular blood flow immediately before rupture (Jones *et al.* 1988). Therefore, the stigma is avascular and its rupture is due to ischemia and necrosis (Jones 1987).

The final number of follicles to be ovulated per clutch (instantaneous fecundity) depends on several factors: number of follicles growing together in a batch, the number of vitellogenic follicles produced from this batch, and the number of follicles that become atretic. Therefore, atresia is one of the mechanisms that govern the final number of ovulated eggs (Jones 1978; Méndez-de la Cruz *et al.* 1993). All of these numbers are determined by circulating hormones and peptides (Jones *et al.* 1982; Jones 2011). The number of follicles growing together in each batch is in turn related to the number of follicles recruited from the GBs. This number is linked to the number of oogonia per GB, which increases along with gonadotropin levels during reproductive recrudescence in seasonal reproductive lizards. Follicle stimulating hormone stimulates oogonial division as well as primary oocytes and primordial follicles (Jones *et al.* 1975, 1976).

The presence of a batch of vitellogenic follicles seems to inhibit the development of another set of follicles unless vitellogenic follicles are ovulated or become atretic (Jones *et al.* 1984; Shanbhag and Prasad 1993). However, in some species with a high annual reproductive output (e.g., *Pogona barbata*, Amey and Wittier 2000; *Calotes versicolor*, Radder *et al.* 2001; *Cnemidophorus lemniscatus*, Mojica *et al.* 2003; Madagascar Ground Gecko, *Paroedura picta*, Weiser *et al.* 2012) the clutches overlap such that the next clutch begins vitellogenesis before the first is laid. For these species hormonal regulation seems to differ. In gravid females of *C. versicolor* with vitellogenic follicles, levels of estradiol are high while plasma progesterone remains low allowing the growth of a new batch of eggs and suggesting that high levels of progesterone are not required for egg retention during late gravidity (Radder *et al.* 2001).

It has been observed that vitellogenesis and ovulation can also be promoted by prostaglandins (PGs), which can be considered regulators of steroidogenesis and are synthesized by the ovarian follicles and corpora lutea (Guillette *et al.* 1988; Gobbetti *et al.* 1993, 1994). Guillette *et al.* (1988) showed that ovarian tissue of a lizard synthesizes PGs, which might act in several ways to support or induce preovulatory degradative changes in the follicle during ovulation; i.e., support follicular contractile activity (Jones *et al.* 1984), yield collagen degradation and cellular dissociation in the follicular wall, regulate preovulatory vascular changes in the follicular wall and cause ischemic necrosis at the point of follicular rupture, as observed in *Anolis carolinensis* (Jones 1987; Jones *et al.* 1990). Gobbetti *et al.* (1993) identified both PGF2a and PGE2 in ovarian tissues of *Podarcis siculus*. Whereas PGE2 was produced during early vitellogenesis suggesting that it could have a stimulatory effect on estradiol release by the growing follicles, preovulatory follicles produced PGF2a, suggesting that it may be involved in ovulation. Ovulation is also correlated with elevated plasma testosterone concentrations in *Tiliqua nigrolutea* (Edwards and Jones 2001).

The processes that regulate ovarian growth, vitellogenesis, and ovulation in lizards are under the control of a complex network of intracellular and extracellular signals including GnRHs and their receptors, GTHs, several peptides, cyclic AMP (Borelli *et al.* 2002), and prostaglandins (Guillette *et al.* 1991; Gobetti *et al.* 1994).

Corpus luteum. All lizards, whether oviparous or viviparous, develop true, transitory endocrine structures after ovulation termed corpora lutea. These structures develop from the postovulatory follicular tissues (the granulosa and thecal tissues) (Browning 1973; Chieffi and Chieffi-Baccari 1999); therefore, the corpus luteum (CL) is the dominant structure in the ovary of gravid/pregnant females when fertilized eggs/embryos are retained in the oviduct for a species-specific duration before oviposition or parturition (Jones and Guillette 1982).

Luteal morphology is relatively consistent among lizards (Guillette *et al.* 1981; Fox and Guillette 1987; Guarino *et al.* 1998; Gómez and Ramírez-Pinilla 2004). After ovulation, the granulosa cells of the collapsed follicle hypertrophy, filling the empty follicle, and forming a central luteal mass (LCM) surrounded by a relatively thin theca (Varma 1970; Fox and Guillette 1987). In most lizard species only the hypertrophied granulosa cells are responsible for the formation of luteal tissue, which possesses a more or less well-defined vascular system; e.g., Italian Three-toed Skink (*Chalcides chalcides*; Guarino *et al.* 1998) and *Sceloporus jarrovii* (Guillette *et al.* 1981). The theca interna and its vascularization contribute in different ways in corpus luteum development. In some species, fibroblasts of the theca interna invade adjacent luteal cells and produce superficial septae and contribute to the composition of the luteal mass; e.g., Imbricate Alligator Lizard (*Barisia imbricata*; Martínez-Torres *et al.* 2003). In *Saara hardwicki* and *Zootoca vivipara* thecal luteal cells are organized in superficial small clusters (Chiaffi and Chiaffi-Baccari 1999). In other species the corpus luteum becomes richly vascularized and the theca interna invades the luteal cell mass forming fibrous dividing septae, but thecal tissue remains distinct from luteal cell mass (as in *Calotes versicolor*; Varma 1970).

The lifespan of a typical corpus luteum may be divided into three stages: luteogenesis, luteal maturity, and luteal regression or luteolysis (Fox and Guillette 1987). In early gravid females newly ruptured postovulatory follicles are observed in the state of luteinization. New formed corpora lutea have a large opening through which the ovulated oocyte was extruded. Corpora lutea are flaccid and contain a large number of erythrocytes in their luteal cavity and the thecae appear poorly developed. During luteal maturity the multilayered granulosa almost completely occupies the follicular cavity; therefore, the CLs are more compact and active. The central luteal mass acquires large lipid droplets and secretory granules in its cytoplasm (Fox and Guillette 1987). The ultrastructure of these luteinized cells shows abundant smooth endoplasmic reticulum, characteristic of steroid-secreting cells (Al-Amri *et al.* 2012). The theca interna is well developed and clearly distinct from the theca externa. At luteolysis in post-oviposition females, LCM cells have a vacuolated cytoplasm with small and pycnotic nuclei. Macrophages, as well as fibrous connective tissue from the thecal layers, invade the LCM, and the theca interna and externa become thinner and more fibrous. Corpora lutea do not fully degenerate in the ovaries but regress into permanent ovarian scars (corpora albicantia). These scars might be used to estimate total reproductive potential of a female (Trauth 1978; Fox and Guillette 1987).

Significant plasma levels of estrogen are found during oocyte growth, whereas progesterone levels are higher during ovulation and egg/embryo retention in the oviducts (Ciarcia *et al.* 1986). Luteal cells are steroidogenic and secrete progesterone (Jones 2011). Progesterone is a quantitatively important product of the corpus luteum and is considered to be the major source of circulating plasma progesterone concentrations (P4). Plasma progesterone begins to increase during the vitellogenic phase but only reaches maximum levels during gestation and, thus, during the luteal phase (Xavier 1982). In those groups in which viviparity has not evolved (e.g., turtles and birds) a predominantly preovulatory pattern of P4 production is observed. By contrast, a predominantly post-ovulatory pattern of P4 production is exhibited in those groups in which viviparity evolved (e.g., squamates: snakes and lizards) (Callard *et al.* 1992). Studies have shown a positive correlation between concentration of progesterone in the plasma and histological features of the CL as well as histochemical activity of Δ5-4–3β-HSD in luteal cell mass, suggesting that the CL is the major source of P4 during pregnancy (Martínez-Torres *et al.* 2003). Progesterone concentration is greatest during luteogenesis and is positively correlated with follicular atresia. Progesterone concentrations are moderate in mature CL and lowest at luteolysis (Fox and Guillette 1987). The steroidogenic enzyme 3β-HSD, which converts pregnenolone to P4, is detectable in luteinized granulosa cells by histochemistry. The intensity of 3β-HSD activity in the corpus luteum has been positively correlated with plasma P4 concentrations in both oviparous and viviparous species; e.g., *Calotes versicolor* (Shanbhag *et al.* 2001) and *Barisia imbricata imbricata* (Martínez-Torres *et al.* 2003).

Studies in several species of lizards have shown that the CLs vary greatly in the duration of their secretion and regression time (Fox and Guillette 1987; Guarino *et al.* 1998; Martinez-Torres *et al.* 2003). There is a strong correlation between luteal life and the period of gravidity or, in viviparous species, of pregnancy (Saidapur 1982; Xavier 1987; Guraya 1989; Villagrán-Santacruz and Méndez de la Cruz 1999). While both oviparous and viviparous species are capable of synthesizing P4, there are differences in the onset, duration and quantity of P4 secretion in animals with different reproductive modes. In oviparous lizards, plasma P4 is reported to peak shortly after ovulation in *Agama atra* (Van Wyk 1984) and *Pogona barbata* (Amey and Whittier 2000) or at mid- or even late gestation (Xavier 1987; Jones and Baxter 1991). In this latter condition, CLs are secretory during most of the gravid period, during egg-shell formation, and then regress around the time of oviposition (Jones and Guillette 1982; Shanbhag *et al.* 2001); therefore, luteal regression is positively correlated with oviposition (Jones 2011). The drop in P4 level during gravidity might facilitate the growth of a second set of vitellogenic follicles in those oviparous species with multiple clutches per reproductive season (Shanbhag *et al.* 2001). The adrenal glands also seem to secrete P4 to help in inhibition of oviposition and egg retention during unfavorable conditions when the lizards are forced to retain eggs in the oviduct (Shanbhag *et al.* 2001).

In most viviparous squamates, the CL have a longer lifespan and P4 production continues for a variable period after ovulation, remaining high throughout pregnancy and then markedly decreases close to delivery (Guillette 1987; Jones *et al.* 1997). The majority of studies have focused on this hormone especially in the maintenance of gestation and evolution of reptilian viviparity (Martinez-Torres *et al.* 2003; Jones and Guillette 1982; Callard *et al.* 1992; Bennet and Jones 2002; Jones 2011). Ovarian progesterone has an important role in the evolution of viviparity, as it inhibits myometrial contractions providing a primary condition for egg retention and viviparity. It also might inhibit estrogen-induced hepatic vitellogenin synthesis as part of both normal oviparous cycles and as a concomitant of placental evolution (Callard *et al.* 1992).

The participation of corpus luteum in the maintenance of gestation in the viviparous Squamata is variable. A CL is not necessary for the maintenance of gestation in some species that show early luteolysis (*Chalcides ocellatus*, Badir 1968; *Mabuya* sp., Gómez and Ramírez-Pinilla 2004; *Barisia i. imbricata*, Martínez-Torres *et al.* 2010); consequently, a secondary extra-ovarian source of P4 capable of maintaining gestation must exist in the absence of luteal tissue. The P4 secretion by adrenal glands has been suggested in *Chalcides chalcides* (Guarino *et al.* 1998) and *Barisia i. imbricata* (Martínez-Torres *et al.* 2003, 2010; Guillette and Fox 1985; Shine and Guillette 1988). However, adrenals produced P4 at approximately one-tenth the rate of production by CLs (Bennet and Jones 2002). In *Sceloporus jarrovii*, the chorioallantoic placenta is a possible source of progesterone during late pregnancy (Guillette *et al.* 1981). *In vitro* production of P4 by the chorioallantoic placenta has been confirmed in the Southern Snow Skink (*Niveoscincus microlepidotus*; Girling and Jones 2003) and in *S. jarrovii* (Painter and Moore 2005). Also, in *C. chalcides* the chorioallantoic placenta produces progesterone and may take over the primary role of P4 production when the CL degenerates in late pregnancy (Guarino *et al.* 1998).

8.2.3 Follicular Atresia

Follicular atresia, the process by which the ovarian follicles degenerate, occurs in all vertebrate ovaries from cyclostomes to mammals (Gouder *et al.* 1979). A large proportion of follicles undergo atresia to form corpora atretica (CA) before completing their development (Guraya 1989). Follicular atresia in the ovaries of lizards can be rare or common, depending upon reproductive state of the female, season, and species (Saidapur 1978). Although atresia can affect follicles at any stage of their development (Varma 1970) at any time of their reproductive cycle, among lizards atresia is most common during the previtellogenic stages (Fox and Guillette 1987) and in follicles with polymorphic granulosa in post-ovulatory ovaries (Guraya and Varma 1976). Low rates of atresia at vitellogenic stages have been documented in the Metallic and Ocellated Cool-skinks (*Niveoscincus metallicus* and *N. ocellatus*; Jones and Swain 1996; Jones *et al.* 1997) and *Mabuya* sp. (Gómez and Ramírez-Pinilla 2004). Thus, as suggested by Jones and Swain (2000) for *Niveoscincus* spp., clutch size is determined early in follicular development. High levels of follicular atresia are also observed during gravidity and pregnancy (Leyton and Valencia 1992; Manriquez-Morán *et al.* 2005; Gómez and Ramírez-Pinilla 2004) and are positively correlated with elevated plasma P4 concentrations (Fox and Guillette 1987), which inhibit the GTH secretion and consequently follicular development. Experimental administration of progesterone during the breeding period in *Podarcis siculus* causes atresia of the vitellogenic follicles (Limatola *et al.* 1989).

Histological characteristics of different stages of follicular atresia have been described in the ovaries of several species of lizards (Corso *et al.* 1978; Saidapur 1978; Guraya and Varma 1976; Trauth 1978; Fox and Guillette

1987; Guraya 1989; Gómez and Ramírez-Pinilla 2004). Atretic follicles are classified as derived from early previtellogenic follicles, advanced or mature previtellogenic follicles, and vitellogenic follicles. During gravidity, even the germinal beds contain atretic primary follicles (Fox and Guillette 1987).

The process of atresia commences with cessation of growth of the oocyte, disintegration of its nucleus, and ooplasm shrinkage by loss of water. The ooplasm appears highly flocculent and vacuolated (Fig. 8.1A). The oolemma is undulated and the ZP appears folded inward toward the ooplasm. The zona radiata and vitelline membrane break up. Granulosa cells became polymorphic, show hypertrophy and hyperplasia, and give rise to leucocytes and macrophages that invade the oocyte and remove the ooplasm by phagocytosis and digestion. The granulosa cells develop many cytoplasmic vacuoles. The theca interna possesses vacuolated spaces and granulocytes, and the fibroblasts of the theca interna appear swollen. Phagocytic leucocytes invade the yolk mass of vitellogenic follicles during atresia. Granulosa cells and phagocytes have vacuolated and granular cytoplasm with lipid droplets, as well as ingested yolk globules. Granulosa cells along with thecal elements evaginate into the ooplasm forming rugose or villi-like folds making inroads into the ooplasm. After removing the ooplasmic contents, granulosa cells diminish in size, show pyknosis, and degenerate. Simultaneously, the thecal tissue becomes relatively more conspicuous and theca interna cells are hypertrophied to form interstitial gland cells in the walls of degenerated follicles (Guraya and Varma 1976). Based on conflicting evidence (Guraya 1976; Gouder *et al.* 1979; Ciarcia *et al.* 1993), atretic follicles at this stage are considered degenerating with active steroidogenesis.

At the end of the process described above, the atretic follicles degenerate. They appear as saclike structures distorted in shape with only a very thin theca that continues to shrink. Thecal elements merge with the ovarian stroma, eventually leaving only a thecal framework that is gradually absorbed (Browning 1973). At this stage, these structures can be confused with regressed corpora lutea or even corpora albicantia (Trauth 1978; Saidapur 1982).

A type of bursting atresia has been described in *Calotes versicolor* and in the Argentine Black and White Tegu (*Salvator merianae*; Gouder *et al.* 1979; García-Valdez *et al.* 2011). It occurs in large-sized follicles that burst and extrude yolk into the ovarian stroma and, consequently, ectopic yolk masses are formed. The atretic follicles of bursting type show rugose granulosa with an orifice through which yolk is extruded.

Atresia significantly decreases the number of eggs to be ovulated. Therefore, follicular atresia has been considered an important factor in the proximate control of clutch size in many lizard species (Guillette and Casas-Andrew 1987; Méndez de la Cruz *et al.* 1993). Jones *et al.* (1982) inferred that the rate of atresia may be relatively fixed within taxa depending on the size of the clutch and the number of germinal beds present. Rates of follicular atresia in preovulatory ovaries are inversely proportional to instantaneous fecundity (Jones *et al.* 1978). In some monoautochronic lizards, each adult ovary contains two germinal beds on the dorsal surface and therefore contains a relatively high number of growing follicles; however, extensive follicular atresia reduces the number of preovulatory follicles to one in each ovary (Jones *et al.* 1978). During each cycle, one follicle becomes atretic and one ovulates. A compensatory follicular hypertrophy occurs, leading to replacement of lost follicles and maintenance of the follicular size hierarchy (Jones and Summers 1984).

Atresia can also control a female's readiness for reproduction. Proximate factors in the local environment influence whether these follicles continue growing or go through atresia. For example, in the Galápagos Marine Iguanas (*Amblyrhynchus cristatus*), a species in which females typically reproduce biennially, nearly all adult-sized females initiated follicular development prior to the lekking period; however, atresia allows for reabsorption of all developing follicles 5–15 days before the start of copulations in about 38% of females. These females are non-receptive and showed significant peaks in both testosterone and progesterone during follicular atresia, suggesting that these hormones may be involved in the inhibition of vitellogenesis (Vitouseck *et al.* 2010). Also, in the teiid lizard *Salvator merianae* a blockade of the ovarian cycle due to a massive previtellogenic follicular atresia has been observed as a result of female isolation from reproductive males (Manes *et al.* 2007; García-Valdez *et al.* 2011). Therefore, the ratio between the numbers of growing versus atretic follicles seems to be influenced by environmental (proximate) factors that control circulating concentrations of gonadotropins and steroids (Jones *et al.* 1976). In fact, stressors such as low nutritional state, high predation pressure, changes in temperature, precipitation or environmental humidity, among others, stimulate a rise in plasma progesterone and/or corticosterone concentrations (Guillette and Fox 1985; Summers 1988; Vitouseck *et al.* 2010; reviewed in Tokarz and Summers 2011), which are capable of modifying clutch size by increasing rates of atresia.

In order to identify and localize the sites of sex hormone production in the lizard ovary the distribution of steroid dehydrogenases has often been examined with histochemistry (Jones *et al.* 1974; Gouder and Nadkarni 1976). Histochemical studies have shown that the atretic follicles are not merely degenerating elements of the ovary but that they synthesize biologically active steroid hormones (Gouder *et al.* 1979; Guillette *et al.* 1981; Hammouche *et al.* 2009). The granulosa cells of the atretic follicles of *Calotes versicolor* exhibited steroid converting enzyme activity (e.g., hydroxysteroiddehydrogenases HSDHs: Δ5-3β-HSDH, 17β-HSDH,

11β-HDSH, G-6-PDH, NADH diaphorase, LDH and ICDH) for a short duration that indicated their transient steroidogenic capacity (Gouder *et al.* 1979). In fact, theca interna cells show an increased activity for these enzymes in the atretic follicles and later persist as interstitial gland cells in the ovarian stroma. In the lizard *Uromastyx acanthinura*, granulosa cells of atretic follicles stained positive for testosterone, 17β-estradiol, and progesterone. β-endorphin was present in the atretic follicles, particularly in the apoptotic granulosa cells (Hammouche *et al.* 2009). Also, localization of GnRH I and GnRH I receptor in the granulosa cells of the atretic follicle in ovary of *C. versicolor* suggests that GnRH I may be regulating follicular selection by promoting atresia (Singh *et al.* 2008).

8.3 FUTURE RESEARCH DIRECTIONS

In lizards, the understanding of the dynamic processes of oogenesis from germ cell development to ovulation is far from complete. This chapter demonstrated that there are several and interesting studies on cellular, ultra-structural, and endocrine aspects of oogenesis and oocyte growth in lizards; however, these usually are focused in only a few lizard species, which can be considered "model" species. In spite of the great diversity of lizards (more than 5000 species), and according to the available literature in common data bases, almost nothing is known about oogenesis of several lizard families; e.g., worm lizards of Amphisbaenia and in the lizard families of Gymnophtalmidae, Gerrhosauridae, Anniellidae, Helodermatidae, Shinisauridae, Xenosauridae, Dibamidae, Lanthanotidae, and Varanidae. Most of the research on lizard oogenesis has been conducted in only five genera of Iguania (*Anolis, Crotaphytus, Ctenosaura, Liolaemus* and *Sceloporus*), seven of Gekkota (*Eublepharis, Gekko, Hemidactylus, Heteronotia, Lepidodactylus, Tarentola* and *Woodworthia*), eight of Scincidae (*Bassiana, Carlia, Chalcides, Eutropis, Mabuya, Niveoscincus, Plestiodon, Tiliqua* and *Trachylepis*), five of Agamidae (*Agama, Calotes, Pogona, Saara,* and *Uromastyx*), three of Lacertidae (*Acanthodacylus, Podarcis* and *Zootoca*), two of Anguidae (*Elgaria* and *Barisia*), two of Teiidae (*Aspidoscelis* and *Salvator*), two of Cordylidae (*Cordylus* and *Platysaurus*), one of Chamaeleonidae (*Chamaeleo*), and one of Xantusidae (*Xantusia*).

Most of the studies on this topic were developed from 1960 to the 1990s. At present only a few research groups in the world are studying the morphology and physiology of lizard ovaries. These groups have been preliminarily revealing structural and physiological changes that occur during lizard oogenesis through genomic and proteomic studies (e.g., Di Stasio *et al.* 1999; Borrelli *et al.* 2001, 2002; Sreenivasulu *et al.* 2002; Endo and Park 2005; Maurizii *et al.* 2009; Tripathi and Raman 2010; Verderame and Limatola 2010; Raucci and Di Fiore 2011).

The understanding of lizard oogenesis could be greatly invigorated with a deeper understanding of molecular aspects incorporated with advanced microscopy; e.g., analysis of gene transcript/protein expression and protein sub-cellular localization trough confocal microscopy within the diverse ovarian cell types during different stages of oogenesis. Furthermore, classic molecular genetic analysis, such as transcript and protein localization by means of *in situ* hybridization and immunohistochemistry can be leveraged by accurate quantification of transcript levels of particular genes at different stages of oogenesis through real-time PCR. More modern tools might have an enormous impact in the field of lizard oogenesis, such as the implementation of massive parallel RNA sequencing using the wide array next generation sequencing platforms and proteomics analysis by mass spectrometry. These analyses have the potential to deliver a more global picture of gene/protein expression during oogenesis, where almost all expressed genes can be interrogated (gene sequence, structure and abundance), shedding light into important molecular aspects of lizard oogenesis; e.g., ovarian gene content (also gene discovery), gene structure and sequence divergence, ovarian specific gene expression and temporal differential gene expression.

Multiple avenues of investigation on lizard oogenesis exist using the modern techniques described above: PGC origin, development and migration; oocyte and follicular recruitment; oocyte maturation processes; mechanisms controlling follicular hierarchy in lizards with different ovulation types; yolk protein synthesis and processing within the developing oocyte; vitellogenin-encoding genes in lecitotrophic and matrotrophic clades; structure and function of the zona pellucida and its role in fertilization; identification of the several circulating endocrine and locally-acting paracrine and autocrine factors regulating oogenesis; understanding of the complex hormonal cross-talk between the oocyte and granulosa and thecal layers at all stages of follicle development, including the identification of pathways regulating the synthesis and response to endocrine and growth factors involved during oocyte maturation and ovulation.

Furthermore, the study of lizard oogenesis can be notably influenced by the analysis of lizard genomes. At the time this chapter was written only three squamate genomes have been sequenced, the green anole lizard, the Burmese python snake and the king cobra snake. Further characterization of other Lepidosauria genomes and more genomic resources might soon develop to address several of the above research avenues in the context of gene structure and regulatory elements, gene duplication, gene degradation (pseudogenization) and the evolution of lizard specific ovarian protein coding genes as well as non-coding RNAs.

8.4 ACKNOWLEDGMENTS

The authors acknowledge the Escuela de Biología, Universidad Industrial de Santander and Departamento de Biología, Universidad Nacional de Colombia for supporting our research work; C. Taddei and G. Maurizii provided images on intercellular bridges. C. Hernández helped us obtain histological slides and capture images; M. Ogielska, D. Siegel, J. P. Ramírez, F. Leal and O. Tarazona, reviewed the manuscript.

8.5 LITERATURE CITED

Al-Amri, I. S., Mahmoud, I. Y., Waring, C. P., Alkindi, A. Y., Khan, T. and Bakheit, C. 2012. Seasonal changes in plasma steroid levels in relation to ovarian steroidogenic ultrastructural features and progesterone receptors in the house gecko, *Hemidactylus flaviviridis*, in Oman. General and Comparative Endocrinology 177: 46–54.

Al-Dokhi, O. A. 1998. Ovarian structure and follicle development in the lizard *Mabuya brevicollis*. Saudi Journal of Biological Sciences 5: 12–23.

Amey, A. P. and Whittier, J. M. 2000. Seasonal patterns of plasma steroid hormones in males and females of the Bearded Dragon Lizard, *Pogona barbata*. General and Comparative Endocrinology 117: 335–342.

Andreuccetti, P. 1992. An ultrastructural study of differentiation of pyriform cells and their contribution to oocyte growth in representative Squamata. Journal of Morphology 212: 1–11.

Andreuccetti, P. and Carrera, M. 1987. The differentiation of the zona pellucida (vitelline envelope) in the lizard *Tarentola mauritanica*. Devolpment, Growth and Differerentiation 29: 113–122.

Andreuccetti, P., Taddei, C. and Filosa, S. 1978. Intercellular bridges between follicle cells and oocyte during the differentiation of follicular epithelium in *Lacerta sicula*. Journal of Cell Science 33: 341–350.

Andreuccetti, P., Motta, C. M. and Filosa, S. 1990. Regulation of oocyte number during oocyte differentiation in the lizard *Podarcis sicula*. Cell Differentiation and Development 29: 129–141.

Andreuccetti, P., Iodice, M., Prisco, M. and Gualtieri, R. 1999. Intercellular bridges between granulose cells and the oocyte in the elasmobranch *Raya asterias*. Anatomical Record 255: 180–187.

Andreuccetti, P., Famularo, C., Gualtieri, R. and Prisco, M. 2001. Pyriform cell differentiation in *Podarcis sicula* is accompanied by the appearance of surface glycoproteins bearing a-GalNAc terminated chains. Anatomical Record 263: 1–9.

Austin, H. B. 1988. Differentiation and development of the reproductive system in the iguanid lizard, *Sceloporus undulatus*. General and Comparative Endocrinology 72: 351–363.

Badir, N. 1968. Structure and function of the corpus luteum during gestation in the viviparous lizard *Chalcides ocellatus*. Anatomischer Anzeiger 122: 1–10.

Bennett, E. J. and Jones, S. M. 2002. Interrelationships among plasma progesterone concentrations, luteal anatomy and function, and placental ontogeny during gestation in a viviparous lizard (*Niveoscincus metallicus*: Scincidae). Comparative Biochemistry and Physiology Part A 131: 647–656.

Blackburn, D. 1992. Convergent evolution of viviparity, matrotrophy, and specializations for fetal nutrition in reptiles and other vertebrates. American Zoologist 32: 313–321.

Borrelli, L., De Stasio, R., Parisi, E. and Filosa, S. 2001. Molecular cloning, sequence, and expression of follicle-stimulating hormone receptor in the lizard *Podarcis sicula*. Gene 275: 149–156.

Borrelli, L., De Stasio, R., Parisi, E. and Filosa, S. 2002. Relationship between adenylate cyclase sensitivity to follitropin and FSH receptor mRNA expression in the ovary of the lizard *Podarcis sicula*. Molecular Reproduction and Development 62: 210–215.

Bou-Resly, M. 1974. Ultrastructural studies on the intercellular bridges between the oocyte and follicle cells in the lizard *Acanthodactylus scutellatus* Hardyn. Zeitschrift für Anatomie und Entwicklungsgeschichte 143: 239–254.

Brawand, D., Whali, W. and Kaessman, H. 2008. Loss of egg yolk genes in mammals and the origin of lactation and placentation. PLoS Biology 6: e63.

Browning, H. C. 1973. The evolutionary history of the corpus luteum. Biology of Reproduction 8: 128–157.

Callard, I. P., Fileti, L. A., Perez, L. E., Sorbera, L. A., Giannoukos, G., Klosterman, L. L., Tsang, P. and McCracken, J. A. 1992. Role of the corpus luteum and progesterone in the evolution of vertebrate viviparity. American Zoologist 32: 264–275.

Carnevali, O., Mosconi, G., Angelini, F., Limatola, E., Ciarcia, G. and Polzonetti-Magni, A. 1991. Plasma vitellogenin and 17β-stradiol levels during the annual reproductive cycle of *Podarcis s. sicula* Raf. General and Comparative Endocrinology 84: 337–343.

Chen, J. S., Sappington, T. W. and Raikhel, A. S. 1997. Extensive sequence conservation among insect, nematode, and vertebrate vitellogenesis reveals ancient common ancestry. Journal of Molecular Evolution 44: 440–451.

Chieffi, G. and Chieffi-Baccari, G. 1999. Corpora Lutea of nonmammalian species. pp. 680–688. In E. Knobil and J. D. Neil (eds.), *Encylopedia of Reproduction*. Academic Press, San Diego.

Ciarcia, G., Paolucci, M. and Di Fiore, M. M. 1993. Changes in ovarian follicles and *in vitro* sex hormone release in the lizard *Podarcis sicula sicula*. Molecular Reproduction and Development 35: 257–60.

Ciarcia, G., Angelini, F., Polzonetti, A., Zexani, M. and Botte, V. 1986. Hormones and reproduction in the lizard *Podarcis s. sicula* Raf. pp. 95–100. In I. Assenmacher and J. Boissin (eds.), *Endocrine Regulations as Adaptive Mechanisms to the Environment*. Editions du CNRS, Paris.

Cohen, R. E. and Wade, J. 2010. Newly deposited maternal hormones can be detected in the yolks of oviductal eggs in the green anole lizard. Journal of Experimental Zoology Part A: Ecological Genetics and Physiology 313: 352–358.

Corso, G., Frau, A. M. L. and Pala, M. 1978. Strutture dell' ovario adulto di *Chalcides ocellatus tiligugu* (Gmelin) durante il corso dell'anno. Archivio Italiano di Anatomiae Embriologia 83: 207–224.

Corso, G. and Pala, M. 1987. Aspetti morfologici del follicolo ovárico previtellogenetico dello Scincide viviparo *Chalcides ocellatus tiligugu* (Gmelin) (Squamata. Scincidae). Archivio Italiano di Anatomia e Embriologia 92: 263–281.

Cree, A., Guillette, L. J., Jr., Brown, M. A., Chambers, G. K., Cockrem, J. F. and Newton, J. D. 1991. Slow estradiol-induced vitellogenesis in the tuatara, *Sphenodon punctatus*. Physiological Zoology 64: 1234–1261.

Cree, A., Cockrem, J. F. and Guillette, L. J., Jr. 1992. Reproductive cycles of male and female tuatara (*Sphenodon punctatus*) on Stephens Islands, New Zealand. Journal of Zoology, London 226: 199–217.

Cuellar, O. 1971. Reproduction and the mechanism of meiotic restitution in the parthenogenetic lizard *Cnemidophorus uniparens*. Journal of Morphology 133: 139–165.

Davenport, I. R., Weaver, A. L. and Wourms, J. P. 2011. A novel Set of structures within the elasmobranch ovarian follicle. Journal of Morphology 272: 557–565.

De Caro, M., Indolfi, P., Iodice, C., Spagnuolo, S., Tammaro, S. and Motta, C. M. 1998. How the ovarian follicle of *Podarcis sicula* recyclesthe DNA of its nurse, regressing follicle cells. Molecular Reproduction and Development 51: 421–429.

Derickson, W. K. 1976. Lipid storage and utilization in reptiles. American Zoologist 16: 711–724.

De Stasio, R., Borrelli, L., Kille, P., Parisi, E. and Filosa, S. 1999. Isolation, characterization and molecular cloning of cathepsin D from lizard ovary: changes in enzyme activity and mRNA expression throughout ovarian cycle. Molecular Reproduction and Development 52: 126–134.

Doddamani, L. S. 1994. Histoenzimological studies on embryonic and posthatching development of the ovary in the tropical oviparous lizard *Calotes versicolor*. Journal of Morphology 222: 1–10.

Edwards, A. and Jones, S. M. 2001. Changes in plasma progesterone, estrogen, and testosterone concentrations throughout the reproductive cycle in female viviparous blue-tongued skinks, *Tiliqua nigrolutea* (Scincidae), in Tasmania. General and Comparative Endocrinolpogy 122: 260–269.

Edwards, A., Jones, S. M. and Davies, N. W. 2002. A possible alternative to 17β-estradiol in a viviparous lizard, *Tiliqua nigrolutea*. General and Comparative Endocrinology 129: 114–121.

Endo, D. and Park, M. K. 2005. Molecular cloning of P450 aromatase from the leopard gecko and its expression in the ovary. Journal of Steroid Biochemistry and Molecular Biology 96: 131–40.

Etches, R. J. and Pettite, J. N. 1990. Reptilian and avian follicular hierarchies: Models for the study of ovarian development. Journal of Experimental Zoology Suppplement 4: 112–122.

Filosa, S. 1973. Biological and citological aspects of the ovarian cycle in *Lacerta sicula sicula* Raf. Monitore Zoologico Italiano 7: 151–165.

Filosa, S. and Taddei, C. 1976. Intercellular bridges in lizard oogenesis. Cell Differentiation 5: 199–206.

Filosa, S., Taddei, C. and Andreuccetti, P. 1979. The differentiation and proliferation of the follicle cells during oocyte growth in *Lacerta sicula*. Journal of Embryology and Experimental Morphology 5: 5–15.

Fox, S. L. and Guillette, L. J., Jr. 1987. Luteal morphology, atresia, and plasma progesterone concentrations during the reproductive cycle of two oviparous lizards, *Crotaphytus collaris* and *Eumeces obsoletus*. American Journal of Anatomy 179: 324–332.

Ganesh, C. B. and Yajurvedi, H. N. 2002. Stress inhibits seasonal and FSH-induced ovarian recrudescence in the lizard, *Mabuya carinata*. Journal of Experimental Zoology 292: 640–648.

García-Valdez, M. V., Chamut, S., Valdez-Jaen, G., Arce, O. E. and Manes, M. E. 2011. Dynamics of ovarian follicles in *Tupinambis merianae* lizards. Acta Herpetologica 6: 303–313.

Gharzi, A., Yari, A. and N. Rastegar-Pouyani. 2012. Morphology and Histology of ovary´s germinal beds in the lacertid lizard *Acanthodactylus boskianus* (Sauria, Lacertidae). Research Journal of Animal Sciences 6: 8–11.

Girling, J. E. and Jones, S. M. 2003. *In vitro* progesterone production by maternal and embryonic tissues during gestation in the southern snow skink (*Niveoscincus microlepidotus*). General and Comparative Endocrinology 133: 100–108.

Gobbetti, A., Zerani, M., Di Fiore M. M. and Botte, V. 1993. Prostaglandins and sex steroids from reptilian (*Podarcis sicula sicula*) ovarian follicles at different developmental stages. Zoological Science 10: 321–328.

Gobbetti, A., Zerani, M. and Di Fiore, M. M. 1994. GnRH and substance P regulate prostaglandins and sex steroids from reptilian (*Podarcis sicula sicula*) ovarian follicles and corpora lutea. General and Comparative Endocrinology 93: 153–162.

Gómez, D. and Ramírez-Pinilla, M. P. 2004. Ovarian histology of the viviparous matrotrophic lizard *Mabuya mabouya* (Squamata, Scincidae). Journal of Morphology 259: 90–105.

Gouder, B. Y. M. and Nadkarni, V. B. 1976. Steroid synthesizing cellular sites in the ovaries of *Calotes versicolor* (Daud.), *Hemidactylus flaviviridis* (Ruppel) and *Chamaeleon calcaratus* (Boulenger): Histochemical study. Indian Journal of Experimental Biology 14: 647–651.

Gouder, B. Y. M., Nadkarni, V. B. and Appaswamy, M. R. 1979. Histological and histochemical studies on follicular atresia in the ovary of the lizard, *Calotes versicolor*. Journal of Herpetology 13: 451–456.

Guarino, F. M., Paulesu, L., Cardone, A., Bellini, L., Ghiara, G. and Angelini, Y. F. 1998. Endocrine activity of the corpus luteum and placenta during pregnancy in *Chalcides chalcides* (Reptilia, Squamata). General and Comparative Endocrinology 11: 261–270.
Guillette Jr., L. J. 1987. The evolution of viviparity in fishes, amphibians and reptiles: An endocrine approach. pp. 523–562. In D. O. Norris and R. E. Jones (eds.), *Hormones and Reproduction in Fishes, Amphibians, and Reptiles*. Plenum Press, New York.
Guillette, L. J., Jr. and Fox, S. L. 1985. Effect of deluteinization on plasma progesterone concentration and gestation in the lizard, *Anolis carolinensis*. Comparative Biochemistry and Physiology 80A: 303–306.
Guillette, L. J., Jr., Spielvogel, S. and Moore, F. L. 1981. Luteal development, placentation and plasma progesterone concentration in the viviparous lizard *Sceloporus jarrovii*. General and Comparative Endocrinology 43: 20–29.
Guillette, L. J., Jr. and Casas-Andreu, G. 1987. The reproductive biology of the high elevation Mexican lizard, *Barisia imbricata*. Herpetologica 43: 29–38.
Guillette, L. J., Jr., Herman, C. A. and Dickey, D. A. 1988. Synthesis of prostaglandins by tissues of the viviparous lizard, *Sceloporus jarrovii*. Journal of Herpetology 22: 180–185.
Guillette, L. J., Dubois, D. H. and Cree, A. 1991. Prostaglandins, oviducal function, and parturient behavior in nonmammalian vertebrates. American Journal of Physiology-Regulatory, Integrative and Comparative Physiology 260: R854–R861.
Guillette, L. J., Jr., Gross, T. S., Masson, G. R., Matter, J. M. Percival, H. F. and Woodward, A. R. 1994. Developmental abnormalities of the gonad and abnormal sex hormone concentrations in juvenile alligators from contaminated and control lakes in Florida. Environmental Health Perspectives 102: 680–688.
Guraya, S. S. 1968. Further morphological and histochemical studies on the yolk nucleus and associated cell components in the developing oocyte of the Indian wall lizard. Journal of Morphology 124: 283–294.
Guraya, S. S. 1989. Ovarian follicles in reptiles and birds. Zoophysiology, Vol. 24, Springer, Berlin, 285 pp.
Guraya, S. S. and Varma, S. K. 1976. Morphology of ovarian changes during the reproductive cycle of the house lizard, *Hemidactylus flaviviridis*. Acta Morphologica Neerlando-Scandinavica 14: 165–192.
Hadley, N. F. and Christie, W. W. 1974. The lipid composition and triglyceride structure of eggs and fat bodies of the lizard *Sceloporus jarrovii*. Comparative Biochemistry Physiology 48: 275–284.
Hamlett, W. C., Jezior, M. and Spieler, R. 1999. Ultrastructural analysis of folliculogenesis in the ovary of the yellow spotted stingray *Urolophus jamaicensis*. Annals of Anatomy 181: 159–172.
Hammouche, S., Gernigon Spychalowicz, T. H. and Exbrayat, J. M. 2007. Immunolocalization of estrogens and progesterone receptors within the ovary of the lizard *Uromastyx acanthinura* from vitellogenesis to rest season. Folia Histochemica Cytobiologica 45(Supp. 1): 23–27.
Hammouche, S., Gernigon, T. and Exbrayat, J. M. 2009. Correlation between ovarian steroidogenesis and β-endorphin in the lizard *Uromastyx acanthinura*: Immunohistochemical approach. Folia Histochemica et Cytobiologica 47: S95–S100.
Hernández-Franyutti, A., Uribe Aranzábal, M. C. and Guillette, L. J., Jr. 2005. Oogenesis in the viviparous matrotrophic lizard *Mabuya brachypoda*. Journal of Morphology 265: 152–164.
Ho, S.-M., Kleis, S., McPherson, R., Heisermann, G. J. and Callard, I. P. 1982. Regulation of vitellogenesis in reptiles. Herpetologica 38: 40–50.
Hubert, J. 1970. Etude cytologique et cytochimique des cellules germinales des reptiles au cours du développement embryonnaire et aprés la naissance. Zeitschrift für Zellforschung und Mikroskopische Anatomie 107: 249–264.
Hubert, J. 1971a. Localisation extra-embryonnaire des gonocytes chez l'embryon d'orvet *Aguis fragilis* L. Archives d'anatomie, d'histologie et d'embryologie normales et expérimentales 60: 261–268.

Hubert, J. 1971b. Étude histologique et ultrastructurale de la granulosa a certains stades de depeloppement du follicule ovarien chez un Lézard *Lacerta vivipara* Jacquin. Zeitschrift für Zellforschung und Mikroskopische Anatomie 115: 46–59.

Hubert, J. 1971c. Aspects ultrastructuraux des relations entre les couches folliculaires et el´ovocyte depuis la formation tu follicule jusq´au debut de la vitellogénese chez le Lézard *Lacerta vivípara* Jacquin. Zeitschrift für Zellforschung und Mikroskopische Anatomie 116: 240–249.

Hubert, J. 1973. Les cellules piriformes du follicule ovarien de certain Reptiles. Archives d'anatomie, d'histologie et d'embryologie normales et expérimentales 56: 5–18.

Hubert, J. 1976. Étude ultrastructurale des cellules piriformes du follicule ovarien chez cinq sauriens. Archives d'anatomie, d'histologie et d'embryologie normales et expérimentales 65: 47–58.

Hubert, J. 1985. Origin and development of oocytes. pp. 41–74. In C. Gans, F. S. Billett and P. F. A. Maderson (eds.), *Biology of the Reptilia, Development A*. John Wiley and Sons, New York.

Ibrahim, M. M. and Wilson, I. B. 1989. Light and electron microscope studies on ovarian follicles in the lizard *Chalcides ocellatus*. Journal of Zoology 218(2): 187–208.

Iguchi, T., Fukazawa, Y., Uesugi, Y. and Takasugi, N. 1990. Polyovular follicles in mouse ovaries exposed neonatally to diethylstilbestrol *in vivo* and *in vitro*. Biology of Reproduction 43: 478–484.

Ikemoto, T. and Park, M. K. 2007. Comparative analysis of the pituitary and ovarian GnRH systems in the leopard gecko: signaling crosstalk between multiple receptor subtypes in ovarian follicles. Journal of Molecular Endocrinology 38: 289–304.

Jones, R. E. 1978. Control of follicular selection. pp. 827–840. In R. E. Jones (ed.), *The Vertebrate Ovary: Comparative Biology and Evolution*. Plenum Press, New York.

Jones, R. E. 1987. Ovulation: insights about the mechanisms based on a comparative approach. pp. 203–240. In D. O. Norris and R. E. Jones (eds.), *Hormones and Reproduction in Fishes, Amphibians, and Reptiles*. Plenum Press, New York.

Jones, R. E. and Baxter, D. C. 1991. Gestation with emphasis on corpus luteum biology, placentation and parturition. pp. 238–244. In P. K. T. Pang and M. P. Schreibman (eds.), *Vertebrate Endocrinology: Fundamentals and Biomedical Implications, vol. 4*. Academic Press, New York.

Jones, R. E. and Summers, C. 1984. Compensatory follicular hypertrophy during the ovarian cycle of the house gecko, *Hemidactylus frenatus*. Anatomical Record 209: 59–65.

Jones, R. E., Sedgley, N. B., Gerrard, A. M. and Roth, J. J. 1974. Endocrine control of clutch size in reptiles III. Δ5-3β-hydroxysteroid dehydrogenase activity in different sized ovarian follicles of *Anolis carolinensis*. General and Comparative Endocrinology 22: 448–453.

Jones, R. E., Tokarz, R., Roth, J., Platt, J. and Collins, A. 1975. Mast cell histamine and ovarian follicular growth in the lizard *Anolis carolinensis*. Journal of Experimental Zoology 193: 343–352.

Jones, R. E., Tokarz, R. R., LaGreek, F. T. and Fitzgerald, K. T. 1976. Endocrine control of clutch size in reptiles. VI. Patterns of FSH-induced ovarian stimulation in adult *Anolis carolinensis*. General and Comparative Endocrinology 30: 101–116.

Jones, R. E., Fitzgerald, K. T. and Duvall, D. 1978. Quantitative analysis of the ovarian cycle of the lizard, *Lepidodactylus lugubris*. General and Comparative Endocrinology 35: 70–76.

Jones, R. E. and Guillette, L. J., Jr. 1982. Hormonal control of parturition and oviposition in lizards. Herpetologica 38: 80–93.

Jones, R. E., Swain, T., Guillette, L. J., Jr. and Fitzgerald, K. T. 1982. The comparative anatomy of lizard ovaries, with emphasis on the number of germinal beds. Journal of Herpetology 16: 248–252.

Jones, R.E., Austin, H.B. and Summers, C. H. 1984. Spontaneous, rhythmic contractions of the ovarian follicular wall of a lizard (*Anolis carolinensis*). General and Comparative Endocrinology 56: 252–257.

Jones, R. E., Austin, H. B., Lopez, K. H., Rand, M. S. and Summers, C. H. 1988. Gonadotropin-induced ovulation in a reptile (*Anolis carolinensis*): histological observations. General Comparative Endocrinology 72: 312–322.
Jones, R. E., Orlicky, D. J., Austin, H. B., Rand, M. S. and Lopez, K. H. 1990. Indomethacin inhibits ovarian PGE secretion and gonadotropin-induced ovulation in a reptile (*Anolis carolinensis*). Journal of Experimental Zoology 255: 57–62.
Jones, S. 2011. Hormonal regulation of ovarian function in reptiles. pp. 89–115. In D. O. Norris and K. H. Lopez (eds.), *Hormones and Reproduction of Vertebrates, Vol. 3. Reptiles.* Academic Press, New York.
Jones, S. M. and Swain, R. 1996. Annual reproductive cycle and annual cycles of reproductive hormones in plasma of female of *Niveoscincus metallicus* (Scincidae) from Tasmania. Journal of Herpetology 30: 140–146.
Jones, S. M. and Swain, R. 2000. Effects of exogenous FSH on follicular recruitment in a viviparous lizard *Niveoscincus metallicus* (Scincidae). Comparative Biochemistry and Physiology 127 A: 487–493.
Jones, S. M. and Swain, R. 1996. Annual reproductive cycle and annual cycles of reproductive hormones in plasma of female of *Niveoscincus metallicus* (Scincidae) from Tasmania. Journal of Herpetology 30: 140–146.
Jones, S. M. and Swain, R. 2000. Effects of exogenous FSH on follicular recruitment in a viviparous lizard *Niveoscincus metallicus* (Scincidae). Comparative Biochemistry and Physiology 127A: 487–493.
Jones, S. M., Wapstra, E. and Swain, R. 1997. Asynchronous male and female gonadal cycles and plasma steroid concentrations in a viviparous lizard, *Niveoscincus ocellatus* (Scincidae), from Tasmania. General and Comparative Endocrinology 108: 271–281.
Klosterman, L. L. 1983. The ultrastructure of germinal beds in the ovary of *Gerrhonotus coeruleus* (Reptilia: Anguidae). Journal of Morphology 178: 247–266.
Klosterman, L. L. 1987. Ultrastructural and quantitative dynamics of the granulosa of ovarian follicles of the lizard *Gerrhonotus coeruleus* (family Anguidae). Journal of Morphology 192: 125–144.
Lance, V. and Lofts, B. 1978. Studies on the annual reproductive cycle of the female cobra, *Naja naja*. IV. Ovarian histology. Journal of Morphology 157: 161–180.
Laughran, L. J., Larsen, J. H. and Schroeder P. C. C. 1981. Ultrastructure of developing ovarian follicles and ovulation in the lizard *Anolis carolinensis* (Reptilia). Zoomorphology 98: 191–208.
Leyton, V. and Valencia, J. 1992. Follicular population dynamics: Its relation to clutch and litter size in Chilean *Liolaemus* lizards. pp. 177–199. In W. Hamlett (ed.), *Reproductive Biology of South American Vertebrates: Aquatic and Terrestrial.* Springer-Verlag, New York.
Limatola, E. and Filosa, S. 1989. Exogenous vitellogenesis and micropinocytosis in the lizard *Podarcis sicula* treated with folliclestimulating hormone. General and Comparative Endocrinology 75: 165–176.
Limatola, E., Angelini, F., Restucci, E. and Boccia, A. 1989. Progesterone effects on oocyte Growth in *Podarcis sicula* Raf. *First World Congress of Herpetology*. University of Kent at Canterbury.
Limatola, E., Manzo, C., Manzo, S., Monti, M. G., Rosanova, P. and Romano, M. 2002. Oocyte growth and follicular hierarchy may be locally controlled by an inhibin-like protein in the lizard *Podarcis sicula*. Journal of Experimental Zoology 292: 96–102.
Lovern, M. B. and Adams, A. L. 2008. The effects of diet on plasma and yolk steroids in lizards (*Anolis carolinensis*). Integrative and Comparative Biology 48: 428–436.
Lovern, M. B. and Wade, J. 2003a. Yolk testosterone varies with sex in eggs of the lizard, *Anolis carolinensis*. Journal of Experimental Zoology (Molecular Development and Evolution) 295A: 206–210.
Lovern, M. B. and Wade, J. 2003b. Sex steroids in green anoles (*Anolis carolinensis*): uncoupled maternal plasma and yolking follicle concentrations, potential embryonic steroidogenesis, and evolutionary implications. General Comparative Endocrinology 134: 109–115.

Magnusson, W. E. 1987. Reproductive cycles of teiid lizards in Amazonian savanna. Journal of Herpetology 21: 307–316.

Manes, M. E., Noriega, T., Campos Casal, F. and Apichela, S. 2007. Ovarian changes during the reproductive cycle of the *Tupinambis merianae* lizard raised in a temperate environment. Cuadernos de Herpetología 21: 21–29.

Manríquez-Morán, N. L., Villagrán-Santa Cruz, M. and Méndez de la Cruz, F. R. 2005. Reproductive biology of the parthenogenetic lizard, *Aspidoscelis cozumela*. Herpetologica 61: 435–439.

Martínez-Torres, M., Hernández-Caballero, M. E., Alvarez-Rodríguez, C., Luis-Díaz, J. A. and Ortiz-López, G. 2003. Luteal development and progesterone levels during pregnancy of the viviparous temperate lizard *Barisia imbricata imbricata* (Reptilia: Anguidae). General Comparative Endocrinology 132: 55–65.

Martínez-Torres, M., Hernández-Caballero, M. E., Luis-Díaz, J. A., Ortiz-López, G., Cárdenas-León, M. and Moreno-Fierros, L. 2010. Effects of luteectomy in early pregnancy on the maintenance of gestation and plasma progesterone concentrations in the viviparous temperate lizard *Barisia imbricata imbricata*. Reproductive Biology and Endocrinology 25: 8–19.

Maurizii, M. G. and Taddei, C. 1996. Immunolocalization of cytoskeletal proteins in the previtellogenic ovarian follicle of the lizard *Podarcis sicula*. Cell Tissue Research 284: 489–493.

Maurizii, M. G. and Taddei. C. 2012. Microtubule organization and nucleation in the differentiating ovarian follicle of the lizard *Podarcis sicula*. Journal of Morphology 273(10): 1089–1095.

Maurizii, M. G., Saverino, O. and Taddei, C. 1997. Cytokeratin cytoskeleton in the differentiating ovarian follicle of the lizard *Podarcis sicula* Raf. Molecular Reproduction and Development 48: 536–542.

Maurizii, M. G., Alibardi, L. and Taddei, C. 2000. Organization and characterization of the keratin cytoskeleton in the previtellogenic ovarian follicle of the lizard *Podarcis sicula* Raf. Molecular Reproduction and Development 57: 159–166.

Maurizii, M. G., Alibardi. L. and Taddei, C. 2004. α-tubulin and acetylated α-tubulin during ovarian follicle differentiation in the lizard *Podarcis sicula* Raf. Journal of Experimental Zoology 301A: 532–541.

Maurizii, M. G., Cavaliere, V., Gamberi, C., Lasko, P., Gargiulo, G. and Taddei, C. 2009. Vasa protein is localized in the germ cells and in the oocyte-associated pyriform follicle cells during early oogenesis in the lizard *Podarcis sicula*. Development Genes and Evolution 219: 361–367.

Méndez de la Cruz, F. R., Guillette, L. J., Jr. and Villagrán-Santa Cruz, M. 1993. Differential atresia of ovarian follicles and its effect on the clutch size of two populations of the viviparous lizard *Sceloporus mucronatus*. Functional Ecology 7: 535–540.

Mojica, B. H., Rey, B. H., Serrano, V. H. and Ramírez-Pinilla, M. P. 2003. Annual reproductive activity of a population of *Cnemidophorus lemniscatus* (Sauria, Teiidae). Journal of Herpetology 37: 35–42.

Moore, M. C. and Johnston, G. I. H. 2008. Toward a dynamic model of deposition and utilization of yolk steroids. Integrative Comparative Biology 48: 411–418.

Morales, M. H., Baerga-Santini, C. and Cordero-López, N. 1996. Synthesis of Vitellogenin polypeptides and deposit of yolk proteins in *Anolis pulchellus*. Comparative Biochemistry and Physiology 114B: 225–231.

Morales, M. H., Pagán, S. M. and Cordero-López, N. 2002. Immunodissection of yolk lipovitellin (LV1) demonstrates the existence of different LV1-domains and suggests a complex family of vitellogenin genes in the lizard *Anolis pulchellus*. Comparative Biochemistry and Physiology Part B 131: 339–348.

Motta, C. M., Castriota-Scanderberg, M., Filosa, S. and Andreuccetti, P. 1995. Role of pyriform cells during the growth of oocytes in the lizard *Podarcis sicula*. Journal of Experimental Zoology 273: 247–256.

Motta, C. M., Filosa, S. and Andreuccetti, P. 1996. Regression of the epithelium in the late previtellogenic follicles of *Podarcis sicula*: a case of apoptosis. Journal of Experimental Zoology 276: 233–241.
Motta, C. M., Tammaro, S., Cicale, A., Indolfi, P., Iodice, C., Spagnuolo, M. S. and Filosa, S. 2001. Storage in the yolk platelets of low MW DNA produced by the regressing follicle cells. Molecular Reproduction and Development 59: 422–430.
Motta, C. M., Tammaro, S., de Stasio, R., Borrelli, L. and Filosa, S. 2004. How follicle number is regulated in the ovary of the lizard *Podarcis sicula*? Italian Journal of Zoology 71: S2,109–111.
Neaves, W. B. 1971. Intercellular bridges between follicle cells and oocyte in the lizard *Anolis carolinensis*. Anatomical Record 170: 285–302.
Neaves, W. B. 1972. The passage of extracellular tracers through the follicular epithelium of lizard ovaries. Journal of Experimental Zoology 179: 339–363.
Neaves, L., Wapstra, E., Birch. D., Girling, J. E. and Joss, J. M. 2006. Embryonic gonadal and sexual organ development in a small viviparous skink, *Niveoscincus ocellatus*. Journal of Experimental Zoology A Comparative Experimental Biology 305: 74–82.
Nijagal, B. S. and Yajurvedi, H. N. 1999. Corticosterone interferes with seasonal recrudescence of germinal bed activity in lizard, *Mabuya carinata*. Indian Journal of Experimental Biology 37: 300–301.
Painter, D. L. and Moore, M. C. 2005. Steroid hormone metabolism by the chorioallantoic placenta of the mountain spiny lizard *Sceloporus jarrovii* as a possible mechanism for buffering maternal–fetal hormone exchange. Physiological and Biochemical Zoology 78: 364–372.
Paolucci, M., Rocco, M. and Varricchio, E. 2001. Leptin presence in plasma, liver and fat bodies in the lizard *Podarcis sicula*. Fluctuations thoughout the reproductive cycle. Life Sciences 69: 2399–2408.
Paolucci, M. and Di Cristo, C. 2002. Progesterone receptor in the liver and oviduct of the lizard, *Podarcis sicula*. Life Sciences 71: 1417–1427.
Pieau, C. 1996. Temperature variation and sex determination in reptiles. Bioessays 18: 19–26.
Radder, R. S. and Shine, R. 2007a.Germinal bed condition in a polyautochronic single-clutched lizard, *Bassiana duperreyi* (Scincidae). Amphibia-Reptilia 28: 159–162.
Radder, R. S. and Shine, R. 2007b. Are the phenotypic traits of hatchling lizards affected by maternal allocation of steroid hormones to the egg? General and Comparative Endocrinology 154: 111–119.
Radder, R. S., Shanbhag, B. A. and Saidapur, S. K. 2001. Pattern of Plasma Sex Steroid Hormone Levels during Reproductive Cycles of Male and Female Tropical Lizard, *Calotes versicolor*. General and Comparative Endocrinology 124: 85–292.
Radder, R. S., Pizzatto, L. and Shine, R. 2008. Morphological correlates of life-history variation: is lizard clutch size related to the number of germinal beds in the ovary? Biological Journal of the Linnean Society 94: 81–88.
Ramírez-Pinilla, M. P., Rueda, E. D. and Stashenko, E. 2011. Transplacental nutrient transfer during gestation in the Andean lizard *Mabuya* sp. (Squamata, Scincidae). Journal of Comparative Physiology B 181: 249–268.
Raucci, F. and Di Fiore, M. M. 2011. Localization of c-kit and stem cell factor (SCF) in ovarian follicular epithelium of a lizard, *Podarcis s. sicula*. Acta Histochemica 113: 647–655.
Raynaud, A. and Pieau, C. 1985. Embryonic development of the genital system. pp. 149–300. In C. Gans and F. Billett (eds.), *Biology of the Reptilia, Vol. 15*. Wiley and Sons, New York.
Rhen, T., Sakata, J. T., Zeller, M. and Crews, D. 2000. Sex steroid levels across the reproductive cycle of female leopard geckos, *Eublepharis macularius*, from different incubation temperatures. General and Comparative Endocrinology 118: 322–331.
Rhen, T., Crews, D., Fivizzani, A. and Elf, P. 2006. Reproductive tradeoffs and yolk steroids in female leopard geckos, *Eublepharis macularius*. Journal of Evolutionary Biology 19: 1819–1829.

Romano, M. and Limatola, E. 2000. Oocyte plasma membrane proteins and the appearance of vitellogenin binding protein during oocyte growth in the lizard *Podarcis sicula*. General and Comparative Endocrinology 118: 383–392.

Saidapur, S. K. 1978. Follicular atresia in the ovaries of non-mammalian vertebrates. International Review of Cytology 54: 225–242.

Saidapur, S. K. 1982. Structure and function of postovulatory follicles (corpora lutea) in the ovaries of nonmammalian vertebrates. International Review of Cytology 75: 243–285.

Serrano-Cardozo, V. H., Ramírez-Pinilla, M. P., Ortega, J. E. and Cortes, L. A. 2007. Annual reproductive activity of *Gonatodes albogularis* (Squamata: Gekkonidae) living in an anthropic area in Santander, Colombia. South American Journal of Herpetology 2: 31–38.

Sica, S., Fierro, D., Iodice, C., Muoio, R., Filosa, S. and Motta, C. M. 2001. Control of oocyte recruitment: regulative role of follicle cells through the release of a diffusible factor. Molecular Reproduction and Development 58: 444–450.

Shanbhag, B. A. and Prasad, B. S. K. 1993. Follicular dynamics and germinal bed activity during the annual ovarian cycle of the lizard, *Calotes versicolor*. Journal of Morphology 216: 1–7.

Shanbhag, B. A., Radder, R. S. and Saidapur, S. K. 2001. Plasma progesterone levels and luteal activity during gestation and prolonged oviductal egg retention in a tropical lizard, *Calotes versicolor*. General and Comparative Endocrinology 123: 73–79.

Shine, R. and Guillette, L. J., Jr. 1988. The evolution of viviparity in reptiles: A physiological model and its ecological consequences. Journal of Theoretical Biology 132: 43–50.

Singh, P., Krishna, A. and Sridaran, R. 2007. Localization of gonadotrophin releasing hormone I, bradykinin and their receptors in the ovaries of non-mammalian vertebrates. Reproduction 133: 969–981.

Singh, P., Krishna, A., Sridaran, R. and Tsutsui, K. 2008. Changes in GnRH I, bradykinin and their receptors and GnIH in the ovary of *Calotes versicolor* during reproductive cycle. General and Comparative Endocrinology 159: 158–169.

Speake, B. K. and Thompson, M. B. 2000. Lipids of the eggs and neonates of oviparous and viviparous lizards. Comparative Biochemestry and Physiology 127A: 453–467.

Sreenivasulu, K., Ganesh, S. and Raman, R. 2002. Evolutionarily conserved, DMRT1, encodes alternatively spliced transcripts and shows dimorphic expression during gonadal differentiation in the lizard, *Calotes versicolor*. Gene Expression Patterns 2: 51–60.

Summers, C. H. 1988. Chronic low humidity-stress in the lizard *Anolis carolinensis*: effects on ovarian and oviductal recrudescence. Journal of Experimental Zoology 248: 192–198.

Taddei, C. 1972. Significance of pyriform cells in the ovarian follicle of *Lacerta sicula*. Experimental Cell Research 72: 562–566.

Tammaro, S., Simoniello, P., Filosa, S. and Motta, C. M. 2007. Block of mitochondrial apoptotic pathways in lizard ovarian follicle cells as an adaptation to their nurse function. Cell and Tissue Research 327: 625–635.

Tokarz, R. R. 1977. An autoradiographic study of the effects of FSH and estradiol-17β on early ovarian follicular maturation in adult *Anolis carolinensis*. General and Comparative Endocrinology 31: 17–28.

Tokarz, R. R. 1978. An autoradiographic study of the effects of mammalian gonadotropin (FSH and LH) and estradiol 17β on 3H-thymidine labeling of surface epithelial cells, prefollicular cells and oogonia in the ovary of the lizard *Anolis carolinensis*. General and Comparative Endocrinology 35: 179–188.

Tokarz, R. R. and Summers, C. H. 2011. Stress and Reproduction in Reptiles. pp. 169–213. In D. O. Norris and K. H. Lopez (eds.), *Hormones and Reproduction of Vertebrates, Vol. 3 Reptiles*. Academic Press, New York.

Trauth, S. E. 1978. Ovarian cycle of *Crotaphytus collaris* (Reptilia, Lacertilia, Iguanidae) from Arkansas with emphasis on corpora albicantia, follicular atresia and reproductive potential. Journal of Herpetology 12: 461–470.

Tripathi, V. and Raman, R. 2010. Identification of Wnt4 as the ovary pathway gene and temporal disparity of its expression vis-a-vis testis genes in the garden lizard, *Calotes versicolor*. Gene 449: 77–84.

Uribe, M. C., Omaña, M. E. M., Quintero, J. E. G. and Guillette, L. J., Jr. 1995. Seasonal variation in ovarian histology of the viviparous lizard *Sceloporus torquatus torquatus*. Journal of Morphology 226: 103–119.
Uribe, M. C., Portales, G. L. and Guillette, L. J., Jr. 1996. Ovarian folliculogenesis in the oviparous Mexican lizard *Ctenosaura pectinata*. Journal of Morphology 230: 99–112.
Van Wyk, J. H. 1984. Ovarian morphological changes during the annual breeding cycle of the rock lizard *Agama atra* (Sauria: Agamidae). Navorsinge van die Nasionale Museum, Bloemfontein 4: 237–275.
Varma, S. K. 1970. Morphology of ovarian changes in the garden lizard *Calotes versicolor*. Journal of Morphology 131: 195–210.
Verderame, M. and Limatola, E. 2010. Molecular identification of estrogen receptors (ERa and ERb) and their differential expression during VTG synthesis in the liver of lizard *Podarcis sicula*. General and Comparative Endocrinology 168: 231–238.
Vieira, S., De Pérez, G. R. and Ramírez-Pinilla, M. P. 2010. Ultrastructure of the ovarian follicles in the placentotrophic andean lizard of the genus *Mabuya* (Squamata: Scincidae). Journal of Morphology 271: 738–749.
Villagrán-Santa Cruz, M. and Méndez de la Cruz, F. R. 1999. Corpus luteum through the gestation of *Sceloporus palaciosi* (Sauria: Phrynosomatidae). Copeia 1999: 214–218.
Vitousek, M. N., Mitchell, M. A., Romero, L. M., Awerman, J. and Wikelski, M. 2010. To breed or not to breed: physiological correlates of reproductive status in a facultatively biennial iguanid. Hormones and Behavior 57: 140–146.
Vitt, L. J. 1982. Sexual dimorphism and reproduction in the microteiid lizard, *Gymnophthalmus multiscutatus*. Journal of Herpetology 16: 325–329.
Xavier, F. 1982. Progesterone in the viviparous lizard *Lacerta vivipara*: Ovarian biosynthesis, plasma levels, and binding to transcortin-type protein during the sexual cycle. Herpetologica 38: 62–70.
Xavier, F. 1987. Functional morphology and the regulation of corpus luteum. pp. 241–282. In R. E. Jones and D. O. Norris (eds.), *Hormones in Fishes, Amphibians and Reptiles*. Plenum Press, New York.
Wilhoft, D. C. 1963. Gonadal histology and seasonal changes in the tropical Australian lizard *Leiolopisma rhomboidalis*. Journal of Morphology 113: 185–204.
Weiser, H., Starostová, Z., Kubi ka, L. and Kratochvíl, L. 2012. Overlap of female reproductive cycles explains shortened interclutch interval in a lizard with invariant clutch size (Squamata: Gekkonidae: *Paroedura picta*). Physiological and Biochemical Zoology 85: 491–498.

CHAPTER 9

Male Reproductive Anatomy: The Gonadoducts, Sexual Segment of the Kidney, and Cloaca

Justin L. Rheubert,[1,*] *David M. Sever,*[2] *Dustin S. Siegel*[3] and *Stanley E. Trauth*[4]

9.1 INTRODUCTION

The male reproductive system in lizards consists of the gonads (testis), gonadoducts (testicular ducts), the sexual segment of the kidney (SSK), and the cloaca (Rheubert *et al.* 2010). In lizards, sperm develop in the germinal epithelium of the seminiferous tubules within the testis (Gribbins 2011) and pass sequentially through the various portions of the gonadoducts; rete testis, ductuli efferentes, epididymis, ductus deferens, and, in some species, an ampulla ductus deferentis (Figs. 9.1 and 9.2; Sever 2010). Sperm then usually mix with secretions from the SSK in a variable ductus deferens-ureter complex, and enter the cloaca through the urogenital papillae (Figs. 9.1 and 9.2). Each of these regions is controlled by hormone fluxes that vary throughout the reproductive and non-reproductive seasons subject to seasonality and, thus, undergo morphological (and physiological) changes throughout an entire calendar year (Prasad and Sanyal 1969; Jones 2002). This seasonality reflects the function of the various portions of the male

[1] College of Sciences, The University of Findlay, Findlay, Ohio, 45840 USA.
[2] Department of Biology, Southeastern Louisiana University, Hammond, Louisiana 70402, USA.
[3] Department of Biology, Southeast Missouri State University, Cape Girardeau, Missouri, USA.
[4] Department of Biology, Arkansas State University, Jonesboro, Arkansas, USA.
* Corresponding author

reproductive system and will thus be discussed in more detail later in this chapter. This chapter will focus on these various regions giving a historical, morphological, ultrastructural, and comparative view. Although the gonadoducts begin intratesticularly, morphological descriptions of the testis are discussed in detail in Gribbins and Rheubert Chapter 11, this volume.

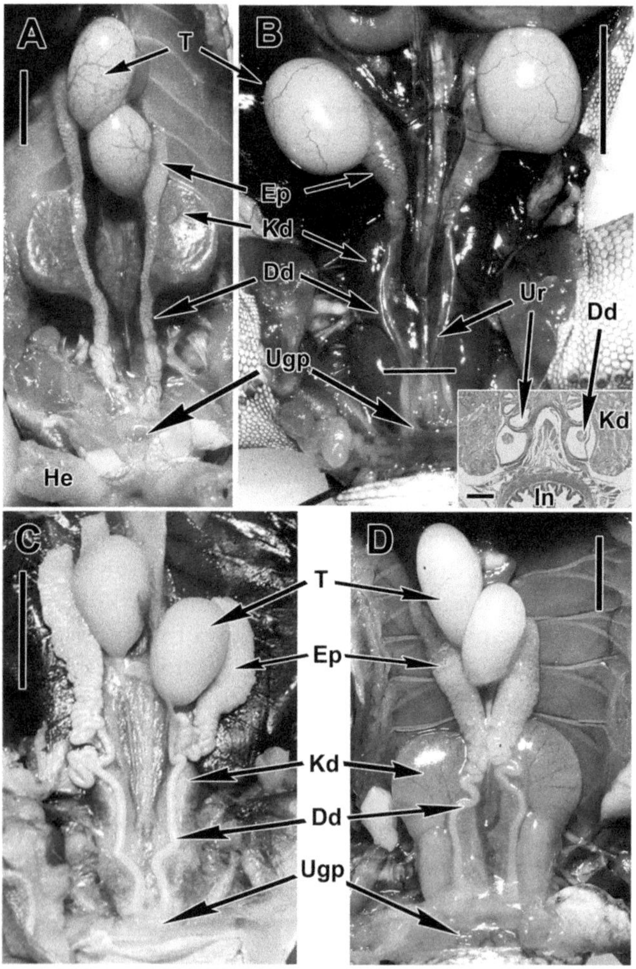

Fig. 9.1 Gross morphology of the urogenital systems in reproductively active representatives of four lizard families. Scale bar = 5 mm for **A-D**. **(A)** *Aspidoscelis sexlineata* (family Teiidae). **(B)** *Crotaphytus collaris* (family Crotaphytidae). Inset: histosection at transverse plane level indicated by bar; inset scale bar = 500 µm. **(C)** *Holbrookia propinqua* (family Phrynosomatidae). **(D)** *Plestiodon fasciatus* (family Scincidae). Ep, ductus epididymis; Dd, ductus deferens; He, hemipenis; In, intestine; Kd, kidney; T, testis; Ugp, urogenital papilla; Ur, ureter.

Color image of this figure appears in the color plate section at the end of the book.

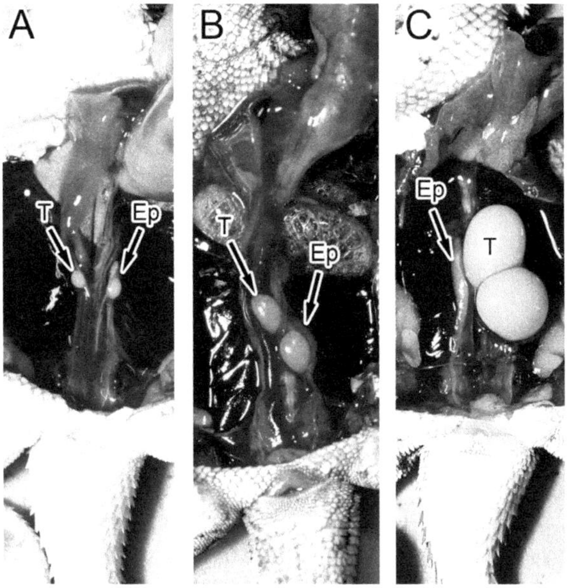

Fig. 9.2 Gross morphology of the reproductive system in **(A)** an immature individual of *Sceloporus undulatus*, **(B)** mature but reproductively inactive individual of *Podarcis muralis*, and **(C)** a mature and reproductively active individual of *Sceloporus virgatus*.

Color image of this figure appears in the color plate section at the end of the book.

9.2 THE GONADODUCTS

9.2.1 Overview

The gonadoducts serve as a corridor for sperm from the testis to the cloaca as well as aiding in sperm maturation. Each region of the testicular ducts is morphologically distinct from one another and based on homologous morphologies follow the same terminology as that found in other amniotes (for review, see Sever 2010).

Few studies exist concerning the morphology of gonadoducts in lizards, a majority of which focus primarily on the epididymis. Lereboullet (1851) and Martin-Saint-Ange (1854) provided the first descriptions of the gonadoducts in the Sand Lizard (*Lacerta agilis*) and the Viviparous Lizard (*Zootica vivipara*), respectively. Henry (1900) followed suit and provided an additional description of the gonadoducts again in *Z. vivipara*. Krause (1922), Alverdes (1926), and Van den Broek (1933) all provided descriptions in *Lacerta agilis*. Alverdes (1928) provided a description of the gonadoducts for the Anguidae with a description in the Slow Worm (*Anguis fragilis*). This was followed by a few studies on snakes before Forbes (1941) provided a description on a phrynosomatid, the Spiny Lizard (*Sceloporus spinosus*). Mulaik (1946) followed up Forbes (1941) description in *Sceloporus* by investigating the Crevice Spiny Lizard (*S. poinsetti*), the Mesquite Graphic Lizard (*S. disparilis*), and the Texas Spiny Lizard (*S. olivaceus*). Reynolds (1943) and Badir (1958) added accounts for the Scincidae with descriptions on the Five-lined Skink (*Plestiodon fasciatus*), the Ocellated Skink (*Chalicides ocellatus*), and the Sandfish (*Scincus scincus*). Dufaure and Saint Girons (1984) provided the most taxonomically extensive study to date by investigating over 40 species in 11 families. However, their study was limited to light microscopy and focused primarily on the epididymis. Daisy *et al.* (2000) investigated the vas deferens and ampulla ductus deferentis of the Indian Garden Lizard (*Calotes versicolor*) and Averal *et al.* (1992) and Meeran *et al.* (2001) extended descriptions in this species by investigating epididymal morphology. Dufaure *et al.* (1986) and Mesure *et al.* (1991) investigated the testosterone content and ultrastructure of the epididymis in *Zootoca vivipara*. Desantis *et al.* (2002) provided histology of the epididymis in the Italian Wall Lizard (*Podarcis sicula*) and Akbarsha *et al.* (2005, 2006a, 2006b) provided descriptions of the testicular ducts in the Fan-Throated Lizard (*Sitana ponticeriana*). Recently, Rheubert *et al.* (2010) added descriptions for the Gekkonidae with a treatise on the Mediterranean Gecko (*Hemidactylus turcicus*), and Sever *et al.* (2013) added the Ground Skink (*Scincella lateralis*). These studies incorporate 13 families of lizards and are summarized in Table 9.1. Within snakes, descriptions have been provided for several species (for review, see Fox 1972; Trauth and Sever 2011) but ultrastructural descriptions have been provided only for a colubrid, the Black Swamp Snake (*Seminatrix pygaea*; Sever 2010) an elapid, the Yellow-Bellied Sea Snake (*Pelamis platurus*; Sever and Freeborn 2011), and viperid, the Cotttomouth (Siegel *et al.* 2009), thus adding 2 more families within the Squamata.

The above studies show that various portions of the gonadoducts can be identified based on their morphology. However, they use a variety of terminology and vary in the amount of details provided. For example, studies prior to that of Akbarsha (2007) on *Sitana ponticeriana* only provided histological descriptions at the light microscopy level and, thus, no ultrastructural or subcellular details were presented. Furthermore, some studies such as Akbarsha *et al.* (2007) and Desantis *et al.* (2002) only focus on a single region of the gonadoducts leaving out descriptions for the other regions. This variation in the level of detail has led to differing terminologies and the inability to provide adequate comparative views. Sever (2010) synonymized nomenclature with all amniotes but the low diversity and sample size still hinders any evolutionary or phylogenetic analyses.

The following is a comparative view that encompasses historical, recent, and new data on the gonadoducts of lizards. Each region of the gonadoducts

Table 9.1 Families that have data concerning the testicular ducts in saurians. EM- electron microscopy study, LM- light microscopy study.

Family	LM or EM	Source
Anguidae	LM	Alverdes 1928
	LM	Dufaure and Saint Girons 1984
Anniellidae	LM	Dufaure and Saint Girons 1984
Agamidae	LM	Dufaure and Saint Girons 1984
	LM and EM	Akbarsha *et al.* 2007
	LM and EM	Akbarsha *et al.* 2007
Chamaeleonidae	LM	Dufaure and Saint Girons 1984
Gekkonidae	LM	Dufaure and Saint Girons 1984
	LM and EM	Rheubert *et al.* 2010
Lacertidae	LM	Martin-Saint-Ange 1854
	LM	Henry 1900
	LM	Dufaure and Saint Girons 1984
Lacertidae	LM	Krause 1922
	LM	Alverdes 1926
	LM	Van den Broek 1933
Teiidae	LM	Dufaure and Saint Girons 1984
	LM and EM	Desantis *et al.* 2002
Phrynosomatidae	LM	Forbes 1941
	LM	Mulaik 1946
		Dufaure and Saint Girons 1984
Polychrotidae	LM	Dufaure and Saint Girons 1984
Scincidae	LM	Reynods 1943
	LM	Badir 1958
	LM	Dufaure and Saint Girons 1984
	LM and EM	Sever *et al.* 2013
Tropiduridae	LM and EM	Ferreira *et al.* 2009
Xantusidae	LM	Dufaure and Saint Girons 1984

will be discussed separately so that comparisons among past and current research can be incorporated in future studies. Finally this chapter will culminate with an overall conclusion and future avenues for research.

9.2.2 Gross Morphology

The testes lie suspended within the pleuroperitoneum posterior to the liver with the left testis slightly more posterior than the right testis. The testes are grossly enlarged in mature individuals and more enlarged during the active season (see Gribbins and Rheubert Chapter 11, this volume). The testicular ducts lie on the lateral side of the testis and are white in color or slightly opaque in immature individuals or during the reproductively inactive season (Fig. 9.2A,B). The rete testis and ductuli efferentes are not visible grossly. The epididymis, which lies juxta-positioned to the testis, appears as a straight opaque tube in immature individuals but becomes coiled in mature individuals and is swollen during the active season as it is filled with sperm (Fig. 9.2). Posteriorly, the epididymis transitions into the ductus deferens, which straightens as it runs posteriorly and ventral to the kidney. The ductus deferens joins the ureter at the posterior to the kidney and anterior to the cloaca, and forms a ductus deferens-ureter complex. This duct enters the cloaca at the dorsal aspect through the urogenital papilla.

9.2.3 The Rete Testis

The rete testis has received the least amount of attention out of the various regions of the gonadoducts. Lereboullet (1851) did not describe the rete testis in his study on *Lacerta agilis* and stated 4 or 5 canals emerged from the testis and dumped into the epididymis. These canals, however, were most likely the ductuli efferentes. Martin-Saint-Ange (1854) was the first to note the rete testis in the lizard *Zootica vivipara* and only referred to them as "tubes" with no informative definition (Fig. 9.3). Alverdes (1926) noticed a single duct emerging from the testis of *Lacerta* sp. and stated, "hier bei den eidechsen erscheint der Langskanal ohne zweifel als ein aquivalent dieser [rete] teile, trotz seiner lage auerhalb des hodens," using the term long canal for the rete testis (Fig. 9.3), and stating it is an equivalent to the rete in mammals despite its location outside of the testis. Alverdes (1928) reiterated that a single duct, the "long canal", emerged from the testis in *Lacerta agilis* (supported again in the study by Van den Broek 1933). Alverdes (1928) found 5–9 ducts emerged from the testis in *Anguis fragilis* but avoided using any anatomical term to describe them.

Forbes (1941) investigated the effects of testosterone and estrone on the gonadoducts in *Sceloporus spinosus* and found that a single tubule exited the testis, which he termed the rete canal. In his study he found that neither testosterone nor estrone had an effect on this duct. However, his study was limited to light microscopy. Reynolds (1943) described the testicular ducts in *Plestiodon fasciatus* and noted an absence of a rete. Instead, Reynolds

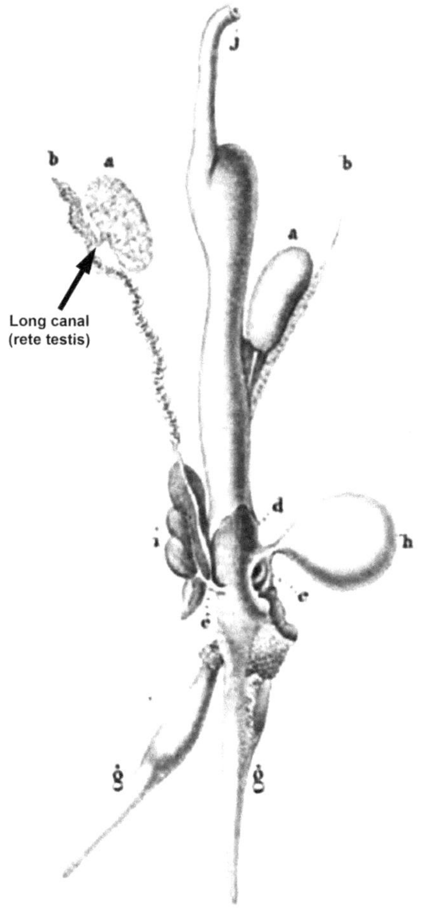

Fig. 9.3 Schematic drawing of the urogenital system in *Zootoca vivipara* adopted from Martin Saint Ange (1854) and adapted from Rheubert *et al.* 2010.

(1943) noted that the seminiferous tubules emptied into a small outlet duct which dumped into a longitudinal canal, which we interpret to represent the intra- and extra-testicular portions of the rete.

Martin-Saint-Ange (1854), Alverdes (1928), Volsøe (1944), and Fox (1952) provided descriptions of the rete testis in snakes. They found that multiple ducts exit from the testis and used the terms petit tubes, hodenhausfuhrgange, ductuli efferentes, and ductuli efferentes, respectively. Volsøe (1944) proposed that snakes exhibited the ancestral condition for lepidosauria with multiple tubules exiting the testis and that lizards represented the derived condition in which only a few ducts remained. He further stated that *Anguis fragilis* represented a transitional stage.

Jones (1998) provided support that the rete ducts that immediately exited the testes were homologous to the rete tubules found in mammals through developmental studies. Akbarsha *et al.* (2007) adopted this terminology in their description of the single testicular duct exiting the testis in the agamid, *Sitana ponticeriana.* Sever (2010) was the first to provide an ultrastructural description of the rete testis in squamates using the snake, *Seminatrix pygaea.* Similar to the other studies on snakes, Sever (2010) found that multiple ducts exited the testis and he elected to term these ducts the rete testis stating they are homologous to the rete tubules in other amniotes based on detailed cellular morphology, further supporting Jones' (1998) hypothesis.

Since the synonymizing of the rete tubule nomenclature in squamates to that of other amniotes, only four studies investigated the morphology of the rete tubules in squamates. One of these studies focused on the snake *Pelamis platurus* (Sever and Freeborn 2011). The other three studies focus on the gekkonid, *Hemidactylus turcicus* (Rheubert *et al.* 2010), the scincid, *Scincella lateralis* (Sever *et al.* 2013), and the phrynosomatid, the Prairie Lizard (*Sceloporus consobrinus*; Rheubert unpublished data). These studies showed that multiple tubules exited the testis in the snake *Pelamis platurus* and a single duct emerged from the testis in the lizards *Scincella lateralis, Hemidactylys turcicus*, and *Sceloporus consobrinus*. Studies on the rete testis and the nomenclature used are listed in Table 9.2.

Sever (2010) was the first to provide an ultrastructural description of the rete testis in squamates in his study on *Seminatrix pygaea*. He found that the rete consisted of a low cuboidal epithelium with irregularly shaped nuclei. Occasional microvilli were found extending into the lumen but no cilia were present (as is found in some amniotes). Intercellular canaliculi were labyrinthine but were narrow in nature and sealed apically by tight junctions and desmosomes. Sever (2010) suggested endocytosis was occurring via vesicular transport and exocytosis was occurring via apocrine blebs.

Following Sever's (2010) study, Rheubert *et al.* (2010) provided the first ultrastructural description of the rete testis in a lizard in their study on *Hemidactylus turcicus.* They found that the rete testis consisted of a simple squamous to low cuboidal epithelium with irregularly shaped nuclei consistent with the results of Sever (2010). The epithelial cells possessed microvilli and few cells possessed cilia. The intercellular canaliculi were labyrinthine, sealed apically by tight junctions, and during the active season were drastically swollen, a finding the authors attributed to the uptake of luminal fluid.

The three most recent studies (Sever and Freeborn 2012; Sever *et al.* in press; Rheubert unpublished data) along with previous ultrastructural studies have all shown that little variation exists in the morphology of the rete testis. These investigations have shown that the rete testis is

Table 9.2 Table showing studies performed on the rete testis of squamates and the terminology used.

Family	Species	terminology used	# of tubules	Reference
Agamidae	*Sitana ponticeriana*	rete tubule	1	Akbarsha *et al.* 2007
Anguidae	*Anguis fragilis*	long canal/Hodenausfuhrgange	5–9	Alverdes 1928
	Anguis fragilis	long canal	5–9	Van den Broek 1933
Phrynosomatidae	*Sceloporus spinosus*	rete tubule	1	Forbes 1941
	Sceloporus consobrinus	rete tubule	1	This study
Gekkonidae	*Hemidactylus turcicus*	rete tubule	1	Rheubert *et al.* 2010
Lacertidae	*Zootoca vivipara*	n/a	1	Martin-Saint-Ange 1854
	Lacerta sp.	long canal/Hodenausfuhrgange	1	Alverdes 1926
	Lacerta sp.	long canal	1	Alverdes 1928
	Lacerta agilis	long canal	1	Van den Broek 1933
Scincidae	*Chalcides ocellatus*	long canal	9	Badir 1958
	Plestiodon fasciatus	longitudinal canal/ testis outlet duct	1	Reynolds 1943
	Scincus scincus	longitudinal canal/ testis outlet duct	6	Badir 1958
	Scincella lateralis	rete tubule	1	Sever *et al.* In press
Colubridae	*Natrix natrix*	petit tubes	numerous	Martin-Saint-Ange 1854
	Thamnophis sp.	ductuli efferentes	numerous	Fox 1952
	Seminatrix pygaea	retetestis	5–7	Sever 2010
Elapidae	*Pelamis platurus*	rete testis	numerous	Sever and Freeborn 2011
Viperidae	*Vipera* sp.	ductuli efferentes	numerous	Volsøe 1944
	Vipera sp.	ductuli efferentes	numerous	Saint Girons 1957
	Agkistrodon piscivorus	ductuli efferentes	numerous	Siegel *et al.* 2009a
	Agkistrodon piscivorus	rete testis	numerous	Trauth and Sever 2011

comprised of a simple squamous or cuboidal epithelium (Fig. 9.4A) with irregularly shaped nuclei (Fig. 9.4B,C, Nu). Cilia (Fig. 9.4B, Ci) are present in *Hemidactylus turcicus* and *Sceloporus consobrinus*, but were not evident in other species studied. Intercellular canaliculi (Fig. 9.4C, Ic) are labyrinthine and swollen during the active season in all species except in the snake *Pelamis platurus*. In the lizards *Hemidactylus turcicus* and *Sceloporus consobrinus*, and the snakes *Seminatrix pygaea* and *Pelamis platurus*, the intercellular canaliculi are sealed apically by tight junctions but tight junctions were not observed in the skink *Scincella lateralis*. Coated vesicles (Fig. 9.4D, Cv) were observed in all species studied to date, which has led to the hypothesis that the rete testis is responsible for absorbing luminal fluids. The results of these studies along with new data can be found summarized in Table 9.3.

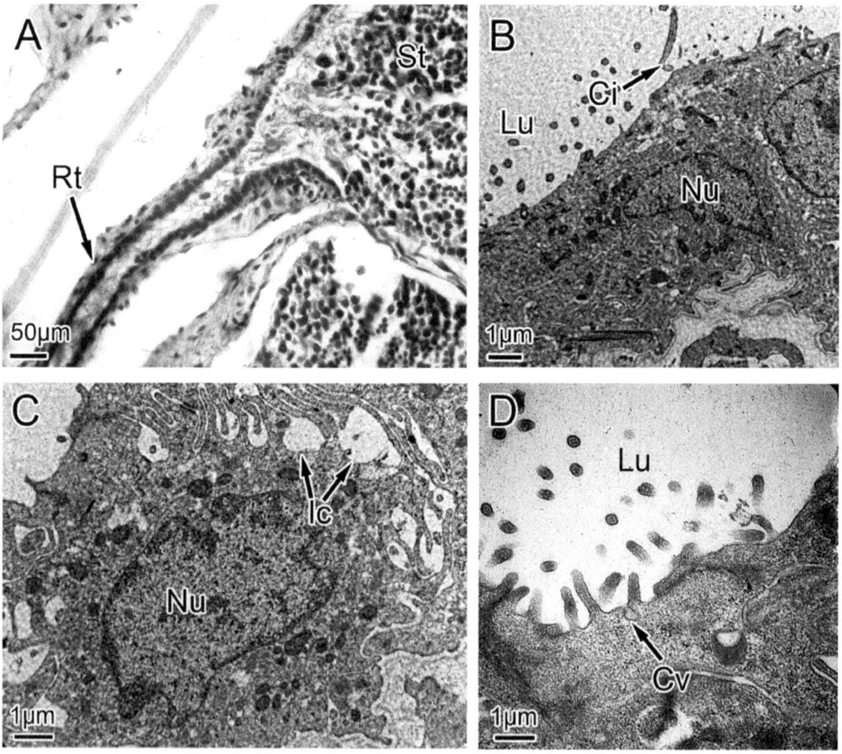

Fig. 9.4 Transmission electron micrographs detailing the fine structure of the rete testis. **(A)** Light micrograph of the rete testis (Rt) in *Sceloporus consobrinus* detailing the simple squamous epithelium. Seminiferous tubule (St). **(B)** Transmission electron micrograph of the rete testis in *Sceloporus consobrinus* detailing the irregular shaped nuclei (Nu), and cilia (Ci) projecting into the lumen (Lu). **(C)** Transmission electron micrograph of the rete testis in *Hemidactylus turcicus* detailing the labyrinthine intercellular canaliculi (Ic). **(D)** Transmission electron micrograph of the rete testis in *Hemidactylus turcicus* detailing the coated vesicles (Cv) at the apical cytoplasm.

Table 9.3 Character summary of studies performed on the rete testis in squamates.

Character	*Hemidactylus turcicus*	*Sceloporus consobrinus*	*Scincella lateralis*	*Seminatrix pygaea*	*Pelamis platurus*
Simple epithelium	Squamous	Squamous	Squamous	Cuboidal	Varies
Cilia	yes	yes	no	no	no
Indented nuclei	yes	yes	yes	yes	yes
Apical tight junctions	yes	yes	no	yes	yes
Coated vesicles	yes	yes	yes	yes	yes
Wide intercellular canaliculi	yes	yes	yes	no	yes
Intercellular spaces	yes	yes	no	no	yes
Labyrinthine membranes	yes	yes	yes	yes	yes
Source	Rheubert *et al.* 2010	This Study	Sever *et al.* 2013	Sever 2010	Sever and Freeborn 2012

Major variation concerning the morphology of the rete testis between taxa is restricted to the number of ducts exiting the testis. The seminiferous tubules merge with one another to either form a single rete duct (Fig. 9.5A) or a series of rete ducts (Fig. 9.5B), with up to 50 ducts present. These ducts exit the testis although they remain within an extension of the tunica albuginea, the epididymal sheath.

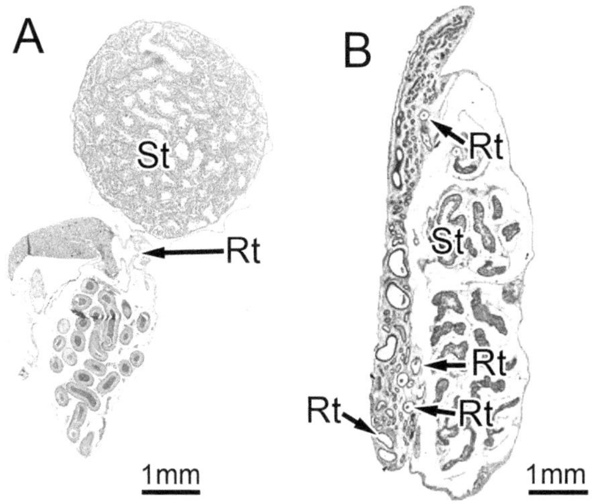

Fig. 9.5 Light microscopy of the reproductive system. **(A)** A single rete testis duct (Rt) emerges from the testis in *Sceloporus consobrinus*. **(B)** Multiple rete testis ducts (Rt) emerge from the testis in *Pelamis platurus*. Seminiferous tubule (St).

Character optimization onto a morphological hypothesis of phylogeny proposed by Conrad (2008) and a molecular hypothesis of phylogeny proposed by Pyron *et al.* (2013) (Fig. 9.6) results in a single rete testis emerging from the testis as the ancestral condition. This contradicts the

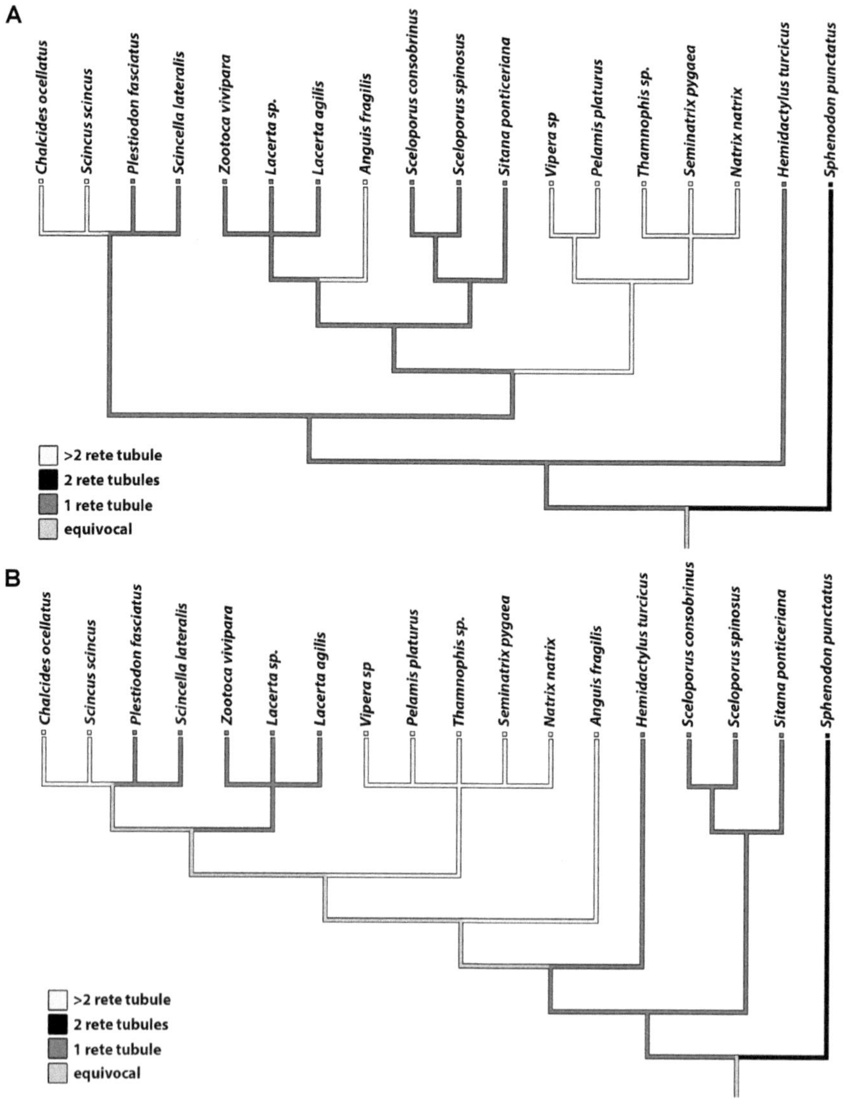

Fig. 9.6 Character optimization of the number of rete tubules present on the **(A)** morphological phylogenetic hypothesis provided by Conrad (2008) and **(B)** the molecular phylogenetic hypothesis provided by Pyron *et al.* (2013).

Color image of this figure appears in the color plate section at the end of the book.

hypothesis proposed by Volsøe (1944) who suggested that snakes exhibited the ancestral condition for lepidosaurs with multiple rete testis tubules. Variation exists within the Lacertoidea + Scincoidea + Anguimorpha clade; thus, any future studies should focus on these clades to fully understand the evolution of the rete complex. Sever and Freeborn (2012) suggested that the number of rete tubules was not dependent on testicular size, but more data concerning that hypothesis is needed for confirmation.

9.2.4 The Ductuli Efferentes

Lereboullet (1851) described the reproductive system in *Lacerta agilis* and stated that 4–5 ducts empty into the epididymis, which he called the efferent canals and we interpret to be the ductuli efferentes. Martin-Saint-Ange (1854) and Henry (1900) referred to these ducts as petit tubes, Morgera (1905) the tubi medii and tubi piccoli, Alverdes (1926, 1928) the Verbinststucke and Hauptstucke (collectively the vasa efferentia), Forbes (1941) the epididymides, and Volsøe (1944) the ductuli epididymides.

Akbarsha *et al.* (2007) adopted the terminology proposed by Jones (1998) and used ductuli efferentes as the homologue to that of the ductuli efferentes in other amniotes. Sever (2010), in his study on *Seminatrix pygaea*, also utilized the term ductuli efferentes in his restructuring of the nomenclature of the testicular ducts in squamates based on morphological similarity. The variation in historical terminology can be found in Table 9.4.

Although the terminology has varied historically, the structure of the ductuli efferentes make them very distinct, as noted by Volsøe (1944); i.e., the ductuli efferentes are "rather distinct and very characteristic." The number of ductuli efferentes seems to be taxa specific and may be dependent on the number of rete tubules that are present. However, the ductuli efferentes are very convoluted, sometimes difficult to count, and thus are often times classified as "numerous."

The epithelium changes from a simple squamous to low cuboidal (found in the rete testis) to a thick cuboidal to low columnar in the ductuli efferentes (Fig. 9.7). Furthermore, the ductuli efferentes have heavily ciliated cells dispersed throughout the epithelium that make the ductuli efferentes unique among the gonadoducts. Akbarsha *et al.* (2007) provided the first ultrastructural description of the ductuli efferentes in the lizard *Sitana ponticeriana*, which provided the first functional hypothesis for these ducts. They found that the epithelium consisted of both ciliated and non-ciliated cells. Based on the presence of coated vesicles and large vacuoles in the apical cytoplasm, the authors suggested that the function of the ductuli efferentes was uptake of luminal fluid while the presence of cilia has led

Table 9.4 Table depicting studies performed on the ductuli efferentes and the terminology used.

Family	Species	Terminology used	Reference
Lacertidae	*Lacerta agilis*	efferent channels	Lereboullet 1851
Agamidae	*Sitana ponticeriana*	ductuli efferentes	Akbarsha *et al.* 2007
Anguidae	*Anguis fragilis*	vasa efferentia	Alverdes 1928
Phrynosomatidae	*Sceloporus spinosus*	epididymides	Forbes 1941
	Sceloporus consobrinus	ductuli efferentes	Rheubert *et al.* in review
Gekkonidae	*Hemidactylus turcicus*	ductuli efferentes	Rheubert *et al.* 2010
Lacertidae	*Lacerta agilis*	efferent channels	Lereboullet 1851
	Lacerta agilis	efferent channels	Martin-Saint-Ange 1854
	Lacerta vivipara	petit tubes	Henry 1900
	Lacerta sp.	vasa efferentia	Alverdes 1926
	Lacerta agilis	vasa efferentia	Van den Broek 1933
	Lacerta sp.	vasa efferentia	Alverdes 1928
Scincidae	*Chalcides ocellatus*	vasa efferentia	Badir 1958
	Plestiodon fasciatus	vasa efferentia	Reynolds 1943
	Scincus scincus	vasa efferentia	Badir 1958
Colubridae	*Natrix natrix*	petit tubes	Martin-Saint-Ange 1854
	Natrix natrix	vasa efferentia	Alverdes 1928
	Natrix natrix	ductuli epididymides	Volsøe 1944
	Natrix natrix	ductuli epididymides	Fox 1952
	Seminatrix pygaea	ductuli efferentes	Sever 2010
Elapidae	*Pelamis platurus*	ductuli efferentes	Sever and Freeborn 2011
Viperidae	*Vipera* sp.	ductuli epididymides	Volsøe 1944
	Vipera sp.	ductuli epididymides	Saint Girons 1957
	Vipera sp.	tubi medii/tubi piccoli	Morgera 1905
	Agkistrodon piscivorus	ductuli epididymides	Siegel *et al.* 2009a
	Agkistrodon piscivorus	ductuli efferentes	Trauth and Sever 2011

to hypotheses that suggest the mixing of seminal fluids and propulsion of spermatozoa. Furthermore, the authors observed evidence of spermiophagic activity by these ducts.

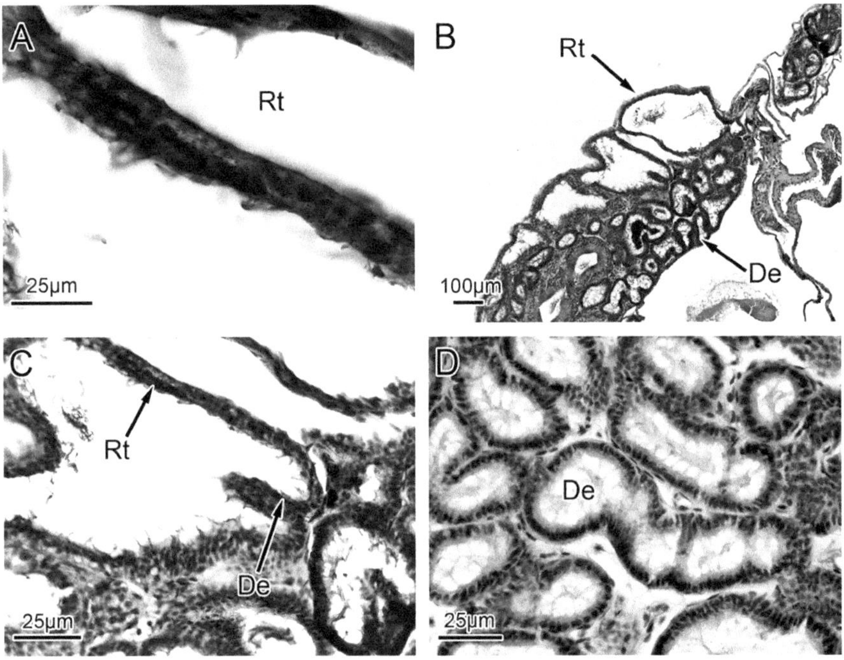

Fig. 9.7 Transition from rete testis (Rt) to ductuli efferentes (De) in *Sceloporus consobrinus*. **(A)** Rete testis. **(B)** Rete testis lying juxtapositioned to the ductuli efferentes. **(C)** Rete testis tubule emptying into the ductuli efferentes. **(D)** Ductuli efferentes.

Sever (2010), Sever and Freeborn (2012), Rheubert *et al.* (2010), and Sever *et al.* (2013) provided ultrastructural descriptions of the efferent ducts for the snakes *Seminatrix pygaea* and *Pelamis platurus* and the lizards *Hemidactylus turcicus* and *Scincella lateralis*, respectively. These studies found similar results to that found in the study by Akbarsha *et al.* (2007). The ductuli efferentes are characterized by a cuboidal to low columnar epithelium (Figs. 9.7D and 9.8A) that is heavily ciliated (Fig. 9.8A, Ci). Two distinct regions were recognized in *Sitana ponticeriana*, but only one region was found in all other species studied to date. In the study on *S. ponticeriana*, four cell types were recognized but only ciliated and non-ciliated cells were recognized in all other species. Apical tight junctions were present in all species as well as coated vesicles. Apocrine blebs were present in all species studied to date except *Scincella lateralis*. The presence of cellular organelles, such as rough endoplasmic reticulum (Fig. 9.8B, Rer) and mitochondria (Fig. 9.8B, Mi), was variable among species. Lysosomes (Fig. 9.8C, Ly) were absent in *S. lateralis* but present in all other species, and spermiophagic activity (Fig. 9.8D, Sp) was noticed in *Sitana ponticeriana* and *Hemidactylus turcicus*.

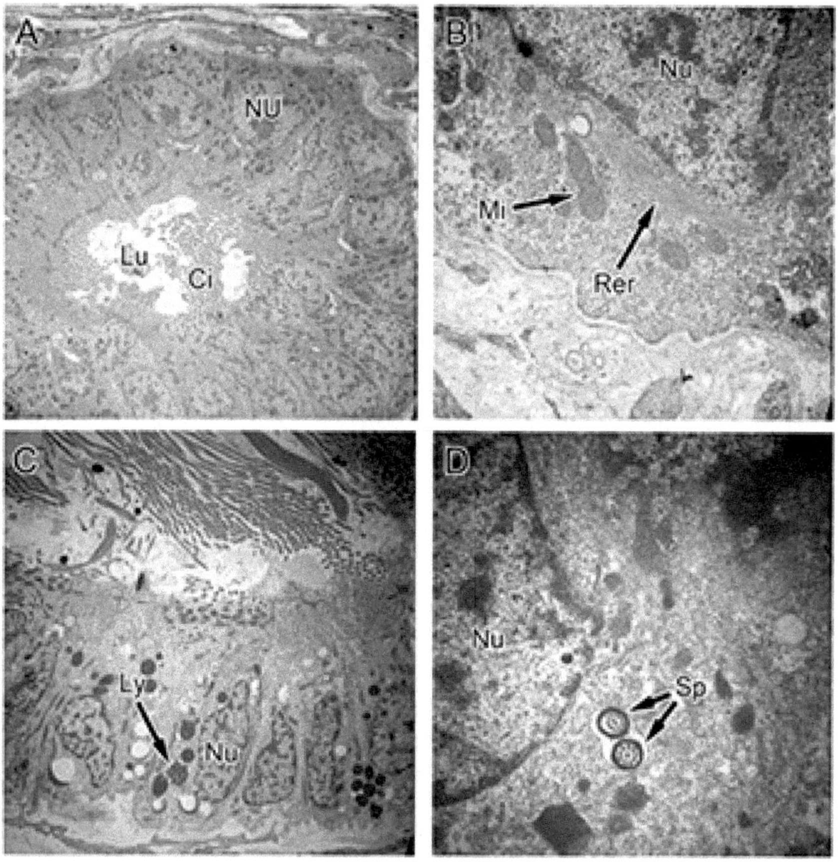

Fig. 9.8 Transmission electron micrographs of the ductuli efferentes in *Hemidactylus turcicus*. **(A)** Overview of the ductuli efferentes detailing cilia (Ci) projecting into the lumen (Lu). **(B)** Enlarged view of the basal portion of a ciliated cell detailing the rough endoplasmic reticulum (Rer) and mitochondria (Mi) basal to the nucleus (Nu). **(C)** Micrograph of the ductuli efferentes detailing the lysosomes (Ly) present juxtapositioned to the nucleus (Nu). **(D)** Evidence of spermiophagic activity in the ductuli efferentes as sperm tails (Sp) are found embedded in the cytoplasm.

These studies have suggested a function of uptake of luminal fluid and spermiophagic activity. The results of these studies along with new data on *Scelporus consobrinus* can be found in Table 9.5.

9.2.5 The Epididymis

The epididymis has been the focus of the majority of studies concerning testicular ducts in lizards. Martin-Saint-Ange (1854), Henry (1900), and Alverdes (1926) provided the first descriptions of the epididymis in lizards,

Table 9.5 Ultrastructural characteristics of the ductuli efferentes.

Character	***Sitana ponticeriana***	***Hemidactylus turcicus***	***Sceloporus consobrinus***	***Scincella lateralis***	***Seminatrix pygaea***	***Pelamis platurus***
Regions	two	one	one	one	one	one
Epithelium	columnar	cuboidal	cuboidal	cuboidal	columnar	columnar
Cell types	four	two	two	two	two	two
Cilia	yes	yes	yes	yes	yes	yes
Apical tight junctions	yes	yes	yes	yes	yes	yes
Coated vesicles	yes	yes	yes	yes	yes	yes
Endosomes	yes	yes	yes	no	yes	yes
Apocrine blebs	yes	yes	yes	no	yes	yes
Wide intercellular canaliculi	no	no	no	yes	no	yes
Rough endoplasmic reticulum	filamentous/saccular	filamentous	?	saccular	filamentous	no
Golgi complexes	yes	no	?	no	no	no
Lysosomes	yes	yes	?	no	yes	yes
Spermiophagy	yes	yes	?	no	no	no
Source	Akbarsha *et al.* 2007	Rheubert *et al.* 2010	This study	Sever *et al.* 2013	Sever 2010	Sever and Freeborn 2012

but Dufaure and Saint Girons (1984) provided the first thorough review of the epididymides in lizards by investigating 51 species encompassing 12 families. They found five types of secretory activity in squamates, but the 5th type characterized by a high cytoplasm devoid of secretory granules was only found in snakes. Type 1 was found only in the Lacertidae and was characterized by secretory granules ranging in size from 10–12 µm. Type 2 found in the Agamidae was characterized by several small granules that rarely exceed 1µm in diameter. Type 3 found in the Iguanidae (except *Anolis*) and the Gekkonidae is similar to Type 2 with small granules, but the granules are smaller than those found in Type 2. Type 4 found in the Gekkonidae, Chamaeleonidae, *Anolis*, Teiidae, and Scincidae was characterized by no secretory activity with dense cytoplasm in the epithelial cells lining the ducts. Anguidae, Anniellidae, Amphisbaenidae, Trogonophidae, and Xantusiidae were all classified as between Types 3 and 4.

Subsequent to the general descriptions of the epididymis, more detailed reports appeared in the literature including ultrastructural descriptions. Haidar and Rai (1987) found 4 distinct regions in the epididymis in the Yellow-Bellied House Gecko (*Hemidactylus flaviviridis*): the initial segment, caput, corpus, and cauda regions. Mesure *et al.* (1991) provided the first ultrastructural description of the epididymis in a lizard, *Zootica vivipara* (Lacertidae). They recognized that secretory principal cells and basal cells were the only two cell types present. Like Haidar and Rai, they recognized four regions of the epididymis, but only minute differences between the regions were observed. Meeran *et al.* (2001) investigated the epididymis of *Calotes versicolor* and also described four epididymal regions, and the basis of the regionality was due to the presence or absence of seven cell types (principal, narrow, apical, clear, dark, basal, and halo). Desantis *et al.* (2002) reported three epididymal regions in *Podarcis sicula*, but only reported two cell types (principal and basal). Akbarsha *et al.* (2006) reported four regions based off presence/absence of six cell types (principal, narrow, apical, clear, basal, and intraepithelial leukocytes). Rheubert *et al.* (2010) reported only two cell types (principal and basal) in *Hemidactylus turcicus* and described no regionalization. Sever and Freeborn (2012) found three cell types (principal, basal, mitochondria rich apical cells) in *Pelamis platurus*, but also did not mention any regionalization. Sever *et al.* (2013) found three cell types (light and dark principal cells and basal cells) in *Scincella lateralis* and differentiated the anterior from the posterior regions based on the amount of secretory material in the lumen. The results of these studies appear to suggest that cellular differentiation beyond the principal and basal cells are only found in the Agamidae and the Scincomorpha. However, this generalization needs further investigation.

From previous detailed ultrastructural investigations, the transition from the ductuli efferentes to the epididymis is sharply delineated as the epithelium changes from a cuboidal ciliated epithelium to a tall columnar secretory epithelium (Fig. 9.9). Two major cell types dominate the epithelium of the epididymis: principal cells and basal cells. Principal cells (Fig. 9.10A, Pcn) are tall columnar cells that have large amounts of secretory material

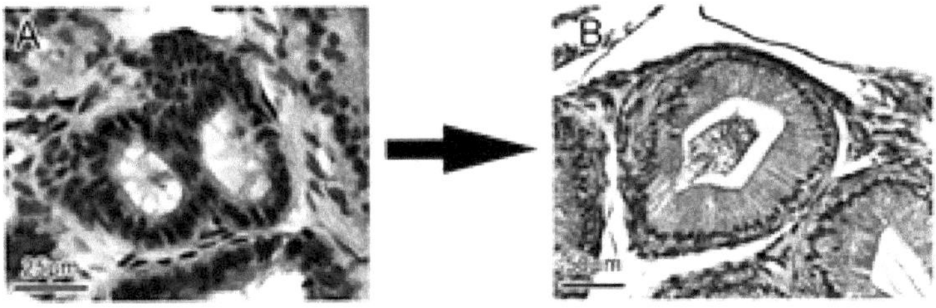

Fig. 9.9 Transition from the **(A)** ductuli efferentes to the **(B)** epididymis in *Sceloporus consobrinus.*

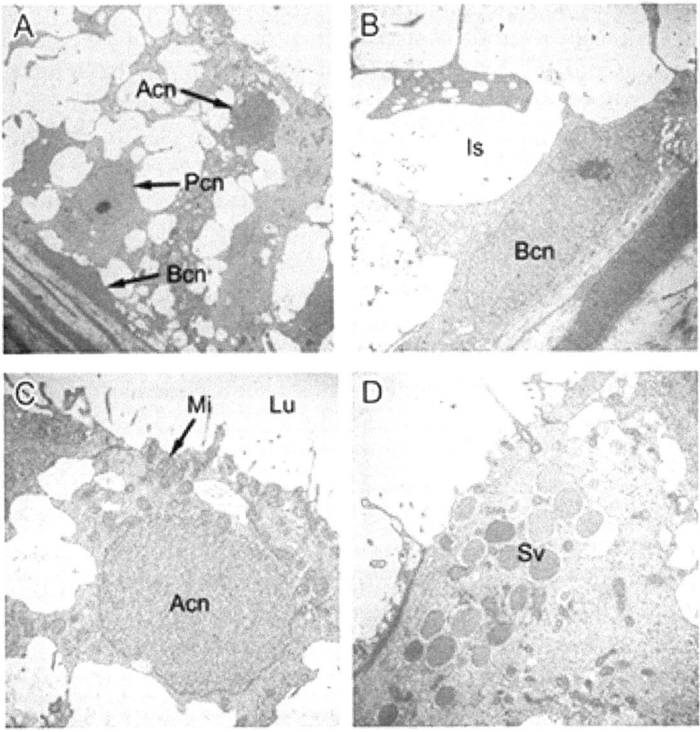

Fig. 9.10 Transmission electron micrographs of the epdidiymis in *Pelamis platurus* detailing different cells types that are recognized. **(A)** Overview of the epididymis detailing an apical cell (Acn), principal cell (Pcn), and basal cell (Bcn). **(B)** Enlarged view of a basal cell (Bcn) and intercellular space (Is). **(C)** Enlarged view of an apical cell (Acn) detailing the numerous mitochondria (Mi) in the cytoplasm and the location of the cell relative to the lumen (Lu). **(D)** Enlarged view of a principal cell detailing the secretory vesicles (Sv).

within them during the reproductively active season. The secretory material can be composed of flocculent material that is released via a merocrine process as seen in *Seminatrix pygaea*, small biphasic secretory granules as seen in *Pelamis platurus*, or large electron dense granules (Fig. 9.10D, Sv) as seen in most species studied to date. Basal cells (Fig. 9.10A,B, Bcn) are short cells that lie on the basal compartment of the epithelium. The role of these cells is largely unknown.

A few other cells types have been described in squamates but the presence of these cell types is extremely variable among taxa. Apical cells (Fig. 9.10A,C, Acn) have been reported in a few species and have slightly different morphologies; however, they all appear to be rich in mitochondria (Fig. 9.10C, Mi). In the snake *Pelamis platurus* these cells are limited to the luminal border, whereas in the lizards *Calotes versicolor* and *Sitana ponticeriana* these cells have a narrow attachment to the basal lamina. Narrow cells, although rare, were found in the lizard *S. ponticeriana*. These cells are similar in nature to mitochondria rich cells but are less wide at the apical portion. Clear cells were also noted in the cauda portion of the epididymis in *S. ponticeriana* and are characterized by a clear cytoplasm that is less electron dense than principal cells. Intraepithelial leukocytes, mainly macrophages, were noted in *S. ponticeriana* and have cellular processes that extended into the intercellular spaces.

Beyond the variability in the presence of cell types, the epididymis is highly variable in other morphological features. Cilia are not present on principal cells, but microvilli (Fig. 9.11A, Mv) are present. Wide intercellular canaliculi (Fig. 9.11B, Ic) occur in the Neotropical Ground Lizard (*Tropidurus itambere*), *Hemidactylus turcicus*, *Scincella lateralis*, and *Pelamis platurus*, but are absent in all other species. Furthermore, the presence of certain cellular machinery, such as filamentous rough endoplasmic reticulum, saccular rough enoplasmic reticulum (Fig. 9.11C, Rer), smooth endoplasmic reticulum, and Golgi complexes (Fig. 9.11D, Gc) is highly variable. These results are summarized in Table 9.6. These results demonstrate that although the rete testis and ductuli efferentes appear to be conserved morphologically among squamates, the epididymis is highly variable between taxa.

Through ultrastructural descriptions of the epididymis in lizards, studies have shown that the primary role of the epididymis is secretion of cellular products into the lumen. This secretory activity is modulated by testosterone and other androgens (Haider and Rai 1986; Depeiges and Dufaure 1977). Although the role of the secretory material from the epididymis in lizards is unknown, secretions from the epididymis in other amniotes function in final maturation of spermatozoa, including motility and fertilization capacity. Furthermore, many studies have suggested that the epididymis is the sperm storage region in lizards as opposed to the ductus deferens, which serves as the sperm storage region in snakes.

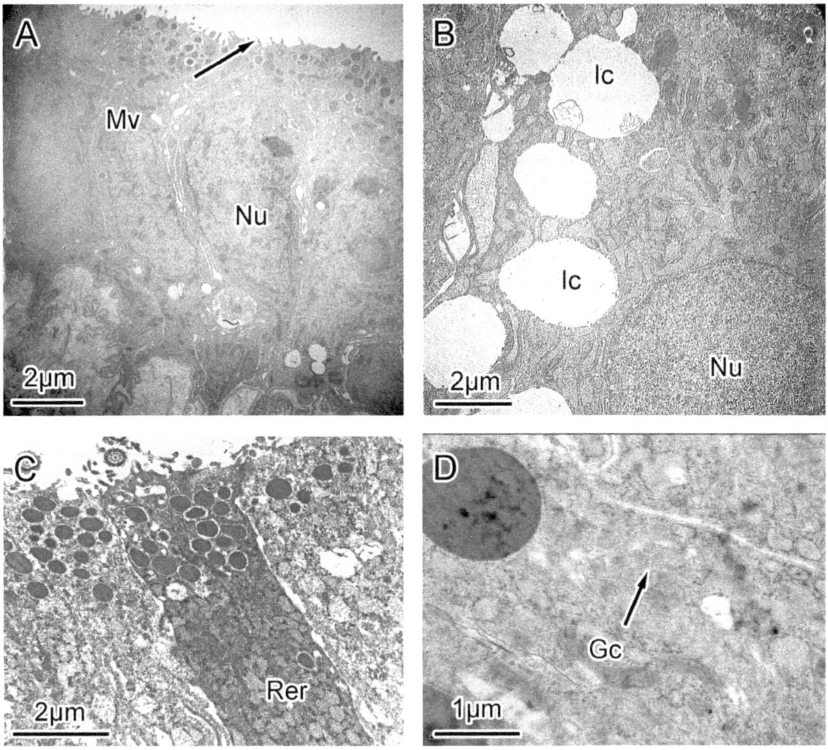

Fig. 9.11 Ultrastructure of the epididymis detailing: **(A)** microvilli (Mv) from *Hemidactylus turcicus*, **(B)** Intercellular canaliculi (Ic) from *Hemidactylus turcicus*, **(C)** Rough endoplasmic reticulum (Rer) from *Scincella lateralis*, and **(D)** Golgi complexes (Gc) in *Sceloporus consobrinus*.

9.2.6 The Ductus Deferens

Grossly, the transition from the epididymis to the ductus deferens in squamates occurs where the duct straightens posterior to the highly coiled epididymis (e.g., see Fig. 9.12B-D). Few studies have been dedicated on this region of the testicular ducts, and most that have were focused on gross morphology and/or light microscopy. Only four studies provided ultrastructural descriptions. Two of these studies were on the lizards *Sitana ponticeriana* and *Scincella lateralis*, and two were on the snakes the Cottonmouth (*Agkistrodon piscivorus*) and *Seminatrix pygaea* (see reviews in Sever *et al.* 2013; Trauth and Sever 2011).

In this section we outline the morphology of the ductus deferens as it extends from its anterior position until it anastomosis with other ducts more posteriorly. In addition, we present novel findings pertaining to distal

Table 9.6 Ultrastructural characteristics of the squamate epididymis.

Character	***Calotes versicolor***	***Sitana ponticeriana***	***Zootoca vivipara***	***Podarcis sicula***
Regions	Four	Four	Four	Three
Cell types	Seven	Six	Two	Two
Cilia	No	No	No	No
Apocrine blebs	Yes	Yes	No	No
Secretory granules	Biphasic	Biphasic	Biphasic	Biphasic
Vesicles	?	Yes	Yes	?
Endosomes	?	Yes	No	?
Wide intercellular canaliculi	?	No	No	No
Flimaentous Rer	?	Yes	Yes	No
Saccular Rer	?	Yes	Yes	Yes
Ser	?	Yes	No	?
Golgi complexes	?	Yes	Yes	?
Lysosomes	?	Yes	Yes	?
Basal junctional complexes	?	?	?	?
Source	Meeran *et al.* 2001	Akbarsha *et al.* 2006	Depeiges *et al.* 1985	Desantis *et al.* 2002

Character	***Tropidurus itambere***	***Hemidactylus turcicus***	***Scincella lateralis***	***Seminatrix pygaea***	***Pelamis platurus***
Regions	One	One	One	One	One
Cell types	Two	Two	Three	Two	Three
Cilia	No	No	No	No	No
Apocrine blebs	No	No	No	No	No
Secretory granules	Dense	Biphasic	Variable	No	Biphasic
Vesicles	?	Yes	Yes	Yes	No
Endosomes	?	No	No	No	No
Wide intercellular canaliculi	Yes	Yes	Yes	No	Yes
Flimaentous Rer	?	Yes	Yes	Yes	Yes
Saccular Rer	?	Yes	Yes	Yes	Yes
Ser	?	No	No	Yes	No
Golgi complexes	?	No	No	Yes	Yes
Lysosomes	?	No	No	Yes	No
Basal junctional complexes	?	No	No	Yes	No
Source	Ferreira and Dolder 2003	Rheubert *et al.* 2010	Sever *et al.* 2013	Sever 2010	Sever and Freeborn 2012

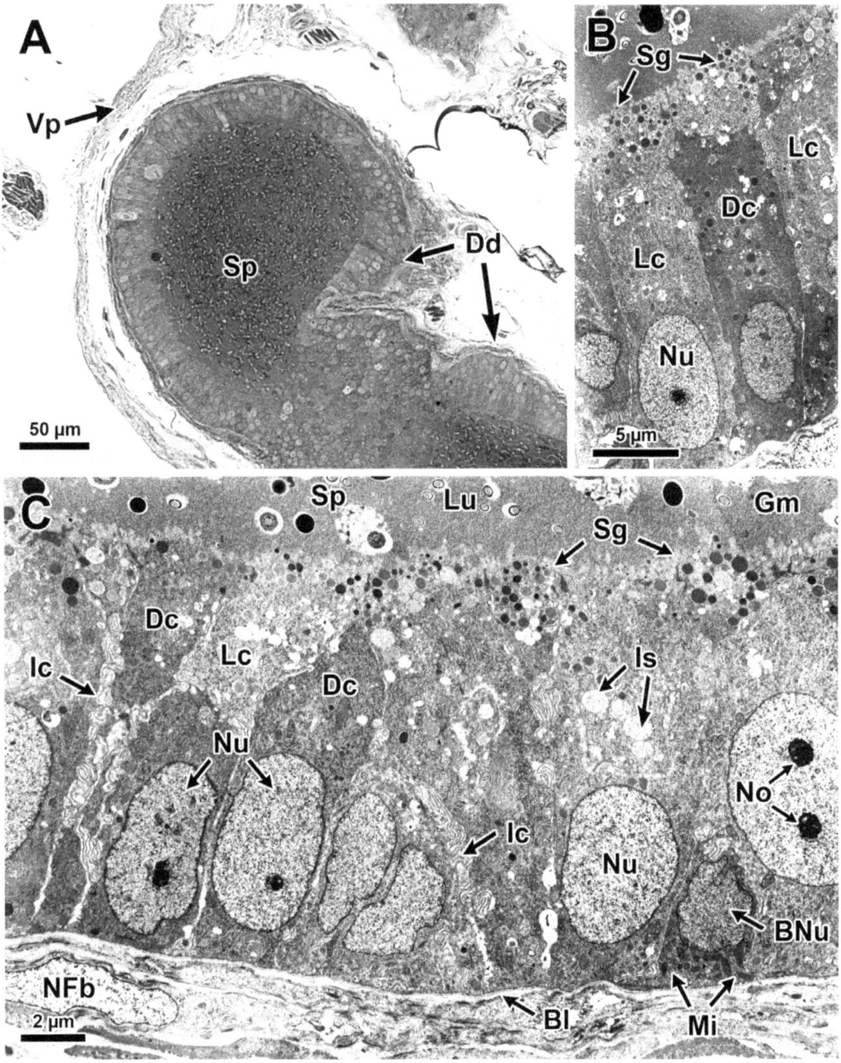

Fig. 9.12 Light micrograph (A, plastic section) and transmission electron micrographs of the anterior ductus deferens (Dd) of *Plestiodon fasciatus*. **(A)** Section plane just prior to duct straightening (see Fig. 2D). Sp, sperm; Vp, visceral pleuroperitoneum. **(B)** A region of tall, simple columnar epithelium as seen in A. Secretory granules (Sg) dominate the apical surfaces of both dark cells (Dc) and light cells (Lc). Nu, nucleus. **(C)** A region of columnar epithelial cells exhibiting diverse nuclei configurations. BNu, basal cell nucleus; Bl, basal lamina; Gm, granular material; Ic, intercellular canaliculi; Is, intercellular space; Lu, lumen; Mi, mitochondria; NFb, nucleus of fibroblast; No, nucleolus.

urogenital anatomy in lizards. Included in our analysis are seven species: 1) a teiid—the Texas Spotted Whiptail (*Aspidoscelis gularis*); 2) two scincids—*Scincella lateralis* and *Plestiodon fasciatus*; 3) and four phrynosomatids—*Sceloporus consobrinus*, the Greater Earless Lizard (*Cophosaurus texanus*), the Keeled Earless Lizard (*Holbrookia propinqua*), and the Texas Horned Lizard (*Phrynosoma cornutum*).

As a starting point in our ductal analysis (i.e., the point at which the ductus deferens straightens; see also Fig. 9.1), we provide a histo-section and two ultrastructural images of the anatomy of the anterior ductus deferens in a reproductively active *Plestiodon fasciatus* (Fig. 9.12). The highly secretory ductal epithelium varies in height and columnar cell stratification, being pseudostratified throughout. Two cell types predominate: dark cells and light cells. Sever *et al.* (2013) also found these two cell types in the posterior ductus epididymis of *Scincella lateralis*. The apical portions of both cell types exhibit an abundance of similar secretory granules that differ only slightly in their size and electron densities. Intercellular canaliculi are interspersed with narrow regions from near the basal lamina to near the cell free surfaces. Although numerous sperm reside near cell apices, none are observed embedded within the cells themselves. A granular material is evident within the duct lumen.

9.2.7 Posterior Ducts and Papillae

Much variation occurs in the posterior urogenital ducts of the lizards examined thus far. *Scincella lateralis* exhibits perhaps the most basic morphology found in squamates (Fig. 9.13). In the posterior trunk, the ductuli deferentia have short serous attachments to the ventral and medial aspects of the kidneys (Fig. 9.13A). As they enter the pelvic floor, each ductus deferens increases two-fold in diameter (Fig. 9.13B,C). The epithelium then becomes highly fluctuated with irregular folds, assuming the appearance of a typical vertebrate ampulla ductus deferentis (Fig. 9.13D; Trauth and Sever 2011).

Collecting ducts of the kidney descend medial to the ampullae (Fig. 9.13E). As the cranial end of the cloaca approaches, the ampullae narrow, but retain the irregular epithelium (Fig. 9.13F). Some collecting ducts empty into the urogenital papilla medial to the ampullae ductus deferentis (Fig. 9.13G), whereas others empty lateral to the ampullae (Fig. 9.13H). The terminal ends of the collecting ducts lose the columnar appearance of more proximal portions and become cuboidal, but we hesitate to call the distal ends "ureters", as this would mean that *Scincella lateralis* has multiple ureters.

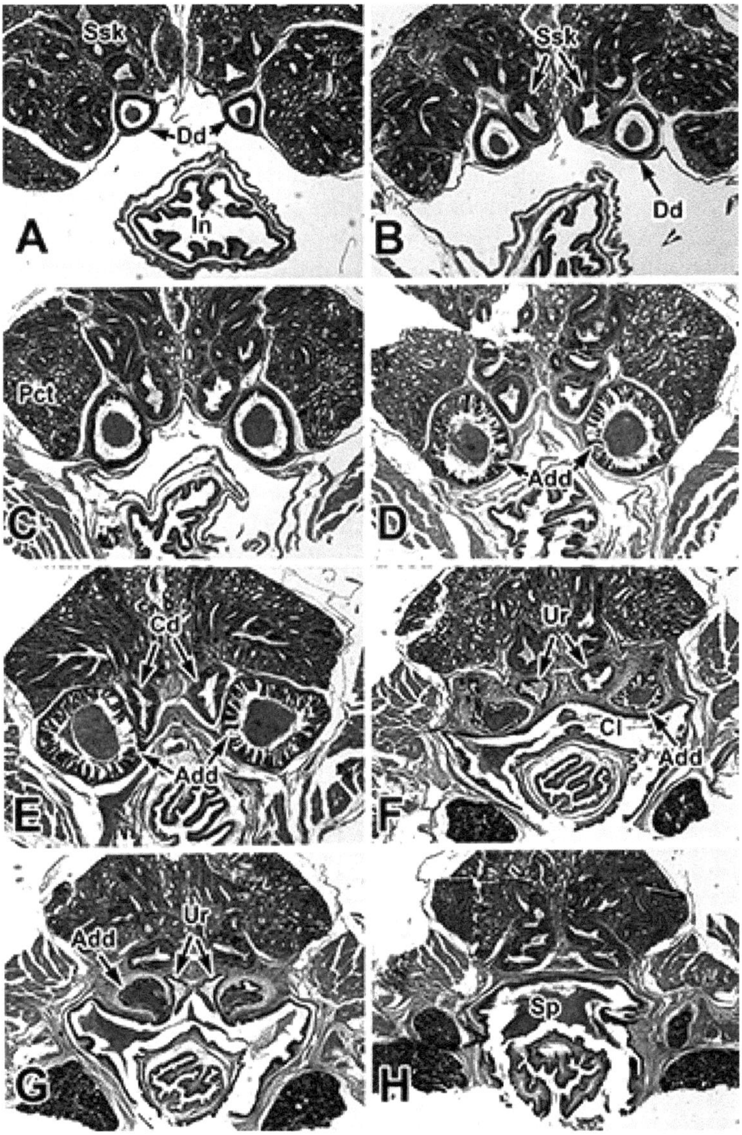

Fig. 9.13 Transverse sections through the posterior ductus deferens (Dd) and anterior cloaca (Cl) of a male *Scincella lateralis* (ASUMZ 32297). Sections stained with hematoxylin-eosin except for **H** treated with periodic acid and Schiff's reagent. (**A-B**) Posterior trunk. **(C)** Entering pelvic floor. **(D-E)** Enlargement into ampulla ductus deferentis (Add). **(F)** Narrowing of Add and descent of collecting ducts (Cd). **(G)** Medial opening of Cd into Up; Ur, ureter. **(H)** Lateral opening of Cd into Up. In, intestine; Pct, proximal convoluted tubules; Sp, sperm; Ssk, sexual segment of the kidney.

The urogenital papillae, therefore, contains the orifice of the ampullae ductus deferentis medially and is surrounded by slits opening from the collecting ducts (Fig. 9.13H; Fig. 9.14). Urinary and genital ducts never merge in *Scincella lateralis*. Histochemically, the distal convoluted tubules and collecting ducts, both of which compose the sexual segment of the kidney (SSK), are periodic acid-Schiff's positive for neutral carbohydrates, whereas the proximal convoluted tubules (Pct) are alcian blue positive at pH 2.5 for carboxylated glycosaminoglycans. The SSK is also positive for proteins with brilliant blue R. Neither the ductus deferens or ampulla ductus deferentis react strongly with any of these stains, although the folds of the ampulla have a weak reaction to proteins.

In addition to its presence in *Scincella lateralis*, we also found an ampulla ductus deferentis as a region of the posterior ductus deferens in several species within the family Phrynosomatidae (Fig. 9.15). Unlike the disproportionally enlarged diameter exhibited by each ampulla in *Scincella lateralis* compared to portions of its anterior ductus deferens, the ampullae of phrynosomatid lizards studied thus far are not greatly expanded segments of the ductus deferens. The tall, convoluted ductal epithelium of the

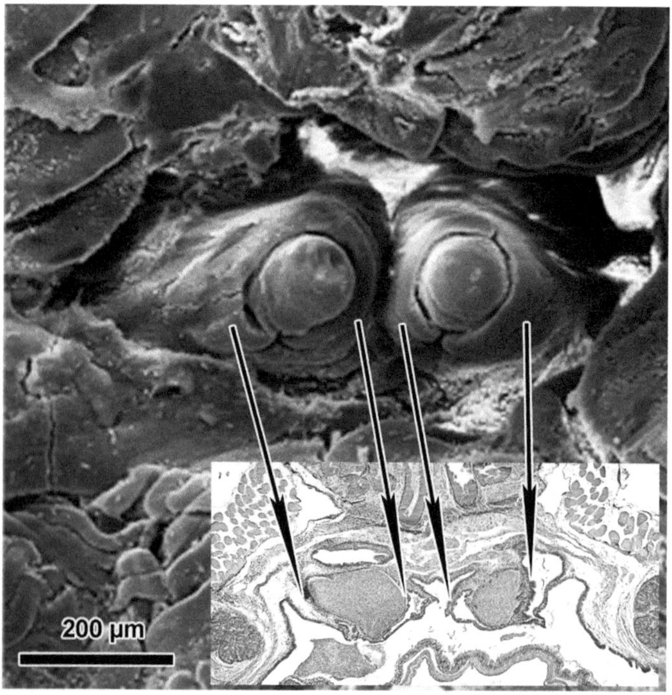

Fig. 9.14 Scanning electron micrograph and light micrograph of the urogenital papillae in *Scincella lateralis* (see also Fig. 9.13G).

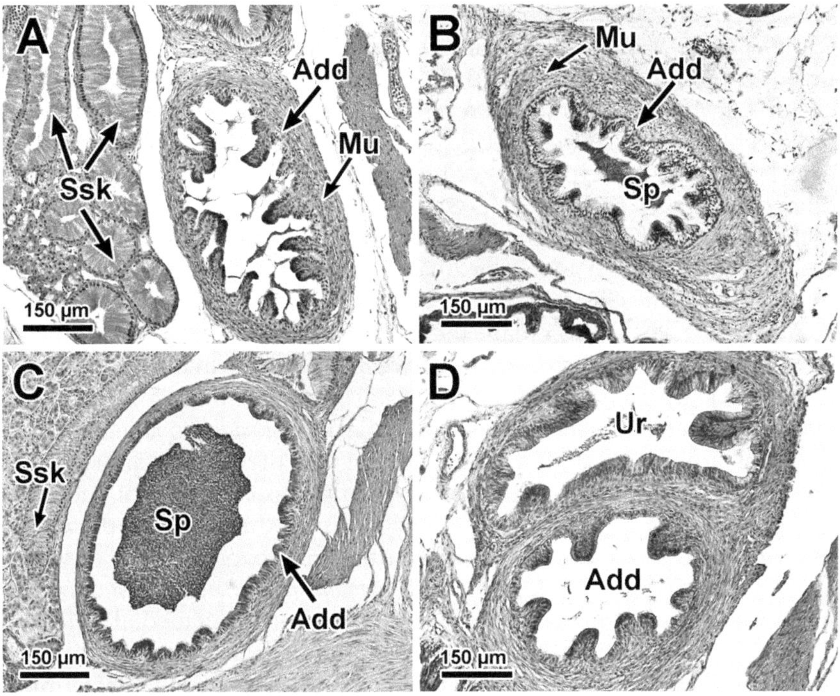

Fig. 9.15 Light micrographs (paraffin sections) of the ampulla ductus deferentis in four phrynosomatid lizards; H and E stains. **(A)** *Cophosaurus texanus*, section through the posterior region of the kidney revealing the ampulla ductus deferentis (Add) and its irregular folds adjacent to sexual segments of the kidney (Ssk); Mu, muscularis. **(B)** *Sceloporus consobrinus,* showing sperm (Sp) within the ampulla, which possesses a folded epithelium similar to the ampulla in **A**. **(C)** *Holbrookia propinqua*, exhibiting an ampulla packed with sperm and a low-folding epithelium. **(D)** *Phrynosoma cornutum*, revealing an ampulla contains tall, widely spaced folds and lying ventral to its ipsilateral ureter (Ur).

ampullae of these phrynosomatid lizards did not exhibit the exceedingly numerous folds or the general folding pattern (folds of mostly equal height) present in *Scincella lateralis*. Instead, irregular folding and festooning of the epithelium is observed in these species. We found interspecific similarities in folding micro-anatomies and epithelial staining properties between *Cophosaurus texanus* (Fig. 9.15A) and *Sceloporus consobrinus* (Fig. 9.15B). In these two species, the apical cell clusters associated with each fold show marked basophilia. This basophilia of the epithelium is also evident in the other two phrynosomatids (*Holbrookia propinqua*, and *Phrynosoma cornutum*), but differed from the former species by being distributed rather evenly among all apical cell surfaces (Fig. 9.15C,D). Another feature found in the ampulla of phrynosomatid species is a thicker muscularis compared relative to that of *Scincella lateralis*.

Our examination of the posterior urogenital anatomy of *Plestiodon fasciatus* (Figs. 9.16–9.18) revealed several anatomical features that differ greatly from the basic design described above for *Scincella lateralis*. For example, paired anterior dorsal recesses of the urodaeum extend anteriorly to lie ventromedial to the ductus deferentia (Fig. 9.16C). These anteriorly directed, pouch-like chambers of the cloacal urodaeum are also found in other lizards, such as *Aspidoscelis gularis* (Figs. 9.19 and 9.20; in other species mentioned elsewhere in this Chapter) and are present in colubrid and viperid snakes (Trauth and Sever 2011); however, these structures are absent in *Scincella lateralis*. The two recesses enlarge posteriorly and eventually combine (Fig. 9.16F) to create the primary cavity of the urodaeum. A second morphological difference between the two skinks is the absence of ampullae ductus deferentia within the ductus deferentia of *Plestiodon fasciatus*. Moreover, as the ductus deferentia of *Plestiodon fasciatus* near their eventual termination within the expanding base of the urogenital papillae (Fig. 9.16H), they merge with the two medial-most concurrent collecting ducts instead of releasing urogenital products directly into the cloaca through the orifices of the individual urogenital papillae, as is observed in *Scincella lateralis* (Fig. 9.13H). In *Plestiodon fasciatus*, a second pair of dorsolateral collecting ducts merge with their concurrent medial collecting ducts (Fig. 9.16J) before all urogenital products exit through individual orifices of the two urogenital papillae; therefore, the arrangement of terminal ducts differs sharply from that seen in *Scincella lateralis*. Finally, the ductal and urodaeal anatomies described above are positioned anterior to the confluence of the urodaeum proper with the coprodaeum at the urodaeal sphincter (Fig. 9.16K).

In *Plestiodon fasciatus*, an individual ductus deferens lies in close proximity to its two concurrent collecting ducts as well as to an anterior dorsal recess throughout most of the entire length of these structures. This aspect allows for a comparative examination of their epithelia at any given point in their pathway, as shown in Figs. 9.17 and 9.18. For example, Fig. 9.17A represents a transverse section of the urogenital anatomy at about the same plane of section as seen in Fig. 9.16E. The three epithelia reveal striking differences in their individual linings (Fig. 9.17B); representative epithelial micro-anatomies and ultrastructures are shown in Fig. 9.17C-E and Fig. 9.18A-C, respectively. The simple columnar epithelium of the collecting ducts reveal the presence of the sexual segment of the kidney (Fig. 9.17C). These tall, narrow columnar cells are laden apically with numerous carbohydrate-rich, periodic acid-Schiff's staining, secretory granules (Fig. 9.18A).

The anterior dorsal recesses of the urodaeum typically possess numerous folds in their walls (Fig. 9.17D). We term the spaces between folds as sperm crypts, due to the presence and orderly arrangement of sperm

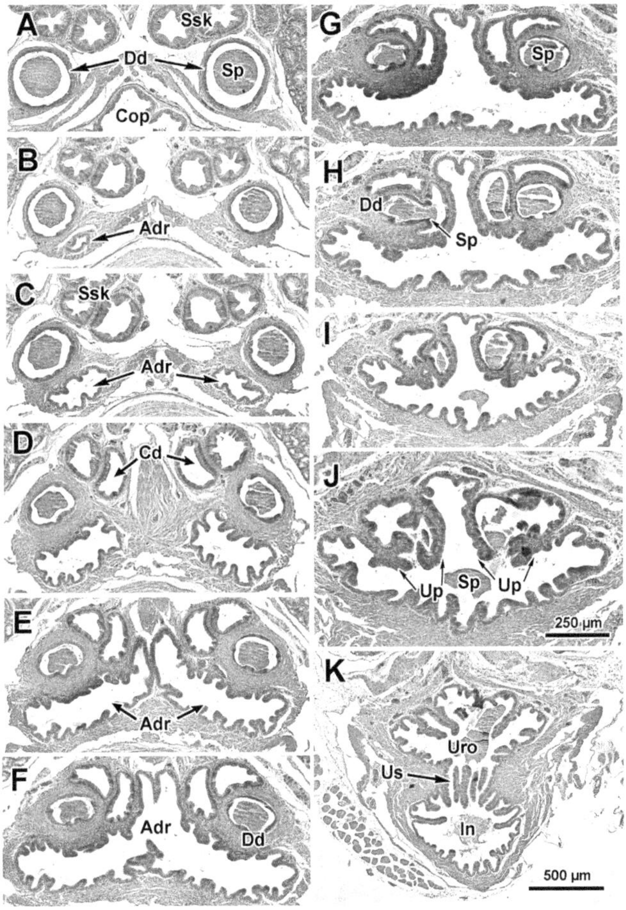

Fig. 9.16 Light micrographs (paraffin sections) of cranial-to-caudal serial histosections of the distal urogenital anatomy of *Plestiodon fasciatus* stained with hematoxylin and eosin. **(A)** Section through the posterior region of the kidney revealing two large collecting ducts on either side and their epithelia comprised of sexual segments (Ssk). The collecting ducts lie dorsal to the straightened ductus deferens (Dd) packed with sperm (Sp). Cop, coprodaeum. **(B)** The anterior dorsal recess of the urodaeum (Adr) emerges ventral to the ductus deferens. **(C)** Pair of medial posteriormost collecting ducts enlarges as Adr expand medially. **(D)** Medial-most collecting duct (Cd) begins to flatten (see Fig. 16 for epithelial morphology of Ssk, Dd, and Adr). **(E)** The Adr expand medially to reside adjacent to medial collecting duct. **(F)** Both extensions of the anterior dorsal recess of the urodaeum (Adr) merge medially forming the urodaeum proper. **(G)** The urodaeum (Uro) exhibits a highly folded epithelium similar to that of the Adr. The two collecting ducts and the Dd, together, form a bulge in the urodaeal wall. **(H)** Sperm within the Dd enter into the medial collecting duct as the lateral collecting duct flattens dorsally. **(I)** The lateral collecting duct shifts from its dorsal position to lie laterally, taking the place vacated by the Dd. The bulging duct complex begins to form the base of the urogenital papilla. **(J)** The lateral collecting duct merges (free arrow) with the medial collecting duct as both communicate with the orifice of the urogenital papilla (Ugp) **(K)** The urodaeum merges with the coprodaeum via the urodaeal sphincter (Us) as the intestine (In) merges ventrally.

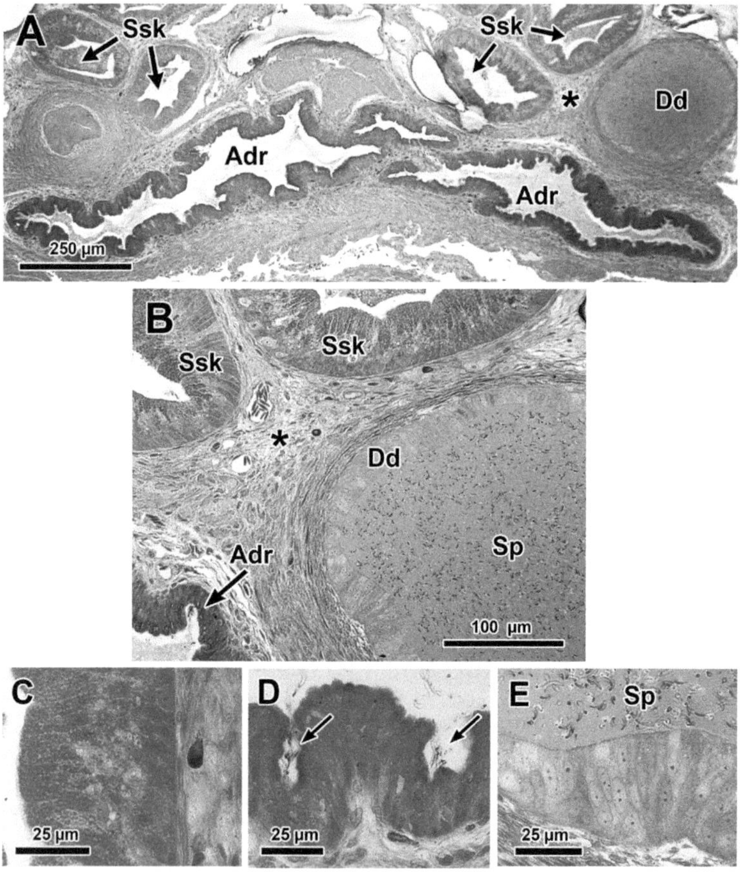

Fig. 9.17 Light micrographs (plastic sections) of the distal urogenital anatomy in *Plestiodon fasciatus* revealing the sexual segment of the kidney (Ssk), the posterior ductus deferens (Dd) and the anterior dorsal recess of the urodaeum (Adr). **(A)** Section through cloacal region similar to the plane shown in Fig. 15D. Asterisk denotes position shown in **B**. **(B)** Magnification of region around asterisk of **A**. Sp, sperm. **(C)** Tall columnar epithelium of the Ssk shown in **B** (upper area). **(D)** Pseudostratified columnar epithelium of the Adr; arrows point to sperm within epithelial folds. **(E)** Pseudostratified columnar epithelium of the Dd shown in **B** (lower middle area).

Color image of this figure appears in the color plate section at the end of the book.

clusters. The apical surfaces of the pseudostratified columnar epithelial cells, which line these crypts, are characterized by large irregular secretory vacuoles, and this constitutes a relatively thick cytoplasmic layer next to the luminal border. The amorphous, electron-lucent secretory material is characteristic of mucoid secretory products (Fig. 9.18C).

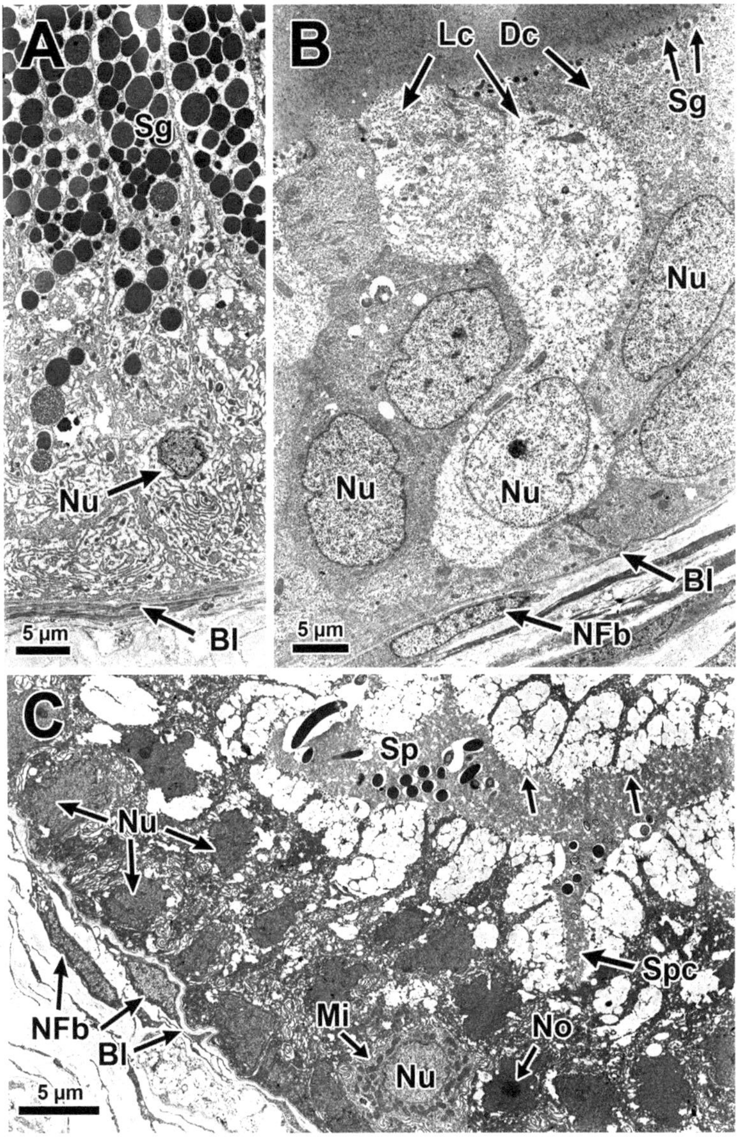

Fig. 9.18 Transmission electron micrographs of epithelia of **A**. the sexual segment of the kidney, **B**. posterior ductus deferens, and **C**. anterior dorsal recess of the urodaeum **C**. in *Plestiodon fasciatus*. Epithelia in **A**, **B**, and **C** extracted from the plane level of tissues shown in Fig. 9.17. **A.** Basal nucleus (Nu) of tall columnar cell; Bl, basal lamina; Sg, secretory granules. **B**. Dark (Dc) and light (Lc) pseudostratified columnar cells. NFb, nucleus of fibroblast cell. **C**. Large masses of opaque mucoid substance (arrows) dominate the apical regions of cells. Sperm are clustered within crypts (Spc). Mi, mitochondria; Sp, sperm; No, nucleolus.

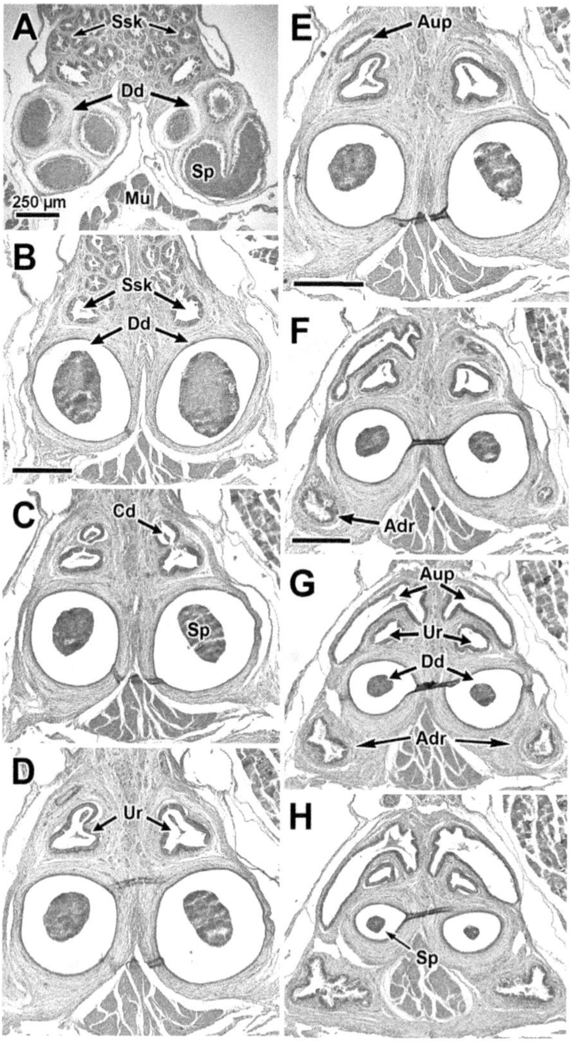

Fig. 9.19 Light micrographs (paraffin sections) of cranial-to-caudal serial histosections of the distal urogenital anatomy of *Aspidoscelis gularis* stained with hematoxylin and eosin. **A**. Section through the posterior region of the kidney revealing numerous tubules and associated sexual segments of the kidney (Ssk) lying dorsal to the ductus deferens (Dd) packed with sperm (Sp). Mu, medial striated muscle. **B**. Kidney tubules appear to merge, adjoining a large ventral collecting duct. The ductus deferens has now become greatly enlarged compared to its condition in **A**. **C**. The posteriormost collecting ducts (Cd) on both sides merge into single ducts. **D**. The ureter (Ur) now drains kidney products. **E**. The ampulla urogenital papilla (Aup) appears dorsolateral to the ureter. **F**. The anterior dorsal recess of the urodaeum (Adr) appears ventrolateral to the ductus deferens. **G-H**. The Aup greatly expands to occupy the dorsolateral region of the distal urogenital anatomy complex.

Color image of this figure appears in the color plate section at the end of the book.

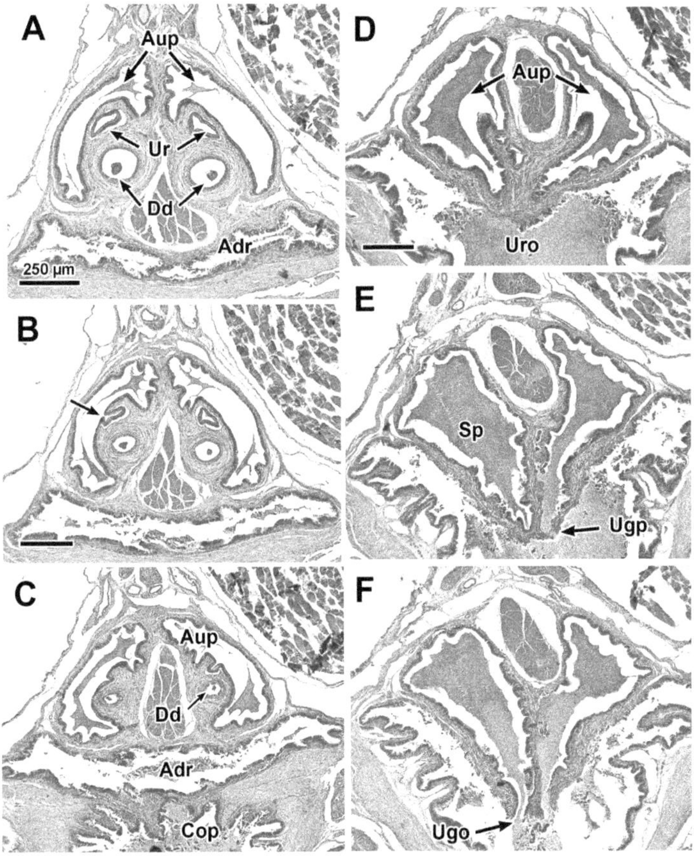

Fig. 9.20 *Aspidoscelis gularis*, continuation of Fig. 9.19. **(A)** Section through the posterior region of the distal urogenital complex; the paired ampulla urogenital papillae (Aup) dominate the dorsolateral region, whereas the paired anterior dorsal recesses (Adr) have greatly expanded within the ventral region. **(B)** A ureter (arrow) is shown merging with an Aup; the diameter of the ductus deferens has become greatly reduced. Both Adr have merged into a single cavity of the anterior urodaeum. **(C)** The Aup have expanded to occupy most of the dorsolateral, lateral, and ventrolateral region around the much-reduced ductus deferens; the Adr of the urodaeum nears its confluence with the coprodaeum (Cop). **(D)** Section at the level of the urodaeum proper (Uro), just posterior to the merging of the ductuli deferentia with the greatly expanded Aup. Sperm (Sp) and other urogenital products fill the Aup. **(E)** The Aup appears elongated in the region of the urogenital papilla (Ugp). **(F)** Release of sperm and urogenital products from the paired orifices (Ugo) of the urogenital papilla.

Color image of this figure appears in the color plate section at the end of the book.

The pseudostratified columnar epithelium of the ductus deferens is variable in thickness and stature (Fig. 9.17E). Both dark cells and light cells are present here, but their morphology and secretory activity differ from the same cells that are characteristic of the anterior ductus deferens (Fig. 9.18B).

New information on the posterior urogenital anatomy of *Aspidoscelis gularis* is possibly the most noteworthy and revealing of the lizards examined thus far. As with the two skinks above, we follow the path of the ductus deferens in a reproductively active *Aspidoscelis gularis*, through a series of cranial-to-caudal serial histo-sections (Figs. 9.19–9.20). The anterior ductus deferens begins following a straightening of the duct (Fig. 9.19B) along with a concomitant diameter enlargement compared to its previous looped morphology (Fig. 9.19A). Along with this straightening and enlargement, the ductal columnar epithelium loses its previous pseudostratification to become a low columnar lining, sparsely populated with cells. The lamina propria of the duct is now rich in scattered elastic fibers, and the muscularis is thin or absent. Several enlarged kidney collecting ducts (modified as a sexual segments) move ventrally to become associated with each ductus deferens; soon thereafter, however, a single tube (Fig. 9.19D) emerges that is triangular in transverse section, being formed by the merging of the last two collecting ducts. At this point, the collecting duct can be called a ureter. The next remarkable event is the appearance of the paired ampullae urogenital papillae, each residing dorsal to the ureters (Fig. 9.19F). These pouch-like sacs are mentioned here for the first time in lizards, but they have been previously reported only in male colubrine snakes (Siegel *et al.* 2011; Trauth and Sever 2011). These urogenital pouches are homologous to the ampullae uriniferous papillae of female colubrine snakes (Siegel *et al.* 2011). As the ampullae urogenital papillae increase dramatically in size during their caudal progression, paired anterior dorsal recesses of the urodaeum appear and likewise enlarge (Fig. 9.19G-H; Fig. 9.20A-C). While these expansions are taking place, a concomitant decrease in size is occurring in the diameters of the ductus deferens and the ureters. In Fig. 9.20B, one ureter can be seen merging with its ipsilateral ampulla urogenital papilla; soon thereafter, the ductuli deferentia merge with their ipsilateral ampullae. At the level of Fig. 9.20D-E, the ampullae are enlarged due to the presence of sperm within the single urogenital papilla. Finally, in Fig. 9.20F, sperm exit through paired orifices of the urogenital papilla.

9.3 THE SEXUAL KIDNEY

9.3.1 Overview

Much of the information on the sexual segment of the kidney (SSK) comes from data on snakes and was thoroughly reviewed by Aldridge *et al.* (2011); thus, data on snakes will not be presented again in this volume. Previous studies have provided morphological descriptions of the SSK as well as performed experimental analyses to determine the function of the SSK in lizards.

The first description of a "sexually" functioning kidney was by Mobius (1885) in stickleback fish and was first noticed in squamates by Gampert (1866). The earliest study on a lizard was by Leydig (1872) in *Lacerta agilis* but Regaud and Policard (1903) coined the term "sex segment" in their study on snakes. Herlant (1933) and Regamey (1935) investigated the seasonality of the sexual segment and found the link between androgens and the hypertrophy of the SSK. Badir (1958) investigated the sexual kidney of *Scincus scincus* and *Chalcides ocellatus*. He noticed that the secretory activity of males switched from a mucoidal secretion to a granular secretion during the active season. Anderson (1960) provided the first ultrastructural descriptions of the SSK in *Phrynosoma cornutum*. Pandha and Thapliyal (1964) followed up on work by Forbes (1941) and Altland (1943) and showed that the development of the SSK was induced by testosterone in *Calotes versicolor*. Sanyal and Prasad (1966) investigated the SSK of *Hemidactylus flaviviridis*, and Fox (1977) noted that the portion of the renal tubule occupied by the SSK differs among taxa. Rheubert *et al.* (2010) examined the location of the SSK through character optimization to squamate phylogenies and found that the location of the SSK was consistent among families, but varied at higher level taxonomic groupings. Furthermore, they found that the ancestral condition was most likely a SSK limited to the collecting duct, but the absence of the SSK in the Tuatara (*Sphenodon punctatus*) rendered this ancestral character state equivocal with parsimony ancestral state reconstruction.

Prasad and Sanyal (1969) followed up on their study of the SSK in *Hemidactylus flaviviridis* with a study on the effects of sex hormones on SSK development, which led to more studies concerning the purpose of the SSK. Cuellar *et al.* (1972) were the first to investigate the function of SSK secretions and found that the SSK secretions increased sperm sustenance in the Green Anole (*Anolis carolinensis*). Other studies concerning the possible functions of the SSK secretions, including blocking renal tubules during copulation to prevent the mixture of sperm and urine, assuring total emptying of the cloaca, forming a copulatory plug, and pheromone production, were proposed from studies on snakes. However, similar studies have been performed on lizards. Olsson (2001) showed that the SSK secretions do not form a copulatory plug in the Mallee Dragon (*Ctenophorus fordi*), whereas Moreira and Birkhead (2003) showed that a plug was formed from the SSK in the Iberian Rock Lizard (*Lacerta monticula*).

Del Conte (1972) and Del Conte and Tamayo (1973) were the first to note that the SSK can also be found in females of the Rainbow Whiptail (*Cnemidophorus lemniscatus*). Subsequent studies showed that a similar hypertrophy was noted in females of *Scincella lateralis* (Sever and Hopkins 2005). From their study they hypothesized that lizard species in which the SSK is hypertrophied in females, the hypertrophy is caused by a low level of natural androgens present in the females. However, this hypothesis has not yet been tested.

9.3.2 Sexual Segment of the Kidney

Previous studies have shown that the portion of the kidney that is occupied by the SSK is variable. The SSK can be found in the intermediate segment (IS), distal convoluted tubule (DCT), collecting duct (CD), ureter (UR), or a combination of those regions (Fig. 9.21). Within snakes the SSK is found

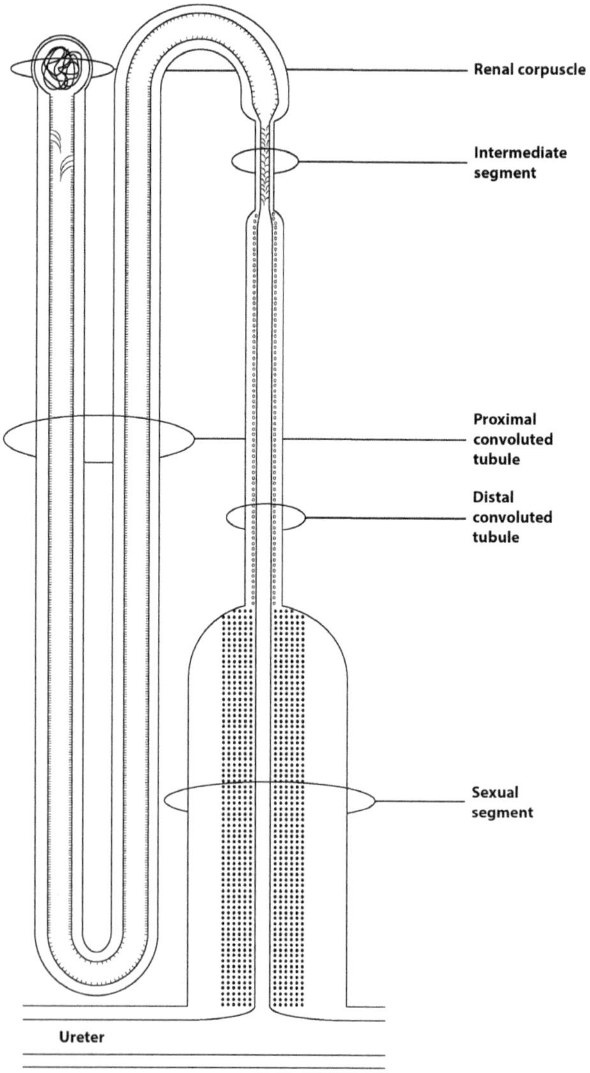

Fig. 9.21 Schematic drawing of the nephron in *Hemidactylus turcicus* from Rheubert *et al.* 2010 detailing the various regions of the nephron and indicating the sexual segment is restricted to the distal convoluted tubule.

in the DCT, CD, and UR within the scolecophidians, but only in the DCT in the Alethinophidia. The SSK is limited to the IS in the Varanidae and the CD in Gekkonidae, Xantusidae, and *Podarcis sicula*. Within the other families of lizards it can either be found in the DCT, CD, and UR or just the CD and UR. This variability is summarized in Table 9.7.

Table 9.7 Table depicting location of the SSK within the nephron in squamates.

Family	Region Occupied by SSK	Source
Sphenodontidae	no SSK present	Gabe and Saint Girons 1965; Rheubert *et al.* 2013
Gekkonidae	collecting duct	Sanyal and Prasad 1966; Saint Girons 1972; Rheubert *et al.* 2011
Xantusiidae	collecting duct	Saint Girons 1972
Scincidae	collecting duct, ureter	Badir 1957; Saint Girons 1972; Sever and Hopkins 2005
Teiidae	distal convoluted tubule, collecting duct, ureter	Del Conte and Tamayo 1973; Saint Girons 1972
Blanidae	collecting duct, ureter	Saint Girons 1972
Trogonophidae	collecting duct, ureter	Saint Girons 1972
Lacertidae	collecting duct, ureter	Saint Girons 1972
Lacertidae	collecting duct	Gabri 1983
Agamidae	distal convoluted tubule, collecting duct, ureter	Saint Girons 1972
Chamaeleonidae	distal convoluted tubule, collecting duct, ureter	Saint Girons 1972
Iguanidae	distal convoluted tubule, collecting duct, ureter	Saint Girons 1972
Polychrotidae	distal convoluted tubule, collecting duct, ureter	Fox 1958; Saint Girons 1972
Phrynosomatidae	distal convoluted tubule, collecting duct, ureter	Saint Girons 1972; this study
Anguidae	collecting duct, ureter	Saint Girons 1972
Varanidae	intermediate segment	Saint Girons 1972
Viperidae	Distal convoluted tubule	Volsøe 1944; Gabe 1959; Saint Girons 1972; Sever *et al.* 2008
Elapidae	Distal convoluted tubule	Saint Girons 1972
Colubridae	Distal convoluted tubule	Saint Girons 1972; Sever *et al.* 2002; Krohmer 2004; Siegel *et al.* 2009b
Xenopeltidae	Distal convoluted tubule	Saint Girons 1972
Boidae	Distal convoluted tubule	Saint Girons 1972
Leptotyphlopidae	distal convoluted tubule, collecting duct, ureter	Fox 1965
Typhlopidae	distal convoluted tubule, collecting duct, ureter	Fox 1965

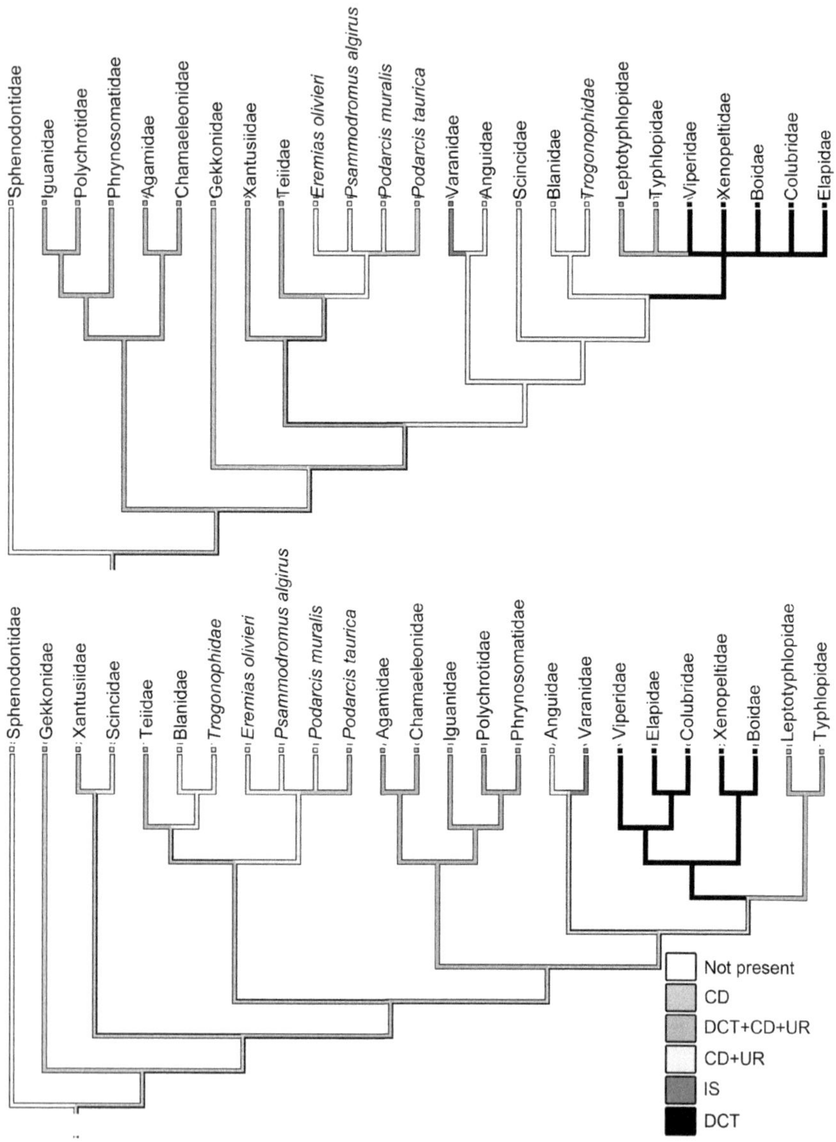

Fig. 9.22 Character mapping of region of kidney occupied by the SSK onto a morphological and molecular phylogenetic hypothesis for squamates. **(Top)** Morphological topology proposed by Conrad (2008). Bulletin of the American Museum of Natural History 2008: 1–182. **(Bottom)** Molecular topology proposed by Pyron *et al.* (2013). BMC Evolutionary Biology 13: 93.

Color image of this figure appears in the color plate section at the end of the book.

Analyses using ancestral state reconstruction for portion of the kidney occupied by the SSK revealed similar results when mapped onto both molecular and morphological phylogenetic hypotheses. On both phylogenetic hypotheses evolution of SSK location required 8 steps; however, we question the character state of the Balkan Wall Lizard (*Podarcis taurica*) proposed by Gabri (1983). He found that the SSK was restricted to the distal convoluted tubule. However, in all other lacertid species studied to date the SSK is found in both the distal convoluted tubule and collecting duct.

Ancestral character states were resolved for the Gekkota (restricted to collecting duct), Iguania (restricted to distal convoluted tubule, collecting duct, and ureter), and Alethinophidia (restricted to distal convoluted tubule, but were unresolved in all other clades. The absence of the SSK in Sphenodontidae (Rheubert *et al.* 2013) further suggests that this is a synapomorphic character within the squamates.

Structurally the SSK is similar in most species studied to date. The SSK is comprised of a tall columnar epithelium (Figs. 9.15 and 9.24) and the nuclei are situated basally (Fig. 9.24B, Nu). However, in *Seminatrix pygaea*, *Pelamis platurus,* and some other water snakes (DSS unpublished data) the nuclei are more centrally located (Fig. 9.24A, Nu). Large electron dense secretory granules are released into the lumen by either a merocrine process (Fig. 9.24B,C) as seen in *Agkistrodon piscivorus*, *Cnemidophorus lemniscatus*, and *Hemidactylus turcicus*, or via an apocrine process as seen in the Grass Snake (*Natrix natrix*), *Seminatrix pygaea*, *Scincella lateralis*, and the Brown Tree Snake (*Boiga irregularis*). Elaborate rough endoplasmic reticulum can be found within the cytoplasm as well as Golgi complexes (Fig. 9.23D). However, the Golgi complexes can be either concentric as observed in *Agkistrodon piscivorus*, *Scincella lateralis*, *Boiga irregularis*, and *Hemidactylus turcicus*, or linear as observed in *Natrix natrix*, *Seminatrix pygaea*, *Cnemidophorus lemniscatus*, *Podarcis sicula*, and *Podarcis taurica*.

9.4 THE CLOACA

Throughout this chapter we have mentioned that the distal urogenital ductal complex has ipsilateral ducts that convey several primary urogenital products in all squamates; these include the following: sperm via the ductuli deferentia, uric acid via either the ureters or collecting ducts of the kidneys, and secretions from the sexual segments of the kidneys via the ureters or collecting ducts. In some species, the ampullae urogenital papillae may also transport all of these materials. Regardless of transport system, all ductal components are released into the cloacal urodaeum through common pores (or orifices) associated with either a single, medial urogenital papilla

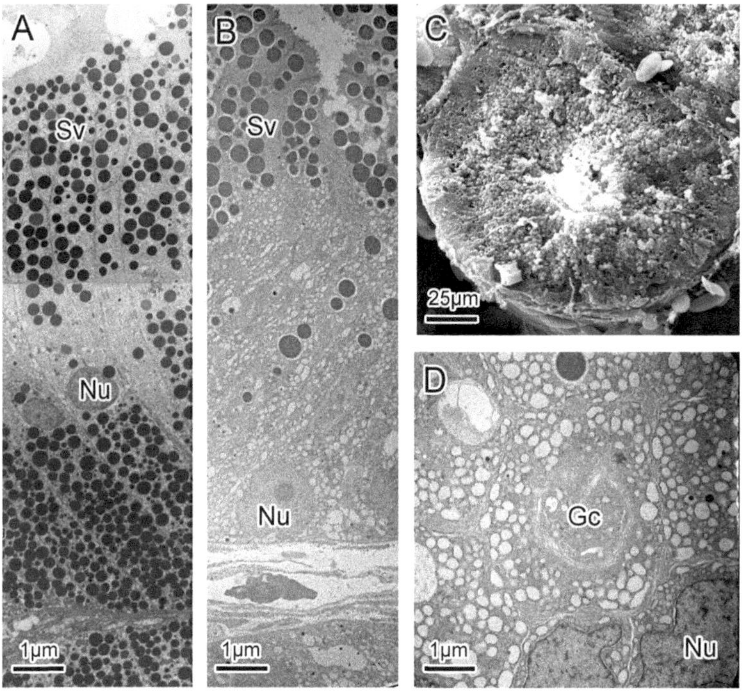

Fig. 9.23 Electron micrographs of the sexual kidney in: **(A)** *Seminatrix pygaea*, **(B)** *Hemidactylus turcicus* **(C)** *Sceloporus consobrinus*, and **(D)** *Hemidactylus turcicus*. The ultramicrographs detail the cellular morphology of the Golgi complex (Gc), nucleus (Nu), and secretory vesicles (Sv).

(Fig. 9.25) or through pores of bilateral urogenital papillae (Figs. 9.14, 9.26 and 9.27). A seminal paper by Trauth and Sever (2011) was the first to describe and illustrate the generalized anatomical features of a snake urogenital papilla. These authors also presented scanning electron micrographs of the micro-anatomies of urogenital papillae in 38 snake species.

In the following paragraphs and in addition to the urogenital papillae of *Scincella lateralis* and *Aspidoscelis sexlineata* (already shown in Figs. 9.14 and 9.25, respectively), we present anatomical features and micro-anatomies of urogenital papillae from lizard species representing 6 families using scanning electron micrographs. In addition, we compare these anatomies with those previously found in snakes using descriptive terminology from Trauth and Sever (2011).

The urogenital papillae of six species within the genus *Aspidoscelis* are illustrated in Figs. 9.25 and 9.26. One species, *Aspidoscelis sexlineata*, is remarkable in that it possesses a variable papillary morphology. Unlike other cnemidophorine species of the family Teiidae examined thus far that possess a single urogenital papilla, *Aspidoscelis sexlineata* was found to exhibit both a single papilla (Figs. 9.25C,D; Fig. 9.26B—note: Figs. 9.25C; Fig. 9.26B are the same specimen) as well as paired papillae (Fig. 9.25A). The specimen with two papillae exhibited separate papillary mounds and paired orifices surrounded by fleshy papillary lips; yet, these papillary mounds projected from a common basal platform, which is similar to the condition found in several conspecifics (Fig. 9.25C,D). In the two specimens examined with a single papilla, the papillary lips are not fleshy or robust. The orifices did project laterally, similar to the specimen in Fig. 9.25A.

The urogenital papillae of the other four teiid lizards shown in Fig. 9.26 are very similar in anatomical structure to one another. In *Aspidoscelis gularis* (Fig. 9.26C), a single prominent, conical papillary mound exists and lends support to the histo-morphology shown in Fig. 9.21E,F. In the Marbled

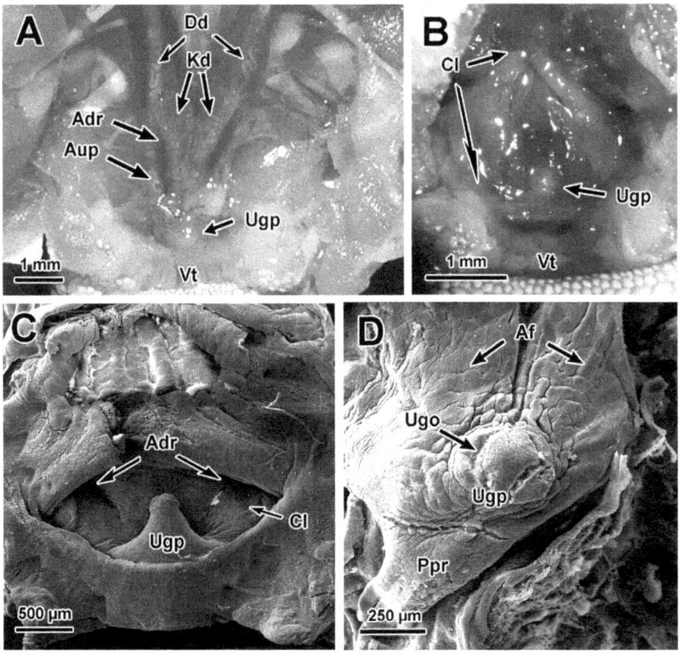

Fig. 9.24 Ventral views of the cloaca and distal urogenital anatomy of *Aspidoscelis sexlineata*. **A**. Macroscopic view of urogenital structures revealing the ductus deferens (Dd) lying ventrolateral to paired posterior extensions of the kidney (Kd), the right ampulla urogenital papilla (Aup) shown lateral to the right anterior dorsal recess (Adr) of the urodaeum (filled with blood on both sides), the urogenital papilla (Ugp), and the vent (Vt). **B**. Macroscopic view of the cloaca (Cl) showing the urogenital papilla (Ugp) situated posteriorly in the cloacal cavity just anterior to the vent (Vt). **C**. Scanning electron micrograph of the cloaca (similar to view as shown in **B**) revealing the paired anterior dorsal recesses (Adr) of the urodaeum and an unpaired urogenital papilla (Ugp). **D**. Scanning electron micrograph of an unpaired urogenital papilla (Ugp) along with its right urogenital orifice (Ugo) as well as the paired anterior ampullary folds (Af) and the single posterior papillary ridge (Ppr).

Color image of this figure appears in the color plate section at the end of the book.

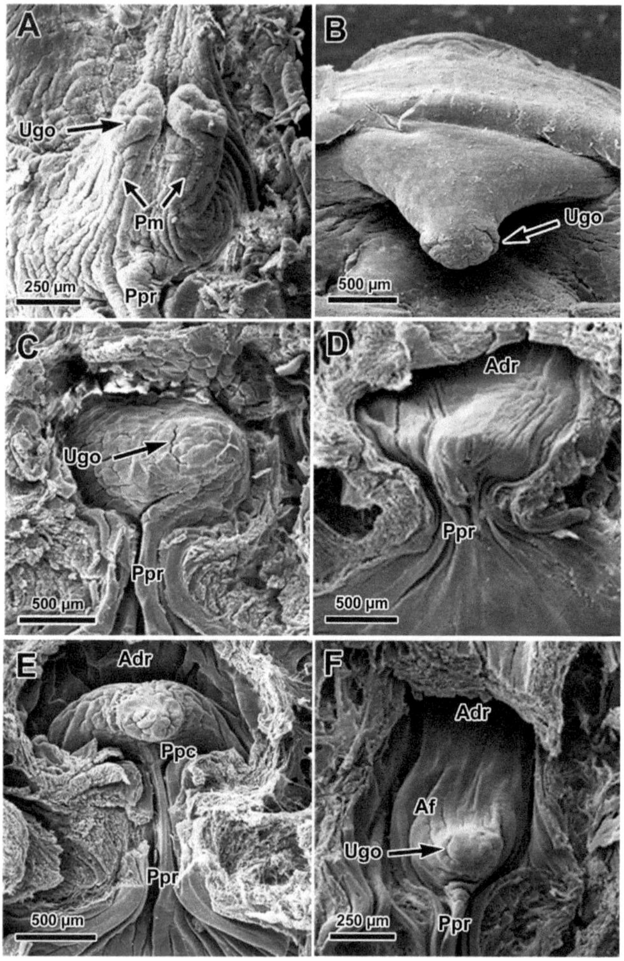

Fig. 9.25 Scanning electron micrographs of the urogenital papilla in five species of whiptail lizards (genus *Aspidoscelis*) of the family Teiidae. **(A)** *A. sexlineata*, a posteroventral view revealing a single posterior papillary ridge (Ppr) and paired papillary mounds (Pm), which supports paired urogenital papillae. The urogenital orifices (Ugo) of each papilla are encircled by irregular fleshy lips and appear to face in opposite directions. **(B)** *A. sexlineata* (same specimen as seen in Fig. 9.24C), an anteroventral view of a single urogenital papilla with laterally facing urogenital orifices (Ugo). **(C)** *A. gularis,* an end-on view of a single, broad-based, conical urogenital papilla occupying most of the cloacal cavity. **(D)** *A. marmorata*, an end-on view of a single, broad-based urogenital papilla along with its expansive anterior dorsal recess (Adr). The papillary mound is supported by a single posterior papillary ridge (Ppr) that extends into the proctodaeum. **(E)** *A. septemvittata*, an end-on view revealing a single, broad-based urogenital papilla similar to those found in **C** and **D**. Well-defined posterior papillary recesses (Ppc) can be seen. **(F)** *A. tigris*, an end-on view of a single, slender, narrow-based urogenital papilla supported by its posterior papillary ridge (Ppr); ampullary folds (Af). As in **A**, the urogenital orifices of **B-F** project laterally and possess irregular fleshy lips.

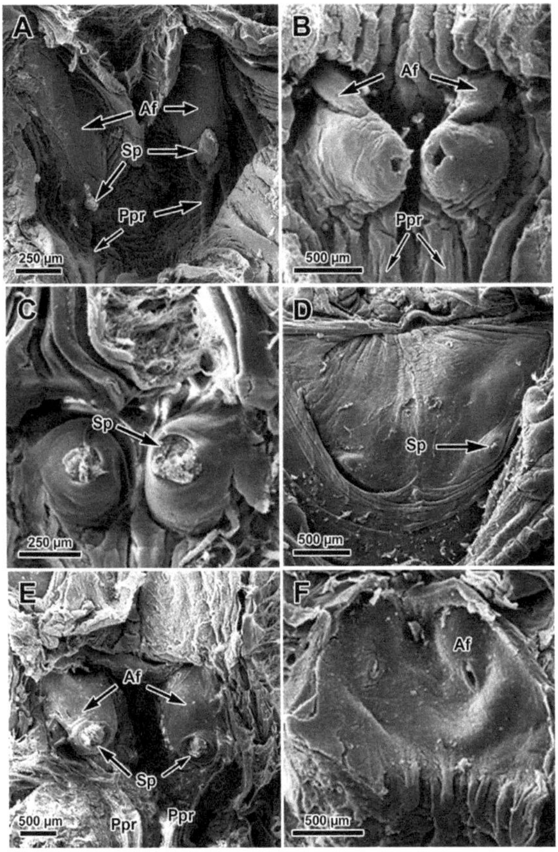

Fig. 9.26 Scanning electron micrographs of the paired urogenital papillae of species from five lizard families. **(A)** *Ophisaurus attenuatus* (family Anguidae), a ventral view revealing long ampullary folds (Af) associated with each urogenital papilla. Each papilla is also supported by well-developed posterior papillary ridges (Ppr). Sperm (Sp) are exiting the urogenital orifices, which are surrounded by smooth, hoop-shaped lips. **(B)** *Crotaphytus collaris* (family Crotaphytidae), an end-on view of tall, well-developed, mostly conical urogenital papillae supported by prominent ampullary folds (Af). The urogenital orifices are mostly narrow pores with poorly-developed lips. **(C)** *Anolis carolinensis* (family Dactyloidae), an end-on view of conical urogenital papillae pack with sperm (Sp). The urogenital orifices are broad and circular with no fleshy lips; posterior papillary ridges are indistinct. **(D)** *Sceloporus consobrinus* (family Phrynosomatidae), a ventral view of the cloacal chamber revealing widely-spaced, diminutive, urogenital papillae situated along its posterolateral margins. Poorly-developed ampullary folds and posterior papillary ridges support each papilla; sperm (Sp) are being released from the left urogenital papilla. **(E)** *Plestiodon laticeps* (family Scincidae), end-on view revealing prominent urogenital papillae releasing sperm (Sp). A tubular expansion of the ampullary folds (Af) appears to accommodate the passage of sperm; the posterior papillary ridges have been torn away from the bases of the papillae. The urogenital orifices are circular and have narrow lips. **(F)** *Plestiodon anthracinus* (family Scincidae), a ventral view of the cloacal chamber revealing poorly-developed urogenital papillae possessing orifices that appear as longitudinal slits surrounded by valve-like lips. No ampullary folds or posterior papillary ridges are present.

Whiptail (*A. marmorata*), the Plateau Spotted Whiptail (*A. septemvittata*), and the Tiger Whiptail (*A. tigris*), the anterior dorsal recess of the urodaeum can be seen extending anterior to the urogenital papilla (Fig. 9.26D-F). A well-developed, posterior papillary ridge secures the papilla to the posterior roof of the cloaca in all teiid lizards; this ridge is a common feature of the papillae in most snakes (Trauth and Sever 2011).

Distinctive micro-anatomies of paired urogenital papillae of representative species of five lizard families are shown in Fig. 9.26. All differed sharply from the paired papillary condition of *Aspidoscelis sexlineata* (Fig. 9.25A) by their possession of papillae whose basal papillary mounds were totally separated from one another. Interfamilial morphological similarities were found between members of several families. For example, the Western Slender Glass Lizard (*Ophisaurus attenuatus*; Fig. 9.26A) and the Broadhead Skink (Plestiodon laticeps; Fig. 9.26E) exhibited elongated ampullary folds. In addition, in *Crotaphytus collaris* (Fig. 9.26B) and *Anolis carolinensis* (Fig. 9.26C) possessed conical, pillar-like papillae similar to those found in Scincella lateralis (Fig. 9.14). Finally, in *Sceloporus consobrinus* (Fig. 9.26D) and the Coal Skink (*Plestiodon anthracinus*; Fig. 9.26F), widely spaced diminutive papillae and slit-like orifices are present. In most of the species displayed in Fig. 9.26, the posterior papillary ridges appear as stabilization structures of the urogenital papillae, and this feature is clearly evident in *Ophisaurus attenuatus* (Fig. 9.26A); however, in *Plestiodon anthracinus* (Fig. 9.26F) these structures are absent.

9.5 CONCLUSIONS

The reproductive system of lizards, consisting of the testis, testicular ducts, sexual kidney, urogenital papillae, and cloaca has recently received much needed attention. In our review we have shown that some regions of the reproductive system are highly conserved, such as the rete testis, whereas other regions show striking variability, such as the epididymis. The contrast between conservatism and variability in the reproductive system can lead to interesting results when investigating the evolution of the reproductive system. However, to date, few studies have compared the morphology of male lizard reproductive characteristics in a phylogenetic context. Therefore, we hope that this chapter stimulates the interest in the morphology of the reproductive system so that new hypotheses concerning the evolution and/or function of these different structures can be made. Furthermore, male repro-morphological characteristics may be beneficial in investigating evolutionary relationships or testing previous morphological and molecular hypotheses on the phylogeny of lizards. Future areas of research are unlimited concerning the reproductive system. There are many intriguing avenues in this fruitful area of research including the variation of the urogenital papilla and its evolution. For example, currently, the presence of a single urogenital papilla only occurs in some teiids and snakes. However, the lack of data on many taxonomic clades, such as the varanids, makes any conclusions on the evolution of a single median papilla in squamates premature. Therefore, more attention must be directed toward increasing the diversity of species / taxonomic clades studied, and focus should be placed on those that currently have no information regarding the morphology of the reproductive system.

9.6 ACKNOWLEDGMENTS

Justin L. Rheubert would like to express extreme gratitude to his mentors, some of whom are coauthors on this chapter: Dr. Robert D. Aldridge, Dr. David M. Sever, Dr. Stanley E. Trauth, Dr. Dustin S. Siegel, and Dr. Kevin M. Gribbins. Without them Justin would not be where he is today, and none of this would have been possible. Justin would also like to thank Dr. Richard Mayden for his discussions on phylogenetic analyses and allowing the use of various equipment. Drs. James M. Walker and James E. Cordes provided the specimens of *Aspidoscelis* species from west Texas for which we are greatly indebted. A special thanks to Christopher M. Murray, Jennifer Lee, and Michael K. Wilder who have all supported Justin throughout his journey. Special thanks go to his mother and father, Diane and Ronald Rheubert, his grandparents, Barbara and Richard Jones, his sister, Amanda Rheubert, and his nephew Trevor Castle, who have supported him both emotionally and financially, were continuously there for him.

9.7 LITERATURE CITED

Akbarsha, M. A., Tamilarasan, V., Kadalmani, B. and Daisy, P. 2005. Ultrastructural evidence for secretion from the epithelium of ampulla ductus deferentis of the fan-throated lizard *Sitana ponticeriana* Cuvier. Journal of Morphology 266: 94–111.

Akbarsha, M. A., Kadalmani, B. and Tamilarasan, V. 2006a. Histological variation along and ultrastructural organization of the epithelium of the ductus epididymides of the fan-throated lizard *Sitana ponticeriana* Cuvier. Acta Zoological (Stockholm) 87: 181–196.

Akbarsha, M. A., Tamilarasa, V. and Kadalmani, B. 2006b. Light and electron microscopic observations of fabrication, release, and fate of biphasic secretion granules produced by epididymal epithelial principal cell of the fan-throated lizard *Sitana ponticeriana* Cuvier. Journal of Morphology 267: 713–729.

Akbarsha, M. A., Kadalmani, B. and Tamilarasan, V. 2007. Efferent ductules of the fan-throated lizard *Stiana ponticeriana* Cuvier. Acta Zoologica 88: 265–274.

Aldridge, R. D., Jellen, B. C., Siegel, D. S. and Wisniewski, S. S. 2011. The sexual segment of the kidney. pp. 477–509. In R.D. Aldridge and D. M. Sever (eds.), *Reproductive Biology and Phylogeny of Snakes*. Boca Raton, FL: CRC Press.

Anderson, E. 1960. The ultramicroscopic structure of a reptilian kidney. Journal of Morphology 106: 205–240.
Alverdes, K. 1926. Die samenableitungswege der eidechsen. Z fur Mikrosk Anat Forsch 6: 420–442.
Alverdes, K. 1928. Die epididymis der sauropsiden im vergleich zu säugetier und mensch Zeitschrift fuer Mikroskopisch-Anatomische Forschung 1928: 405–471.
Averal, H. I., Manimekalai, M. and Akbarsha, M. A. 1992. Differentiation along the ductus epididymis of the Indian garden lizard *Calotes versicolor* (Daudin). Biological Structures and Morphogenesis 4: 53–57.
Badir, N. 1958. Seasonal variation of the male urogenital organs of *Scincus scincus* L. and *Chalcides ocellatus*. Forsk. Zeitschrift fuer Wissenschaftliche Zoologie 160: 290–351.
Cuellar, H. S., Roth, J. J., Fawcett, J. D. and Jones, R. E. 1972. Evidence for sperm sustenance by secretions of the renal sexual segment of male lizards, *Anolis carolinensis*. Herpetologica 28: 53–57.
Conrad, J. L. 2008. Phylogeny and systematics of Squamata (Reptilia) based on morphology. Bulletin of the American Museum of Natural History 2008: 1–182.
Daisy, P., Meeran, M. M. and Akbarsha, M. A. 2000. Histological variation along the vas deferens of the Indian garden lizard *Calotes versicolor* Daudin with special reference to the ampulla ductus deferentis. Journal of Endocrinology and Reproduction 4: 74–88.
Del Conte, E. 1972. Granular secretion in the kidney sexual segments of female lizards, *Cnemidophorus l. lemniscatus* (Sauria, Teiidae). Journal of Morphology 137: 181–192.
Del Conte, E. and Tamayo, J. G. 1973. Ultrastructure of the sexual segments of the kidney in male and female lizards, *Cnemidophorus l. lemniscatus* (L.). Zeitschrift fuer Zellforschung und Mikroskopische Anatomie 144: 325–327.
Desantis, S., Labate, M., Labate, G. M. and Cirillo, F. 2002. Evidence of regional differences in the lectin histochemistry along the ductus epididymis of the lizards, *Podarcis sicula* Raf. Histochemistry Journal 34: 123–130.
Depeiges, A. and Dufaure, J. P. 1977. Secretory activity of the lizard epididymis and its control by testosterone. General and Comparative Endocrinology 33: 473–479.
Dufaure, J. and Saint Girons, H. 1984. Histologie compare de l'epididyme et de ses secretions chez les reptiles (lezards et serpents). Arch d'Anato Micros 73: 15–26.
Dufaure, J. P., Courty, Y. V., Depeiges, A., Mesure, M. and Chevalier, M. 1986. Evolution and testosterone content of the epididymis during the annual cycle of the lizard *Lacerta vivipara*. Biology of Reproduction 35: 667–675.
Ferreira, A., Silva, D. N., Van Sluys, M. and Dolder, H. 2009. Seasonal changes in testicular and epididymal histology of the tropical lizard, *Tropidurus itambere* (Rodrigues 1987), during its reproductive cycle. Brazilian Journal of Biology 69: 429–435.
Forbes, T. R. 1941. Observations on the urogenital anatomy of the adult male lizard, *Sceloporus*, and on the action of implanted pellets of testosterone and of estrone. Journal of Morphology 68: 31–69.
Fox, H. 1977. The urogenital system of reptiles. pp. 1–157. In C. Gans and T. S. Parsons (eds.), *Biology of the Reptilia, vol. 6*. Academic Press, New York.
Fox, W. 1952. Seasonal variation in the male reproductive system of Pacific coast garter snakes. Journal of Morphology 90: 481–553.
Fox, W. 1958. Sexual cycle of the male lizard, *Anolis carolinensis*. Copeia 1958: 22–29.
Fox, W. 1965. A comparison of the male urogenital systems of blind snakes, Leptotyphlopidae and typhlopidae. Herpetologica 21: 241–265.
Gabe, M. 1959. Données histochimiques sur le rein de *Vipera aspis* (L.). Ann Histochim 4: 23–31.
Gabe, M. and Saint Girons, H. 1964. Contribution a l'histologie de *Sphenodon punctatus* Gray. Paris: Editions du Centre National de la Recherche Scientifique.
Gabri, M. S. 1983. Seasonal changes in the ultrastructure of the kidney collecting tubule in the lizard *Podarcis (=Lacerta) taurica*. Journal of Morphology 175: 143–151.
Gampert, O. 1866. Ueber di niere von *Tropidonotus natrix* und der Cyprinoiden. Zeitschrift für Wissenschaftliche Zoologie 16: 369–373.

Gribbins, K. M. 2011. Reptilian spermatogenesis. Spermatogenesis 1: 250–269.
Haider, S. and Rai, U. 1986. Effects of cyproterone acetate and flutamide on the testis and epididymis of the Indian wall lizard, *Hemidactylus flaviviridis* (Ruppell). General and comparative endocrinology 64: 321–329.
Haider, S. and Rai, U. 1987. Epididymis of the Indian wall lizards (*Hemidactylus flaviviridis*) during the sexual cycle and in response to mammalian pituitary gonadotropins and testosterone. Journal of Morphology 191: 151–160.
Henry, P. A. 1900. Étude histologique de la function sécrétoire de l'épididyme chez les vertébrés supérieurs. Archives d'anatomie microscopique 3: 229–292.
Herlant, M. 1933. Recherches histologiques et expérimentales sur les variations cycliques du testicule et des caractéres sexuels secondaires chez les reptiles. Archive de Biologie 44: 347–468.
Jones, R. C. 1998. Evolution of the vertebrate epididymis. Journal of Reproduction and Fertility Supplements 53: 163–181.
Krause, R. 1922. Mikroskopische anatomie der wirbeltiere. II. Vogel und reptilien. Vereinigung wissenschaftlicher Verleger., Berlin und Leipzig, pp. 268.
Krohmer, R. W., Martinez, D. and Mason, R. T. 2004. Development of the renal sexual segment in immature snakes: effect of sex steroid hormones. Comparative Biochemistry and Physiology, Part A 139: 55–64.
Lereboullet. 1851. *Recherches Sur L'Anatomie des organes genitaux des animaux vertebres*. Kessinger Publishing Company, pp. 272.
Leydig, F. 1872. Die in Deutschland lebenden Arten der Saurier. Laupp'schen Buchhandlung, Tübingen, Germany.
Martin Saint-Ange, G. J. 1854. *Étude de la'appareil reproducteur dan les cinq d'animaux vertébrés, au point de vue anatomique, physiologique et zoologique*. Paris, France: Imprimeri Impériale, pp. 281.
Meeran, M. M., Daisy, P. and Akbarsha, M. A. 2001. Histological differentiation along the ductus epididymis of the lizard *Calotes versicolor* Daudin. Journal of Animal Morphology and Physiology 48: 85–96.
Mesure, M., Chevalier, M., Depeiges, A., Faure, J. and Dufaure, J. P. 1991. Structures and ultrastructure of the epididymis of the viviparous lizard during the annual hormonal cycle: Changes of the epithelium related to secretory activity. Journal of Morphology 210: 133–145.
Mobius, K. 1885. Ueber die eigenschaften und den ursprund der schleimfaden des seestichlingnestes. Archiv fuer Mikroskopische Anatomie 25: 554–563.
Moreira, P. L. and Birkhead, T. R. 2003. Copulatory plugs in the Iberian rock lizard do not prevent insemination by rival males. Functional Ecology 17: 796–802.
Morgera, A. 1905. La relaZione tra it testiculo ed il deferente di alcuni Rerrili. Boll. Soc. Nat. Napoli 18: 114–128.
Mulaik, D. M. 1946. A comparative study of the urinogenital systems of an oviparous and two ovoviviparous species of the lizard genus *Sceloporus*. Bulletin of the University of Utah 37: 3–24.
Olsson, M. 2001. 'Voyeurism' prolongs copulation in the dragon lizard *Ctenophorus fordi*. Behavioral Ecology and Sociobiology 50: 378–381.
Pandha, S. K. and Thapliyal, J. P. 1964. Effect of male hormone on the renal sex segment of castrated males of the lizard *Calotes versicolor*. Copeia 1964: 579–581.
Prasad, M. R. N. and Sanyal, M. K. 1969. Effect of sex hormones on the sexual segment of kidney an other accessory reproductive organs of the Indian house lizard *Hemidactylus flavlviridis* Ruppell. General and Comparative Endocrinology 12: 110–118.
Pyron, R. A., Burbrink, F. T. and Wiens, J. J. 2013. A phylogeny and revised classification of Squamata, including 4161 species of lizards and snakes. BMC Evolutionary Biology 13: 93.
Regamey. 1935. Les caractéres sexuels du lezards (*Lacerta agilis* L.). Revue Suisse de Zoologi 42: 87–168.

Regaud, C. and Policard, A. 1903. Recherches sur la structure du rein de quelques ophidians. Archives d'Anatomie Microscopique et de Morphologie Experimentale 6: 191–282.
Reynolds, A. E. 1943. The normal seasonal reproductive cycle in the male *Eumeces fasciatus* together with some observations on the effect of castration and hormone administrations. Journal of Morphology 72: 331–377.
Rheubert, J. L., Sever, D. M., Geheber, A. D. and Siegel, D. S. 2010. Proximal testicular ducts of the Mediterranean Gecko (*Hemidactylus turcicus*). Anatomical Record 293: 2176–2192.
Rheubert, J. L., Murray, C. M., Siegel, D. S., Babin, J. and Sever, D. M. 2011. The sexual segment of *Hemidactylus turcicus* and the evolution of sexual segment location in Squamata. Journal of Morphology 272: 802–813.
Rheubert, J. L., Cree, A., Downes, M. and Sever, D. M. 2013. Reproductive morphology of the male Tuatara, *Sphenodon punctatus*. Acta Zoologica 94: 454–461.
Saint Girons, H. 1957. The cycle sexuel chez *Vipera aspis* (L.) dans l'ouest de la France. Bull Biol. France Belgique 91: 284–350.
Saint Girons, H. 1972. Morphologie compare du segment sexuel du rein des Squamates (Reptilia). Archives d'Anatomie Microscopique et de Morphologie Expérimentale 61: 243–266.
Saint Girons, H. 1984. Lez cycles sexuels des lézards males et leurs rapports avec le climat et les cycles reproducteurs des femelles. Annales des sciences naturelles. Zoologie et biologie animale. 6: 221–243.
Sanyal, M. K. and Prasad, M. R. N. 1966. Sexual segment of the kidney of the Indian house lizard, *Hemidactylus flaviviridis* Ruppell. Journal of Morphology 118: 511–527.
Sever, D. M. 2010. Ultrastructure of the reproductive system of the black swamp snake (*Seminatrix pygaea*). VI. Anterior testicular ducts and their nomenclature. Journal of Morphology 271: 104–115.
Sever, D. M., Stevens, R. A., Ryan, T. J. and Hamlett, W. C. 2002. Ultrastructure of the reproductive system of the black swamp snake (*Seminatrix pygaea*). III. Sexual segment of the male kidney. Journal of Morphology 252: 238–254.
Sever, D. M. and Hopkins, W. A. 2005. Renal sexual segment of the ground skink, *Scincella laterale* (Reptilia, Squamata, Scincidae). Journal of Morphology 266: 46–59.
Sever, D. M., Siegel, D. S., Bagwill, A., Eckstut, M. E., Alexander, L., Camus, A. and Morgan, C. 2008. Renal sexual segment of the Cottonmouth Snake, *Agkistrodon piscivorous* (Reptilia, Squamata, Viperidae). Journal of Morphology 269: 640–653.
Sever, D. M. and Freeborn, L. R. 2012. Observations on the anterior testicular ducts in snakes with emphasis on sea snakes and ultrastructure in the yellow-bellied sea snake, *Pelamis platurus*. Journal of Morphology 273: 324–336.
Sever, D. M., Rheubert, J. L., Hill, T. A. and Siegel, D. S. 2013. Observations on variation in the ultrastructure of the proximal testicular ducts of the ground skink, *Scincella lateralis* (reptilian: squamata). Journal of Morphology 274: 429–446.
Siegel, D. S., Sever, D. M., Rheubert, J. L. and Gribbins, K. M. 2009a. Reproductive biology of *Agkistrodon piscivorus* Lacépède (Squamata, Serpentes, Viperidae, Crotalinae). Herpetological Monographs 23: 74–107.
Siegel, D. S., Aldridge, R. D., Clark, C. S., Poldemann, E. H. and Gribbins, K. M. 2009b. Stress and reproduction in the brown treesnake (*Boiga irregularis*) with notes on the ultrastructure of the sexual segment of the kidney in squamates. Canadian Journal of Zoology 87: 1138–1146.
Siegel, D. S., Trauth, S.E., Sever, D. M. and Aldridge, R. D. 2011. The phylogenetic distribution of the ampulla ureter and ampulla urogenital/urrinferous papilla in the Serpentes. Journal of Zoological Systematics and Evolutionary Research 49: 160–168.
Trauth, S. E. and Sever, D. M. 2011. Male urogenital ductus and cloacal anatomy. pp. 411–475. In R. D. Aldridge and D. M. Sever (eds.), *Reproductive Biology and Phylogeny of Snakes*. Boca Raton, FL: CRC Press.

Van den Broek, A. J. P. 1933. Goaden und susführungsgänge. pp. 1–154. In L. Bolk, E. Göpper, E. Kallius and W. Lubosh (eds.), *Vergleichenden Anatomie der Wirbelthiere, Vol. 6.* Berlin: Urban and Schwarzenberg.
Volsøe, H. 1944. Structure and seasonal variation of the male reproductive organs of *Vipera berus* (L.). Spolia Zoologica Musei Hauniensis 5: 1–157.

CHAPTER 10

Male Reproductive Cycles in Lizards

Fausto R. Méndez de la Cruz,[1,*] *Norma L. Manríquez Morán,*[2] *Edith Arenas Ríos*[3] and *Nora Ibargüengoytía*[4]

10.1 INTRODUCTION

Previous studies of reproductive ecology in lizards have shown that the most important traits in the life history of reptiles are: age and size at sexual maturity, clutch size, frequency of clutches, reproductive phenology and the period of the reproductive activity (Duvall *et al.* 1982; Licht 1984; Galán 1997; Ramírez-Pinilla *et al.* 2009). In addition, the differences in the reproductive season between species or even among populations of the same species depend heavily on environmental conditions (Galán 1997). Seasonal variation in environmental factors such as temperature (Cruz 1994; Abu-Zinadah 2008), precipitation (Colli 1991; Cruz *et al.* 1999) and photoperiod (Rodriguez-Ramírez and Lewis 1991; Censky 1995; Manríquez-Morán *et al.* 2005; Zaldívar-Rae *et al.* 2008) play a fundamental role in the timing of the reproductive events such as gametogenesis, mating, gestation, egg laying, hatching or birth (Litch 1984; Medina and Ibargüengoytía 2010; Boretto and Ibargüengoytía 2006; Ortiz 1981; Olivares *et al.* 1987; Pough *et al.* 1998).

[1] Departamento de Zoología, Instituto de Biología, Universidad Nacional Autónoma de México, México.
[2] Laboratorio de Sistemática Molecular, Centro de Investigaciones Biológicas, Universidad Autónoma del Estado de Hidalgo, México.
[3] Departamento de Biología de la Reproducción, División de Ciencias Biológicas y de la Salud, Universidad Autónoma Metropolitana–Iztapalapa, México.
[4] Instituto de Biodiversidad y Medio Ambiente INIBIOMA Universidad Nacional del Comahue, Bariloche, Argentina.
* Corresponding author

The reproductive cycle starts with the gametogenesis (Fig. 10.1), which goes from the proliferation of spermatogonia, spermatocytes I and II, spermatids, and finally the differentiation into spermatozoa (Pudney 1995; Villagrán-Santa Cruz *et al.* 2009; Gribbins and Rheubert Chapter 11, this volume).

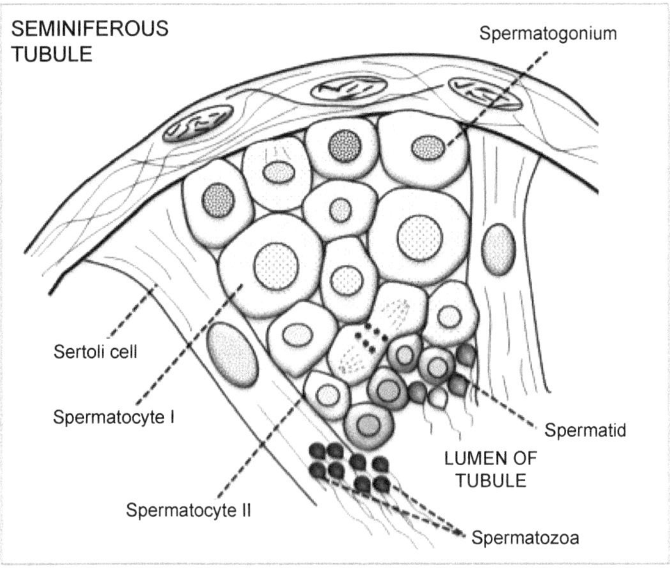

Fig. 10.1 Diagrammatical representation of the differentiation of germ cells in the seminiferous tubule during the spermatogenesis.

10.2 STAGES OF REPRODUCTIVE ACTIVITY IN THE TESTIS

Lizard testes consist of tubules lined by a seminiferous epithelium integrated by germ cells and nourishing Sertoli cells (Gribbins 2011; Gribbins and Rheubert Chapter 11, this volume). In males, the gonads show a cycle characterized by two main phases, one of activity with germinal cells undergoing development within the tubules during spermatogenesis, and the other characterized by a period of testicular inactivity or quiescence only when spermatogonia and Sertoli cells remain in the tubules (Licht 1971).

The male gonad cycle can be divided into four stages: quiescence, reactivation or recrudescence, spermiogenesis (maximum activity) and regression, which have been characterized based on the external morphology and the histological analysis of testis (Miller 1948; Fitch 1970; Fitch 1982; Villagrán-Santa Cruz *et al.* 2009; Gribbins 2011). Quiescence is characterized by testicular inactivity and during this time, the testis generally has a lower volume and weight (Licht 1971). During quiescence only Sertoli cells and some spermatogonia near the basal membrane of the seminiferous tubules are observed (Villagrán-Santa Cruz *et al.* 2009; Gribbins 2010; Gribbins 2011). During the recrudescence (reactivation), gonads can increase in size and volume as a result of the expanded activity inside of the seminiferous tubules. Within the tubules the proliferation of the germinal cells is observed, which begins when the spermatogonia have rapid consecutive mitotic divisions before meiosis (Gribbins 2011). Recrudescence also includes the meiotic phase, in which spermatogonia division produces secondary spermatocytes. Finally, spermatids appear after completing the second division of meiosis (Zaldivar-Rae *et al.* 2008; Gribbins 2011). During spermiogenesis (maximum activity), testes usually reach their maximum volumes and are widely vascularized showing an opaque color. At the microscopic level, one can observe spermiogenesis, the stage when the spermatids differentiate into spermatozoa. Spermatids can be seen distributed on the luminal edge of the seminiferous tubules with the sperm tails orientated towards the lumen. Similarly, in this stage, the first sparse spermatozoa in the lumen can be seen, which will be transported to the excurrent ducts: efferent tubules, epididymides and ductus deferens, and copulation occurs (Villagrán-Santa Cruz *et al.* 2009; Gribbins 2010; Gribbins 2011).

Controversy arises in determining the phase of maximum activity because testicular size, volume, and mass do not always allow accurate inference about spermatogenesis since they are the result not only of spermatogenic activity in the tubules but also the result of other reproductive events such as the hormonal activity of the Leydig cells in the interstitial compartment (Ibargüengoytía *et al.* 1999; Boretto *et al.* 2007, 2012). In fact, to determine the male cycle, it is necessary to understand the progress of spermatogenesis and the presence or absence of sperm in the epididymis by histology of the gonads and the reproductive tract (Mayhew and Wright 1970) as has been done in several genera such as *Liolaemus* (Ortiz 1981; Leyton *et al.* 1982; Ibargüengoytía and Cussac 1999; Medina and Ibargüengoytía 2010); *Phymaturus* (Boretto and Ibargüengoytía 2006; Boretto *et al.* 2007; Boretto and Ibargüengoytía 2009) and *Microlophus* (Goldberg and Rodríguez 1986; Olivares *et al.* 1987).

Finally, the last stage of testicular activity, regression, occurs when spermatids have finished the differentiation process and spermatozoa are transported to the excurrent ducts. Microscopically, there is a reduction in the number of cell layers that constitute the germinal epithelium. Only spermatogonia and Sertoli cells with phagocytes consuming the residual cytoplasm during spermatogenesis can be seen during advanced regression (Pudney 1995). There is also an increment of the lumen of the seminiferous tubules (Gribbins 2010; Gribbins 2011), where only sparse spermatozoa remain (Mayhew and Wright 1970; Ortiz 1981; Leyton *et al.* 1982; Goldberg and Rodríguez 1986; Olivares *et al.* 1987). Nevertheless, the efferent ducts exhibit a great quantity of spermatozoa and, at least during the early part of this stage, copulation can still occur (González-Espinoza 2006; Villagrán-Santa Cruz *et al.* 2009).

The events related to the reproductive activity of lizards (quiescence, recrudescence, spermiogenesis and regression) may occur at different times of the year and determine the different reproductive patterns, such as continuous (acyclic) or seasonal (cyclic) (Wen-San 2007).

10.3 REPRODUCTIVE CYCLES

The timing of reproductive activity is widely variable in lizards and several studies have shown that reproductive cycles are influenced by both the environment (Manríquez-Morán *et al.* 2005; Zaldívar-Rae *et al.* 2008) and phylogeny (Craig and Shine 1985; Dunham *et al.* 1988).

Continuous patterns are common in species inhabiting tropical areas where the environmental conditions are less variable throughout the year and can occur at individual or population levels (Licht 1984). When continuous activity is present at the individual level, males can breed throughout their lifetime once they reach sexual maturity. Microscopically, the testes show constant cycles of spermatogenesis-spermiogenesis for the production of gametes (Wen-San 2007; Gribbins 2011). The continuous cycles in males have been associated with the presence of vitellogenic females throughout the year in certain species (Hernández-Gallegos *et al.* 2002).

Continuous reproductive activity is common in several genera of lizards, *Gonatodes* (Serrano-Cardozo *et al.* 2007), *Sceloporus* (García-Collazo *et al.* 1993), *Aspidoscelis* (Fitch 1973), *Ameiva* (Vitt 1982; Censky 1995), *Cnemidophorus* (León and Cová 1973; Mojica *et al.* 2003), *Tropidurus* (Vieira *et al.* 2001; Ortiz, Boretto, Piantoni, Álvarez and Ibargüengoytía, in press), *Anolis* (Gribbins *et al.* 2009) and *Microlophus* (Leyton *et al.* 1982). Nevertheless, even in areas near the equator some species cease or reduce their reproductive effort and instead of having a single event (one large clutch in females), they have many small clutches throughout the year (Colli *et al.* 1997; Mesquita and Colli 2010).

On the other hand, most lizards inhabiting high latitudes (temperate and cold areas) exhibit seasonal patterns (Fitch 1970; Fitch 1972), characterized by alternating periods of gonadal activity (recrudescence-spermiogenesis) with periods of regression and quiescence. The most common pattern in both hemispheres around the world is one in which both females and males exhibit gametogenesis, courtship and mating during spring, with the oviposition in early summer and hatchlings or birth at mid-summer

(Fitch 1970; Vitt 1973; Stewart 1979; Xavier 1982; Guillette and Méndez de la Cruz 1993; Boretto and Ibargüengoytía 2009; Medina and Ibargüengoytía 2010; Méndez de la Cruz *et al.* 2013). This appears to be the ancestral pattern in lizards and is common in the oviparous species of temperate areas of several genera and even families of lizards. Fortunately, several groups of lizards have been studied with enough detail to define geographical or phylogenetical trends.

The genus *Sceloporus* is one of the most diverse groups of lizards in North America. It is integrated by almost 100 species of oviparous and viviparous lizards. The oviparous species of this genus are spring breeders, but depending on the latitude and altitude, the timing of reproductive activity may vary from a few months in spring in species such as the southern bunchgrass lizard, *Sceloporus aeneus* (Manríquez-Morán *et al.* 2013), the eastern spiny lizard, *S. spinosus* (Méndez de la Cruz *et al.* 2013) and the rosebelly lizard, *S. variabilis* (Benabib 1994) to several months during spring and summer in other species such as the jalapa spiny lizard, *S. jalapae* (González-Espinoza 2006), the queretaran spiny lizard, *S. ochoterenae* (Bustos-Zagal *et al.* 2011), the yellow-backed spiny lizard, *S. uniformis* (Goldberg 2012) and *S. variabilis* (Benabib 1994). This variation has been observed even in populations of the same species (Benabib 1994).

Another well-studied genus in North America is *Aspidoscelis* (Fig. 10.2), a group integrated by more than 80 unisexual and gonochoristic taxa, all of them oviparous. Except for some populations of the blackbelly racerunner, *A. deppii*, in Costa Rica (Fitch 1973), in the south limit of the genus distribution, all gonochoristic species exhibit seasonal patterns with maximum activity in spring and summer (Vitt and Breitenbach 1993). As it occurs in other groups of lizards, as latitude or altitude increases, the period of reproductive activity decreases. Species from temperate areas (the western whiptail, *A. tigris*, Goldberg 1973, 1976; the six-lined racerunner, *A. sexlineata*, Etheridge *et al.* 1986; Belding's orange-throated whiptail, *A. hyperythra*, Bostic 1966) exhibit a short reproductive season during spring and summer (four to five months), while tropical species show more extended seasons (6 to 8 months) as in the western Mexico whiptail, *A. costata* (Zaldivar-Rae *et al.* 2008; Granados-González *et al.* 2013), the many-lined whiptail, *A. lineatissima* (Ramírez-Bautista *et al.* 2000) and the eastern spotted whiptail, *A. gularis* (Ramírez Bautista *et al.* 2009). One of the most important differences between lizards of low and high latitudes is the moment when reproductive activity starts. In low latitude species of *Aspidoscelis*, testicular recrudescence begins in the last part of winter or beginning of spring, while in species of high latitude the reproductive activity initiates at the end of spring, when environmental conditions are hotter (Vitt and Breitenbach 1993). In these species, gonads exhibit a rapid growth during the reproductive season. Nevertheless, there are a number of

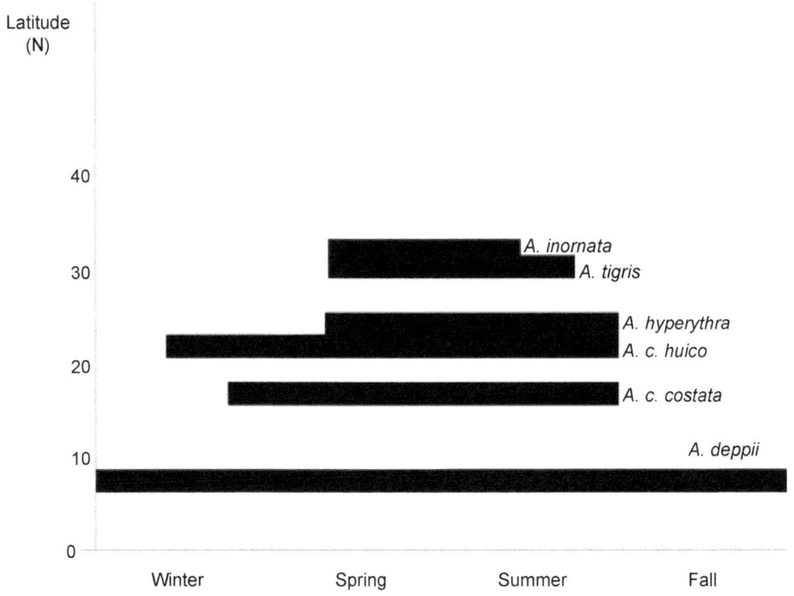

Fig. 10.2 Schematic representation of the variation in the reproductive season along a latitudinal gradient of different species of *Aspidoscelis*.

studies that suggest testicular activity may be during hibernation (Goldberg 1976; Vitt and Ohmarth 1977), but histological analyses are necessary to clarify the spermatogenic activity during winter in the genus.

The other genera of the Teiidae family, distributed from Central to South America, exhibit the same pattern observed in *Aspidoscelis*. The southernmost species (*Teius oculatus*, Cappellari *et al*. 2011; the red tegu, *Salvator rufescens*, Fitzgerald *et al*. 1993) show shorter periods of reproductive activity than other species or populations closer to the Equator (Spix's whiptail, *Ameivula ocellifera* and the rainbow lizard, *Cnemidophorus lemniscatus*, Mezquita and Colli 2003).

Many of the extant Squamata of southern South America belong to the family Lioleamidae with three genera. The monospecific genus *Ctenoblepharys*, with *Ctenoblepharys adspersa* an oviparous and *Psamnophilus* species, the entirely viviparous and saxicolous genus *Phymaturus* and the genus *Liolaemus*, one of the most versatile and rich in terms of the number of the species (Cei 1986; Donoso-Barros 1966). The genus *Liolaemus*, with at least 300 known species, is distributed from the highlands of Perú and Bolivia to Tierra del Fuego, and from the Pacific islands to Brazil (Cei 1986).

Oviparous species of *Liolaemus* of "*nigromaculatus* group" inhabiting the coasts of Chile, around 30° S, have an annual reproductive activity except the many-spotted tree iguana, *L. nigromaculatus*, which has a shorter reproductive cycle, with two clutches per year (Ortiz 1981). In these species, the association between temperature and spermatogenesis is evident since testicular development reduces in the summer months, when temperatures are higher (Ortiz 1981). Males of Phillipi's tree iguana, *L. bisignatus*, Kuhlman's tree iguana (*L. kuhlmanni*), *L. nigromaculatus*, the Zapallaren tree iguana (*L. zapallarensis*), and Muller's tree iguana, *L. copiapensis* inhabit sandy areas with sparse vegetation and their food is predominantly insectivorous (Ortiz and Riveros 1976; Ortiz 1981). These species present recrudescence in autumn, stop spermatogenesis during dormancy (March–August) and reach its peak in spring, when sperm can be found in the seminiferous tubules and epididymis, and show quiescence the rest of the summer (Ortiz 1981). Conversely, males of *L. nigromaculatus*, have maximum activity and presence of sperm until late summer, coinciding with the presence of ovigerous females at this time (Ortiz 1981).

In the north of Argentina *Liolaemus ramirezae*, the striped tree iguana, *L. bitaeniatus,* the shoulder tree iguana, *L. scapularis*, and *L. pacha* (Juárez-Heredia *et al.* 2013) spermatogenesis correlates inversely with the photoperiod and the maximum gonadal activity occurs in autumn and males show the maximum spermatogenesis and spermiogenesis in fall (Pinilla 1991). In contrast, male cycles of the high altitude species but at the same latitude as *L. huacahuasicus* have the peak of spermiogenesis at the end of spring (Pinilla 1991).

Liolaemids, can show prenuptial (Fig. 10.3, Type I) or postnuptial cycles (Fig. 10.3, Type II). In prenuptial cycles spermatogenesis occurs from spring to mid-summer, in synchrony with the peak of vitellogensis in females. This is the case of *Phymaturus vociferator* that inhabits Laguna del Laja National Park in severe weather conditions, with long periods of dormancy (Habit and Ortiz 1996) and Cei's mountain lizard, *P. punae* from the highlands of the Andes in San Guillermo Provincial Reserve in the province of San Juan, in the north of Argentina (Boretto *et al.* 2007). In contrast, postnuptial cycles have been found in lizards from steppe environments in west and central Patagonia and are characterized by spermatogenesis in mid-summer and spermiogenesis from early autumn to the next spring, when mating and ovulation occur. These males have an annual and seasonal cycle, with the onset of spermatogenesis in spring, ending in mid-summer with plenty sperm in the epidydimis (i.e., *Phymaturus zapalensis*, Boretto and Ibargüengoytía 2009; Bibron's tree iguana (*L. bibronii*), Boulenger's tree iguana (*L. boulengeri*), and the decorated tree iguana (*L. lineomaculatus;* Medina and Ibargüengoytía 2010) and *L. elongatus* (Ibargüengoytía and Cussac 1998).

Other liolaemids show continuous cycles with spermatozoa in the seminiferous tubules and/or sperm in the epididymides throughout the

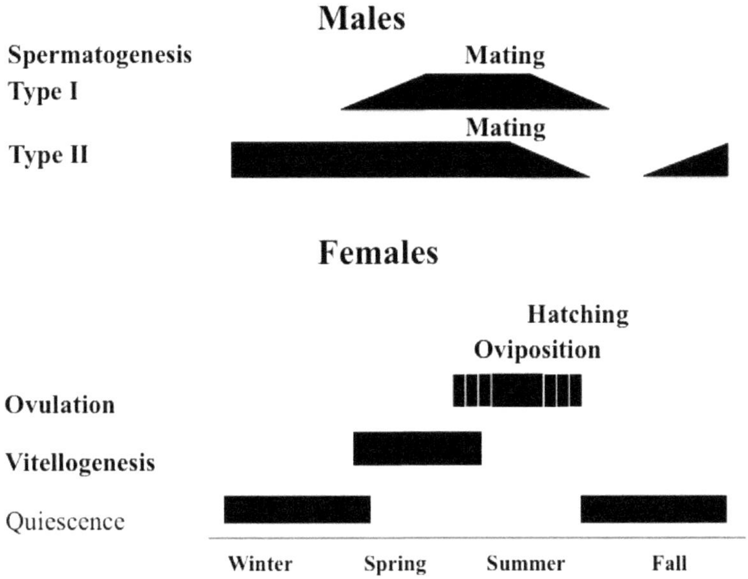

Fig. 10.3 Diagrammatical representation of the different events of the summer-fall reproductive cycle. The males may present two different types of spermatogenesis: Type I: summer fall, or Type II: fall-summer. In both cases the dormancy may interrupt the process.

whole year (*P. antofagastensis*; Boretto and Ibargüengoytía 2006; *P. aguanegra*, Cabezas Cartes *et al.* 2010); the painted tree iguana, *L. pictus* (Ibargüengoytía and Cussac 1999; *P. tenebrosus*, Ibargüengoytía 2004) and/or sperm in the epididymides throughout the activity season.

The family Lacertidae is a group of lizards integrated by more than 300 oviparous and viviparous species from Europe, Asia and Africa. Males of different genus exhibit reproductive cycles linked with female activity and influenced by photoperiod and temperature (Carretero 2006). Several studies have shown that thermal seasonality changes over the years and in different locations affecting the length of the reproductive season (Gavaud 1991), even in populations of the same species (the sand lizard, *Lacerta agilis*, Jackson 1978; Glandt 1993; Olsson and Madsen 1996; Amat *et al.* 2000). Nevertheless, most of the species studied to date exhibit seasonal reproductive patterns with maximum activity in spring and summer (Galán 1996, 1997; Amat *et al.* 2000; Carretero 2006). As in other families of lizards, patterns with short periods of reproductive activity are found in species that inhabit areas of high latitudes and/or altitudes. The males of a Pyrenean population of *Lacerta agilis* exhibit one of the shortest periods of reproductive activity, limited to two months (May and June) at the end of spring (Amat *et al.* 2000).

The family Agamidae is a group of more than 400 species of lizards, mainly oviparous, from Africa, Asia, Australia and some parts of Europe. Reproductive patterns in these lizards are influenced by temperature, precipitation and availability of resources and in most species the reproductive activity is restricted to spring and summer, as in the short-legged japalure, *Japalura brevipes* (Huang 1997). Some studies have shown that species of desert areas (the central netted dragon, *Ctenophorus nuchalis* and the ring-tile dragon, *C. caudicinctus*) exhibit different reproductive cycles that change annually in response to environmental variations (Bradshaw *et al.* 1991).

Reproductive patterns with maximum activity in spring are common in oviparous lizards, but have been observed in viviparous species at high latitudes (the northern alligator lizard, *Elgaria coerulea principis*, Vitt 1973; *Elgaria coerulea*, Stewart 1979; the common lizard, *Zootoca vivipara*, Xavier 1982; the elongate tree iguana, *Liolaemus elongatus*, *L. pictus*, Ibargüengoytía and Cussac 1999; *L. lineomaculatus*, Medina and Ibargüengoytía 2010; *Phymaturus tenebrosus*, Ibargüengoytía 2004; *P. zapalensis*, Boretto and Ibargüengoytía 2009), as environmental factors restrict the activity of lizards to some months in spring and summer.

Reptile spermatogenetic cycles have been classified according to the period in which the sexual cells become activated, which could precede or follow the reproductive period (Saint-Girons 1963, 1984; Carretero 2006). There are species such as *Phymaturus zapalensis* and *P. tenebrosus* (Boretto and Ibargüengoytía 2009) which produce spermatozoa immediately after the breeding season (postnuptial spermatogenesis), but delay this process along a variable period of time, thus distributing the energetic costs associated with spermatogenesis (Olsson *et al.* 1997; Roig *et al.* 2000). In other species, spermatogonia develop into spermatocytes and spermatids immediately after breeding and then the maturation to spermatozoa extends until the following season (mixed type). This is the case of *Psammodromus*, *Podarcis*, and *Lacerta* (Carretero 2006) species. In contrast, the whole maturation from spermatogonia to spermatozoa can be produced immediately before or during breeding (prenuptial or vernal type) in *Acanthodactylus*, *Mesalina* and *Phymaturus* (Bons and Saint-Girons 1982; Boretto and Ibargüengoytía 2009). These patterns have been observed in most of the known spring breeders.

The cessation of activity at high altitudes or latitudes is a common phenomenon due to the low temperatures occurring in these areas. In tropical latitudes, the period of dormancy is shorter (Galán 1996) and only the adults of some species go into dormancy, while the young remain active year-round (Zaldivar-Rae *et al.* 2008).

Moreover, there are viviparous species inhabiting tropical-temperate latitudes that show patterns in which the gametogenesis, courtship and mating occur in fall (Fig. 10.4, Type II), gestation is carried out during winter and the birth of offspring occurs in the following spring (*S. torquatus*, Guillette and Méndez de la Cruz 1993; Méndez de la Cruz *et al.* 1998), or a variation where all events occurs during the same time but mating take place during summer months (Fig. 10.4, Type I), as in the mesquite graphic lizard, *Sceloporus grammicus* (Guillette and Casas-Andreu 1980); the central cleft lizard, *S. mucronatus* (Méndez-de la Cruz *et al.* 1988). In South America, at high altitudes in the north of Chile (18° S and >3000 m) the viviparous species of *Liolaemus*: the zodiac tree iguana, *L. signifer* (= *multiformis*); James' tree iguana, *L. jamesi*; *L. aymararum*; and the brilliant tree iguana, *L. alticolor* (Leyton *et al.* 1982), show cyclic reproduction synchronized with female reproductive activity (Ibargüengoytía 2010). These species exhibit a fall pattern very similar to what it is observed in many viviparous species of *Sceloporus* at similar latitudes and altitudes in the northern hemisphere (e.g., *S. torquatus* 19° N from 2250 to 2600 m; the crevice spiny lizard, *S. poinsettii* 26° N). These species present a slow gametogenesis from winter through fall and mating in autumn with gestation periods ranging from 5 to 10 months during fall and winter, and births occur the following spring or summer (Méndez de la Cruz *et al.* 1998; Gadsden *et al.* 2005).

Dormancy is not present in lizards with fall reproductive activity, as the females become pregnant and should bask to develop the embryos,

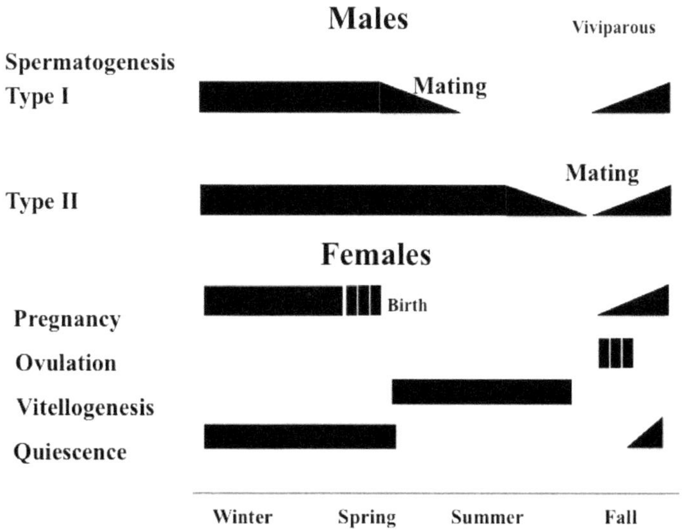

Fig. 10.4 Diagrammatical representation of the different events of the fall reproductive cycle. The males may present two different types of spermatogenesis: Type I (dissociated cycle): fall-summer, sperm storage occurs in the oviducts until the ovulation (that occurs from 3 to 4 months later), or Type II (associated cycle): fall-fall, with slow differentiation of the spermatic cells. In both cases the lizards do not hibernate.

although there may be a decrease in the hours of activity regardless of the altitude or latitude (Sinervo *et al.* 2011). This winter pregnancy surely limits the distribution of species that present fall activity to areas with tropical or subtropical influence (Méndez de la Cruz *et al.* 1998), apparently because the low winter temperatures force lizards living at high latitudes to dormancy, impeding uterine embryo development.

In some areas, at high altitudes or latitudes, environmental conditions are unfavorable for lizards, consequently, their reproduction is limited and several species of both hemispheres exhibit biennial patterns of reproductive activity (Capula *et al.* 1992; Boretto and Ibargüengoytía 2009; Cabezas-Cartes *et al.* 2010; Ibargüengoytía and Casalins 2010). These patterns are more common in females because the unfavorable conditions promote periods of gestation up to 14 months and limited resources, with a slow recovery of lost energy reserves (Boretto and Ibargüengoytía 2009). Under cold temperate environments in Argentina and Chile, liolaemid males have developed a variety of reproductive cycles that allow them to coordinate with the female cycles, pointing out the existence of the following chain of causal events: environmental conditions—female cycle-male cycle and male dimorphic traits (Ibargüengoytía and Cussac 1999). This is especially notorious under harsh environments at high altitudes in the Andes or in the Patagonian steppe, in which female cycles are constrained by the length of the activity season. Species from cold temperate environments in South America live under severe weather conditions, with long periods of dormancy which restricts the reproductive activity period up to 6 months resulting in prolonged biennial or triennial female cycles (Ibargüengoytía 2010). Thereby, the low frequency of reproductive females affects the male cycles and the development of male dimorphic traits (Ibargüengoytía and Cussac 1999; Ortiz 1981; Boretto and Ibargüengoytía 2009).

10.4 ASYNCHRONY AND SPERM STORAGE

Most of the lizards exhibit associated reproductive cycles (also called synchronic), i.e., reproductive phenologies in which males and females reach the maximum activity during the same period, thus sperm production, mating, and ovulation occur at the same time, for example in *Sceloporus jalapae* (González-Espinoza 2006), *S. poinsettii* (Gadsden *et al.* 2005), *Aspidoscelis costata* (Zadívar-Rae *et al.* 2008), *A. inornata* (Christiansen 1971), *Teius teyou* (Cruz *et al.* 1999) and *Sceloporus torquatus* (Feria-Ortíz *et al.* 2001). However, there are species that exhibit dissociated reproductive patterns (also called asynchronous), i.e., the males and females show reproductive phenologies with maximum activity at different times. In these species (e.g., *Sceloporus grammicus*, *S. mucronatus*, the Palacio's bunchgrass lizard, *S. palaciosi*,

Guillette and Casas 1980; Méndez de la Cruz *et al.* 1988, 1999; Estrada-Flores *et al.* 1990; Villagrán-Santa Cruz *et al.* 1994; Méndez-de la Cruz and Villagrán-Santa Cruz 1998), males reach the maximum reproductive activity earlier than females. These species display patterns with reproductive activity in spring-summer, while the females reach the maximum activity in fall (Méndez de la Cruz and Villagrán-Santa Cruz 1998). In populations with dissociated reproductive cycles sperm retention by females is required, as mating takes place when the males are in maximum testicular activity or during early regression (Méndez-de la Cruz *et al.* 1988, 1999; Villagrán-Santa Cruz *et al.* 1994). Therefore, male reproductive success in many cases depends on the viability of sperm stored in the reproductive tract of the females (Girling 2002; Kast 2007; Olsson *et al.* 2007).

Retention of sperm by females is common among lizards. The sperm is stored in the vaginal region within the lumen and folds or epithelial outgrowths (Blackburn 1998; Girling 2002; Sever and Hamlett 2002; Siegel *et al.* Chapter 6, this volume). The period of sperm retention by females is widely variable in lizards, but has been recorded to be viable a maximum of seven years (Birkhead and Møller 1993). However, experimental studies have demonstrated that the sperm retention enables the production of viable clutches after the isolation periods with similar success as those of the females with no retention (Murphy *et al.* 1996; Chun-Fu *et al.* 2004; Yamamoto and Ota 2006; Kast 2007; Ortega León *et al.* 2009). The sperm retention has allowed the existence of disassociation in the reproductive cycles of these males and females (Smyth and Smyth 1968; Guillette and Casas-Andreu 1980; Guillette and Sullivan 1985; Méndez de la Cruz *et al.* 1988; Villagrán-Santa Cruz *et al.* 1992).

Two different hypotheses have been suggested to explain the evolution of sperm retention (Birkhead and Moller 1993): 1) Election of male quality, with two options, the election may occur as a precopulatory mechanism after a male-male competition for territory, or a postcopulatory mechanism, involving the insemination by more than one male and the posterior sperm competition in the tract of the female, and 2) season optimality, when the timing for fertilization or hatching/birth is incompatible with time for copulation, the sperm storage result in the mechanism to solve the differences in the timing for copulation in asynchronic reproduction between males and females.

10.5 PARTICIPATION OF THE EPIDYDIMIS IN THE REPRODUCTIVE CYCLES

The description of gametic activity based only on the maximum gonadal size has been shown to be inaccurate. The gonadal size may be due to a testicular edema following the accumulation of water, which is not related to spermatogenic activity (Vitt 1986; El-rajam *et al.* 2009). Gonadal size has also been used to describe the dissociated cycles and consequently, the protandry. Nevertheless, in the green anole (*Anolis carolinensis*), one of the preferred models to study the protandry, after considering several evidences (sex ratio, social behavior, stomach content, sex steroids and gonadal condition), showed that in fact the cycles of both sexes were associated (Jenssen *et al.* 2001). Some studies based solely on testicular activity have led to consider dissociated cycles (Villagrán Santa Cruz *et al.* 2009), when they were indeed associated, by missing the fact that males continued the copulas during the early regression, when the epididymides were full of sperm (Méndez de la Cruz *et al.* 1988; Villagrán-Santa Cruz *et al.* 1994; Méndez-Juárez 2003). This evidence indicated that the epididymis activity should be incorporated to the description of the reproductive cycle, in particular during the maximum reproductive activity. Actually, the epididymis should be the focus of detailed studies addressing, among several issues, the sperm maturation as an important process of the reproduction.

During sperm maturation, the acquisition of proteins and reconfiguration of the plasmatic membrane (Aitken *et al.* 2007) is of vital importance for sperm motility (Mohri and Yanagimachi 1980; Ishijima and Witman 1987; Clulow *et al.* 1992) in mammals, even though there are no studies about the acquisition of proteins or reconfiguration of the plasmatic membrane in lizards. However, there are evidences that indicate that sperm motility changes positively from regions proximal to distal of the epididymis, similarly to mammals (Akbarsha *et al.* 2006; Nirmal and Rai 1997).

During the sperm maturation in mammals, one of the processes that occur is the migration of the cytoplasmic drop from the basis of the sperm head toward the medium part before it is totally detached. This process is concomitant with the passing in the epididymis from the head to the body region (Gatti *et al.* 2004). This migration may affect the lipid mitochondrial dynamic (see Cooper 2011), but there are no evidences of this, even though several pathologies are related with the migration of the cytoplasmic drop process (Bonet *et al.* 1992), and the sperm should discard the cytoplasmic drop to acquire the progressive motility and to be able to fertilize the ova. Recently, some evidences in our laboratory (EAR) with *Sceloporus mucronatus* indicate that the sperm from the proximal and distal part of the epididymis present no migration of the cytoplasmic drop (unpublished information; Fig. 10.5), in contrast to evidence found in mammals, where the passing through the epididymis is related with greater motility unless prevented by high temperatures.

The appropriate temperature is important to eliminate the cytoplasmic drop, the increase of temperature (in humans, scrotal temperature rises due to tight clothes) prevent the migration of this drop which stays attached to the medium part of the spermatozoa (Bedford *et al.* 1982). On the other

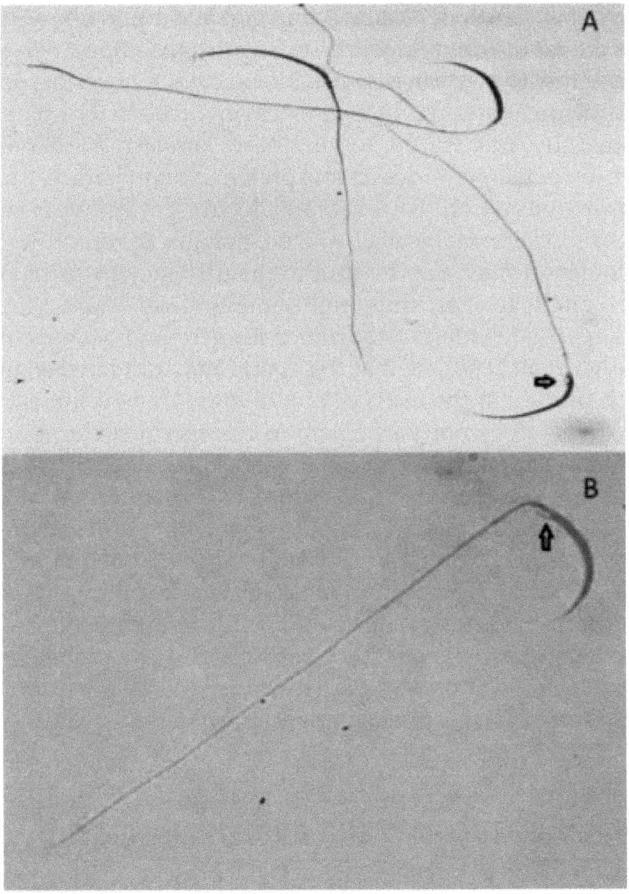

Fig. 10.5 *Sceloporus mucronatus* sperm from **(A)** proximal region of the epididymis; **(B)** Distal region of the epididymis.

hand, the exposition to fluctuant temperatures in reptiles is important to complete the testis cycle (Gavaud 1991). Unfortunately, the exposure to low temperatures has been poorly studied in reptiles, in contrast with the studies carried out in other groups. In mammals, evidence indicates that a temperature lower than the body core is necessary to attain an appropriate spermatogenesis (Setchell 1998), or to store the sperm in the distal segment of the epididymis (Foldesy and Bedford 1982). When testis cannot attain lower temperature in mammals, their weight decreases (Bartlett and Sharpe 1987), inducing apoptosis of the germinal cells (Allan *et al.* 1987) as well as a reduction in testosterone (Larsson *et al.* 1983), or they become disable to produce viable sperm in cryptorchid testis (Crew 1922).

Sinervo *et al.* (2010) determined that global warming limits the time of activity of the reptiles and may even extinguish the population that do not have a possibility to perform essential biological activities. It is possible that within the incipient findings of the process involved in the epididymis, the reproductive structures of the reptiles need, similarly to mammals, to be exposed to lower temperatures (below preferred temperatures), to complete spermatozoa maturation. For example, during a warm year some lizard populations (*Sceloporus mucronatus*) were not able to reproduce; however, this phenomenon may also be due to insufficiency to generate enough energy for vitellogenesis, thus high temperatures retard vitellogenesis as well as spermatogenesis (Rodríguez-Romero and Méndez de la Cruz 2004; Sinervo *et al.* 2010), or that the sperm may not have completed the maturation process in the testis or epididymis. However, it is considered that epididymis does not participate in the sperm maturation (Sever *et al.* 2002; Marinho *et al.* 2009) in Squamata (Jones 1998; Sever *et al.* 2002), suggesting that the sperm should complete its maturation in the ductus deferens (Tourmente *et al.* 2007). Some evidence in reptiles suggest that the first three zones of the epididymis (initial, head and body) are associated with the secretion and absorption of fluids, whereas the epididymis tail is in charge of the cytoplasmic absorption and sperm storage (Akbarsha *et al.* 2006). Many questions are still unanswered. How global temperature increases will affect lizards is still to be answered, evidence from our studies indicate that sperm motility stay longer at low temperatures, similarly to mammals.

10.6 REPRODUCTIVE PATTERN EVOLUTION: A STUDY CASE IN PHRYNOSOMATIDS

Some iguanian groups have experienced a big diversification in reproductive patterns such as the phrynosomatids. The lizards of the genus *Sceloporus* present a wide distribution, from Canada to Panamá, they are therefore exposed to diverse environmental conditions (Sites *et al.* 1992) and consequently adopt different reproductive and life history strategies.

10.6.1 Continuous Reproductive Patterns

According to the reconstruction of the ancestral states (Madisson and Madisson 2011) of the reproductive patterns (Fig. 10.6), the ancestral reproductive cycle in *Sceloporus* is seasonal (with maximum activity during spring-summer), which is the common pattern in the genus. One of the derived patterns is continuous reproduction, which has evolved independently two times. Two species present continuous spermatogenesis, *Sceloporus variabilis*

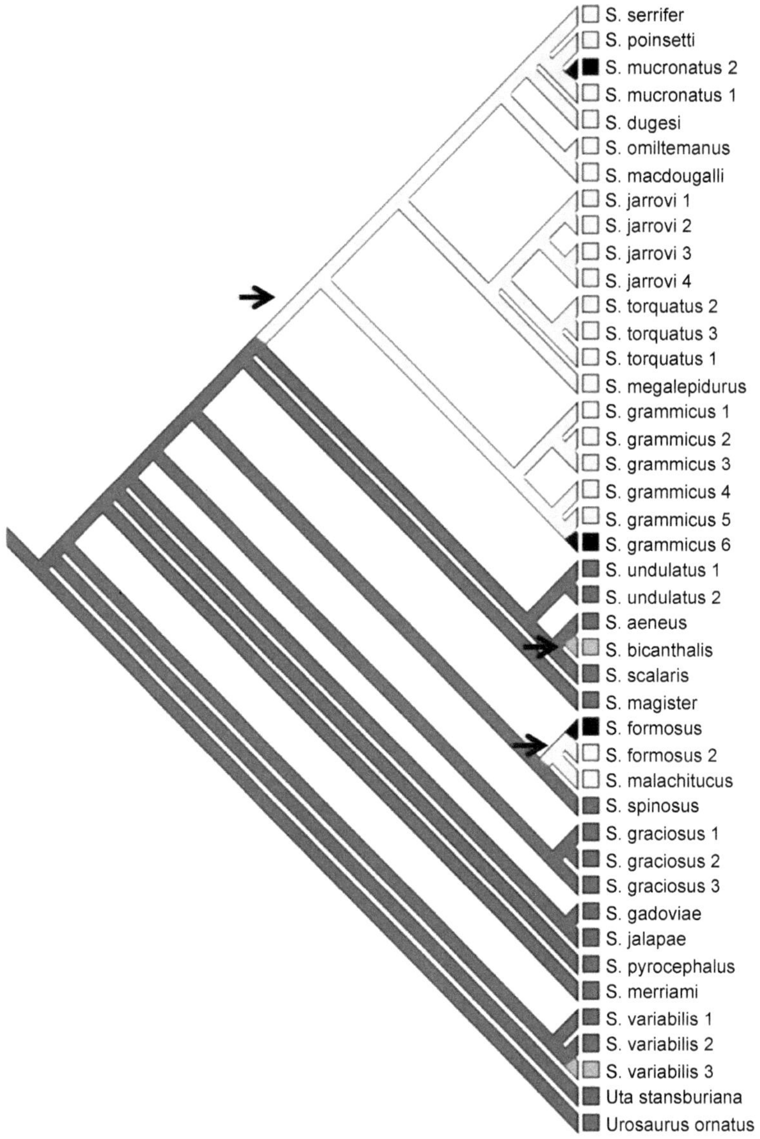

Fig. 10.6 Evolution of reproductive cycles in the species of the genus *Sceloporus*. **Blue**—Associated cycles with spring-summer reproductive activity. **Green**—Continuous reproduction. **Yellow**—Associated cycles with fall reproductive activity. **Black**—Dissociated cycles with fall reproductive activity of females and spring-summer of males. **Arrows** indicate events of evolution of viviparity. Reconstruction of ancestral states using parsimony. Phylogenetic information was based on Wiens *et al.* 2010.

Color image of this figure appears in the color plate section at the end of the book.

(García-Collazo *et al.* 1993) and *Sceloporus bicanthalis* (Hernández-Gallegos *et al.* 2002). Nevertheless, the continuous reproductive pattern in these species of *Sceloporus* evolved under different scenarios, as these species occur in contrasting environmental conditions. The populations of *S. variabilis* in Alvarado, Veracruz, México inhabit a tropical environment at sea level and environmental conditions allow for continuous spermatogenesis. Pregnant females are present throughout the year although its number varies per month. However, other populations of the same species present a defined pattern of seasonality (García Collazo *et al.* 1993). The populations of *S. variabilis* studied in Los Tuxtlas, Veracruz, México (Benabib 1994) at different altitudes (45 and 1000 meters) show a seasonal reproductive cycle with a tendency to increase the months of reproduction from high elevation (8 months at 1000 m) to low (9.5 months at 45 m).

On the other hand, the second species that presents continuous reproduction belongs to the light-bellied bunch grass lizard (*S. scalaris*) group and occurs in an entirely different environment from *S. variabilis*. The viviparous trans-volcanic bunchgrass lizard (*Sceloporus bicanthalis*) is closely related to the oviparous southern bunchgrass lizard (*S. aeneus*), which shows a spring-summer reproductive pattern (Hernández-Gallegos *et al.* 2002; Manríquez-Morán *et al.* 2013), and the highest elevation where it occurs is 3100 m (García Collazo *et al.* 2012). Whereas, the viviparous species *S. bicanthalis* lives in a pine area in the mountains of central México at 3200 (but may also reach 4200) meters and reproduce continually. *S. bicanthalis* is unique presenting a continuous reproduction, unlike all the other species that inhabit the same area (*S. mucronatus*, *S. grammicus*, the Cope's skink *Plestiodon copei* and the imbricate alligator lizard *Barisia imbricata*) which reproduce seasonally. There are two traits that seem to influence this particularity, one of which is short life expectancy, which in the males is 15 months; therefore, they reach sexual maturation very fast (close to five months; Rodríguez Romero *et al.* 2011). Another important trait is that *S. bicanthalis* can compensate the low thermal quality of the environment (11.8°C) with an active thermoregulatory behavior and can reach 27.8°C, i.e., 12°C above the environmental temperature (Lara Resendiz 2008).

10.6.2 Reproductive Fall Patterns

One of the events that promote the evolution of the reproductive phenology (continuous or seasonal) was during the acquisition of viviparity in the genus *Sceloporus* within the groups *S. torquatus* and the Mexican emerald spiny lizard, *S. formosus* (Méndez de la Cruz *et al.* 1998). In these groups, the viviparity is associated with changes in the phenology of the reproduction.

The changes involve a delay in vitellogenesis that occur in spring and summer in the oviparous sceloporines, to summer and fall (in mentioned viviparous groups), and pregnancy now moves from summer (in oviparous species) to winter (in the species of these viviparous groups).

This shift in the reproductive season shown by fall activity and winter pregnancy could be originated by the thermal quality of the environment. In the central mountains of México, winter is less cloudy and rainy than summer, which provides more basking days. Therefore, the influence of the tropical climate promotes better thermal quality during the winter months versus less suitable basking days during summer. Nevertheless, the fall reproductive pattern is present also in lizards that inhabit subtropical areas, some of which occur in desert environment and also in some species that inhabit low elevation localities as the rough-scaled lizard *S. serrifer* (in the Yucatán Peninsula, México) and MacDougall's spiny lizard *S. macdougalli* (in the coastal areas of Oaxaca, México) and in southern populations of the formosus group (the green spiny lizard, *S. malachitucus*). Therefore, once viviparity was acquired, the species kept the reproductive phenology independently of the habitat. In addition, the fall reproductive activity seems to limit to the distribution of the species to environments where winter allows activity (Méndez de la Cruz *et al.* 1998) and thus they cannot be present if winter is too harsh for winter viviparity, whereas oviparous species are present at northern areas (Fig. 11.7). It is important to mention that food resources are available during winter months for fall reproductive lizards (Méndez de la Cruz *et al.* 1992) and do not seem to limit their activity.

On the other hand, this change in the phenology of the females seems not to have affected the first stages of the spermatogenesis in the males. In the fall associated cycles (males and females with maximum activity during the fall months) and also in the dissociated cycles (maximum activity of males during summer and females during fall), the first stages of sperm differentiation (proliferation of spermatogonia and their transformation to primary spermatocytes) occur during winter and spring (as well as in the oviparous reproductive cycles of spring-summer). However, the phases of differentiation to secondary spermatocytes and spermatids occur immediately in the dissociated cycles and become slower in the associated ones (Fig. 10.8), and the maximum peaks of these cells are delayed and are thus presented at the end of summer, so that during fall spermiation events lead to a large sperm population in the epididymis (Villagrán Santa Cruz *et al.* 1994; Méndez de la Cruz *et al.* 1994). Consequently, copulation occurs at the end of fall or the beginning of winter so that males synchronize with the ovulation of the females (Méndez de la Cruz *et al.* 1988). One phenomenon that appears as a derived event is the reproductive dissociation between both sexes that is present in the high mountain viviparous species of the groups *S. torquatus*, *S. grammicus* and *S. formosus* (Fig. 10.6). These

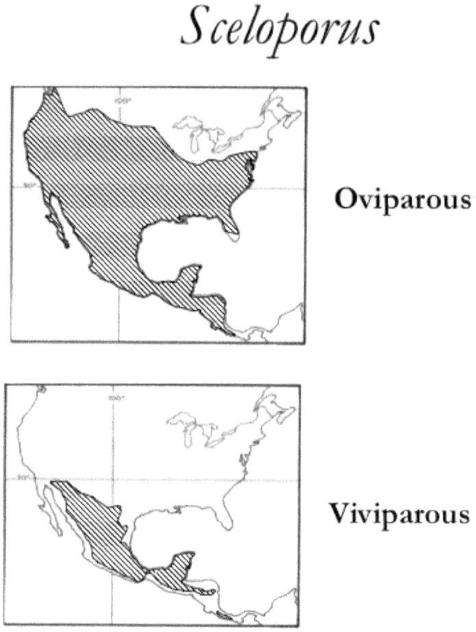

Fig. 10.7 Distribution of the oviparous and viviparous species of the genus *Sceloporus*. Note that the viviparous species are more restricted to tropical and subtropical weather.

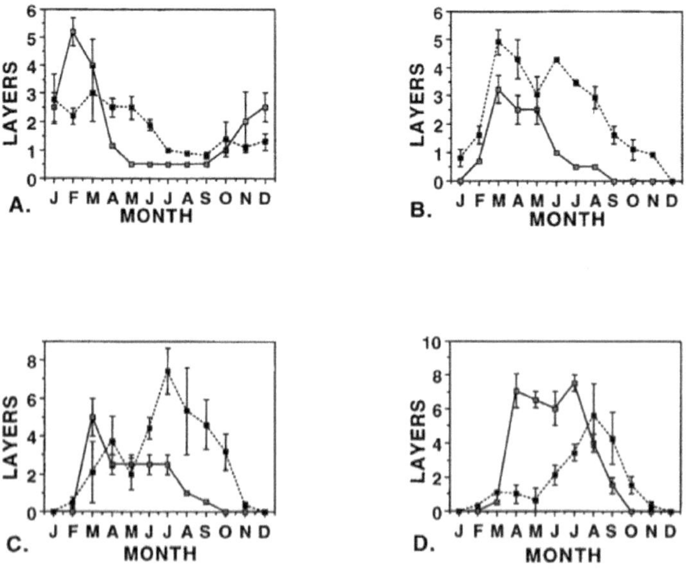

Fig. 10.8 Monthly variation in the number of cell layers of **(A)** Primary spermatocytes, **(B)** Secondary spermatocytes, **(C)** Secondary spermatocytes and **(D)** spermatids in populations of *Sceloporus mucronatus* from low (2500 m, dotted line) and high elevation (3200 m, continuous line). Note the slow differentiation of the spermatic cells in the low elevation population.

dissociated cycles are present in species that occupy altitudes of 2500 m or higher, for example: in the mountain population: *S. grammicus* to 3200 m and *S. mucronatus* to 3400 m (Guillette and Casas-Andreu 1980; Méndez de la Cruz *et al.* 1988). Meanwhile, the fall associated reproductive cycles are present in populations at lower altitudes (Martínez-Isac 1985; Ortega and Barbault 1984). Even the species that appear in the base of the viviparous clade (large scale spiny lizard, *S. megalepidurus*, *S. grammicus* and *S. torquatus*), exhibit a reproductive pattern associated with the fall season (Godínez-Cano 1985). A different explanation was suggested by Martin (1958), who assumed that the populations from the High Central Plateau of México were derived from the mountain species (Martin 1958; Méndez de la Cruz *et al.* 1994; Gadsden *et al.* 2005). In this proposal, it has been suggested that the drought conditions and the low abundance of food that prevail at lower elevation promote a mating season at the end of the rainy season providing enough energy for males to defend their territories (Méndez de la Cruz *et al.* 1994).

The fall reproductive activity is not exclusive to the viviparous species of the genus *Sceloporus*. Some other species of lizards, not closely related, that inhabit the same environment (mountains of México and Central America) have also developed the same strategy, i.e., fall reproductive activity and winter pregnancy. Some of these lizards are *Plestiodon copei* (Guillette 1983), the mountain horned lizard, *Phrynosoma orbiculare* (Hodges 2002), the Mexican horned lizard, *P. taurus*, and the short-tail horned lizard, *P. braconnieri* (Zamudio and Parra-Olea 2000), *Barisia imbricata* (Guillette and Casas-Andreu 1987) and Gadow's alligator lizard *Mesaspis gadovii* (Ramírez Pinilla *et al.* 2009). Therefore, considering that the same strategy was developed independently more than once, the environmental conditions are a major influence in the colonization and expansion of this reproductive strategy, and once acquired the regression to the spring-summer activity could be possible only if lizards can afford a shorter pregnancy period (see below the evolution of the *Phrynosoma* reproductive cycles).

10.6.3 Dissociated Reproductive Cycles and the Effect of the Synchrony/Asynchrony

As previously mentioned, Jensen *et al.* (2001) using different eco-physiological traits such as sex ratio, social behavior, stomach contents, steroid hormones, and gonadal condition, found that what was considered as protandry before, is in fact synchronic reproductive cycles of males and females. In fact, the involvement of the epididymis has been shelved because studies have failed to substantiate whether there is asynchrony between testicular and epididymal function in lizards (Villagrán-Santa Cruz 2009). In this regard, several studies involving the reproductive cycles first obtained the weight of organs throughout their annual cycle (Fig. 10.9), then confirmed them by histological changes of testes and epididymis and verified the presence of prolonged storage of spermatozoa in the epididymis, especially in the absence of testicular involvement (Boretto *et al.* 2012; see the same in bats Arenas-Ríos *et al.* 2005). In lizards, in addition to what has already been mentioned, it is necessary to test epididymal sperm maturation, perhaps following the physiological and biochemical changes of sperm during the passage through epidydimis in order to better understand the epididymal function in sperm maturation (Cooper 1995, 2007).

Asynchronous or dissociated cycles (sensu Pough *et al.* 1998) between males and females can be found when the reproductive conditions change for females due to climate or feeding resources restrictions, but these restrains not necessarily affect males in the same way (Saint Girons 1985; Olivares *et al.* 1987). These cycles are characteristic of the species that

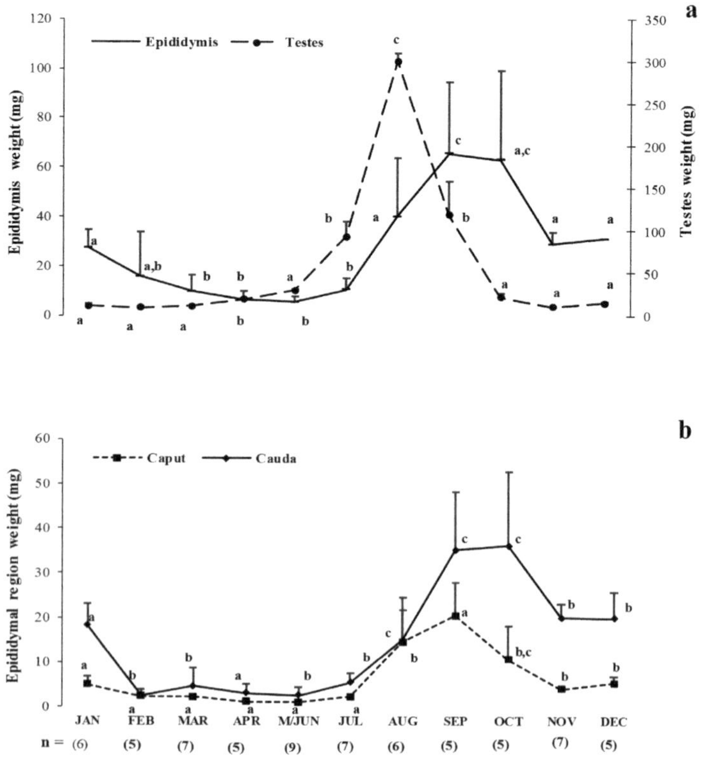

Fig. 10.9 Monthly variation (average and DS) of the weight of the: **(A)** Testis and epididymis; **(B)** Caput and cauda of the epididymis from the bat throughout its annual cycle. Different letters define statistical differences (P < 0.05). Modified from Arenas-Ríos *et al.* 2011.

live in habitats where the breeding season is short (e.g., at high latitudes and altitudes) and the best time for gonadal activity may not be the best time for egg laying or births (Crews and Garstka 1982). In such cycles, the spermatozoa are stored in uterus or epididymides until fertilization (Pough *et al.* 1998) and copulation occurs when there is low spermatogenic activity. This is the case of the lizards from the north of Chile where the climate is desert with low rainfall and a uniform temperature ranging from 22°C in January to 15°C in June (di Castri 1968), where cycles tend to be extensive and some species reproduce once or twice per year (Leyton *et al.* 1982; Goldberg and Rodríguez 1986; Olivares *et al.* 1987). For example, the northernmost population of the four-banded Pacific iguana, *Microlophus quadrivittatus,* from Arica, Chile (18 ° 10′ S, Leyton *et al.* 1982) has continuous reproduction in both males and females (see a review in Ibargüengoytía 2010). However, *M. quadrivittatus* (Goldberg and Rodríguez 1986) and *M. atacamensis* (Olivares *et al.* 1987) in the southern distribution maintain a continuous reproduction in males with spermiogenesis throughout most of the year, while females have an annual reproductive cycle (Goldberg and Rodríguez 1986). It has been observed in *Microlophus* that the female cycle is positively related to temperature, photoperiod and rainfall, probably because these factors favor a greater abundance of insects (Olivares *et al.* 1987). These observations are consistent with females requiring a large amount of energy to perform the vitellogenesis, which is mainly from fat bodies or by having the possibility of feeding (Saint Girons 1985). But the male cycle relates positively with the photoperiod, while the relationship with temperature is reversed, since spermatogenesis starts at the end of summer, continues during fall and winter, and peaks again in spring (Olivares *et al.* 1987).

The possibility of storing sperm of *Microlophus* males appears as an alternative to continuous reproduction observed in *M. quadrivittatus* from Arica (Leyton *et al.* 1982), since males can achieve some independence from the environmental temperature and copulate with annual but asynchronic females (Ibargüengoytía 2010). Greater availability of males for reproduction matches the character polygyny (Heisig 1993) and in the case of *M. atacamensis,* it is reinforced by the existence of sexual dimorphism typical of intrasexual competition in this species (Vidal *et al.* 2002).

10.6.3.1 Phrynosoma

In the genus *Phrynosoma,* there are two different patterns of reproductive activity (Fig. 10.10). Although patterns of all species of the genus are not known, the information to date indicates that as in other lizard groups, the ancestral pattern is spring reproductive activity, in which males and females

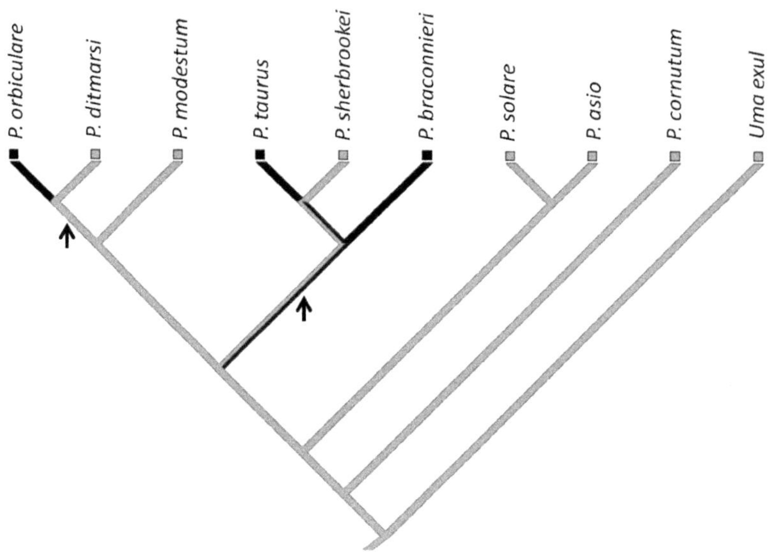

Fig. 10.10 Evolution of reproductive cycles in the species of the genus *Phrynosoma*. **Green**—Associated cycles with spring-summer reproductive activity. **Black**—Associated cycles with fall reproductive activity. **Arrows** indicate events of evolution of viviparity. Phylogenetic information was based on Nieto Montes de Oca *et al.* 2014.

Color image of this figure appears in the color plate section at the end of the book.

have associated cycles with maximum gonadal activity during spring and summer. This pattern occurs in species such as the Texas horned lizard, *P. cornutum* (Ballinger 1974), the giant horned lizard, *P. asio* (Hodges 2002), and the regal horned lizard, *P. solare* (Parker 1971). The fall reproductive pattern appears in clades composed by viviparous species and appears to have evolved independently two or three times in the genus: once in the clade that includes the mountain horned lizard, *P. orbiculare* and a second time in the group of *P. taurus* and *P. braconnieri* (Zamudio and Parra-Olea 2000). This, however, would imply the existence of plasticity in the reproductive season, considering that the rock horned lizard, *P. ditmarsi* and *P. sherbrookei* present a spring pattern.

The shift from viviparity with fall reproductive activity and winter pregnancy to viviparity with spring-summer reproductive activity with summer pregnancy can be afforded only for lizard species that may have a short pregnancy period (from two to three months of pregnancy), as in *Phynosoma ditmarsi* and *P. sherbrooki* (Table 10.1). Oviparous species that present summer reproduction can be present in the northern or southern limits of the distribution, whereas viviparous in the southern distribution may present winter or summer gestation, but only populations with summer

Table 10.1 Information used for the reconstruction of ancestral states character for the Fig. 11.6.

Species	Reproductive mode	Type of Cycle	Reference
Urosaurus ornatus	Oviparous	Associated	Martin 1977; Asplund and Lowe 1964
Uta sturburiana	Oviparous	Associated	Asplund and Lowe 1964
Sceloporus variabilis 3	Oviparous	Continuum	García-Collazo *et al.* 1993
Sceloporus variabilis 2	Oviparous	Associated	Benabib 1994
Sceloporus variabilis 1	Oviparous	Associated	Benabib 1994
Sceloporus merriami	Oviparous	Associated	Dunham 1981
Sceloporus pyrocephalus	Oviparous	Associated	Ramírez-Bautista and Olvera Becerril 2009
Sceloporus jalapae	Oviparous	Associated	Ramírez-Bautista *et al.* 2005
Sceloporus gadoviae	Ovíparous	Associated	Lemos-Espinal *et al.* 1999
Sceloporus graciosus 1	Oviparous	Associated	Goldberg 1975
Sceloporus graciosus 2	Oviparous	Associated	Tinkle 1993
Sceloporus graciosus 3	Oviparous	Associated	Parker 1973
Sceloporus spinosus	Oviparous	Associated	Méndez de la Cruz *et al.* 2013
Sceloporus malachiticus	Viviparous	Associated	Marion and Sexton 1971
Sceloporus formosus 2	Viviparous	Associated	Ramírez-Pinilla *et al.* 2009
Sceloporus formosus	Viviparous	Dissociated	Guillette and Sullivan 1985
Sceloporus magister	Oviparous	Associated	Tanner *et al.* 2010
Sceloporus scalaris	Ovipara-R	Associated	Newlin 1976
Sceloporus bicanthalis	Viviparous	Continuum	Hernández-Gallegos *et al.* 2002
Sceloporus aeneus	Oviparous	Associated	Manriquez-Morán *et al.* 2013
Sceloporus undulatus 1	Oviparous	Associated	McKinney and Marion 1985
Sceloporus undulatus 2	Oviparous	Associated	Jones and Ballinger 1987
Sceloporus grammicus 6	Viviparous	Dissociated	Guillete and Casas-Andreu 1980
Sceloporus grammicus 5	Viviparous	Associated	Guillette and Bearce 1986
Sceloporus grammicus 4	Viviparous	Associated	Jiménez-Cruz *et al.* 2005
Sceloporus grammicus 3	Viviparous	Associated	Hernández-Salinas *et al.* 2010
Sceloporus grammicus 2	Viviparous	Associated	Martínez-Isac 1985
Scelporus grammicus 1	Viviparous	Associated	Ortega and Barbault 1984
Sceloporus megalepidurus	Viviparous	Associated	Godínez-Cano 1985

Table 10.1 contd....

Table 10.1 contd....

Species	Reproductive mode	Type of Cycle	Reference
Sceloporus torquatus 1	Viviparous	Associated	Guillete and Mendéz-de la Cruz 1993
Sceporus torquatus 2	Viviparous	Associated	Feria *et al.* 2001
Sceloporus torquatus 3	Viviparous	Associated	Ramírez-Bautista *et al.* 2002
Sceloporus jarrovi 1	Viviparous	Associated	Ballinger 1973
Sceloporus jarrovi 2	Viviparous	Associated	Ballinger 1979
Sceloporus jarrovi 3	Viviparous	Associated	Tinkle and Hardley 1973
Sceloporus jarrovi 4	Viviparous	Associated	DeMarco 1992
Sceloporus macdougalli	Viviparous	Associated	Martínez-Bernal 2004
Sceloporus omiltemanus	Viviparous	Associated	Ramírez-Pinilla *et al.* 2009
Sceloporus dugesi	Viviparous	Associated	Ramírez-Bautista and Dávila-Ulloa 2009
Sceloporus mucronatus 2	Viviparous	Dissociated	Méndez de la Cruz *et al.* 1988; Estrada *et al.* 1990
Sceloporus mucronatus 1	Viviparous	Associated	Villagrán-Santa Cruz *et al.* 2009
Sceloporus poinsetti	Viviparous	Associated	Ballinger 1972
Sceloporus serrifer	Viviparous	Associated	Morales-Rivera 2001

Table 10.2 Information used for the reconstruction of ancestral states character for the Fig. 11.10.

Species	Reproductive mode	Reproductive season	Reference
Uma exul	Oviparous	Spring-Summer	Gadsden *et al.* 2006
Phrynosoma cornutum	Oviparous	Spring-Summer	Ballinger 1974
Phrynosoma asio	Oviparous	Spring-Summer	Hodges 2002
Phrynosoma solare	Oviparous	Spring-Summer	Parker 1971
Phrynosoma braconnieri	Viviparous	Fall-Winter	Hodges 2002
Phrynosoma sherbrookei	Viviparous	Spring-Summer	Santos Bibiano 2014
Phrynosoma taurus	Viviparous	Fall-Winter	Hodges 2002
Phrynosoma modestum	Oviparous	Spring-Summer	Hernández-Ibarra y Ramírez-Bautista 2002
Phrynosoma ditmarsi	Viviparous	Spring-Summer	Hodges 2002
Phrynosoma orbiculare	Viviparous	Fall-Winter	Hodges 2002

pregnancy are present in the high latitudinal areas, as they would not have pregnancies longer than 3 months because the growing and basking season is shorter than in the tropics (Fig. 10.11). Lizards that occur in latitudes higher than 30° N face harsh environments and have to hibernate to survive winter conditions (Méndez de la Cruz *et al.* 1998).

These lizards may be active only during 6 to 7 months of the year. The species of the genus *Sceloporus* cannot afford a period shorter than 5 months of pregnancy even in the tropical localities. For example, *Sceloporus serrifer* occurs in low elevation tropical areas, but as the females are not able to bask longer than 4 hours per day because the operative temperatures are too hot for embryos (Beauchot 1986), and pregnant females should get to their retreats in order not to expose the embryos to fatal temperatures. Therefore, high temperatures limit the shortage of the pregnancy period to 5 months, which is too long to survive in a habitat with 5 or 6 months of activity, as those from latitudes higher than 32° N, the mountain spiny lizard, *Sceloporus jarrovi,* and *S. poinsetti,* which also present 5 months of pregnancy so they are not able to colonize northern latitudes that seem too harsh and too cold during winter or too hot during summer. Therefore, the distribution of viviparous species of the *S. torquatus* group is limited toward the North by frozen temperatures because of winter pregnancy. In contrast, the greater short horned lizard *Phrynosoma hernandezi* presents a pregnancy period of less than 3 months (Goldberg 1971) or 4 months in the Rock horned lizard,

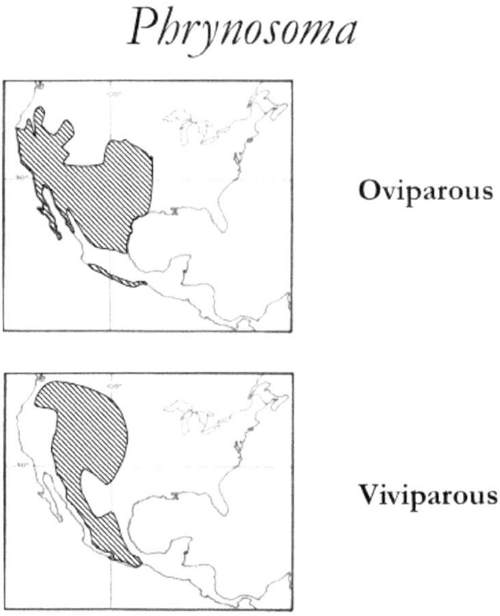

Fig. 10.11 Distribution of the oviparous and viviparous species of the genus *Phrynosoma*.

P. ditmarsi (Hodges 2002). The length of the pregnancy period present in the *P. tapaja* group (8 to 9 months in the Mountain horned lizard, *P. orbiculare* or less than 3 months in *P. hernandesi*), seems to be related with the thermal quality of the environment: when it is low, allows a maximum of 4 hours of activity in *P. orbiculare* in the Central mountains of México (Urzúa Vazquez 2008); when is high, permits until 9 hours of activity in Texas populations of *P. hernandesi* (Christian 1998). This possibility to shorten the gestation does not seems to occur in the viviparous *Sceloporus* due that females of *Phrynosoma* may afford temperatures higher than 35° C with no damage to embryos (Christiansen 1998; Urzua-Vázquez 2008; Lara-Resendiz 2013).

10.7 CONCLUSION

Reproductive patterns in lizards are highly variable. Most groups are able to reproduce throughout spring and summer, however, in some cases the reproduction is continuous regardless the environment they inhabit. The evolution of viviparity has led to the evolution of reproductive activity in both sexes during fall (associated pattern) or males in summer and females in autumn (dissociated pattern).

In general, the reproductive cycle has been defined considering only the size of the testis. However, the incipient evidence in lizards and many studies in mammals indicate that epididymal involvement is essential to determine the maximum phase of activity (spermatic maturation) process, which has not been defined well in lizards yet. However, the body temperature can affect both spermatogenesis and the maturation phases, so that lizards must be exposed to temperatures lower than the preferred temperatures to complete these processes.

Even when a considerable amount of information on the reproductive ecology of lizards is available, allowing, among other things, to describe the different reproductive patterns, studies on the reproductive biology have progressed very slowly in relation to the urge needed to support species that are in danger of disappearing.

Temperatures are rising too fast due to global warming (Sinervo *et al.* 2010) which has caused the extinction of many species of lizards since many of them are especially sensitive to temperature increase and, in particular, the reproductive physiology of the male. Specifically, processes such as sperm production, maturation, and sperm storage are insufficient to enable us to perform conservation studies well supported by the literature. Even though many studies are required to understand the reproductive biology of reptiles, and lizards in particular, the studies in the future should be cautious about the massive sacrifice of lizards to understand their reproductive cycle, therefore, we strongly encourage to sacrifice lizards only when it is strictly needed to work with more specific issues, and researchers should use some low invasive techniques to obtain information related with the reproductive season or clutch size, among others.

10.8 ACKNOWLEDGMENTS

Financial support was provided by the PAPIIT-UNAM 215011-3, Consejo Nacional de Investigaciones Científicas y Técnicas (CONICET, PIP 100271), by MINCYT-CONACYT MX10/11, NSF award EF 1241885 ("Collaborative Research: Quantifying Climate-forced Extinction Risks for Lizards, Amphibians, and Plants").

10.9 LITERATURE CITED

Abu-Zinadah, O. A. 2008. Variation in testicular histology of the spiny tailed lizard *Uromastyx aegyptius microlepis* during hibernation and active periods. Pakistan Journal of Biological Sciences 11: 1615–1619.

Aitken, R. J., Nixon, B., Lin, M., Koppers, A. J., Lee, Y. H. and Baker, M. A. 2007. Proteomic changes in mammalian spermatozoa during epididymal maturation. Asian Journal of Andrology 9: 554–564.

Akbarsha, M. A., Tamilarasan, V. and Kadalmani, B. 2006. Light and electron microscopic observations of fabrication, release, and fate of biphasic secretion granules produced by epididymal epithelial principal cells of the fan-throated lizard *Sitana ponticeriana* cuvier. Journal of Morphology 267: 713–729.

Allan, D. J., Harmon, B. V. and Kerr, J. F. 1987. Cell death in spermatogenesis. pp. 229–258. In C. S. Potten (ed.), *Perspectives on mammalian cell death*. London. Oxford University Press.

Amat, F., Llorente, G. A. and Carretero, M. A. 2000. Reproductive cycle of the sand lizard (*Lacerta agilis*) in its southestern range. Amphibia-Reptilia 21: 463–476.

Arenas-Ríos, E., León-Galván, M. A., Mercado, P. E. and Rosado, G. A. 2005. Superoxide dismutase, catalase, and glutathione peroxidase during epididymal maturation and prolonged storage of spermatozoa in the mexican big-eared bat (Corynorhinus mexicanus). Canadian Journal Zoology 83: 1556–1565.

Asplund, K. K. and Lowe, C. H. 1964. Reproductive cycles of the iguanid lizards *Urosaurus ornatus* and *Uta stansburiana* in Southeastern Arizona. Journal of Morphology 115: 27–34.

Ballinger, R. E. 1973. Comparative demography of two viviparous iguanid lizards (*Sceloporus jarrovi* and *Sceloporus poinsetti*). Ecology 54: 269–283.

Ballinger, R. E. 1974. Reproduction of the Texas horned lizard, *Phrynosoma cornutum*. Herpetologica 27: 321–327.

Ballinger, R. E. 1979. Intraspecific variation in demography and life history of the lizard, *Sceloporus jarrovi*, along an altitudinal gradient in southeastern Arizona. Ecology 60: 901–909.

Barlett, J. M. S. and Sharpe, R. M. 1987. Effect of local heating of the rat testis on the levels in interstitial fluid of a putative paracrine regulator of the Leydig cells and its relationship to changes in Sertoli cell secretory function. Journal of Reproduction and Fertility 80: 279–287.

Beauchot, C. A. 1986. Reproductive influence on the thermoregulatory behavior of a live bearing lizard. Copeia 1986: 971–979.

Bedford, J. M., Berrios, M. and Dryden, G. L. 1982. Biology of the scrotum. IV. Testis location and temperature sensitivity. Journal Experimental of Zoology 224: 379–388.

Benabib, M. 1994. Reproduction and lipid utilization of the tropical populations of *Sceloporus variabilis*. Herpetological Monographs 8: 160–180.

Birkhead, T. R. and Møller, A. P. 1993. Sexual selection and temporal separation of reproductive events: sperm storage data from reptiles, birds and mammals. Biological Journal of the Linnean Society 50: 295–311.

Blackburn, D. G. 1998. Structure, function and evolution of the oviducts of squamate reptiles, with special reference to viviparity and placentation. Journal of Experimental Zoology 282: 560–617.

Bonet, S., Briz, M., Fradera, A. and Egozcue, J. 1992. Origin, development and ultrastructure of boar spermatozoa with folded tails and with two tails. Human Reproduction 7: 523–528.

Bons, J. and Saint-Girons, H. 1982. Le cycle sexuel des reptiles au Maroc et ses rapports avec la répartition géographique et le climat. Bulletin de la Société Zoologique de France 107: 71–86.

Boretto, J. M. and Ibargüengoytía, N. R. 2006. Asynchronous spermatogenesis and biennial female cycle of the viviparous lizard *Phymaturus antofagastensis* (Liolaemidae): reproductive responses to high altitudes and temperate climate of Camarca, Argentina. Amphibia-Reptilia 27: 25–36.

Boretto, J. M., Ibargüengoytía, N. R., Acosta, J. C., Blanco, G. M., Villavicencio, H. J. and Marinero, J. A. 2007. Reproductive biology and sexual dimorphism of a high-altitude population of the viviparous lizard *Phymaturus punae* from the Andes in Argentina. Amphibia-Reptilia 28: 1–7.

Boretto, J.M. and Irbargüengoytía, N. R. 2009. *Phymaturus* of Patagonia, Argentina: reproductive biology of *Phymaturus zapalensis* (Liolaemidae) and a comparison of sexual dimorphism within the genus. Journal of Herpetology 43: 96–104.

Boretto, J. M., Ibargüengoytía, N. R., Jahn, G. A., Acosta, J. C., Vincenti, M. W. and Fornés, M. W. 2010. Asynchronic steroids activity of Leydig and Sertoli cells related to spermatogenic and testosterone cycle in *Phymaturus antofafastensis*. General and Comparative Endocrinology 166: 556–564.

Boretto, J., Jahn, G. A., Fornés, M. W., Cussac, V. E. and Ibargüengoytía, N. R. 2012. How males cope with the scarcity of reproductive females: An endocrinal and ultrastructural study of *Phymaturus zapalensis* lizards (Liolaemidae). Herpetological Journal 22: 33–42.

Bostic, D. L. 1966. A preliminary report of reproduction in the teiid lizard *Cnemidophorus hyperythrus beldingi*. Herpetologica 22: 81–90.

Bradshaw, S. D., Saint-Girons, H. and Bradshaw, F. J. 1991. Patterns of breeding in two species of agamid lizards in the arid subtropical Pilbara region of western Australia. General and Comparative Endocrinology 82: 407–424.

Bustos-Zagal, G., Méndez de la Cruz, F. R., Castro-Franco, R. and Villagrán-Santa Cruz, M. 2011. Ciclo reproductor de Sceloporus ochoterenae en el estado de Morelos, México. Revista Mexicana de Biodiversidad 82: 589–597.

Cabezas-Cartes, F., Boretto, J., Aosta, J. C., Jahn, G., Blanco, G., Laspiur, A. and Ibargüengoytía, N. 2010. Reproductive biology of *Phymaturus* cf. *palluma*: A vulnerable lizard from the highlands of the Andes, San Juan, Argentina. Herpetological Conservation and Biology 5: 430–440.

Cappellari, L. H., Balestrin, R. L., de Lema, T. and Rocha, C. F. D. 2011. Reproductive biology of *Teius oculatus* (Sauria, Teiidae) in Southern Brazil (Dom Feliciano, Rio Grande do Sul). North Western Journal of Zoology 7: 270–276.

Capula, M., Luiselli, L. and Anibald, C. 1992. Biennial reproduction and clutch parameters in an alpine population of the Slow Worm, *Anguis fragilis* Linnaeus 1758 (Squamata: Sauria: Angyidae). Hepetozoa 5: 95–98.

Cei, J. M. 1986. Reptiles del centro, centro-oeste y sur de la Argentina. Herpetofauna de las zonas áridas y semiáridas. Museo Regionale di Scienze Naturali, Torino, Monografie IV: 527.

Censky, E. J. 1995. Reproduction in two lesser Antillean populations of *Ameiva plei* (Teiidae). Journal of Herpetology 29: 553–560.

Carretero, M. A. 2006. Reproductive cycles in Mediterranean lacertids: plasticity and constraints. pp. 33–54. In C. Corti, P. Lo Cascio and M. Biaggini (eds.), *Mainland and Insular Lacertids Lizards: A Mediterranean Perspective*. Firenze University Press, Fire.
Colli, G. R. 1991. Reproductive ecology of *Ameiva ameiva* (Sauria, Teiidae) in the Cerrado of Central Brazil. Copeia 1991: 1002–1012.
Colli, G. R., Péres, A. K., Jr. and Zatz, M. G. 1997. Foraging mode and reproductive seasonality in tropical lizards. Journal of Herpetology 31: 490–499.
Cooper, T. G. 1995. Role of the epididymis in mediating changes in the male gamete during maturation. Advances in Experimental Medicine and Biology 377: 87–101.
Cooper, T. G. 2007. Sperm maturation in the epididymis: a new look at an old problem. Asian Journal of Andrology 9: 533–539.
Cooper, T. G. 2011. The epididymis, cytoplasmic droplets and male fertility. Asian Journal of Andrology 13: 130–138.
Craig, J. and Shine, R. 1985. The seasonal timing of reproduction. A tropical-temperate comparison in Australian lizards. Oecologia (Berlin) 67: 464–474.
Crew, F. A. E. 1992. A suggestion as to the cause of the aspermatic condition of the imperfectly descendent testis. Journal of Anatomy 56: 98–106.
Crews, D. and Garstka, W. 1982. The ecological physiology of a garter snake. Scientific American 247: 158–168.
Cruz, F. B. 1994. Actividad reproductora en *Vanzosaura rubricauda* (Sauria: Teiidae) del Chaco occidental en Argentina. Cuadernos de Herpetología 8: 112–118.
Cruz, F. B., Teisaire, E., Nieto, L. and Roldán, A. 1999. Reproductive biology of *Teius teyou* in the semiarid Chaco of Salta, Argentina. Journal of Herpetology 33: 420–429.
Christian, A. K. 1998. Thermorregulation by the short-horned lizard (*Phrynosoma douglassi*) at high elevation. Journal of Thermal Biology 23: 395–399.
Christiansen, J. L. 1971. Reproduction of *Cnemidophorus inornatus* and *Cnemidophorus neomexicanus* (Sauria, Teiidae) in northern New Mexico. American Museum Novitates 2442: 1–48.
Chun-Fu, L., Yen-Long, C. and Ya-Fen, T. 2004. A production of four successive clutches of eggs by a female grass lizard (*Takydromus stejnegeri* van Denburgh) in captivity. Endemic Species Research 6: 35–40.
Demarco, V. and Guillette, L. J. 1992. Physiological cost of pregnancy in a viviparous lizard (*Sceloporus jarrovi*). Journal of Experimental Zoology 262: 383–390.
Di Castri, F. 1968. Esquisse écologique du Chili. pp. 7–52. In D. Debouttevílle and C. Rapoport (eds.), Biologie de L'Amérique Australe. Editions du Centre National de la Researche Scientific, París, France.
Donoso-Barros, R. 1966. Reptiles de Chile. Ediciones Universidad de Chile, Santiago. cxliv + 458 pp.
Dunham, A. E. 1981. Populations in a fluctuate environment; The comparative population ecology of *Sceloporus merriami* and *Urosaurus ornatus*. Miscellaneous Publications, Museum of Zoology, University of Michigan 158: 1–62.
Dunham, A. E., Miles, D. B. and Reznick, D. N. 1988. Life history patterns in squamate reptiles. pp. 441–522. In C. Gans and R. B. Huey (eds.), *Biology of the Reptilia. Vol. 16, Ecology B. Defense and Life History*. Alan R. Liss, Inc., New York.
Duvall, D., Guillette, L. J., Jr. and Jones, R. E. 1982. Environmental control of reptilian reproductive cycles. pp. 201–231. In C. Gans and F. H. Pough (eds.), *Biology of the Reptilia. Volume 3*. Academic Press, New York.
El-Rahman, A., Gabr, A., Abdel-Raheim, M. and El-Saed, M. 2009. Altered testicular morphology and oxidative stress induced by cadmium in experimental rats and protective effect of simultaneous green tea extract. Inernational Journal of Morphology 27: 757–764.
Estrada-Flores, E., Villagrán-Santa Cruz, M., Méndez-de la Cruz, F. R. and Casas-Andreu, G. 1990. Gonadal changes throughout the reproductive cycle of the viviparous lizard *Sceloporus mucronatus* (Sauria: Iguanidae). Herpetologica 46: 43–50.

Etheridge, K., Wit, L. C., Sellers, J. C. and Trauth, S. E. 1986. Seasonal changes in reproductive condition and energy stores in *Cnemidophorus sexlineatus*. Journal of Herpetology 20: 554–559.
Feria-Ortíz, M., Nieto-Montes de Oca, A. and Salgado Ugarte, I. H. 2001. Diet and reproductive biology of the viviparous lizard *Sceloporus torquatus torquatus* (Squamata: Phrynosomatidae). Journal of Herpetology 35: 104–112.
Fitch, H. S. 1970. Reproductive cycles of lizards and snakes. University of Kansas Publications of Museum of Natural History, Miscelanean Publications 52: 1–247.
Fitch, H. S. 1973. A field study of Costa Rican lizards. University of Kansas Science Bulletin 50: 39–126.
Fitch, H. S. 1982. Reproductive cycles in tropical reptiles. Occasional Papers of the Museum of Natural History, the University of Kansas 96: 1–53.
Fitzgerald, L. A., Cruz, F. B. and Perotti, G. 1993. The reproductive cycle and the size at maturity of *Tupinanbis rufescens* (Sauria: Teiidae) in the dry Chaco of Argentina. Journal of Herpetology 27: 70–78.
Flemming, A. F. 1993. The male reproductive cycle of the lizard *Pseudocordylus m. melanotus* (Sauria: Cordylidae). Journal of Herpetology 27: 473–478.
Foldesy, R. G. and Bedford, J. M. 1982. Biology of the scrotum I. Temperature and androgen as determinants of the sperm storage capacity of the rat cauda epididymidis. Biology of Reproduction 26: 673–682.
Gadsden, H., Rodríguez-Romero, F. J., Méndez de la Cruz, F. R. and Gil-Martínez, R. 2005. Ciclo reproductor de *Sceloporus poinsettii* Baird y Girrard, 1852 (Squamata: Phrynosomatidae) en el centro del Desierto Chihuahuense, México. Acta Zoológica Mexicana 21: 93–107.
Gadsden, H., Dávila-Carrazco, M. D. L. L. and Gil-Martínez, R. 2006. Reproduction in the arenicolous mexican lizard *Uma exsul*. Journal of Herpetology 40: 117–122.
Galán, P. 1996. Reproductive and fat body cycles of the lacertid lizard *Podarcis bocagei*. Herpetological Journal 6: 20–25.
Galán, P. 1997. Reproductive ecology of the lacertid lizard *Podarcis bocagei*. Ecography 20: 197–209.
Galán, P. 2009. Ecología de la reproducción de los saurios ibéricos. Boletín Asociación Herpetológica Española 20: 2–34.
García-Collazo, R., Alatamirano-Álvarez, T. and Gómez-Soto, M. 1993. Reproducción continua en *Sceloporus variabilis variabilis* (Sauria: Phrynosomatidae) en Alvarado, Veracruz, México. Boletin de la Sociedad Herpetológica Mexicana 5: 51–59.
García-Collazo, R., Meza-Lázaro, R. N., Morales Guillaumin, E., Villagrán-Santa Cruz, M. and Méndez de la Cruz, F. R. 2012. Egg retention and intrauterine embryonic development in *Sceloporus aeneus* (Reptilia: Phrynosomatidae): implications for the evolution of viviparity. Revista Mexicana de Biodiversidad 83: 802–808.
Gatti, J. L., Castella, S., Dacheux, F., Ecroyd, H., Métayer, S., Thimon, V. and Dacheux, J. L. 2004. Post-testicular sperm environment and fertility. Animal Reproduction Science 82: 321–339.
Gavaud, J. 1991. Role of cryophase temperature and thermophase duration in thermoperiodic regulation of testicular cycle in the lizard *Lacerta vivipara*. Journal of Experimental Zoology 260: 239–246.
Girling, J. E. 2002. The reptilian oviduct: a review of structure and function and directions for future research. Journal of Experimental Zoology 293: 141–170.
Glandt, D. 1993. Seasonal activity of the sand lizard (*Lacerta agilis*) and the common lizard (*Lacerta vivipara*) in an experimental outdoor enclosure. pp. 229–231. In G. A. Llorente, A. Montori, X. Santos and M. A. Carretero (eds.), *Scientia Herpetologica*. Asociación Herpetologica Española.
Godínez-Cano, E. 1985. Ciclo reproductivo de *Sceloporus megalepidurus* Smith (Reptilia: Sauria: Iguanidae), en la parte oriental de Tlaxcala, México. Bch. Thesis, ENEP Iztacala, Universidad Nacional Autónoma de México, México.
Goldberg, S. R. 1973. Ovarian cycle of the western fence lizard, *Sceloporus occidentalis*. Herpetologica 29: 284–289.

Goldberg, S. R. 1975. Reproduction in the sagebrush lizard, *Sceloporus graciosus*. American Midland Naturalist 93: 177–187.

Goldberg, S. R. 1976. Reproduction in a mountain population of the coastal whiptail lizard, *Cnemidophorus tigris multiscutatus*. Copeia 1976: 260–266.

Goldberg, S. R. 2006. Reproductive cycle of the bushveld lizard *Heliobolus lugubris* (Squamata: Lacertidae) from southern Africa. Salamandra 42: 151–154.

Goldberg, S. R. 2012. Reproduction of the Yellow-backed spiny lizard, *Sceloporus uniformis* (Squamata: Phrynosomatidae) from California. Bulletin of the Southern California Academy of Sciences 11: 25–28.

Goldberg, S. R. and Rodriguez, E. 1986. Reproductive cycles of two iguanid lizards from northern Chile, *Tropidurus quadrivittatus* and *T. theresioides*. Journal of Arid Environments 10: 147–51.

González-Espinoza, J. E. 2006. Ecología reproductiva de *Sceloporus jalapae* (Reptilia: Phrynosomatidae) en Zapotitlán de Salinas, Puebla. M. S. Thesis. FES-Iztacala, Universidad Nacional Autónoma de México.

Granados-González, G., Rheubert, J. L., Villagrán-Santa Cruz, M., González-Herrera, M. E., Dávila-Cedillo, J. V., Gribbins, K. M. and Hernández-Gallegos, O. 2013. Male reproductive cycle in *Aspidoscelis costata costata* (Squamata: Teiidae) from Tonatico, Estado de México, México. Acta Zoologica 1–9.

Gribbins, K. M. 2010. Temperate reptilian spermatogenesis: A new amniotic mode of germ cell development. pp. 137–167. In O. Hernández-Gallegos, F. R. Méndez-de la Cruz and J. F. Méndez-Sánchez (eds.), Reproducción en Reptiles: Morfología, Ecología y Evolución.

Gribbins, K. M. 2011. Reptilian spermatogenesis: A histological and ultrastructural perspective. Spermatogenesis 1: 250–269.

Gribbins, K. M., Rheubert, J. L., Poldemann, E. H., Collier, M. H., Wilson, B. and Wolf, K. 2009. Continuous spermatogenesis and the germ cell development strategy within the testis of the Jamaican Gray Anole, *Anolis lineatopus*. Theriogenology 72: 484–492.

Guillette, L. J., Jr. 1983. Notes concerning reproduction of the montane skinks, *Eumeces copei*. Journal of Herpetology 17: 144–148.

Guillette, L. J. and Casas-Andreu, G. 1980. Fall reproductive activity in the high altitude Mexican lizard *Sceloporus grammicus microlepidotus*. Journal of Herpetology14: 143–147.

Guillette, L. J., Jr. and Sullivan, W. P. 1985. The reproductive and fat body cycles of the lizard *Sceloporus formosus*. Journal of Herpetology 19: 474–480.

Guillette Jr., L. J. and Bearce, D. A. 1986. The reproductive and fat body cycles of the lizard, *Sceloporus grammicus disparilis*. Transactions of the Kansas Academy of Science 1903: 31–39.

Guillette, L. J. and Casas-Andreu, G. 1987. The reproductive biology of the high elevation Mexican lizard, *Barisia imbricata imbricata*, with notes on the other *imbricata* subspecies. Herpetologica 43: 29–38

Guillette, L. J., Jr. and Méndez de la Cruz, F. R. 1993. The reproductive cycle of the viviparous Mexican lizard *Sceloporus torquatus*. Journal of Herpetology 27: 168–174.

Habit, E. M. and Ortiz, J. C. 1996. Ciclo reproductivo de *Phymaturus flagellifer* (Reptilia, Tropiduridae). Boletín de la Sociedad de Biología de Concepción 67: 7–14.

Heisig, M. 1993. An etho-ecological study of an island population of *Tropidurus atacamensis*. Salamandra 29: 65–81.

Hernández-Gallegos, O., Méndez-de la Cruz, F. R., Villagrán-Santa Cruz, M. and Andrews, R. M. 2002. Continuous spermatogenesis in the lizard *Sceloporus bicanthalis* (Sauria: Phrynosomatidae) from high elevation habitat of central México. Herpetologica 58: 415–421.

Hernández-Ibarra, X. and Ramírez-Bautista, A. 2002. Reproductive characteristics of the roundtail horned lizard, *Phrynosoma modestum* (Phrynosomatidae), from the Chihuahuan Desert of Mexico. The Southwestern Naturalist 47: 138–141.

Hernández-Salinas, U., Ramírez-Bautista, A., Leyte-Manrique, A. and Smith, G. R. 2010. Reproduction and sexual dimorphism in two populations of *Sceloporus grammicus* (Sauria: Phrynosomatidae) from Hidalgo, México. Herpetologica 66: 12–22.

Hodges, W. L. 2002. *Phrynosoma* systematics, comparative reproductive ecology, and conservation of a Texas Native. PhD Dissertation, pp. 209.
Holmes, K. M. and Cree, A. 2006. Annual reproduction in females of a viviparous skink (*Oligosoma maccanni*) in a subalpine environment. Journal of Herpetology 40: 141–151.
Huang, W. S. 1997. Reproductive cycle of the oviparous lizard *Japalura brevipes* (Agamidae: Reptilia) in Taiwan, Republic of China. Journal of Herpetology 31: 22–29.
Ishijima, S. and Witman, G. B. 1987. Flagellar movement of intact and demembranated, reactivated ram spermatozoa. Cell Motility and the Cytoeskeleton 8: 375–391.
Ibargüengoytía, N. R. 2004. Prolonged cycles as a common reproductive pattern in viviparous lizards from Patagonia, Argentina. Reproductive cycle of *Phymaturus patagonicus*. Journal of Herpetology 38: 73–79.
Ibargüengoytía, N. R. and Cussac, V. E. 1996. Reproductive biology of the viviparous lizard *Liolaemus pictus* (Tropiduridae): biennial female reproductive cycle? Herpetological Journal 6: 137–143.
Ibargüengoytía, N. R. and Cussac, V. E. 1998. Reproduction of the viviparous lizard *Liolaemus elongatus* in the highlands of southern South America: plastic cycles in response to climate? Herpetological Journal 8: 99–105.
Ibargüengoytía, N. R., Pastor García, L. M. and Pallares, J. 1999. Histological and ultrastructural study of the testes of *Testudo graeca* (Testudinidae). Journal of Submicroscopic Cytology and Pathology 31: 221–230.
Jenssen, T. A., Lovern, M. B. and Congdon, J. D. 2001. Field-testing the protandry-based mating system for the lizard, *Anolis carolinesis*: does the model organism have the right model? Behavioral Ecology and Sociobiology 50: 162–172.
Jackson, H. C. 1978. Low May sunshine as a possible factor in the decline of sand lizard (*Lacerta agilis*) in northwest England. Biological Conservation 13: 1–12.
Jiménez-Cruz, E., Ramírez-Bautista, A., Marshall, J. C., Lizana-Avia, M. and Nieto-Montes De Oca, A. 2005. Reproductive cycle of *Sceloporus grammicus* (Squamata: Phrynosomatidae) from Teotihuacan, Mexico. The Southwestern Naturalist 50: 178–187.
Jones, R. C. 1998. Evolution of the vertebrate epididymis. Journal of Reproduction and Fertility 53: 163–181.
Jones, S. M. and Ballinger, R. E. 1987. Comparative life histories of Holbrookia maculata and *Sceloporus undulatus* in western Nebraska. Ecology 68: 1828–1838.
Kast, J. 2007. Prolonged sperm storage in the Asian wáter dragon Physignathus cocincinus. Herpetological Review 38: 172–173.
Lara-Resendiz, R. A. 2008. Comparación y modelación de algunos aspectos del nicho ecológico de dos especies de *Sceloporus* con diferente modo reproductor. MSc Thesis. Posgrado en Ciencias Biológicas, UNAM.
Lara-Resendiz, R. A. 2013. Ecología térmica de lacertilios mexicanos: implicaciones de su distribución y modo reproductor. PhD Thesis. Instituto de Biología, Posgrado en Ciencias Biológicas, UNAM.
Larsson, K., Einarsson, S., Lundstöm and Hakkarainen, J. 1983. Endocrine effects of heat stress in boars. Acta Veterinaria Scandinavica 24: 305–314.
Lemos-Espinal, J. A., Smith, G. R. and Ballinger, R. E. 1999. Reproduction in Gadow's Spiny Lizard, *Sceloporus gadovae* (Phrynosomatidae), from arid tropical México. Southwestern Naturalist 44: 57–63.
León, J. R. and Cova, L. J. 1973. Reproducción de *Cnemidophorus lemniscatus* (Sauria: Teiidae) en Cumana, Venezuela. Caribbean Journal of Science 13: 63–73.
Leyton, V. C., Veloso, A. and Bustos-Obregón, E. 1982. Modalidad reproductiva y actividad cíclica gonadal en lagartos iguánidos de distintos pisos altitudinales del interior de Arica (lat. 18° 10′ S), pp. 293–315. En: El Hombre y los Ecosistemas de Montaña: 1. La Vegetación y los Vertebrados Ectodérmicos del Transecto Arica-Lago Chungará. Veloso, A. and Bustos-Obregón, E. (eds.). Contribución sistemática al conocimiento de la herpetofauna del extremo norte de Chile. Oficina Regional de Ciencias y Tecnología de la UNESCO para América Latina y el Caribe, Montevideo, Uruguay.

Licht, P. 1971. Regulation of the annual testis cycle by photoperiod and temperature in the lizard *Anolis carolinensis*. Ecology 52: 240–252.

Licht, P. 1984. Reptiles. Pages 206–282 in Marshall's physiology of reproduction. Volume 1. Reproductive cycles of vertebrates (G. E. Lamming, editor). Churchill Livingstone, Elsevier Ltd., Edinburg, Scotland.

Maddison, W. P. and Maddison, D. R. 2011. Mesquite: a modular system for evolutionary analysis. Version 2.75 http://mesquiteproject.org.

Manríquez-Morán, N. L., Villagrán-Santa Cruz, M. and Méndez-de la Cruz, F. R. 2005. Reproductive biology of the parthenogenetic lizard, *Aspidoscelis cozumela*. Herpetologica 61: 435–439.

Manríquez-Morán, N. L., Villagrán-Santa Cruz, M. and Méndez-de la Cruz, F. R. 2013. Reproductive activity in females of the oviparous lizard *Sceloporus aeneus*. The Southwestern Naturalist 58: 325–329.

Marinho, C. E., Almeida-Santos, S. M., Yamasaki, S. C. and Silveira, P. F. 2009. Seasonal variation of peptidase activities in the reproductive tract of *Crotalus durissus terrificus*. General and Comparative Endocrinology 160: 84–89.

Marion, K. R. and Sexton, O. J. 1971. The reproductive cycle of the lizard *Sceloporus malachiticus* in Costa Rica. Copeia 1971: 517–526.

Martin, R. F. 1973. Reproduction in the tree lizard (*Urosaurus ornatus*) in Central Texas: drought conditions. Herpetologica 29: 27–32.

Martin, P. S. 1958. A biogeography of reptiles and amphibians in the Gomez Farías region, Tamaulipas, México. Miscellaneous Publications. Museum of Zoology. University of Michigan 101: 1–102.

Martínez Bernal, R. L. 2004. Contribución al conocimiento de la biología de la especie endémica *Sceloporus macdougalli* en el Istmo de Tehuantepec, Oaxaca México. Bch Thesis, FES Iztacala, UNAM. México.

Martínez-Isac, R. 1985. Estudio comparativo de dos poblaciones de la lagartija *Sceloporus grammicus microlepidutus* en el Ajusco y Pedregal de San Angel, D. F. Bch. Thesis, ENEP Iztacala, Universidad Nacional Autónoma de México, Mexico.

Maythew, W. W. and Wright, S. J. 1970. Seasonal changes in testicular histology of three species of lizard genus *Uma*. Journal of Morphology 130: 163–168.

McKinney, R. B. and Marion, K. R. 1985. Reproductive and fat body cycles in the male lizard, *Sceloporus undulatus*, from Alabama, with comparisons of geographic variation. Journal of Herpetology 19: 208–217.

Medina, M. and Ibargüengoytía, N. R. 2010. How do viviparous and oviparous lizards reproduce in Patagonia? A comparative study of three species of *Liolaemus*. Journal of Arid Environments 74: 1024–1032.

Méndez Juárez, B. H. 2003. Retención de esperma y edad a la madurez sexual en las hembras de la especie vivípara *Sceloporus serrifer*. Bch. Thesis, FES Iztacala, UNAM.

Méndez de la Cruz, F. R., Casas-Andreu, G. and Villagrán-Santa Cruz, M. 1992. Variación en la alimentación y condición física a lo largo del año en *Sceloporus mucronatus* (Sauria: Iguanidae). The Southwestern Naturalist 37: 349–355.

Méndez de la Cruz, F. R., Villagrán-Santa Cruz, M. and Cuellar, O. 1994. Geographic variation of spermatogenesis in the Mexican viviparous lizard *Sceloporus mucronatus*. Biogeographica 70: 59–67.

Méndez de la Cruz, F. R., Guillette, L. J., Villagrán-Santa Cruz, M. and Casas-Andreu, G. 1988. Reproductive and fat body cycles of the viviparous lizard, *Sceloporus mucronatus* (Sauria: Iguanidae). Journal of Herpetology 22: 1–12.

Méndez de la Cruz, F. R. and Villagrán-Santa Cruz, M. 1998. Reproducción asincrónica en *Sceloporus palaciosi* (Sauria: Phrynosomatidae), con comentarios sobre sus ventajas y regulación. Revista de Biología Tropical 46: 1159–1161.

Méndez de la Cruz, F. R., Villagrán-Santa Cruz, M., López-Ortíz, L. and Hernández-Gallegos, O. 2013. Reproductive cycle of a high-elevation, oviparous lizard (*Sceloporus spinosus*: Reptilia: Phrynosomatidae). The Southwestern Naturalist 58: 54–63.

Mesquita, D. O. and Colli, G. R. 2003. Geographical variation in the ecology of populations of some Brazilian species of *Cnemidophorus* (Squamata, Teiidae). Copeia 2003: 285–298.

Mesquita, D. O. and Colli, G. R. 2010. Life history patterns in tropical South America lizards. In O. Hernández-Gallegos, F. R. Méndez-de la Cruz and J. F. Méndez-Sánchez (eds.), Reproducción en Reptiles: Morfología, Ecología y Evolución. Universidad Autónoma del Estado de México.

Miller, M. R. 1948. The seasonal histological changes occurring in the ovary, corpus luteum, and testis of the viviparous lizard, *Xantusia vigilis*. University of California Publications in Zoology 47: 197–224.

Mohri, H. and Yanagimachi, R. 1980. Characteristics of motor apparatus in testicular, epididymal and ejaculated spermatozoa. A study using demembranated sperm models. Experimental Cell Research 127: 191–196.

Mojica, B. H., Rey, B. H., Serrano, V. H. and Ramírez-Pinilla, M. P. 2003. Annual Reproductive Activity of a Population of *Cnemidophorus lemniscatus* (Squamata: Teiidae). Journal of Herpetology 37: 35–42.

Murphy-Walker, S. and Haley, S. R. 1996. Functional sperm storage duration in female *Hemidactylus frenatus* (family Gekkonidae). Herpetologica 52: 365–373.

Newlin, M. E. 1976. Reproduction in the bunch grass lizard, *Sceloporus scalaris*. Herpetologica 322: 171–184.

Nieto-Montes de Oca, A., Arenas-Moreno, D., Beltrán Sánchez, E. and Leache, A. 2014. A new species of horned lizard (genus *Phrynosoma*) from Guerrero, México, with an updated multilocus phylogeny. Herpetologica 70: 241–257.

Nirmal, B. K. and Rai, U. 1997. Epididymal Influence on Acquisition of Sperm Motility in the Gekkonid Lizard *Hemidactylus flaviviridis*. Systems Biology in Reproductive Medicine 39: 105–110.

Nonecke, S., Sage-Ciocca, D., Wollnik, F. and Pévet, P. 2013. Photoperiod can entrain circannual rhythms in pinealectomized European hamsters. Journal of Biological Rhythms 28: 278–290.

Olivares, A., Tapia, L., Estica, O., Henriquez, R. and Bustos-Obregon, E. 1987. Reproductive cycle of two coastal *Tropidurus* lizards. Microscopía Electrónica and Biología Celular 11: 107–117.

Olsson, M. and Madsen, T. 1996. Cost of mating with infertile males selects for late emergence in female sand lizard (*Lacerta agilis*). Copeia 1996: 462–464.

Olsson, M., Madsen, T. and Shine, R. 1997. Is sperm really so cheap? Costs of reproduction in male adders, *Vipera berus*. Proceedings of the Royal Society of London (B) 264: 455–459.

Olsson, M., Schwartz, T., Uller, T. and Healey, M. 2007. Sons are made from old stores: sperm storage effects on sex ratio in a lizard. Biological Letters 3: 491–493.

Ortega, A. and Barbault, H. 1984. Reproductive cycles in the Mesquite lizard, *Sceloporus grammicus*. Journal of Herpetology 18: 168–175.

Ortega León, A. M., Villagrán Santa Cruz, M., Zúñiga Vega, J. J., Cueva del Castillo, R. and Méndez de la Cruz, F. R. 2009. Sperm viability in the reproductive tract of females in a population of *Sceloporus mucronatus* exhibiting asynchronous reproduction. Western North American Naturalist 69: 96–104.

Ortiz, J. C. 1981. Révision taxonomique et biologie des *Liolaemus* du groupe *nigromaculatus* (Squamata, Iguanidae). Thèse de Doctorat d'Étatès Sciences Naturelles, Université Paris VII, 438 pp.

Parker, W. S. 1971. Ecological observations on the regal horned lizard (*Phrynosoma solare*) in Arizona. Herpetologica 27: 333–338.

Parker, W. S. 1973. Notes on reproduction of some lizards from Arizona, New Mexico, Texas, and Utah. Herpetologica 29: 258–264.

Pough, F. H., Andrews, R. M., Cadle, J. E., Crump, M. L., Savitzky, A. H. and Wells, K. D. 1998. Herpetology. Upper Saddle River, NewJersey: Prentice Hall.

Pudney, J. 1995. Spermatogenesis in nonmammalian vertebrates. Microscopy Research and Technique 32: 459–497.

Ramírez Bautista, A. and Becerril-Olvera, V. 2004. Reproduction in the boulder spiny lizard, *Sceloporus pyrocephalus* (Sauria: Phrynosomatidae), from a tropical dry forest of México. Journal of Herpetology 38: 225–231.

Ramírez-Bautista, A. and Dávila-Ulloa, E. G. 2009. Reproductive characteristics of a population of *Sceloporus dugesii* (Squamata: Phrynosomatidae) from Michoacán, Mexico. The Southwestern Naturalist 54: 400–408.

Ramírez-Bautista, A. and González-Romero, A. 2002. Some reproductive and feeding characteristics of the viviparous Mexican lizard *Sceloporus torquatus* (Phrynosomatidae). The Southwestern Naturalist 47: 98–102.

Ramírez-Bautista, A., Balderas-Valdivia, C., Vitt, L. J. and Price, A. H. 2000. Reproductive ecology of the whiptail lizard *Cnemidophorus lineatissimus* (Squamata: Teiidae) in a tropical dry forest. Copeia 2000: 712–722.

Ramírez-Bautista, A., Ortíz-Cruz, A. L., Arizmendi, M. C. and Campos, J. 2005. Reproductive characteristics of two syntopic lizard species, *Sceloporus gadoviae* and *Sceloporus jalapae* (Squamata:Phrynosomatidae) from Tehuacán Valley, Puebla, México. Western North American Naturalist 65: 202–209.

Ramírez-Bautista, A., Smith, G. R. and Hernández-Ibarra, X. 2009. Reproduction and sexual dimorphism in the whiptail lizard *Aspidoscelis gularis* (Squamata: Teiidae) in Guadalcázar, San Luis Potosí, Mexico. The Southwestern Naturalist 54: 453–460.

Ramírez-Pinilla, M. P., Calderón-Espinosa, M. L., Flores-Villela, O., Muñóz-Alonso, A. and Méndez-de la Cruz, F. R. 2009. Reproductive activity on three sympatric viviparous lizards at Omiltemi, Guerrero, Sierra Madre del Sur, Mexico. Journal of Herpetology 43: 409–420.

Rivera Morales, F. 2001. Estrategia reproductora de la lagartija tropical *Sceloporus serrifer* en el estado de Yucatán. Bch Thesis FES Iztacala, Universidad Nacional Autónoma de México, México.

Rodríguez-Ramírez, J. and Lewis, A. R. 1991. Reproduction in the Puerto Rican teiids *Ameiva exsul* and *Ameiva wetmorei*. Herpetologica 47: 395–403.

Rodríguez-Romero, F. J. and Méndez de la Cruz, F. R. 2004. Reproductive arrest in *Sceloporus mucronatus* (Sauria: Prhynosomatidae), due to "El niño" climate effects. Herpetological Review 35: 121–123.

Rodríguez-Romero, F., Smith, G. R., Méndez-Sánchez, F., Hernández-Gallegos, O., Fox, S. S. and Méndez de la Cruz, F. R. 2011. Demography of a semelparous high elevation population of *Sceloporus bicanthalis* (Lacertilia: Phrynosomatidae) from the Nevado de Toluca Volcano, México. The Southwestern Naturalist 56: 71–77.

Roig, J. M., Carretero, M. A. and Llorente, G. A. 2000. Reproductive cycle in a Pyrenean oviparous population of the common lizard (*Zootoca vivipara*). Netherlands Journal of Zoology 50: 15–27.

Saint-Girons, H. 1963. Spermatogenèse et évolution cyclique des carècteres sexuels secondaires chez les Squamata. Annales des Sciences Naturelles 5: 461–476.

Saint-Girons, H. 1984. Les cycles sexuels des lezards mâles et leurs rapports avec le climat et les cycles reproducteur des famelles. Annales des Sciences Naturelles. 6: 221–243.

Saint Girons, H. 1985. Comparative data on Lepidosaurian reproduction and some time tables. En: C. Carl Gans (ed.), Biology of the Reptilia. John Wiley & Sons. New York 15: 35–58.

Santos-Bibiano, R. 2014. Aspectos reproductores de *Phrynosoma sherbrookei* (Sauria: Phrynosomatidae) en la localidad de Tenexatlajco, Chilpa de Alvarez, Guerrero, México. Bch. Thesis. Universidad de Guerrero, México.

Serrano-Cardozo, V. H., Ramírez-Pinilla, M. P., Ortega, J. E. and Cortés, L. A. 2007. Annual reproductive activity of *Gonatodes albogularis* (Squamata: Gekkonidae) living in an anthropic area in Santander, Colombia. South American Journal of Herpetology 2: 31–38.

Setchell, B. P. 1998. The Parkes Lecture Heat and the testis. Journal Reproduction and Fertility 114: 179–194.

Sever, D. M., Stevens, R. A., Ryan, T. J. and Hamlett, W. C. 2002. Ultrastructure of the reproductive system of the black swamp snake (*Seminatrix pygaea*): III. Sexual segment of the male kidney. Journal of Morphology 252: 238–254.

Sever, D. M. and Hamlett, W. C. 2002. Female sperm storage in reptiles. Journal of Experimental Zoology 292: 187–199.
Sinervo, B., Méndez-de-la-Cruz, F., Miles, D. B., Heulin, B., Bastiaans, E., Villagrán-Santa Cruz, M., Lara-Reséndiz, R., Martínez-Méndez, N., Lucía Calderón-Espinosa, M., Nelsi Meza-Lázaro, R., Gadsden, H., Avila, L. J., Morando, M., De la Riva, I. J., Victoriano Sepulveda, P., Duarte Rocha, C. F., Ibargüengoytía, N., Aguilar Puntriano, C., Massot, M., Lepetz, V., Oksanen, T. A., Chapple, D. G., Bauer, A. M., Branch, W. R., Clobert, J. and Sites Jr., J. W. 2010. Erosion of lizard diversity by climate change and altered thermal niches. Science 328: 894–899.
Sinervo, B., Miles, D. B., Martínez-Méndez, N., Lara-Reséndiz, R. and Méndez-De la Cruz. F. R. 2011. Response to Comment on "Erosion of lizard diversity by climate change and altered thermal niches". Science 332: 537.
Sites, J. W., Jr., Archie, J. W., Cole, C. J. and Villela, O. F. 1992. A review of phylogenetic hypothesis for the lizards of the genus *Sceloporus* (Phrynosomatidae): Implications for ecological and evolutionary studies. Bulletin of the American Museum of Natural History 213: 1–110.
Smyth, M. and Smyth, M. J. 1968. Obligatory sperm storage in the skink *Hemiergis peronii*. Science 161: 575–576.
Stewart, J. R. 1979. The balance between number and size of young in the live bearing lizards *Gerrhonotus coeruleus*. Herpetologica 35: 342–350.
Tanner, W. W. and Krogh, J. E. 2010. Ecology of *Sceloporus magister* at the Nevada test site, Nye County, Nevada. Western North American Naturalist 33: 133–146.
Tinkle, D. W. and Hadley, N. F. 1973. Reproductive effort and winter activity in the viviparous montane lizard *Sceloporus jarrovi*. Copeia 272–277.
Tinkle, D. W., Dunham, A. E. and Congdon, J. D. 1993. Life history and demographic variation in the lizard *Sceloporus graciosus*: a long-term study. Ecology 74: 2413–2429.
Tourmente, M., Cardozo, G. A., Guidobaldi, H. A., Giojalas, L. C., Bertona, M. and Chiaraviglio, M. 2007. Sperm motility parameters to evaluate the seminal quality of *Boa constrictor occidentalis*, a threatened snake species. Research in Veterinary Science 82: 93–98.
Urzúa-Vázquez, U. E. 2008. Termorregulación de la lagartija vivípara *Phrynosoma orbiculare* (Phrynosomatidae) en zonas de alta montaña. Posgrado en Ciencias Biológicas, IBUNAM. México.
Vieira, G. H. C., Wiederhecker, H. C., Colli, G. R. and Báo, S. N. 2001. Spermiogenesis and testicular cycle of the lizard *Tropidurus torquatus* (Squamata, Tropiduridae) in the Cerrado of central Brazil. Amphibia-Reptilia 22: 217–233.
Vidal, M., Ortiz, J. C. and Labra, A. 2002. Sexual and age differences in ecological variables of the lizard *Microlophus atacamensis* (Tropiduridae), from northern Chile. Revista Chilena Historia Natural 75: 283–292.
Villagrán-Santa Cruz, M., Méndez de la Cruz, F. R. and Cuellar, O. 1992. Obligatory sperm storage in the lizard *Sceloporus grammicus*. Acta Zoológica Mexicana 49: 23–31.
Villagrán-Santa Cruz, M., Méndez de la Cruz, F. R. and Parra, L. 1994. Ciclo espermatogénico del lacertilio *Sceloporus mucronatus* (Reptilia, Sauria, Phrynosomatidae). Revista de Biología Tropical 42: 289–296.
Villagrán-Santa Cruz, M., Hernández-Gallegos, O. and Méndez de la Cruz, F. R. 2009. Reproductive cycle of the lizard *Sceloporus mucronatus* with comments on intraspecific geographic variation. Western North American Naturalist 69: 437–446.
Vitt, L. J. 1973. Reproductive biology of the anguid lizard *Gerrhonotus coeruleus principis*. Herpetologica 29: 176–184.
Vitt, L. J. 1982. Reproductive tactics of *Ameiva ameiva* (Lacertilia: Teiidae) in a seasonally fluctuating tropical habitat. Canadian Journal of Zoology 60: 3113–3120.
Vitt, L. J. 1986. Reproductive tactics of sympatric gekonid lizards with comments on the evolutionary and ecological consequences of invariant clutch size. Copeia 1986: 773–786.
Vitt, L. J. and Breitenbach, L. 1993. Life histories and reproductive tactics among lizards in the genus *Cnemidophorus* (Sauria: Teiidae). pp. 211–243. In J. W. Wright and L. J. Vitt (eds.),

Biology of Whiptail Lizards (Genus *Cnemidophorus*). Oklahoma Museum of Natural History, Norman, Oklahoma, U.S.A.

Vitt, L. J. and Ohmart, R. D. 1977. Ecology and reproduction of lower Colorado river lizards: II. *Cnemidophorus tigris* (Teiidae), with comparisons. Herpetologica 33: 223–234.

Wen-San, H. 2007. Ecology and reproductive patterns of the agamid lizard *Japalura swinhonis* on an East Asian Island, with comments on the small clutch sizes of island lizards. Zoological Science 24: 81–188.

Wiens, J. J., Kuczynski, C. A., Arif, S. and Reeder, T. W. 2010. Phylogenetic relationships of phrynosomatid lizards based on nuclear and mitochondrial data, and a revised phylogeny for *Sceloporus*. Molecular Phylogenetics and Evolution 54: 150–161.

Xavier, F. 1982. Progesterone in the viviparous lizard *Lacerta vivipara*: ovarian biosynthesis, plasma levels, and briding to transcorticon-type protein during the sexual cycle. Herpetologica 38: 62–70.

Yamamoto, Y. and Ota, H. 2006. Long-term functional sperm storage by a female common house gecko, *Hemidactylus frenatus*, from the Ryukyu Archipelago, Japan. Current Herpetology 25: 39–40.

Zaldívar-Rae, J., Drummond, H., Ancona-Martínez, S., Manríquez-Morán, N. L. and Méndez de la Cruz, F. R. 2008. Seasonal breeding in the Western Mexican whiptail lizard, *Aspidoscelis costata* on Isla Isabel, Nayarit, México. The Southwestern Naturalist 53: 175–184.

Zamudio, K. R. and Parra-Olea, G. 2000. Reproductive mode and female reproductive cycles of two endemic Mexican horned lizards (*Phrynosoma taurus* and *Phrynosoma braconnieri*). Copeia 2000: 222–229.

CHAPTER 11

The Architecture of the Testis, Spermatogenesis, and Mature Spermatozoa

Kevin M. Gribbins[1,*] and *Justin L. Rheubert*[2]

11.1 INTRODUCTION

The intent of this chapter is to review the major morphological data on the architecture of the testis, the histological ontogeny of spermatogenesis, and the ultrastructure of the mature spermatozoa in lizards. There is a great deal of information on reproduction in male lizards and the above task might seem impossible to encompass in a book chapter. However, much of what has been done in male lizard reproduction focuses on reproductive cycles, mating, the common anatomy/histology of the reproductive tract, or specific parts of spermatogenesis. In the last 15 years, there has been an effort to provide ultrastructural information for the spermatozoa in many species of squamates (including both snakes and lizards), which has led to preliminary constructions of phylogenetic relationships between certain families within Squamata (Jamieson 1995, 1999; Oliver *et al.* 1996; Teixeira *et al.* 1999; Tavares-Bastos *et al.* 2002, 2008; Tourmente *et al.* 2008; Gribbins and Rheubert 2011).

This chapter will be limited to general data that detail the morphology and ultrastructure of the testis, completed data on the entire germ cell development strategy of spermatogenesis, and ultrastructural information on the products or process of spermiogenesis. Although these data are

[1] Department of Biology, University of Indianapolis, Indianapolis, IN 46227.
[2] College of Sciences, The University of Findlay, Findlay, OH 45840, USA.
* Corresponding author

limited in many regards, they suffice for a preliminary analysis of the comparative histology of the testis and spermatogenic cycle/phases in lizards. Reproductive or spermatogenic cycles will not be a focus in this chapter, for considerations of these cycles refer to Mendez de la Cruz (Chapter 10, this volume). The information provided here will be more than a simple review of what is known but, when appropriate (sperm ultrastructure for example), the morphology will be placed in an evolutionary context. Providing morphological characteristics during spermatogenesis, for the mature spermatozoa, and for the testis of lizards will offer non-traditional data that can be useful for cladistic analyses within lizards and the Reptilia. We hope that this chapter stimulates interest in encouraging research efforts in the histological analysis of the male reproductive tracts in reptiles.

11.2 THE TESTIS

The male reproductive tract in lizards consists of the testes and the excurrent ducts (see Rheubert *et al.* Chapter 9, this volume), which drain testicular products, the spermatozoa, and seminal fluids to the exterior. Lizards are amniotes and thus the testis is enclosed in a connective tissue capsule with the testicular interior subdivided into seminiferous tubules and the interstitial compartment similar to that described for birds and mammals (Russell *et al.* 1990).

As in birds, the testes of lizards and other squamates are intra-abdominal and develop embryonically in close association with the kidneys (Raynaud and Peiau 1985). Most lizard testes are round to slightly oval in shape (Fig. 11.1A), except in legless species that have more elongated snake-like testes (Fox 1975). They are often white to cream in color, but can be pigmented as seen in chameleons (Bouix and Bourgat 1970). Each testis is held together and is encased with a thin fibrous tunica albuginea (Fig. 11.1B) and connected to the body wall by a mesochorium. Because of the restricted width of the tunica (roughly 3–15 μm) the testes in lizards can be soft to the touch (especially in quiescent periods), and one can typically visualize the coiled seminiferous tubules just under this exterior connective tissue layer (Fig. 11.1A). However, it should be noted that in times of maximum spermatogenic activity the testis can easily double its size (especially in some *Sceloporus* species) and produce high intratesticular pressure that gives the testis an added firmness. In most lizards studied to date, the tunica sends a few to many septa (Fig. 11.1B) interiorly, which sometimes separates the testis into distinct compartments or lobes. The lizard testis, as in all vertebrates, is composed of a coiled mass of seminiferous tubules where germ cells develop through spermatogenesis (Figs. 11.1, 11.2, 11.3A). The intertubular space between tubules is occupied by interstitial (Leydig)

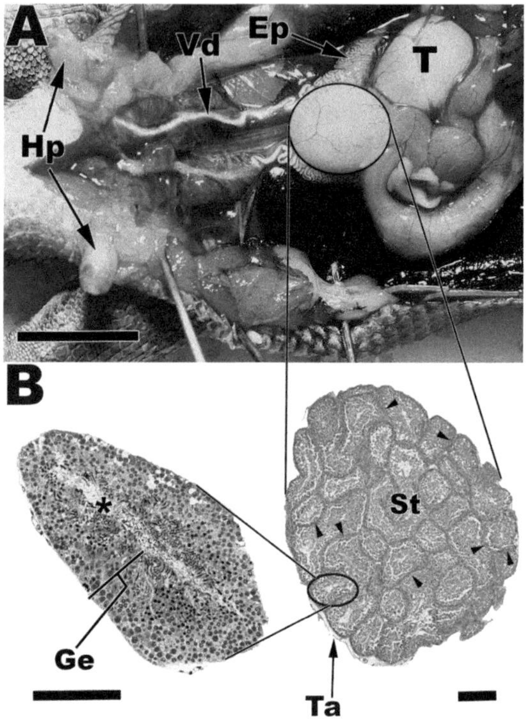

Fig. 11.1 (A) Gross dissection of the urogenital tract of *Sceloporus bicanthalis*. Scale = 10 mm. Note the location of the testes (T) and excurrent ducts (Ep, Vd) of the reproductive system. Ep, epididymis; Vd, vas deferens; Hp, hemipenis. **(B)** Cross sections of the entire testis (*Sceloporus consobrinus*), right, and of an individual seminiferous tubule (left) within *Barisia imbricata*. Bar (testis) = 10 mm, Bar (tubule) = 75 µm. A thin tunica albuginea (Ta) encases the highly convoluted seminiferous tubules (ST). Septa (black arrows) originating from the tunica albuginea and penetrate the testis providing the framework of the interstitial space interiorly. Individual seminiferous tubules have a thick germinal epithelium (GE) and a lumen (*) where spermatozoa are released once spermatogenesis is complete. Photo A: Oswaldo Hernández-Gallegos, Photo B: (CS Testis) Justin Rheubert (CS Tubule), Kevin Gribbins.

Color image of this figure appears in the color plate section at the end of the book.

cells (Fig. 11.3B), a few lymph and blood vessels (Fig. 11.3B), leukocytes, and myofibroblasts (Fig. 11.3A,B), and is surrounded by an extracellular matrix that consists of multiple collagen fibers (Fig. 11.2B).

11.2.1 Seminiferous Tubules

The seminiferous tubules of lizards form a constricted mass of highly convoluted tubules that are continuous with the anterior testicular ducts

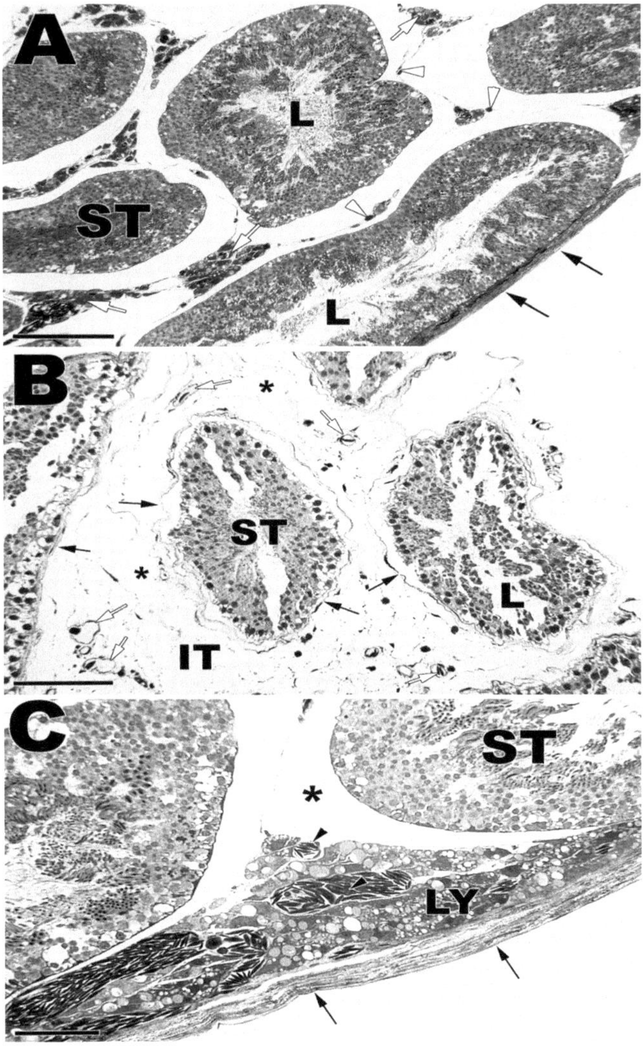

Fig. 11.2 (A) Cross/Sagittal sectional views of seminiferous tubules within the testis of *Phrynosoma cornutum*. Bar = 100 μm. Lumen of tubules (L), tunica albuginea (black arrows), seminiferous tubules (ST), Leydig cells (white arrows), myofibroblasts (white arrowheads). **(B)** Transverse view of two seminiferous tubules (ST) within the testis of *Anolis sagrei*. Bar = 50 μm. The basement membrane (black arrows) is visible and a thick seminiferous epithelium lines the tubules with prominent lumina (L). Within the interstitial space (IT) there are blood vessels (white arrows) and a thick collagenous matrix (*) visible at this magnification. **(C)** Superficial section close to the tunica albuginea (black arrowhead) in *P. cornutum*. Bar = 50 μm. The interstitial space (*) is thick with blood vessels (black arrowheads) and what appears to be a subtunica layer of Leydig cells (LY) just under the tunica albuginea (black arrows).

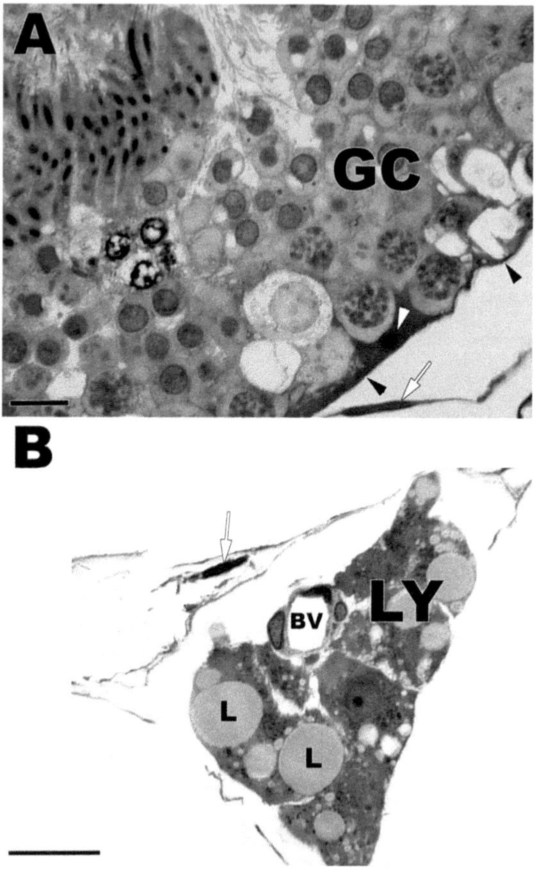

Fig. 11.3 *Phrynosoma cornutum.* **(A)** A July seminiferous tubule under high magnification within the testis. Bar = 20 μm. Note the prominent basement membrane (black arrowheads), a myofibroblast nucleus (white arrow), and the layers of germ cells (GC) within the epithelium. Sertoli cell nucleus (white arrowhead). **(B)** A high power view of Leydig cells (LY) heavy with lipid droplets (L) and occupying the interstitial space in July. Bar = 25 μm. Blood vessel (BV), myofibroblast nucleus (white arrowhead).

(see Rheubert *et al.* Chapter 9, this volume). The mature spermatozoa are evacuated from the seminiferous epithelia into the lumina of these tubules and flushed to the efferent ductuli, which empty to the epididymis where spermatozoa will then reside until ejaculation. The seminiferous tubules are hollow (Fig. 11.2A-C) and typically have a well-defined lumen that varies in size due to seasonal activities of the testis. The lumen is surrounded by a thick seminiferous epithelium (Figs. 11.2, 11.3A, 11.4), which is made up of complex differentiated Sertoli cells and developing germ cells (Fig. 11.4). This epithelium resides on a thin basal lamina (~between 2–10 μm)

(Figs. 11.4, 11.5) produced by this epithelium, which is embedded into the boundary layer (Figs. 11.4, 11.5A) of the interstitial compartment of the testis. The Sertoli cells are highly branched cells that invest and nurture the developing germ cells during spermatogenesis (Figs. 11.5A, 11.6). The width of the seminiferous epithelium is also contingent on the seasonal activity of the testis, and on the stage of the spermatogenic cycle of the testis.

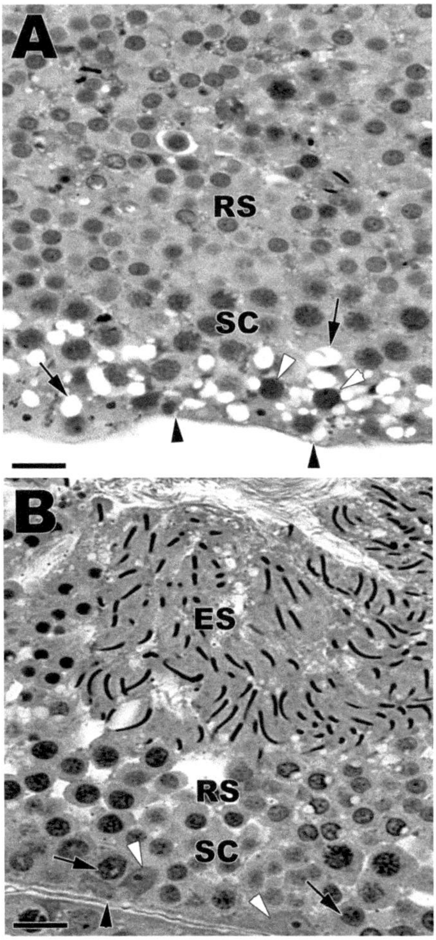

Fig. 11.4 *Podarcis muralis.* **(A)** Early August seminiferous tubule within the testis, showing the early stages of spermatogenesis with a large population of lipid inclusions (black arrows) amongst spermatogonia (white arrowheads). Bar = 20 μm. Spermatocytes (SC), round spermatids (RS), Basal lamina (black arrowheads). **(B)** May seminiferous tubule that has late elongating spermatids (ES), spermatocytes (SC), some round spermatids (RS), and no lipid droplets within the spermatogonia basal layer (black arrows) and a prominent boundary layer (black arrowhead). Sertoli cell nuclei (white arrowheads). Bar = 20 μm.

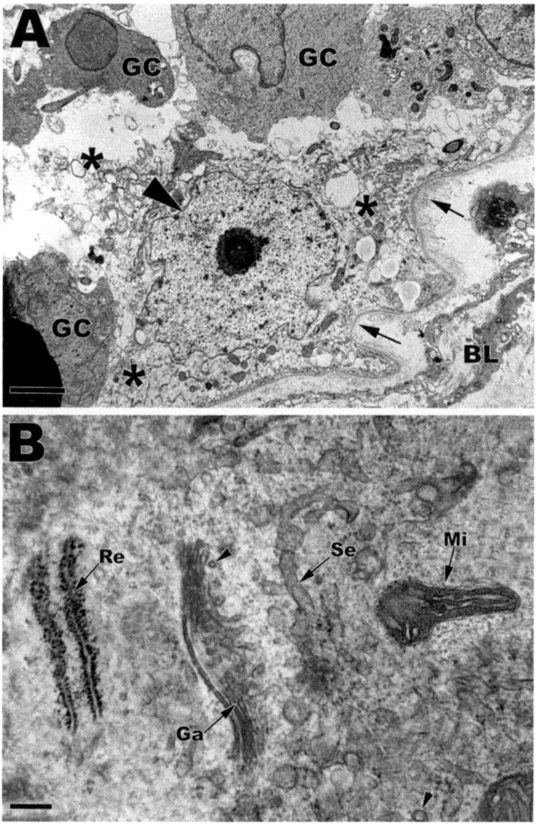

Fig. 11.5 *Podarcis muralis* **(A)** Sertoli cell nucleus (black arrowhead) resting on a prominent basal lamina (black arrows) of the boundary layer (BL). The Sertoli cell cytoplasmic processes (*) wrap around the dark staining developing germ cells (GC). Bar = 15 μm. **(B)** Typical organelles within the Sertoli cytoplasm, which includes mitochondria (Mi), abundant smooth (Se) and rough (Re) ER, Golgi apparatus (Ga), and vesicles (black arrowheads). Bar = 3 μm.

11.2.2 Sertoli Cells

The Sertoli (sustentacular) cells in lizards are a permanent component of the germinal epithelium irrespective of seasonal activity of the testis. Sertoli cell nuclei and their epithelial extensions are usually recognizable within thin plastic sections of the testis. Their cytoplasm and the nucleus of the Sertoli cell tend to stain darker than the germ cell cytoplasm in epon sections that are 1 μm or thinner (Fig. 11.3A). It is not surprising that Sertoli cells in squamates (Gribbins and Rheubert 2011) and other amniotes (Cooksey and Rothwell 1973; Pudney 1993) are often referred to as germ cell nurturing cells. These specialized epithelial cells support, synthesize,

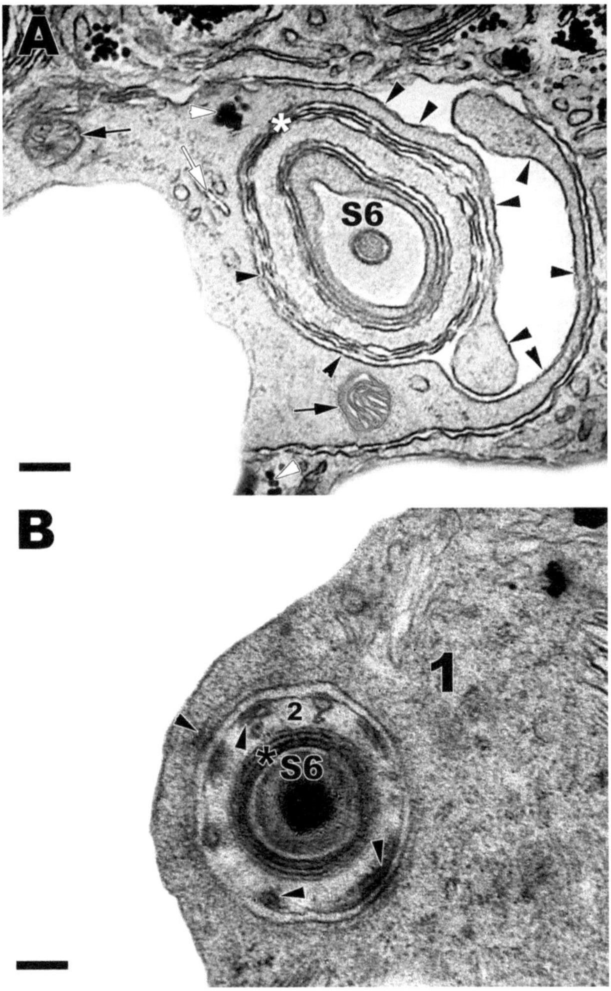

Fig. 11.6 (A) A Sertoli cell process (black arrowheads) wrapping around a developing elongating spermatid (S6) in the seminiferous epithelium of *Anolis lineatopus*. Bar = 0.2 μm. Note the enveloping process produces several cell membrane laminae (white *) around the spermatid. There are also mitochondria (black arrows), smooth ER (white arrow), and intermediate filaments in cross section (white arrowheads) within the cytoplasm of the Sertoli cell process. **(B)** Two Sertoli cells (1,2) surrounding the nuclear head of an elongating spermatids (S6) in *Sceloporus variabilis*. Multiple Sertoli cell membrane laminae (*) surround the developing spermatid and numerous adhering junctions are found between the associated Sertoli cells. These desmosome like junctions have thick subsurface densities that are very thick on the cytoplasmic side of Sertoli cell 2 (Black arrowheads).

and store steroids/lipids, form ectoplasmic specializations, and provide nutritional provisions to the germ cells during their development.

Remarkably, Sertoli cells in the Western Fence Lizard (*Sceloporus occidentalis*; Wilhoft and Quay 1961) and the Tiger Whiptail (*Cnemidorphorous tigris*; Goldberg and Lowe 1966) have been reported to be shed to the lumina of seminiferous tubules at the end of the breeding season. But these studies were only performed at the light microscope level and have not been confirmed via electron microscopy. In other seasonal lizard species, such as the Horned Lizard (*Phrynosoma cornutum*; Lofts *et al.* 1968) and the Golden Skink (*Eutropis carinata*; Shivanandappa *et al.* 1979; Aranha *et al.* 2006), Sertoli cell changes have been linked or synchronized to the reproductive cycle. One good example of this is lipid depletion and accumulation within the germinal epithelium of lizards and other squamates. Depending on the species, lipid droplets are typically smallest during active spermatogenesis (through spermiogenesis; Fig. 11.4B). Conversely, during spermatogenic quiescent and early recrudescence periods, there is a rapid accumulation of lipid rich droplets within the lizard seminiferous epithelium (Fig. 11.4A). Once spermatogenesis is well underway through the events of mitosis, meiosis, and spermiogenesis, lipid content slowly relapses to its lowest levels during late spermiogenesis.

There are only a few fine structural studies on Sertoli cells in squamates, with only a handful focusing on lizard sustentacular cells. These saurian studies include the Common Wall Lizard (*Podarcis muralis*), the Italian Wall Lizard (*Podarcis sicula*; Baccetti *et al.* 1983), and the Common Lizard (*Zootoca vivipara*; Dufaure 1971). One other study provides some ultrastructural information; however, the authors focus on the double nuclei and their relationship to cell volumes within the Sertoli cells of the Green Anole (*Anolis carolinensis*) a condition apparently also observed in another lizard, the Side-blotched Lizard (*Uta stansburiana*) (Hahn 1964).

General characteristics of Sertoli cells in lizards will be given here with the caveat that more ultrastructural information on the Sertoli cell needs to be gathered not only in lizards, but also within other reptiles. Most of the data (photographic) presented in this section are unpublished. The cytology of Sertoli cells from the scarce number of species studied to date appears similar in ultrastructural detail between species. The irregular nucleus is typically basally located juxtaposed to the basal lamina (Fig. 11.5A). Each nucleus also normally has one to three nucleoli. The cytoplasm contains numerous organelles associated/correlated with the complex functions of these cells; they include small to large lipid droplets (Fig. 11.4A), abundant mitochondria, well-developed rough and smooth endoplasmic reticulum, Golgi apparatus, many variable sized vesicles, and ribosomes (Fig. 11.5B). Bundles of intermediate filaments, actin filaments, and moderate amounts of glycogen granules are also represented within the Sertoli cell cytoplasm during different times of the year in lizards and other squamates (Gribbins and Rheubert 2011).

Relatively detailed information on the ultrastructure of Sertoli/germ cell and Sertoli/Sertoli junctions have been described for some reptiles, including a small number of lizards. The most comprehensive description of ectoplasmic specializations in squamates (Ophidia; Gribbins and Rheubert 2011), for which lizards belong, corroborates the ultrasructural information provided here for saurian species. Sertoli cell to cell contacts with spermatids have been documented for the Rainbow Whiptail (*Cnemoidophorus lemniscatus*; Del Conte 1976), *Anolis carolinensis* (Clark 1967), and the Yellow-bellied House Gecko (*Hemidactylus flaviviridis*; Khan and Rai 2004). These junctions are plentiful between the Sertoli cell and the elongating spermatids because Sertoli cells envelop these spermatids with many layers of cell membrane or laminae (Fig. 11.6A), increasing the surface area contacts between germ cells and sustentacular cells. This increase in surface area provides more interaction and thus plentiful junctions between elongating spermatids, which are hypothesized to prevent these spermatids from being pulled away from the epithelium during spermiation (Baccetti *et al.* 1983; Gribbins and Rheubert 2011). Desomosome-like junctions (adhering) are seen between spermatogonia, spermatocytes, and most often between spermatids and juxtaposed Sertoli cells. All three major cell types show the same morphology for ectoplasmic specializations, which are similar for those in snakes (Gribbins and Rheubert 2011). The desmosome-like junctions in lizards are very unassuming in structure and may be described as areas of juxtaposed plasma membranes separated by a roughly 20 nm space (Fig. 11.6B) (Gribbins unpublished). The junctions are flanked on the germ cell and Sertoli cell sides of the membranes by thick dark staining subsurface densities (Figs. 11.6B, 11.7), which are particularly robust on the germ cell side of the inner Sertoli cell when they are present (Fig. 11.6B). Many species of lizard, as in the Imbricate Alligator Lizard (*Barisia imbricate*; Gribbins *et al.* 2013a) and the Trans Volcanic Bunchgrass Lizard (*Sceloporus bicanthalis*; Rheubert *et al.* 2012), have two morphologically distinct Sertoli cells that spiral around each other and envelop elongating spermatids during the end events of spermiogenesis. The exact reason for this relationship between Sertoli cells and elongating spermatids is unknown. In some cases, the subsurface densities are so thick they appear to cover the intercellular space between the opposing membranes (Fig. 11.7C). Smooth endoplasmic reticulum (SER) is associated with the germ cell cytoplasmic and Sertoli sides of the cell membranes (Figs. 11.7C,D). The SER is located between 8 and 60 nm away from the plasma membrane and lies either in sagittal or cross sectional positions to the desmosome-like junction (Gribbins unpublished). Some of the desmosome-like junctions, especially those associated with spermatogonia and spermatids, show the presence of 7 nm filaments near the subsurface densities (Fig. 11.7C).

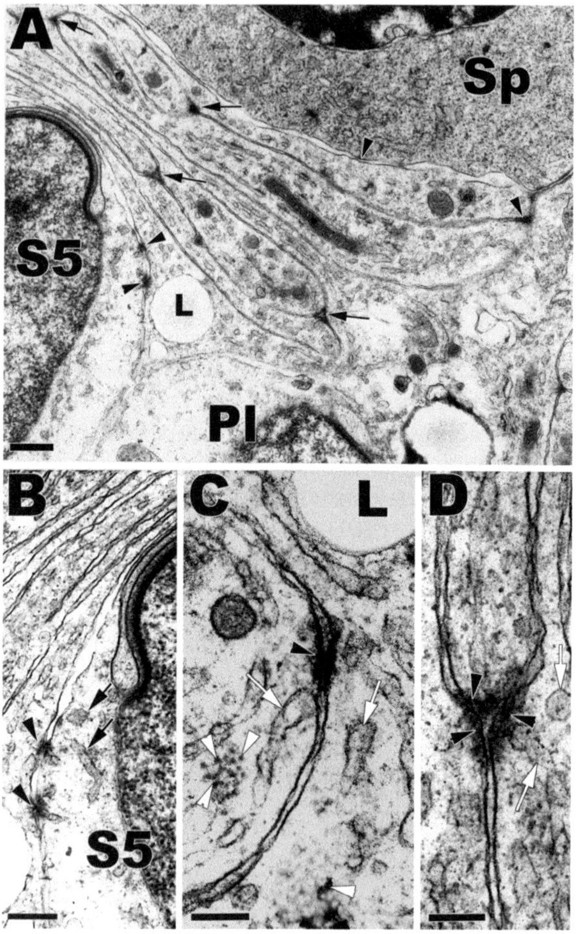

Fig. 11.7 *Podarcis muralis.* **(A)** Adhering junctions between Sertoli cells and germ cells within the seminiferous epithelium. Bars = 1 μm. Note the triad-like junctional combinations (black arrows) between Sertoli cell interdigitations versus single desmosomes (black arrowheads) between Sertoli cells and germ cells. Lipids (L), spermatid (S5), spermatogonia (Sp), preleptotene spermatocyte (Pl). **(B)** Desmosome-like junction (black arrowhead) between a Sertoli cell and step 5 spermatid (S5). The subsurface densities (black arrowheads) are on both sides of the cell membranes of the Sertoli and germ cells. Smooth endoplasmic reticulum (black arrows). Bar = 1 μm. **(C)** Higher Magnification of a single desmosome, showing simple ectoplasmic specializations that includes smooth endoplasmic reticulum (white arrows), subsurface densities (black arrowhead), and 7 nm filaments (white arrowheads). Bar = 0.2 μm. **(D)** Shows the details of a triad combination (black arrowheads) of desmosome-like junctional complexes between interdigitations of the Sertoli cells. Smooth ER (white arrow). Bar = 0.2 μm.

Tight (Occluding) junctions between adjacent Sertoli cells are easily seen within lizard germinal epithelia with transmission electron microscopy and, in at least *Podarcis muralis*, they appear similar in structure. They are often observed in the basal compartment of the testis and separate spermatogonia from more advanced spermatocytes and spermatids intercellularly (Fig. 11.8A). These junctions are often found grouped together between adjacent Sertoli cells and zip-lock their adjacent cell membranes together. They can be numerous and may be found between juxtaposed Sertoli cells even in the absence of spermatogonia. Morphologically, they show focal contacts between opposing Sertoli cell membranes that are fused together. Subsurface

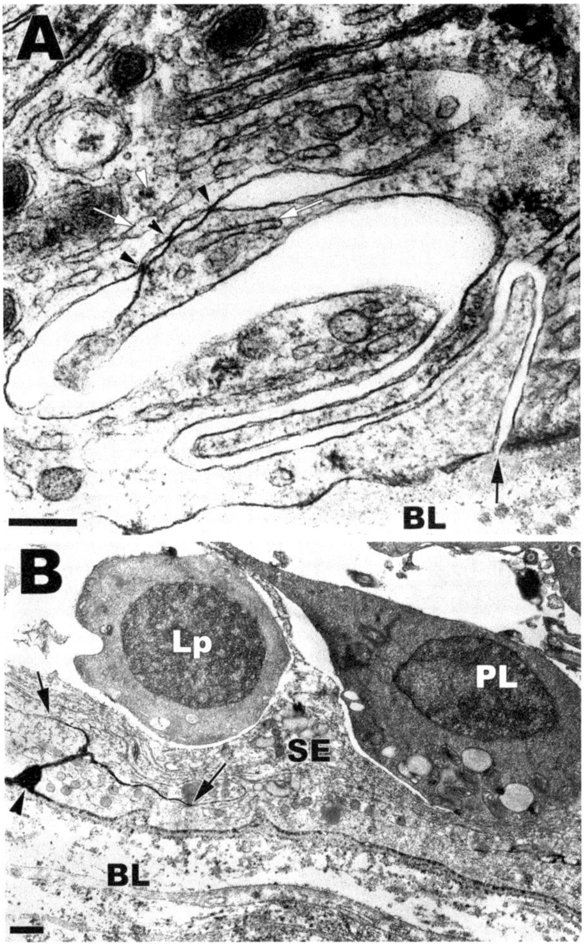

Fig. 11.8 *Podarcis muralis.* **(A)** Three tight junctions (black arrowheads) near the basal lamina (BL) between adjacent Sertoli cells of the germinal epithelium. Bar = 0.5 μm. Entrance to an intercellular pathway between Sertoli cells (black arrow). Focal subsurface densities (black arrowheads), smooth ER (white arrows), 7 nm filaments (white arrowhead). **(B)** Lanthanum nitrate penetration within an intercellular pathway (black arrowhead) is stopped (black arrows) just below leptotene spermatocytes (Lp) in the germinal epithelium (SE). Boundary layer (BL), preleptotene spermatocyte (PL). Bar = 0.5 μm.

densities are found on both sides of the juxtaposed Sertoli cell membranes but in a more restricted manner than that seen in adhering junctions (Fig. 11.8A). The subsurface densities are concentrated at the contact points between opposing membranes. Frequently, smooth endoplasmic reticulum is found associated with the tight junctions on both sides of the juxtaposed membranes (typically between 13–60 nm away from the junctional complex) (Fig. 11.8A) in *Podarcis muralis* (Gribbins unpublished).

A blood testis barrier has been confirmed in squamates (Bacetti *et al.* 1983; Bergmann *et al.* 1984; Gribbins and Rheubert 2011) using electron-dense intercellular tracers, including lanthanum nitrate. Only a small number of studies have looked at the blood testis barrier in detail within lizards; these studies include *Anolis carolinensis* (Bergman *et al.* 1984), *Podarcis muralis* (Gribbins unpublished), *P. sicula* (Baccetti *et al.* 1983), Bibron's Tree Iguana (*Liolaemus bibroni*), Ruibal's Tree Iguana (*L. ruibali*), and the High Mountain Lizard (*Phymaturus palluma*) (Cavicchia and Miranda 1988). *Podarcis muralis* has a functional blood testis barrier similar to that depicted in other reptiles, birds (Osman *et al.* 1980; Baccetti *et al.* 1983; Bergmann and Schindelmeiser 1987), and mammals (see Pudney 1993). Testes from active (April, May) and inactive (July) months in *P. muralis* were treated with lanthanum nitrate during fixation for this chapter's study of the blood testis barrier. In active and inactive months, tight junctions are intact and show similar morphological characteristics (Fig. 11.8A). These tight junctions resemble those found in the Japanese Rat Snake (*Elaphe climacophora*; Hondo *et al.* 1996), Cottonmouth Snake (*Agkistrodon piscivorus*; Gribbins and Rheubert 2011), and the lizards mentioned above.

Like the other lizards studied to date, lanthanum penetration in the Wall Lizard, *Podarcis muralis*, is stopped (Figs. 11.8A, 11.9) above (apically) spermatogonia before the intercellular tracer reaches the level of the spermatocytes and spermatids within the germinal epithelium (Gribbins unpublished). When spermatogonia and spermatocytes were not present within the seminiferous epithelium the tracer would penetrate just below the developing spermatids (picture not shown). In contrast, the lanthanum nitrate diffused easily through the basement membrane and into intercellular pathways around spermatogonia (Fig. 11.9) even before active spermatogenesis starts. Thus, as proposed by Pudney (1993) for other reptiles, the lizards also have an intact blood testis barrier before meiosis and spermiogenesis begin. Since the seminiferous epithelium is a permanent component of the testis and intact tight junctions are seen throughout the year in all lizards studied to date, it is hypothesized that the blood testis barrier is functional year around.

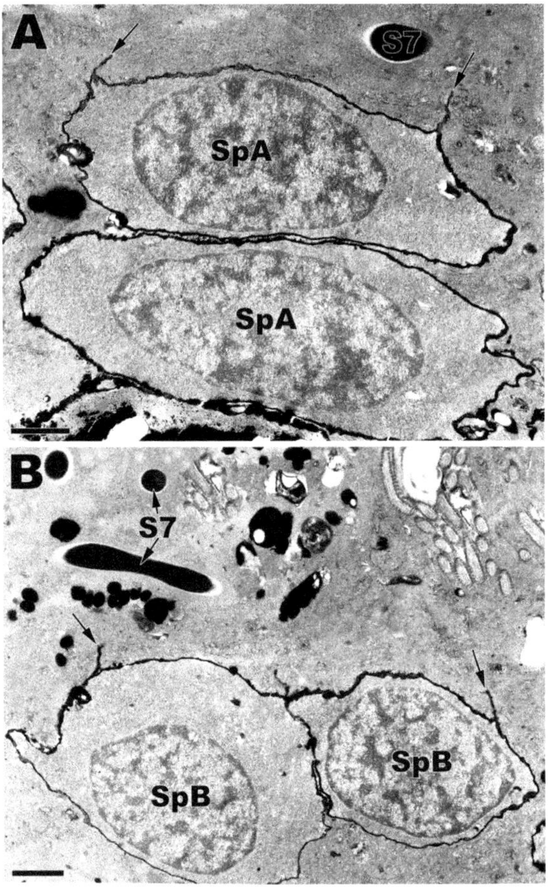

Fig. 11.9 *Podarcis muralis.* Lanthanum nitrate penetrates freely around spermatogonia A (SpA) and B (SpB) but penetration of lanthanum intercellularly is stopped (black arrows) just below step 7 spermatids (S7) in both active (**A**, May) and inactive (**B**, Late July) testes. Bars = 2 μm.

11.2.3 Interstitial Tissue of the Testis

The interstitial tissue of the testis within lizards is composed of two key layers. The first layer is a compact boundary layer (Fig. 11.10), which includes the basal lamina, and is composed of acellular and cellular strata that are roughly 2 to 6 μm in thickness. The other, an inner layer, consists of loose connective tissue, which lies between the boundary layers of adjacent seminiferous tubules and is called the interstitium (Fig. 11.10A). The interstitium is variable in size and is the space where Leydig cells, fibroblasts, blood and lymphatic vessels, and nerves can be found. Typically

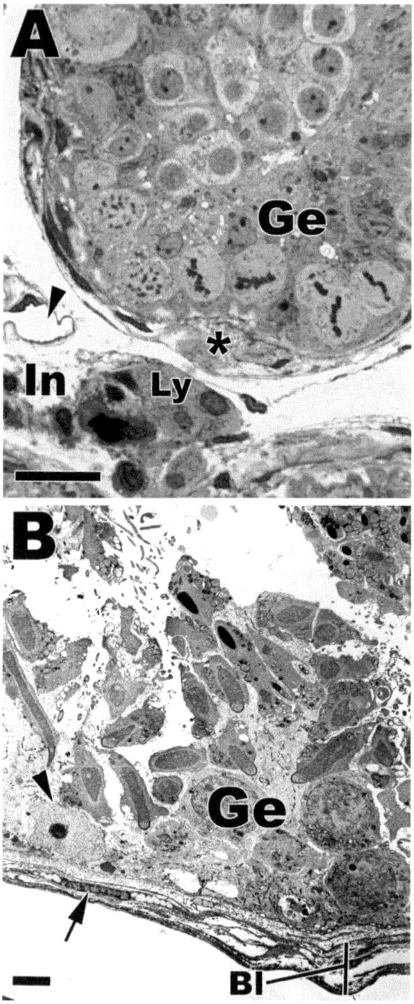

Fig. 11.10 (A) *Podarcis muralis*. Light micrograph showing the boundary layer (*) and interstitial tissue (In) outside of the germinal epithelium (Ge) of a seminiferous tubule within the August testis. Bar = 20 μm. Leydig cells (Ly), blood vessel (black arrowhead). **(B)** Low power electron micrograph depicting the boundary layer (Bl) below the germinal epithelium (Ge) in a late April testis. Bar = 5 μm. Myofibroblasts (black arrow), Sertoli cell nucleus (black arrowhead).

Color image of this figure appears in the color plate section at the end of the book.

during active spermatogenesis in lizards, adjacent seminiferous tubules can be juxtaposed in such a way that the interstitium is hard to discern from the boundary layer.

The boundary or peritubular layer (Figs. 11.10B, 11.11) is the subepithelial layer that is stratified into sequential acellular connective

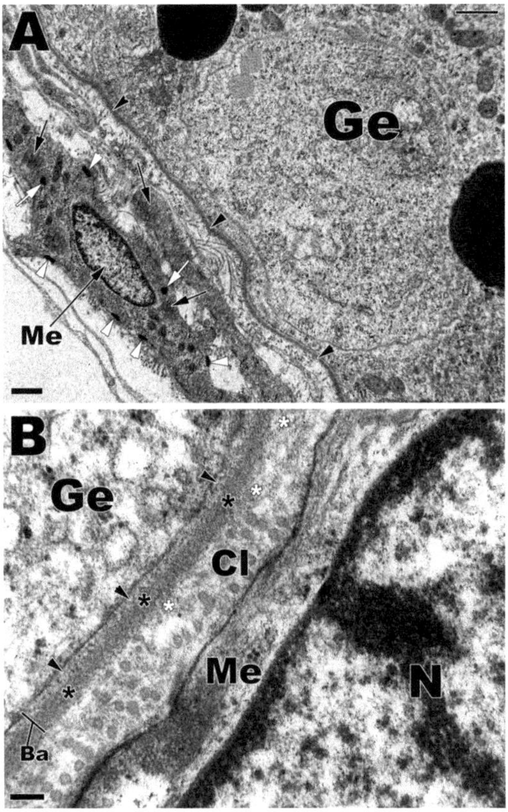

Fig. 11.11 *Hemidactylus turcicus.* **(A)** The boundary layer in located just under the June germinal epithelial (Ge) basal compartment. Bar = 1 μm. The majority of boundary layer cells appear as myofibroblasts (Me). Basal lamina (black arrowheads), thin filaments (black arrows), membrane bound dense bodies (white arrowheads), cytoplasmic dense bodies (white arrows). **(B)** Basal lamina (Ba) and fibroblastic-like cell (Me) details within the October boundary layer of the testis. Bar = 100 nm. The basal lamina is made up of two layers: the lamina lucida (black arrowheads) and the lamina densa (black *). The disorganized lamina reticularis (white *) anchors the basal lamina layers to the 1st tier of the acellular collagenous boundary layer (Cl). Nucleus (N) of the fibroblastic-like boundary layer cell (Me), germinal epithelium (Ge).

tissue and elongated cells and their interconnecting amorphous extracellular matrix. Only one comprehensive study exists in reptiles in which the authors describe the boundary layer in several reptiles including the Madeira Wall Lizard (*Teira dugesi*; Unsicker and Burnstock 1975). There are no present studies that compile data on the testicular interstitia in lizards. Thus, the interstitial and boundary layer data presented here represent new ultrastructural information collected from *Podarcis muralis,* the Mediterranean House Gecko (*Hemidactylus turcicus*), and the Texas Horned Lizard (*Phyronsoma cornutum*) (Gribbins unpublished), which will be compared to the findings observed in *T. dugesi.*

The interstitia among seminiferous tubules within *Podarcis muralis, Hemidactylus turcicus,* and *Phryonsoma cornutum* all show incredible similarity in morphology and ultrastructure. Slightly different from snakes (Gribbins and Rheubert 2011), these seasonally breeding lizards all show differences between fibroblast-like and myoid-like cells during spermatogenically active (Fig. 11.11A) and inactive months (Fig. 11.11B). The fibroblast-like cells (Figs. 11.11B, 11.12A) are more slender than myoid cells

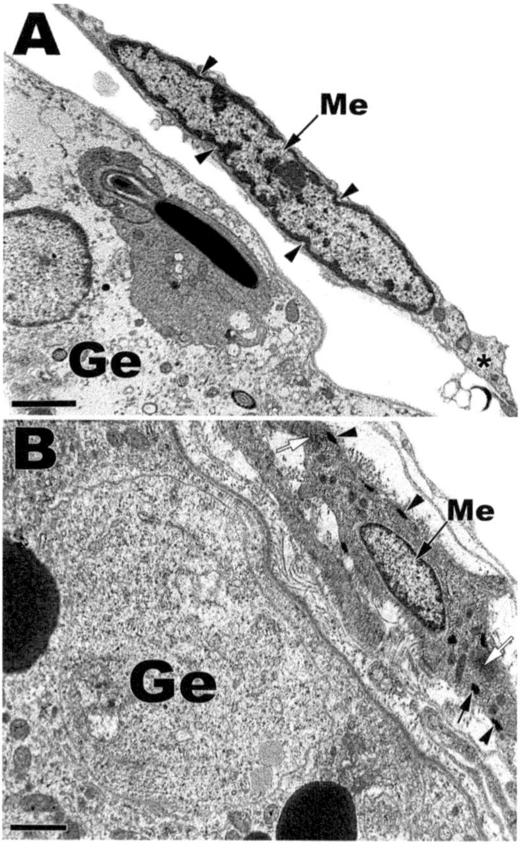

Fig. 11.12 *Phrynosoma cornutum.* **(A)** Large myofibroblast (Me) within the very thin boundary layer of the July testis. Bar = 2 μm. Germinal epithelium (Ge), scalloped nuclear envelop (black arrowheads), light fibroblast-like cytoplasm (*). **(B)** A myofibroblast (Me) near the germinal epithelium (Ge) that expresses characteristics (thin filaments, white arrows; membrane bound dense bodies, black arrowheads; cytoplasmic dense bodies, black arrow) of a typical amniotic myoid testicular cell. Bar = 1 μm. There are normally 1–5 alternating strata within the repeating layers of the boundary region of the lizard testis.

(Figs. 11.11A, 11.12B) and contain very elongated nuclei that typically have a thicker heterochromatin stratum associated with the nuclear envelope. The nuclei of the fibroblast-like cells also have a scalloped nuclear outline (Figs. 11.12A, 11.13B) that is mostly absent in the more round to oval shaped nuclei (Figs. 11.12B, 11.13A) of the myoid-like cells. Although these differences can be seen between the two cell types during the spermatogenic cycle, the cells often appear similar to each other and hard to differentiate in some sections. Within *Teira dugesi* (Unsicker and Burnstock 1975), there were distinct differences between fibroblasts and myoid cells, with myoid cells being more prominent during May and thus the beginning of the breeding season. Cells characteristic of fibroblast were more abundant in early fall outside of the breeding season in *Teira dugesi*. The cells within the boundary layer of the lizards presented here show similar trends to this lacertid lizard. Boundary cells further from the basal lamina during presumed breeding months of *Podarcis muralis* (May; Gribbins *et al.* 2003), *Phrynosoma* (July; Cavazos 1951), and *Hemidactylus turcicus* (June; Rheubert *et al.* 2009a) were almost always myoid-like in ultrastructure. However, because both cell types could be seen within any cross section of testis (for example in *Phrynosoma cornutum*, Fig. 11.12) in the lizards studied below, we simplify the discussion in this chapter by adopting the term myofibroblast for all cell types seen within the boundary layer. Also, the myofibroblasts of lizards studied to date are comparable to ophidian (Gribbins and Rheubert 2011) and avian (see Aire 2007) boundary layer cells.

11.2.3.1 The boundary (peritubular) layer

The boundary layers within *Hemidactylus turcicus* (Fig. 11.11), *Phrynosoma cornutum* (Fig. 11.12), and *Podarcis muralis* (Figs. 11.13, 11.14) contain an inner fibrous layer and outer cellular layers of myofibroblasts. The fibrous layer is located directly under the seminiferous epithelium and varies in thickness and staining efficiency. It consists of a homogeneous and moderately dense basal lamina (roughly 100 nm to 500 nm thick). The basal lamina is composed of two major layers (Fig. 11.11B), the lamina lucida that is in direct contact with the basal Sertoli cell membrane and the lamina densa, which is the thickest layer. The lamina densa is attached to the juxtaposed acellular collagen layer by an irregular lamina reticularis, which is thought to be produced by the underlying myofibroblasts (Bloom and Jansh 2002). The collagen fibers that occupy the alternating bands of acellular regions (Fig. 11.13A) of the boundary layer have multi-directional collagen fibrils (Figs. 11.12B, 11.13B). In rare cases, the collagen fibrils organize into bundles that loosely resemble mature collagen fibers (Fig. 11.13B).

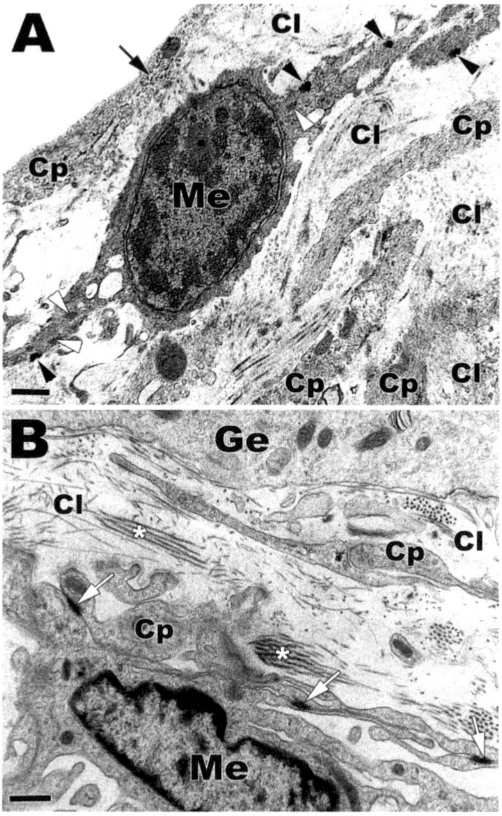

Fig. 11.13 *Podarcis muralis.* **(A)** A myoid-like myofibroblast (Me) typical of the boundary layer within the May testis. Bar = 0.5 μm. These cells express ultrastructural characters within their dark staining cytoplasm that are similar to those of smooth muscle. Thin filaments (white arrowheads), acellular collagen layers (Cl), myofibroblast cytoplasmic processes (Cp), glycogen granules (black arrow), cytoplasmic dense bodies (black arrowheads). **(B)** A fibroblast-like myofibroblast (Me) of the August testis. Bar = 0.5 μm. Their lighter cytoplasm lacks the thin filaments and dense bodies common in boundary layer cells of spermatogenically active months. Desmosomes (white arrows) are common among the distal cellular processes (Cp) of the myofibroblasts. In rare cases, extracellular collagen fibrils form a bundle similar to a mature collagen fiber (white *) within the acellular layers (Cl). Germinal epithelium (Ge).

The outer mesenchymal myofibroblast layer fluctuates in thickness according to the number of concentric, alternating, and overlapping cellular layers. The number of layers observed in the boundary tissue of the three lizard species discussed here are similar to that of the other squamates studied to date, which include *Agkistrodon piscivorus*, the Black Swamp Snake (*Seminatrix pygaea*; Gribbins and Rheubert 2011), the Grass Snake (*Natrix natrix*) and *Teira dugesi* (Unsicker and Burnstock 1975). These cells

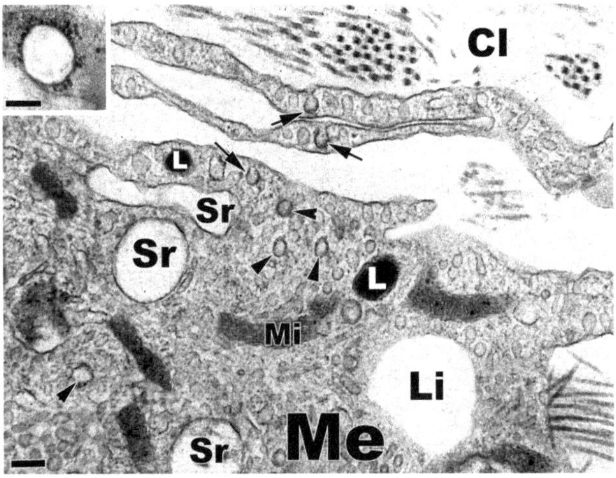

Fig. 11.14 *Podarcis muralis*. Myofibroblast (Me) cytoplasm in August. Several organelles are observed; including dilated smooth ER (Sr), dilated rough ER (inset, Bar = 0.1 μm), lysosomes (L), lipids (Li), and mitochondria (M). There are also many coated pinocytotic vesicles (black arrowheads) arising from pinocytosis (black arrows) that is occurring all over the cell surface of these myofibroblasts. Bar = 0.2 μm. Acellular collagen layer (Cl).

number 1–5 layers thick and contain either elongated nuclei like those of fibroblasts (Figs. 11.12A, 11.13B) or round, robust nuclei like those of myoid cells (Figs. 11.12B, 11.13A) that display uniformly granular chromatin with thin heterochromatin associated with the nuclear envelope. One or two prominently staining nucleoli are visible in most sagittal sections of these nuclei. In some cases, Leydig cells can be found within the boundary layer of spermatogenically active lizard testes.

The myofibroblast cytoplasm in both cell types continues past the thick cell body as very slender processes that wrap around the seminiferous tubules. These processes within the different cellular layers often overlap each other and form desmosomes (Fig. 11.13B), which are responsible for holding these boundary cells together as an intact layer. The myofibroblasts often take on a dark appearance around the nuclei (Fig. 11.13A) in spermatogenically active months in all three lizard species discussed in this chapter. Dark staining filaments, dense bodies, and glycogen granules are what appear to cause the dark staining to the cytoplasm in this myoid-like cell (Figs. 11.12B, 11.13A). These filaments are the same size and show similar characteristics to actin described in other squamate boundary myoid cells (Unsicker and Burnstock 1975; Gribbins and Rheubert 2011). The thin filaments and dense bodies are most abundant, and the myofibroblasts stain the darkest during the end of spermatogenesis and spermiation in these seasonally breeding lizards. The dense bodies (Fig. 11.13A) are most often associated with the cytoplasmic leaflet of the cell membrane. These dense bodies and the presence of actin filaments suggest that these myofibroblasts have the ability to contract because of their similarity to smooth muscle (Unsicker and Burnstock 1975). This hypothesized function has also been proposed for birds (Aire 2007) and other reptiles (Unsicker and Burnstock 1975; Gribbins and Rheubert 2011). The thin filaments are most numerous during the climax of spermatogenesis, when spermatozoa are released into the lumina of the seminiferous tubules. Thus, we suggest that these myofibroblasts contract and help drive sperm through the seminiferous tubules to the anterior ducts at the end of the spermatogenic cycle in lizards. Also, boundary layer cells most likely play a role in production of the extracellular matrix and collagen of this layer.

Lizard myofibroblasts also have numerous elongated mitochondria, some lipid inclusions, a few lysosomes, and a large number of coated and uncoated pinocytotic vesicles that are seen forming at the cell membrane and within the cytoplasm (Fig. 11.14). There are also grossly distended cisternae of both smooth (Fig. 11.14) and rough endoplasmic reticula (Fig. 11.14 inset). The lizard myofibroblast organelle and cytoplasm composition, like that of snakes (Gribbins and Rheubert 2011), is comparable to that described in birds (Aire 2007) and mammals (Kurohmaru *et al.* 1990) and seems to be a combination of the organelles described for the myoid cells and fibroblasts of *Natrix natrix* and *Teira dugesi* (Unsicker and Burnstock 1975). The exact function of the specific (coated) and nonspecific pinocytotic ability of these cells is still unknown (Maekawa *et al.* 1996). The amount of pinocytosis at least in the seasonal reproducing lizards focused on here increases in spermatogenically quiescent months. Further studies on the morphology and physiology of these complex boundary cells need to be performed to further elucidate their role in testicular function.

11.2.3.2 The interstitium and Leydig cells

The interstitium in lizards is normally organized as a loose connective tissue (Gribbins unpublished; Fig. 11.15A). It is most prominent in areas where three seminiferous tubules meet forming a wedge of interstitial tissue. The middle portion of the wedge is where most of the interstitium is located (Figs. 11.15A, 11.16A). Between two seminiferous tubules there is typically little to no interstitial space particularly during spermatogentically active months (Fig. 11.16A). In this case, Leydig cells may be found within the boundary layer (Gribbins unpublished; Gribbins and Rheubert 2011). Some lizards (Lowe and Goldberg 1966; Granados-González *et al.* 2013; Gribbins unpublished, for *Phryonsoma cornutum*) also have a subtunic layer of Leydig cells (Fig. 11.17) that can be quite large in size and undergoes seasonal changes roughly paralleling variations observed within the seminiferous tubules.

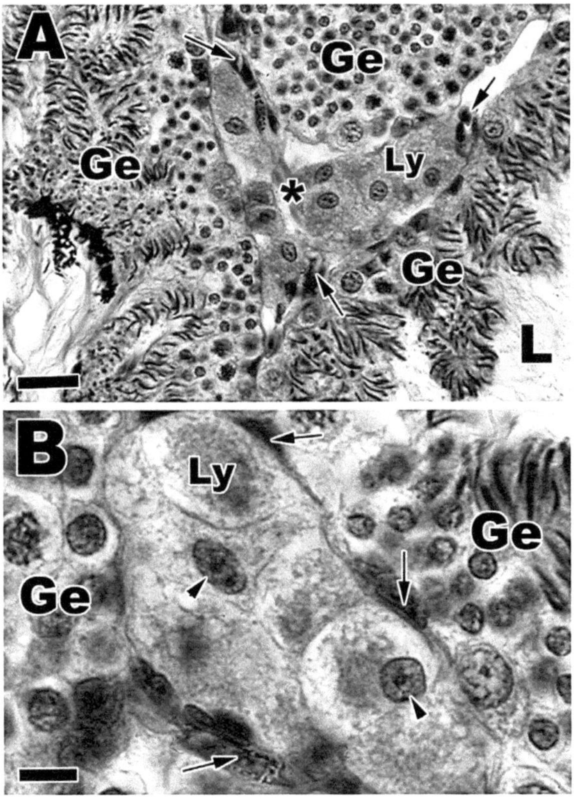

Fig. 11.15 *Sceloporus mucronatus.* **(A)** Shows where three seminiferous tubules meet to form a wedge of interstitium (*) in October within the testis. Leydig cells (Ly) and blood vessels (black arrows) occupy most of the interstitial space. Bar = 25 μm. Germinal epithelium (Ge), lumen of seminiferous tubules (L). **(B)** Higher power of Leydig cells (Ly) within the October interstitium. Bar = 10 μm. Leydig cell nuclei (black arrowheads), germinal epithelium (Ge), blood vessels (black arrows). The cytoplasm at this time has a flocculent look and has evenly distributed lipid droplets.

The components present within the lizard interstitium are a loosely aggregated connective tissue with a number of blood vessels, lymphatic vessels (rare), a few macrophages, and either single or loosely associated Leydig cells (Fig. 11.15). The Leydig cells are typically located close to the blood vessels within the interstitium (Fig. 11.18). Leydig cell cytology presented here will be from the lizard species *Podarcis muralis* (Gribbins unpublished), the Cleft Lizard (*Sceloporus mucronatus*; modified from Villagrán *et al.* 2013), and the Western Mexico Whiptail (*Aspidoscelis costata*; Modified from Granados-González *et al.* 2013).

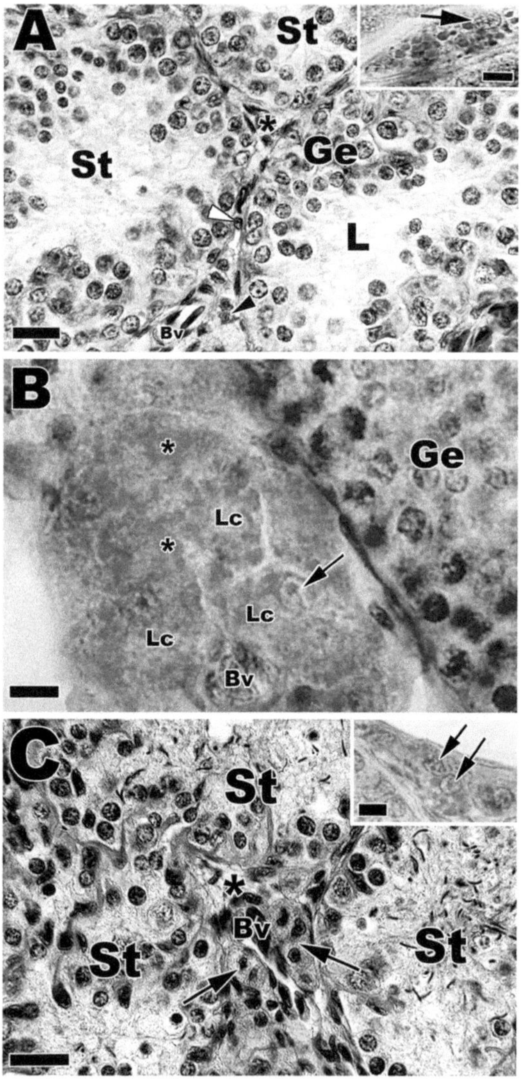

Fig. 11.16 *Sceloporus mucronatus.* The interstitium (*) between seminiferous tubules (St) within the February **(A)** and September **(B)** testes showing Leydig cells (Lc) that are located both within the boundary layer (white arrowhead) and the interstitial space (black arrowhead). The inset in A and C show sudanophilic lipid inclusions that are moderate in February versus hypertrophied lipid droplets (* in B) within Leydig cells (Lc) of September lizards. Blood vessels (Bv), germinal epithelium (Ge), lumen (L), Leydig cell nucleus (black arrow). **(C)** Leydig cells (black arrows) located between November seminiferous tubules (St). Inset reveals smaller sudanophilic lipid droplets around Leydig cell nuclei (black arrows) than during recrudescence in February. Blood vessel (Bv), Interstitium (*). Bar in A/C = 25 μm; Insets A/C = 10 μm and Bar in B = 10 μm.

Color image of this figure appears in the color plate section at the end of the book.

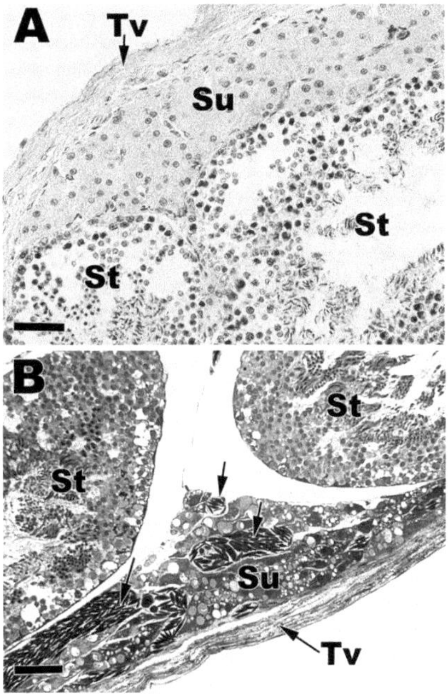

Fig. 11.17 Light micrographs of the subtunic layer (Su) of Leydig cells in *Aspidoscelis costata* (A) and *Phrynosoma cornutum* (B). These are very large layers of Leydig cells just under the tunica vaginialis (Tv) that have ample vascularization (black arrows). Bar = 50 μm. Seminiferous tubules (St).

Color image of this figure appears in the color plate section at the end of the book.

Within these three lizards, Leydig cells demonstrate a cycle (Fig. 11.16) of lipid accumulation and use (Gribbins unpublished). The Leydig cell lipid cycle in lizards is well represented at the light microscope level (for a basic review, see Engel and Callard 2007) and tracks that reported for snakes and other squamates (Lofts 1968; Guraya 1973; Unsicker and Burnstock 1975; Callard *et al.* 1978; Leceta *et al.* 1982; Kan and Rai 2004; Boretto *et al.* 2010). Although asynchrony can be found between Leydig cell morphology and Sertoli cell activity in some lizards (i.e., the Mountain Lizard, *Phymaturus antofagastensis*, from the Andes, Boretto *et al.* 2010), most seasonal lizards examined to date display a building trend of increasing activity (hypertrophy) in Leydig cells (Fig. 11.16A,B) during the progression of spermatogenesis. As spermatogenesis climaxes and spermiation occurs there is usually a rapid loss of lipid content and hypotrophy (Fig. 11.16C; inset) in the Leydig cells of lizards. Often seasonal lizards are prenuptial (Gribbins 2011) in sperm production (Gribbins and Gist 2003; Gribbins 2011; Gribbins and Rheubert 2011) and secondary sex characteristics often develop during spermiation along with rapid lipid loss in Leydig cells (Leceta *et al.*

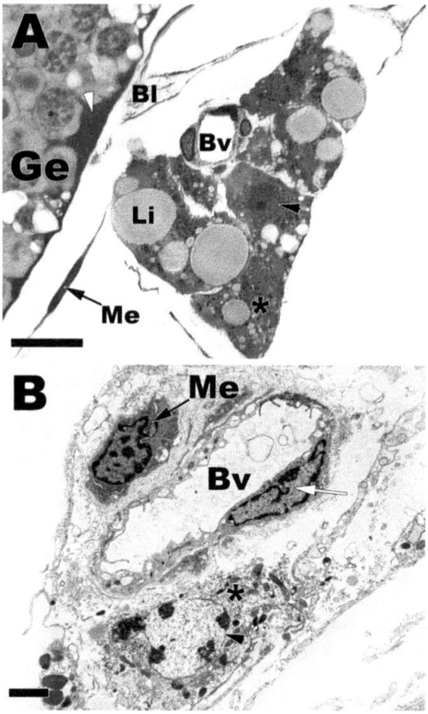

Fig. 11.18 *Podarcis muralis.* Leydig cells (*) near blood vessels (Bv) in April **(A)** and August **(B)**. Note the numerous lipid droplets (Li) found in the April Leydig cells, which are lacking in August (B, *). Bar in A = 25 μm and in B = 5 μm. Germinal epithelium (Ge), Sertoli cell nucleus (white arrowhead), boundary layer (Bl), myofibroblasts (Me), Leydig cell nucleus (black arrowheads), endothelial cell (white arrow).

Color image of this figure appears in the color plate section at the end of the book.

1982; Aranha *et al.* 2006; Villgrán-Santa Cruz *et al.* 2013). Thus, we agree with recent studies that suggest this loss of lipid content and the ultrastructural changes to organelles of the Leydig cells of the lizards and development of coinciding breeding events and secondary sex characteristics in males parallel high androgen production in their Leydig cells. This trend also holds true for many snakes studied to date (Gribbins and Rheubert 2011).

Seasonal lizard Leydig cells in most of the investigated species accumulate lipids from recrudescence through late spermiogenesis (no conclusive data on Leydig cell morphology and ultrastructure during the year exist for continual breeding reptiles). Leydig cells early in spermatogenesis (August) have small evenly distributed lipid droplets (Fig. 11.19A) of varying size in *Podarcis muralis* (Gribbins unpublished), which is similar to what is seen during recrudescence (February) in *Sceloporus mucronatus* (Fig. 11.19A, inset) (Villagrán-SantaCruz 2013). These cells also

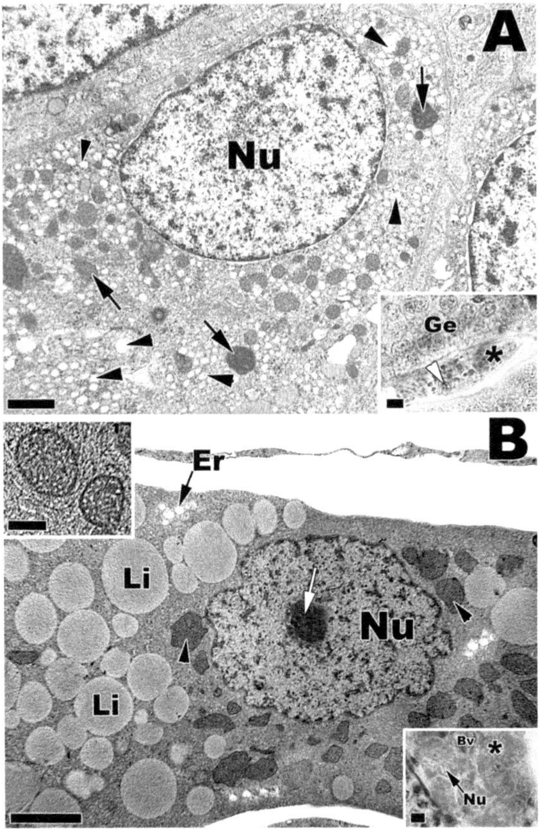

Fig. 11.19 (A) Ultrastructure of a Leydig cell in the August testis of *Podarcis muralis*. Lipid inclusions (black arrowheads) are small, numerous, variable in size, and evenly distributed. Bar = 2 μm. Mitochondria with small tubular cristae (black arrows) Inset: Shows a similar trend in early recrudescence in February in *Sceloporus mucronatus* with small sudanophilic lipid droplets (*) and heterochromatic nucleus (white arrowhead). Bar = 10 μm. Germinal epithelium (Ge). **(B)** Ultrastructure of a Leydig cell near the climax of spermatogenesis (July) in *Phrynosoma cornutum*. Bar = 1 μm Lipid droplets (Li) are large as are the mitochondria (black arrowheads), which have dilated tubular cristae (upper left corner inset, Bar = 0.2 μm). Dialated smooth ER (Er), Nuceolus (white arrow). Lower right corner inset: November Leydig cells in *Sceloporus mucronatus* have large sudanophilic lipid droplets (*) with large nuclei (Nu). Bar = 2 μm. Blood vessel (Bv).

Color image of this figure appears in the color plate section at the end of the book.

have inconspicuous smooth ER and mitochondria with reduced tubular cristae. In contrast, during late spermatogenesis in *Phrynosoma cornutum* (Fig. 11.19B) and *S. mucronatus* (Fig. 11.19B, inset-light microscopy), the Leydig cells have large lipid inclusions, large nucleoli, and appear more granular with enlarged mitochondria with dilated tubular cristae (Fig. 11.19B and inset-TEM). This indicates that these cells are using the lipids as spermatogenesis progresses and climaxes, presumably for steroid synthesis, which is required for completion of spermiogenesis, spermiation, and breeding behavior in reptiles (Engel and Callard 2007). Leydig cells in the lipid accumulation stage also have very large prominent smooth endoplasmic reticula (Fig. 11.19B).

11.3 GERM CELL DEVELOPMENT STRATEGY

Within the last ten years a significant amount of data have been reported that concentrate on the light and electron microscopy of the development, sequence, morphology, germ cell cycle, and ontogenic strategies of spermatogenesis within the class Reptilia (i.e., see Gribbins 2011). Spermatogenesis is required of all vertebrates whose main goal through this process is to package genetic material into a motile cell for use during sexual reproduction. The beginning of this chapter focused on the testicular architecture within reptiles. Among vertebrates there is a distinct shift in morphology of the testis (Fig. 11.20) between anamniotes (fish and amphibians) and amniotes (reptiles, birds, and mammals). Although they

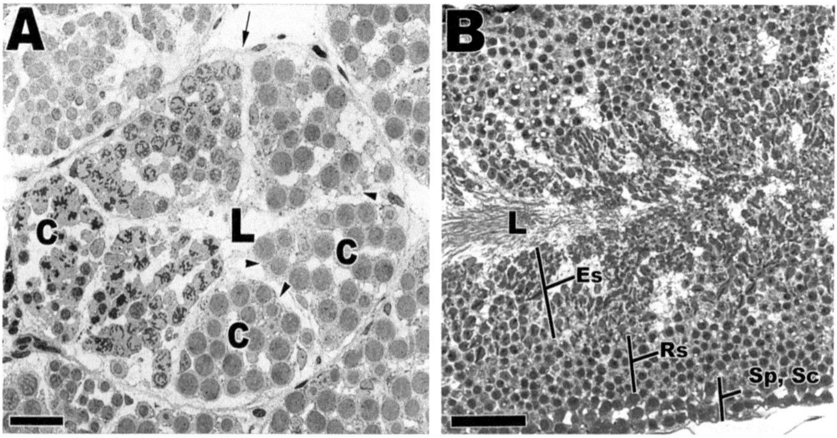

Fig. 11.20 (A) A seminiferous tubule (basal lamina, black arrow) in *Notophthalamus viridescens* (anamniotic cystic testis) showing several early cysts (C) with germ cells in the same phase of development within each cyst. The cyst walls are made up of Sertoli cells (black arrowheads) and when mature they will open to the lumen (L) of the tubule. Bar = 50 μm. **(B)** Partial sagittal section through a seminiferous tubule in *Sceloporus bicanthalis* (amniotic tubular testis). The germ cells develop in successive generations within an epithelium made up of Sertoli cells. The youngest germ cells, spermatogonia (Sp) and spermatocytes (Sc) are positioned on the periphery of the tubule near the basement membrane. The older cell types are found closer to the lumen, i.e., round spermatids (Rs) and elongating spermatids (Es). Once maturation is complete they are shed to the lumen of each tubule (L). Bar = 100 μm.

are paraphyletic, reptiles, which include lizards, and their ancestors (basal amniotes), hold an important phylogenetic position between amphibians, birds, and mammals; little is still known about the progression of germ cells through the phases of spermatogenesis including the cytological events and germ cell development strategies employed during spermatogenesis in these nonmammalian amniotes. The implications of which are most likely linked to the evolutionary adaptations of the amniotic egg and complete reliance of amniotes on internal fertilization, which freed them completely from water for breeding (Pudney 1993; Gribbins and Gist 2003). The differences cytologically and ultrastructurally within the spermatogenic cycle and the spermatozoa between reptilian taxa and other amniotic clades may in time provide influential nontraditional evidence that could be important in understanding the phylogenetics within this basal taxon of amniotes, which at best is considered highly controversial at this time. Furthermore, these data also offer a fundamental model that could be utilized for histopathological studies of the testis in amniotes. In order to understand how toxins or heavy metals disrupt sperm development, a thorough understanding of the histology of spermatogenesis has to be in place (Russell *et al.* 1990). Reptiles provide the perfect model for these types of studies because of their temporal development of germ cells (explained later in this chapter) as they progress through spermatogenesis, which provides a basic and easy to follow series of germ cell morphologies that can be readily defined during the major phases of spermatogenesis. This should make it easier to discern abnormalities among different germ cell generations within the testis and the absence or interruptions of particular stages of sperm development. Also, many reptiles are semi aquatic and relatively abundant and thus are sentinel species for the studies on aquatic pesticides and their toxicological impacts on germ cell development. Reptiles could impact these studies even further as they have a testis more similar to the human testis than is the morphologically very different anamniotic testis, such as that of the Red Spotted Newt (*Notophthalamus viridescens*) (Fig. 11.20A).

Much of the material presently found within the literature for reptilian spermatogenesis still focuses on reproductive cycles, environmental, breeding, and hormonal cues, and these studies only describe some of the germ cell types (see for example Chamut 2012) or organize spermatogenesis in a historical stage scheme similar to that of Marion (1982). Although these studies provide very important data on reproduction as a whole, they do not present much insight into the process of spermatogenesis and thus a way to compare sperm development histologically to that of other amniotic and anamniotic taxa. Germ cells are in layers within the reptilian germinal epithelium suggesting that spatial stages like those seen in mammals and birds may exist. However, when observing germ cells within the testis of most temperate or continuous breeding species of lizards, one can observe up to five layers of spermatids within the seminiferous epithelium (Figs. 11.20, 11.21; Gribbins *et al.* 2008).

Germ cells entering the spermatogenic cycle in continually or seasonally breeding mammals and birds progress through a number of consistent cytological changes during their development (see Roosen-Runge 1972; Russell *et al.* 1990). As new generations of spermatogonia proliferate they shift older generations of germ cells centrally toward the lumen. Thus, the seminiferous epithelium at any point in time contains three to five generations of germ cells that are reliably located together within the germinal epithelium. These consistent spatial relationships between multiple generations of germ cells are termed stages and a complete series of stages has been termed the spermatogenic cycle (Russell *et al.* 1990). This cycle results in multiple waves of spermiation during spermatogenesis.

Although squamates (including lizards) have achieved the amniotic organization of the germinal epithelium, they can tend to have up to eight or nine generations of germ cells at any one time within a cross section of seminiferous tubule (Fig. 11.21). This is many more representative generations of germ cells than their endothermic counterparts. Thus, it is very important to look more closely at the germ cell development strategy within squamates and other reptiles to compare their spermatogenic

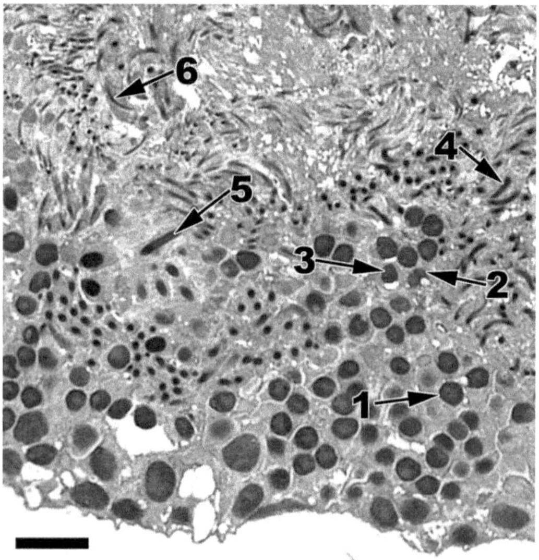

Fig. 11.21 *Sceloporus bicanthalis*. A light micrograph of the germinal epithelium. Bar = 25 μm. There are six different generations of spermatids (1–6). Typically in a spatial type of germ cell development found in mammals and birds, no more than two to three generations of spermatids are found within the epithelium at any one time during development.

cycle to that of mammals and birds. Spermatogenesis in most temperate lizards is highly seasonal. Historically, two types of seasonal cycles exist in seasonal lizards; a few produce sperm after mating (postnuptial) and the majority of saurians produce sperm before mating (prenuptial) (Fig. 11.22; Modified from Gribbins and Rheubert 2011; Licht 1984). How this seasonality influences germ cell development and subsequent maturation of sperm is not well understood in lizards or other reptiles. To date only a few lizard species (*Podarcis muralis*, Gribbins and Gist 2003; *Anolis lineatopus*-Jamaican Anole, Gribbins *et al.* 2009; *Scincella lateralis*-Ground Skink/ *Hemidactylus turcicus*, Rheubert *et al.* 2009a,b; *Sceloporus bicanthalis*, Gribbins *et al.* 2011; multiple species, Gribbins 2011; *Sceloporus consobrinus*-Praire Lizard, Rheubert *et al.* 2013; *Sceloporus mucronatus*, Villagrán-SantaCruz *et al.* 2013; *Aspidoscelis costata*, Granados-González *et al.* 2013) have been studied histologically to determine their germ cell development strategies

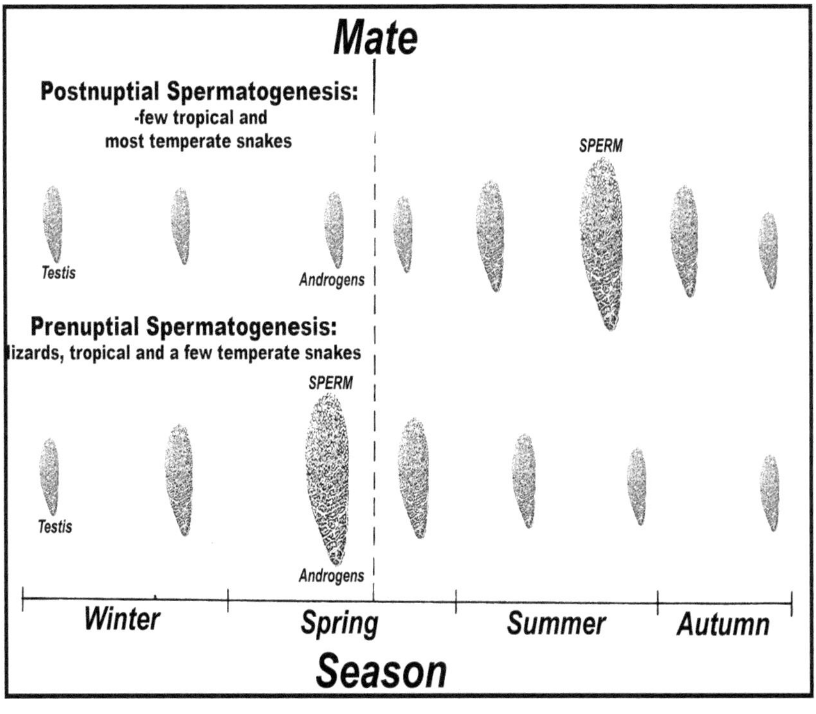

Fig. 11.22 A visual representation of the two major seasonal gonadal activities in temperate reptiles relative to spermatogenesis and mating. Sperm production is completed before or after mating. Hypothesized androgen and sperm presence are also represented for each type of seasonal spermatogenic cycle. Within both types, variations may occur in the rate of recrudescence, spermatogenesis, testis growth, and duration of the breeding season (taken from Gribbins and Rheubert 2011).

as compared to mammals and birds. Thus, not enough species have been studied to confirm commonality to the cytology and the germ cell strategy of spermatogenesis in prenuptial versus postnuptial reproductive cycles within lizard and other reptiles.

Understanding how germ cells are arranged within the reptilian seminiferous epithelium might also afford some insight on lizard testis physiology during seasonal cycles. For instance, in mammals it has been well established that FSH stimulates type A spermatogonia recrudescence during spermatogenesis (Waits and Setchell 1990). FSH has been discovered in reptiles (Licht 1979; Licht *et al.* 1979) and all current data indicates that it may jump start spermatogenesis in temperate reptiles either before or after mating (Licht *et al.* 1989; Masson and Guillette Jr. 2005). However, it is not known whether type A spermatogonia with the ability to undergo mitosis or recognize FSH are located within the germinal epithelium at all times of the year. Temporary or absent populations of dividing or endocrinologically competent spermatogonia during the year may be responsible for refractory periods seen in some reptiles for example (see Licht 1984), and thus these gonial cells are only available for FSH activation within a small window during yearly spermatogenesis in lizards and snakes (Gribbins and Rheubert 2011). Therefore, a basic comprehension of the cytological progression involved with spermatogenesis in lizards may help resolve the existing controversies with regard to refractory periods, temperature effects, and hormonal stimulation within the testicular cycles of reptiles.

Although current data on germ cell development strategies in lizards are limited, these studies do suggest that all reptiles studied, irrespective of the type of reproductive cycle employed (pre- or postnuptial, biannual, or continuous), maintain the same temporal germ cell development strategy (Gribbins and Gist 2003; Gribbins *et al.* 2003, 2005, 2006, 2008, 2009; Rheubert *et al.* 2009, 2013). Again it is obvious that more squamate species need to be studied, especially lizards, as they show much diversity in their reproductive biology (Licht 1984).

In order to determine whether consistent cellular associations are seen between germ cell generations within the lizard germinal epithelium one must be able to recognize the different morphologies of germ cells as they progress through the phases of mitosis (spermatogonia), meiosis (spermatocytes), and spermiogenesis (spermatids) (acrosome formation, elongation and chromatin condensation) (Volsøe 1944; Hess 1990; Russell *et al.* 1990; Gribbins 2011). Three species of lizards will be compared here in order to discuss the temporal germ cell development employed by all lizards and other reptiles studied to date (see Gribbins 2011; Gribbins and Rheubert 2011). The representative lizards are one from Phrynosomatidae: *Sceloporus bicanthalis* (continuous spermatogenesis; Gribbins *et al.* 2008), one from Lacertidae: *Podarcis muralis* (mixed/prenuptial spermatogenesis; Gribbins and Gist 2003), and one from Gekkonidae: *Hemidactylus turcicus* (mixed/prenuptial spermatogenesis; Rheubert *et al.* 2009b).

11.3.1 Germ Cell Cycle

The lizard germinal epithelium has two morphologies of pre-meiotic cells, spermatogonia A and B (Fig. 11.23B,C) during all months of the year in the lizards studied to date. These cells are characterized by nuclei with random clumps of heterochromatin. The major morphological differences between these spermatogonia are the ovoid shape of the A type compared to the round profile of the B type, which also differs in lacking a prominent nucleolus. Also, there may be a difference in the amount of heterochromatin associated with the nuclear membrane between these two cells but that is more inconsistent than the basic shape of the nucleus in the basal compartment of the seminiferous epithelium. Both spermatogonial types are generally located near the basement membrane within the basal compartment formed by Sertoli cells. During spermatogenesis, both types

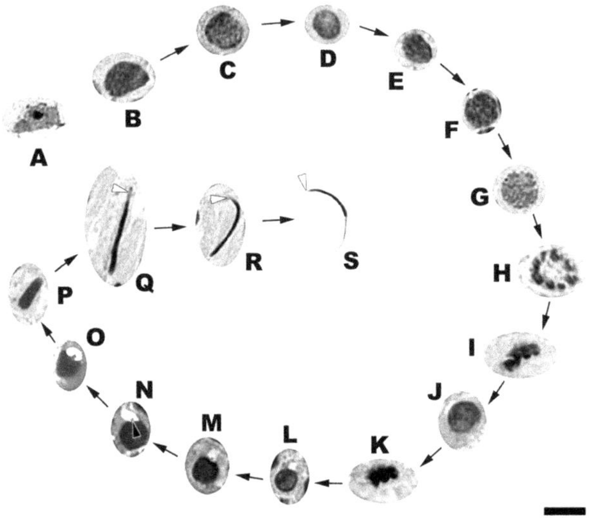

Fig. 11.23 *Sceloporus bicanthalis.* Cell types found within the seminiferous epithelia. Scale bar = 10 μm. Sertoli cell nucleus **(A)**, Type A spermatogonia **(B)**, type B spermatogonia **(C)**, pre-leptotene spermatocytes **(D)**, leptotene spermatocytes **(E)**, zygotene spermatocytes **(F)**, pachytene spermatocytes **(G)**, diplotene spermatocytes **(H)**, meiosis I **(I)**, secondary spermatocytes **(J)**, meiosis II **(K)**, step 1 spermatid **(L)**, step 2 spermatid **(M)**, step 3 spermatid **(N)**, (black arrow: acrosome granule), step 4 spermatid **(O)**, step 5 spermatid **(P)**, step 6 spermatid **(Q)**, step 7 spermatid **(R)**, mature sperm **(S)**. White arrowheads in Q,R,S, apical nuclear extension with acrosome.

of spermatogonia undergo mitosis and maintain a robust spermatogonial population that will then enter the meiotic phase of germ cell development as pre-leptotene spermatocytes (Fig. 11.23D). These spermatogonia are most active mitotically immediately before the onset of meiosis and spermiogenesis in most squamates studied to date (Gribbins and Rheubert 2011). The majority of spermatogonia B undergo mitotic divisions at different points of the reproductive cycle within lizards to produce preleptotene cells. Preleptotene spermatocytes (Fig. 11.23D) have nuclei with prominent nucleoli and fine granular chromatin. Preleptotene cells are typically the smallest germ cells (including nucleus and cytoplasm) within the basal compartment of the germinal epithelium of lizards and other squamates. Their small size easily distinguishes them from the larger spermatogonia A and B within the seminiferous epithelium.

Leptotene (Fig. 11.23E) and zygotene (Fig. 11.23F) spermatocytes are slightly larger and stain more intensely than pre-leptotene cells and their chromatin fibers pack their nuclei. The major cytological distinction between the two is that zygotene cells have more condensed globular chromatin compared to the fine filamentous chromatin found in leptotene cells. They appear together, along with pachytene spermatocytes, and typically make up a large percentage of the primary spermatocyte population when they are observed within the saurian germinal epithelium.

Pachytene cells (Fig. 11.23G) will remain in the seminiferous epithelium of most lizards throughout active spermatogenesis and are the most common spermatocyte seen within the reptilian testis. These germ cells experience an extensive size increase over their ontogeny and their nuclei include very thick chromatin fibers that are scattered among zones of open nucleoplasm. Diakinesis (Fig. 11.23H), metaphase I and II (Fig. 11.23I,K), and secondary spermatocytes (Fig. 11.23J) are transitional cells that are typically located together within the lizard seminiferous epithelium.

Diakinesis cells are characterized by thickly packed chromosomal fibers that are interspersed with large open areas of nucleoplasm. The nuclear membrane is indistinguishable and their spoke-like chromosomal fibers are arranged in a circular pattern. Metaphase I and II cells contain clumps of chromosomes that are positioned along the metaphaseal plate with no apparent nuclear boundaries. Metaphase II cells are smaller in size than metaphase I cells and visually have roughly half the amount of chromatin when compared to metaphase I cells. Secondary spermatocytes usually are bundled with metaphase I and II cells within the seminiferous epithelium. Secondary spermatocytes have centrally located nuclei that are usually 15% bigger than subsequent step 1 spermatids.

Spermiogenesis is divided into steps based on the terminology introduced by Russell *et al.* (1990) for mammalian species and includes the development of the acrosome complex, elongation, and chromatin condensation of the nucleus. Spermiogenesis is the longest phase of germ cell development and there are typically seven or eight recognizable steps of spermiogenesis within the recently studied lizards. The round spermatids (Fig. 11.23L-O) are the cohort of haploid cells that start to develop the differentiated morphology of the mature spermatozoa. The round spermatids undergo the development of the acrosome complex, which includes the production of the acrosome vesicle and granule (Fig. 11.23N) that rest on the apex of the nuclear head of developing spermatids.

Step 1 (Fig. 11.23L) and 2 (Fig. 11.23M) spermatids in the saurian testis mark the beginning of spermiogenesis. Their small size and lightly stained Golgi complexes and acrosomes juxtaposed to the nuclei characterize these spermatids. Their spherical nuclei are centrally located and contain one or more chromatin bodies and light diffuse chromatin. The rest of the ontogeny of the round spermatids involves further contact and expansion of the acrosome vesicle, which routinely forms an indentation on the apex of the nucleus (Fig. 11.23O).

Once the acrosome complex has terminated development, round spermatids begin a transitional step where the nuclear body begins to elongate (Fig. 11.23P). This elongation terminates when the rod shaped nucleus reaches a length of 30 μm or more (Fig. 11.23Q). During early elongation, a flagellum is often seen extending from the spermatids out into the lumen of the seminiferous tubule. The acrosome, in many instances, is still observable and is associated with a thin extension of the apical nucleus (Fig. 11.23Q,R), which to date is a unique feature found in some reptiles at least at the light microscopic level (Gribbins and Rheubert 2011). Once elongation has finished, chromatin material has completely condensed and cytoplasmic material has been eliminated from the developing elongate to form a more hydrodynamic cell with a flagellum that allows movement through the female reproductive tract. During the last step of spermiogenesis in lizards and other squamates, it is common for nuclei to become slightly curved (Fig. 11.23R), which provides the reptilian spermatozoa with its filiform shape. These elongating spermatids are often found in bundles within large columns of seminiferous epithelium and develop together as cohorts. Once spermiogenesis is complete, mature spermatozoa (Fig. 11.23S) are spermiated from the seminiferous epithelium into the lumina of the seminiferous tubules within the lizard testis.

11.3.2 The Mode of Germ Cell Development

After all germ cell morphologies are determined for the phases of spermatogenesis, an examination can be performed to delineate whether germ cells are consistently layered together in the lizard seminiferous epithelium (spatial development) or if the germ cells develop as a single population (temporal development). In *Podarcis muralis* (Gribbins and Gist 2003), before every month of the year was sampled for germ cell development. Spermatogenesis begins in late July/early August right after most mating events are completed, where spermatogonia and early primary spermatocytes (Fig. 11.24A) dominate the germ cell population. No spermatids are seen at this time of the year in this Wall Lizard from Southwest Ohio. Spermatogenesis continues through spermiogenesis (Fig. 11.24B) into late fall and then is suspended over the winter months during hibernation. In this regard, the testis acts as a reservoir for developing sperm over winter. Samples from the early months of the year thus show similar cell types to those of late winter (Fig. 11.24B,C). As temperatures increase into spring and summer, spermatogenesis picks up again and spermiation events start in April and gradually increase through June (Fig. 11.25). This is the time of year when mating behavior is seen in this population of Wall Lizards (Gribbins unpublished). By late July, the seminiferous epithelium is spent and most spermatozoa have moved into the excurrent ducts, thus leaving spermatogonia A and B cells as the dominating germ cell in the seminiferous epithelium at this time (Fig. 11.25C).

In the *Podcaris muralis* example, the majority of germ cells are moving as a major cohort through proliferation (July/August), meiosis (Nov.–March), and spermiogenesis/spermiation (Nov.–June). This temporal development is very similar to the development seen in amphibians and other reptile taxa studied to date and varies greatly from what has been described for all avian and mammalian species, which practice spatial germ cell development irrespective of their seasonal cycle. This type of spermatogenesis within the testis of *P. muralis* is also very comparable to the mixed spermatogenesis and germ cell strategy seen in the male *Hemidactylus turcicus* (Rheubert *et al.* 2009a) from Louisiana and the continuous germ cell development of *Sceloporus bicanthalis* (Gribbins *et al.* 2011) from Mexico.

Specifically in *Hemidactylus turcicus,* the testis is quiescent in early fall (September) (Fig. 11.26A) in Louisiana populations. Subsequently, spermatogenic recrudescence begins in October and November (Rheubert *et al.* 2009a) with division of spermatogonia A and B and the presence of early spermatocytes (Fig. 11.26B). Meiosis 1 predominates through the colder months of Dec. and Jan. and then in late Jan. meiosis 2 and early round spermatids (Fig. 11.26C) are the major cell type of the germinal epithelium. By March, spermiogenesis is well underway with small pockets of spermatozoa observed within the lumina of the gecko seminiferous tubules (Fig. 11.26D). The majority of the germ cell population then moves through the late steps of spermiogenesis (Fig. 11.26E,F) from May–August where the heaviest spermiation take place. These months are when this population is thought to have greatest breeding activity (Eckstut *et al.* 2009). *Podarcis*

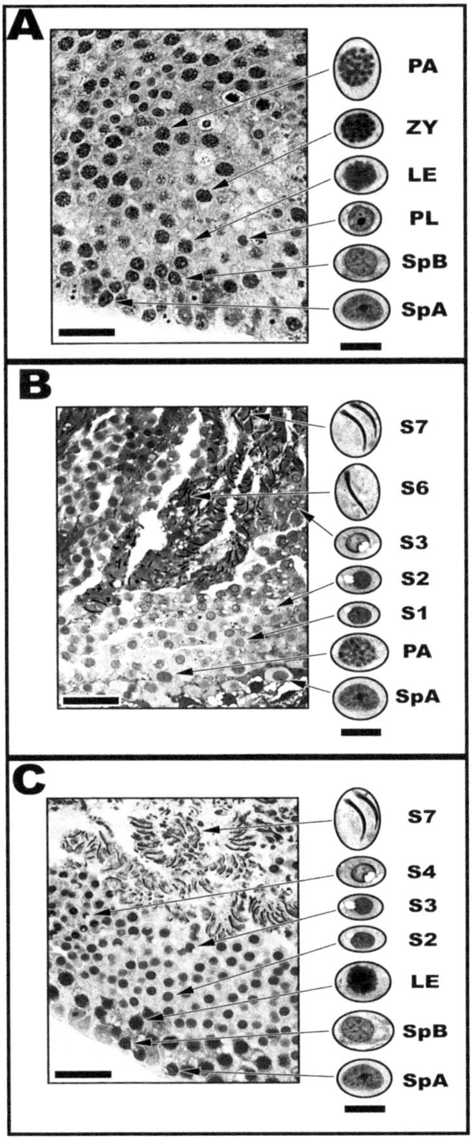

Fig. 11.24 *Podarcis muralis*. Histology of the seminiferous epithelium of represented months within the testis. **(A)** Section of the late July seminiferous epithelium and represented cell types from top to bottom: pachytene (PA), zygotene (ZY), leptotene (LE), pre-leptotene (PL), spermatogonia B (SpB) and A (SpA). **(B)** Section of the November seminiferous epithelium and represented cell types from top to bottom: Step 7 spermatids (S7), step 6 (S6), step 3 (S3), step 2 (S2), step 1 (S1), pachytene (PA), spermatogonia A (SpA). **(C)** Section of the February seminiferous epithelium and represented cell types from top to bottom: Step 7 spermatid (S7), step 4 (S4), step 3 (S3), step 2 (S2), leptotene (LE), spermatogonia B (SpB) and A (SpA). Bar of GE = 25 μm and bar of all types = 10 μm.

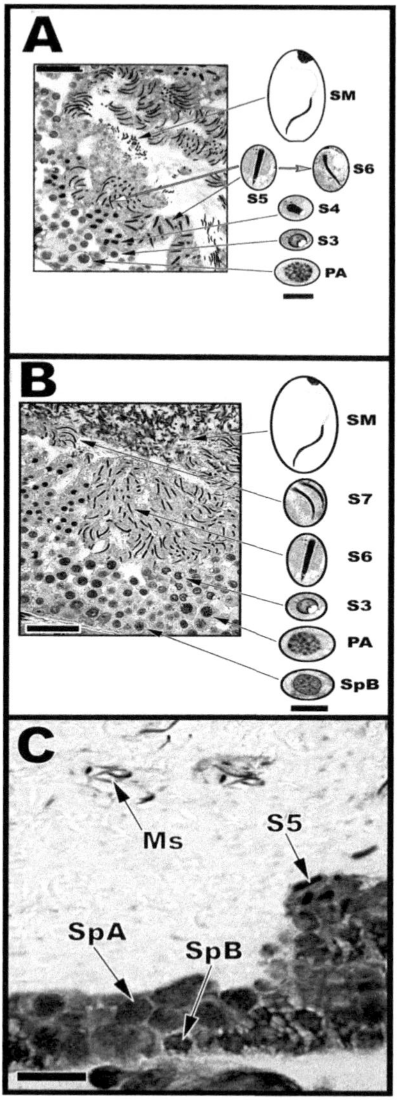

Fig. 11.25 *Podarcis muralis*. Histology of the seminiferous epithelium of represented months within the testis. **(A)** Section of the April seminiferous epithelium and represented cell types from top to bottom: Mature sperm (SM), step 6 spermatids (S6), step 5 (S5), step 4 (S4), step 3 (S3), pachytene (PA). **(B)** Section of the early June seminiferous epithelium and represented cell types from top to bottom: Mature sperm (SM), step 7 spermatid (S7), step 6 spermatid (S6), step 3 (S3), pachytene (PA), spermatogonia B (SpB). Note large number of mature sperm in lumen. **(C)** Higher power view of early July seminiferous tubule showing that spermiation is complete. Remnant spermatids (S5) and spermatozoa in the lumen (Ms). Spermatogonia A (SpA) and B (SpB) predominate in the reduced germinal epithelium. Bars in A and B =100 μm for tubules and 20 μm for cell types. Bar in C = 25 μm.

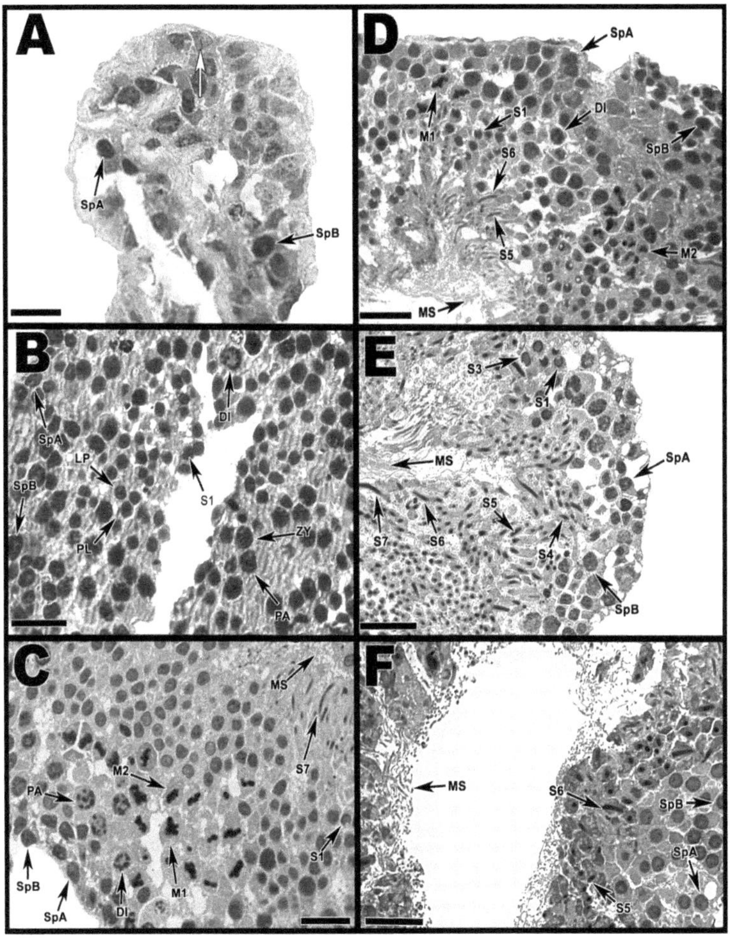

Fig. 11.26 *Hemidactylus turcicus.* Light micrographs of represented months of seminiferous epithelia within the testis. **(A)** The cell types observed in the September seminiferous epithelia. Bar = 20 µm. Spermatogonia A (SpA), spermatogonia B (SpB). **(B)** The cell types observed in the November seminiferous epithelia. Bar = 20 µm. Spermatogonia A (SpA), spermatogonia B (SpB), pre-leptotene spermatocyte (PL), leptotene spermatocyte (LP), pachytene spermatocyte (PA), diplotene spermatocyte (DI), zygotene spermatocyte (ZY), step 1 spermatid (S1). **(C)** The cell types observed in the January seminiferous epithelia. Bar = 20 µm. Spermatogonia A (SpA), spermatogonia B (SpB), pachytene spermatocyte (PA), diplotene spermatocyte (DI), meiosis I (M1), meiosis II (M2), step 1 spermatid (S1), step 7 spermatid (S7), mature sperm (MS). **(D)** The cell types observed in the March seminiferous epithelia. Bar = 20 µm. Spermatogonia A (SpA), spermatogonia B (SpB), diplotene spermatocyte (DI), meiosis I (M1), meiosis II (M2), step 1 spermatid (S1), step 5 spermatid (S5), step 6 spermatid (S6), mature sperm (MS). **(E)** The cell types observed in the May seminiferous epithelia. Bar = 20 µm. Spermatogonia A (SpA), spermatogonia B (SpB), step 1 spermatid (S1), step 3 spermatid (S3), step 5 spermatid (S5), step 6 spermatid (S6), step 7 spermatid (S7), mature sperm (MS). **(F)** The cell types observed in the August seminiferous epithelia. Bar = 20 µm. Spermatogonia A (SpA), spermatogonia B (SpB), step 5 spermatid (S5), step 6 spermatid (S6), mature sperm (MS).

muralis and *Hemidactylus turcicus* exhibit a similar mixed spermatogenesis with the same temporal germ cell development pattern detected in other temperate reptiles (Gribbins and Gist 2003; Gribbins *et al.* 2003; Gribbins *et al.* 2006; Rheubert *et al.* 2009a). These two lizards along with most saurians in temperate regions (see Fig. 11.22) studied to date exhibit prenuptial breeding. The next logical hypothesis was to test if a lack of seasonality in spermatogenesis or breeding could affect the type of germ cell development strategy employed by reptiles, as all continually breeding mammals and birds have a spatial germ cell development strategy (Russell *et al.* 1990).

Two continually spermatogenic lizards have been studied to date as far as germ cell development strategy, *Anolis lineatopus* (Gribbins *et al.* 2009) and *Sceloporus bicanthalis* (Gribbins *et al.* 2011). The two lizards occupy different habitats (the anole is a tropical lowland lizard and the bunchgrass lizard is a high elevation temperate lizard) but both show continuous spermatogenesis and a comparable temporal germ cell development strategy as seasonally breeding reptiles. We will discuss the spermatogenic cycle (Figs. 11.27, 11.28) of *Sceloporus bicanthalis* in this chapter.

Although spermiogenesis and spermiation (Fig. 11.27) ensue during every month of the year within *Sceloporus bicanthalis*, a distinct cycle is visible within the morphometric (Fig. 11.28) and histological data (Figs. 11.27, 11.28). The spermiogenic and spermiation events are observed during every other month. The seminiferous epithelium is continually cycling between early and late spermatids from month to month. Also, the incidents of spermiogenesis are slower than the mitotic and meiotic divisions (relatively consistent between all months of the year), which produced layerings of 3–5 spermatids within the germinal epithelia (Fig. 11.27) during this bimonthly seminiferous epithelial cycle. The two-month spermatogenic cycle or trend within the germinal epithelia is easily seen within the seminiferous epithelial height and tubular diameter. As the seminiferous tubule diameter increases (Fig. 11.28, upper figure), the germinal epithelial height (Fig. 11.28, lower figure) decreases. This phenomenon within the seminiferous tubules is evident even though there is no overall significant difference between SEH means monthly (Kruskal-Wallis; $H=8.71$, $df=11$, $P=0.649$, Fig. 2; Gribbins *et al.* 2011). In contrast, seminiferous tubule diameter (STD) did express a significant monthly trend (Kruskal-Wallis; $H=52.77$, $df=11$, $P=0.000$, Fig. 3; Gribbins *et al.* 2011). Post-hoc Dunn-Sidak data revealed that there was a seasonal component to STD. That was the spring (March–May), fall (Sept.–Oct.), and winter (Nov.–Feb.) months had similar lower STD measurements, while the summer (June–Aug.) STD data were significantly higher than those of spring, fall, and winter (Fig. 11.28, upper figure, subsets A and B; Gribbins *et al.* 2011). The cause is apparent from histological analysis (Fig. 11.28, upper figure micrographs); cross sections of summer seminiferous

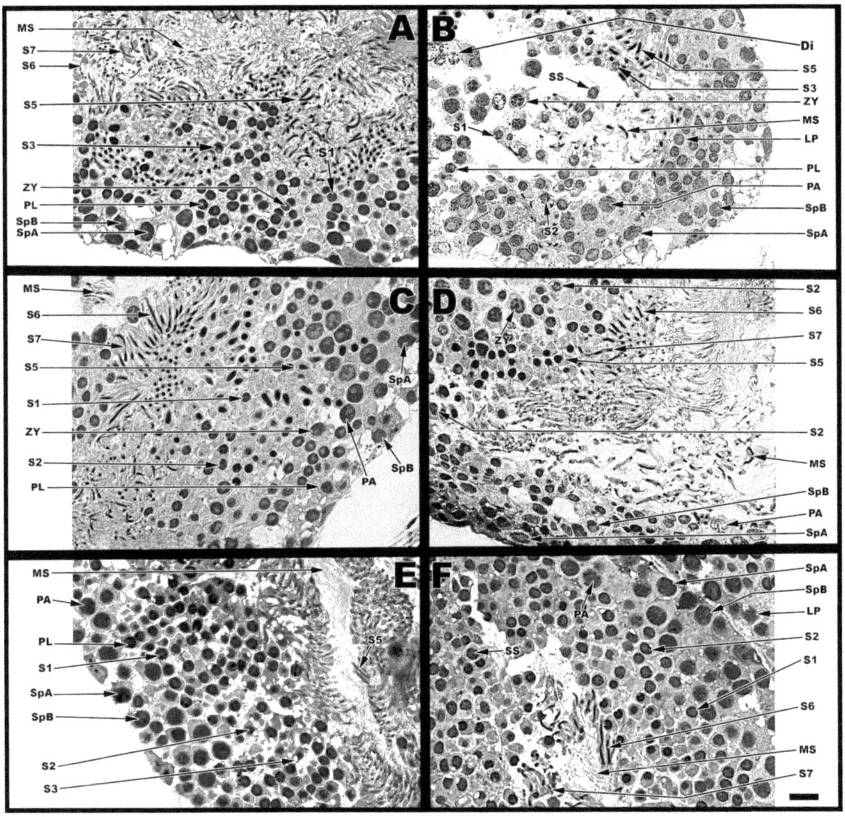

Fig. 11.27 *Sceloporus bicanthalis*. **(A)**. Transverse views of seminiferous tubules under x 1000 magnification from testes collected in: July **(A)**, August **(B)**, September **(C)**, October **(D)**, November **(E)**, and December **(F)**. Bar in F is representative for all cross sections and equals 20 μm. Present cell types: SpA, type A Spermatogonia; SpB, type B Spermatogonia; PL, pre-leptotene spermatocyte; LP, leptotene spermatocyte; ZY, zygotene spermatocyte; PA, pachytene spermatocyte; DI, diplotene spermatocyte; M1, meiosis I; SS, secondary spermatocyte; M2, meiosis II; S1, step 1 spermatid; S2, step 2 spermatid; S3, step 3 spermatid; S4, step 4 spermatid; S5, step 5 spermatid; S6, step 6 spermatid; S7, step 7 spermatid; MS, mature spermatozoa.

tubules reveal that the lumen is much wider, more noticeable, and has diffuse spermatozoa concentrations compared to spring, fall, and winter months.

The most likely cause for the significant difference seen in tubular diameter in the summer months is not epithelial height, as it showed no statistically significant differences. The enlarged and more visible lumen within summer tubules suggests fluid accumulation has increased, as has been shown in mammals (Zhou *et al.* 2001). One plausible hypothesis as to why summer tubules show this fluid accumulation versus the other months

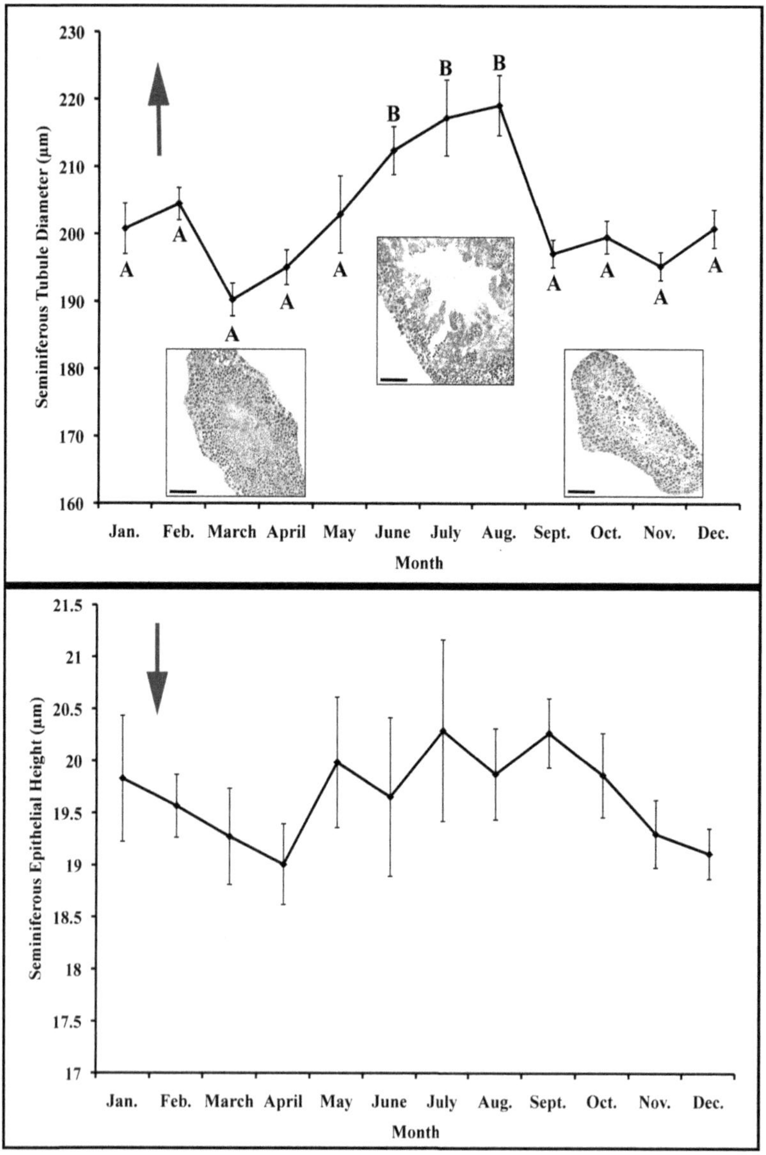

Fig. 11.28 *Sceloporus bicanthalis*. **Upper Figure**. Variation in seminiferous tubule diameter across months of the year in the testes. Values represented on this graph are means ± 1 standard error. Different super and subscripts (A and B) indicate significant differences ($P \leq 0.05$; Dunn-Sidak multiple range test). Micrographs from left to right represent seminiferous tubules from March, July, and November. Note the enlarged lumen of the July tubule compared to the other months. Bars = 40 µm. **Lower Figure**. Variation in seminiferous epithelial height during every month of the year in the testes. Values represented on this graph are means ± 1 standard error (Modified from Gribbins *et al.* 2011).

of the year is increasing precipitation, which occurs where this population of Bunchgrass Lizards is found (Gribbins *et al.* 2011). Summer rains could trigger a spike in hormone levels (such as testosterone and/or luteinizing hormone), which in turn causes the accumulation of fluid in the lumina and an increase in overall tubular diameter in the testis.

The temporal-like germ cell development strategy of these three lizards is similar even though their reproductive patterns are quite different, continuous vs. mixed spermatogenesis. These data in turn support similar findings in snakes (Gribbins and Rheubert 2011) and other reptiles (Gribbins 2011). This temporal germ cell development differs greatly from the spatial germ cell development seen in seasonally and continually breeding birds and mammals (Yamamoto 1967; Roosen-Runge 1972; Tait and Johnson 1982; Tsubota and Kanagawa 1989; Tiba and Kita 1990; Foreman 1997). The reptilian mode of spermatogenesis is also like the temporal germ cell development strategy of derived amphibians such as anurans (Lofts 1964; Van Oordt and Brands 1970). Although anuran amphibians have been called a provisional taxon between the anamniotes and amniotes in terms of testicular organization (Van Oordt 1955), amphibian seminiferous tubules are lined with cysts and do not maintain the permanent germinal epithelium seen in all amniotes. The ancestors to the extant reptiles are presumably considered the most basal amniotes phylogenetically and most likely had testes that were structurally similar to birds and mammals. Thus, reptiles might represent a better transitional intermediary in terms of testicular architecture and germ cell development strategy between anurans and the derived amniotic taxa. However, caution must be used when thinking of the evolution of testis structure within reptiles because of the paraphyly of Reptilia (Pough *et al.* 2001). This makes it difficult to confirm such hypotheses about the evolution of germ cell development strategy and testis function within reptiles. As has been stated in earlier publications on squamates (Rheubert *et al.* 2009a) the most parsimonious hypothesis for the germ cell development strategy in amniotes would involve maintenance of the anamniote germ cell development in basal amniotes and modern reptiles and thus convergence of the spatial development now seen in modern birds and mammals. In an alternative hypothesis, the spatial development of mammals could have developed in an ancestor to Synapsida, but this would necessitate reversal to the plesiomorphic state (temporal strategy) in the crocodilians, lepidosaurians (including lizards), and possibly chelonians.

11.4 SPERMIOGENESIS

Reptiles have been the focus of reproductive ultrastructural studies for more than 30 years. However, for the testis, most of these studies focus on the mature spermatozoa (Jamieson 1999; Gribbins *et al.* 2013a,b). Because of the wealth of data on the microscopic anatomy of spermatozoa at least within many of the major families across orders within the class Reptilia, the majority of recent studies have focused of the phylogenetic significance of this growing data-set of ultrastructural characters (Oliver *et al.* 1996; Jamieson *et al.* 1996; Vieira *et al.* 2005; Tournmente *et al.* 2006; Tavares-Bastos *et al.* 2008; Rheubert *et al.* 2010; Gribbins and Rheubert 2011). It has also been noted that the spermatozoa of Squamata has more variation across species (apomorphies) than any other amniotic taxa (Jamieson 1995; Jamieson 1999; Gribbins *et al.* 2013a). This may make sense in light of how speciose the squamate taxon is at over 9,400 known species (Gribbins *et al.* 2013a). Unfortunately, the number of ultrastructural studies completed on different species for spermatozoa is still lacking for reptiles as a whole. Even more discouraging, there are relatively few studies that exist on spermiogenic characterization during spermatogenesis within reptiles and particularly for squamates. Spermiogenic ultrastructure could provide a new suite of characters that could be added to spermatozoa data to increase the robustness of phylogenetic analyses in squamates and other reptiles. This is particularly true because most structures that are observed within the mature spermatozoa arose ontogenically during spermiogenesis. The major hope of this chapter is to stimulate and encourage other interested researchers to add data to this growing field by providing an ultrastructural model to follow in order to describe spermiogenic and spermatozoal characters within lizards.

There is a large pool of data that spotlights certain aspects or that mentions ultrastructural features of certain spermatid steps within lizards and other squamates (i.e., Clark 1967; Butler and Gabri 1984; Hondo *et al.* 1994; Al-Dokhi 2004a,b; Mubarak 2006; Al-Dokhi 2013). They are too numerous to list or to focus on in this chapter. The focus presently will be on studies that have complete ultrastructural descriptions of the ontogeny of spermatid formation during spermiogenesis within squamates, with an emphasis on lizards. In Squamata (snakes, lizards, worm-lizards), only 12 studies (of which 10 are for lizards) have comprehensive ultrastructural descriptions of spermiogenesis, including the Common Lizard (*Zootoca vivipara*; Courtens and Depeiges 1985), the Oscillated Skink (*Chalcides ocellatus*; Carcupino *et al.* 1989), the Amazon Lava Lizard (*Tropidurus torquatus*; Verira *et al.* 2001), the Green Iguana (*Iguana iguana*; Ferreira and Dolder 2002), *Scincella lateralis* (Gribbins *et al.* 2007), *Anolis lineatopus* (Rheubert *et al.* 2010), *Hemidactylus turcicus* (Rheubert *et al.* 2011), multiple species (Gribbins 2011), *Sceloporus bicanthalis* (Rheubert *et al.* 2012), the Imbricate Alligator Lizard (*Barisia Imbricata*; Gribbins *et al.* 2013a), and the Rosebelly Lizard (*Sceloporus variabilis*; Gribbins *et al.* 2013b).

Lizard ultrastructural studies during spermiogenesis make up the largest pool of data for sperm maturation in squamates. Unfortunately, there are only 10 species with extensive data, thus representing a limited number of families overall (of the more than 30 lizard families only 7 families have detailed spermiogenic data). Furthermore, when combining incomplete and complete studies on characteristics of spermiogenesis in lizards and reptiles in general, few researchers use consistent terms to describe structures common or uncommon to reptilian spermatids. Thus, this inconsistency in terminology makes comparative studies of spermiogenesis in lizards and reptiles a daunting task. Hopefully within this chapter we can provide consistent terms that can be used by other researchers to minimize confusion and aid in future comparative studies that could be useful, with sperm morphology, as non-traditional data for phylogenetic analyses of reptiles. Whenever possible within the present data, we choose to use the original descriptive terms to maintain historical precedence and provide a solid, universal backbone for nomenclature of saurian spermatid structures.

Most researchers agree that within Squamata most features or the process of spermiogenesis are somewhat conserved (see Jamieson 1999; Gribbins 2011; Gribbins and Rheubert 2011). Thus, there are common characteristics to spermatid development within squamates and lizards that have been observed within many of the studies to date. There is almost always a series of microtubules called the manchette associated with lizard elongating spermatids except in *Anolis lineatopus* (Rheubert *et al.* 2010), which is one of the few amniotes investigated to date that does not possess this structure. The acrosome forms close to the nucleus and then indents its apical surface. There is the presence of an acrosome granule within the acrosome vesicle during some point of the round spermatid stage in lizards. The acrosome complex in squamates is stratified and contains multiple structures. An extranuclear perforatorium is also always present in lizards. The perforatorium outside the nucleus and the stratification of the subacrosomal cone are considered by most to be synapomorphies for squamates (Jamieson 1995; Oliver *et al.* 1996; Teixeira *et al.* 1999a,b; Gribbins *et al.* 2013a,b). The acrosomal lucent ridge that separates the subacrosomal cone into two layers is seen in almost every micrograph of lizard elongating spermatids that have been studied to date (possible synapomorphy for squamates). Most researchers neglect, ignore its presence, or misidentify it as the epinuclear lucent zone within the acrosome complex of lizards. Rheubert *et al.* (2012) coined the term acrosomal lucent ridge to universally recognize this seemingly common structure in lizard spermatids. The last structure common in the subacrosome space of squamates is the epinuclear lucent zone, which sits onto the tip of the nuclear rostrum. This zone is found in all lizards studied to date except Scincomorpha (Gribbins *et al.*

2007; Rheubert *et al.* 2011) and the gekkonid, the Prickly Gecko, *Heteronotia binoei* (Jamieson *et al.* 1996).

The condensation of chromatin in elongating spermatids almost always results in filamentous chromatin fibers within the nucleus during the mid/late stages of spermiogenesis in squamates. However, in many lizard species chromatin starts condensing early in nuclear elongation in a granular form and then takes on its filamentous form as spermiogenesis continues (Clark 1967; Ferreira and Dolder 2002; Vieria *et al.* 2004; Rheubert *et al.* 2011; Rheubert *et al.* 2012). However, in at least three lizards studied (Carcupino *et al.* 1989; Gribbins *et al.* 2013a,b), chromatin condensation commences with chromatin filaments and maintains this form throughout maturation. In most previous studies, an annulus is seen migrating distally with the developing axoneme and marks the end of the midpiece. The fibrous sheath of the developing lizard flagellum starts within the midpiece, resulting in a shortened distal centriole. This unique feature of saurian squamates is considered another synapomorphy for this taxon (Jamieson 1995, 1999).

The inconsistencies within the description of spermatids and the lack of relatively complete information for each phase of spermiogenesis in the present lizard literature, lends cause to the idea that a universal model is needed so that dependable data can be obtained in future studies, especially if these data are to be used in phylogenetic analyses. The study of spermatid ultrastructure during lizard spermiogenesis may provide robust data that could deliver many new character differences (possibly derived in nature) between closely related species (see Fig. 11.29). Thus, in this chapter three recently studied examples of lizard spermiogenesis will be presented for comparison, which include data for the species *Sceloporus bicanthalis* (Rheubert *et al.* 2012), *S. variabilis* (Gribbins *et al.* 2013b), and *Barisia imbricata* (Gribbins *et al.* 2013a). Again the following discussion is a preliminary one and comes with the caution that too few lizard genera have been studied to date to make firm phylogenetic assumptions.

11.4.1 Spermatid Ultrastructure During Spermiogenesis

The ultrastructure of the spermatids as they progress through spermiogenesis within the germinal epithelia of *Sceloporus bicanthalis*, *S. variabilis*, and *Barisia imbricata* resembles that of most other squamates (Gribbins 2011). During acrosome formation in the round spermatid stage (Figs. 11.30, 11.31) within these lizards, the growing acrosome vesicle forms from transport vesicles budding from the Golgi apparatus (Figs. 11.30, 11.33D). The acrosome grows in size even before it reaches the nuclear surface in the sceloporine lizards (Fig. 11.30). This is similar to what has been described in *Zootoca vivipara* (Courtens and Depeiges 1985), *Scincella lateralis* (Gribbins *et al.*

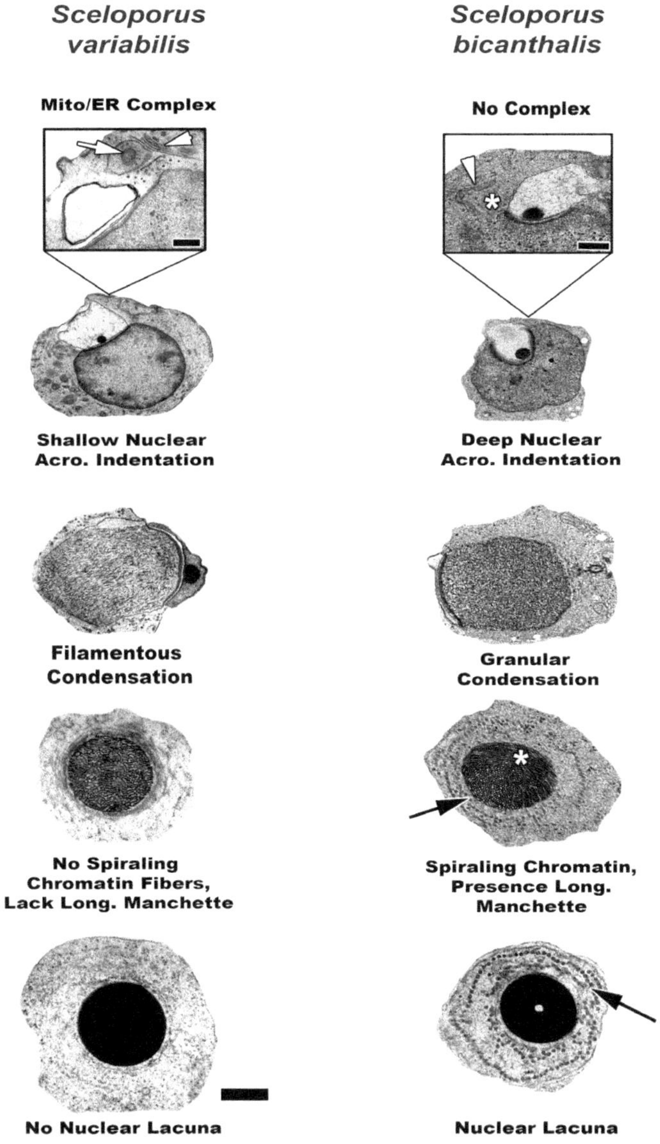

Fig. 11.29 Comparison of ultrastructural characters observed during spermiogenesis between *Sceloporus variabilis* and *S. bicanthalis*. There are six main character differences seen during spermiogenesis between these two species within the same genus, *Sceloporus*. These include: 1) presence (white arrow)/absence (*) of a mitochondrial/endoplasmic reticulum complex near Golgi complex (white arrowhead), 2) deep or shallow nuclear depression accommodating the acrosome, 3) filamentous vs. granular condensation of chromatin, 4) spiraling (*) or lack of spiraling during condensation of chromatin, 5) presence (black arrows) or absence of the longitudinal manchette microtubules, 6) presence or absence of nuclear lacunae. Bars = 1 μm.

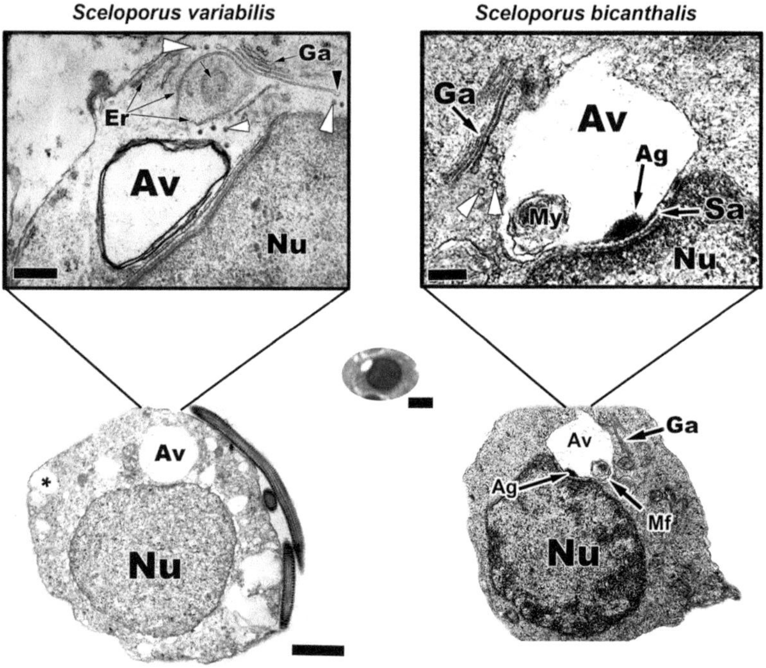

Fig. 11.30 Early round spermatids of *Sceloporus variabilis* and *S. bicanthalis*. Bars = 2 μm for whole cell TEM, 0.5 μm for insets, and 5 μm for (central) light micrograph. These two lizards have very similar ontogeny to their early acrosome vesicle (Av) formation. The only two differences are the presence of myelin figures (Mf, My) within the *S. bicanthalis* acrosome vesicle and the ER/Mitochondria (black arrow, Er) complex between the Golgi complex (Ga) and the acrosome vesicle of *S. variabilis*. Dark staining proteins (the acrosome granule, Ag) accumulate within the acrosome vesicle and rest on the inner acrosome membrane. Transport vesicles (white arrowheads), nucleus (Nu), lipid droplets (*), transport vesicles (white arrowheads), budding transport vesicle (black arrowhead), subacrosome space (Sa).

2007), and *Chalcides ocellatus* (Carcupino *et al.* 1989). However, the acrosome vesicle grows much longer away from the nuclear membrane in *Sceloporus variabilis* than that of other lizards studied to date and leads to a much shallower acrosome indentation on the nuclear surface compared to *S. bicanthalis*, *Barisia imbricata*, and other lizards (Figs. 11.31, 11.33D, Table 11.1), which may be an autapomorphy for *S. variabilis*. In *B. imbricata*, the acrosome vesicle makes nuclear contact immediately after its initial growth phase. The acrosome granule is not seen before making contact with the nuclear envelope in these lizards as it is in *Scincella lateralis* and *Agkistrodon piscivorus* (the only snake with comprehensive ultrastructural data; Gribbins *et al.* 2010). Subsequent features of acrosome development (Figs. 11.30, 11.31, 11.33, 11.36A,B) within these lizards are similar to that described for other squamates such as: transport vesicles from the Golgi,

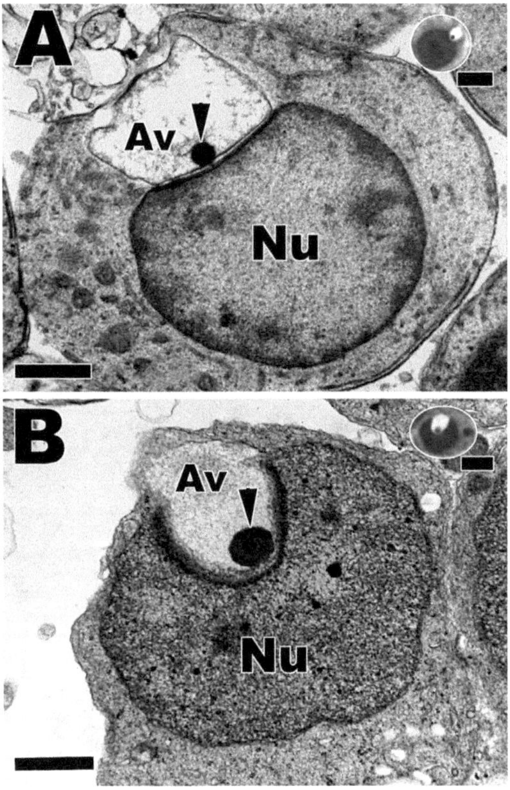

Fig. 11.31 Comparison of acrosome vesicle (AV) indentation depth into the apical nucleus (NU) of round spermatids in *Sceloporus variabilis* **(A)** and *S. bicanthalis* **(B)**. Bar = 2 μm for TEM and 5 μm for light micrographs (insets). Note how much deeper *S. bicanthalis* indentation, which is typical of the lizards studied to date, is compared to *S. variabilis*. Acrosome granule (black arrowhead).

Color image of this figure appears in the color plate section at the end of the book.

prominent subacrosome space, multilaminar Sertoli cell membranes, and lateral foldings (Clark 1967; Da Cruz-Landim and Da Druz-Hofling 1977; Butler and Gabri 1984; Dehlawi *et al.* 1992; Ferreira and Dolder 2002, 2003; Gribbins *et al.* 2007).

Both *Sceloporus variabilis* and *S. bicanthalis* also possess clear well-developed subacrosomal granules that have been reported in every lizard studied to date excluding *Iguana iguana* and *Chalcides ocellatus* (Table 11.1). There, however, has been some misunderstanding about the fate of the granules within mature spermatozoa. In early studies of spermiogenesis, the perforatorium was thought to arise from the subacrosomal granule (Humphreys 1975; Del Conte 1976) or the acrosomal granule (Adelina *et*

Table 11.1 Comparison of spermatid ultrastructural characters for the ten species of lizards within the literature that have extensive ontogenetic details of spermatid development. Red colored letters represent unique characters of a species compared with the other lizards within the table. Sources: *Barisia imbricata*, Gribbins *et al.* 2013b; *Lacerta vivipara*, Courtens and depeiges 1985; *Chalcides ocellatus*, Carcupino *et al.* 1989; *Tropidurus torquatus*, Vieira *et al.* 2001; *Sceloporus bicanthalis*, Rheubert *et al.* 2012; *Sceloporus variabilis*, Gribbins *et al.* 2013a; *Scincella lateralis*, Gribbins *et al.* 2007; *Anolis lineatopus*, Rheubert *et al.* 2010; *Iguana iguana*, Ferreira and Dolder 2002; *Hemidactylus turcicus*, Rheubert *et al.* 2011.

	Acrosome Complex												
	ALR	PE	BP	ELZ	ER/MI COM	NUCR	EMY	LB	ER	SGA	NR	AI	PV
Barisia Imbricata	1	B	1	B	0	1	1	1	0	1	0	D	0
Zootoca vivipara	1	N	NR	N	0	1	S	1	0	Poss	1	D	0
Chalcides ocellatus	1	NR	NR	N	0	1	0	1	0	0	0	D	0
Tropidurus torquatus	1	N	NR	N	0	1	0	1	0	1	0	D	0
Sceloporus bicanthalis	1	N	1	N	0	1	S	1	0	1	0	D	0
Sceloporus variabilis	1	N	1	N	1	1	0	1	0	1	0	S	0
Scincella lateralis	1	S	1	0	0	1	0	1	0	1	0	D	0
Iguana Iguana	1	N	1	N	0	1	0	NR	1	NR	0	D	1
Anolis Lineatopus	1	N	1	N	0	1	0	1	0	1	0	D	0
Hemidactylus turcicus	1	N	1	N	0	1	S	1	0	1	0	D	0

	Nucleus						
	CH	MA	LIP	LA	NP	NF	CHS
Barisia Imbricata	F	L/C	1	0	NS	C	0
Zootoca vivipara	G/F	L/C	1	Poss	1/NS	N	1
Chalcides ocellatus	F	L/C	NR	0	NS	N	0
Tropidurus torquatus	G/F	L/C	Poss	NR	1/NS	N	NR
Sceloporus bicanthalis	G/F	L/C	1	1	1/NS	N	1
Sceloporus variabilis	F	l/C	0	0	1/NS	N	0
Scincella lateralis	G/F	L/c	0	1	1/Poss NS	N	0
Iguana Iguana	G/F	L/C	NR	Poss	Poss Both	N	1
Anolis Lineatopus	G/F	0	1	0	IR/NS	N	0
Hemidactylus turcicus	G/F	L/C	0	0	1/NS	N	0

	Flagellum						
	PC/CP	P3/8	MIL	MIR	DB	AN	MFS
Barisia Imbricata	1	M/e	4 to 5	5 to 7	1(4/3-5L)	1	3
Zootoca vivipara	Poss	NR	3	NR	1 (NR/2L)	1	NR
Chalcides ocellatus	1	NR	4	NR	1 (3/4L)	1	Poss 2
Tropidurus torquatus	1	NR	NR	NR	1 (NR)	1	NR
Sceloporus bicanthalis	1	M/p	4 to 6	4 to 6	1 (5-7/2-3L)	1	2
Sceloporus variabilis	1	M/p	4 to 6	4 to 6	1(3-5/3-4L)	1	2
Scincella lateralis	1H	M/p	4-6L	12	rare	1	Poss 1 (longNeck)
Iguana Iguana	1	M	4	NR	1 (NR)	1	poss 2
Anolis Lineatopus	1	M	4 to 5	NR	1(2/3L)	1	2
Hemidactylus turcicus	1	M	12	10 to 12	1 (8-10T/L)	1	3

ALR	Acrosomal Lucent Ridge	1: Present; 0: Absent
PE	Peforatorium	N: Normal; B: Bulbous; S: Short
BP	Basal Plate	1: Present; 0: Absent
ELZ	Epinuclear Lucent Zone	N: Normal; B: Bulbous; 0: Asent
ER/MI COM	ER/Mitochondria Complex	1: Present; 0: Absent
NUCR	Nuclear Rostrum	1: Present; 0: Absent
EMY	Extra Myelin Acrosomal Material	1: Present; 0: Absent; S: Some
LB	Lamellar Body (ER)	1: Present; 0: Absent
ER	ER Aid in Acrosome Dev.	1: Present; 0: Absent
SGA	Subacrosomal Granule	1: Present; 0: Absent
NR	Nuclear Ribbon	1: Present; 0: Absent
AI	Acrosomal Nuclear Indentation	D: Deep; S: Shallow
PV	Proacrosomal Vesicles	1: Present; 0: Absent
		****NR: Not Reported**
		**** Poss: Possibly Found (Not reported but can observe in micrographs)**
CH	Chromatin Condensation	F: Filamentous; G/F: Granular then Filamentous
MA	Manchette Microtubule Arrangement	L/C: Longitudinal/Circular; l/c: small longitudinal/circular; 0: Absent
LIP	Lipids surrounding late elongates	1: Present; 0: Absent
LA	Nuclear Lacuna	1: Present; 0: Absent
NP	Nuclear Membrane Pouches	1: Present; NS: Nuclear Shoulders present ; IR: Irregular pouches; 0: Absent
NF	Nuclear Implatation Fossa Shape	C: Clover Shaped; N: Cylindrical
CHS	Significant Chromatin Spiraling	1: Present; 0: Absent
AN	Annulus	1: Present; 0: Absent
P3/8	Peripheral Fibers 3 and 8	M/p: Enlarged MidP/slightly enlarged Principal P; M: Enlarged MP; e: endP
MIL	# Mitochondria in Longitudinal Plane	L: oriented elongated only
DB	Dense Bodies Present	1: Present; 0: Absent; #/#L: # in Transverse/# in Longitudinal
PC	Pericentriolar Material Present/Conn. Piece	1: Present; 0: Absent; 1H: Heavy Accumulation
MIR	# of Mitochondria in individual rings	
MFS	Tier of Mito. Where Fiberous Sheath Forms	Long Neck: Longer connecting piece than other lizards

Color Table appears in the color plate section at the end of the book.

al. 2006). More recently in *Barisia imbricata* (Gribbins *et al.* 2013a) and in *S. bicanthalis* (Rheubert *et al.* 2012) the same granule appears to develop either during the late round or early elongation phases of spermiogenesis. However, these two studies call the granule within the subacrosomal space the developing basal plate of the perforatorium. In both previous studies, it seemed difficult to distinguish whether the subacrosome granule originated from the acrosomal granule or was a new protein accumulation within the subacrosomal space (see Fig. 2C in *B. imbricata*, Gribbins *et al.* 2013a). At least in *S. variabilis* it is apparent that this granule arises independently of the acrosomal granule and most likely gives rise to the epinuclear lucent zone based on its close association with this zone during mid to late elongation (see Fig.11.36, *). Thus, we concur with Ferreira *et al.* (2006) in that the perforatorium most likely originates from the acrosomal granule and the epinuclear lucent zone is hypothesized to arise from the subacrosomal granule established on its location and spatial organization within the developing acrosome complex.

There are several morphological differences between these two species of sceloporines (Rheubert *et al.* 2012; Gribbins *et al.* 2013a,b) and *Barisia imbricata* during early acrosome formation. The ER/Mitochondria complex (Fig. 11.30) forms early in acrosome development of *Sceloporus variabilis* spermatids and is a unique feature (Table 11.1) that has not been observed in any other squamate to date. Whether this complex is an autapomorphy of *S. variabilis* or a synapomorphy for the *variabilis* group within the genus *Sceloporus* remains to be seen. Mitochondria and ER have been reported near the acrosome vesicle but not together and not between the Golgi complex and the developing acrosome vesicle as observed in *S. variabilis* round spermatids. Endoplasmic reticulum has even been described as playing a role in early round spermatid development in *Iguana iguana*, an exclusive feature (Table 11.1) of this lizard species. The other differences between these two sceloporines are the presence of myelin figures within the acrosome vesicles of *S. bicanthalis* spermatids and the shallow depth of the acrosome vesicle as it indents the nucleus during the late round spermatid stage in *S. variabilis*. Myelin figures (Figs. 11.30, 11.33A,B) are described in other lizard species (Table 11.1), but are most commonly seen in large numbers in *Barisia imbricata*. The high frequency of myelin figures in *Barisia* spermatids provided evidence that they have common origins and attachments to the acrosomal membrane (Fig. 11.33A). These membrane structures seem to wind up into layers from the developing cell membrane of the acrosomal vesicle, which has its largest growth phase at this time. The authors hypothesize this is an effective mechanism to rid the acrosomal vesicle of excess membrane that is delivered by the multiple transport vesicles supplied to the acrosomal vesicle via the Golgi complex. As the acrosome continues to develop during the round stage of spermiogenesis, visual evidence of the exocytosis of these degrading myelin figures is apparent and the newly formed cytoplasmic residual bodies accumulate near the apex of the spermatid nucleus (Fig. 11.33B,C, white arrows). The shallow acrosome fossa (Fig. 11.31) of the apical nucleus is another possible autapomorphy for *S. variabilis* or maybe a synapomorphy again for the *variabilis* group of the *Sceloporus* clade. The significance of the large number of myelin figures phylogenetically will have to be explored in other lizard groups before any attempt at the evolutionary or physiological implications of these strange acrosome formations can be deduced.

Two additional features that seem to be found in only two of the ten lizards studied to date during early acrosome development are the nuclear ribbon and more than one proacrosomal vesicle (Table 11.1). The nuclear ribbon has only been described for *Zootoca vivipara* and could be either unique to this species or a shared apomorphy of the *Zootoca* genus or the Lacertidae family (Table 11.1). The ribbon consists of a series or parallel array of microtubules (7 or more total) just under the acrosome vesicle initially during its development (Courtens and Depeiges 1985). Two proacrosomal vesicles have only been delineated in *Iguana iguana* but those authors make no attempt to suggest their significance as far as function or phylogenetics; only time will tell as more species are studied within the family Iguanidae.

As early elongation (Fig. 11.32) begins in *Sceloporus variabilis*, *S. bicanthalis*, and *Barisia imbricata*, the acrosome complex starts to compartmentalize, which is common in lepidosaurians (Healy and Jamieson 1994; Harding *et al.* 1995; Jamieson and Scheltinga 1994; Jamieson *et al.* 1996; Tavares-Bastos *et al.* 2007). As the acrosome collapses against the cell membrane of the elongating spermatid (Fig. 11.32), a subacrosomal space that is rich in dark paracrystalline protein deposits develops between the acrosome vesicle and the apical nuclear surface (Figs. 11.33B, 11.34B, 11.36C). The cytoplasmic shift posteriorly that caused the acrosome to collapse and migrate laterally along the nuclear apex (Figs. 11.32, 11.36C; Clark 1967; Butler and Gabri 1984), also results in relocation of cellular organelles (Sprando and Russell 1988; Soley 1997; Lin and Jones 1993; Lin *et al.* 1997; Ventela *et al.* 2003). The squeezing of the apical nuclear head by the differentiating acrosome complex causes the acrosome to rest on thin nuclear shoulders (Figs. 11.36C,D, 11.37A) in all species of lizard studied to date. The manchette (Fig. 11.32) can be visualized in both transverse and sagittal sections early in elongating spermatids of both sceloporine species and *Barisia imbricata* and is thought to aid in the elongation of the nucleus (Russell *et al.* 1990), although many have suggested its role is minor or not involved at all in nuclear elongation (Fawcett *et al.* 1971; Cole *et al.* 1988). There are two arrays of microtubule scaffolding that make up the manchette in squamates. Parallel (longitudinal) microtubules run along the long axis of the elongating spermatids and circum-cylindrical tubules wrap around

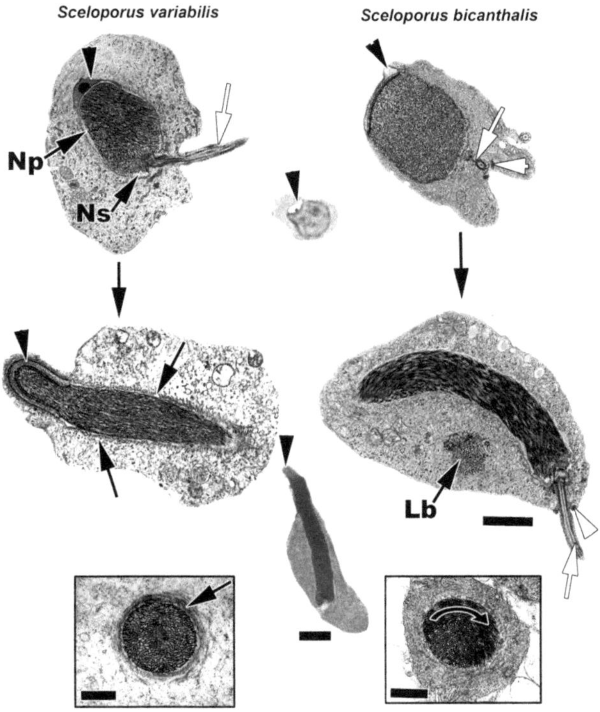

Fig. 11.32 Early and mid-elongation in *Sceloporus variabilis* and *S. bicanthalis*. Bar for TEM of sagittal elongates = 1 μm, bar for transverse elongates 0.5 μm, and bar for light micrographs 2 μm. Condensation involves spiraling of chromatin and granular condensation and then filamentous chromatin in *S. bicanthalis*. Both the spiraling and granular chromatin are absent in *S. variabilis*. Acrosome (black arrowheads), developing flagellum (white arrows), developing annulus (white arrowheads), nuclear pouches (Np), nuclear shoulders (Ns), manchette (black arrows), lamellar body (Lb), spiraling chromatin (curved black arrow).

Color image of this figure appears in the color plate section at the end of the book.

the circumference of the nuclear body. Most species of lizard (Table 11.1), including *Sceloporus bicanthalis* and *Barisia imbricata* (Fig. 11.34D), have fairly equal numbers of both types of microtubule scaffolding as part of their manchettes. Notable exceptions to equal manchette microtubule distribution in lizards occur in three of the ten species studied. Although *S. bicanthalis* and *S. variabilis* have both microtubule scaffoldings, *S. variabilis* has few of the longitudinal manchette microtubules, while *S. bicanthalis* has less of the circum-cylindrical tubules (Rheubert *et al.* 2012; Gribbins *et al.* 2013a,b). Also, it has been noted that there is an absence of circum-cylindrical microtubules in *Scincella lateralis* (Gribbins *et al.* 2007), which have thicker bodied spermatozoa than those of other squamate species (Jamieson and Scheltinga 1994). This result of fewer circum-cylindrical fibers may also

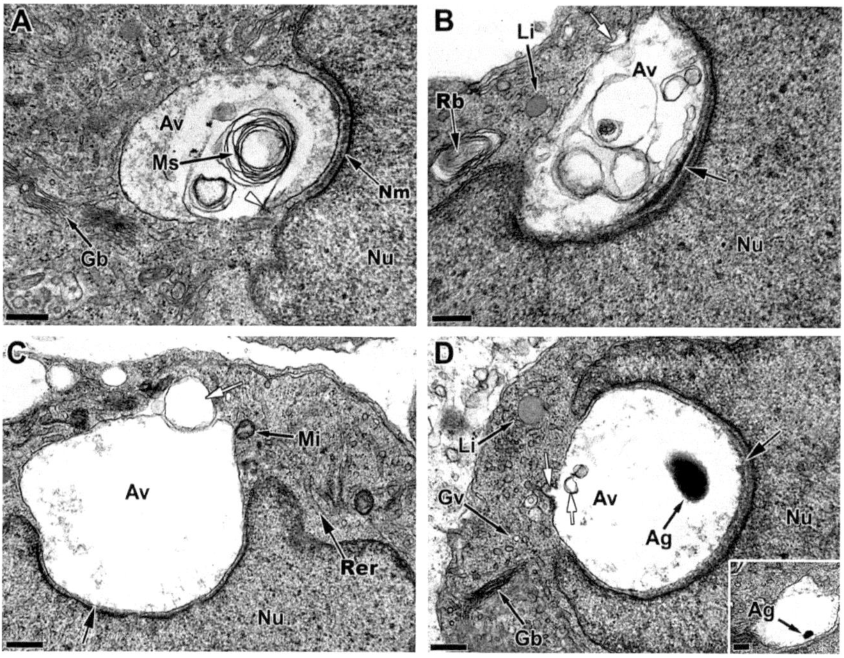

Fig. 11.33 *Barisia imbricata*. Beginning stages of spermiogenesis. **(A)** The acrosomal vesicle (Av) develops at the apex of the nucleus (Nu) from vesicles budding from the Golgi body (Gb). Multiple membranous structures (Myelin figures-Ms) can be seen within the developing vesicle. Nuclear membrane (Nm). **(B)** The acrosome vesicle (Av) increases in size causing an indentation in the nucleus (Nu) to increase in size. Lipid inclusion (Li), residual bodies (Rb), subacrosomal space (black arrow), myelin figure exocytosis (white arrow). **(C)** The myelin figures continue to undergo exocytosis (white arrow) as the acrosome vesicle (Av) is cleared of debris. Mitochondria (Mi), nucleus (Nu), subacrosomal space (black arrow), rough endoplasmic reticulum (Rer). **(D)** The acrosomal granule (Ag) becomes evident centrally within the acrosomal vesicle (Av) before migrating to its basal position (inset, Ag; Bar = 300 nm). Some myelin figure debris can be seen in the acrosome and also undergoing exocytosis from the acrosome vesicle (white arrows). Lipid inclusion (Li), Golgi apparatus (Gb), transport vesicle (Gv). Bars = 0.5 μm.

result in more robust spermatozoa in *Agkistrodon piscivorus* (Gribbins *et al.* 2010). *Anolis lineatopus* lacks a manchette altogether and is the only lizard or reptile that has this condition to date, a conceivable autapomorphy for this species.

Throughout the events of elongation and initial condensation, uniform translucent nuclear shoulders (Fig. 11.32) are present in all lizards studied to date except possibly *Scincella lateralis* (Table 11.1) and are detected on either side of the caudally located nuclear fossa (Fig. 11.34A, inset) where the flagellum attaches to the nuclear body of the developing spermatid. Also, translucent nuclear pouches seem to be visible in all lizards studied

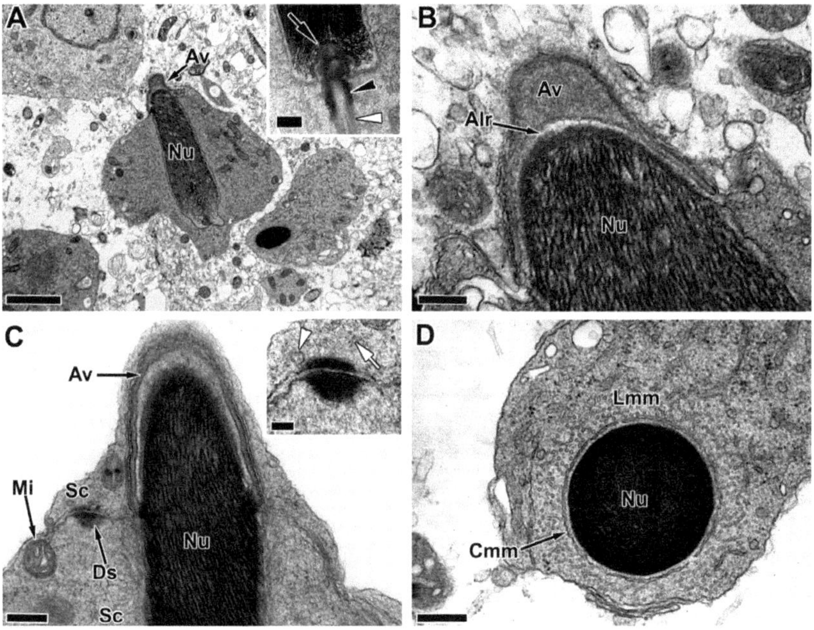

Fig. 11.34 *Barisia imbricata*. Elongation steps of spermiogenesis. **(A)** The germ cell nucleus (Nu) migrates to the apical portion of the cytoplasm causing the acrosomal vesicle (Av) to push against the cell membrane. Bar = 2 μm. Inset: Caudal end of the nucleus where distal centriole (black arrowhead) attaches to the nucleus via the nuclear fossa (black arrow). Extending axoneme of flagellum (white arrow). Bar = 200 nm. **(B)** The acrosomal lucent ridge (Alr) becomes evident separating the acrosomal vesicle (Av) from the nucleus (Nu). Bar = 200 nm. **(C)** The acrosomal vesicle (Av) and apex of the germ cell become surrounded by Sertoli cells (Sc) that are adhered by desmosomes (Ds). Inset: Bar = 50 nm. Enlarged view of the desmosome cellular junction showing cisterna of endoplasmic reticulum (white arrowhead) and intermediate filaments (white arrow) in CS. Mitochondria (Mi), nucleus (Nu). Bar = 0.5 μm. **(D)** The nucleus (Nu) in transverse section has both circum-cylindrical (Cmm) and longitudinal (Lmm) microtubules of the manchette present. Bar = 0.5 μm.

to date, except possibly *Iguana iguana* (Table 11.1), along the lateral sides of developing elongates in the early to mid-stage of development (Fig. 11.32). It should be noted that these structures have been completely ignored within the literature and were first described by Courtens and Depeiges (1985) for *Zootoca vivipara*. A quick search of the literature and observation of micrograph data shows relatively clearly that these translucent characters are common in almost all lizards studied to date and thus may be suspected synapomorphies for Squamata or even plesiomorphies for reptiles, which would require studying these pouches and shoulders outside of squamates.

Most lizards start the condensation of their chromatin in a granular form similar to that of *Sceloporus bicanthalis* (Fig. 11.32) and then later in development condensation continues in a filamentous fashion. However, three of the ten lizards studied to date immediately start condensation with filamentous fibers (Figs. 11.32, 11.34), skipping the granular mechanism, and this includes *S. variabilis*, *Barisia imbricata*, and *Chalcides ocellatus* (Table 11.1). About half the number of lizards practice twisting or spiraling of DNA (including *S. bicanthalis*) during the condensation of chromatin (Fig. 11.32), while the other half does not (*S. variabilis* and *B. imbricata* lack this spiraling phenomenon). During this process of condensation, *S. bicanthalis* and *Scincella lateralis* both possess nuclear lacunae within their nucleoplasm. These are the only two lizards (besides lizards from the genus *Tupinambis*; Tavares-Bastos *et al.* 2002) at present known to exhibit these spaces within the nucleus during elongation, a condition similar to that seen in other non-squamate reptiles such as turtles (Ibarguengoytia *et al.* 1999). One other feature that is common to almost every lizard and squamate studied to date is the presence of a lamellar body within the cytoplasm of developing spermatids during the round and elongation phases of spermiogenesis. This body of dark staining endoplasmic reticulum (Carcupino *et al.* 1989) has been given many names within the literature. Here we give historical precedence to the nomenclature of this structure and use lamellar body after Clark (1967) who first described it in spermatids during spermiogenesis. The last major difference between *Sceloporus variabilis*, *S. bicanthalis*, and *B. imbricata* is the presences of lipid inclusions around the nucleus during elongation in spermatids. The significance of these lipids is unknown and two other lizards within our data set also exhibit these droplets, *Zootoca vivipara* and *Anolis lineatopus*.

As the flagellum develops during late elongation (Figs. 11.35, 11.38), a prominent nuclear fossa houses the proximal part of the distal centriole and large numbers of mitochondria become present in the posterior portion of the sceloporine and the Imbricate Alligator Lizard, *Barisia imbricata*, spermatid cytoplasm near the axoneme, which is consistent with other amniotes (McIntosh and Porter 1967; Lin and Jones 1993; Ferreira and Dolder 2002; Gribbins *et al.* 2007; Cunha *et al.* 2008). The nuclear fossae of all the lizards in the present comparison have a deep round shape except B. *imbricata*, which has a clover shaped caudal nuclear depression (Fig. 11.34A, Inset) during elongation. This unique feature of the caudal nucleus is a possible autapomorphy of *B. imbricata* or a potential synapomorphy for either the *Barisia* genus or the Anguidae family. Within or around the nuclear fossa in all lizards studied to date (Table 11.1), there exists a connecting piece or dense dark staining protein plaques that often are reported as becoming the connecting piece. These plaques are termed pericentriolar material (Fig. 11.38B) by most authors. In *Scincella lateralis* (Gribbins *et al.* 2007), the distal centriole/neck region is elongated compared to the other nine lizards within this comparison. This skink also has larger accumulations of the

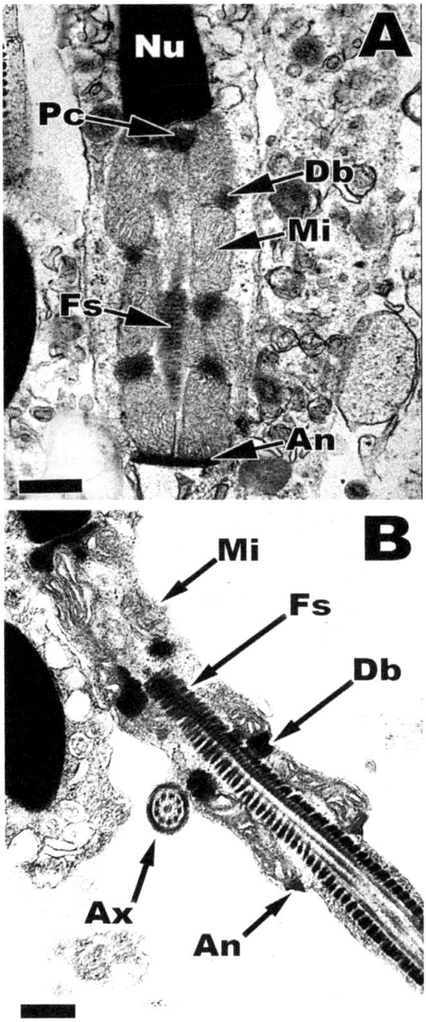

Fig. 11.35 (A) Midpiece in sagittal section late in the development in *Sceloporus variabilis*. The oblong mitochondria (Mi) and dense bodies (Db) are associated with the axoneme of the flagellum; the fibrous sheath (Fs) of the principal piece begins at mitochondria tier 2. The annulus (An) appears as a perfect ring in sagittal section. Nucleus, Nu. Proximal centriole (Pc). Bar = 2 μm. **(B)** The flagellum becomes differentiated as mitochondria (Mi) and dense bodies (Db) surround the midpiece in *S. bicanthalis*. A fibrous sheath (Fs) begins in the midpiece at mitochondrial tier 2 and continues past the annulus (An) as the principle piece. Axoneme (Ax). Bar = 1 μm.

pericentriolar material than that reported for other lizards. It again is hard to judge the importance of these two characters associated with the neck region with so little lizard data available.

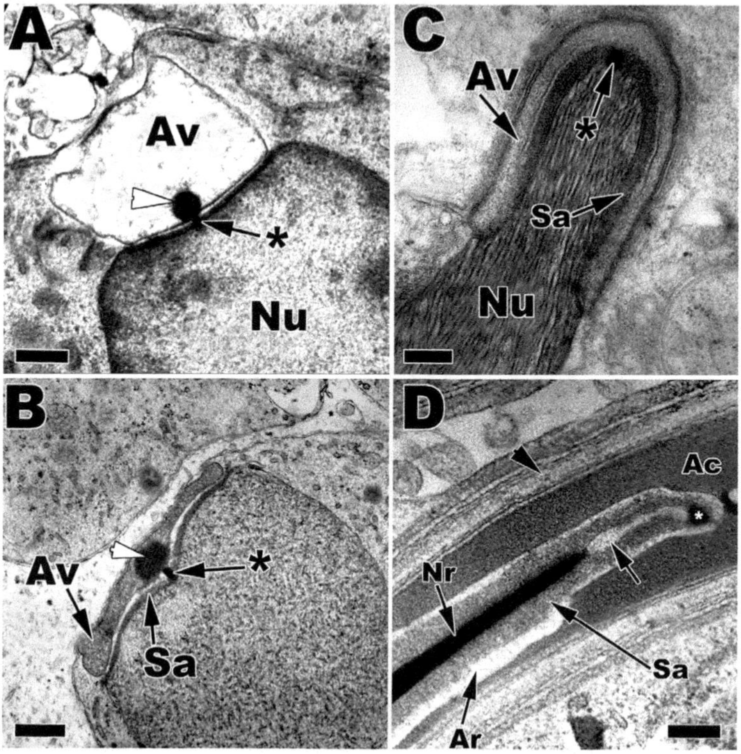

Fig. 11.36 *Sceloporus variabilis.* Showing the progression and position of the subacrosomal granule (*) through the ontogeny of spermiogenesis in spermatids. The granule is located within the subacrosome space (Sa) and just under the acrosome granule (white arrowhead), which is in the acrosome vesicle (Av), during the round spermatid stages **(A, B)** of development. In mid-elongation **(C)** it is still located within the subacrosomal space just above the nucleus (Nu). The remnants (white *) of the subacrosomal granule **(D)** still persist at the top of the epinuclear lucent zone (black arrow), which has its base resting on the edge of the nuclear rostrum (Nr). Sertoli cell membrane laminae (black arrowhead), Subacrosome lucent ridge, Acrosome Vesicle (Ar). Bars for A,B,C = 1 μm, Bar for D = 0.5 μm.

Lizards in general have short midpieces (4 to 6 mitochondrial tiers deep; Table 11.1; Figs. 11.35, 11.38) when compared to their counterparts the snakes, within Squamata. Snakes tend to have more than 12 tiers of mitochondria and thus very long midpieces in sagittal section (Furieri 1970; Oliver *et al.* 1996; Gribbins 2011; Gribbins and Rheubert 2011). *Hemidactylus turcicus* (Rheubert *et al.* 2011) has a midpiece more similar to snakes (12 tiers deep) and at least as many tiers of dense body between mitochondrial rings in sagittal section in this gecko. Both the very large numbers of mitochondria and dense bodies are a result of this gecko's elongated midpiece and are definitely potential autapomorphies for the Mediterranean Gecko. Dense bodies are common within the rings of mitochondria of the midpiece of lizards and have been thought to be products of degenerating mitochondria as dense bodies only appear in the midpiece during development when mitochondria are present (Hamilton and Fawcett 1968; Vieira *et al.* 2007); the origin from mitochondria has been demonstrated ontogenetically in some squamates (Oliver *et al.* 1996). The sperm of *Scincella lateralis* does not exhibit or only rarely shows dense bodies within the midpiece (Gribbins *et al.* 2007), absence being a similarity with non-squamate reptiles (Gribbins 2011). The midpiece ends at the annulus (Figs. 11.35, 11.38), a protein rich ring that is located juxtaposed caudally to the last tier of mitochondria in all squamates and amniotes studied to date (Jamieson *et al.* 1996). The last common feature of all squamates studied, including lizards, is the presence of the fibrous sheath around the flagellar axoneme starting within the midpiece (Figs. 11.35, 11.38), a commencement which is considered a synapomorphy for squamates (Jamieson 1999). The fibrous sheath in the lizards included in this chapter typically begins at either mitochondrial tier 2 or 3 except in *Scincella lateralis*, whose long distal centriolar neck predetermines its fibrous sheath origin at mitochondrial tier 1 (Table 11.1).

The axoneme displays (Fig. 11.37I) the conserved 9+2 microtubule arrangement seen in other amniotes. The previously described enlarged peripheral fibers (Fig. 11.37G) of the distal centriole are also common in lizard and other squamate flagella (Gribbins and Rheubert 2011; Gribbins *et al.* 2013a,b). The peripheral fibers that are associated with microtubule doublets 3 and 8 continue into the midpiece of lizards and other squamates (Table 11.1; Healy and Jamieson 1994; Ferreira and Dolder 2003; Cunha *et al.* 2008; Tourmente *et al.* 2008) and this characteristic is considered a synapomorphy for Squamata and Lepidosauria. In some lizards (*Sceloporus bicanthalis, S. variabilis, Scincella lateralis*) these fibers extend into the principle piece and in at least one lizard, *Barisia imbricata,* they reach all the way into the endpiece (Fig. 11.37J). The fibrous sheath (Fig. 11.35) in lizards is composed of dark staining protein blocks that sequentially run along parts of the midpiece and principle piece (Gribbins *et al.* 2013a,b). The sheath terminates at the most caudal end of the flagellum, which is called the endpiece and thus has a naked axoneme surrounded by cell membrane (Fig. 11.28E).

The acrosome complex at the completion of elongation and condensation resembles that of the mature spermatozoa in lizards. The acrosome cap is typically rounded and can have stratification of diffuse material within it leading to an acrosome medulla and cortex (Fig. 11.39). The acrosomes in all lizards and squamates studied are surrounded by either one or sometimes two Sertoli cell processes that wrap the maturing elongates in multiple cell membrane lamina (Figs. 11.37, 11.39, 11.40). These multiple membranes around the head of a developing elongate allow for numerous desomosome-

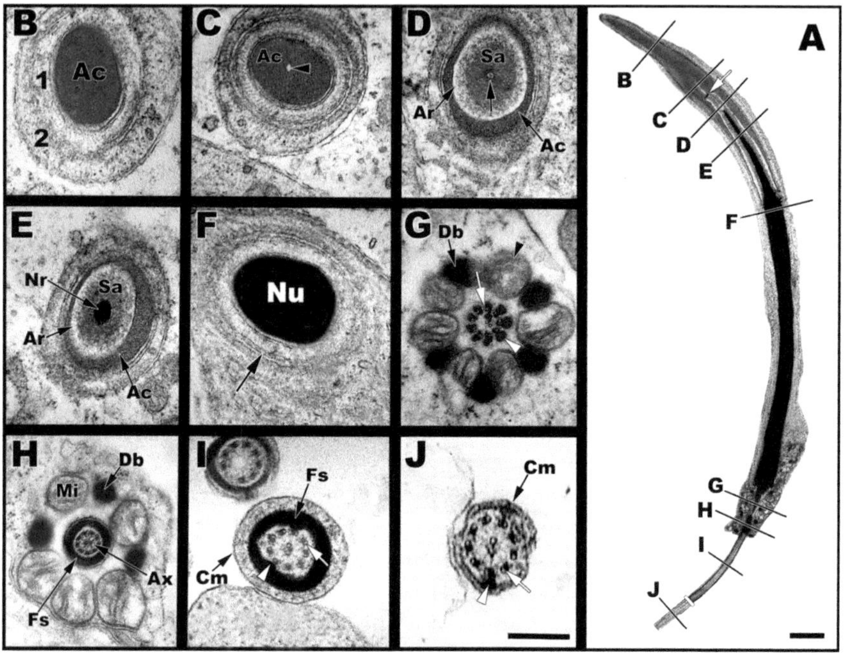

Fig. 11.37 *Sceloporus variabilis.* Late elongation and condensation during spermiogenesis. **(A)** Sagittal section of the mature spermatid showing the acrosome, basal perforatorial plate (white arrow), nucleus, and flagellum. Lettered perpendicular lines correspond to cross sections B-J. Bar = 1 μm. **(B)** Cross section through the apical acrosome vesicle (Ac) and its Sertoli membrane (1,2) cap. **(C)** Cross section through the acrosome vesicle (Ac) and the perforatorium (black arrowhead). **(D)** Cross section through the epinuclear lucent zone (black arrowhead) of the subacrosome cone (Sa). Subacrosome lucent ridge, Ar; acrosome vesicle, Ac. **(E)** Cross section through the nuclear rostrum (Nr). Subacrosome space, (Sa); Subacrosome lucent ridge, Ar; acrosome vesicle, Ac. **(F)** Cross section through the nucleus proper (Nu). Manchette microtubules, black arrow. **(G)** Cross section through the midpiece and distal centriole. Mitochondria, black arrowhead; dense body, Db; microtubule triplet, white arrow; peripheral fiber, white arrowhead. **(H)** Cross section through the midpiece and the principal piece. Mitochondria, Mi; dense body, Db; axoneme, Ax; fibrous sheath, Fs. **(I)** Cross section through principal piece. Fibrous sheath, Fs; axoneme, white arrow; cell membrane, Cm. **(J)** Cross section through the endpiece. Cell membrane, Cm; peripheral fiber number 3, white arrowhead; axoneme, white arrowhead. Bars = 1 μm.

like (Fig. 11.39E,F) junctions to exist between the Sertoli cells and the spermatid, preventing its premature release to the lumina of seminiferous tubules. All lizards examined to date, like their other reptilian kin, have acrosome vesicles that envelop a compartmentalized subacrosome space (Figs. 11.37, 11.39,11.40), which has been described as a plesiomorphy for Reptilia (Oliver *et al.* 1996; Scheltinga *et al.* 2000). All lizards, like squamates in general, have the same structures within the acrosome. These include an extranuclear perforatorium (Figs. 11.37, 11.39, 11.40) resting on a prominent

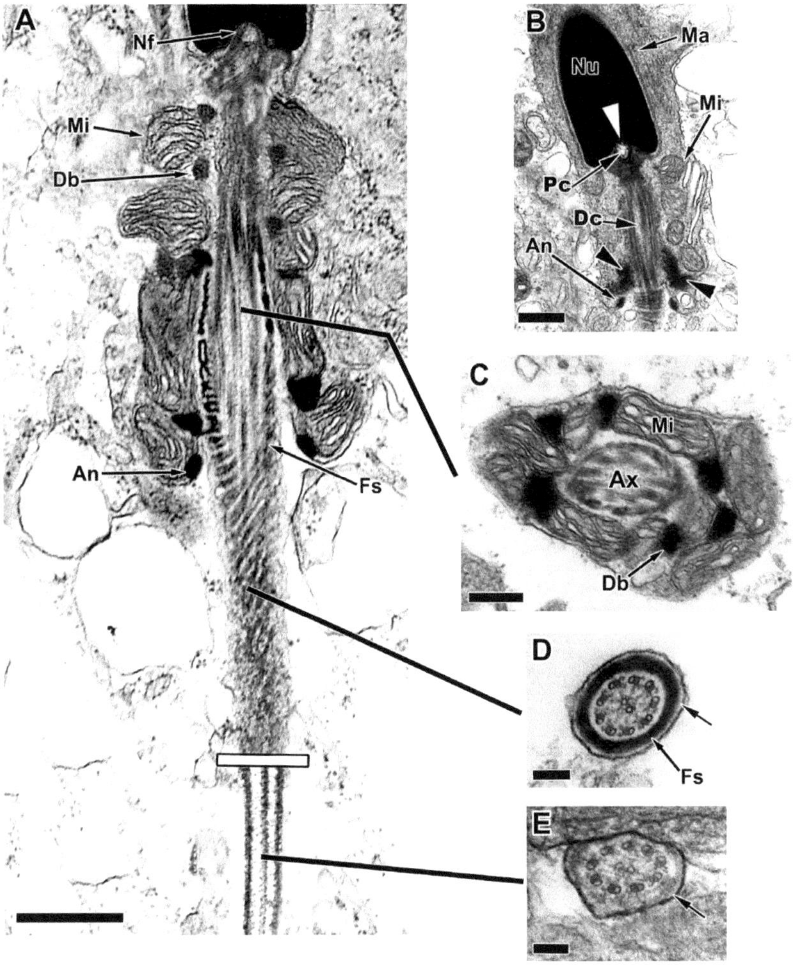

Fig. 11.38 *Barisia imbricata*. Caudal end of a spermatid prior to spermiation. **(A)** Sagittal section detailing the nuclear fossa (Nf) and the developing flagella. Annulus (An), dense bodies (Db), mitochondria (Mi), fibrous sheath (Fs). **(B)** Sagittal view the proximal centriole (Pc) situated within the nuclear fossa (white arrowhead). Distal centriole (Dc), annulus (An), mitochondria (Mi), nucleus (Nu), manchette (Ma), pericentriolar material (black arrowheads). **(C)** Cross sectional view through the midpiece of the flagellum detailing the mitochondria ring (Mi) and dense bodies (Db) surround the axoneme (Ax). **(D)** Cross sectional view through the principal piece of the flagellum detailing the fibrous sheath (Fs). Cell membrane (black arrow). **(E)** Cross sectional view through the endpiece of the flagellum showing the axoneme and cell membrane (black arrow). Bar for sagittal section = 1 μm, Bars for B,C = 0.25 μm, Bars for D,E = 0.1 μm.

basal plate. Presence of the plate is a synapomorphic condition (Jamieson 1999) of Squamata; it is thought to help propel the acrosome forward during fertilization because actin is found within it (Ferreira *et al.* 2006). The

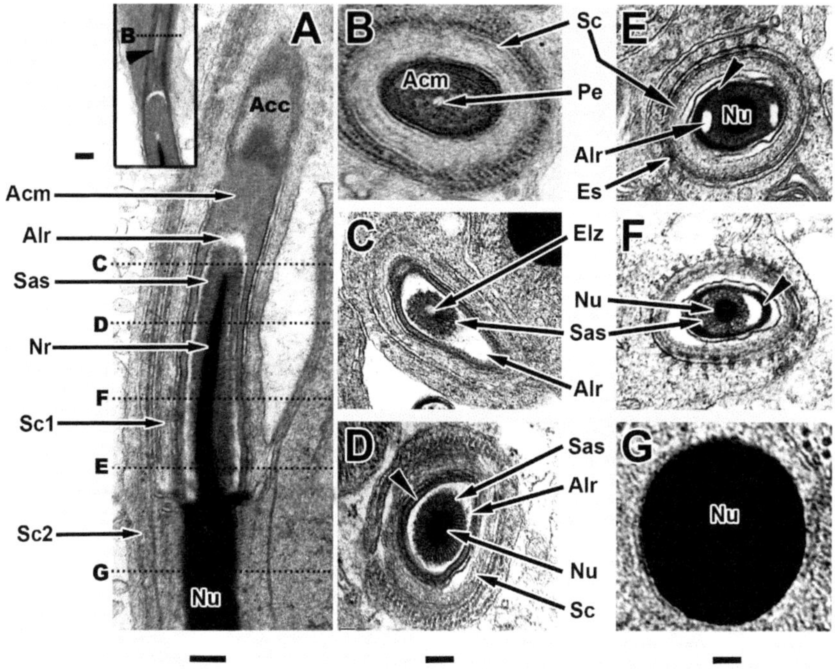

Fig. 11.39 *Sceloporus bicanthalis*. Nuclear head of a spermatid close to completing spermiogenesis. **(A)** Sagittal view of the apical germ cell demonstrating the organization of the acrosome complex with the acrosome cortex (Acc), medulla (Acm), acrosome lucent ridge (Alr), subacrosome space (Sas), nuclear rostrum (Nr), and surrounding Sertoli cell processes (Sc1,Sc2). Inset: showing the perforatorium (black arrowhead). **(B)** Cross section through the apex of the acrosome showing the perforatorium (Pe) extending into the acrosome medulla (Acm). Sertoli cell laminae (Sc). **(C)** Cross sectional view of the epinuclear lucent zone (Elz) within the subacrosome space (Sas). Subacrosome lucent ridge (Alr). **(D)** Cross section through the nuclear rostrum (Nu) within the subacrosome space (Sas). Subacrosome lucent ridge (Alr). **(E)** Cross sectional view of the nucleus caudally below the acrosome. Notice the small pockets of the acrosome lucent ridge (Alr). Sertoli cell laminae (Sc), Nucleus (Nu), Ectoplasmic specializations of the desmosomes (Es). **(F)** Nuclear (Nu) cross section through acrosome (black arrowhead) and subacrosome cone (Sas). **(G)** Cross section through the nucleus proper (Nu). Bars = 200 nm.

perforatorium is approximately cylindrical in all species of lizard studied to date (Table 11.1) except for the shortened perforatorium of *Scincella lateralis* and the wide bulbous version found in *Barisia imbricata* (Fig. 11.40A). Another synapomorphy of squamates (Jamieson 1995; Jamieson *et al.* 1996; Oliver *et al.* 1996; Jamieson 1999; Gribbins and Rheubert 2011; Gribbins *et al.* 2013a), including lizards, is the presence of an epinuclear lucent zone within the subacrosome cone. It rests atop the nuclear rostrum, which penetrates the acrosome complex. Again in most lizard species this lucent

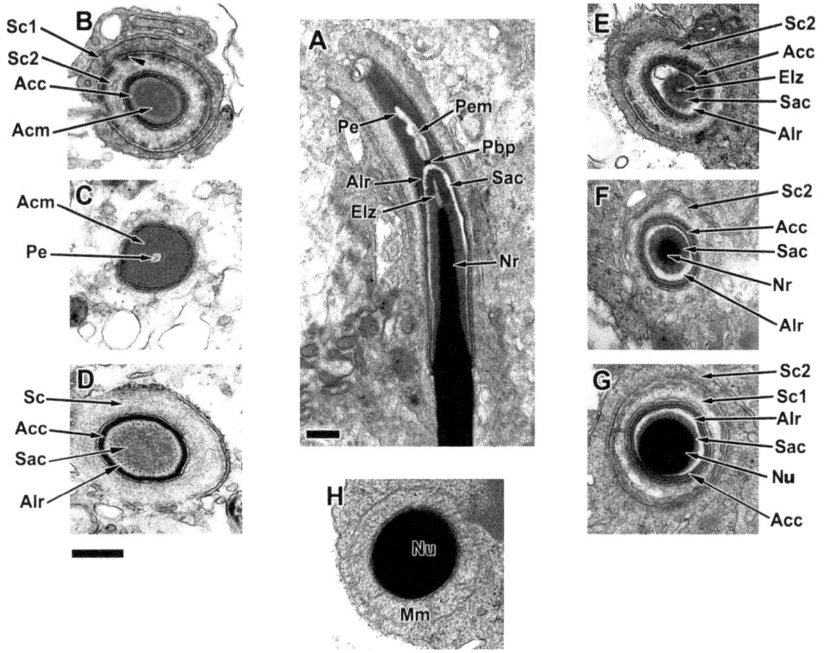

Fig. 11.40 *Barisia imbricata*. Acrosome complex and apical nucleus in a spermatid just prior to spermiation. **(A)** Sagittal section through the acrosomal complex showing the compartmentalization of the acrosome with the perforatorium (Pe), perforatorial microfilaments (Pem), perforatorial base plate (Pbp), acrosomal lucent ridge (Alr), subacrosomal cone (Sac), epinuclear lucent zone (Elz), and nuclear rostrum (Nr). Bar = 0.2 μm. **(B)** Cross sectional view through the apex of the acrosome showing Sertoli cells (Sc1 and Sc2) surrounding the acrosome. The acrosomal cap is divided into a cortex (Acc) and medulla (Acm). Desomosomes (black arrowhead). **(C)** Cross sectional view through the acrosomal apex (Sas) detailing the perforatorium (Pe) and acrosomal medulla (Acm). **(D)** Cross sectional view through the subacrosomal cone (Sac) and the acrosomal shoulders (Acc). Sertoli cell (Sc), acrosomal lucent ridge (Alr). **(E)** Cross sectional view through the epinuclear lucent zone (Elz) within the subacrosomal cone (Sac). The acrosomal lucent ridge (Alr) separates the subacrosomal cone from the acrosomal shoulders (Acc). The Sertoli cells (Sc2) surrounds the entire acrosome. **(F)** Cross sectional view through the nuclear rostrum (Nr) which extends into the subacrosomal cone (Sac). The acrosomal lucent ridge (Alr) separates the subacrosomal cone and the acrosomal shoulders (Acc). The entire acrosome is surrounded by the Sertoli cells (Sc2). **(G)** Cross sectional view through the nucleus proper (Nu) and acrosomal shoulders (Acc). Acrosomal lucent ridge (Alr), Serotli cells (Sc1,Sc2), Subacrosomal cone (Sac) acrosomal lucent ridge (Alr). **(H)** Cross sectional view through the nucleus (Nu) detailing the microtubules of the manchette (Mm). Bar for all transverse sections = 0.2 μm.

zone is cylindrical with the exception of its absence in *Scincella lateralis* and it wide bulbous shape within the acrosome of *Barisia imbricata* (Fig. 11.40A; Table 11.1).

11.5 THE SAURIAN SPERMATOZOON

Spermatozoa morphology has received a vast amount of attention historically with over 75 species of squamates within 26 families studied. Furthermore, spermatozoon descriptions for the squamate sister taxon, Sphenodontida, have also been provided and are presented in Jamieson (Chapter 17, this volume) which also outlines general descriptions of the ancestral lepidosaurian spermatozoon and amniotic spermatozoon and therefore will not be presented here.

11.5.1 Introduction to Squamate Sperm

The mature spermatozoa of squamates follows the general bauplan of spermatozoa for amniotes. They are highly elongated, filiform in shape and can be broken into three main regions (see Fig. 11.41): the acrosome, nucleus, midpiece and flagellum. Each of the regions will be discussed separately. Following the descriptions and variations observed in each region, an evolutionary analysis will be performed in order to outline ancestral characters for major taxonomic groupings.

The acrosome. As seen in other amniotes, the acrosome is an elongated structure at the apex of the sperm that contains proteins and enzymes that are responsible for the degradation of oocyte extracellular layers during fertilization. The acrosome complex (Fig. 11.42) is as a cone shaped-structure that drapes over the apical portion of the nucleus. In most species studied, such as *Sceloporus consobrinus*, the acrosome complex is laterally compressed and appears "spatula" shaped (Fig. 11.42B) and thus appears laterally flattened in cross section. A single unilateral ridge (Fig 11.42C, black arrow bracket) may be present on the lateral surface; however, this morphological arrangement is highly variable among the squamates. The outer portion of the acrosome can be divided into the acrosome cortex (Figs. 11.42C, 11.43A,B,C) and acrosome medulla (Figs. 11.42A,C, 11.43A) which are easily distinguishable based on electron density with the inner acrosome medulla being more electron dense than the acrosome cortex. Within the acrosome medulla resides a single perforatorium (Fig. 11.42A, Pe), which houses microfilaments (Actin-Fererria *et al.* 2006). In some species, the perforatorium rests on an electron dense perforatorial base plate (Fig. 11.42A, Pbp) but this structure is absent in other taxa such as *Tropidurus* (Teixeira *et al.* 1999a,b; Ferreira and Dolder 2003). If present, the perforatorial base plate can be stopper shaped as observed in *Barisia imbricata* (Gribbins *et al.* 2013a) or knob shaped as seen in *Iguana iguana* (Vieira *et al.* 2004).

The outer acrosome vesicle is separated from the inner subacrosome cone (Figs. 11.42A, 11.43B,C, Sac), sometimes referred to as the subacrosome

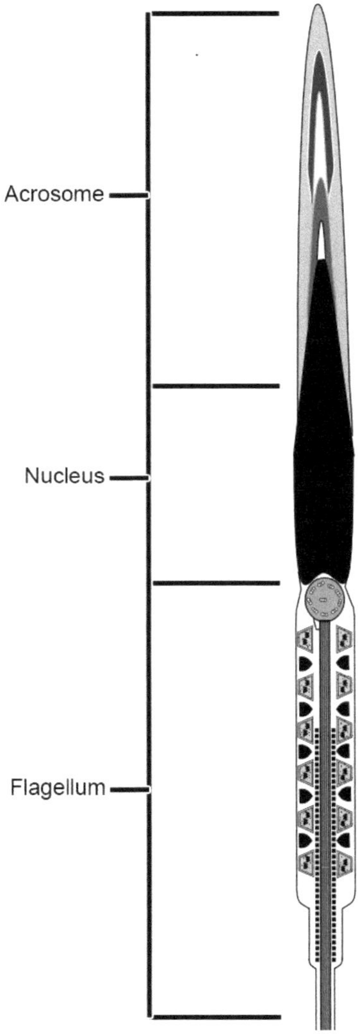

Fig. 11.41 Schematic drawing of a generalized squamate spermatozoon detailing the acrosome, nucleus, and flagellum.

space, by a thin electron lucent area, which Rheubert *et al.* (2012) termed the acrosome lucent ridge (Figs. 11.42A, 11.43A,C,D, Alr). This structure has been observed in all species studied (JLR personal observation) but has sometimes been overlooked or inaccurately described as the acrosome vesicle, subacrosome clear zone, or epinuclear lucent zone. The subacrosome cone is paracrystalline in nature and is in direct contact with the apical portion of the nucleus, the nuclear rostrum (Figs. 11.42A, 11.43C,D, Nr).

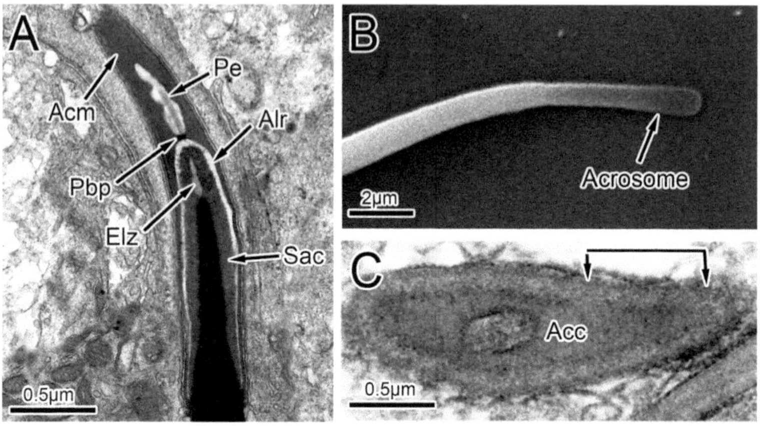

Fig. 11.42 Electron micrographs of the acrosome in *Barisia imbricata* **(A)** and *Sceloporus consobrinus* (**B** [SEM] and **C**) detailing the acrosome complex (Acc), acrosome medulla (Acm), acrosome perforatorium (Pe), Perforatorial base plate (Pbp), acrosome lucent ridge (Alr), epinuclear lucent zone (Elz), subacrosome cone (Sac), and acrosome ridge (arrowhead bracket).

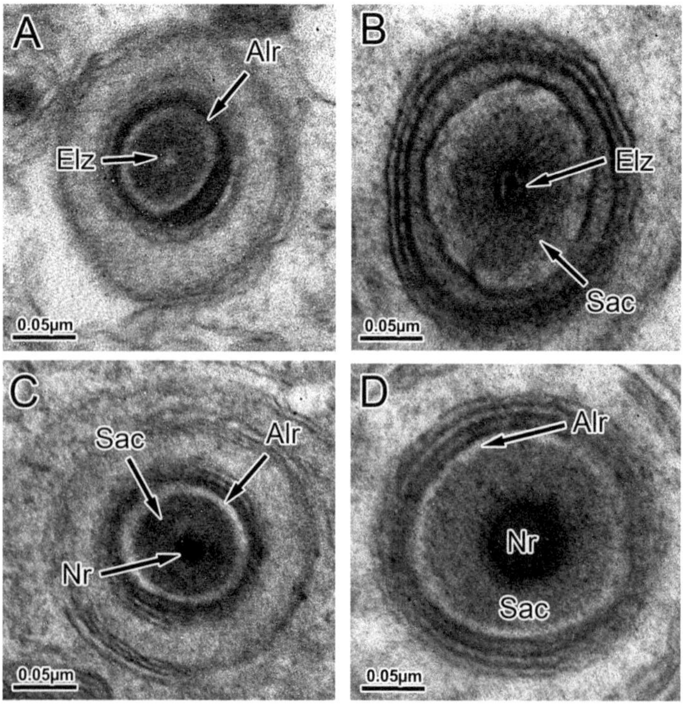

Fig. 11.43 *Hemidactylus turcicus*. **A-D**. Electron micrographs of the acrosome region detailing the acrosome complex, acrosome lucent ridge (Alr), epinuclear lucent zone (Elz), subacrosome cone (Sac), and nuclear rostrum (Nr).

An electron lucent region extending off the nuclear rostrum, termed the epinuclear lucent zone (Figs. 11.42A, 11.43A,B, Elz), is present in all taxa studied except the Scincoidea and the gekkonid, *Heteronotia binoei*.

The nucleus. The nucleus (Fig. 11.44) is an elongated electron-dense structure composed of fully condensed chromatin. The apical portion of the nucleus, the nuclear rostrum (Fig. 11.44A, Nr), is tapered and extends into the acrosome complex. In some species, an electron lucent lacuna (Fig. 11.44B, La) may be observed within the nucleus. This structure does not extend the length of the nucleus (Fig. 11.44C, Nu) and thus may have been missed in studies in which no lacuna is mentioned. The nucleus is indented at the most distal (basal) aspect forming the nuclear fossa (Fig. 11.44D, Nf), which houses the proximal and distal centrioles (Fig. 11.44D, Pc and Dc) of the flagellar axoneme, a long initial portion of which is contained within the midpiece.

The flagellum. The flagellum can be broken into four distinct regions: the neck region, the midpiece, the principal piece, and the endpiece. The neck region is made up of the proximal centriole and anterior portion of the distal centriole. The proximal centriole (Fig. 11.44D, Pc) resides within the nuclear fossa (Fig. 11.44D, Nf). Lateral projections may be observed off the proximal centriole but in some species such as Prickly Gecko (*Heteronotia*

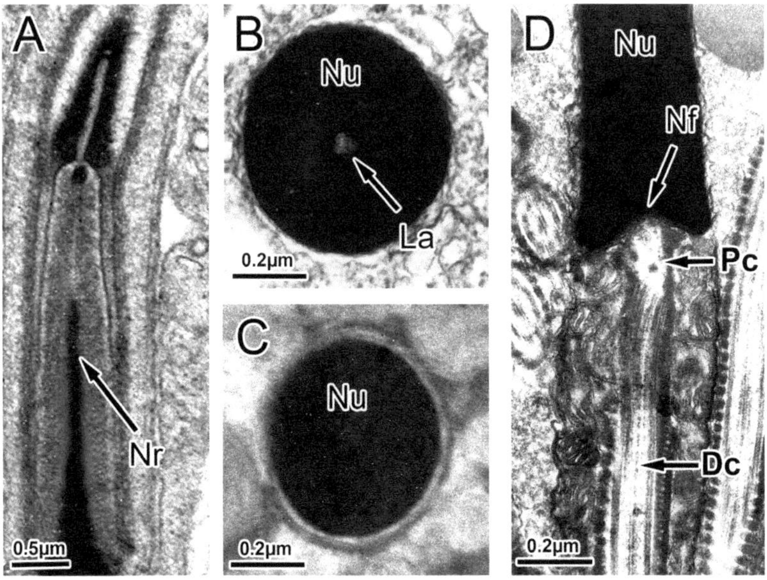

Fig. 11.44 *Sceloporus consobrinus*. Electron micrographs of the nucleus, detailing the nuclear rostrum (Nr), nuclear lacuna (La), and nuclear fossa (Nf). Nucleus (Nu), Proximal centriole (Pc), Distal Centriole (Dc).

binoei; Jamieson *et al.* 1996) they are absent. If present, the projection can be unilateral, only projecting off one side such as in *Sceloporus consobrinus* (Rheubert unpublished data), or bilateral as seen in *Anolis carolinensis* (Scheltinga *et al.* 2001). Lateral to the proximal centriole are electron dense structures that extend posteriorly to the distal centriole as a dense collar surrounding the connecting piece. This structure has been observed in several taxa but has also been absent in certain taxa such as *Tropidurus torquatus* (Teixeira *et al.* 1999c).

The midpiece of the flagellum (Fig. 11.45A) is easily identifiable as it is the region of the flagellum surrounded by mitochondria (Fig. 11.45A,

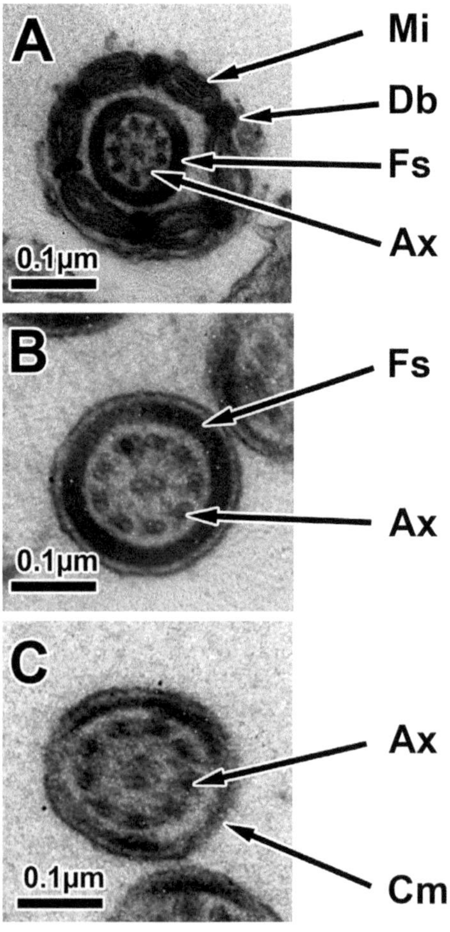

Fig. 11.45 Transmission electron micrograph of the midpiece **(A)**, principal piece **(B)**, and endpiece **(C)** detailing the mitochondria (Mi), dense bodies (Db), fibrous sheath (Fs), axoneme (Ax), and cell membrane (Cm).

Mi). Within the centrioles are 9 sets of triplet microtubules surrounding a central pair of microtubules, which then continues on in the typical amniotic organization of the 9+2 microtubule arrangement with the axoneme (Fig. 11.45A, Ax). Electron dense structures, termed peripheral fibers, associated with microtubule doublets 3 and 8 are enlarged in all species studied to date except the Brazilian Bush Anole (*Polychrus acutirostris*; Teixeira *et al.* 1999b). The axoneme is enveloped by an electron dense fibrous sheath (Fig. 11.45A, Fs) that begins shortly after the neck region. However, the starting location of the fibrous sheath is highly variable within squamates. Surrounding the centriole exterior to the fibrous sheath are mitochondria (Fig. 11.45A, Mi) with linear cristae that in cross section can appear round in shape as seen in the Great Basin Collared Lizard (*Crotaphytus bicintores*; Scheltinga *et al.* 2001), trapezoidal as seen in *Iguana iguana* (Vieira *et al.* 2004), or irregular in shape as seen in *Tropidurus torquatus* (Teixeira *et al.* 1999c). In longitudinal section, the mitochondria can either be columnar as seen in *Iguana iguana* (Vieira *et al.* 2004), have rounded ends as seen in *Anolis carolinensis* (Scheltinga *et al.* 2001), or trapezoidal as seen in the Four-striped Tegu (*Tupinambis quadrilineatus*; Tavares-Bastos *et al.* 2002). Electron dense structures known as dense bodies (Fig. 11.45A, Db) are observed within the midpiece and are typically juxtaposed to the fibrous sheath. However, in Mixamilian's Blue-tailed Microteiid, *Micrablepharus maximilliani* (Teixeira *et al.* 1999a) they are separated from the fibrous sheath by the mitochondria. The midpiece is terminated by the electron dense annulus, which is irregularly shaped in squamates as opposed to the scythe shape in other amniotes Fawcett and Bloom (1970).

The principal piece (Fig. 11.45B) is devoid of mitochondria but the fibrous sheath (Fig. 11.45B, Fs) persists. In some species, such as the Red Worm Lizard, *Amphisbaenia alba* (Teixeira *et al.* 1999d), the grossly enlarged peripheral fibers associated with microtubule doublets 3 and 8 are greatly enlarged at the anterior most region of the principal piece. The terminal portion of the flagellum, the endpiece (Fig. 11.45C), consists only of the 9+2 microtubule arrangement of the axoneme (Fig. 11.45C, Ax) and the cellular membrane (Fig. 11.45C, Cm). Mitochondria and the fibrous sheath are not found in this region.

11.5.2 Insights into Sperm Evolution

Previous works have investigated sperm characters in an evolutionary and phylogenetic context (Jamieson 1995; Oliver *et al.* 1996; Tavares-Bastos *et al.* 2008; Gribbins and Rheubert 2011) but no study to date has investigated the evolution of these characters across all squamates. Although the data provided by such analyses are preliminary since the number and diversity of taxa studied in terms of sperm morphology are limited they still provide valuable insights into the evolution of the spermatozoa. In order to understand the evolution of the spermatozoa in squamates 28 morphological characters (same characters utilized by Gribbins and Rheubert 2011 for snakes) were coded by a single individual (JLR) for 80 species representing 26 families from previously published works (For species utilized see Table 11.2). The characters and character states can be found in Table 11.3. Furthermore, characters were mapped onto the phylogenetic hypothesis proposed by Pyron *et al.* (2013) to try to reconcile ancestral character states under a parsimonious framework using Mesquite. Many characters are highly variable among species and thus analyses were limited to major taxonomic levels above family level designation (Fig. 11.46). More profound analyses in the future may lead to a more rigorous testing of character optimization and/or phylogenetic analyses if and when other squamate species are studied for spermatozoal characters.

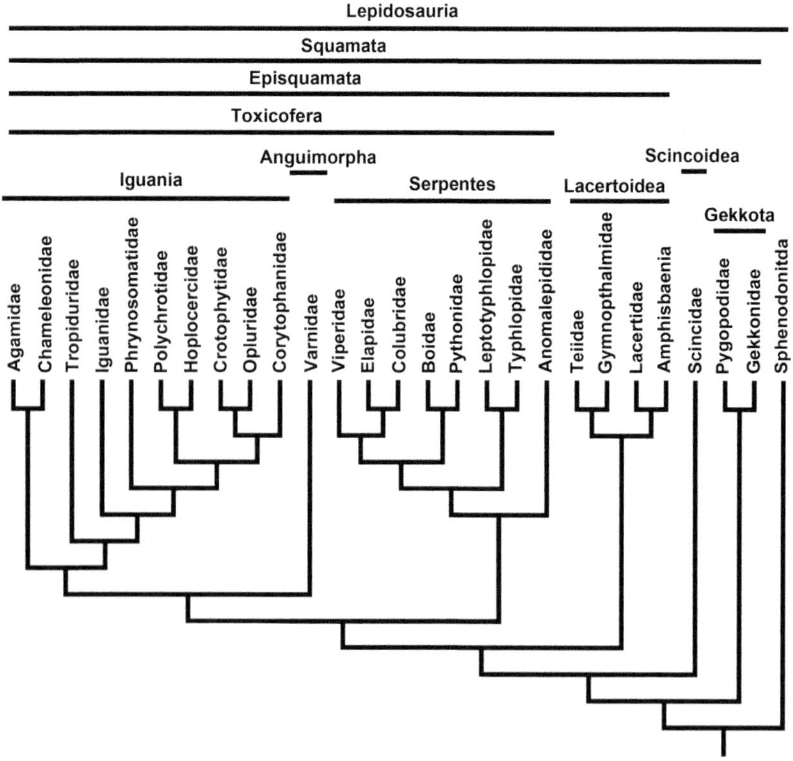

Fig. 11.46 Phylogenetic hypothesis of squamates proposed by Pyron *et al.* (2013) showing family level relationships and higher taxonomic classifications.

Table 11.2 List of species with ultrastructural sperm data.

species	**Family**	**source**
Sphenodon punctatus	Sphenodontidae	Healy and Jamieson 1994
Pogona barbata	Agamidae	Oliver *et al.* 1996
Bradypodion karrooicum	Chameleonidae	Jamieson 1995
Enyalioides laticeps	Hoplocercidae	Vieira *et al.* 2007
Hoplocercrus spinosus	Hoplocercidae	Vieira *et al.* 2007
Anolis carolinensis	Polychrotidae	Scheltinga *et al.* 2001
Polychrus acutirostris	Polychrotidae	Teixeira *et al.* 1999
Corytonphanes cristatus	Corytophanidae	Vieira *et al.* 2005
Basiliscus vittatus	Corytophanidae	Vieira *et al.* 2005
Laemanctus longipes	Corytophanidae	Vieira *et al.* 2005
Iguana iguana	Iguanidae	Vieira *et al.* 2004
Crotaphytus bicinctores	Crotaphytidae	Scheltinga *et al.* 2001
Gambelia wislizenii	Crotaphytidae	Scheltinga *et al.* 2001
Oplurus cyclurus	Opluridae	Vieira *et al.* 2007
Tropidurus semitaeniatus	Tropiduridae	Teixeira *et al.* 1999
Tropidurus torquatus	Tropiduridae	Teixeira *et al.* 1999
Tropidurus itambre	Tropiduridae	Ferreira and Dolder 2003
Urosaurus ornatus	Phrynosomatidae	Scheltinga *et al.* 2000
Uta stansburiana	Phrynosomatidae	Scheltinga *et al.* 2000
Sceloporus undulatus	Phrynosomatidae	Rheubert unpublished
Heteronotia binoei	Gekkonidae	Jamieson *et al.* 1995
Hemidactylus frenatus	Gekkonidae	Furieri 1970
Lygodactylus picturatus	Gekkonidae	Furieri 1970
Lialis burtonis	Pygopodidae	Jamieson *et al.* 1995
Ameiva Ameiva	Teiidae	Giuliano *et al.* 2002
Tupinambis merianae	Teiidae	Tavares-Bastos *et al.* 2002
Tupinambis quadrilineatus	Teiidae	Tavares-Bastos *et al.* 2002
Tupinambis duseni	Teiidae	Tavares-Bastos *et al.* 2002
Tupinambis teguixin	Teiidae	Tavares-Bastos *et al.* 2002
Callopistes flavipunctatus	Teiidae	Colli *et al.* 2007
Crocodilurus amazonicus	Teiidae	Colli *et al.* 2007

Table 11.2 contd....

Table 11.2 contd.

species	Family	source
Dicrodon guttulatum	Teiidae	Colli *et al.* 2007
Draceaena guianensis	Teiidae	Colli *et al.* 2007
Cnemidophorus gularis	Teiidae	Teixeira *et al.* 2002
Cnemidophorus ocellifer	Teiidae	Teixeira *et al.* 2002
Cnemidophorus sexlineatus	Teiidae	Newton and Trauth 1992
Kentropyx altamazonica	Teiidae	Teixeira *et al.* 2002
Teius oculatus	Teiidae	Colli *et al.* 2007
Cercosaura ocellata	Gymnophthalmidae	Colli *et al.* 2007
Micrablepharus maximiliani	Gymnophthalmidae	Teixeira *et al.* 1999
Takydromus septentrionalis	Lacertidae	Yong-pu *et al.* 2005
Eumeces elegans	Scincidae	Yong-Zhang and Yong-pu 2004
Chalcides ocellatus	Scincidae	Furieri 1970
Ctenotus robustus	Scincidae	Jamieson *et al.* 1995
Tiliqua scincoides	Scincidae	Jamieson *et al.* 1995
Nangura spinosa	Scincidae	Jamieson *et al.* 1995
Ctenotus taeniolatus	Scincidae	Jamieson *et al.* 1995
Anomalopus verreauxii	Scincidae	Jamieson *et al.* 1995
Carlia pectoralis	Scincidae	Jamieson *et al.* 1995
Cryptoblepharus virgatus	Scincidae	Jamieson *et al.* 1995
Lampropholis delicata	Scincidae	Jamieson *et al.* 1995
Scincimorphus indicus	Scincidae	Yong-Pu *et al.* 2009
Mabuya multifasciata	Scincidae	Yong-Pu *et al.* 2009
Mabuya nigropunctata	Scincidae	Mandel *et al.* 2009
Mabuya frenata	Scincidae	Mandel *et al.* 2009
Mabuya guaporicola	Scincidae	Mandel *et al.* 2009
Mabuya dorsivittata	Scincidae	Mandel *et al.* 2009
Amphisbaenia alba	Amphisbaenia	Teixeira *et al.* 1999
Eryx jayakari	Boidae	Al-Dokhi *et al.* 2007
Boa constrictor amarali	Boidae	Tavares-Bastos *et al.* 2008
Boa constrictor occidentalis	Boidae	Tourmente *et al.* 2006
Corallus hortulanus	Boidae	Tavares-Bastos *et al.* 2008

Table 11.2 contd....

Table 11.2 contd.

species	Family	source
Epicrates cenchria	Boidae	Tavares-Bastos *et al.* 2008
Aspidites melancephalus	Pythonidae	Oliver *et al.* 1996
Crotallus durissus	Viperidae	Cunha *et al.* 2008
Bothrops alternatus	Viperidae	Tourmente *et al.* 2008
Bothrops diporus	Viperidae	Tourmente *et al.* 2008
Vipera aspis	Viperidae	Furiera 1965
Oxyuranus microlepidotus	Elapidae	Oliver *et al.* 1996
Typhlops reticulatus	Typhlopidae	Tavares-Bastos *et al.* 2007
Ramphotyphlops waitii	Typhlopidae	Harding *et al.* 1995
Ramphotyphlops endoterus	Typhlopidae	Harding *et al.* 1995
Leptotyphlops koppesi	Leptotyphlops koppesi	Tavares-Bastos *et al.* 2007
Liotyphlops beui	Anomalepididae	Tavares-Bastos *et al.* 2007
Nerodia sipedon	Colubridae	Jamieson and Koehler 1994
Boiga irregularis	Colubridae	Oliver *et al.* 1996
Elaphe scalaris	Elapidae	Camps and Bargallo 1977
Stegonotus cucullatus	Colubridae	Oliver *et al.* 1996
Seminatrix pygaea	Colubridae	Rheubert *et al.* 2010
Varanus gouldii	varanidae	Oliver *et al.* 1996

11.5.3 Acrosomal Ancestral States

An acrosomal ridge was resolved as absent in the Anguimorpha, Scincoidea, Gekkota, and Squmata and Lepidosauria, equivocal in the Episquamata, Toxicofera, and present in the Lacertoidea and Iguania. The acrosomal shape in cross section was resolved as circular in Lepidosauria, Squamata, Gekkota, and Serpentes but depressed in the Scincoidea, Episquamata, Toxicofera, and Iguania. An acrosomal vesicle subdivision is resolved as present in all major clades of the Squmata but equivocal in the Lepidosauria due to its absence in *Sphenodon*. A subacrosomal cone that is paracrystalline is resolved as ancestral for the Squamata but equivocal in the Lepidosauria. An acrosomal vacuity subdivision was resolved as absent in all major clades of Lepidosauria but present in *Ramphotyphlops*. The epinuclear lucent zone is resolved as present in all major clades of Squamata but equivocal in the Lepidosauria due to its absence in *Sphenodon*. A single perforatorium was resolved as the ancestral condition to the Squamata

Table 11.3 Characters and character states used in evolutionary analyses (from Gribbins and Rheubert 2011).

Character	*States*
1. Acrosome Complex Ridge	0-absent; 1-present
2. Acrosome CS Shape	0-circular; 1-depressed
3. Acrosome Subdivisions	0-absent; 1-present
4. Subacrosome Cone	0-paracrystalline; 1-not paracrystalline
5. Acrosome: Vacuity Subdivision	0-absent; 1-present
6. Epinuclear lucent Zone	0-absent; 1-present
7. Perforatorium #	0-2; 1-1
8. Perforatorium Tip	0-pointed; 1-rounded
9. Perforatorium Basal Plate	0-absent; 1-present
10. Perfor. Basal Plate Shape	0-knob-like; 1-stopper-like; 2-N/A
11. Nuclear Lacunae	0-absent; 1-present
12. Neck: Stratified Laminae	0-absent; 1-present
13. Neck: Laminar Projections	0-unilateral; 1-bilateral; 2-N/A
14. Dense Material: Prox. Centriole	0-absent; 1-present
15. Dense Collar (Cylinder)	0-absent; 1-present
16. Midpiece 3 and 8 Fibers	0-grossly enlarged; 1 not grossly enlarged
17. Midpiece: Mito. Cristae	0-concentric; 1-linear
18. Mito. Shape OS	0-columnar; 1-slightly curved; 2-sinuous tubes
19. Mito. Shape LS	0-Trapezoidal; 1-Oval; 2-Columnar sq. ends
20. Mito. Shape CS	0-round/oval; 1- irregular; 2-trapezoidal
21. Midpiece Dense Bodies	0-solid; 1-granular; 2-N/A
22. Dense Bodies CS	0-separated fibrous sh.; 1-juxta fibrous sh.; 2-N/A
23. Midpiece Fibrous Sheath	0-t4; 1-t3; 2-t2; 3-t1; 4-before t1; 5-at annulus level; 6-N/A
24. Midpiece Annulus	0-absent; 1-present
25. Midpiece Annulus Shape	0-scythe; 1-irregular; 2-N/A
26. Principal Piece 3 and 8 Fibers	0-absent; 1-present
27. Multilaminar Membranes	0-absent; 1-present
28. Extracellular Microtubules	0-absent; 1-present

but equivocal in the Lepidosauria as *Sphenodon* posses 2 perforatoria. A rounded perforatorium tip was resolved as the ancestral condition in all major clades of the Lepidosauria except in the Iguania in which it is resolved as pointed. A perforatorial base plate was resolved as present for all major clades of the Lepidosauria except the Scincoidea in which it was resolved as equivocal. The perforatorial base plate was resolved as knob shaped in the Anguimorpha, stopper shaped in the Polychrotidae + Cortyophanidae + Opluridae clade and the Typhlopidae, and not available in the remaining major clades of the Lepidosauria. The exceptions for *Sphenodon*, above, add weight to the contention (Jamieson and Healy 1992) that it is not a squamate (see also Jamieson Chapter 17, this volume).

11.5.4 Nucleus Ancestral States

A nuclear lacuna is resolved as absent in all major clades of the Squamata but equivocal in the Lepidosauria due to its presence in *Sphenodon*.

11.5.5 Flagellum Ancestral States

A stratified laminar structure in the neck region is resolved as present in the Iguania, Toxicofera, and Serpentes, equivocal in the Episquamata, and absent in the Squamata and Lepidosauria. The stratified laminar structure projecting off the neck piece is resolved as being bilateral for the Iguania, Anguidae, Toxicofera, and Serpentes, equivocal in the Lacertoidea and Episquamata, and not available in the Squamata and Lepidosauria. Electron dense structures inside the proximal centriole are resolved as present in all major clades of the Lepidosauria. A dense collar in the neck region is resolved as equivocal in the Iguania and Lepidosauria but present in all other major clades of the Squamata. Peripheral fibers in the midpiece are resolved as grossly enlarged in all major clades of the Squamata but equivocal in the Lepidosauria. Midpiece mitochondrial cristae are resolved as linear in all major clades of the Squamata. The shape of midpiece mitochondria in oblique section is resolved as equivocal for the Iguania, Lacertoidea, and Lepidosauria and sinuous tube shaped in the Toxicofera, Episquamata, and Squamata. The shape of the midpiece mitochondria in longitudinal section was resolved as equivocal for the Iguania, Toxicofera, and Lepidosauria, and columnar for the Episquamata and Squamata. The mitochondrial shape in cross section was resolved as round/oval for all major clades of the Lepidosauria. Midpiece dense bodies were resolved as granular for all major clades of the Lepidosauria. The midpiece dense bodies were resolved as juxtaposed to the fibrous sheath for all major clades of the Lepidosauria. The beginning of the fibrous sheath in the midpiece was resolved as originating at mitochondrial tier 2 in the Iguania and Lacertoidea, equivocal in the Toxicofera and Episquamata and Lepidosauria, and starting at mitochondrial tier 1 in the Squamata. A midpiece annulus that is irregular in shape is resolved as present for all the Lepidosauria.

Enlarged peripheral fibers in the principal piece is resolved as absent in all major clades of the Lepidosauria except the Scincoidea in which the ancestral state is resolved as present. Multilaminar membranes and extracellular membranes are resolved as absent in all clades of Lepidosauria but are resolved as equivocal in the Serpentes.

11.6 CONCLUSIONS

The authors hope the information in this chapter will stimulate the general interests of the reproductive and herpetological communities for the reptilian testis and spermatozoa. There is much that needs to be accomplished as far as ultrastructural studies on testicular architecture, particularly the general anatomy and function of the Sertoli cell and the morphology of the spermatozoon. Spermiogenesis and sperm ultrastructure need to be studied in more taxa within lizards and other squamates. Although the data to date are very incomplete, there are many stimulating preliminary results especially with the ontogenic features of spermiogenesis and spermatozoa. For example, Fig. 11.29 and Table 11.1 reveal potential characters that could provide some insight into the phylogeny of certain groups of lizards (such as the 6 differences between *Sceloporus bicanthalis* and *S. variabilis*). This would provide at least six differences that may be important in phylogenetic outcomes if they were mapped onto phylogenies with sperm data. The major hurdle at the present time for such phylogenetic analysis is the lack of spermiogenic information for other closely related lizard families. This is true also for all of Ophidia and Squamata and thus does not allow for much resolution in determining ancestral conditions. Before definitive support can be given to such spermiogenic ultrastructural differences many more species will have to be studied. Hopefully, the information provided here will stimulate research efforts in the area of spermiogenic and spermatogenic histology and ultrastructure within lizards and other squamate taxa. These data, however, do provide a solid morphological sequence for the development of spermatids during spermiogenesis and fine structure of the sperm and may provide baseline ultrastructural information important to histopathological studies regarding spermatogenesis for species of reptiles, whose distribution and large numbers make them potential sentinel species for reproductive toxicant studies.

From the *Sceloporus* comparison, there were six character state differences between two species within the same genus. If this trend holds true for other clades then spermiogenesis could be a powerful tool to help resolve phylogenetic inferences below the family level. Barrie G. M. Jamieson has opined for more than 20 years that squamate spermatozoa data seem to be highly conserved. It is exciting to consider that if there is as much diversity in the development of sperm (i.e., large character differences during spermiogenesis) then why do we not see more differences in spermatozoal ultrastructure? These and other interesting questions need to be answered and this chapter will at least provide some direction to reproductive anatomists interested in pursuing these issues in lizards and other reptiles.

From the detailed description of the spermatozoa multiple ancestral states are recovered in the evolutionary analyses. However, these evolutionary analyses are based off phylogenetic hypotheses that are controversial and new hypotheses are being put forth frequently. Data concerning spermatozoa morphology, which seems to be variable at even the species level, may help to relieve some of the controversy regarding lizard phylogenetics, and/or reptilian phylogenetics. Therefore, in order to accurately answer questions regarding the use of spermatozoal characters in phylogenetic/evolutionary analyses, hypothesis driven questions need to be the focus in the future.

11.7 ACKNOWLEDGMENTS

Kevin Gribbins: I wish to thank Wittenberg University for their support and resources, which aided in the completion of much of the research that appears in this chapter. Without the support of the university, such as maintaining my aging microscopes and other histological equipment, I would not be able to perform research with my very eager students. I would also like to acknowledge the tremendous Wittenberg undergraduates who enhanced my research experiences, especially my coauthor Justin Rheubert. Their hard work and dedication to this type of research rivals what is produced at the graduate school level. Lastly, I acknowledge with my sincerest gratitude and adoration to my wife, Sara, who has always supported my efforts and has endured many evenings without my presence.

Justin Rheubert: I wish to thank my co-author Kevin for all his hardwork and time he spent in my training on spermatozoal biology. Without his assistance this would not have been possible. Furthermore I would like to thank Dr. David Sever, Dr. Dustin Siegel, Dr. Stanley Trauth and Dr. Robert Aldridge for their continuous support. Last but not least I would like to thank my friends and family especially my parents, Ron and Diane Rheubert, who have supported me 100%.

11.8 LITERATURE CITED

Aire, T. A. 2007. Anatomy of the testis and male reproductive tract. pp. 37–113. In B. G. M. Jamieson (ed.), *Reproductive Biology and Phylogeny of Birds, Volume 6A*. Science Publishers, Enfield, NH.

Adelina, F., Mehanna, M. and Dolder, H. 2006. Ultrastructural immunocytochemical evidence for actin in the acrosomal complex during spermiogenesis of the lizard *Tropidurus itambere* (Rodrigues 1987) (Reptilia: Tropiduridae). Caryologia-Firenze 59: 213–220.

Al-Dokhi, O. A. 2004a. Electron microscopic study of sperm head differentiation in the Arabian Horned Viper *Cerastes cerastes* (Squamata, Reptilia). Journal of Biological Sciences 4: 111–116.

Al-Dokhi, O. A., Al-Onazee, Y. Z. and Mubarak, M. 2004b. Light and electron microscopy of the testicular tissue of the snake *Eryx jayakari* (Squamata, Reptilia) with reference to the dividing germ cells. Journal of Biological Sciences 4: 345–351.

Al-Dokhi, O.A., Al-Onazee, Y. Z. and Mubarak, M. 2007. Fine structure of the epididymal sperm of the snake *Eryx jayakari* (Squamata, Reptilia). International Journal of Zoological Research 3: 1–13.

Al-Dokhi, O., Ahmed, M., Al-Dosary, A. and Al-Sadoon, M. K. 2013. Ultrastructural study of spermiogenesis in a rare desert amphisbaenian *Diplometopon zarudnyi*. Comptes rendus biologies 336: 473–478.

Aranha, I., Bhagya, M. and Yajurvedi H. N. 2006. Testis of the lizard *Mabuya carinata*: a light microscopic and ultrastructural seasonal study. Journal of Submicroscopic Cytology and Pathology 38: 93–102.

Aldridge, R. D. 2002. Evolution of mating season in the pitvipers of North America. Herpetological Monographs 16: 1–25.

Austin, C. R. 1965. Fine structure of the snake tail. Journal of Ultrastructure Research 12: 452–462.

Baccetti, B., Bigliardi, E., Talluri, M. V. and Burrini, A. G. 1983. The Sertoli cell in lizards. Journal of Ultrastructure Research 168: 268–275.

Bergmann, M. and Schindelmeiser, J. 1987. Development of the blood testis barrier in the domestic fowl (*Gallus domesticus*). International Journal of Andrology 10: 481–488.

Bergmann, M., Schindelmeiser, J. and Greven, H. 1984. The blood-testis barrier in vertebrates having different testicular organization. Cell and Tissue Research 238: 145–150.

Bloom, D. W. and Jensh, R. P. 2002. Bloom and Fawcett: Concise Histology, 2nd ed. Arnold Publishing, New York NY.

Boisson, C. and Mattei, X. 1965. Sur la spermiogenese de *Python sebae* (Gmelin) étudiée au microscope électronique. Comptes Rendus des Séances de la Societé de Biologia et de ses Filiales 159: 1192–1194.

Boisson, C. and Mattei, X. 1966. La spermiogenese de *Python sebae* (Gmelin) observée au microscope électronique. Annales des Sciences Naturelles. Zoologie et Biologie Animale 8: 363–390.

Boretto, J. M., Ibargüengoytía, N. R., Jahn, G. A., Acosta, J. C., Vincenti, A. E. and Fornés, M. W. 2010. Asynchronic steroid activity of Leydig and Sertoli cells related to spermatogenic and testosterone cycle in *Phymaturus antofagastensis*. General and Comparative Endocrinology 166(3): 556–564.

Bouix, G. and Bourgat, R.M. 1970. Cycle spermatogenetique de Chamaeleopardalis CUV. de l'ile de la Reunion. Ann. Univ. Madagascar 7: 307–315.

Butler, R. D. and Gabri, M. S. 1984. Structure and development of the sperm head in the lizard *Podarcis* (*Lacerta*) *taurica*. Journal of Ultrastructure Research 88: 261–274.

Callard, I. P., Callard, G. V., Lance, V., Bolaffi, J. L. and Rosset, J. S. 1978. Testicular regulation in nonmammalian vertebrates. Biology of Reproduction 18: 16–43.

Camps, J. L. and Bargallo, R. 1977. Espermatogenesis de reptiles. I. Ultraestructura dose 1 espermatozoides de *Elaphe scalaris* (Schinz). Boletin de la Royal Sociedad Espanola Historia Naturelles 75: 429–446.
Carcupino, M., Corso, G. and Pala, M. 1989. Spermiogenesis in *Chalcides ocellatus tiligugu* (Gmelin) (Squamata: Scincidae): an electron microscope study. Bollettino di Zoologia 56: 119–124.
Cavicchia, J. C. and Miranda, J. C. 1988. The blood testis barrier in lizards during annual spermatogenic cycle. Microscopy and electron biology of the cell 12: 73–88.
Cavazos, L. F. 1951. Spermatogenesis of the horned lizard *Phrynosoma cornutum*. American Naturalist 373–379.
Chamut, S., Jahn, G. A., Arce, O. E. and Manes, M. E. 2012. Testosterone and reproductive activity in the male tegu lizard, *Tupinambis merianae*. Herpetological Conservation and Biology 7(3): 299–305.
Clark, A. Q. 1967. Some aspects of spermiogenesis in a lizard. American Journal of Anatomy 121: 369–400.
Cole, A., Meistrich, M. L., Cherry, L. M. and Trostle-Weige, P. K. 1988. Nuclear and Manchette development in spermatids of normal and azh/azh mutant mice. Biology of Reproduction 38: 385–401.
Colli, G. R., Teizeira, R. D., Scheltinga, D. M., Mequita, D. O., Wiederhecker, H. C. and Báo, S. N. 2007. Comparative study of sperm ultrastructure of five speices of teiid lizards (Teiidae, Squamata), and *Cercosaura ocellata* (Gymnophthalmidae, Squamata). Tissue & Cell 39: 59–78.
Cooksey, E. J. and Rothwell, B. 1973. The ultrastructure of the Sertoli cell and its differentiation in the domestic fowl (*Gallus domesticus*). Journal of Anatomy 114: 329–345.
Courty, Y. and Dufaure, J. P. 1979. Levels of testosterone in the plasma and testis of the viviparous lizard (*Lacerta vivipara* jacquin) during the annual cycle. General and Comparative Endocrinology 39: 336–342.
Courtens, J. L. and Depeiges, A. 1985. Spermiogenesis of *Lacerta vivipara*. Journal of Ultrastructure Research 90: 203–220.
Cunha, L. D., Tavares-Bastos, L. and Bao, S. N. 2008. Ultrastructural description and cytochemical study of the spermatozoa of *Crotallus durissus* (Squamata, Serpentes). Micron 39: 915–925.
Da Cruz-Landim, C. and Da Cruz-Hofling, M. A. 1977. Electron microscope study of lizard spermiogenesis in *Tropidurus torquatus* (Lacertilia). Caryologia 30: 151–162.
Dehlawi, G. Y., Ismail, M. F., Hamdi, S. A. and Jamjoom, M. B. 1992. Ultrastructure of spermiogenesis of a Saudian reptile. The sperm head differentiation in *Agama adramitana*. Archives of Andrology 28: 223–234.
Del Conte, E. 1976. The subacrosomal granule and its evolution during spermiogenesis in a lizard. Cell and Tissue Research 171(4): 483–498.
Defuare, J. P. 1971. L'ultrastructure du testicule de lézard vivipara (Reptile, Lacertilien) II. Les cellules de Sertoli. Etude du glycogene. Zeitschrift fur Zellforschung 115: 565–578.
Eckstut, M. E., Sever, D. M., White, M. E. and Crother, B. I. 2009. Phylogenetic analysis of sperm storage in female squamates. pp. 185–218. In L. T. Dahnof (ed.), *Animal Reproduction: New Research Developments*. Nova Science Publishers, Inc., Hauppauge, NY.
Engel, K. B. and Callard, G. V. 2007. Endocrinology of Leydig cells in nonmammalian vertebrates. pp. 207–224. In A. H. Payne and M. P. Hardy (eds.), *Contemporary Endocrinology: The Leydig cell in health and disease*. Humana Press Inc., Totowa, NY.
Fawcett, D. W., Anderson, W. A. and Phillips, D. M. 1971. Morphogenetic factors influencing the shape of the sperm head. Developmental Biology 26(2): 220–251.
Fawcett, D. W. and Bloom, J. R. 2002. Fawcett's concise histology. Arnold Publishing, London.
Ferreira, A. 2009. Formation of the spermatozoon during and after spermiogenesis of *Iguana iguana* Linnaeus, 1758 (Reptilia, Iguanidae), with special attention to the middle piece and axonemic complex. Bioscience Journal 25: 181–187.

Ferreira, A. and Dolder, H. 2002. Ultrastructural analysis of spermiogenesis in *Iguana iguana* (Reptilia: Sauria: Iguanidae). European Journal of Morphology 40: 89–99.
Ferreira, A. and Dolder, H. 2003. Sperm ultrastructure and spermatogenesis in the lizard, *Tropidurus itambre*. Biocell 27: 353–362.
Ferreira, A., Mehanna, M. and Dolder, H. 2006. Ultrastructural immunocytochemical evidence for actin in the acrosomal complex during spermiogenesis of the lizard *Tropidurus itambere* (Rodrigues 1987) (Reptilia: Tropiduridae). Caryologia 59: 213–9.
Fox, W. 1952. Seasonal variation in the male reproductive system of he Pacific Garter Snake. Journal of Morphology 90: 481–553.
Fox, S. 1975. Natural selection on morphological phenotypes of the lizard *Uta stansburiana*. Evolution 29: 95–107.
Foreman, D. 1997. Seminiferous tubule stages in the prairie dog (*Cynomys ludovicianus*) during the annual breeding cycle. Anatomical Record 247: 355–367.
Furieri, P. 1965. Prime osservazioni al microscopio elettronico sulla ultrastruttura dello spermatozoo di Vipera aspis *aspis*. Bollettino della Societa Italiana di Biologia Spermimentale 41: 478–480.
Furieri, P. 1970. Sperm morphology in some reptiles: Squamata and Chelonia. pp. 115–132. In B. Baccetti (ed.), *Comparative Spermatology*. Academic Press, New York, NY.
Giugliano, L. G., Teizeira, R. D., Colli, G. R. and Báo, S. N. 2002. Ultrastructure of spermatozoa of the lizard *Ameiva ameiva*, with considerations on polymorphism within the family Teiidae (Squamata). Journal of Morphology 253: 264–271.
Goldberg, S. R. and Lowe, C. H. 1966. The reproductive cycle of the Western Whiptail Lizard (*Cnemidophorous tigris*) in southern Arizona. Journal of Morphology 118: 543–548.
Granados-González, G., Rheubert, J. L., Villagrán-SantaCruz, M., González-Herrera, M. E., Dávila-Cedillo, J. V., Gribbins, K. M. and Hernández-Gallegos, O. 2013. Male reproductive cycle in *Aspidoscelis costata costata* (Squamata: Teiidae) from Tonatico, Estado de México, México. Acta Zoologica Early View.
Gribbins, K. and Gist, D. 2003. Cytological evaluation of the germinal epithelium and the germ cell cycle in an introduced population of European Wall Lizards, *Podarcis muralis*. Journal of Morphology 256: 296–306.
Gribbins, K., Gist, D. and Congdon, J. 2003. Cytological evaluation of spermatogenesis in the Red-Eared Slider, *Trachemys scripta*. Journal of Morphology 255: 337–346.
Gribbins, K., Happ, C. S. and Sever, D. M. 2005. Ultrastructure of the reproductive system of the Black Swamp Snake (*Seminatrix pygaea*). V. The temporal germ cell development strategy of the testis. Acta Zoologica 86: 223–230.
Gribbins, K., Elsey, R. and Gist, D. 2006. Cytological evaluation of spermatogenesis in the germinal epithelium of the American Alligator, *Alligator mississippiensis*. Acta Zoologica 87: 59–69.
Gribbins, K., Mills, E. and Sever, D. 2007. The ultrastructure of spermiogenesis within the testis of the ground skink, *Scincella laterale* (Scincidae, Squamata). Journal of Morphology 268: 181–192.
Gribbins, K., Rheubert, J., Collier, M., Siegel, D. and Sever, D. 2008. Histological analysis of spermatogenesis and the germ cell development strategy within the testis of the male Western Cottonmouth Snake, *Agkistrodon piscivorus*. Annals of Anatomy 190: 461–476.
Gribbins, K. M., Rheubert, J. L., Poldemann, E. H., Collier, M. H., Wilson, B. and Wolf, K. 2009. Continuous spermatogenesis and the germ cell development strategy within the testis of the Jamaican Gray Anole, *Anolis lineatopus*. Theriogenology 72: 484–492.
Gribbins, K. M., Rheubert, J. L., Anzalone, M. L., Siegel, D. S. and Sever, D. M. 2010. Ultrastructure of spermiogenesis in the cottonmouth, *Agkistrodon piscivorus* (Squamata: Viperidae: Crotalinae). Journal of Morphology 271: 293–304.
Gribbins, K. M. 2011. Reptilian spermatogenesis: A histological and ultrastructural perspective. Spermatogenesis 1: 250–269.

Gribbins, K. M. and Rheubert, J. L. 2011. The Ophidian testis, spermatogenesis, and mature spermatozoa. pp. 183–264. In R. D. Aldridge and D. M. Sever (eds.), *Reproductive biology and phylogeny of snakes.* Science Publishes, Enfield, NH.

Gribbins, K., Anzalone, M., Collier, M., Granados-González, G., Villagrán-Santa Cruz, M. and Hernández-Gallegos, O. 2011. Temporal germ cell development strategy during continuous spermatogenesis within the montane lizard, *Sceloporus bicanthalis* (Squamata; Phrynosomatidae). Theriogenology 76: 1090–1099.

Gribbins, K. M., Rheubert, J. L., Touzinsky, K., Hanover, J., Matchett, C. L., Granados-González, G. and Hernández-Gallegos, O. 2013a. Spermiogenesis in the Imbricate Alligator Lizard, *Barisia imbricata* (reptilia, squamata, anguidae). Journal of Morphology 274(6): 603–614.

Gribbins, K., Matchett, C. L., DelBello, K. A., Rheubert, J., Villagrán-SantaCruz, M., Grandados-González, G. and Hernández-Gallegos, O. 2013b. The ultrastructure of spermatid development during spermiogenesis within the Rosebelly Lizard, *Sceloporus variabilis* (Reptilia, Squamata, Phrynosomatidae). Journal of Morphology. Early View.

Guraya, S. S. 1973. Histochemical observations on the interstitial (Leydig) cells of the snake. Acta Morphologica Hungarica 21: 1–12.

Furieri, P. 1970. Sperm morphology of some reptiles. Squamata and Chelonia. pp. 115–131. In B. Baccetti (ed.), *Comparative Spermatology.* Accademia Nazionale dei Lincei, Rome, and Academic Press, New York.

Hahn, W. E. 1964. Seasonal changes in testicular and epididymal histology and spermatogenic rate in the lizard *Uta stansburiana stejnegeri.* Journal of Morphology 115: 447–459.

Hamilton, D. W. and Fawcett, D. W. 1968. Unusual features of the neck and middle-piece of snake spermatozoa. Journal of Ultrastructure Research 23(1): 81–97.

Harding, H. R., Aplin, K. P. and Mazur, M. 1995. Ultrastructure of spermatozoa of Australian Blindsnakes, *Ramphotyphlops* spp. (Typhlopidae, Squamata): first observations on the mature spermatozoon of scolecophidian snakes. pp. 385–396. In B. G. M. Jamieson, J. Ausio and J. L. Justine (eds.), *Advances in spermatozoal phylogeny and taxonomy. vol. 166.* Mémoires du Musséum National de Histoire Naturelle. Editions du Museum, Paris.

Healy, J. M. and Jamieson, B. G. M. 1994. The ultrastructure of spermatogenesis and epididymal spermatozoa of the Tuatara *Sphenodon punctatus* (Sphenodontidae, Amniota). Philosophical Transactions: Biological Sciences 344: 187–199.

Hess, R. 1990. Quantative and qualitative characteristics of the stages and transition in the cycle of the rat seminiferous epithelium: light microscopic observations of perfusion-fixed and plastic-embedded testes. Biological Reproduction 43: 525–542.

Hondo, E., Kurohmaru, M., Toriba, M. and Hayashi, Y. 1994. Seasonal changes in spermatogenesis and ultrastructure of developing spermatids in the Japanese Rat Snake, *Elaphe climacophora.* Journal of Veterinary Medical Sciences 56: 836–40.

Honda, E., Kitamura, N., Toriba, M., Kurohmaru, M. and Hayashi, Y. 1996. Histological study of the seminiferous epithelium in the Japanese Rat Snake, *Elaphe climacophora*: Identification of the spermatogonia. Journal of Veterinary Medical Science 58: 23–29.

Humphreys, P. N. 1975. The differentiation of the acrosome in the spermatid of the budgerigar (*Melopsittacus undulatus*). Cell and Tissue Research 156(3): 411–416.

Ibargüengoytía, N. R., Pastor, L. M. and Pallares, J. 1999. A light microscopy and ultrastructural study of the testes of tortoise *Testudo graeca* (Testudinidae). Journal of Submicroscopic Cytology and Pathology 31(2): 221–231.

Jamieson, B. G. M. 1995. The ultrastructure of spermatozoa of the Squamata (Reptilia) with phylogenetic considerations. pp. 359–383. In B. G. M. Jamieson, J. Ausio and J. L. Justine (eds.), *Advances in spermatozoal phylogeny and taxonomy. Vol. 166.* Mémoires du Musséum National de Histoire Naturelle. Editions du Museum, Paris.

Jamieson, B. G. M. 1999. Spermatozoal phylogeny of the Vertebrata. pp. 303–331. In C. Gagnon (ed.), *The Male Gamete. From Basic Science to Clinical Applications.* Cache River Press, Clearwater, Fl.

Jamieson, B. G. M. 2007. Avian spermatozoa: Structure and phylogeny. pp. 349–511. In B. G. M. Jamieson (ed.), *Reproductive Biology and Phylogeny of Birds, Volume 6A*. Science Publishers, Enfield, NH.
Jamieson, B. G. M. and Healy, J. M. 1992. The phylogenetic position of the tuatara, *Sphenodon* (Sphenodontida, Amniota) as indicated by cladistic analysis of the ultrastructure of spermatozoa. Philosophical Transactions of the Royal Society of London B Biological Sciences 335: 207–219.
Jamieson, B. G. M. and Koehler, L. 1994. The ultrastructure of the spermatozoa of the Northern Water Snake, *Nerodia sipedon* (Colubridae, Serpentes) with phylogenetic considerations. Canadian Jouranl of Zoology 72: 1648–1652.
Jamieson, B. G. M. and Scheltinga, D. M. 1993. The ultrastructure of spermatozoa of *Nangura spinosa* (Scincidae, Reptilia). Memoirs of the Queenlsand Museum 34: 169–179.
Jamieson, B. G. M. and Scheltinga, D. M. 1994. The ultrastructure of spermatozoa of the Australian skinks, *Ctenotus taeniolatus*, *Carlia pectoralis*, and *Tiliqua scincoides scincoides* (Scincidae, Reptilia). Memoirs of the Queensland Museum 37: 181–193.
Jamieson, B. G. M., Oliver, S. C. and Scheltinga, D. M. 1996. The ultrastructure of spermatozoa of Squamata. I. Scincidae, Gekkonidae, and Pygonidae (Reptilia). Acta Zoologica 77: 85–100.
Khan, U. W. and Rai, U. 2004. *In vitro* effect of FSH and testosterone on Sertoli cell nursing function in wall lizard *Hemidactylus flaviviridis* (Rüppell). General and Comparative Endocrinology 136: 225–231.
Khan, U. W. and Rai, U. 2005. Endocrine and paracrine control of Leydig cell steroidogenesis and proliferation in the wall lizard: an *in vitro* study. General and Comparative Endocrinology 140: 109–115.
Kurohmaru, M., Hikim, A. P. S., Mayerhofer, A., Bartke, A. and Russell, L. D. 1990. Golden hamster myoid cells during active and inactive states of spermatogenesis: correlation of testosterone levels with structure. American Journal of Anatomy 188: 319–327.
Leceta, J., Barrutia, M. G. and Fernandez, J. 1982. Seasonal ultrastructural changes of Leydig cells in *Lacerta muralis*. Acta Zoologica 63: 33–38.
Lee, M. S. Y., Hugall, A. F., Lawson, R. and Scanlon, J. D. 2007. Snake phylogeny combining morphological and molecular data in likelihood, bayesian and parsimony analyses. Systematic Biodiversity 5: 371–389.
Licht, P. 1979. Reproductive endocrinology of reptiles and amphibians: Gonadotropins. Annual Review of Physiology 41: 337–351.
Licht, P. 1984. Reptiles. pp. 206–282. In G. E. Lamming (ed.), *Marshall's Physiology of Reproduction Vol. 1. Reproductive Cycles of Vertebrates*. Chruchhill Livingstone, New York, NY.
Licht, P., Farmer, S. W., Bona-Gallo, A. and Papkoff, H. 1979. Pituitary gonadotropins in snakes. General and Comparative Endocrinology 39: 34–52.
Licht, P., Denver, R. J. and Pavgi, S. 1989. Temperature dependence of *in vitro* pituitary, testis, and thyroid secretion in a turtle, *Pseudemys scripta*. General Comparative Endocrinology 76: 274–285.
Lin, M. and Jones, R. C. 1993. Spermiogenesis and spermiation in the Japanese Quail (*Coturnix coturnix japonica*). Journal of Anatomy 183: 525–535.
Lin, M., Harman, A. and Rodger, J. C. 1997. Spermiogenesis and spermiation in a marsupial, the Tammar Wallaby (*Macropus eugenii*). Journal of Anatomy 190: 377–395.
Lofts, B. 1964. Seasonal changes in the functional activity of the interstitial and spermatogenetic tissues of the green frog, *Rana esculenta*. General and Comparative Endocrinology 4: 550–562.
Lofts, B. 1968. Patterns of testicular activity. pp. 239–245. In E. J. Barrington and C. B. Jorgensen (eds.), *Perspectives in Endocrinology: Hormones in the Lives of Lower Vertebrates*. Academic Press, New York, NY.
Lofts, B., Phillips, J. G. and Tam, W. H. 1966. Seasonal changes in the testis of the cobra, *Naja naja* (Linn). General and Comparative Endocrinology 6: 466–475.
Lowe, C. H. and Goldberg, S. R. 1966. Variation in the circumtesticular Leydig cell tunic of teiid lizards (*Cnemidophorus* and *Ameiva*). Journal of Morphology 119: 277–282.

Maddison, D. R. and Maddison, W. P. 2005. MacClade version 4.08. Sinauer Associates, Inc. Sunderland, MS USA.
Maekawa, M., Kamimura, K. and Nagano, T. 1996. Peritubular myoid cells in the testis: their structure and function. Archives of Histology and Cytology 59: 1–13.
Mandel, S. M., Cunha, L. D., Brandão, J. C., Colli, G. R. and Báo, S. N. 2009. Ultrastucture of spermatozoa of lizards in the genus *Mabuya* from Central Brazil. Acta Zoologica 90: 68–74.
Marion, K. R. 1982. Reproductive cues for gonadal development in temperate reptiles: temperature and photoperiod effects on the testicular cycle of the lizard *Sceloporus undulatus*. Herpetologica 26–39.
Marshall, A. J. and Woolf, F. M. 1957. Seasonal lipid changes in the sexual elements of a male snake *Vipera berus*. Quarterly Journal of Microscopy Science 98: 89–100.
Masson, G. R. and Guillette Jr., L. J. 2005. FSH-induced gonadal development in juvenile lizards, *Eumeces obsoletus*. Journal of Experimental Zoology 236: 343–351.
McIntosh, J. R. and Porter, K. R. 1967. Microtubules in the spermatids of the domestic fowl. Journal of Cell Biology 35: 153–173.
Mubarak, M. 2006. Ultrastructure of sperm tail differentiation of the lizard *Stenodactylus dorie* (Squamata, Reptilia). Journal of Biological Sciences 6: 187–192.
Newton, W. D. and Trauth, S. E. 1992. Ultrastructure of the spermatozoon of the lizard *Cnemidophorus sexlineatus* (Sauria: Teiidae). Herpetologica 48: 330–343.
Oliver, S. C., Jamieson, B. G. M. and Scheltinga, D. M. 1996. The ultrastructure of the spermatozoa of Squamata. II. Agamidae, Varanidae, Colubridae, Elapidae, and Boidae (Reptilia). Herpetologica 52: 216–241.
Osman, D. I., Ekwall, H. and Ploen, L. 1980. Specialized cell contacts and the blood-testis barrier in the seminiferous tubules of the domestic fowl (*Gallus domesticus*). International Journal of Andrology 3: 553–562.
Pough, F. H., Andrews, R. M., Cadle, J. E., Crump, M. L., Savitzky, A. H. and Wells, K. D. 2001. Herpetology, 2nd Edition. Prentice Hall, New Jersey, pp. 124–125.
Pudney, J. 1993. Comparative Cytology of the non-mammalian vertebrate Sertoli cell. pp. 611–657. In L. D. Russell and M. D. Griswold (eds.), *The Sertoli Cell*. Cache River Press, Clearwater, FL.
Pyron, R. A., Burbrink, F. T. and Wiens, J. J. 2013. A phylogeny and revised classification of Squamata, including 4161 species of lizards and snakes. BMC Evolutionary Biology 13: 93.
Raynaud, A. and Pieau, C. 1985. Embryonic development of the genital system. pp. 149–300. In C. Gans and F. Billett (eds.), *Biology of Reptilia Vol. 15, Development B*. John Wiley and Sons, New York, NY.
Rheubert, J. L., Poldemann, E. H., Eckstut, M. E., Collier, M. H., Sever, D. M. and Gribbins, K. M. 2009a. Temporal germ cell development strategy during mixed spermatogenesis within the Mediterranean Gecko, *Hemidactylus turcicus* (Reptilia: Gekkonidae). Copeia.
Rheubert, J. L., McHugh, H., Collier, M. H., Sever, D. M. and Gribbins, K. M. 2009b. Temporal germ cell development strategy during spermatogenesis within the testis of the ground skink, Scincella laterale (Sauria: Scincidae). Theriogenology 72: 54–61.
Rheubert, J. L., Wilson, B. S., Wolf, K. W. and Gribbins, K. M. 2010. Ultrastructural study of spermiogenesis in the Jamaican Gray Anole, *Anolis lineatopus* (Reptilia: Polychrotidae). Acta Zoologica 91: 484–494.
Rheubert, J. L., Siegel, D. S., Venable, K. J., Sever, D. M. and Gribbins, K. M. 2011. Ultrastructural description of spermiogenesis within the Mediterranean Gecko, *Hemidactylus turcicus* (Squamata: Gekkonidae). Micron 42: 680–690.
Rheubert, J., Touzinsky, K., Hernández-Gallegos, O., Granados-González, G. and Gribbins, K. 2012. Ontogenic development of spermatids during spermiogenesis in the high altitude bunchgrass lizard (*Sceloporus bicanthalis*). Spermatogenesis 2: 94–103.
Rheubert, J. L., Touzinsky, K., Sever, D. M., Aldridge, R. D., Wilmes, A. J., Siegel, D. S. and Gribbins, K. M. 2013. Reproductive Biology of *Sceloporus consobrinus* (Phrynosomatidae): Male Germ Cell Development and Reproductive Cycle Comparisons within Spiny Lizards. Journal of Herpetology. Early View.

Rheubert, J. L., McMahan, C. D., Sever, D. M., Bundy, M. R., Siegel, D. S. and Gribbins, K. M. 2010. Ultrastructure of the reproductive system of the black swamp snake (*Seminatrix pygaea*). VII. spermatozoon morphology and evolutionary trens of sperm characters in snakes. Journal of Zoological Systematics and Evolutionary Research 48: 366–375.
Robb, J. 1960. The internal anatomy of *Typhlops schneider* (Reptilia). Australian Journal of Zoology 8: 181–216.
Roosen-Runge, E. C. 1972. The process of spermatogenesis in animals. University Press, Cambridge, UK, pp. 1–150.
Russell, L. D., Hikim, S. A. P., Ettlin, R. A. and Legg, E. D. 1990. Histological and Histopathological Evaluation of the Testes. Cache River Press, Clearwater, FL, pp. 211–275.
Saita, A., Comazzi, M. and Perrotta, E. 1988. Ulteriori osservazioni at M. E. sulla spermiogenesi di un serpente: Coluber viridiflavus (Lacepede) in reiferimento ad elementi comparativi nella spermiogenesi dei rettili. Atti dell a Accademia Nazionale dei Lincei. Rendiconti. Classe di Scienze Fisiche, Matematiche e Naturali LXXXII: 137–143.
Scheltinga, D. M., Jamieson, B. G. M., Trauth, S. E. and McAllister, C. T. 2000. Morphology of the spermatozoa of the iguanian lizards *Uta stansburiana* and *Urosaurus ornatus* (Squamta, Phrynosomatidae). Journal of Submicroscopic Cytology and Pathology 32: 261–271.
Scheltinga, D. M., Jamieson, B. G. M., Espinoza, R. E. and Orrell, K. S. 2001. Descriptions of the mature spermatozoa of the lizards *Crotaphytus bicinctores*, *Gambelia wislizenii* (Crotaphytidae), and *Anolis carolinensis* (Polychrotidae) (Reptilia, Squamata, Iguania). Journal of Morphology 247: 160–171.
Schuett, G. W., Harlow, H. J., Rose, J. D., Van Kirk, E. A. and Murdoch, W. J. 1997. Annual cycle of plasma testosterone in male copperheads, *Agkistrodon contortrix* (Serpentes, Vperidae): Relationship to timing of spermatogenesis, mating, and agonistic behavior. General and Comparative Endocrinology 105: 417–427.
Scott, D. E., Fishcer, R. U., Congdon, J. D. and Buss, S. A. 1995. Whole body lipid dynamics and reproduction in the Eastern Cottonmouth, *Agkistrodon piscivorus*. Herpetologica 51: 472–487.
Sever, D. M. 2004. Ultrastructure of the reproductive system of the Black Swamp Snake (*Seminatrix pygaea*). IV. Occurrence of an ampulla ductus deferentis. Journal of Morphology 262: 714–730.
Sever, D. M., Ryan, T. J., Stephens, R. and Hamlett, W. C. 2002. Ultrastructure of the reproductive system of the Black Swamp Snake (*Seminatrix pygaea*). III. Sexual segment of the male kidney. Journal of Morphology 252: 238–254.
Siegel, D. S. and Sever, D. M. 2008. Sperm aggregations in female *Agkistrodon piscivorous* (Reptilia: Squamata): A histological and ultrastructural investigation. Journal of Morphology 269: 189–206.
Shea, G. M. 2001. Spermatogenic cycle, sperm storage, and Sertoli cell size in Scolecophidian (*Ramphotyphlops nigrescens*) from Australia. Journal of Herpetology 35: 85–91.
Shivanandappa, T. and Sarkar, H. B. 1979. Seasonal lipid cycle and steroidogenic activity in the testis of the tropical skink, *Mabuya carinata* (Schneider): A histochemical study. General and Comparative Endocrinology 38(4): 491–495.
Soley, J. T. 1997. Nuclear morphogenesis and the role of the manchette during spermiogenesis in the ostrich (*Struthio camelus*). Journal of Anatomy 190: 563–576.
Sprando, R. L. and Russell, L. D. 1988. Spermiogenesis in the red-eared turtle (*Pseudemys scripta*) and the domestic fowl (*Gallus domesticus*): A study of cytoplasmic events including cell volume changes and cytoplasmic elimination. Journal of Morphology 198: 95–118.
Tait, A. J. and Johnson, E. 1982. Spermatogenesis in the Grey Squirrel (*Sciurus carolinensis*) and changes during sexual regression. Journal of Reproduction and Fertility 65: 53–58.
Tavares-Bastos, L., Teixeira, R. D., Colli, G. R. and Báo, S. N. 2002. Polymorphism in the sperm ultrastructure among gour species of lizards in the genus *Tupinambis* (Squamata: Teiidae). Acta Zoologica 83: 297–307.
Tavares-Bastos, L., Cunha, G. R., Colli, G. R. and Bao, S. N. 2007. Ultrastructure of spermatozoa of scolecophidian shakes (Lepidosuaria, Squamata). Acta Zoologica 88: 189–197.

Tavares-Bastos, L., Colli, G. R. and Bao, S. N. 2008. The evolution of sperm ultrastructure among Boidae (Serpentes). Zoomorphology 8: 62–68.

Teixeira, R. D., Colli, G. R. and Báo, S. N. 1999a. The ultrastructure of the spermatozoa of the lizard *Micrablepharus maximiliani* (Squamata, Gymnophthalmidae), with considerations on the use of sperm ultrastructure characters in phylogenetic reconstruction. Acta Zoologica 80: 47–59.

Teixeira, R. D., Colli, G. R. and Báo, S. N. 1999b. The ultrasturcutre of the spermatozoa of the lizard, *Polychrus acutirostris* (Squamata, Polychrotidae). Journal of Submicroscopic Cytology and Pathology 31: 387–395.

Teixeira, R. D., Vieira, G. H. C., Colli, G. R. and Báo, S. N. 1999c. Ultrastructural study of spermatozoa of the neotropical lizards, *Tropidurus semitaeniatus* and *Tropidurus toquatus* (Squamata, Tropiduridae). Tissue & Cell 31: 308–317.

Teixeira, R. D., Colli, G. R. and Báo, S. N. 1999d. The ultrastructure of the spermatozoa of the worm lizard *Amphisbaena alba* (Squamata, Amphisbaenidae) and the phylogenetic relationship of amphibaenians. Canadian Journal of Zoology 77: 1254–1264.

Teixeira, R. D., Scheltinga, D. M., Trauth, S. E., Colli, G. R. and Báo, S. N. 2002. A comparative ultrastructual study of spermatozoa of the teiid lizards *Cnemidophorus gularis gularis*, *Cnemidophorus ocellifer*, and *Kentropyx altamazonica* (Reptilia, Squamata, Teiidae). Tissue & Cell 34: 135–142.

Tibia, T. and Kita, I. 1990. Undifferentiated spermatogonia and their role in seasonally fluctuating spermatogenesis in the ferret, *Mustela putorius furo* (Mammalia). Zoology of Australia and New Zealand 224: 140–155.

Tourmente, M., Cardozo, G., Bertona, M., Guidobaldi, A., Giojalas, L. and Chiaraviglio, M. 2006. The ultrastructure of the spermatozoa of *Boa constrictor occidentalis*, with considerations on its mating system and sperm competition theories. Acta Zoologica 87: 25–32.

Tourmente, M., Giojalas, L. and Chiaraviglio, M. 2008. Sperm ultrastructure of *Bothrops alternatus* and *Bothrops diporus* (Viperidae, Serpentes), and its possible relation to the reproductive features of the species. Zoomorphology 127: 241–248.

Tsubota, T. and Kanagawa, H. 1989. Annual changes in serum testosterone levels and spermatogenesis in the Hokkaido Brown Bear, *Ursus arctos yesoensis*. Journal of Mammalian Society of Japan 14: 11–17.

Unsicker, K. and Burnstock, G. 1975. Myoid cells in the peritublar tissue (lamina propria) of the reptilian testis. Cell and Tissue Research 163: 545–560.

Van Oordt, P. G. W. J. 1955. Regulation of the spermatogenetic cycle in the frog. Memoires of the Society of Endocrinology 4: 25–38.

Van Oordt, P. G. W. J. and Brands, F. 1970. The Sertoli cell in the testis of the common frog, Rana temporaria. Proceedings of the Society of Endocrinology 119th Meeting Journal of Endocrinology 48, Abstract 100.

Ventela, S., Toppari, J. and Parvinen, M. 2003. Intercellular organelle traffic through cytoplasmic bridges in early spermatids of the rat: Mechanism of haploid gene product sharing. Molecular Biology and Cell 14: 2768–2780.

Vieira, G. H. C., Colli, G. R. and Báo, S. N. 2001. Spermiogenesis and testicular cycle of the lizard *Tropidurus torquatus* (Squamata, Tropiduridae) in the Cerrado of central Brazil. Amphibia-Reptilia 22(2): 217–233.

Vieira, G. H. C., Colli, G. R. and Báo, S. N. 2004. The ultrastructure of the spermatozoon of the lizard *Iguana iguana* (Reptilia, Squamata, Iguanidae) and the variability of sperm morphology among iguanian lizards. Journal of Anatomy 204: 451–464.

Vieira, G. H. C., Colli, G. R. and Báo, S. N. 2005. Phylogenetic relationship of corytophanid lizards (Iguania, Squamata, Reptilia) based on partitioned and total evidence analyses of sperm morphology, gross morphology, and DNA data. Zoologica Scripta 34: 605–625.

Vieira, G. H. C., Cunha, L. D., Scheltinga, D. M., Glaw, F., Colli, G. R. and Báo, S. N. 2007. Sperm ultrastructure of hoplocercid and oplurid lizards (Sauropsidae, Squamata, Iguania) and the phylogeny of Iguania. Journal of Zoological Systematics and Evolutionary Research 45: 230–241.

Villagrán-SantaCruz, M., Hernández-Aguacaliente, M. J., Méndez-de la Cruz, F. R., Gribbins, K. M., Granados-González, G. and Hernández-Gallegos, O. 2013. The germ cell development strategy and seasonal changes in spermatogenesis and Leydig cell morphologies of the spiny lizard *Sceloporus mucronatus* (Squamata: Phrynosomatidae). Zoomorphology: Early View.

Volsøe, H. 1944. Structure and seasonal variation of the male reproductive organs of Vipera *berus* (L.). Spolia Zoologica Musei Hauniensis V. Skrifter, Universitetets Zoologiske Museum, København, pp. 157.

Waits, G. M. H. and Setchell, B. P. 1990. Physiology of the mammalian testis. pp. 1–105. In G. E. Lemming (ed.), *Marshall's Physiology of Reproduction. Vol. 2. Reproduction in the Male. 4th Ed.* Churchill Livingstone, London.

Weins, J. J. 2004. The role of morphological data in phylogeny reconstruction. Systemic Biology 53: 653–661.

Werner, Y. L. and Drook, K. 1967. The multipartite testis of the snake *Leptotyphlops phillipsi*. Copeia 1967: 159–163.

Wharton, C. H. 1966. Reproduction and growth in Cottonmouths, *Agkistrodon piscivorus* (Lacepede) of Cedar Key, Florida. Copeia 1966: 149–161.

Wilhoft, D. C. and Quay, W. B. 1961. Testicular histology and seasonal changes in the lizard, *Sceloporus occidentalis*. Journal of Morphology 108: 95–106.

Yamamoto, S., Tamate, H. and Itikawa, O. 1967. Morphological studies on the sexual maturation in the male Japanese Quail (*Coturnix coturnix japonica*). Tohuku Jorunal of Agricultural Research 18: 27–37.

Yong-pu, Z., Xue-ping, Y. and Xiang, J. 2005. Ultrastructure of the spermatozoon of the Northern Grass Lizard (*Takydromus septentrionalis*) with comments on the variability of sperm morphology among lizard taxa. Zoological Research 26: 518–526.

Yong-Zhang and Yong-pu. 2004. Blue-tailed skink spermatozoa ultrastructure. Zoological Research 25: 429–435.

Zhou, Q., Clarke, L., Nie, R., Carnes, K., Lai, L. W., Lien, Y. H. H., Verkman, A. and Hess, R. A. 2001. Estrogen action and male fertility: roles of the sodium/hydrogen exchanger-3 and fluid reabsorption in reproductive tract function. Proceedings of the National Academy of Sciences 98: 14132–14137.

CHAPTER 12

The Evolutionary Ecology of Reproductive Investment in Lizards

Tobias Uller[1,2,*] and *Geoffrey M. While*[3]

12.1 INTRODUCTION

All animals face the decision of when, where, and how much to invest in reproduction. This investment can vary as a function of clutch interval, clutch size, and egg size. Consequently, variation in reproductive investment can refer to variation in any of these characters, which can be observed to different degree among species, among populations within species, and among individuals within populations. The processes that shape this variation are of great interest to evolutionary ecologists. This is not only because reproduction is one of the most fundamental aspects of organismal biology, but also because variation in reproductive investment enables tests of adaptive and non-adaptive models of phenotypic evolution and provides insights into the interplay between demographic, ecological, and evolutionary change.

Lizards have a long history as model organisms for studies of reproductive investment and, more generally, life history biology. Classic long-term studies of small and abundant lizards in the United States provided some of the first data on individual variation in life history (e.g., Blair 1960; Tinkle 1967, 1969). Following these initial works, lizards became

[1] Edward Grey Institute, Department of Zoology, University of Oxford, UK.
[2] Department of Biology, University of Lund, Sweden.
[3] School of Zoology, University of Tasmania, Australia.
* Corresponding author

widely used as study organisms for developing, testing, and refining life history models, in particular for reproductive traits. Seminal work by, among others, Ballinger, Pianka, Shine, and Tinkle outlined the theories and empirical evidence as it applied within and across species. These studies significantly enhanced our understanding of the trade-off between offspring size and number, reproductive effort, costs of reproduction, and the evolution of reproductive mode (the transition from oviparity to viviparity) (e.g., Tinkle *et al.* 1970; Andrews and Rand 1974; Shine 1980, 1985; Huey and Pianka 1981; Ballinger 1983; Dunham and Miles 1985; Vitt and Seigel 1985). Many of these original insights remain valid despite the fact that rigorous data on the relationships between lizard taxa and statistics for comparative studies were not developed until later.

Research on reproductive investment continues to rely heavily on observations from natural populations. Although the number of long-term studies of marked lizards in natural populations is still limited, they have contributed substantially to furthering our understanding of the processes that underlie individual variation in reproductive investment. These observational studies are increasingly being combined with a broad range of contemporary techniques that have allowed for greater insights into factors responsible for the evolutionary diversification of reproductive investment for future research. For example, there has been a surge in experimental tests of adaptive hypotheses and development of methods aimed at establishing links between individual variation and population-level processes (e.g., allometric engineering: Sinervo and Huey 1990; Sinervo and Licht 1991; semi-natural enclosures: Le Galliard *et al.* 2005; Warner and Shine 2008). Modern comparative analysis has brought a focus on linking species and population variation to climatic variables (e.g., Meiri *et al.* 2012), which has been facilitated by the use of publicly available climate data.

Our aim in this chapter is to give an overview of research on reproductive investment in lizards, focusing on recent contributions and future directions. In doing so we do not aim to provide a comprehensive review of all aspects of reproductive investment, as this is clearly beyond the scope of a single chapter. Instead we will single out some of the key aspects of contemporary research on the evolutionary ecology of reproductive investment. Some of these are topics for which lizards have been extensively used as models, others are those we feel deserve greater attention by herpetologists. We begin by discussing the comparative biology of reproductive investment, followed by a discussion of some key aspects of the causes and consequences of individual variation in female reproductive investment within populations. We then discuss how one could connect within-population processes and population and species divergence. Finally, we suggest some potential directions.

12.2 DIVERSITY IN REPRODUCTIVE INVESTMENT AMONG SPECIES AND POPULATIONS

With approximately 5600 species currently described, it is not surprising that lizards exhibit considerable variation in reproductive investment. However, it appears that species tend to allocate a similar fraction of their metabolic expenditure to reproduction (Warne and Charnov 2008; Meiri *et al.* 2012). In other words, species with low metabolic turnover, such as those in cool climates or many fossorial lizards, reproduce less in absolute terms but not necessarily in relative terms. Because metabolic rate scales with body size, it is not surprising that the best predictor of annual reproductive investment across species is body mass (Meiri *et al.* 2012). Nevertheless, there remains much variation that cannot simply be explained by metabolic scaling, and this variation forms the basis for most of the work discussed in this chapter.

Total annual reproductive investment can be divided into the number of clutches produced, clutch size, and the size of eggs or offspring. Each of these varies phylogenetically and geographically. For example, while *Anolis* lizards produce a single egg at a time, they reproduce frequently and have a long reproductive season (Andrews and Rand 1974). In contrast, snow skinks (*Niveoscincus*) in temperate Tasmania reproduce at a low rate with some species completing a 14 month gestation, retaining embryos over winter and giving birth to one to five offspring the following spring (Olsson and Shine 1999). Other species, including some chameleons, may reproduce more frequently despite the fact that they lay tens of eggs in a single clutch. How individuals acquire energy for reproduction also varies considerably between species. Lizards, being ectotherms, exhibit a relatively low energy life style. Unlike endothermic animals, this facilitates storage and retention of resources, allowing for the accumulation of nutrients during periods of resource abundance and their use even when resource availability is low (i.e., capital breeding; Bonnet *et al.* 1998). To what extent species rely on stored resources or resources obtained during the breeding season is variable and spans a continuum that ranges from pure income to pure capital breeders (Bonnet *et al.* 1998; Warner *et al.* 2007; Telemeco and Baird 2011; Warne *et al.* 2012). The processes underlying this diversity in reproductive investment between species represent a combination of extrinsic (e.g., climatic) and intrinsic (e.g., body size) constraints on reproductive output, the trade-off between number and size of offspring, genetic adaptation to local conditions, and the contingency of the evolutionary process that makes closely related lineages more similar than less closely related species.

Lizards also exhibit substantial diversity in reproductive investment within species. The processes that shape variation in reproductive investment between populations within species should be similar to those that shape variation between species, with one important caveat; populations within species are (potentially) connected by gene flow. This implies that some populations, and in particular those in marginal habitats that represent sinks, may be unable to adapt to local environmental conditions because of the constant influx of genes from other populations (Kawecki 2008). However, gene flow in lizards is often restricted because of limited dispersal ability. It is therefore not surprising that many species show pronounced local adaptation of life histories across their range. For example, the common lizard (*Zootoca vivipara*) has evolved viviparity relatively recently with the ancestral oviparous condition being restricted to two regions in southern Europe (Surget-Groba *et al*. 2006). Furthermore, viviparous populations of this species exhibit divergence in life history traits across climatic regimes, including total reproductive investment, clutch size, and offspring size, which parallel differences found among species (discussed further below).

Lizards have been extensively used to address the relative importance of extrinsic and intrinsic factors and the role of local adaptation in causing life history variation between species and populations. Here we first briefly cover the evidence for phylogenetic conservatism, discuss comparative patterns of reproductive investment from the perspective of life history theory, and then identify climatic effects and other potential causes of divergence in reproductive investment between populations.

12.2.1 Phylogenetic Conservatism and Constraints

Phylogenetic signals are strong for all three aspects of reproductive investment; number of clutches produced (or reproductive interval), clutch size and egg size. Because a phylogenetic effect partly reflects that closely related species tend to inhabit similar environments, this does not necessarily imply a 'phylogenetic constraint' on reproductive investment that prevents local adaptation nor does it imply that investment is independent of current environmental conditions such as food availability. Yet, there is some evidence that reproductive investment may be constrained by body shape, which itself shows both phylogenetic inertia and adaptive match to ecological conditions (Oufiero *et al.* 2011). Thus, stabilizing selection or phylogenetic inertia in morphology can give rise to phylogenetically conserved patterns of reproductive investment.

Mode of reproduction can also place significant constraints on reproductive investment. Viviparous species naturally have longer reproductive intervals than oviparous species. Females of some viviparous species appear to only breed every other year, possibly because of limited ability to replenished energy stores following gestation. Viviparity is also associated with changes in clutch size and egg size, but the extent to which this represents physical constraints is still debated. Experimental evidence from *Takydromus septentrionalis* shows that reducing abdominal space reduces clutch size (but not egg size; Du and Lu 2010) but even experimental lizards did not use up the entire space available (75% compared to control females 55%). Furthermore, comparisons of oviparous and viviparous species suggest that body shape remains unaltered despite an increase in the space required for offspring development following the evolution of viviparity (Sun *et al.* 2012; Yang *et al.* 2012), which may suggest that body shape itself has limited explanatory power for phylogenetic patterns of investment. However, comparison of snakes and lizards suggests that elongated body shape may account for at least some variation in relative clutch mass as elongated bodies have larger abdominal volumes (Shine 1992). In contrast, more recent studies of skinks of different body shapes have failed to find evidence that a flattened body shape results in reduced reproductive output (Goodman *et al.* 2009). Eggs in flat lizards simply fill up a greater proportion of the available space. The lack of physical constraints may be a general phenomenon as it is possible to increase clutch size and egg size beyond that observed in natural populations under high-food conditions in captivity (e.g., Olsson and Shine 1997; Uller and Olsson 2005). The available evidence therefore suggests that abdominal cavity space is unlikely to be a widespread limiting factor for reproductive output. Instead, females may adjust their investment to balance the cost of carrying a clutch, such as impaired locomotor ability or organ functioning.

Maximum egg size can also be constrained by the size of the pelvic girdle (Congdon and Gibbons 1987). This may imply that the egg size—clutch size trade-off in small lizards will be shifted towards a smaller egg size than would be optimal in terms of offspring recruitment. Species with invariant low clutch sizes, often a single egg, may be particularly prone to such morphological constraints (Doughty 1997; Michaud and Echternacht 1995). However, direct evidence that morphology is a severe constraint on reproductive investment is overall weak (Oufiero *et al.* 2007; but see Sinervo 1994).

12.2.2 Reproductive Investment in Life History Theory

Reproductive investment is central to life history theory. Life history theory is defined as the body of theory that predicts how natural selection should influence the allocation of resources towards different functions and the consequences this has for traits such as age at maturity and lifespan (Stearns 1976; Roff 1992, 2002). Life history theory commonly analyzes maximization of lifetime reproductive success through optimization of age-specific investment into, for example, growth and reproduction. The various models included in life history theory make certain assumptions regarding the nature of costs, trade-offs and constraints. For example, low extrinsic mortality (e.g., due to low predation rate) can select for delayed maturity at an increased body size with relatively low and infrequent reproduction (Roff 1992).

Predictions regarding co-variation among body size, age at maturity, and different reproductive parameters depend on the assumptions of survival of adults and offspring and the cost of reproduction (reviewed in Roff 1992, 2002). Thus, it is important to document costs and trade-offs. Lizards have been extensively used to this end. For example, studies have indicated that reproductive costs should not be considered only in terms of a loss of energy or the need to acquire more resources before or following reproduction but also in terms of reduced performance during pregnancy (Schwartzkopf 1994). The trade-off between the size and number of offspring is a classic problem in life history biology (e.g., Lack 1947) for which lizards have been extensively used as models and we treat this separately in Box 12.1.

Testing life history theory is compromised by the typically poor quantitative data on actual survival or life expectancy in the wild, but lizards tend to follow qualitative predictions reasonably well. Small, short-lived lizards have short reproductive intervals and produce clutches of relatively small eggs compared to long-lived species that reproduce infrequently and produce larger clutches (Tinkle *et al.* 1970; Dunham *et al.* 1994). Species that occupy islands offer opportunities to test these general predictions. Population densities and hence intraspecific competition are typically higher on islands than on the mainland, whereas predation levels are lower (Adler and Levins 1994; Novosolov *et al.* 2013). All else being equal, this should result in selection for late onset and infrequent reproduction of a relatively small number of large offspring that are competitive and run a lower risk of cannibalism (which can be a common cause of mortality in island populations). A recent comparative study found that island lizard endemics do indeed produce small clutches of large offspring, but did not find support for increased clutch interval or reduced annual reproductive investment (Novosolov *et al.* 2013; see also Fitch 1985; Meiri

BOX 12.1 OFFSPRING SIZE VERSUS NUMBER.

The trade-off between offspring size and offspring number is at the heart of optimality models of reproductive investment. Lizards have been extensively used to test both the assumptions of these models and their predictions. Despite that a trade-off should occur almost by necessity, it has proven difficult to establish empirically. It appears that this is largely due to among-individual variation in total reproductive effort. The logic behind this was described by van Noordwijk and de Jong (1986; see also Roff 2002 for a detailed discussion). In brief, when individuals differ in how much resource they have available for reproduction, those with large amounts may produce both larger clutches and larger eggs than individuals with limited resources. The fitted slope to the data may therefore not be negative even if there is a trade-off between size and number of young. There is abundant evidence from within populations of lizards that individuals vary in total reproductive investment and hence this is likely to explain the inability to detect phenotypic trade-offs. Another possible problem is that what constitutes investment into a single offspring is not always obvious. Not only is egg size at oviposition not necessarily a very good proxy for the nutrients available in the egg, different macro- and micro-nutrients may come with different costs and have different effects on offspring growth and development (Thompson and Speake 2003). Finally, eggs take up water as they develop and a study of snakes suggest that larger clutches are more efficient at this, resulting in increased size of hatchlings from experimentally enlarged clutches (Brown and Shine 2009).

Dividing resources among offspring into a single optimal size is only possible when clutch sizes are large. At a small clutch size, there is a range of resource availabilities for which adding resources to produce another egg of optimal size is not possible (Ricklefs 1968). The consequences of this will depend on a number of factors, including constraints on within-clutch variation in offspring size and whether or not there is an upper or lower limit on egg size (Ebert 1994; Charnov and Downhower 1995). Under assumptions regarding these and a number of additional aspects (listed in Uller and Olsson 2009; Rollinson *et al.* 2013), models predict that the ranges of investment per offspring should be inversely proportional to clutch size (Charnov and Downhower 1995). The maximum and minimum per-offspring investment at each clutch size should therefore converge upon the optimal level of investment per offspring as clutch size increases. This model has been tested on twelve lizard species (Uller *et al.* 2009). Overall the fit to model predictions was relatively poor, suggesting that one or several assumptions do not apply to lizards. For example, there could be a (species-specific) minimum offspring size below which fitness is zero, an upper limit on egg or offspring size, or a mechanistic coupling of total reproductive investment and offspring size such that that they are not independently and hierarchically controlled (Uller *et al.* 2009; Rollinson *et al.* 2013).

Given these complexities, is offspring size optimized in the wild? The best evidence that it is comes from a study of side-blotched lizards (*Uta stansburiana*) by Sinervo and colleagues that was published in a series of papers in the early 1990's. By removing yolk from eggs and ablating follicles during oogenesis Sinervo and colleagues were able to increase and reduce egg size (Sinervo and Huey 1990; Sinervo and Licht 1991). They subsequently studied survival of those offspring in the wild. The results provide strong evidence that offspring size is maximized given the size-number trade-off (Sinervo and Huey 1990). Similar egg size manipulations have been conducted in other species and provide additional support for the general theory (e.g., Warner and Andrews 2002). However, egg size manipulations remain to be done in the context of variable selection on offspring size within and between populations. Furthermore, not all studies detect positive selection on offspring size (Uller and Olsson 2010), suggesting that selection on egg size above a certain minimum for offspring viability may be weak in some species or populations.

et al. 2012). Mainland populations of the same species also commonly differ in reproductive investment, and this has sometimes been attributed to differences in predation or food availability (Iraeta *et al.* 2013).

Co-variation between low mortality rate, longevity and reproductive output has similarly been suggested to influence the relative importance of capital vs. income breeding. Short-lived lizard should rely more heavily on capital resources in order to maintain high annual reproduction. In contrast, longer lived species have been suggested to conserve a greater proportion of their capital resources to ensure survival and future reproduction (Schwartzkopf 1994; Doughty and Shine 1998). Despite this, studies have failed to identify a relationship between annual mortality rate and the proportion of capital resources allocated to reproduction across species (Warne *et al.* 2012). More generally, the fact that the predictions so strongly depend on specific assumptions that often are poorly understood makes it difficult to assess to what extent life history theory for reproductive investment in lizards is well supported.

12.2.3 Climatic Effects on Species and Population Divergence in Reproductive Investment

Climate is an important determinant of life history in general and thus also for comparative patterns of reproductive investment. Cool climates reduce activity periods, may limit body temperature and metabolic rate, and are often associated with low food availability. As a result, lizards in cooler climates should grow slowly, mature late, and have a low reproductive frequency. This may be further reinforced by the evolution of viviparity in cool climates, which tends to result in lower reproductive output (both in terms of offspring size or number and clutch frequency), due to demands imposed by carrying the clutch to full term. These predictions are largely confirmed by comparative data variation between species and populations, including patterns of latitudinal population divergence in North American *Sceloporus* lizards (Niewairowski 1994; Niewiarowski *et al.* 2004) and between high and low altitude populations of Tasmanian snow skinks (Wapstra and Swain 2001). Although ambient temperature directly contributes to the slow life history of cool climate lizards, recent evidence suggests that the length of the activity season may be equally or more important than temperature *per se* in explaining climatic effects on reproductive investment (Meiri *et al.* 2012; Horvathova *et al.* 2013). By limiting activity and hence reducing annual growth and mortality rates, selection favors shifts towards late onset of reproduction and low annual productivity (Adolph and Porter 1996; Sears and Angilletta 2004). In the common lizard, air temperatures during the activity period did not correlate with reproductive investment across populations whereas the length of the activity period was a good predictor of body size and reproductive output (Horvathova *et al.* 2013).

Whether or not activity period or temperature affect the balance between number and size of offspring is poorly understood. On the one hand, prolonged hibernation may select for large offspring to enhance survival. On the other hand low population densities in extreme environments could result in low intraspecific competition and therefore relaxed selection on offspring size. Studies that have examined variation in the size of offspring have produced conflicting results; some studies have found larger hatchlings produced at warmer sites (Sinervo and Huey 1990; Diaz *et al.* 2007) while others have found larger hatchlings produced in colder environments (Niewiarowski and Angilletta 2008; Du *et al.* 2010). The causes of inter-population variation in reproductive output have been particularly well studied in the Eastern fence lizard, *Sceloporus undulatus*. Angilletta and co-workers have shown that population divergence in clutch size and egg size is likely to be an indirect consequence of temperature effects on female body size (Angilletta *et al.* 2006). Thus, egg size clines in this species are largely an effect of clines in female body size, itself affected by a number of factors that covary with climate. There may also be stronger selection on offspring size in cooler environments which contribute to divergence in the size-number trade-off between *Sceloporus* populations.

Differences in the frequency of reproduction (e.g., clutch interval) have not been studied in the same detail but are well known to occur. For example, clutch numbers range from two to four in *Sceloporus undulatus* (Niewairowski *et al.* 2004), and from one to five in the Mongolia racerunner, *Eremias argus* (Wang *et al.* 2011). This variation in the frequency of reproduction is most likely due to variation in activity season, which limits the extent to which later clutches will produce viable offspring (Meiri *et al.* 2012; see also Wang *et al.* 2011). However, at least in *Sceloporus undulatus*, which show a complex phylogeography (Leache and Reeder 2002), it is possible that part of the variation may be explained by lineage identity rather than by geographic-specific features of the environment (e.g., Niewairowski *et al.* 2004).

Other climatic factors that could contribute to variation in reproductive investment across species include those that affect food availability. Experimental increase in food availability has been shown to increase clutch size in the cool-temperate water skink, *Eulamprus tympanum* (Doughty and Shine 1998). Furthermore, increased food availability during the breeding season can allow species to compensate for poor food conditions during the rest of the year (Warne *et al.* 2012). However, it is important to note that most lizards appear to rely to some extent on stored resources for reproduction and therefore that investment may be less affected by variation in food availability immediately before or during reproduction (e.g., Bleu *et al.*

2013). This emphasizes that the main problem in linking climate and life history is that the relationships between climate and the main components of optimality models are typically assumed rather than well established. A prediction from a life history model is only as good as its assumptions.

12.2.4 Finding the Causes of Divergence in Reproductive Investment Between Populations

Despite that optimality models predict the reproductive investment that maximizes fitness under different environmental and life history scenarios, it is not easy to assess if population divergence is due to natural selection. This is because several predictions about adaptive shifts in resource allocation are similar to those expected based on plastic, perhaps inevitable, responses to climate. Short activity season may select for delayed age and increased size at maturity, reduced annual reproductive investment or large offspring size, but this is also the expected result of direct physiological effects of temperature or opportunity for thermoregulation. Indeed, few studies have unequivocally demonstrated that inter-population divergence in reproductive investment is a result of adaptive genetic divergence. A recent study of the common lizard (*Z. vivipara*) argued that when predictions differ for local genetic adaptation and (non-adaptive) plasticity they were in the direction favoring plasticity (Roitberg *et al.* 2013). However, as there are many life history models and predictions can vary depending on assumptions of relationships that are typically poorly documented, it is difficult to evaluate the evidence from comparative studies.

The most straightforward way to test for genetic divergence is to conduct common garden or reciprocal transplant experiments. When the trait of interest is reproduction, this requires a substantial time for animals to mature. As a consequence the number of studies that have investigated local adaptation in reproductive investment is very limited. However, studies that have reared animals to maturity typically find that plasticity accounts for a large proportion of population divergence in reproductive output (e.g., Ferguson and Talent 1993). This is not surprising considering the strong environmental effects on growth and hence female size in lizards. Studies that have tested for a within-species phylogenetic signal in reproductive traits in lizards have generally found them to be weak (Niewiarowski *et al.* 2004; Horvathova *et al.* 2013; see also Diaz *et al.* 2012; Roitberg *et al.* 2013). This is in marked contrast to the strong phylogenetic effect in among-species comparative studies and may support the idea that most within-species variation is due to plasticity rather than genetic divergence. We return to the role of plasticity versus genetic divergence at the end of this chapter.

12.3 WITHIN-POPULATION VARIATION IN REPRODUCTIVE INVESTMENT

The phylogeographic patterns in reproductive investment detailed above ultimately originate from variation within populations. Understanding the origin of variation can therefore be informative for understanding how populations and species diverge. In addition, variation can also be maintained within populations because of selection. Within-population variation, adaptive or non-adaptive, therefore occupies a central role in evolutionary biology.

Lizard populations typically show variation among females in how much they invest in reproduction at any given time (e.g., Tinkle 1967; Olsson and Shine 1997; Bleu *et al.* 2013). Most obviously, reproductive output is typically affected by body size, which varies among individuals due to both genetic and environmental factors. The latter includes the timing of hatching, food availability, and microhabitat variation in thermal opportunity. For example, in two short-lived lizards (*Uta stansburiana* and *Ctenophorus fordi*), late-hatching females are smaller as adults and show reduced reproductive output (Sinervo and Doughty 1996; Uller and Olsson 2010). Rather than describe the causes of variation in reproductive investment in detail, we here focus on two different aspects that are of particular interest for linking variation within populations with divergence among populations: (i) the extent to which variation in reproductive investment is heritable and hence can respond to selection; (ii) whether or not plasticity in reproductive investment within populations is adaptive.

12.3.1 Genetic Variation in Reproductive Investment

Heritable variation is the raw material for long-term evolutionary responses to selection and adaptive divergence between populations. Thus, it is important to establish the degree to which reproductive investment exhibits genetic variation in the wild. There are currently only a few studies of the quantitative genetics of reproductive investment in lizards and none that has gone further than establishing the quantitative genetic basis of (co) variation of phenotypic traits (e.g., no mapping of quantitative trait loci or genome-wide association studies; see Santure *et al.* 2013 for an example in wild birds). Egg size is heritable in side-blotched lizards and offspring size is heritable in the viviparous common lizard, whereas clutch size shows no or very low heritability in both species (Sinervo and Doughty 1996; Le Galliard *et al.* 2006). Furthermore, egg size appears to vary less within populations and show lower covariation with environmental variables than does clutch mass or clutch size (e.g., Ferguson and Talent 1993; Meiri *et al.* 2013). This is perhaps somewhat surprising from a mechanistic perspective since it is tempting to think about reproductive investment as a hierarchical set of decisions, where egg size is an outcome of upstream regulation of total allocation of resources towards reproduction and division of those resources into a discrete number of eggs. However, a high heritability of egg size is entirely consistent with the more extensive data on birds (Christians 2002). The underlying mechanisms of regulation of egg size remain unknown, however. In side-blotched lizards, it gives rise to genetic correlations between total investment and egg size and number (Sinervo and Doughty 1996), which is manifested in the phenotypic integration of these traits through shared hormonal regulation (Sinervo and Licht 1991). Whether or not such integration represents a constraint or an adaptive mechanism that can facilitate a coordinated shift in reproductive investment is poorly understood (see also Lancaster *et al.* 2010).

Sometimes within-population genetic variation is not just a source for selection to act upon but may be maintained by selection. The most famous cases are perhaps genetically determined alternative male mating strategies, such as the ones observed in side-blotched lizards (Sinervo and Lively 1996). These can evolve because of frequency-dependent selection where each strategy does well when rare (Sinervo and Calsbeek 2009). Female side-blotched lizards also have color morphs, which are associated with differences in reproductive output (Sinervo and Zamudio 2001). Yellow-throated females lay small clutches of large eggs whereas orange-throated females lay large clutches of small eggs. As a consequence, fitness of the two morphs is reversed across environmental conditions according to the r- versus K-selection paradigm (Pianka 1970). Fluctuations in environmental conditions or density cycles may, therefore, contribute to the maintenance of genetic variation in egg size in this population. A similar scenario of frequency-dependent selection, driven by social interactions, has been argued to occur in some French populations of common lizards (Vercken *et al.* 2007, 2010); however, the discreteness of color morphs has been questioned in this species (Cote *et al.* 2008). It is possible that fluctuating or frequency-dependent selection on reproductive investment is common in lizards as populations tend to vary substantially in density and demography, partly as a result of annual variation in egg survival (e.g., Tinkle *et al.* 1993). This means that individuals that reproduce fast and produce many small offspring are favored when populations grow, whereas individuals that invest into large offspring do better during periods of high juvenile and adult density. Whether or not this contributes to maintenance of genetic variation in reproductive traits in natural populations is poorly understood but further studies on the link between demography, density, and selection on reproductive investment in lizards are likely to be rewarding.

12.3.2 Adaptive and Non-adaptive Plasticity in Reproductive Investment

Rather than maintaining genetic variation, populations may respond to variable selection by evolving phenotypic plasticity. Variation in reproductive investment (e.g., differential allocation to clutch size versus egg size) may therefore reflect adaptive plasticity that maximizes fitness in a fluctuating environment. Predictions for adaptive plasticity in response to abiotic factors partly follow those for population or species divergence. For example, individuals that find themselves in thermally challenging microhabitats should adjust their growth and reproductive investment in ways that mirror adaptive population divergence along temperature clines (e.g., a poor thermal habitat may delay maturation). Indeed, many experimental studies have demonstrated plasticity in clutch size or offspring size in response to thermal opportunity, food availability, and so on (e.g., Shine and Downes 1999; Wapstra 2000; Warner *et al.* 2007). However, there is scant evidence that individual responses to, for example, microhabitat variation actually are adaptive within wild populations (Uller *et al.* 2011). Demonstrating this would require assessment of the fitness consequences of investment across different environmental conditions (e.g., Ferguson *et al.* 1982; Ferguson and Fox 1984), which is challenging to establish experimentally. Furthermore, documentation of behavioral responses to microhabitat profiles in natural settings is also difficult, but increasingly made feasible by technological development (e.g., small thermal loggers). As a consequence the adaptive significance of plasticity in reproductive investment within populations remains surprisingly poorly understood. A recent study comparing offspring size variation in response to thermal differences suggested that variation in offspring size between years is better explained by weak selection rather than adaptive plasticity (Uller *et al.* 2011; but see Goodman *et al.* 2013). Furthermore, the same response (e.g., increased investment) can be the result of different allocation decisions (e.g., stored versus income resources). The adaptive value of reproductive investment may therefore only be detected by considering their long-term consequences for individual growth or survival. Nevertheless, since lizards are able to make flexible use of stored resources when the income during egg production varies (Bonnet *et al.* 1998; Warne *et al.* 2012), individual-level optimization of reproductive effort is likely to be widespread. Recent work making use of stable isotopes are promising for establishing the extent of capital breeding within populations (Warner *et al.* 2007; Warne *et al.* 2012), and it may provide opportunities to establish relationships to individual characteristics and features of individual home ranges or territories.

The expected return on investment into size and number of offspring may also vary as a result of plastic responses to conditions that are specific to particular population structures or individual interactions. We discussed the link between density and offspring size arising from the hypothesized adaptive significance of the island syndrome above. A similar scenario is expected to apply within populations, as densities often vary substantially across spatial and temporal scales. An experimental manipulation of density in the common lizard found that reducing density increased clutch size, but the effect on offspring size was marginal at best (Meylan *et al.* 2007). This could reflect limited plasticity in terms of offspring size, with an increased clutch size being an outcome of increased reproductive effort when competition is weaker. However, a problem with an adaptive hypothesis for local (as opposed to population-specific) variation in density is that density also affects dispersal tendencies of offspring, which itself may put different demands on components of reproductive investment (e.g., offspring size; Burgess *et al.* 2013; Kuijper and Johnstone 2013). Thus, it is not clear that the increased offspring size predicted in high-density island populations will apply to the typical meta-population scenario of mainland lizard populations that vary in density among sites and among years. Similar arguments could be made for other variables that vary on small scales (e.g., predation); in contrast to large-scale geographic variation lizards may have evolved reproductive investment responses that facilitate offspring escaping from poor conditions rather than tolerating those conditions.

In many species with paternal care, such as mammals, the quality of the partner is one of the most important predictors of how reproductive investment is transformed into fitness. Females may therefore be selected to adjust their investment in response to the quality of their partners (Burley 1986; Sheldon 2000). Whether females should invest more or less when mated to high-quality partners depend on a number of biological features that influence how male and female characters influence the fitness returns of current versus future reproduction (Harris and Uller 2009; Ratikainen and Kokko 2010). Although male lizards rarely exhibit care (see While *et al.* Chapter 15, this volume), an analogous argument applies also to genetic differences among males that affect the fitness of offspring (Sheldon 2000). Many male lizards display colors or other forms of ornaments that appear to signal some form of 'male quality'. Thus, we may expect that females should adjust their investment in response to such indicators of quality. This hypothesis has received support from a study of sand lizards (*Lacerta agilis*) (Anderholm *et al.* 2004). Male sand lizards with enlarged green lateral patches were more successful in acquiring females in the field. Furthermore, females mated to those males showed increased relative clutch mass, suggesting that females that paired up with these 'enhanced' males increased their reproductive effort. Similar effects may occur in other species as well. In territorial species, females that pair up with attractive males may invest more not only because of genetic benefits but also because male ornamentation signals high resource availability and hence an opportunity to reduce the cost of reproduction, for example, by increased food intake post-oviposition. However, some species have limited ability to adjust their reproductive investment in response to male traits as mating occurs close to ovulation, at which time females already have committed to a certain clutch size and even egg size. Understanding the reproductive physiology and ecology of the species is essential to designing appropriate tests of the differential allocation hypothesis (Harris and Uller 2009).

In addition to the potential for adaptive adjustment of offspring size and number, many environmental effects on reproductive investment can, of course, reduce fitness. For example, low food availability in the previous year limits the ability to store energy for reproduction the following year (e.g., Bleu *et al.* 2013). Similarly, high density and intra-specific competition or social interference can reduce energy acquisition and hence reproductive output. For example, an experimental study of common lizards found that male-biased sex ratio was associated with reduced female reproductive success (Le Galliard *et al.* 2005). This may be a direct effect of male harassment and injury incurred during mating, which is particularly long in this species (over 1 h compared to the more typical 1–10 minutes in lacertids; Olsson and Madsen 1998). As a consequence, a male-biased population sex ratio should reduce recruitment in this species and may lead to interesting population dynamics via direct and indirect effects on reproductive investment (Le Galliard *et al.* 2005; Fitze *et al.* 2005).

More generally, population growth and stability is affected by conditions that individuals experience early in life and that have delayed effects on reproductive performance. In lizards, conditions following hatching or birth not only affect survival and hence recruitment into the breeding population, but also the size and age at maturity. This in turn can influence reproductive output, suggesting there may be a substantial lag between, for example, poor environmental condition and its effect on the breeding population. However, a recent study of common lizards showed that cohort effects primarily exercise their effect through juvenile survival (Le Galliard *et al.* 2010). The effect on reproductive traits in this species may be buffered as a result of catch-up growth or flexible adjustment of stored resources to maintain reproductive output. Nevertheless, we anticipate that further studies of different species living in more unstable habitats will show strong cohort effects that are manifested in terms of reproductive output and that this will feed back on changes in population density through time and space. Long-term studies such as the one of common lizards in France (e.g., Le Galliard *et al.* 2010) are crucial for testing theory and establishing patterns of phenotypic variation across contexts. Hopefully, it will be possible to secure sufficient funding to maintain ongoing studies of populations of individually marked animals as well as setting up new studies on systems that may provide particularly interesting insights into the links between ecology, population demography and reproductive investment.

12.4 FROM INDIVIDUAL VARIATION TO EVOLUTIONARY DIVERSIFICATION

Variation within populations is the raw material for divergence between populations and species. This raises an interesting question regarding whether or not the variation observed within populations is consistent with population divergence. For example, does plasticity to locally variable environments match adaptive divergence among populations? As suggested above, there is currently limited evidence that within-population responses to abiotic conditions are adaptive. Nevertheless, plasticity is clearly ubiquitous. This suggests that population divergence in reproductive investment across environmental gradients may occur in two steps (West-Eberhard 2003). The first is a plastic response to changes in the abiotic or biotic environments. This response will partly reflect evolved mechanisms that enabled organisms to respond appropriately in their ancestral environment. This step is followed by selection on reaction norms such that individuals that respond to the new environment in functional ways do better. If there is heritable variation in reaction norms, the population may evolve to genetically diverge from the ancestral population, perhaps also with respect to the degree of plasticity (Waddington 1942; Suzuki and Nijhout 2006; see Lande 2009 for a quantitative genetic model). In the metaphor of the adaptive landscape, the population moves towards a new fitness peak initially via those individuals that persist via plasticity and only subsequently accumulate genetic changes that enable the population to reach the peak (Price *et al.* 2003). In some cases plasticity may be sufficient for populations to move to new fitness peaks without genetic evolution. Thus, intermediate levels of plasticity should be associated with the most rapid adaptive evolution at the genetic level (Price *et al.* 2003), whereas divergence through genetic evolution only would take longer and perhaps require more extensive genetic divergence.

Reproductive investment exhibits substantial plasticity, and the evidence reviewed above suggests that plasticity is greater for total investment and clutch size than it is for offspring size. Furthermore, population divergence can often be attributed to plastic responses with more or less additional divergence due to genetic differences (Schlichting and Wund 2014). Thus, diversification in reproductive investment provides an opportunity to test the evolutionary scenario that plasticity shapes the direction of evolutionary diversification. For example, divergence in the mechanisms and patterns of investment across climatic regimes should mirror the plasticity exhibited by the ancestral population in response to annual variation in, for example, temperature and rainfall. If ancestral plasticity facilitated persistence in novel climates, selection may have fine-tuned the responses such that the population now exhibit increased, or more specific, responses to the local environment. If the ancestral responses were non-adaptive, we predict evolution of reduced plasticity in derived environments (West-Eberhard 2003; Grether 2005). Finally, the difference within population in the degree of plasticity and genetic variation between egg size and clutch size suggest that clutch size is more likely to evolve via plasticity than egg size. Thus, population divergence in egg size may be more likely to be associated with substantial genetic difference and respond slower to new conditions compared to clutch size.

To our knowledge, no study has so far evaluated these, or similar, scenarios of evolutionary change in reproductive investment in lizards. However, the data on common lizards suggest that population divergence in clutch mass and relative clutch mass is largely due to plasticity, but that hatchling mass varies across lineages and with geographic distance, suggesting genetic divergence (Roitberg *et al.* 2013). Furthermore, co-variation between reproductive traits and body size across common lizard populations parallels that occurring in response to climatic variation within populations (Sorci and Clobert 1999; Lorenzon *et al.* 2001; Le Galliard *et al.* 2010). Experimental tests of responses to environmental gradients for lizards from different populations could allow the reconstruction of the adaptive scenarios described above and evaluation of the hypothesis that selection has modified reaction norms across climatic regimes.

Moving from studies of patterns of reproductive investment towards dissecting the evolutionary process by which such patterns arise will require increased attention to the mechanisms of reproductive investment. Unfortunately, we have a limited understanding of the regulation of reproductive investment in lizards, the genetic basis of reproductive traits, and the extent to which reaction norms show heritable variation (although this is likely given genetic variation in growth plasticity). Studies that combine assessment of the mechanistic basis of plasticity (or lack thereof) with studies of population divergence in reproductive investment would be able to go beyond studying the pattern of adaptive variation and target the process of adaptation itself. Such an integrated view on evolution has recently shown how house finches are able to rapidly adapt to novel climatic conditions via adjustment of oocyte growth and the onset of laying (Badyaev 2009). Lizards that inhabit highly variable environments, that occur across steep environmental gradients, that show substantial demographic lability and that have been introduced into novel environments are likely to be particularly useful for establishing the mechanisms by which populations adapt.

One of the most exciting aspects of modern biology is the ability to study the developmental genetic basis of evolution. Genome sequencing of species that exhibit individual and geographic variation would open up opportunities to also address the genetic basis of phenotypic variation within- and between populations in detail, providing further opportunities to address to what extent ancestral mechanisms facilitate or constrain evolutionary divergence. *Anolis carolinensis* is currently the only available whole genome sequence for a lizard, which limits the ability to investigate functional divergence between populations (but more genomes are expected to soon be available). The common lizard is an obvious candidate for a genomic approach as its recent evolution of viviparity makes it a model for understanding transitions in reproductive mode. Viviparity is a highly evolutionarily labile reproductive trait in squamates (Shine 1985) and hence an outstanding character for studying the genomics of convergent evolution (Stern 2013). However, many more lizard species exhibit a remarkable variation in reproductive traits, which is why lizards facilitated much of the early development of life history theory. Perhaps they can also contribute to the new wave of interest in the role of plasticity for evolutionary diversification of life histories and the evolution of development.

12.5 ACKNOWLEDGMENTS

We are grateful to Justin Rheubert, Dustin Siegel, Stanley Trauth and Barrie Jamieson for the invitation to contribute to this volume. TU is supported by the Royal Society of London and the Knut and Alice Wallenberg Foundation. GW is supported by a Marie Curie research fellowship. Reviewers provided valuable comments on a draft chapter.

12.6 LITERATURE CITED

Adler, G. H. and Levins, R. 1994. The island syndrome in rodent populations. Quarterly Review of Biology 69: 473–490.

Adolph, S. C. and Porter, W. P. 1996. Growth, seasonality, and lizard life histories: Age and size at maturity. Oikos 77: 267–278.

Anderholm, S., Olsson, M., Wapstra, E. and Ryberg, K. 2004. Fit and fat from enlarged badges: a field experiment on male sand lizards. Proceedings of the Royal Society B-Biological Sciences 271: S142–S144.

Andrews, R. M. and Rand, A. S. 1974. Reproductive effort in anoline lizards. Ecology 55: 1317–1327.

Angilletta, M. J., Oufiero, C. E. and Leach, A. D. 2006. Direct and indirect effects of environmental temperature on the evolution of reproductive strategies: an information —theoretic approach. American Naturalist 168: E123–E135.

Badyaev, A. V. 2009. Evolutionary significance of phenotypic accommodation in novel environments: an empirical test of the Baldwin effect. Philosophical Transactions of the Royal Society B: Biological Sciences 364: 1125–1141.

Ballinger, R. E. 1983. Life history variations. pp. 241–260. In R. B. Huey, E. R. Pianka and T. W. Schoener (eds.), *Lizard ecology: studies of a model organism*. Harvard University Press, Cambridge, MA.

Blair, W. F. 1960. *The Rusty Lizard*. University of Texas Press, Austin.

Bleu, J., Le Galliard, J. -F., Fitze, P. S., Meylan, S., Clobert, J. and Massot, M. 2013. Reproductive allocation strategies: a long term study on proximate factors and temporal adjustments in a viviparous lizard. Oecologia 171:141–151.

Bonnet, X., Bradshaw, D. and Shine, R. 1998. Capital versus income breeding: an ectothermic perspective. Oikos 107: 333–342.

Brown, G. P. and Shine, R. 2009. Beyond size-number trade-offs: clutch size as a maternal effect. Philosophical Transactions of the Royal Society B: Biological Sciences 364: 1097–1106.

Burgess, S.C., Bode, M. and Marshall, D. J. 2013. Costs of dispersal alter optimal offspring size in patchy habitats: combining theory and data for a marine invertebrate. Functional Ecology 27: 757–765.

Burley, N. 1986. Sex-ratio manipulation in color-banded populations of zebra finches. Evolution 40: 1191–1206.

Calsbeek, R. and Sinervo, B. 2008. Alternative reproductive tactics in reptiles. pp. 518. In J. Brockmann, L. Oliveira and M. Taborsky (eds.), *Alternative reproductive tactics*. Cambridge University Press.

Charnov, E. L. and Downhower, J. F. 1995. A trade-off-invariant life-history rule for optimal offspring size. Nature 376: 418–419.

Christians, J. K. 2002. Avian egg size: variation within species and inflexibility within individuals. Biological Reviews 77: 1–26.

Congdon, J. D. and Gibbons, J. W. 1987. Morphological constraint on egg size—a challenge to optimal egg size theory. Proceedings of the National Academy of Sciences of the United States of America 84: 4145–4147.

Cote, J., Le Galliard, J. F., Rossi, J. M. and Fitze, P. S. 2008. Environmentally induced changes in carotenoid-based coloration of female lizards: a comment on Vercken *et al*. Journal of Evolutionary Biology 21: 1165–1172.

Diaz, J. A., Perez-Tris, J., Bauwens, D., Perez-Aranda, D., Carbonell, R., Santos, T. and Telleria, J. L. 2007. Reproductive performance of a lacertid lizard at the core and the periphery of the species' range. Biological Journal of the Linnean Society 92: 87–96.

Diaz, J. A., Iraeta, P., Verdu-Ricoy, J., Siliceo, I. and Salvador, A. 2012. Intraspecific Variation of Reproductive Traits in a Mediterranean Lizard: Clutch, Population, and Lineage Effects. Evolutionary Biology 39: 106–115.

Doughty, P. 1997. The effects of "fixed" clutch sizes on lizard life-histories: Reproduction in the Australian velvet gecko, *Oedura lesueurii*. Journal of Herpetology 31: 266–272.

Doughty, P. and Shine, R. 1998. Reproductive energy allocation and long-term energy stores in a viviparous lizard (*Eulamprus tympanum*). Ecology 79: 1073–1083.

Du, W.- G. and Lue, D. 2010. An experimental test of body volume constraint on female reproductive output. Journal of Experimental Zoology A 313A: 123–128.

Du, W. G., Warner, Daniel, A., Langkilde, T., Robbins, T. and Shine, R. 2010. The physiological basis of geographic variation in rates of embryonic development within a widespread lizard species. American Naturalist 176: 522–528.

Dunham, A. E. and Miles, D. B. 1985. Patterns of covariation in life-history traits of squamate reptiles: the effects of size and phylogeny reconsidered. American Naturalist 126: 231–257.

Dunham, A. E., Miles, D. B. and Reznick, D. N. 1994. In *Life history patterns in squamate reptiles*. pp. 441–522. In C. Gans and R. B. Huey (eds.), *Biology of the Reptilia. Vol 16*, Ecology B. Defense and life history, Branta Books, Ann Arbor, MI.

Ebert, D. 1994. Fractional resource-allocation into few eggs—Daphnia as an example. Ecology 75: 568–571.

Ferguson, G. and Talent, L. 1993. Life-history traits of the lizard *Sceloporus undulatus* from two different populations raised in a common laboratory environment. Oecologia 93: 88–94.

Fitch, H. S. 1985. Variation in clutch and litter size in New World reptiles. Museum of Natural History University of Kansas, Miscellaneous Publications 76: 1–76.
Fitze, P. S., Le Galliard, J. -F., Federici, P., Richard, M. and Clobert, J. 2005. Conflict over multiple-partner mating between males and females of the polygynandrous common lizards. Evolution 59: 2451–2459.
Goodman, B. A., Hudson, S. C., Isaac, J. L. and Schwarzkopf, L. 2009. The evolution of body shape in response to habitat: is reproductive output reduced in flat lizards? Evolution 63: 1279–1291.
Goodman, B. A., Schwarzkopf, L. and Krockenberger, A. K. 2013. Phenotypic integration in response to incubation environment adaptively influences habitat choice in a tropical lizard. American Naturalist 182: 666–673.
Grether, G. F. 2005. Environmental change, phenotypic plasticity, and genetic compensation. American Naturalist 166: E115–E123.
Harris, W. E. and Uller, T. 2009. Reproductive investment when mate quality varies: differential allocation versus reproductive compensation. Philosophical Transactions of the Royal Society B: Biological Sciences 364: 1039–1048.
Horvathova, T., Cooney, C. R., Fitze, P. S., Oksanen, T. A., Jelic, D., Ghira, I. and Jandzik, D. 2013. Length of activity season drives geographic variation in body size of a widely distributed lizard. Ecology and Evolution 3: 2424–2442.
Huey, R. B. and Pianka, E. R. 1981. Ecological consequences of foraging mode. Ecology 62: 991–999.
Iraeta, P., Salavador, A. and Diaz, J. A. 2013. Life history traits of two Mediterranean lizard populations: a possible example of countergradient covariation. Oecologia 172: 167–176.
Kawecki, T. J. 2008. Adaptation to Marginal Habitats, Pages 321–342. Annual Review of Ecology Evolution and Systematics 20: 321–342.
Kuijper, B. and Johnstone, R. A. 2013. How should parents adjust the size of their young in response to local environmental cues? Journal of Evolutionary Biology 26: 1488–1498.
Lack, D. 1947. The significance of clutch size. Ibis 89: 302–352.
Lancaster, L. T., McAdam, A. G. and Sinervo, B. 2010. Maternal adjustment of egg size organizes alternative escape behaviors, promoting adaptive phenotypic integration. Evolution 64: 1607–1621.
Lande, R. 2009. Adaptation to an extraordinary environment by evolution of phenotypic plasticity and genetic assimilation. Journal of Evolutionary Biology 22: 1435–1446.
Leache, A. D. and Reeder, T. W. 2002. Molecular systematic of the eastern fence lizards (*Sceloporus undulatus*): A comparative of parsimony, likelihood, and Bayesian approaches. Systematic Biology 51: 44–68.
Le Galliard, J. F., Fitze, P. S., Ferriere, R. and Clobert, J. 2005. Sex ratio bias, male aggression, and population collapse in lizards. Proceedings of the National Academy of Sciences of the United States of America 102: 18231–18236.
Le Galliard, J. F., Massot, M., Landys, M. M., Meylan, S. and Clobert, J. 2006. Ontogenic sources of variation in sexual size dimorphism in a viviparous lizard. Journal of Evolutionary Biology 19: 690–704.
Le Galliard, J.- F., Marquis, O. and Massot, M. 2010. Cohort variation, climate effects and population dynamics in a short-lived lizard. Journal of Animal Ecology 79: 1296–1307.
Lorenzon, P., Clobert, J. and Massot, M. 2001. The contribution of phenotypic plasticity to adaptation in *Lacerta vivipara*. Evolution 55: 392–404.
Meiri, S. 2008. Evolution and ecology of lizard body sizes. Global Ecology and Biogeography 17: 724–734.
Meiri, S., Brown, J. H. and Sibly, R. 2012. The ecology of lizard reproductive output. Global Ecology and Biogeography 21: 592–602.
Meylan, S., Clobert, J. and Sinervo, B. 2007. Adaptive significance of maternal induction of density-dependent phenotypes. Oikos 116: 650–661.

Michaud, E. J. and Echternacht, A. C. 1995. Geographic variation in the life-history of the lizard *Anolis carolinensis* and support for the pelvic constraint model. Journal of Herpetology 29: 86–97.

Niewiarowski, P. H. and Dunham, A. E. 1994. The Evolution of Reproductive Effort in Squamate Reptiles—Costs, Trade-Offs, and Assumptions Reconsidered. Evolution 48: 137–145.

Niewiarowski, P. H. 1994. Understanding geographic life-history variation in lizards. pp. 31–50. In L. J. Vitt and E. R. Pianka (eds.), *Lizard ecology. Historical and experimental approaches.* Princeton University Press, Princeton, NJ.

Niewiarowski, P. H., Angilletta, M. J. and Leache, A. D. 2004. Phylogenetic comparative analysis of life-history variation among populations of the lizard *Sceloporus undulatus*: An example and prognosis. Evolution 58: 619–633.

Niewiarowski, P. H. and Angilletta, M. J. 2008. Countergradient variation in embryonic growth and development: do embryonic and juvenile performances trade off? Functional Ecology 22: 895–901.

van Noordwijk, A. and de Jong, G. 1986. Acquisition and allocation of resources: the influence on variation in life history tactics. American Naturalist 128: 137–142.

Novosolov, M., Raia, P. and Meiri, S. 2013. The island syndrome in lizards. Global Ecology and Biogeography 22: 184–191.

Olsson, M. and Shine, R. 1997. The limits to reproductive output: offspring size versus number in the sand lizard. American Naturalist 149: 179–188.

Olsson, M. and Madsen, T. 1998. Sexual selection and sperm competition in reptiles. In T. R. Birkhead and A. P. Møller (eds.), *Sperm competition and sexual selection*. Academic Press.

Olsson, M. and Shine, R. 1999. Plasticity in frequency of reproduction in an alpine lizard, *Niveoscincus microlepidotus*. Copeia 1999: 794–796.

Oufiero, C. E., Smith, A. J. and Angilletta, M. J., Jr. 2007. The importance of energetic versus pelvic constraints on reproductive allocation by the eastern fence lizard (*Sceloporus undulatus*). Biological Journal of the Linnean Society 91: 513–521.

Oufiero, C. E., Gartner, G. E. A., Adolph, S. C. and Garland, T. 2011. Latitudinal and climatic variation in body size and dorsal scale counts in *Sceloporus* lizards: A phylogenetic perspective. Evolution 65: 3590–3607.

Pianka, E. R. 1970. On r and K selection. American Naturalist 104: 592–597.

Price, T. D., Qvarnstrom, A. and Irwin, D. E. 2003. The role of phenotypic plasticity in driving genetic evolution. Proceedings of the Royal Society of London B 270: 1433–440.

Ratikainen, I. I. and Kokko, H. 2010. Differential allocation and compensation: who deserves the silver spoon? Behavioral Ecology 21: 195–200.

Ricklefs, R. 1968. A note on the evolution of clutch size in altricial birds.

Roff, D. A. 1992. The evolution of life histories. Chapman and Hall Publishers.

Roff, D. A. 2002. Life history evolution. Sinauer Associates Inc.

Roitberg, E. S., Kuranova, V. N., Bulakhova, N. A., Orlova, V. F., Eplanova, G. V., Zinenko, O. I., Eplanova, G. V., Zinekos, O. I., Shamgunova, R. R., Hofmann, S. and Yakovlev, V. A. 2013. Variation of Reproductive Traits and Female Body Size in the Most Widely-Ranging Terrestrial Reptile: Testing the Effects of Reproductive Mode, Lineage, and Climate. Evolutionary Biology 40: 420–438.

Rollinson, N., Edge, C. B. and Brooks, R. J. 2013. Recurrent violations of invariant rules for offspring size: evidence from turtles and the implications for small clutch size models. Oecologia 172: 973–982.

Santure, A. W., De Cauwer, I., Robinson, M. R., Poissant, J., Sheldon, B. C. and Slate, J. 2013. Genomic dissection of variation in clutch size and egg mass in a wild great tit (*Parus major*) population. Molecular Ecology 22: 3949–3962.

Schlichting, C. D. and M. A. Wund. 2014. Phenotypic plasticity and epigenetic marking: an assessment of evidence for genetic accommodation. Evolution 68: 656–672.

Schwartzkopf, L. 1994. Measuring trade-offs: a review of studies of costs of reproduction in lizards. pp. 7–30. In L. J. Vitt and E. R. Pianka (eds.), *Lizard Ecology: historical and experimental perspectives*. Princeton University Press, Princeton, NJ.

Sears, M. W. and Angilletta, M. J. 2004. Body size clines in *Sceloporus* lizards: Proximate mechanisms and demographic constraints. Integrative and Comparative Biology 44: 433–442.

Sheldon, B. 2000. Differential allocation: tests, mechanisms and implications. Trends in Ecology and Evolution 15: 397–402.

Shine, R. 1980. Costs of reproduction in reptiles. Oecologia 46: 92–100.

Shine, R. 1985. The evolution of viviparity in reptiles: an ecological analysis. pp. 605–694. In C. Gans and F. Billet (eds.), *Biology of the Reptilia.* . Volume 15 Development B. Wiley—Interscience, John Wiley & Sons.

Shine, R. 1992. Relative Clutch Mass and Body Shape in Lizards and Snakes—Is reproductive investment constrained or optimized? Evolution 46: 828–833.

Shine, R. and Downes, S. J. 1999. Can pregnant lizards adjust their offspring phenotypes to environmental conditions? Oecologia 119: 1–8.

Sinervo, B. 1994. Experimental tests of reproductive allocation paradigms. Chapter 4. pp. 73–90. In L. J. Vitt and E. R. Pianka (eds.), *Lizard Ecology: Historical and Experimental Perspectives.* Princeton University Press.

Sinervo, B. and Huey, R. B. 1990. Allometric engineering—an experimental test of the causes of interpopulational differences in performance. Science 248: 1106–1109.

Sinervo, B. and Licht, P. 1991. Proximate constraints on the evolution of egg size, number, and total clutch mass in lizards. Science 252: 1300–1302.

Sinervo, B. and Doughty, P. 1996. Interactive effects of offspring size and timing of reproduction on offspring reproduction: Experimental, maternal, and quantitative genetic aspects. Evolution 50: 1314–1327.

Sinervo, B. and Lively, C. 1996. The rock-paper-scissors game and the evolution of alternative mating strategies. Nature 380: 240.

Sinervo, B. and Zamudio, K. R. 2001. The evolution of alternative reproductive strategies: fitness differential, heritability, and genetic correlations between the sexes. Journal of Heredity 2: 198–205.

Sorci, G. and Clobert, J. 1999. Natural selection on hatchling body size and mass in two environments in the common lizard (*Lacerta vivipara*). Evolutionary Ecology Research 1: 303–316.

Stearns, S.C. 1976. Life history tactics: a review of the ideas. Quarterly Review of Biology 51: 3–47.

Stern, D. L. 2013. The genetic causes of convergent evolution. Nature Reviews Genetics 14: 751–764.

Sun, Y.- Y., Du, Y., Yang, J., Fu, T.- B., Lin, C.- X. and Ji, X. 2012. Is the Evolution of Viviparity Accompanied by a Relative Increase in Maternal Abdomen Size in Lizards? Evolutionary Biology 39: 388–399.

Surget-Groba, Y., Heulin, B., Guillaume, C. P., Puky, M., Semenov, D., Orlova, V., Kupriyanova, L. *et al.* 2006. Multiple origins of viviparity, or reversal from viviparity to oviparity? The European common lizard (*Zootoca vivipara*, Lacertidae) and the evolution of parity. Biological Journal of the Linnaean Society 87: 1–11.

Suzuki, Y. and Nijhout, H. F. 2006. Evolution of a polyphenism by genetic accommodation. Science 311: 650–652.

Telemeco, R. S. and Baird, T. A. 2011. Capital energy drives production of multiple clutches whereas income energy fuels growth in female collared lizards *Crotaphytus collaris.* Oikos 120: 915–921.

Thompson, M. B. and Speake, B. K. 2003. Energy and nutrient utilisation by embryonic reptiles. Comparative Biochemistry and Physiology A 133: 529–538.

Tinkle, D. W. 1967. The life and demography of the side-blotched lizard, *Uta stansburiana.* Miscellaneous Publications of the Museum of Zoology of the University of Michigan 132: 1–182.

Tinkle, D. W. 1969. The concept of reproductive effort and its relation to the evolution of life histories in lizards. American Naturalist 103: 501–516.

Tinkle, D. W. and Wilbur, H. M. 1970. Evolutionary strategies in lizard reproduction. Evolution 24: 55–74.
Tinkle, D. W., Dunham, A. E. and Congdon, J. D. 1993. Life history and demographic variation in the lizard *Sceloporus graciosus*—a long term study. Ecology 74: 2413–2429.
Uller, T. and Olsson, M. 2005. Trade-offs between offspring size and number in the lizard *Lacerta vivipara*: a comparison between field and laboratory conditions. Journal of Zoology 265: 295–299.
Uller, T. and Olsson, M. 2009. Offspring size-number trade-off in a lizard with small clutch sizes: tests of invariants and potential implications. Evolutionary Ecology 23: 363–372.
Uller, T. and Olsson, M. 2010. Offspring size and timing of hatching determine survival and reproductive output in a lizard. Oecologia 162: 663–671.
Uller, T., While, G. M., Wapstra, E., Warner, D. A., Goodman, B. A., Schwarzkopf, L. and Olsson, M. 2009. Evaluation of offspring size-number invariants in 12 species of lizard. Journal of Evolutionary Biology 22: 143–151.
Uller, T., While, G. M., Cadby, C. D., Harts, A., O'Connor, K., Pen, I. and Wapstra, E. 2011. Altitudinal divergence in maternal thermoregulatory behavior may be driven by differences in selection on offspring survival in a viviparous lizard. Evolution 65: 2313–2324.
Vercken, E., Massot, M., Sinervo, B. and Clobert, J. 2007. Colour variation and alternative reproductive strategies in females of the common lizard *Lacerta vivipara*. Journal of Evolutionary Biology 20: 221–232.
Vercken, E., Clobert, J. and Sinervo, B. 2010. Frequency Dependent reproductive success in female common lizards: a real-life hawk-dove-bully game? Oecologia 162: 49–58.
Vitt, L. J. and Seigel, R. A. 1985. Life history traits of lizards and snakes. American Naturalist 125: 480–484.
Waddington, C. H. 1942. Canalization of development and the inheritance of acquired characteristics. Nature 3811: 563–565.
Wang, Z., Xia, Y. and Ji, X. 2011. Clutch frequency affects the offspring size-number trade-off in lizards. PLoS ONE 6: 1.
Wapstra, E. 2000. Maternal basking opportunities affects juvenile phenotype in a viviparous lizard. Functional Ecology 14: 345–352.
Wapstra, E. and Swain, R. 2001. Geographic and annual variation in life-history traits in a temperate zone Australian skink. Journal of Herpetology 35: 194–203.
Warne, R. W. and Charnov, E. L. 2008. Reproductive allometry and the size-number trade-off for lizards. American Naturalist 172: E80–E98.
Warne, R. W., Gilman, C. A., Garcia, D. A. and Wolf, B. O. 2012. Capital breeding and allocation to life-history demands are highly plastic in lizards. American Naturalist 180: 130–141.
Warner, D. A. and Andrews, R. M. 2002. Laboratory and field experiments identify sources of variation in phenotypes and survival of hatchling lizards. Biological Journal of the Linnean Society 76: 105–124.
Warner, D. A. and Shine, R. 2008. The adaptive significance of temperature-dependent sex determination in a reptile. Nature 451: 566–571.
Warner, D. A., Lovern, M. B. and Shine, R. 2007. Maternal nutrition affects reproductive output and sex allocation in a lizard with environmental sex determination. Proceedings of the Royal Society of London B 274: 883–890.
West-Eberhard, M.- J. 2003. *Developmental plasticity and evolution*. Oxford University Press, New York, pp. 797.
Yang, J., Sun, Y.- Y., Fu, T.- B., Xu, D.- D. and Ji, X. 2012. Selection for increased maternal body volume does not differ between two *Scincella* lizards with different reproductive modes. Zoology 115: 199–206.

CHAPTER 13

Viviparity and Placentation in Lizards

James R. Stewart[1,*] and *Daniel G. Blackburn*[2]

13.1 INTRODUCTION

Viviparity (live-bearing reproduction) is a widespread reproductive pattern of considerable biological significance and a subject of intensive ongoing investigation. Viviparity in squamate reptiles is a particular focus of such research, with particular regard to its significance, the mechanisms by which it is accomplished, and how and why it has so frequently evolved (Blackburn 2000a).

Because the embryos of viviparous lizards and snakes complete their development inside the maternal reproductive tract, they are protected from environmental threats and maintained under maternally-controlled conditions. However, oviductal gestation poses critical physiological problems for the embryo that must be addressed. At minimum, embryos need ways to exchange respiratory gases, to obtain water, and to replace calcium lost through evolutionary reduction of the eggshell. These functions are accomplished in viviparous squamates by means of placentas–organs formed through a close association of the oviductal lining to the fetal (extraembryonic) membranes. Such placentas also can provide nutrients to embryos that supplement those contained in the ovulated yolks.

[1] Department of Biological Sciences, East Tennessee State University, Johnson City, TN 37614 USA.
[2] Department of Biology and Electron Microscopy Facility, Trinity College, Hartford, CT 06106 USA.
* Corresponding author

For several reasons, squamate reptiles are ideal organisms for the study of viviparity and placentation. First, more than 1100 squamate species are viviparous (Blackburn 2006), offering a large and diverse collection of organisms for investigation. In fact, the number of viviparous squamate species exceeds that of all other non-mammalian vertebrates combined (Blackburn 1999a, 2014). Second, squamates are thought to have evolved viviparity on at least 115 separate occasions, constituting 75% of the ≥150 origins that are known to have occurred in vertebrate history (Blackburn 2014). Most of these squamate origins have occurred among lizards (Blackburn 1985; Shine 1985, 1999b). Third, these multiple origins of viviparity allow for quantitative analyses of selective pressures, exaptations, and constraints, as well as exploration of the divergent means by which this homoplastic pattern has been achieved. Fourth, most of the known origins of squamate viviparity have occurred at low (sub-generic) taxonomic levels (Shine 1985; Blackburn 1999b) making it feasible to reconstruct historical details of how and why this pattern has arisen. Fifth, studies on squamates have the potential to provide valuable information relevant to other viviparous vertebrates. Viviparous lizards are extremely diverse reproductively, with modes of fetal nutrition that range from relatively lecithotrophic to extremely placentotrophic (Blackburn 1994a; Stewart and Thompson 2000; Thompson and Speake 2006). This range of fetal nutritional modes parallels that of fishes (Wourms *et al.* 1988) and exceeds that of viviparous mammals. Sixth, viviparous lizards and snakes form placentas from the same basic components as other amniotes, and may therefore offer (as well as inform) insights applicable to therian mammals. At their simplest, squamate placentas provide useful information about early stages in placental evolution, stages no longer present among extant mammals. Further, at their most complex, placentas of viviparous lizards show specializations that have converged on features of eutherian mammals.

In this review chapter, we have two interrelated goals. One is to summarize aspects of viviparity in lizards from descriptive, functional, and evolutionary standpoints. The other goal, the main focus of this chapter, is to explore lizard placentation in terms of development, structure, function, and evolution. Although the taxonomic focus of this chapter is on lizards, phylogenetically saurians are a broad, paraphyletic group of which snakes are a single monophyletic clade. Consequently, lizards are not distinct from snakes except insofar as the latter may exhibit derived specializations. The overall range of reproductive diversity of viviparous lizards encompasses that of snakes (see Section 13.2.2, this volume). Features of snake viviparity and placentation have recently been reviewed elsewhere (Blackburn and Stewart 2011).

13.2 VIVIPARITY

13.2.1 Reproductive Modes in Squamates

"Viviparity" is the vertebrate reproductive pattern in which females maintain developing eggs in their reproductive tracts and give birth to their offspring, a pattern also known as "live-bearing reproduction". In contrast, "oviparity" (or "egg-laying reproduction") is the pattern in which females deposit eggs with intact shells or jelly coats, either fertilized and developing (as in birds, chelonians, and most squamates and chondrichthyans) or unfertilized (as in most teleosts and anurans).

This classification system of reproductive patterns is universally used for reptiles and has been widely adopted for other vertebrates as well (Wourms 1981; Blackburn 1999a,b, 2000b, 2014; Wake 2002). Literature sources that pre-date the 1970s sometimes reflect a reluctance to consider live-bearing squamates as fully "viviparous", a category traditionally (in a form of phylogenetic chauvinism) reserved for mammals (Lamarck 1835). Accordingly, the now-archaic term "ovoviviparity" was often applied to live-bearing lizards and snakes under the mistaken notion that the females of these species simply retained eggs in their oviducts that hatched at the time of birth. In contrast, in other older literature, "ovoviviparity" was applied to egg-laying squamates that deposited shelled eggs with visible embryos—a category that arguably includes all oviparous Reptilia including birds (Blackburn 1994b, 2000b). The use of "ovoviviparity" in discrepant (and mutually exclusive) ways for reptiles produced much confusion, and the term has been abandoned in the primary literature. The fact that most oviparous squamates deposit eggs with embryos in the limb bud stage (at about 30% of the way through development) (Shine 1983a; Blackburn 1995) upholds the oviparity-viviparity dichotomy. However, the distinction is less easily applied to the few lizard species in which "oviparous" females deposit eggs with advanced embryos that hatch within days of being laid (see Smith and Shine 1997). Such species can be viewed as being on the verge of viviparity. A hierarchy of criteria has been established for recognition of oviparity, viviparity, and patterns of fetal nutrition (Blackburn 1993a, 1994a).

13.2.2 Distribution and Nature of Viviparity in Lizards

Viviparity is broadly distributed phylogenetically among lizards. It occurs in 13 nominal families (Table 13.1) and an estimated 19% of the extant species (Blackburn 1982; Shine 1985). Distribution of this reproductive pattern varies at a wide range of taxonomic levels. Three small families (Xantusiidae, Xenosauridae, and the monotypic Shinisauridae) are entirely

Table 13.1 Lizard families that include viviparous species. Taxonomy draws on Sites et al. (2011). Distributions refer to viviparous species, as do common names (not all available names are listed). Information on reproduction is summarized in many reviews, including Fitch (1970), Blackburn (1982, 1985a, 1999b), Shine (1984, 1985), Vitt and Caldwell (2013).

Family	Common names	Distribution
Diplodactylidae [1, 2]	common-, Pacific-, green-, tree-, striped-, cloudy-, & forest- geckoes	New Zealand
Xantusiidae	night lizards	s. w. US, n. Mexico
Cordylidae [1]	girdled-, grass-, spiny tailed lizards	s. & e. Africa
Scincidae [1]	"skinks" (variable)	worldwide
Lacertidae [1]	"viviparous lizard"; Gobi racerunner	Europe; c. & e. Asia
Anguidae [1, 3]	glass-, alligator-, & legless lizards	N. S. Am, Eur, Asia
Xenosauridae	knob- scaled lizards	Cent. Amer., Mexico
Shinisauridae	Chinese crocodile lizard	southeastern Asia
Chamaeleonidae [1]	dwarf-, 2 lined-, side-striped chameleons	s. & e.- central Africa
Agamidae [1]	Ceylon deaf agama; toad-headed agamas	Sri Lanka; Asia
Phrynosomatidae [1]	horned-, fence-, & spiny lizards, etc.	N. through Cent. Am.
Corytophanidae [1]	helmeted iguanas, helmeted basilisk	s. Mex to n. S. Am.
Liolaemidae [1]	tree iguanas, mountain lizards	South America

[1] also includes oviparous species
[2] traditionally placed in the family Gekkonidae
[3] including "Anniellidae"

viviparous, whereas 10 other families contain both viviparous and oviparous representatives (Table 13.1). Many genera contain both viviparous and oviparous species, including *Phrynocephalus* (Agamidae), *Trioceros* (*sensu* Tilbury and Tolley 2009; Chameleonidae), *Liolaemus* (Liolaemidae), *Corytophanes* (Corytophanidae), *Sceloporus* (Phrynosomatidae), *Phrynosoma* (Phrynosomatidae), *Eremias* (Lacertidae), *Diploglossus* (Anguidae), *Elgaria* (Anguidae), and several among the Scincidae, e.g., *Anomalopus, Chioninia, Eumeces, Eutropis, Leiolopisma, Lerista, Lioscincus, Lipinia, Lobulia, Lygosoma, Oligosoma, Prasinohaema, Scincella, Sphenomorphus, Tribolonotus,* and *Trachylepis* (Tinkle and Gibbons 1977; Blackburn 1982, 1985a, 1999b; Shine 1984, 1985). Furthermore, a few species contain both oviparous and viviparous populations, notably the European Common Lizard, *Zootoca vivipara* (Braña and Bea 1987; Heulin *et al.* 1993), the Three-toed Skink, *Saiphos equalis* (Smith and Shine 1997; Smith *et al.* 2001), the Southern Lerista, *Lerista bougainvillii* (Qualls *et al.* 1995; Fairbairn *et al.* 1998), and reportedly, the Cape Skink, *Trachylepis capensis* (Brown Wessels 1989).

Viviparous lizards exhibit a worldwide distribution that extends to every habitable continent (Table 13.1); they also occupy at least as wide a range of habitats as their oviparous counterparts. For example, viviparous lizards can be arboreal (as in chameleons and various liolaemids, agamids, and skinks), aquatic (as in Mesozoic mosasauroids), fossorial (e.g., the

Checkerboard Worm Lizard, *Trogonophis wiegmanni*, the California Legless Lizard, *Anniella pulchra*, and acontine and feylinine skinks), or nocturnal (xantusiids and New Zealand geckos of the genus *Woodworthia*), as well as diurnal and terrestrial (most viviparous lizards) (Fitch 1970; Vitt and Caldwell 2013). Viviparous species are represented disproportionately among lizards of high latitudes and high altitudes (Weekes 1933; Duellman 1965; Tinkle and Gibbons 1977; Shine and Berry 1978). However, they are widespread elsewhere in cold, temperate, and tropical climates; in latitudes that range from sub-arctic (in the northern ranges of *Zootoca vivipara* and the Slow Worm *Anguis fragilis*) to equatorial; in humid through xeric conditions; and in deserts, tropical rainforests, and montane habitats.

Litter sizes in viviparous lizards show moderate interspecific variation. A complicating factor is that they also can vary within a species according to maternal body size, age, and locality (Fitch 1970). At the low end of the range, litter sizes of only one offspring have been reported in the Cape Girdled Lizard, *Cordylus cordylus* (range = 1–3) (FitzSimons 1943), the Lined Lance Skink, *Acontias* ("*Typhlosaurus*") *lineatus* (Fitch 1970), and the Chonburi Snake Skink, *Isopachys roulei* (Taylor 1963). Litter sizes of ≤2 neonates are routinely produced by New Zealand geckos of the genera *Naultinus*, *Hoplodactylus*, and *Woodworthia* (Boyd 1942; Cree 1994), African *Scelotes* (Scincidae) (FitzSimons 1943), and North American *Xantusia* (Brattstrom 1951; Miller 1951, 1954; Goldberg and Bezy 1974). Small litter sizes also characterize the Grand Skink, *Oligosoma grande* (mean = 2.4) (Cree 1994) and the Knob-scaled Lizard, *Xenosaurus grandis* (mean = 3.2) (Lemos-Espinal *et al.* 2003). At the other end of the range of litter sizes lies the Pigmy Short-horned Lizard, *Phrynosoma douglasii*, which averages about 15 young but can produce as many as 31 (Stebbins 1954; Fitch 1970). Although viviparous females in a few other genera (e.g., *Anguis, Diploglossus, Sceloporus*, and *Trioceros*) can occasionally produce more than 20 offspring, litter sizes of 3 to 10 offspring are typical of viviparus squamates (Fitch 1970; Dunham and Miles 1985; Vitt and Caldwell 2013).

Gestation commonly lasts 2 to 4 months in viviparous lizards, depending on the species (Fitch 1970; Tinkle and Gibbons 1977; Vitt and Caldwell 2013). In seasonal climates, pregnancy usually is timed to correspond with the warmer summer months. Longer gestation periods (≥6 months) characterize various live-bearing *Sceloporus*, which undergo vitellogenesis in the autumn, overwinter while pregnant, and give birth in the spring or early summer (Goldberg 1971; Ballinger 1973; Guillette and Casas-Andreu 1980; Ortega and Barbault 1984; Guillette and Méndez de la Cruz 1993; also see Marion and Sexton 1971). This pattern facilitates reproduction in the seasonally cold climates of montane environments (Méndez de la Cruz *et al.* 1998). A similar pattern occurs in various other viviparous species of montane habitats, such as the Keeled Helmeted Iguana, *Corytophanes pericarinatus*, of Guatemala

(McCoy 1968), the Imbricate Alligator Lizard, *Barisia imbricata* (Guillette and Casas-Andreu 1987) of Mexico, and Cope's Skink, *Plestiodon copei* also from Mexico (Guillette 1983). Extended gestation lengths of 12 months or more have been documented in a few other lizards in thermally challenging environments, such as Magellan's Tree Iguana, *Liolaemus magellanicus* (Jaksić and Schwenk 1983), the anguid *Mesaspis monticola* (Vial and Stewart 1985), and the Common Gecko, *Woodworthia maculatus* (Cree and Guillette 1995); the last of these holds the known record among squamates, with its 14 month gestation. However, extended gestation periods are not confined to lizards of cold climates. For example, in *Xenosaurus grandis* of temperate montane rainforests, gestation lengths of 9–10 months are reported (Ballinger *et al.* 2000). Likewise, in *Mabuya* of tropical Brazil, gestation can last up to 12 months, a pattern reflected in their extreme placentotrophy (Vitt and Blackburn 1983, 1991).

Although viviparous lizards typically reproduce once per year, long gestation lengths can preclude reproduction on an annual basis (Tinkle 1967; Dunham *et al.* 1988), sometimes requiring a pattern termed "low frequency reproduction" (LFR) (Bull and Shine 1979). Accordingly, biennial reproduction occurs in several distantly-related species with long gestations including certain *Liolaemus* (Ibargüengoytía and Cussac 1996; Medina and Ibargüengoytía 2010), *Xenosaurus* (Ballinger *et al.* 2000; Goldberg 2009), *Phymaturus* (Ibargüengoytía 2004), *Tiliqua* (Edwards *et al.* 2000) and *Woodworthia* (Cree and Guillette 1995). However, an association of long gestations with LFR is not invariant. In the Brazilian Mabuya, *Mabuya heathi,* despite a gestation length of nearly 12 months all females in the population become pregnant each year. As a result, beyond the age of 3–4 months, the females spend virtually their entire lives pregnant (Vitt and Blackburn 1983). Furthermore, biennial reproduction also occurs in the Island Night Lizard, *Xantusia riversiana* (Goldberg and Bezy 1974), a viviparous species with a conventional gestation length. In at least some cases, as in Tasmania's Southern Snow Skink, *Niveoscincus microlepidotus,* LFR has a facultative component; whereas females in nature reproduce on a 2 or 3 year cycle, those housed under warmer temperatures revert to an annual cycle (Olsson and Shine 1999).

A useful question to consider is how reproductive parameters compare between viviparous and oviparous lizards. Overall figures for total development time (fertilization through hatching/parturition) in oviparous and viviparous species broadly overlap (Tinkle and Gibbons 1977). Any interspecific differences in development time attributable to reproductive mode are obscured by the wide ranges within each mode. Litter/clutch sizes also vary broadly and equivalently in lizards of each reproductive mode (Fitch 1970; Tinkle and Gibbons 1977). Although viviparous lizards have been said to have smaller clutch sizes than oviparous lizards of equivalent size (Tinkle 1967), phylogenetic variation (including the small clutch sizes of anoles and geckoes) challenges such generalizations. The very small litter sizes of some viviparous lizards tend to be shared with their oviparous congeners and confamilials (Greer 1976; Cree 1994), reflecting phylogenetic features unrelated to reproductive mode. Although the largest clutch sizes are found among oviparous lizards, these are large- bodied forms of the genera *Iguana, Tupinambis*, and *Varanus* (Fitch 1970) from families that lack viviparous members, again making meaningful comparisons difficult. One potential difference between the two reproductive modes lies in the frequency of reproduction. Whereas oviparous females often produce multiple clutches per year, a single clutch is standard for viviparous lizards (Tinkle 1967; Tinkle *et al.* 1970; Fitch 1970; Vitt and Caldwell 2013). However, even the viviparous pattern of LFR with biennial cycles has an equivalent among oviparous lizards, notably Darwin's Marked Gecko, *Homonota darwini*, of Patagonia (Ibargüengoytía and Casalins 2007).

Comparison to snakes. Another useful question to consider is how aspects of viviparity compare between lizards and snakes. Information on snake reproduction was summarized by Blackburn and Stewart (2011), from which the following generalizations are taken. Like that of lizards, snake viviparity is widely distributed phylogenetically; it is represented in 14 ophidian families and about 20% of the species. Viviparous snakes have a worldwide distribution and occupy an extensive array of habitats. The range of gestation lengths in viviparous lizards encompasses those of snakes, which typically are on the order of 2–3 months, with some attaining 6–8 months. However, no known snakes exhibit the extended 10–14 month gestation periods of some lizards. The wide range of average litter sizes of snake species (2 to 47 offspring) parallels those of lizards, although the very large litters (70–100 offspring) of some individual female snakes (Fitch 1970) exceeds maximum values for lizards. Like live-bearing lizards, viviparous snakes typically reproduce once per year, with some species showing biennial cycles. In certain rattlesnakes and vipers, females in the northern part of their ranges reproduce at 3 or 4 year intervals (Saint Girons 1957; Brown 1991). This is an extreme manifestation of the LFR pattern of some cold climate lizards (Ibargüengoytía 2004). The degree of overlap of saurian and ophidian reproductive parameters emphasizes the point that lizard viviparity is not distinct from that of snakes.

13.2.3 Historical Overview of Research on Lizard Viviparity

References to viviparous habits in snakes date back to ancient Hebrew writings of more than 2700 years ago (Yaron 1985; Blackburn and Stewart 2011). Four centuries later, Aristotle's *Historia Animalium* (circa 343 BCE) noted that vipers differed from other snakes in giving birth to their young, a pattern that was said to be shared with sharks and our own species. References to lizard viviparity date at least to the late 1700s (Jacquin 1787; Fig. 13.1A), and were a common feature of works on natural history in the 1800s and beyond (Cuvier 1829; Duméril and Bibron 1834–1854; Bell 1839) (Fig. 13.1B). During the 20th century, many thousands of literature sources documented viviparous habits in particular lizard species, along with information on litter size, gestation length, breeding season, and other parameters. Unfortunately, the older literature is not entirely reliable. Not only have sources used the terms "ovoviviparity" and "viviparity" in discrepant ways (Blackburn 1994b) but identity of many of the species is uncertain, and reproductive information about them is often anecdotal or based on unspecified criteria (Blackburn 1985a). Nevertheless, Fitch's (1970) comprehensive monograph is unsurpassed as a guide to the older literature. No subsequent work has tried to compile the large amount of information that has accumulated since that time.

In addition to primary works on lizard reproduction, analyses from the early through mid-20th century sought to explain the evolution of viviparity with reference to its ecological distribution (Rollinat 1904; Gadow 1910; Mell 1929; Weekes 1933; Sergeev 1940; Neill 1964). Such analyses generally concluded that viviparity was advantageous in cold climates. However, being based on correlations of reproductive modes and habitats, such works were unable to distinguish between current distributions of viviparous species and habitats in which their reproductive mode had arisen (see Section 13.2.5).

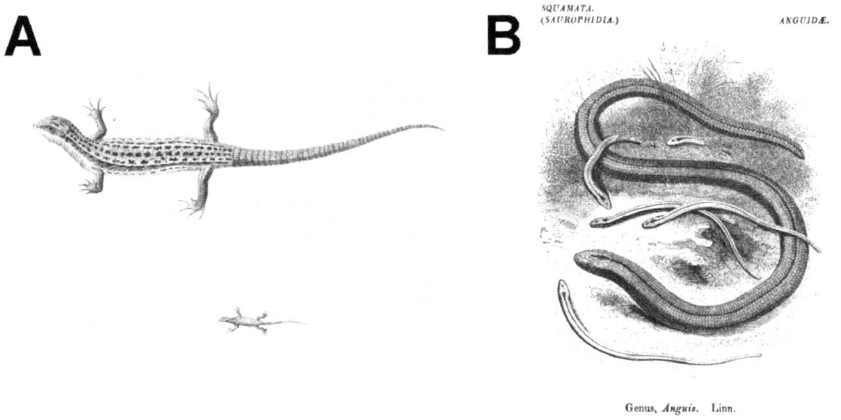

Fig. 13.1 Early illustrations of viviparous lizards. **(A)** The European Common Lizard, *Zootoca vivipara*, showing an adult and neonate; from Jacquin (1787). **(B)** The Slow Worm, *Anguis fragilis;* from Thomas Bell's *A History of British Reptiles* (1839).

Meanwhile, research on lizard viviparity proceeded on several other fronts. Early studies of placental histology and development offered insight into how pregnant lizards sustained their developing embryos (Haacke 1885; Giacomini 1891, 1893, 1906; ten Cate-Hoedemaker 1933; Harrison and Weekes 1925; Weekes 1927a, 1930; Boyd 1940; Panigel 1951). Such research also revealed that placental structure and function varied markedly between species, especially with regard to specializations for maternal-fetal nutrient transfer (Giacomini 1891, 1906; Weekes 1935). In addition, physiological and anatomical studies sought to clarify mechanisms through which reproductive cycles, gestation, and parturition were controlled (Weekes 1934; Rahn 1939; Boyd 1940; Clausen 1940; Bragdon 1951; Miller 1951; Panigel 1953, 1956).

Two conceptual advances stimulated new approaches to the study of squamate viviparity. One was the application of life history theory to lizards (Tinkle 1967, 1969; Tinkle *et al.* 1970; Tinkle and Hadley 1975; Vitt and Congdon 1978; Vitt 1981; Vitt and Price 1982). With it came recognition that viviparity was a pattern with benefits as well as costs, and the product of selective pressures and constraints (Packard *et al.* 1977; Tinkle and Gibbons 1977; Shine and Bull 1979). The influence of life history theory on studies of viviparity can hardly be overstated; it has spawned four decades of descriptive and experimental research on squamates. A second conceptual advance was the rise of cladistics, which stimulated and facilitated phylogenetic approaches to squamate viviparity. Application of cladistic principles revealed that viviparity had evolved convergently in more than 100 squamate lineages (Blackburn 1982, 1985a,b, 1999b; Shine 1985). Definition of these separate viviparous clades provided a database for analysis of selective pressures, constraints, and ecological correlates (Shine 1985), as well as morphological and physiological specializations associated with viviparity. It also provided a database for comparisons between lineages, and potentially, for quantification of convergent and divergent trends. Through phylogenetic approaches, broad correlational studies of past decades have given way to work that focuses on individual lineages to reconstruct their evolutionary history (e.g., Braña *et al.* 1991; Stewart and Thompson 2003, 2009a; Surget-Groba *et al.* 2006).

Over the past 25 years, research on lizard viviparity has intensified and diversified. Contemporary studies on lizard viviparity are inherently integrative and draw on the methods of physiology, anatomy, biochemistry, molecular biology, genomics, ethology, ecology, and phylogenetic analysis (Blackburn 2000a; Thompson and Blackburn 2006; Thompson *et al.* 2010; Murphy and Thompson 2011; Brandley *et al.* 2012; Van Dyke *et al.* 2014). Life history theory and principles of phylogenetic systematics have continued to provide a theoretical and methodological backdrop to hypothesis-testing. Furthermore, technological advances have contributed in a major way to contemporary research. For example, studies on physiological ecology were boosted by development of radiotelemetric devices, thermal data loggers, sophisticated incubators, and laboratory racetracks for measuring sprint speed of offspring and pregnant females (Shine 2014). Likewise, studies of placentation have increasingly drawn upon sophisticated techniques of electron microscopy, confocal microscopy, immunocytochemistry, and quantitative analyses of placental transport (Blackburn 2000a; Blackburn and Stewart 2011; Section 13.3).

13.2.4 Viviparity as a "Reproductive Strategy"

Advantages. In accord with the principles of life history theory, viviparity is best viewed as a pattern that confers both advantages and disadvantages (Tinkle and Gibbons 1977; Shine and Bull 1979; Shine 1985; Blackburn and Stewart 2011). Numerous factors that have been hypothesized in lizards are listed in Table 13.2. Several of the postulated benefits relate to how pregnant females might protect their developing embryos, e.g., from environmental threats, temperature extremes, and predators (Fitch 1970; Packard *et al.* 1977; Tinkle and Gibbons 1977; Shine and Bull 1979; Shine 1985; Andrews 2000). Other potential benefits reflect ways that females enhance development of their offspring, through thermoregulation of embryos at optimal temperatures (Shine 1983b, 2004; Mathies and Andrews 1995), through maternal provision of small quantities of nutrients (Stewart 1989, 1992), and by optimally timing the birth of offspring (Olsson and Shine 1998; Shine and Olsson 2003; Wapstra *et al.* 2010). Still others focus on maternal benefits, i.e., ability to provide nutrients to embryos facultatively (Swain and Jones 2000; Atkins *et al.* 2003; Itonaga *et al.* 2012a) or at times that are optimal for the pregnant female (Blackburn and Vitt 1992), and postponement of nutrient investment until late in gestation as a means of "bet hedging" (Vitt and Blackburn 1983, 1991). Yet another proposed benefit of viviparity is that it may allow females to resorb abortive eggs and embryos (Mingazzini 1892; Domini 1928; Bustard 1966). However, empirical studies and examination of the literature have revealed no evidence that the squamate oviducts have this ability (Blackburn 1998a,b; Blackburn *et al.* 1998, 2003a); instead, squamate oviducts extrude abortive eggs. A further potential benefit of viviparity is that it may permit a live-bearing species to inhabit environments that lack suitable nest sites for oviposition, including arboreal and aquatic habitats (Neill 1964).

Many studies on lizards have documented thermoregulatory benefits of viviparity. Early researchers noted that viviparous squamates are disproportionately represented in the cold climates of high altitudes and latitudes (Gadow 1910; Mell 1929; Weekes 1933, 1935; Sergeev 1940). This

Table 13.2 Potential benefits and costs of viviparity in lizards. A given benefit did not necessarily act as a selective advantage during evolution of viviparity.

Potential benefits	Associated factors
Protects eggs from environmental hazards	temperature extremes; hydric extremes; highly variable temperatures; predators; microbial and fungal infection
Accelerates development, speeds offspring emergence	maternal thermoregulation of embryos
Frees females from having to find suitable nest sites	environments without nest sites maternal energy expenditure and exposure risk
Allows use of a range of environments	arboreal or aquatic habitats; xeric conditions; cold climates; fluctuating temperatures
Permits maternal thermoregulation of embryos	optimal mean temperatures stable temperatures
Permits maternal oxygen supply to embryos	high altitudes and microenvironments with low oxygen tensions
Lets females predict environmental conditions	unpredictable climates
Lets females optimize offspring phenotypes	maternal manipulation of incubation
Permits sex allocation	temperature-dependent sex determination
Allows resorption of eggs	recycling of nutrients from abortive eggs
Allows extra-vitelline nutrient provision by the female	nutrient supplementation of yolk facultative nutrient supply
Allows maternal "bet hedging" in nutrient investment	extreme placentotrophy
Allows precocial pregnancy at juvenile body sizes	extreme placentotrophy
Potential costs	
Decreased locomotor performance	physical burden of the litter
Anorexia/reduced feeding during pregnancy	abdominal space constraints; decreased foraging ability; inability to replenish lipid stores
Decreased maternal activity during pregnancy	altered thermoregulatory behavior; decreased locomotor ability
Increased predation on reproducing females	decreased locomotion; increased exposure to predators due to thermoregulatory needs
Metabolic costs	maintenance of litter; burden of pregnancy
Physiological debilitation of reproducing females	physiological costs of pregnancy coupled with decreased feeding
Decreased litter size	abdominal space constraints; female mobility
Decreased offspring size	abdominal space constraints; female mobility

correlation was supported in subsequent studies on particular fauna (Fitch 1935; Smith 1939; Duellman 1965; McCoy 1968; Greer 1968; Greene 1970; Tinkle and Gibbons 1977; Shine and Berry 1978). The rationale traditionally offered for this correlation was that because viviparous female squamates can thermoregulate their embryos, they can thrive in cold environments that make oviparous reproduction difficult (Shine and Bull 1979; Shine 1985). This explanation came to be known as the "cold climate hypothesis" for the evolution of viviparity. Early researchers proposed that pregnant female squamates (a) can protect embryos from freezing night-time temperatures (Mell 1929); (b) can maintain embryos at optimal temperatures for development (Duellman 1965; Greer 1968); and (c) can accelerate rates of development, thereby allowing offspring to emerge before the end of a short breeding season (Sergeev 1940) (For an analysis of the development of these hypotheses, see Shine 2014). As evidence accumulated in favor of the thermal benefits of viviparity, the "cold-climate" hypothesis was subsumed into a broader "maternal manipulation hypothesis." Under this explanation, pregnant female squamates can manipulate incubation conditions, not only affecting developmental rates and embryo survival but also affecting a variety of phenotypic traits (Shine 1995, 2004, 2012; Li *et al.* 2009; see Schwarzkopf and Andrews 2012a for a comprehensive summary). Accordingly, while viviparity can be advantageous in cold temperatures, its advantages also extend to tropical climates (Shine 2003; Ji *et al.* 2007; Webb *et al.* 2006).

Interestingly, studies on viviparous squamate species have revealed no consistent pattern of thermoregulation. Compared to males and non-pregnant females, pregnant lizards can regulate at temperatures that are higher, lower, more stable, more variable, or identical, depending on the species (Beuchat 1986; Mathies and Andrews 1997; Andrews *et al.* 1999; Blackburn 2000a; Robert and Thompson 2009). Likewise, the duration of basking may be increased or decreased during pregnancy (Stewart 1984; Schwarzkopf and Shine 1991; Labra and Bozinovic 2002). Similar variation is evident among viviparous snakes (Blackburn and Stewart 2011).

A significant point to note is that advantages of viviparity vary between species. For example, the potential that viviparity offers to control sex allocation of offspring reportedly applies only to temperature-dependent sex determination (Robert and Thompson 2001, 2009; Wapstra *et al.* 2004), a very rare feature in squamates. Likewise, advantages of substantial placentotrophy (Blackburn and Vitt 1992; Itonaga *et al.* 2012a) apply only to the small percentage of viviparous lizards with this fetal nutritional pattern. Another example is the interspecific variation in thermoregulatory patterns of pregnant females, variation that probably reflects interspecific differences in benefits and costs of pregnancy (Table 13.2), compromises between maternal and embryonic thermal needs (Beuchat and Ellner 1987), varying maternal effects on offspring phenotype, and differential tactics that maximize offspring fitness vs. maternal reproductive potential (e.g., Schwarzkopf and Andrews 2012a,b; DeNardo *et al.* 2012). In a group as diverse as lizards, it is understandable that the manifestations of live-bearing reproduction would vary along with other reproductive parameters, especially given the diversity of selection pressures and constraints, and the multiple origins of this pattern from ancestors that themselves were reproductively diverse.

Disadvantages. In addition to its advantages, viviparity also has been shown to entail significant disadvantages in lizards. Empirically-demonstrated costs of viviparity include the negative effects of pregnancy on maternal locomotion, feeding, thermoregulation, fat reserves, metabolism, and predator avoidance (Shine 1980, 2003; Bauwens and Thoen 1981; Birchard *et al.* 1984a; Van Damme *et al.* 1989; Beuchat and Vleck 1990; DeMarco and Guillette 1992; Schwarzkopf 1996; Olsson *et al.* 2000; Lin *et al.* 2008; Bleu *et al.* 2012a). As noted above, viviparity also can limit a species to reproducing but once per year (Fitch 1970). This potential cost is evident in some species from comparisons of closely related oviparous and viviparous forms (Heulin *et al.* 1997; Bleu *et al.* 2012b). In addition, viviparity may limit litter size or offspring size due to constraints imposed by maternal body volume (Qualls and Shine 1995; Qualls and Andrews 1999a; Sun *et al.* 2012). In theory, each of the above costs might be reflected in decreased offspring size, quantity, quality, or a decreased frequency of reproduction, if not addressed through modifications to the maternal phenotype (e.g., Qualls and Shine 1998; Bleu *et al.* 2012b; Sun *et al.* 2012).

Like the benefits of viviparity, costs are species-specific (Shine 1980). For example, viviparity would not necessarily affect clutch frequency in squamates that reproduce once per year (excluding certain cold-climate snakes: Blackburn and Stewart 2011). As another example, such pregnancy-associated costs as decreased maternal mobility and increased vulnerability to predation would not apply to species in which females are large with defensive capabilities (Blackburn and Stewart 2011). Thus, unlike many viviparous lizards and snakes, female rattlesnakes and vipers continue to forage and feed during pregnancy (Lorioux *et al.* 2013; Schuett *et al.* 2013).

13.2.5 Evolution of Viviparity

Until a few decades ago, biologists tended to assume that advantages of viviparity (Table 13.2) served as selective pressures during its evolution. Thus, if viviparity were beneficial in aquatic, subterranean, or arboreal habitats, those benefits were assumed to have selected for that pattern (Neill 1964; Fitch 1970). However, as subsequent researchers noted, selective pressures leading towards viviparity must be distinguished from benefits that only accrue after its evolution, including the potential for invasion of new habitats where viviparity may be advantageous (Greene 1970; Packard *et al.* 1977; Shine and Bull 1979; Blackburn 1982; Shine 1985). In addition, it is important to consider whether alleged advantages could have selected for intermediate stages in the evolution of viviparity (Weekes 1933; Packard 1966). These complications underscore the importance of taking historical and phylogenetic approaches towards viviparity, in order to distinguish temporal sequences as well as the role of causal influences (selective pressures and constraints) during the evolutionary transition towards that pattern.

Origins of viviparity. In principle, precise definition of each of the phylogenetic origins of viviparity makes it possible to analyze selective pressures, ecological habitats, and evolutionary transitions. Two phylogenetic analyses published in the 1980s defined the separate evolutionary origins of viviparity among reptiles (Blackburn 1982, 1985; Shine 1985). A follow-up compilation of these analyses revealed a total of more than 100 well-established origins of viviparity, including at least 64 origins in lizards and 39 origins in snakes (Blackburn 1999b). These origins are discontinuously distributed among squamate families. For example, while numerous origins of viviparity have occurred among scincid lizards, in several other speciose lizard groups, viviparity is absent (Dactyloidae, Gymnophthalmidae, Teiidae) or scarce (Gekkota, Lacertidae) (taxonomy of Sites *et al.* 2011). Subsequent studies on a few taxa (Méndez de la Cruz 1998; Zamudio *et al.* 2000; Lynch 2010; Lynch and Wagner 2010; Stanley *et al.* 2011; Lambert and Wiens 2013) have either corroborated or modestly altered the discernible origins of viviparity.

Evidence for additional origins of squamate viviparity has continued to accumulate. For example, an analysis of the genus *Liolaemus* has revealed six independent origins of viviparity (Schulte *et al.* 2000), an increase over previous estimates. A more extensive analysis suggests additional origins of viviparity within the genus (Pincheira-Donoso *et al.* 2013). Likewise, fossil evidence indicates that viviparity was present in a terrestrial Mesozoic lizard (Wang and Evans 2011), as well as in giant aquatic mosasaurs of the Cretaceous (Caldwell and Lee 2001). Many of the ~115 origins of viviparity that can now be recognized have occurred at subgeneric levels; in fact viviparity has evolved multiple times in some lizard genera (e.g., *Phrynosoma, Sceloporus, Liolaemus,* and *Eumeces*). Likewise, at least three saurian origins of viviparity have occurred at (subspecific) populational levels (Heulin *et al.* 1993; Qualls *et al.* 1995; Smith and Shine 1997; Fairbairn *et al.* 1998; Surget-Groba *et al.* 2001, 2006). Several origins of viviparity have occurred in geologically recent times, such as the Pleistocene (Heulin and Guillaume 1989; Surget-Groba *et al.* 2001; Heulin *et al.* 2011) or late Pliocene (Calderón-Espinosa *et al.* 2006). Other origins are scattered throughout the Cenozoic (Lynch 2010; Schulte and Moreno-Roark 2010), in addition to the two Mesozoic origins mentioned above.

Selective advantages. Many factors are proposed to have acted as selective advantages and constraints in the evolution of squamate viviparity (Table 13.3). Until the development of experimental methods and the use of phylogenetic analyses, these were judged mainly on the basis of their plausibility. Indeed, several factors (the hypothetical constraints in particular) remain difficult to define in testable terms.

Table 13.3 Factors postulated to affect the evolutionary origin of viviparity in lizards. Sources include: Sergeev 1940; Fitch 1970; Packard *et al.* 1977; Tinkle and Gibbons 1977; Shine and Bull 1979; Bull 1980; Blackburn 1982, 1985a; Shine 1983b, 1985, 2014; Qualls and Andrews 1999; Andrews and Mathies 2000; Andrews 2002; Shine *et al.* 2003. Among the selective pressures listed, empirical evidence strongly supports temperature–related factors (see text). Several others fail to consider stages in the evolutionary transition to viviparity.

Selective pressures	Constraints
Cold climates	Decreased clutch size
Short breeding season	Decreased offspring size
Freezing temperatures	Decreased clutch frequency
Lethally high temperatures	Decreased maternal survival
Fluctuating temperatures	Mode of sex determination
Environmental unpredictability	Highly calcified eggshells
Soil aridity	Limited oxygen availability to oviductal embryos
Aquatic habitat	Egg guarding behavior
Arboreal habitat	
Predation on eggs	
Low oxygen levels	

A comprehensive analysis that compared closely related viviparous and oviparous clades concluded that most of the origins of viviparity have occurred in relatively cold climates (Shine 1985). Subsequent analyses focusing on particular taxa (e.g., Schulte *et al.* 2000; Hodges *et al.* 2004; Lynch 2010; Lambert and Wiens 2013; Pincheira-Donoso *et al.* 2013) have reached similar conclusions. The overall implication of these conclusions is that squamate viviparity commonly has evolved due to its thermal benefits. However, as noted above (Section 13.2.4), such benefits are not restricted to cold environments.

Numerous studies on lizards have tested the hypothesis that maternal thermoregulation of oviductal eggs enhances survival and viability of viviparous offspring. For example, experimental studies have shown that through basking, viviparous females can enhance overall quality of offspring (Qualls and Andrews 1999b; Shine and Harlow 1993; Wapstra 2000; Webb *et al.* 2006); improve their locomotor performance (Shine and Olsson 2003; Ji *et al.* 2006, 2007; Li *et al.* 2009; Yan *et al.* 2011); increase their rate of development (Schwarzkopf and Shine 1991; Wapstra 2000); and facilitate offspring survival (Rock and Cree 2003). From this broad range of studies, the overall thermal benefits of viviparity are well established. However, perspectives differ (Schwarzkopf and Andrews 2012a,b; Shine 2012) on aspects of their adaptive nature.

Other relevant research has been conducted on oviparous species to identify costs and benefits of egg thermoregulation, constraints on egg retention, thermal effects on hatchling phenotypes, and differences between oviparous species in ability to retain developing eggs (Andrews and Rose 1994; Mathies and Andrews 1996, 2000; Shine and Harlow 1996; Andrews 1997, 2000, 2002; Andrews *et al.* 1997; Qualls and Andrews 1999a,b; Andrews and Mathies 2000; Shine 2002; Shine *et al.* 2003; Warner and Andrews 2003; Calderón-Espinosa *et al.* 2006; Rodríguez-Díaz *et al.* 2012; Wang *et al.* 2013). This work includes studies designed to determine whether specializations that facilitate viviparity (e.g., thinning of the eggshell [Heulin 1990; Mathies and Andrews 2000; Calderón-Espinosa *et al.* 2006] and altered maternal thermoregulation [Shine 2006]) arise during evolution of that pattern or constitute pre-adaptations in oviparous lineages. Still other research has focused on lizard species with reproductive bimodality. Such research includes analytical comparisons between oviparous and viviparous populations (Heulin *et al.* 1991, 2002; Qualls and Shine 1995, 1996, 1998, 2002; Qualls 1996; Stewart *et al.* 2004; Voituron *et al.* 2004; Bleu *et al.* 2012b), hybridization experiments (Heulin *et al.* 1989; Arrayago *et al.* 1996; Lindtke *et al.* 2010), and tests of evolutionary assumptions (Rodríguez-Díaz and Braña 2011a,b, 2012; Rodríguez-Díaz *et al.* 2010; Roitberg 2013).

Functional mechanisms. Viviparity in squamates is accomplished by a diversity of mechanisms that vary interspecifically (Blackburn 2000a). For example, in various species, fetal blood has a higher affinity for oxygen than maternal blood, a feature that facilitates gas exchange *in utero* (Grigg and Harlow 1981; Birchard *et al.* 1984b). This is achieved in some species via a specialized fetal hemoglobin, whereas in others derived from a different origin of viviparity, by altered maternal nucleoside triphosphate levels (Ingermann *et al.* 1991; Ragsdale and Ingermann 1993; Ragsdale *et al.* 1993).

A complex set of mechanisms maintain pregnancy, which at minimum requires a prolonged delay in the timing of parition (oviposition/birth), reduced shell gland activity, an increase in uterine vascularity, maternal behavior, and a suspension of vitellogenesis and ovulation. Several studies have implicated progesterone and/or the corpus luteum (CL) in stimulating some of these responses. However, species vary greatly in progesterone profiles during gestation, sources of progesterone, responses to progesterone, and in whether or not a CL is necessary to maintain pregnancy (Yaron 1985; Guillette 1987, 1989; Jones and Baxter 1991; Blackburn 2000a; Martínez-Torres *et al.* 2010; Albergotti and Guillette 2011; Lovern

2011). Viviparous lizards also differ in sources of progesterone. Although this hormone generally is secreted by the CL, in some species it also is synthesized and secreted by the placental membranes themselves (Guarino *et al.* 1998; Girling and Jones 2003; Painter and Moore 2005). The variation in progesterone sources during gestation is understandable in the context of the multiple origins of viviparity. However, fetal membranes can synthesize steroids in turtles, birds, and alligators (Albergotti *et al.* 2009; Cruze *et al.* 2012, 2013). Thus, this capacity may predate viviparity in squamates, with the specific functions that progesterone controls in viviparous clades being independently derived.

Another mechanism of viviparity is the physiological control of parturition. A variety of factors are involved through stimulatory and inhibitory effects on the oviduct (Blackburn 1998b, 2000a), including neurohypophysial hormones (Jones and Guillette 1982; Cree and Guillette 1991; Fergusson and Bradshaw 1992), prostaglandins and autonomic neurons (Guillette *et al.* 1991, 1992), as well as maternal behavior and environmental stimuli. Parturition clearly differs from oviposition in its timing relative to embryonic development. However, the extent to which the mechanisms of parition and their control differ between oviparous and viviparous species has not been sufficiently investigated.

Other mechanisms involved in viviparity include specializations for placental nutrient transfer, maternal provision of calcium (to compensate for eggshell loss: Stewart *et al.* 2009a,b; Stewart and Ecay 2010), and in placentotrophic forms, the inhibition of vitellogenesis and yolk deposition. Diversity in structure and function of the oviduct and placental membranes (including increased vascularity: Parker *et al.* 2010a,b) is explored in detail below (Section 13.3). Overall, the diversity of mechanisms involved in viviparous reproduction and the structural features by which they are accomplished reflects the multiple evolutionary origins of squamate viviparity (Blackburn 2000a). A recently published paper is entitled "Different mechanisms lead to convergence of reproductive strategies in… lizards" (Sun *et al.* 2013). Although this paper deals with oviparous lacertids, its title articulates a pattern that broadly characterizes viviparous squamates and their evolution.

Evolutionary transitions. Questions about how viviparity has evolved in squamates have stimulated research for more than a century. A significant conceptual advance came with Weekes' (1933, 1935) recognition that scenarios for the origins of viviparity must take into account transitional evolutionary stages. Accordingly, Packard *et al.* (1977) suggested that over the generations, gravid female squamates retained their developing eggs for progressively longer periods of time, laying them in later and later stages of development. At the culmination of this trend, females would retain their eggs for the entire period of gestation and give birth to their young. A corollary of this scenario is that thinning of the eggshell would parallel the trend towards increased egg retention, to allow for gas exchange *in utero.* One implication is that viviparity would more likely evolve in cold climates than in xeric conditions; the thinned eggshells necessary for extended egg retention would be incompatible with development of those eggs after oviposition in dry climates (Packard 1966).

Traditionally, evolution of squamate reproductive modes has been suggested to evolve through four successive transformations: (a) incremental, progressive increases in egg retention, culminating in (b) viviparity; (c) origin of placentas that facilitate embryonic exchange; (d) placental provision of small quantities of nutrients; and (e) placental supply of substantial quantities of nutrients. Implicit tests of elements of this scenario have falsified several elements (Blackburn 1992, 1995, 2006). As noted by Packard *et al.* (1977), placentas are a natural consequence of eggshell thinning and arise along with viviparity. Furthermore, as subsequent study has shown, such placentas transfer small quantities of nutrients, including calcium, sodium, and amino acids. Such nutrient transfer has been found in every viviparous squamate studied, including lizards (Thompson 1977a,b, 1981, 1982; Stewart and Castillo 1984; Jones and Swain 2010; Itonaga *et al.* 2012a,b) as well as snakes (Stewart 1989; Stewart *et al.* 1990; Van Dyke and Beaupre 2012; Van Dyke *et al.* 2012). Consequently, available evidence supports a modified scenario in which viviparity, placentation, and placental supply of nutrients evolve simultaneously rather than as three successive stages (Blackburn 1995, 2005, 2006; Stewart 2013).

Another debatable feature of the traditional scenario is the assumption that viviparity evolves as the culmination of incremental evolutionary increases in egg retention, during which females lay eggs with increasingly advanced embryos. This scenario has long been supported by the untested assumption that oviparous squamates lie along a broad continuum in degree of development of their eggs at oviposition. In marked contrast to this assumption, a survey of the literature has shown that the great majority of squamates follow one of two patterns: they either lay eggs in the limb bud stage, at about 30% of the way through development (Shine 1983a; Blackburn 1995) or give birth to their young. In very few species do females lay eggs with markedly advanced embryos (Blackburn 1995, 1998c; Mathies and Andrews 1995). This bimodal distribution of species has been cited in support of a punctuated equilibrium model of change, in which the transition from oviparity to viviparity occurs relatively quickly through an intermediate phenotype that is evolutionarily unstable (Blackburn 1995, 1998c; *contra* Qualls *et al.* 1997). The rationale is that deposition of eggs with advanced embryos places conflicting demands on eggshell structure. Oviparous retention of eggs into advanced stages requires eggshells thin enough to permit gas exchange, eggshells that may be too thin to protect eggs after oviposition. Accordingly, prolonged oviductal retention of developing eggs may be an evolutionarily unstable strategy that is traversed relatively quickly, in favor of the relatively stable strategies of viviparity and typical oviparity (Blackburn 1995). Neither physical characteristics of eggshells, nor physiological attributes of embryos, that facilitate intrauterine egg retention are known but may vary interspecifically. For example, experimental manipulation of length of intrauterine egg retention in species of the genus *Sceloporus* found that eggshell thickness did not predict the capacity to retain eggs beyond the mean stage of oviposition for the species (Mathies and Andrews 2000). However, eggshell thickness is correlated inversely with length of intrauterine egg retention in populations of two lizard species with geographic variation in reproductive mode (Heulin 1990; Qualls 1996; Heulin *et al.* 2002). Notably, there is a dramatic difference in thickness of oviparous eggshells between these two species (Table 13.9). Understanding the physiology of intrauterine egg retention is critical to the formulation of models for the evolutionary transition from oviparity to viviparity and should be a fruitful area for future research. Evidence that has accumulated over the past 15 years has corroborated the bimodal distribution of squamate species, having revealed very few additional species in which females routinely lay eggs in advanced stages. These findings are consistent with predictions from the punctuated equilibrium model. However, questions over the pace and tempo of the origin of viviparity remain, and continue to stimulate research.

Reversibility? Phylogenetic analyses of squamate reproductive modes commonly have assumed that viviparity evolves irreversibly from oviparity (Blackburn 1982; Shine 1985). While an evolutionary reversal from viviparity to oviparity remains a theoretical possibility (Fenwick *et al.* 2012), detailed analyses have concluded that such reversals have played little role (if any) in the history of squamate reproduction (Lee and Shine 1998; Blackburn 1999b; Shine and Lee 1999; Pincheira-Donoso *et al.* 2013; *contra* de Fraipont *et al.* 1996). In fact, several putative cases of reversals in lizards have been shown to be no more parsimonious than the alternatives. For example, a phylogenetic analysis of the speciose genus *Liolaemus* yielded two scenarios that are equally conservative—six origins of viviparity or three origins of that mode followed by three reversals (Schulte *et al.* 2000; also see Pincheira-Donoso *et al.* 2013). Another tentative case of reversal is in the reproductively bimodal lacertid *Zootoca vivipara*. A recent analysis suggests two alternative interpretations–three (subspecific) origins of viviparity, vs. one origin and one reversal (Surget-Groba *et al.* 2006). Thus, the possibility of a reversal in this species is only marginally more parsimonious than the alternative.

Furthermore, other evidence suggests multiple origins of viviparity in this species (Odierna *et al.* 2004; Kupriyanova *et al.* 2006).

Limited evidence is available for other possible cases of reversal, i.e., when an oviparous taxon is nested within a larger viviparous group (Lynch and Wagner 2010; Chapple *et al.* 2009). However, phylogenetic information alone may be insufficient to conclude occurrence of a viviparity-oviparity reversal. The alternative interpretation of multiple origins of viviparity may be nominally less parsimonious in a very limited context but retain a higher degree of probability, especially in light of the indisputably large number of transitions in squamate reproductive modes. A recent analysis has proposed that viviparity may be an ancestral rather than a derived pattern, having evolved in a basal squamate stock; thus, oviparity is said to have evolved from viviparity as many as 59 times (Pyron and Burbrink 2013). This interpretation offers an interesting challenge to widespread assumptions about reproductive evolution. However, in considering the possibility of reversals, close attention should be given to what we know about squamate reproductive biology, information that bears directly on questions of how (and how readily) such reversals could be accomplished.

13.3 PLACENTATION

Viviparity is a homoplastic trait, a phenotypic similarity that has evolved independently from oviparous ancestors as a novel relationship between embryos and mothers (Blackburn 2014; Wake 2014). The morphology of this relationship is expressed in multiple forms among vertebrates, and although little is known of the underlying mechanisms, genetic and developmental pathways are undoubtedly diverse as well. The evolution of viviparity and placentation in squamates are intertwined (Blackburn 2005, 2006; Stewart 2013). Viviparity is dependent on physiological exchange between mothers and embryos and placentas provide functional support for embryonic development. Embryos of modern viviparous amniotes—lizards, snakes and mammals—develop within the maternal reproductive tract and placentas arise through an intimate relationship between the uterus and fetal membranes (Mossman 1937, 1987; Luckett 1977). Like viviparity itself, placentas have evolved over 100 times in lizards and snakes (Blackburn 2005, 2014), as a novel relationship between tissues, each of which shares a genetic and developmental history (Wake 1985; Blackburn 1992; Blackburn and Flemming 2009). The maternal oviduct, derived from the embryonic Müllerian duct, is an ancestral, plesiomorphic, trait for amniotes (Wake 1985). The extraembryonic membranes that participate in placentation, principally the yolk sac and chorioallantoic membrane, are shared amniote structures (Stewart 1997; Blackburn and Flemming 2009; Ferner and Mess

2011). Given the similarity between species in the particular tissues from which placentas are derived, it is not surprising that squamate placentas exhibit considerable structural similarity interspecifically (Blackburn 1993b; Stewart and Thompson 2000, 2003; Blackburn and Stewart 2011).

Squamate viviparity is the outcome of multiple natural experiments in the evolution of phenotypic similarity. Thus, we have an excellent opportunity to explore the patterns that have led to the transitions between traits. Homoplasy is likely to be promoted in evolutionary transitions involving similar ancestral structures exposed to novel, yet similar, environments (Wake 1991) and phenotypic similarity among viviparous squamates may result from adaptive responses to similar selection pressures. However, homoplasy may also arise through constraints on morphology integrated through genetics and development (Wake *et al.* 2011). Thus, similar phenotypes may arise through either shared or different developmental pathways (Parra-Olea and Wake 2001; Hall 2003; Wake 2003). Therefore, distinguishing patterns of similarity and understanding the mechanisms that produce them require knowledge of functional attributes and developmental sequences of ancestral and derived conditions (Wake *et al.* 2011). Squamates generally, and lizards specifically, offer an opportunity to understand conditions that promote the evolution of homoplasy through research programs aimed at determining the patterns of similarity across lineages and the genetic and developmental mechanisms that contribute to those patterns.

Although similarity in placental structure is a prominent feature of interspecific comparisons among squamates, some variations reflect phylogenetic history and unique specializations for placental function (Giacomini 1891; ten Cate-Hoedemaker 1933; Weekes 1935; Blackburn and Vitt 1992; Blackburn 1993c, 2000a; Flemming and Branch 2001; Jerez and Ramírez-Pinilla 2001; Stewart and Thompson 2003, 2009a; Ramírez-Pinilla *et al.* 2006; Blackburn and Flemming 2009, 2012; Blackburn and Stewart 2011). Such specializations provide an opportunity to understand how maternal–fetal interactions shape placental evolution. The evolution of placentation involves co-evolutionary events in which the fitness of one individual, the offspring, is intertwined with the fitness of a second individual, the mother. Placentas are sites of nutritive exchange that potentially promote resource competition between mothers and embryos; a circumstance that could promote rapid evolutionary change and lead to divergent evolutionary pathways among lineages (Crespi and Semeniuk 2004).

Studies of viviparity and placentation in lizards and snakes provide insight into both the evolution of homoplasy and the subsequent evolution of specializations leading to divergent structures. The past 30 years have seen a dramatic increase in our understanding of placental diversity, yet we are in the early stages of uncovering patterns of similarity in development and morphology. The increasing accessibility of genomic data offers exciting opportunities for comparative analyses of the genetics of reproductive mode and insight into the mechanisms underlying patterns of reproductive evolution (Van Dyke *et al.* 2014). However, the genomic approach will be limited in the absence of further research aimed at understanding the development of maternal–fetal functional interactions.

Our discussion of placentation begins with overviews of the uterus, the extraembryonic membranes, placental categories, and patterns of embryonic nutrition. We then focus on patterns of placental diversity. A final section considers patterns of placental evolution. A primary goal of our treatment of placentation is to explore what is known regarding placental diversity and evolution and to suggest a research program for future emphasis.

13.3.1 The Oviduct

Anatomy of the lizard oviduct is discussed in detail in Siegel *et al.* (Chapter 6, this volume). Our emphasis will be on structure and function of the uterus (glandular uterus) relevant to comparisons between modes of parity. Oviductal structure and function in oviparous reptiles provide a basis for understanding the specialized roles of the oviduct in viviparous squamates and how they are accomplished. Reptilian oviducts are paired, elongate tubes that develop from the embryonic Müllerian ducts, like their homologues in other gnathostomes (for a comprehensive review of squamate oviducts, see Blackburn 1998b; Fox [1977] and Girling [2002] also consider other reptiles). Orientated lengthwise in the abdominal cavity, each oviduct is associated with the ovary anteriorly and opens into the cloaca posteriorly. Each oviduct is suspended from the dorsal wall of the abdominal cavity by means of a mesentery. A single, longitudinal artery (the "secondary oviductal artery") lies along the dorsal aspect of each oviduct; in the uterine region, it is called the "uterine artery". Multiple vessels ("primary oviductal arteries") supply the secondary oviductal artery with blood from the aorta. A "secondary oviductal vein" runs adjacent to the corresponding artery. It is drained via multiple vessels into the renal vein, renal portal vein, and caudal vena cava.

Oviducts of oviparous reptiles serve several reproductive functions, including the following: sperm transport and storage, egg transport, fertilization, synthesis and secretion of albumen, deposition of the eggshell (including its calcium), physiological maintenance of the early embryo (through provision of water and respiratory gases), and oviposition (Fox 1977; Blackburn 1998b; Girling 2002).

Unlike those of chelonians and crocodilians, squamate oviducts do not secrete a layer of albumen around the yolk (Packard and Packard 1988;

Cordero-López and Morales 1995; Stewart 1997; for a review of the evidence, see Blackburn 1998b). In addition, whereas non-squamate reptiles deposit eggs very early in development (i.e., the primitive streak or gastrula stage), oviducts of most oviparous squamates retain developing eggs through the limb bud stage or beyond (Shine 1983; Blackburn 1995). These two specializations of oviparous squamates represent the ancestral condition from which viviparity has arisen, and for the following reasons, may well have facilitated the evolution of viviparity. In viviparous squamates, the duration of embryo maintenance within the oviduct is prolonged, representing a quantitative extension of the ancestral squamate pattern of oviparous egg retention. In addition, the absence of an albumenous layer in squamate eggs permits the contact between fetal and uterine tissues necessary for prolonged embryo maintenance under conditions of viviparity. In chelonians, in contrast, oviductal secretions arrest development of the egg *in utero* by limiting oxygen diffusion between maternal and embryonic tissues (Rafferty *et al.* 2013).

Interspecific variation. Oviducts of squamate species vary in several macroscopic features, including the following: number of oviducts (one vs. two), the extent and nature of regional differentiation, vascular supply, topographic asymmetry, vaginal pouches, and relationship to the cloaca (Blackburn 1998b). Characteristics of these features in ancestral oviparous clades must be recognized if adaptations for viviparity are to be distinguished. For example, given that space constraints can limit litter and offspring size in live-bearing forms, one might be tempted to interpret presence of but a single oviduct in Schmidt's Helmet Skink *Tribolonotus schmidti* (Greer 1976) as a specialization for viviparity. However, reduction to a single oviduct characterizes oviparous *Tribolonotus* as well as other oviparous Australo-Pacific skinks, where it correlates not with reproductive mode but with an unusually small clutch size (Greer 1976). Another variable feature is an exaggerated degree of asymmetry that minimizes overlap between egg-bearing regions of left and right oviducts (Owen 1866; Das 1960; Perkins and Palmer 1996). Although a common feature of viviparous snakes and readily interpreted as beneficial for elongated viviparous forms, this feature also is widespread (if not universal) among oviparous snakes (Blackburn 1998b). Thus, while an exaggerated oviductal asymmetry may have facilitated viviparity, it predates viviparity and is not an adaptation for that reproductive mode in snakes.

A broad survey of oviductal morphology involving representatives from 14 squamate families has revealed no macroscopic specializations for viviparity (Blackburn 1998b). Similarities between the oviducts of closely related oviparous and viviparous squamates (e.g., Mulaik 1946; Blackburn 1998b) indicate that the evolution of viviparity generally has not accompanied macroscopic modifications of the reproductive tract. Specializations for viviparity at microscopic levels are outlined below in sections on the uterus and the placentae.

Regional differentiation. Three major regions of the squamate oviduct can be recognized histologically, each of which serves specific functions. These regions are the "infundibulum" (which receives the ovulated egg and provides a site of fertilization), the "uterus" (which houses the developing egg and deposits the eggshell), and the muscular "vagina" ("non-glandular uterus") (which facilitates expulsion of the egg or fetus at parition) (Blackburn 1998b; also see Siegel *et al.* 2011). Distinctive characteristics of the anatomy of the three regions of the lizard oviduct are discussed in Siegel *et al.* (Chapter 6, this volume).

Uterus. The oviductal uterus in squamates is a very thin-walled tube that is greatly distended during gravidity to accommodate each of the eggs. Accordingly, the uterus becomes expanded into multiple "incubation chambers" that alternate with constricted "inter-embryonic" regions. These expanded incubation chambers can persist long after gravidity/pregnancy in some species (Giacomini 1891; Weekes 1927a), although whether the same sites are used in subsequent reproductive episodes is unknown. During gravidity, small arteries and veins communicate with the afore-mentioned uterine artery and pass circumferentially around each incubation chamber. There they connect with tiny blood vessels that supply and drain the sub-epithelial capillary networks located in the lamina propria.

As the site of eggshell deposition, the oviductal uterus contains multicellular glands that synthesize and secrete the proteinaceous eggshell fibers (Giersberg 1922; Botte 1973; Ortiz and Morales 1974; King 1977; Packard *et al.* 1988; Guillette *et al.* 1989; Picariello 1989; Packard and DeMarco 1991; Heulin *et al.* 2005; Stewart *et al.* 2010). A study of the oviparous lizard *Sceloporus woodi* showed that elongated fibers are extruded from the openings of the shell glands and become wrapped around the egg (Palmer *et al.* 1993). Through an unknown mechanism, the fibers polymerize or combine extracellularly from the glandular secretions. These observations are consistent with those on other species (Giersberg 1922; Raynaud 1962; Packard and DeMarco 1991). A study of the oviparous skink *Lampropholis guichenoti*, the Grass Sun-Skink, using immunofluorescence labeling implicated both uterine shell glands and surface epithelium in deposition of eggshell calcium (Herbert *et al.* 2006; Thompson *et al.* 2007).

From the standpoint of viviparity, the uterus of squamates is of particular interest because it houses the developing embryos throughout gestation and contributes to placentas that sustain them until parturition. Maintenance of the developing fetuses requires that the uterus meet embryonic needs for respiratory gas exchange, needs that undergo a dramatic increase as development proceeds (Packard *et al.* 1977; Guillette 1982a; Birchard *et al.* 1995; Thompson and Stewart 1997). In addition, the pregnant uterus provides water to the embryo as well as at least small quantities of organic and inorganic nutrients (Stewart and Thompson 2000; Thompson and Speake 2006; Blackburn and Stewart 2011). However, the uterus of oviparous squamates is not well-suited to meet the physiological requirements of advanced embryos, especially the need for gas exchange. Indeed, low oxygen levels may represent a significant constraint on the evolution of viviparity (Andrews 2002; Parker and Andrews 2006). Accordingly, the evolution of viviparity is associated with modifications in uterine structure and function that have evolved convergently.

The uterus of viviparous squamates exhibits three features that contribute to the capacity for maintenance of advanced fetuses: a reduction in shell gland activity (and deposition of correspondingly reduced eggshell), an increase in vascularity, and a decrease in epithelial height. A reduction in uterine shell glands is widespread in viviparous squamates (Giersberg 1922; Jacobi 1936; Hoffman 1970; Angelini and Ghiara 1984; Picariello *et al.* 1989; Girling *et al.* 1998), and is probably universal. As compared to uterine glands of oviparous squamates, those of viviparous species are reduced in size, quantity, or both (Blackburn 1998b; Heulin *et al.* 2005; Stewart *et al.* 2010; see Section 13.3.5.1). Gland reduction is associated with deposition of a proteinaceous eggshell membrane that is much thinner than its oviparous counterpart. The thinned eggshell enhances gas exchange by decreasing the diffusion distance between fetal and maternal capillaries.

The increased uterine vascularity during pregnancy occurs through proliferation of capillaries of the incubation chambers. Hyper-vascularization of the uterus has been noted in viviparous lizards (Giersberg 1922; Weekes 1927a, 1930; Jacobi 1936; Boyd 1942; Heimlich and Heimlich 1950; Panigel 1951; Blackburn *et al.* 2010; Murphy *et al.* 2010), as well as snakes (Rahn 1942; Matthews 1955; Parameswaran 1962; Hoffman 1970; Gerrard 1974; Blackburn and Lorenz 2003a). An increase in uterine vascularity during gravidity also has been reported in a few oviparous lizards (Bhatia and Dayal 1933; Masson and Guillette 1987; Picariello *et al.* 1989). However, histological comparisons suggest that the uterus is more vascularized in viviparous than in oviparous squamates (Giacomini 1906; Jacobi 1936; Guillette and Jones 1985), an inference supported by quantitative analyses using confocal microscopy (Parker *et al.* 2010a,b).

A third feature that contributes to intrauterine maintenance of the embryo is attenuation of the uterine epithelium. During pregnancy, the uterine epithelium that lines the dorsal (embryonic) hemisphere of the incubation chamber is reduced in height from cuboidal or columnar cells (of the non-pregnant state) to thin, squamous cells (Blackburn 1993b, 1998b; Heulin *et al.* 2005). Transmission electron microscopy has revealed that where epithelium overlies uterine capillaries, it can be reduced in thickness to 0.5 to 1 μm, less than the width of an erythrocyte (Ibrahim 1977; Ghiara *et al.* 1987; Blackburn and Lorenz 2003a; Adams *et al.* 2005; Blackburn *et al.* 2010). The fact that epithelial reduction is a local effect is revealed by a comparison to inter-embryonic segments of the uterus, which retain a cuboidal epithelium (e.g., Blackburn 1993c; Heulin *et al.* 2005; Uribe *et al.* 2006). Although distension of the uterus by the egg possibly contributes to thinning of the uterine epithelium, it chiefly appears to be caused by migration of uterine capillaries towards the uterine lumen with displacement of epithelial cell cytoplasm (Blackburn 1998b; Blackburn and Lorenz 2003a). Like reduction of the eggshell and hyper-vascularization of the uterus, epithelial attenuation enhances maternal–fetal gas exchange during pregnancy.

13.3.2 Development of the Extraembryonic Membranes

Embryonic amniotes share a suite of tissues, termed extraembryonic membranes, which serve as an interface with the external environment and a conduit between compartments of the developing egg. Development of these tissues—the amnion, chorion, allantois and several tissues collectively called yolk sac—is interactive and produces a series of structural–functional complexes as the embryo matures. During late developmental stages of oviparous amniotes, the embryo is surrounded by the amnion, whereas a vascular yolk sac membrane surrounds the yolk mass, and a vascular chorioallantoic membrane contacts the eggshell. Although unique attributes are associated with variation in egg physiology, development of the extraembryonic membranes is highly conserved across the major amniote lineages (Luckett 1977; Stewart 1997).

Extraembryonic membranes first appear at the perimeter of the blastodisc, which rests on the surface of a mass of yolk, as two sheets of cells derived from ectodermal and endodermal germ layers. This structure, the bilaminar omphalopleure, expands to surround the yolk as the primary yolk sac. Mesoderm from the blastodisc contributes a middle layer between ectoderm and endoderm, converting the bilaminar omphalopleure to a trilaminar omphalopleure, the secondary yolk sac. Blood vessels, and blood cells, arise within the mesodermal sheet and grow inward to the embryo and outward concentrically over the surface of the yolk. The resulting structure is the vascular trilaminar omphalopleure, or choriovitelline membrane, which is the tertiary yolk sac. During the advance of the choriovitelline membrane, a vascularized yolk sac is adjacent to the eggshell. In monotremes, turtles, and birds, the choriovitelline membrane grows over the surface of the yolk around to the abembryonic pole of the egg (Fig. 13.2A) (Agassiz 1857; Semon 1894; Hamilton 1952; Romanoff 1960; Luckett 1977; Hughes 1993).

Development of the yolk sac of lizards and snakes differs from other amniotes because of a unique pattern of growth of extraembryonic mesoderm that limits the distribution of the choriovitelline membrane (Stewart 1997). The sheet of mesoderm initially grows between the endodermal and ectodermal layers of the bilaminar omphalopleure, as it does in other amniotes but then alters course to grow into the yolk mass parallel to the bilaminar omphalopleure (Hrabowski 1926; Weekes 1927b; Stewart and Blackburn 1988; Stewart 1993). This intravitelline mesoderm eventually forms a continuous sheet parallel to the bilaminar omphalopleure across the abembryonic region and separates an outer layer of yolk (the "isolated yolk mass" or IYM) from the main body of the yolk (the vitellus) (Fig. 13.2B). The isolated yolk mass is bordered externally by the bilaminar omphalopleure at the perimeter of the egg and by the sheet of intravitelline mesoderm on its inner aspect. The distribution of mesoderm ultimately determines the distribution of blood vessels. Thus the outer perimeter of the egg in the region of the isolated yolk mass, which is not supplied by mesoderm, lacks vascular support.

In all reptiles, the final stage of yolk sac development begins with splitting of the mesodermal sheet of the choriovitelline membrane (Luckett

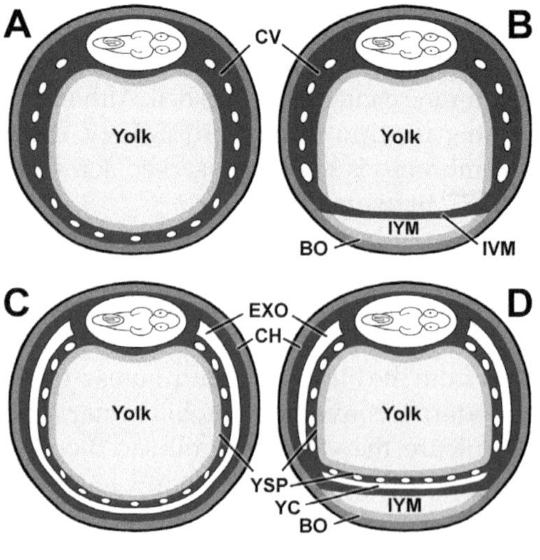

Fig.13.2 Topology of the two forms of vascular yolk sacs of turtles, crocodilians and birds **(A, C)** in contrast to squamates **(B, D)**. The diagrams emphasize the yolk sac at the expense of other structures. Blood vessels are indicated by small ovals. BO, bilaminar omphalopleure; CH, chorion; CV, choriovitelline membrane; EXO, exocoelom; IVM, intravitelline mesoderm; IYM, isolated yolk mass; YC, yolk cleft; YSP, yolk splanchnopleure.

1977; Stewart 1997). The inner layers, which contain the vascular network, remain adjacent to the yolk and constitute the yolk splanchnopleure (Figs. 13.2C,D). The outer two layers, the chorion, line the eggshell and lack vascular support. The cavity between the yolk splanchnopleure and chorion, an extraembryonic coelom, initially appears in the choriovitelline membrane in the vicinity of the embryo. In turtles and birds, the yolk splanchnopleure, which replaces the choriovitelline membrane, gradually expands around the entire yolk mass (Agassiz 1857; Hamilton 1952; Romanoff 1960) (Fig. 13.2C). In contrast, the bilaminar omphalopleure and isolated yolk mass of squamates cover the abembryonic region of the egg and the choriovitelline membrane does not extend beyond the margin of the isolated yolk mass (Stewart and Blackburn 1988; Stewart 1997; Stewart and Thompson 2000) (Fig. 13.2D). The yolk sac of the abembryonic surface of the vitellus (main body of yolk) of squamate reptiles is yolk splanchnopleure, as it is in other reptiles but the development of this membrane is unique among amniotes. Layers of intravitelline mesoderm split to form a cavity, the yolk cleft (Fig. 13.2D). The epithelium of the yolk cleft, the intravitelline mesoderm, in association with the underlying yolk endoderm, forms splanchnopleure. Omphalomesenteric blood vessels vascularize the yolk splanchnopleure at the base of the vitellus, i.e., the inner wall of the yolk cleft. This vascular yolk splanchnopleure is continuous with the remainder of the yolk sac. The intravitelline mesoderm plus endoderm of the isolated yolk mass, forming the outer wall of the yolk cleft is not vascularized.

Although not the first to describe the phenomenon, Boyd (1942) introduced the terms "intravitelline mesoderm" and "yolk cleft," in a study of the gekkonid lizard *Woodworthia maculatus*. This terminology has been conventionally adopted. However, Boyd's (1942) interpretation of the origin of cells that contribute to this complex differs from subsequent authors. Boyd (1942) proposed that the sheet of intravitelline cells remained linear and contributed to the yolk splanchnopleure lining the inner wall of the yolk cleft. In her view, the outer wall was endoderm; thus, the yolk cleft was not an extraembryonic coelom. This interpretation contradicted Hrabowski's (1926) original description of the yolk cleft (= distaler *Dotterspalt*) as a cavity lined by mesoderm. Most studies of development of the yolk cleft have concluded that it forms within sheets of mesoderm, as described above (Stewart 1993; Stewart and Thompson 2000). Boyd (1942) also followed Harrison and Weekes (1925) and Weekes (1930, 1935) in interpreting the omphalopleure of the isolated yolk mass to be trilaminar (ectoderm, mesoderm, and endoderm). More recent investigations using transmission electron microscopy to study lizards and various natricine snakes reported that the omphalopleure of the isolated yolk mass is developmentally bilaminar (ectoderm, endoderm) (Hoffman 1970; Ibrahim 1977; Blackburn *et al.* 2002, 2009, 2010; Blackburn and Lorenz 2003b; Stewart and Brasch

2003; Anderson *et al.* 2011). The yolk cleft–isolated yolk mass complex is structurally similar in all squamate lineages and it is likely to also have a common developmental derivation. However, it is a defining characteristic of squamate yolk sac development and should be investigated further in gekkonid lizards as a possible variant on the general pattern.

The yolk cleft is a derived squamate characteristic (Stewart and Blackburn 1988; Stewart 1997). There are no descriptions of a yolk cleft for amniote lineages other than squamate reptiles. Accordingly, a choriovitelline membrane extends over the abembryonic surface of the yolk mass in monotremes, turtles, and birds (Agassiz 1857; Semon 1894; Hamilton 1952; Romanoff 1960; Luckett 1977; Hughes 1993). Descriptions of development of the extraembryonic membranes of squamates are heavily biased toward viviparous species (reviews in Yaron 1985; Blackburn 1993b, 2000a; Stewart 1993; Blackburn and Stewart 2011), and with the exception of specialized placentotrophic forms with substantial reduction in yolk mass (Jerez and Ramírez-Pinilla 2001, 2003; Flemming and Blackburn 2003; Blackburn and Flemming 2012), all viviparous species have a yolk cleft–isolated yolk mass complex. The yolk cleft has been described in viviparous species of each of the six major lineages of squamate reptiles and in oviparous species of four of these lineages (Fig. 13.3; Table 13.4). This phylogenetic distribution indicates that the yolk cleft is a shared, derived structure of squamate yolk sac development.

There have been few studies of yolk sac development in reptiles and, with the exception of the chicken, structure and function of the yolk sac and yolk sac cavity are largely unknown. Initial stages of yolk cleft formation are similar for both oviparous and viviparous squamates but subsequent development of the yolk cleft–isolated yolk mass complex

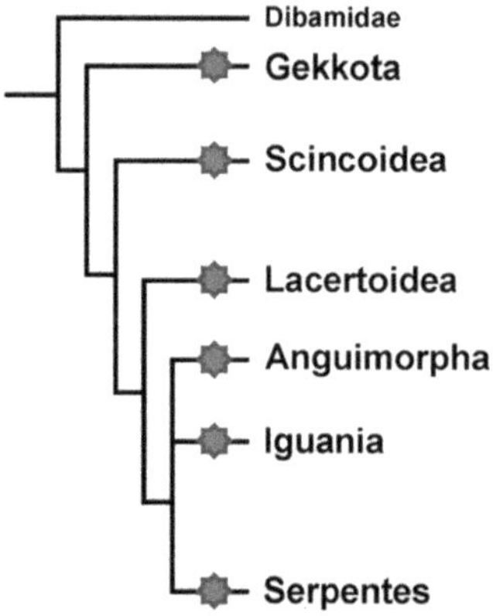

Fig. 13.3 Phylogenetic distribution of the yolk cleft of squamate reptiles. Phylogeny after Sites *et al.* 2011. Annual Review of Ecology, Evolution, and Systematics 42: 227–244.

Table 13.4 Squamate groups in which the yolk cleft has been described microscopically. Taxa include representatives of the six major squamate lineages. "O" refers to oviparous species and "V" to viviparous species.

Taxon	Reproductive mode	Source
Gekkota		
Diplodactylidae	V	Boyd 1942
Scincoidea		
Xantusiidae	V	Yaron 1977,1985
Scincidae	O, V	ten Cate-Hoedemaker 1933; Weekes 1927a,b, 1930, 1935; Blackburn 1993c; Stewart and Thompson 1994, 1996, 1998, 2004, 2009b; Blackburn and Callard 1997; Stewart and Florian 2000; Murphy *et al.* 2012; Stewart *et al.* 2012
Lacertoidea		
Lacertidae	O, V	Hrabowski 1926; Stewart *et al.* 2004a
Anguimorpha		
Anguidae	O, V	Stewart 1985
Iguania		
Phrynosomatidae	V	Villagran *et al.* 2005; Blackburn *et al.* 2010
Colubroidea		
Homalopsidae	V	Parameswaran 1962
Elapidae	V	Weekes 1929; Kasturirangan 1951a,b
Colubridae	O, V	Hoffman 1970; Stewart 1990; Blackburn *et al.* 2003b, 2009; Blackburn and Lorenz 2003b

often differs between reproductive modes. These differences are likely to impart different functional attributes to the yolk sac system. The ontogeny of the yolk cleft has been described histologically in three species of oviparous lizards, *Zootoca vivipara* (Lacertidae) (Hrabowski 1926; Stewart *et al.* 2004), Five-lined Skink, *Plestiodon fasciatus* (Scincidae) (Stewart and Florian 2000) and Lord Howe and Norfolk Islands Skink, *Oligosoma lichenigerum* (Scincidae) (Stewart *et al.* 2012). The lining of the yolk cleft in these species is a germinal epithelium that releases cells into the yolk cleft. The yolk cleft of *Z. vivipara* is intact throughout incubation, whereas the yolk clefts of the two scincid species are modified into secondary cavities that persist to hatching. In each of these cases, the cavities fill with cells during late stages of development (Fig. 13.4). The function of these cells is unknown but the presence of clusters of similar cells among the larger endodermal cells and yolk platelets within the main yolk sac cavity of all three species points to a role in yolk degradation and/or immunoprotection (Stewart and Florian 2000; Stewart *et al.* 2004, 2012). Development of the yolk cleft–isolated yolk mass complex is similar for both oviparous and viviparous populations of the reproductively bimodal lacertid *Zootoca*

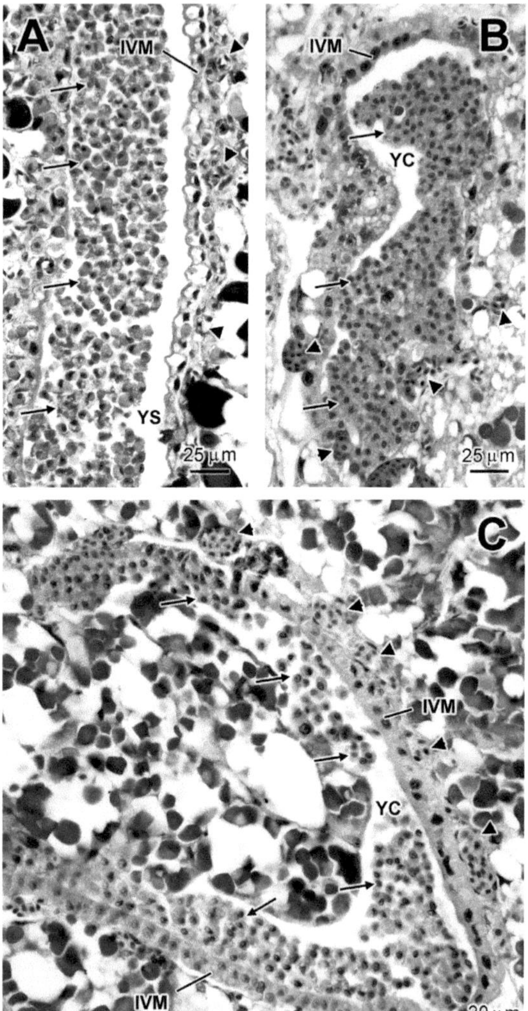

Fig. 13.4 Yolk cleft and associated structures in three species of oviparous lizards. **(A)** *Plestiodon fasciatus*, embryonic stage 38. **(B)** *Zootoca vivipara*, embryonic stage 37. **(C)** *Oligosoma lichenigerum*, embryonic stage 39. IVM, intravitelline mesoderm; YC, yolk cleft; YS, yolk sinus, derivative of the yolk cleft. Arrows, cells in yolk cleft/yolk sinus. Arrowheads, omphalomesenteric blood vessels.

Color image of this figure appears in the color plate section at the end of the book.

vivipara but differs between oviparous and viviparous species of skinks. These embryonic tissues contribute to placental structures that are unique to squamates and innovations of viviparous species are likely to enhance placental exchange. In addition, the yolk cleft of viviparous species of skinks is not transformed into a site of cell proliferation, indicating a different functional relationship with the contents of the yolk sac cavity. These differences suggest the possibility that the process of yolk mobilization differs between reproductive modes.

Whatever its function, the yolk cleft–isolated yolk mass complex influences growth of the allantois and topology of the vascular chorioallantoic membrane. The allantois is an extension of the hindgut that initially appears as a small bulb-like vesicle. The allantoic vesicle grows into the extraembryonic coelom as it expands outward from the embryo, positioning the outer allantoic membrane in contact with the chorion (Luckett 1977; Mossman 1987). These two tissues subsequently fuse to form the chorioallantoic membrane, which is vascularized by allantoic blood vessels. In turtles, crocodilians and birds, the chorioallantoic membrane follows the distribution of the extraembryonic coelom and its predecessor the choriovitelline membrane and surrounds the entire yolk mass (Agassiz 1857; Hamilton 1952; Romanoff 1960; Ferguson 1982, 1985). Thus, the vascular chorioallantoic membrane is adjacent to the eggshell and the vascular yolk splanchnopleure surrounds the yolk (Fig. 13.5A).

The limited growth of the choriovitelline membrane in squamates constrains the expansion of the allantois, which is initially restricted to the perimeter of the egg above the bilaminar omphalopleure of the isolated yolk mass (Fig. 13.5B). Subsequent distribution of the chorioallantoic membrane is dependent on interaction between the allantois and further development of the yolk cleft–isolated yolk mass complex. Several developmental patterns have been described but there are two primary structural alignments for the final disposition of extraembryonic membranes in the abembryonic hemisphere of the egg; an omphalopleure persists over some or all of the region originally occupied by the isolated yolk mass or a chorioallantoic membrane replaces the omphalopleure. Extension of a

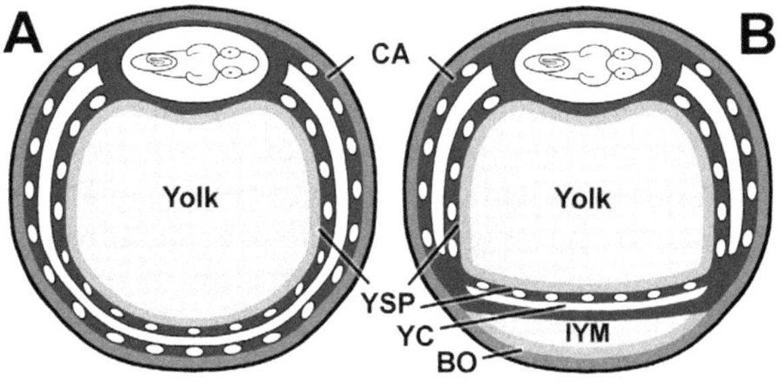

Fig. 13.5 Topology of the yolk sac membranes and chorioallantoic membranes of turtles, crocodilians and birds **(A)** in contrast to squamates **(B)**. Blood vessels are indicated by small ovals. BO, bilaminar omphalopleure; CA, chorioallantoic membrane; IYM, isolated yolk mass; YC, yolk cleft; YSP, yolk splanchnopleure.

chorioallantoic membrane around the entire perimeter of the egg occurs in the four oviparous lizards that have been studied, three scincids and a lacertid (Stewart and Thompson 1996; Stewart and Florian 2000; Stewart *et al.* 2004, 2012), a viviparous gekkonid (Boyd 1942), a viviparous lacertid (Stewart *et al.* 2003), species of the viviparous scincid genus *Eulamprus* (Weekes 1927a; Murphy *et al.* 2012) and an oviparous snake (Blackburn *et al.* 2003b). There are differences among these species in the alignment of tissues during formation of the chorioallantoic membrane in the abembryonic hemisphere but in all cases the pattern of development differs from that of the embryonic hemisphere of the egg.

Persistence of the omphalopleure in the abembryonic region of the egg is preceded by one of two developmental events. In one pattern, the allantois expands into the yolk cleft, and in the second, the allantois remains positioned at the periphery of the omphalopleure. Growth of the allantois into the yolk cleft occurs in two viviparous phrynosomatid lizards (Villagran *et al.* 2005; Blackburn *et al.* 2010; Anderson *et al.* 2011) and in several viviparous natricine snakes (Stewart 1990; Blackburn and Lorenz 2003b; Blackburn and Stewart 2011). In species with this pattern, the isolated yolk mass may be reduced considerably in thickness and extent, being represented in some species by sparse, isolated droplets (Blackburn and Lorenz 2003b; Blackburn *et al.* 2010; Anderson *et al.* 2011). Nevertheless, the allantois is separated from the outer perimeter of the egg by the omphalopleure and any remnants of the isolated yolk mass (Stewart 1990; Villagran *et al.* 2005; Blackburn and Stewart 2011). In the second pattern, the yolk cleft is not breached by the allantois and the omphalopleure remains intact in a viviparous anguid (Stewart 1985), the viviparous scincine skinks *Chalcides chalcides* (ten Cate-Hoedemaker 1926; Blackburn 1993c; Blackburn and Callard 1997) and *C. ocellatus* (Ibrahim 1977; Corso *et al.* 2000) and in several viviparous lygosomine skinks of the *Eugongylus* species group (Stewart and Thompson 2009, 2010). Details of the developmental pattern resulting in this extraembryonic membrane topology vary among species but the common outcome is the absence of a vascular chorioallantoic membrane in at least a portion of the abembryonic region that was initially occupied by the bilaminar omphalopleure of the isolated yolk mass.

In summary, development of the yolk sac of squamate reptiles differs from other amniotes in the pattern of growth of extraembryonic mesoderm in the abembryonic hemisphere of the egg. Growth of the allantois and formation of the chorioallantoic membrane is influenced by growth of extraembryonic mesoderm. Consequently, formation of the chorioallantoic membrane in the embryonic hemisphere of the egg is a shared derived amniote trait, while formation of the chorioallantoic membrane in the abembryonic hemisphere of the egg of squamates, if it occurs, is a novel pattern of development.

Notably, the presence of the yolk cleft–solated yolk mass complex influences extraembryonic membrane development with a diversity of specific outcomes. Some features of the diverse patterns of development among squamates may reflect phylogenetic relationships but there are too few data to analyze. In addition, functional attributes of the yolk cleft–isolated yolk mass complex are unknown. However, variation in the development of these tissues among oviparous antecedents and subsequent specializations for placentation contribute significantly to squamate placental diversity.

13.3.3 Placental Categories

As noted above, embryos of viviparous amniotes develop within the uterine region of the maternal oviduct. Amniote placentas are sites of physiological exchange between the uterus and embryonic tissues that contact the perimeter of the egg. Placentas are designated by the embryonic tissues that participate and lizard placentas include a chorioplacenta, three types of yolk sac placentas and a chorioallantoic placenta (Table 13.5). Of the yolk sac placentas, the choriovitelline placenta is a transitional structure; the choriovitelline membrane is the first vascular yolk sac to develop and it is replaced by the yolk splanchnopleure. The omphaloplacenta and omphalallantoic placentas are associated with the yolk cleft–isolated yolk mass complex. The omphaloplacenta is the terminal form of yolk sac placentation for some lizard species, whereas in other species it is transformed into the omphalallantoic placenta by a contribution from the allantois.

Table 13.5 Categories of placentas in viviparous lizards. Terminology follows Stewart and Blackburn (1988), Stewart (1993, 1997), and Blackburn and Stewart (2011). The chorioplacenta and choriovitelline placenta are transitory. The chorioallantoic placenta persists until birth in all species. The omphaloplacenta, when present, may persist until birth unless it is supplanted by the omphalallantoic placenta.

Placental type	Components
Chorioplacenta	Chorion + uterine lining
Choriovitelline placenta	Choriovitelline membrane + uterine lining
Chorioallantoic placenta	Chorioallantoic membrane + uterine lining
Omphaloplacenta	Bilaminar omphalopleure/IYM complex + uterine lining
Omphalallantoic placenta	Omphalallantoic membrane complex + uterine lining

13.3.4 Pattern of Embryonic Nutrition

Pattern of embryonic nutrition refers to the proximate source of nutrients available to developing embryos. Vertebrate embryos are nourished from two primary sources, yolk and maternal sources other than yolk. Lecithotrophy denotes nutrients derived from yolk; matrotrophy refers to maternal nutrient provision other than yolk (Wourms 1981; Blackburn 1992, 2000b). The temporal and spatial separation of the two sources has physiological implications for both mothers and embryos. Yolk is deposited in the developing oocyte within the ovary and is available to embryos as they develop mechanisms to mobilize nutrients stored as yolk. The source of matrotrophic nutrients for squamate reptiles is the female reproductive tract, the oviduct. The embryonic tissues that are primarily responsible for transport of oviductal secretions are the yolk sac and chorioallantoic membrane (Table 13.5). This pattern of matrotrophy, which is shared with other amniote vertebrates, is termed placentotrophy (Blackburn *et al.* 1985; Blackburn 1992, 2014; Stewart 1989, 1992). Mobilization of nutrients directly from the oviduct to the embryo requires coordination between maternal and embryonic tissues such that embryonic mechanisms for nutrient uptake are functionally capable of transporting and metabolizing these molecules. Whereas mobilization of yolk nutrients extends throughout most of the incubation period, placental nutrient transfer in many species occurs primarily during later developmental stages when the yolk sac and chorioallantoic membrane are functionally mature (Stewart *et al.* 2009a,b; Linville *et al.* 2010). In highly placentotrophic species with little yolk, embryonic and uterine specializations occur early in development and accordingly nutrient transfer is extended throughout most of intrauterine gestation (Ramirez-Pinilla 2006; Ramirez-Pinilla *et al.* 2011b).

Categories of vertebrate embryonic nutritional patterns originally were based on sources of organic molecules (Wourms 1981; Blackburn *et al.* 1985) but were later made more inclusive by recognition of the importance of inorganic molecules as well (Stewart 1989; Blackburn 1992, 1994a). Inorganic molecules are an important constituent of yolk and also are transported directly to embryos across the placenta. Thus, when applied generally, the terms lecithotrophy and placentotrophy refer to all nutrients, inorganic and organic. In reality, patterns of embryonic nutrition are likely to be a mosaic in which some molecules originate entirely from one source, while others arise from more than one source. For example, female oviparous lizards deposit calcium in two compartments of the egg, yolk and eggshell, and embryos mobilize calcium from both sources (Packard 1994; Stewart and Ecay 2010). This property of squamate embryonic nutrition challenges the dichotomy implied by early distinctions between lecithotrophy and matrotrophy (Wourms 1981; see Blackburn 2000b).

Oviparous lizards are primarily lecithotrophic because yolk provides all organic and inorganic nutrients for developing embryos except for calcium derived from the eggshell (Packard *et al.* 1985; Shadrix *et al.* 1994; Stewart and Thompson 1993; Stewart *et al.* 2009a,b; Linville *et al.* 2010). In contrast, a few viviparous species are primarily placentotrophic and obtain all but a small quantity of nutrients from placental transfer (Blackburn *et al.* 1984; Blackburn and Vitt 1992; Ramírez-Pinilla 2006; Blackburn and Flemming 2009, 2012; Ramírez-Pinilla *et al.* 2011a). This range of functional characteristics represents the hypothetical ends of a continuum in an evolutionary transformation series. However, detailed studies of placental function are scarce and the range of diversity of patterns of embryonic nutrition is unknown. In the absence of information on how different placentas process specific molecules (i.e., what the actual variation in placental function looks like), the pattern of embryonic nutrition is commonly assessed by using a matrotrophic index, computed as the ratio of hatchling to egg dry mass (Blackburn 1994a; Stewart and Thompson 2000; Thompson *et al.* 2000) (Table 13.6). This statistic provides a rough estimate of conversion of yolk to hatchling in oviparous species and a useful comparison across lineages with the assumption that metabolic costs of development are equivalent. When applied to comparisons between oviparous and viviparous species, most viviparous species have a matrotrophic index similar to oviparous species, implying that most viviparous species are primarily lecithotrophic (Stewart and Thompson 2000; Thompson and Stewart 2000). There are a few species with a matrotrophic index greater or equal to 1.0, which is interpreted as indicating placental transport of organic nutrients but this pattern is uncommon. Using this measure of placental function, the transition from oviparous lecithotrophy to viviparous placentotrophy may, like intrauterine egg retention, be more of a step function than a continuum.

13.3.5 Placental Diversity

The large number of independent origins of viviparity among squamates generally and lizards specifically promise a wealth of information on evolutionary patterns and the underlying mechanisms. Our understanding of placental diversity has grown substantially in recent years with discovery of remarkable structural specializations in multiple lineages (Blackburn and Vitt 1992; Blackburn 1993c; Flemming and Branch 2001; Jerez and Ramírez-Pinilla 2001, 2003; Ramírez-Pinilla *et al.* 2006; Blackburn and Flemming 2012), variation in structures among multiple species within genera (Stewart and Thompson 1998, 2009b; Murphy *et al.* 2011, 2012) and structural and functional characteristics of reproductively bimodal species (Smith and Shine 1997; Heulin *et al.* 2002, 2005; Stewart *et al.* 2004, 2009b, 2010, 2011;

Table 13.6 Dry mass (mg) and calcium content (mg) of recently ovulated egg yolks and hatchlings/neonates of lizards. Values are reported as means. For mode of parity, "O" = oviparous, and "V" = viviparous. "Hatch" = hatchling or neonate.

Species	Parity Mode	Yolk Dry	Yolk Ca	Hatch Dry	Hatch Ca	Hatch Dry/ Yolk Dry	Yolk Ca/ Hatch Ca
Geographic variation in reproductive mode (text Section 13.3.5.1)							
Zootoca vivipara[1]	O	54.4	0.17	46.8	0.89*	0.86	0.19
Zootoca vivipara[1]	V	57.6	0.14	40.2	0.65	0.70	0.22
Saiphos equalis[2]	O	71.7	0.47	52.0	0.89	0.73	0.53
Saiphos equalis[2]	V	107	0.74	77.1	1.03	0.72	0.72
Oviparous Scincidae (text Section 13.3.5.2)							
Eumeces chinensis[3]	O	300	2.07	200	2.65*	0.67	0.78
Plestiodon fasciatus[4]	O	84.1	0.72	72.3	1.19*	0.86	0.61
Bassiana duperreyi[5]	O	79.8	0.44	68.0	1.01	0.85	0.44
Lampropholis delicata[6]	O	26.6	0.19	20.7	0.44	0.78	0.43
Lampropholis guichenoti[6]	O	41.6	0.30	31.2	0.64	0.75	0.47
Saproscincus mustelinus[7]	O	44.9	0.33	31.1	0.60	0.69	0.55
Placentotrophic scincid lineages (text Section 13.3.5.2)							
Niveoscincus coventryi[8]	V	31.9	0.16	25.6	0.52	0.80	0.31
Niveoscincus ocellatus[9]	V	59.4	2.19	100.0	4.76	1.68	0.46
Niveoscincus metallicus[10]	V	45.7	0.39	34.5	0.85	0.75	0.46
Niveoscincus metallicus[11]	V	41.8	0.66	37.9	0.72	1.00***	1.00**
Pseudemoia entrecasteauxii[5]	V	32.5	0.30	54.6	1.09	1.68	0.28
Pseudemoia pagenstecheri[7]	V	18.0	0.16	50.2	1.10	2.79	0.15
Pseudemoia spenceri[12]	V	55.8	0.71	68.6	1.72	1.23	0.41
Placentotrophic scincid species with extremely small eggs (text Section 13.3.5.3)							
Mabuya bistriata[13]	V	0.47	ND	222.4	ND	473	ND
Mabuya heathi[14]	V	0.40	ND	154	ND	385	ND
Mabuya sp.[15]	V	0.40	<0.1	191.8	6.71	480	<0.01
Lecithotrophic species (text Section 13.3.5.4)							
Elgaria coerulea[16]	V	202	ND	165	ND	0.82	ND
Eulamprus quoyii[17]	V	286	ND	240	ND	0.84	ND
Eulamprus tympanum[18]	V	187	1.63	157	4.77	0.84	0.34

Table 13.6 contd....

Table 13.6 contd.

Species	Parity Mode	Yolk Dry	Yolk Ca	Hatch Dry	Hatch Ca	Hatch Dry/ Yolk Dry	Yolk Ca/ Hatch Ca
Oviparous non-scincid							
Iguana iguana[19]	O	ND	59	ND	101*	ND	0.58
Pogona barbata[20]	O	710	9.42	500	15.8*	0.70	0.60
Podarcis muralis[21]	O	93.4	0.64	70.7	1.40	0.76	0.46
Sceloporus aeneus[22]	O	79.4		32.3	ND	0.41	ND
Viviparous non-scincid							
Barisia imbricata[23]	V	220	ND	223	ND	1.01	ND
Sceloporus bicanthalis[22]	V	90.6	ND	41.5	ND	0.46	ND

*Includes internal yolk.
**Yolk and hatchling calcium values do not differ significantly.
***Yolk and hatchling dry mass values do not differ significantly.

[1]Stewart *et al.* (2009b); [2]Linville *et al.* (2010); [3]Ji *et al.* (1996); [4]Shadrix *et al.* (1994); [5]Stewart and Thompson (1993); [6]Thompson *et al.* (2001c); [7]Stewart *et al.* (2009a); [8]Thompson *et al.* (2001b); [9]Thompson *et al.* (2001a); [10]Thompson *et al.* (2000); [11]Thompson *et al.* (1999a); [12]Thompson *et al.* (1999c); [13]Vitt and Blackburn (1991); [14]Blackburn *et al.* (1984); [15]Ramírez-Pinilla (2006); [16]Stewart and Castillo (1984); [17]Thompson (1977b); [18]Thompson *et al.* (2001d); [19]Packard *et al.* (1992); [20]Packard *et al.* (1985); [21]Ji and Braña (1999); [22]Guillette (1981); [23]Guillette and Casas-Andreu (1987).

Linville *et al.* 2010). These studies provide a contrast between species that have evolved viviparity relatively recently and those for which selection has favored complete dependence on placental nutrients for embryonic development. We thus have some insight into the extremes of placental evolution, yet we have barely scratched the surface in exploring placental diversity among the ~1100 viviparous species of lizards (Tables 13.7, 13.8). A greater understanding of placental diversity will not only uncover evolutionary patterns but also identify lineages that will be most informative for predicting characteristics of the transition from oviparity to viviparity.

Reproductive characteristics relevant to understanding development and evolution of placentation include: 1) mode of reproduction, 2) eggshell structure, 3) pattern of embryonic nutrition, 4) ontogeny of the extraembryonic membranes, 5) structure and function of the uterus, and 6) structure and function of the maternal–fetal placental interface. Variation in these features determines attributes of the maternal–fetal interaction. For the following discussion of placental diversity, we will focus on these specific reproductive characteristics of: 1) species for which reproductive mode varies geographically, 2) viviparous placentotrophic lineages with interspecific variation in placental structure and/or pattern

Table 13.7 Viviparous lizards in which placental morphology has been studied in detail. "Methods" abbreviations: CM = confocal microscopy; LM = light microscopy; SEM = scanning electron microscopy; TEM = transmission electron microscopy.

Taxon	Methods	Sources
Geographic variation in reproductive mode (text Section 13.3.5.1)		
Lacertidae *Zootoca vivipara*	LM	Hrabowski 1926; Stewart *et al.* 2004, 2009b
Placentotrophic lineages (text Section 13.3.5.2)		
Scincidae – *Chalcides*		
C. chalcides	LM	Giacomini 1891; ten Cate-Hoedemaker 1933; Blackburn 1993c; Blackburn and Callard 1997
C. ocellatus	LM, TEM, SEM	Giacomini 1906; Ibrahim 1977; Corso *et al.* 2000
Scincidae–*Niveoscincus*		
N. coventryi	CM, LM	Stewart and Thompson 1998; Ramírez-Pinilla *et al.* 2011b
N. greeni	LM	Stewart and Thompson 2009b
N. metallicus	LM	Stewart and Thompson 1994
N. microlepidotus	LM	Stewart and Thompson 2009b
N. ocellatus	LM	Stewart and Thompson 2004
Scincidae - *Pseudemoia*		
P. entrecasteauxii	LM, TEM, SEM	Weekes 1930; Stewart and Thompson 1996, 2009a; Adams *et al.* 2005; Stewart *et al.* 2006
P. pagenstecheri	LM	Harrison and Weekes 1925; Weekes 1930
P. spenceri		Weekes 1929; Stewart and Thompson 1998
Placentotrophic species with extremely small eggs (text Section 13.3.5.3)		
Scincidae		
Eumecia anchietae	LM	Flemming and Branch 2001
Mabuya heathi	LM, SEM	Blackburn *et al.* 1984; Blackburn and Vitt 2002
Mabuya mabouya	LM	Jerez and Ramírez-Pinilla 2001, 2003
Mabuya sp.	LM, TEM	Ramírez-Pinilla 2006; Ramírez-Pinilla *et al.* 2006; Leal and Ramírez-Pinilla 2008; Vieira *et al.* 2007
Trachylepis ivensi	LM	Blackburn and Flemming 2010, 2012
Lecithotrophic species (text Section 13.3.5.4)		
Gekkonidae		
Woodworthia maculatus	LM	Boyd 1942

Table 13.7 contd....

Table 13.7 contd....

Taxon	Methods	Sources
Phrynosomatidae		
Sceloporus jarrovi	LM, TEM, SEM	Blackburn *et al.* 2010; Anderson *et al.* 2011
Sceloporus mucronatus	LM	Villagran *et al.* 2005
Anguidae		
Elgaria coerulea	LM	Stewart 1985
Scincidae		
Eulamprus sp.	LM	Weekes 1927a; Murphy *et al.* 2012
Eulamprus quoyii	LM, CM	Murphy *et al.* 2010, 2011
Eulamprus tympanum	TEM, SEM	Hosie *et al.* 2003; Adams *et al.* 2007
Xantusiidae		
Xantusia vigilis	LM	Heimlich and Heimlich 1950; Yaron 1977, 1985
Species not covered in text		
Tropiduridae		
Liolaemus elongatus	LM	Crocco *et al.* 2008
Liolaemus gravenhorsti	LM	Lemus and Badinez 1967

of embryonic nutrition, 3) viviparous placentotrophic species with a highly modified maternal–embryonic interface, and 4) viviparous species that are predominantly lecithotrophic.

13.3.5.1 Intraspecific geographic variation in mode of reproduction

Reproductive mode varies geographically in three species of lizards (Table 13.9). Each of these species has populations in which females give birth to free living young, i.e., are viviparous. Other populations of each species are oviparous and retain intrauterine eggs for some interval of time but the stage at oviposition varies among populations and between species. There are no records of variation in mode of reproduction among females within populations. All lizard species, with rare exceptions, retain eggs in the oviduct for a considerable interval of embryonic development (Shine 1983a; DeMarco 1993; Blackburn 1995; Andrews and Mathies 2000). The modal embryonic stage at oviposition for a large sample of oviparous lizard species is stage 30 of the 40-stage Dufaure and Hubert (1961) staging system; egg-laying lizards rarely oviposit later than stage 33 (Blackburn 1995). Oviparous populations of *Zootoca vivipara* and the Eyre Peninsula population of *Lerista bougainvillii* oviposit eggs within the range that is typical for oviparous lizards, i.e., with embryos at stage 33 or younger (Heulin 1990; Heulin *et al.* 1991, 2002; Qualls 1996; Roitberg *et al.* 2013). The

Table 13.8 Functional studies of extraembryonic membranes and/or uterus of viviparous lizards. "Methods" abbreviations: CA = composition analysis; EHC = enzyme histochemistry; IHC = immunohistochemistry; MU = uptake of macromolecules; RI transfer = radioisotope transfer.

Taxon	Methods	Sources
Geographic variation in reproductive mode (text Section 13.3.5.1)		
Lacertidae *Zootoca vivipara*	CA, IHC, RI	Panigel 1951, 1956; Paulesu *et al.* 2005; Stewart *et al.* 2009b, 2011
Scincidae *Lerista bougainvillii*	IHC, EHC	Biazik *et al.* 2008, 2010a,b; Wu *et al.* 2011
Saiphos equalis	IHC, EHC	Biazik *et al.* 2007, 2010a,b; Murphy *et al.* 2009
Placentotrophic lineages (text Section 13.3.5.2)		
Scincidae–*Chalcides*		
C. chalcides	IHC	Paulesu *et al.* 1995; Jones *et al.* 2003
C. ocellatus	EHC	Corso *et al.* 2000
Scincidae - *Niveoscincus*		
N. coventryi	CA	Thompson *et al.* 2001b
N. metallicus	CA, IHC, RI	Jones *et al.* 1998; Thompson *et al.* 1999a; Swain and Jones 1997; Wu *et al.* 2011
N. microlepidotus	CA	Atkins *et al.* 2006
N. ocellatus	CA, IHC	Thompson *et al.* 2001a; Wu *et al.* 2011
Scincidae - *Pseudemoia*		
P. entrecasteauxii	CA, EHC, IHC, MU	Stewart and Thompson 1993; Herbert *et al.* 2006, 2010; Stewart *et al.* 2006; Biazik *et al.* 2007, 2008, 2009, 2010a,b, 2012
P. pagenstecheri	CA, IHC	Thompson and Stewart 1994; Thompson *et al.* 1999b; Stewart *et al.* 2009a; Stinnett *et al.* 2012
P. spenceri	CA, EHC, IHC	Thompson *et al.* 1999c; Herbert *et al.* 2006, 2010; Biazik *et al.* 2007, 2008, 2009, 2010a,b
Placentotrophic species with extremely small eggs (text Section 13.3.5.3)		
Scincidae		
Mabuya sp.	CA, IHC	Ramírez-Pinilla 2006; Wooding *et al.* 2010; Ramírez-Pinilla *et al.* 2011
Lecithotrophic species (text Section 13.3.5.4)		
Anguidae		
Elgaria coerulea	CA	Stewart and Castillo 1984
Scincidae		
Eulamprus quoyii	CA, RI	Thompson 1977a,b, 1981, 1982
Eulamprus tympanum	EHC, IHC	Biazik *et al.* 2007, 2008, 2010b
Xantusiidae		
Xantusia vigilis	RI	Yaron 1977, 1985

Table 13.9 Reproductive mode and eggshell characteristics for lizard species with geographic variation in reproductive mode. Stages at parition follow Dufaure and Hubert (1961), in which stage 40 is the stage at hatching or birth. ND = no data.

Species	Population	Stage at oviposition	Post-oviposition incubation time (days)	Eggshell thickness (µm)	Eggshell calcium (mg)
Lerista bougainvillii[1]	Kangaroo Island	40 (39.5–40)	0	6.3	ND
	Gippsland	35.8 (35–37)	19.3 (29°C)	13.9	ND
	Eyre Peninsula	32.6 (32–33)	28.7 (29°C)	18.3	ND
Zootoca vivipara[2,3,4]	Paimpont	40	0	9.0	0.1
	France	33 (30–35)	33.3 (22.5°C)	46.3	1.2
	Sloveno-Italian	31 (30–32)	31.2 (22.5°C)	72.0	ND
Saiphos equalis[5,6,7]	NE NSW	40	1.5 (23°C)	18.4	0.19
	SE NSW	38–39	5.5 (23°C)	37.4	0.37

[1]Qualls 1996; [2]Heulin 1990; [3]Heulin *et al.* 2002; [4]Stewart *et al.* 2009b; [5]Smith and Shine 1997; [6]Linville *et al.* 2010; [7]Stewart *et al.* 2010.

population of *L. bougainvillii* that oviposits at embryonic stage 36 retains eggs beyond what is common for lizards generally. Although stage 36 embryos are larger and with greater metabolic demands than stage 33 embryos, the greatest increase in mass for lizard embryos occurs later in development (Thompson and Stewart 1997; Stewart *et al.* 2009a,b). Thus, oviparous eggs of *L. bougainvillii* and *Z. vivipara* are oviposited prior to the developmental interval of greatest embryonic metabolism and growth. *Saiphos equalis* differs because intrauterine egg retention is prolonged in all populations and embryos undergo considerable increase in mass within the uterus (Smith and Shine 1997; Linville *et al.* 2010). The diversity in embryonic stage at oviposition among these three species includes "typical" oviparous eggs, that is, eggs oviposited immediately prior to the embryonic growth phase, eggs oviposited late within the embryonic growth phase and viviparous eggs (Table 13.9). These species provide an unusual opportunity to study functional characteristics of eggs in relation to reproductive mode.

Eggshell structure. A reproductive characteristic that has been explored for all three species is the relationship between egg retention and eggshell thickness. Thickness of the eggshell is correlated inversely with length of intrauterine egg retention within species but there is considerable variation in eggshell thickness between species (Table 13.9). Eggshells are a potential impediment to exchange of respiratory gases, water and other molecules between the embryo and its surroundings. Reduction in thickness of the eggshell is predicted to be a necessary correlate with intrauterine egg retention and viviparity, in order to allow the required exchange of respiratory gases to maintain embryonic metabolism and growth in the uterine environment (Packard *et al.* 1977; Guillette 1982a; Blackburn 1995; Qualls 1996; but see Andrews and Mathies 2000). The three species support the prediction, yet also reveal that thickness of the eggshell of oviparous species may be a poor predictor of the propensity to evolve viviparity within a specific lineage. For example, the eggshell of oviparous *Zootoca vivipara* is considerably thicker than that of oviparous *Lerista bougainvillii*, yet eggshells of viviparous populations are comparable in thickness.

The eggshell of squamates consists of an inner boundary layer of glycoprotein, multiple layers of fibrous protein, and an outer calcareous layer (Packard and Demarco 1991). Protein fibers are secreted by uterine shell glands that undergo a cycle of activity paralleling the ovarian cycle (Heulin *et al.* 2005; Stewart *et al.* 2010). The protein component of the eggshell is fully formed early in intrauterine gestation. Uterine shell glands fill with secretory material during vitellogenesis and empty shortly after ovulation. Reduction of thickness of the eggshell is associated with smaller size of the shell glands in *Zootoca vivipara* (Heulin *et al.* 2005) and decreased density of shell glands in *Saiphos equalis* (Stewart *et al.* 2010). Calcium is added to the eggshell subsequent to formation of the protein matrix. The quantity of calcium in the eggshell varies with reproductive mode (Table 13.9). Eggshells of oviparous *Z. vivipara* contain a substantial amount of calcium, whereas eggshells of viviparous populations of both *Z. vivipara* and *S. equalis* contain much less calcium. Although thickness of the eggshells of *S. equalis* females with prolonged egg retention do not differ substantially from eggshells of some oviparous populations of *Z. vivipara*, they contain much less calcium. This comparison suggests that the mechanisms of calcium secretion are independent of those for protein secretion and that reduction in eggshell calcium content is a critical factor promoting a suitable environment for intrauterine egg retention.

Pattern of embryonic nutrition. Matrotrophic indices (offspring dry mass/egg dry mass) indicate that yolk provides most of the nutrients for embryonic development for both reproductive modes of *Zootoca vivipara* and *Saiphos equalis* (Table 13.6). Embryonic gains in dry mass are gradual and commensurate with yolk loss during early stages of development but increase dramatically in late embryonic stages (Fig. 13.6A,B). However, although there is no difference between reproductive modes in dry mass of recently ovulated eggs of *Z. vivipara*, viviparous neonates have less dry mass than oviparous hatchlings. This difference is reflected in the matrotrophic indices and results from a greater increase in dry mass of oviparous embryos

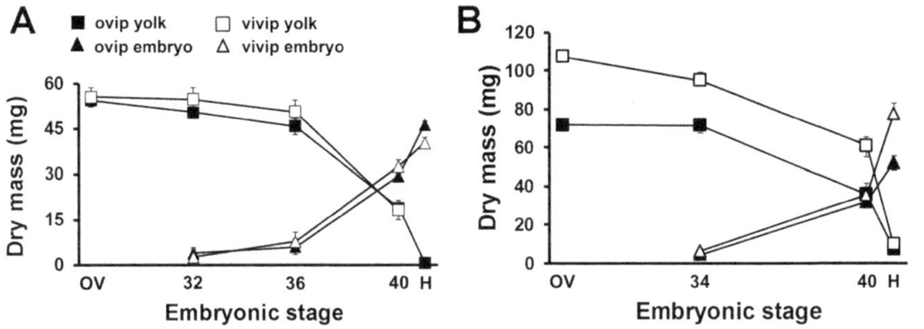

Fig. 13.6 Ontogeny of embryonic uptake and yolk loss of dry mass for oviparous and viviparous populations of **(A)** *Zootoca vivipara* and **(B)** *Saiphos equalis*. Embryonic staging based on Dufaure and Hubert (1961). H, hatchling/neonate; OV, recently ovulated egg.

during the ultimate embryonic stage. Why oviparous embryos grow larger than viviparous embryos on the same amount of yolk is unknown.

Yolk is the primary source of organic nutrients for embryonic development, yet both species have an additional source of calcium (Table 13.6). Oviparous *Zootoca vivipara* are similar to other oviparous lizards in that embryos extract a substantial amount of calcium from the eggshell; 81% of hatchling calcium is mobilized from the eggshell. Eggshell calcium is deposited prior to oviposition at embryonic stage 33 and is extracted by the embryo late in incubation. Loss of calcium from the eggshell is pronounced during the ultimate embryonic stage (Fig. 13.7A). Viviparous embryos also mobilize calcium late in development but this calcium is taken up directly from uterine secretions, i.e., placental transport (Fig. 13.7B). The quantity of calcium in ovulated eggs does not differ between reproductive modes but oviparous hatchlings contain more calcium than viviparous neonates (Table 13.6). This difference is not surprising because oviparous hatchlings also have greater dry mass; however, the concentration of calcium (calcium mass relative to dry mass) in oviparous hatchlings is also greater. Oviparous embryos obtain a greater mass of calcium from the eggshell than the placenta provides to viviparous embryos and the difference is independent of offspring size.

Saiphos equalis with prolonged intrauterine egg retention have a unique pattern of calcium provision commensurate with their intermediate pattern of reproductive mode. Calcium deposited in the eggshell is extracted during the ultimate embryonic stage following oviposition of the egg at stage 38–40 (Fig. 13.7C). Yolk provides 53% of offspring calcium and 27% is extracted from the eggshell. The remaining 20%, which is mobilized directly from uterine secretion prior to oviposition, is placental transport. Thus, uterine calcium secretion contributes to the eggshell following ovulation, as in other oviparous lizards but also occurs during the prolonged egg

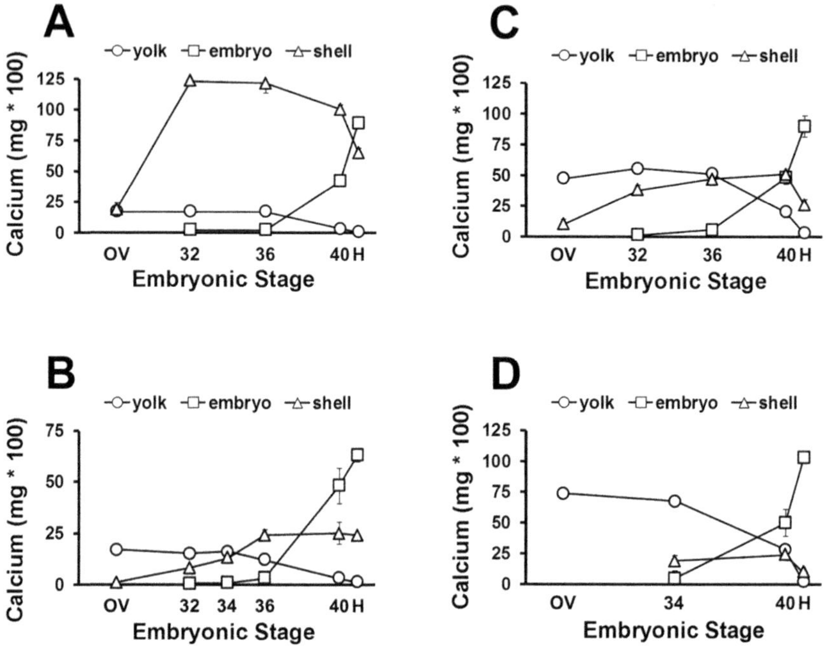

Fig.13.7 Ontogeny of distribution of calcium in egg components for **(A)** oviparous and **(B)** viviparous populations of *Zootoca vivipara* and for **(C)** oviparous and **(D)** viviparous populations of *Saiphos equalis*. Embryonic staging is based on Dufaure and Hubert (1961). H, hatchling/ neonate; OV, recently ovulated egg.

retention. Viviparous females also deposit calcium in the eggshell but less than oviparous females (Fig. 13.7D). Placental secretion of calcium late in development provides 28% of calcium in neonates. As with *Z. vivipara*, offspring of viviparous female *S. equalis* contain relatively less calcium than those of oviparous females (Linville *et al.* 2010).

Structure and function of the extraembryonic membranes and uterus. Of the three reproductively bimodal lizard species mentioned above, detailed information on placental membranes is available only for *Zootoca vivipara*. Development and morphology of the extraembryonic membranes does not differ between reproductive modes of this species (Stewart *et al.* 2004). Initial stages of development of the extraembryonic membranes and formation of the yolk cleft are consistent with the general description in Section 13.3.2 (above) but the development of a vascular yolk sac splanchnopleure in the abembryonic hemisphere differs from all other species. The growth of intravitelline mesoderm into the yolk mass is followed by a sheet of vascular tissue positioned in a parallel plane internal to the intravitelline mesoderm

(Fig. 13.8A). Blood vessels from this membrane provide vascular support for the inner epithelium of the yolk cleft and supply blood vessels to the main yolk mass.

During mid-gestation (embryonic stages 34–36), a chorioallantoic membrane covers the embryonic hemisphere and extends beyond the equator of the egg. The bilaminar omphalopleure occupies the remaining area of the abembryonic hemisphere of the egg. During late developmental stages, the bilaminar omphalopleure is supplanted by the chorioallantoic membrane but the yolk cleft remains intact throughout embryonic development. The mesodermal contribution to the chorioallantoic membrane in this region comes solely from the allantois because the allantois intrudes between the omphalopleure and the isolated yolk mass (Stewart *et al.* 2004). During the final stages of embryonic development, the chorioallantoic membrane is in contact with the eggshell except for a small area at the embryonic pole of the egg (Figs. 13.8B, 13.9).

Uterine morphology of viviparous females is the same throughout; a single layer of low cuboidal cells makes up the uterine epithelium and blood vessels and inactive glands are prominent. The epithelia of the omphalopleure and chorion are squamous in both reproductive modes. Allantoic blood vessels provide support for the chorionic epithelium but the isolated yolk mass is not vascularized and the nearest blood vessels to the omphalopleure lie on the inside of the yolk cleft. As the embryo matures the chorioallantoic membrane advances toward the abembryonic pole of the egg and the uterine epithelium of viviparous females becomes squamous. The chorionic epithelium of the chorioallantoic membrane of both reproductive modes is a bilayer of thin, squamous cells supported by a rich vascular network of allantoic blood vessels (Fig. 13.10A, B). The mature

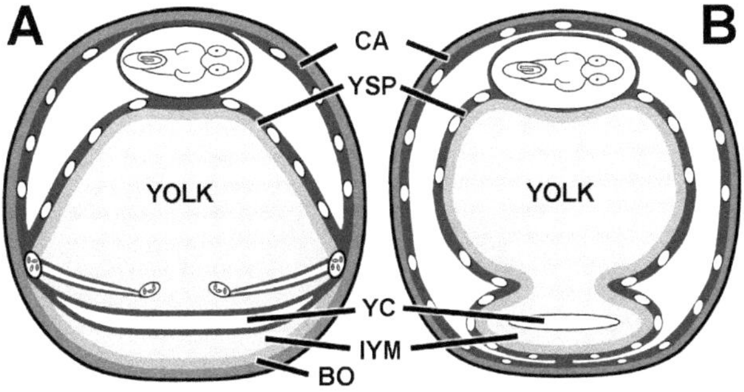

Fig. 13.8 Topology of the yolk sac membranes and chorioallantoic membranes of two embryonic stages (**A** = stage 33; **B** = stage 36) of *Zootoca vivipara*. Oviparous and viviparous populations do not differ. Blood vessels are indicated by small ovals. BO, bilaminar omphalopleure; CA, chorioallantoic membrane; IYM, isolated yolk mass; YC, yolk cleft; YSP, yolk splanchnopleure.

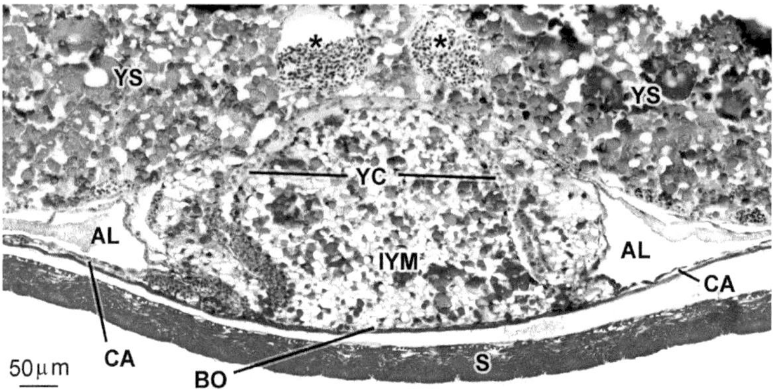

Fig. 13.9 Yolk cleft–isolated yolk mass complex of a stage 34 embryo from an oviparous population of *Zootoca vivipara*. With the exception of the thick eggshell (S), the structures are similar in viviparous populations. AL, allantois; BO, bilaminar omphalopleure; CA, chorioallantoic membrane; IYM, isolated yolk mass; YC, yolk cleft; YS, yolk sac cavity.

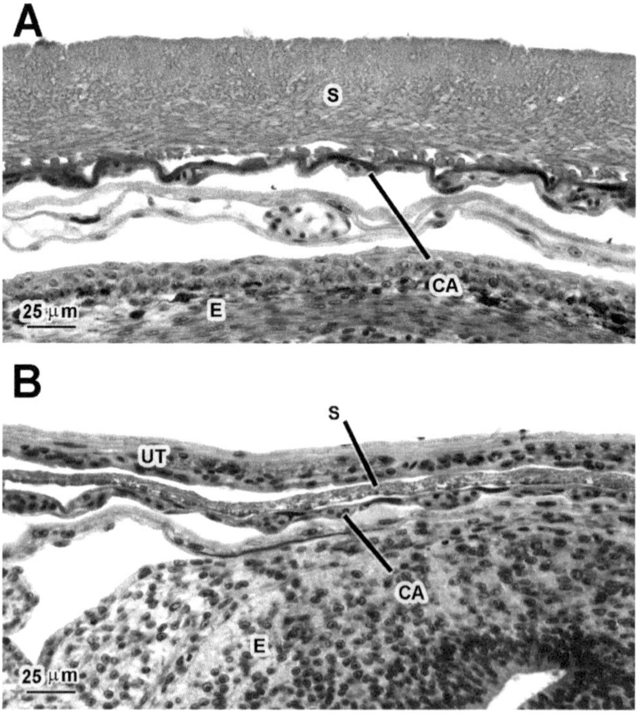

Fig. 13.10 Chorioallantoic membrane of **(A)** oviparous (embryonic stage 37) and **(B)** viviparous (embryonic stage 34) forms of *Zootoca vivipara*. CA, chorioallantoic membrane; E, embryo; S, eggshell; UT, uterus.

Color image of this figure appears in the color plate section at the end of the book.

chorioallantoic placenta also has thin uterine epithelial cells and numerous blood vessels (Fig. 13.10B). An eggshell surrounds the egg throughout incubation in both reproductive modes but is much thinner in viviparous females, as described above (Fig. 13.10A,B).

Our understanding of the function of the extraembryonic membranes and placenta of *Zootoca vivipara* is primarily based on inference from structure (Stewart *et al.* 2004). There is a dramatic reduction in yolk dry mass and increase in embryonic dry mass in both reproductive modes between embryonic stage 36 and parition (Fig. 13.6A). Transport of yolk nutrients to the embryo occurs via omphalomesenteric blood vessels, which are located within the yolk splanchnopleure and also course throughout the yolk sac cavity. Large endodermal cells line the yolk splanchnopleure and likely transport nutrients to the vascular system. Endodermal cells also accompany blood vessels throughout the yolk sac cavity and are in position to mobilize yolk to these vessels as well. There are also cells with a granular cytoplasm scattered among the yolk platelets within the yolk sac cavity. As with the large endodermal cells, these cells frequently are associated with blood vessels and may have a role in yolk degradation. The granulocyte-like cells are morphologically distinct from the endodermal cells and their origin is unknown. One of the hypotheses for the origin of these cells is that they arise from a germinal epithelium lining the yolk cleft and form aggregations within the yolk cleft (Fig. 13.4) before migrating into the yolk sac cavity (Hrabowski 1926; Stewart *et al.* 2004).

The highly vascular chorioallantoic membrane contacts the eggshell around most of the perimeter of the egg by embryonic stage 34 for both reproductive modes of *Zootoca vivipara*. The chorioallantoic membrane is the only structure in position to mediate respiratory exchange and the high density of allantoic blood vessels and their close proximity to the eggshell are attributes that would enhance diffusion of gases. The chorioallantoic membrane is also the most likely site of water exchange during late stages of development when embryos are growing rapidly. Increases in embryonic mass are greatest following embryonic stage 36 (Stewart *et al.* 2009b). The chorioallantoic membrane contacts the eggshell over all but a small area of the perimeter of the egg in stage 36 embryos and increases in area of distribution in subsequent embryonic stages (Stewart *et al.* 2004).

In addition to functioning in gas and water exchange, the chorioallantoic membrane of *Zootoca vivipara* is the major site of calcium transport from either eggshell or placenta depending on reproductive mode. The remaining calcium is mobilized from yolk by cells in the yolk sac splanchnopleure. Calcium transport is facilitated by calcium transporting proteins in the chorionic epithelium and in yolk splanchnopleure (Stewart *et al.* 2011).

Summary. Studies of lizards with geographic variation in reproductive mode reveal few structural differences in embryonic development (Stewart *et al.* 2004). There are, however, differences in uterine structure and function. Uterine glands responsible for secreting the eggshell are either smaller or in lower density in viviparous females (Heulin *et al.* 2005; Stewart *et al.* 2010). As a result, viviparous females secrete thinner eggshells that contain less calcium than oviparous eggs of conspecific females (Heulin 1990; Heulin *et al.* 2002; Stewart *et al.* 2009b). The uterus is a major source of calcium for embryos, irrespective of reproductive mode but the timing of uterine calcium secretion occurs later in embryonic development in viviparous females. Placental transport of calcium thus is highest during the developmental interval when embryonic growth is high.

13.3.5.2 Interspecific variation in placentotrophic lineages

The family Scincidae, the most speciose lizard family, has both the largest number of viviparous species and the most independent origins of viviparity of any squamate family (Blackburn 1982, 1985b, 1999b; Shine 1985). Six scincid genera include the only squamate species that are known to be substantially placentotrophic (Blackburn 2014). These species offer a view of the structural diversity associated with functionally specialized placentas. Scincidae was partitioned taxonomically into four subfamilies by Greer (1970), although recent phylogenetic analyses recognize three primary lineages (Skinner *et al.* 2011; Brandley *et al.* 2012; Pyron *et al.* 2013). The monophyletic Lygosominae, a major radiation with multiple origins of viviparity, is the subject of classical studies on viviparity and placentation (Harrison and Weekes 1925; Weekes 1927a,b, 1929, 1930, 1933, 1935) and continues to be the focus of recent research (Thompson *et al.* 2002; Stewart and Thompson 2003, 2009a; Murphy *et al.* 2012). A second lineage, which includes a large number of species assigned to the subfamily Scincinae (Greer 1970), contains species that were the subject of some of the earliest records of placentation in reptiles (Giacomini 1891, 1906).

Oviparous Scincidae. Comparisons between viviparous lineages and oviparous sister taxa provide an important platform for hypotheses for the relationship between reproductive characteristics and pattern of reproductive mode. This approach cannot be applied rigorously for scincid lizards because of limited data on embryonic development of oviparous species and, in some cases, lack of certainty of phylogenetic relationships. Embryonic development has been described for three oviparous species representing the two largest scincid lineages, the Eastern Three-lined Skink, *Bassiana duperreyi* (Stewart and Thompson 1996) and *Oligosoma lichenigerum* (Stewart *et al.* 2012) of the *Eugongylus* species group of Lygosominae, and

Plestiodon fasciatus (Stewart and Florian 2000) of the Scincinae. Pattern of embryonic nutrition and major features of embryonic development are similar among the three species and, given that they represent two large radiations within Scincidae, they may portray a basic oviparous scincid pattern. With that assumption, we will consider reproductive and developmental patterns of these species as characters for comparison with placentotrophic viviparous scincid lineages.

Embryos of oviparous scincid lizards obtain organic nutrients from yolk but mobilize calcium from both yolk and eggshell (Table 13.6). Embryonic uptake of organic nutrients and calcium is greatest during the latest stages of development in the two species that have been studied (Shadrix *et al.* 1994; Stewart *et al.* 2009a). Thickness of the eggshell has rarely been measured (Qualls 1996; Stewart and Florian 2000; Stewart *et al.* 2010) but the amount of calcium extracted by embryos (Table 13.6) indicates that the eggshell is a substantial structure throughout development in oviparous skinks.

Topology of the extraembryonic membranes during the latest stages of development of *Plestiodon fasciatus* is similar to that of two oviparous skinks of the *Eugongylus* species group, *Bassiana duperreyi* and *Oligosoma lichenigerum* (Stewart and Thompson 1996; Stewart and Florian 2000; Stewart *et al.* 2012) (Fig. 13.11). The chorioallantoic membrane, supplied by the allantoic vascular system, surrounds the living compartments of the egg and contacts the eggshell. This structural alignment positions the chorioallantoic membrane to mediate exchange between the eggshell and the embryo. The chorioallantoic membrane of these oviparous skinks is structurally well suited to effect exchange between the embryonic vascular system and the external environment. The chorionic epithelium is a stratified squamous epithelium of two layers in close apposition to the underlying allantoic blood vessels (Fig. 13.12). Chorionic epithelial cells are frequently thinner where they extend over vessels than in the intervening

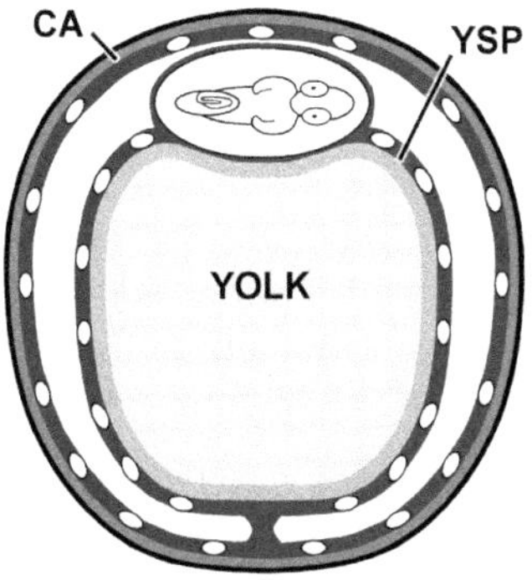

Fig. 13.11 Topology of the extraembryonic membranes of a stage 40 embryo of an oviparous scincid lizard. CA, chorioallantoic membrane; YSP, yolk sac splanchnopleure.

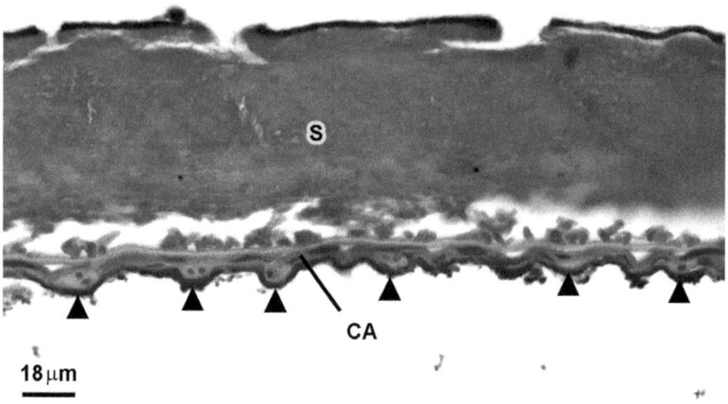

Fig. 13.12 Chorioallantoic membrane of a stage 40 embryo of *Oligosoma lichenigerum*. CA, chorioallantoic membrane; S, eggshell. Arrowheads, allantoic blood vessels.

regions. The primary functions of this membrane are respiratory exchange and transport of calcium from the eggshell to the embryo via the allantoic circulation (Stewart and Thompson 1993; Shadrix *et al.* 1994). The yolk sac is splanchnopleure supplied by omphalomesenteric blood vessels. These vessels, which surround and course throughout the yolk sac cavity, transport yolk nutrients to the embryo.

Early development of the yolk sac is similar to other squamates. A choriovitelline membrane extends outward from the embryo over the surface of the yolk to the region of the equator of the egg where the sheet of mesoderm is diverted into the yolk mass. This intravitelline mesoderm invades the yolk and forms the yolk cleft–isolated yolk mass complex, resulting in regional diversification of the yolk sac. The bilaminar omphalopleure (ectoderm, endoderm) remains over the abembryonic surface of the yolk and the choriovitelline membrane (ectoderm, mesoderm, endoderm) confines the yolk in the embryonic region. The chorioallantoic membrane attains its position around the circumference of the egg during late embryonic stages by replacing the choriovitelline membrane in the embryonic hemisphere and the bilaminar omphalopleure in the abembryonic hemisphere of the egg. As the allantois expands around the abembryonic hemisphere, intravitelline cells surrounding the yolk cleft of *Plestiodon fasciatus* (Stewart and Florian 2000) and *Oligosoma lichenigerum* (Stewart *et al.* 2012) grow into the yolk to transform the yolk cleft into a cavity that persists to the time of hatching (Fig. 13.4A,C).

Viviparous Scincidae. Whereas species with geographic variation in reproductive mode offer insight into the evolution of viviparity, viviparous lineages exhibiting interspecific variation in pattern of embryonic nutrition contribute importantly to understanding the evolution of placental specializations. Three viviparous lineages of scincid lizards have been studied sufficiently to suggest patterns of interspecific variation in placental structure and pattern of embryonic nutrition.

Chalcides. The genus *Chalcides*, which consists of approximately 28 viviparous species, is morphologically diverse; for example elongate body shape and reduced limbs have evolved independently in multiple lineages (Carranza *et al.* 2008), and reproductive features, i.e., clutch size and placental structure, vary among species (Ghiara *et al.* 1987; Caputo *et al.* 2000). Reproductive characteristics indicate that the lineage could contribute importantly to understanding placental evolution but relevant data are available for only two species. Early studies of placentation in *Chalcides chalcides*, the Italian Three-Toed Skink, and *C. ocellatus*, the Ocellated Skink (Giacomini 1891, 1906), which appeared during an era of the emergence of greater understanding of placental diversity among marsupial and eutherian mammals, established that a comprehensive understanding of the evolution of placentation among amniotes must include reptiles. Subsequent studies of placental development and morphology provided further details of placental specializations and verified that *C. chalcides* is highly placentotrophic (ten Cate-Hoedemaker 1933; Ghiara *et al.* 1987; Blackburn 1993c; Blackburn and Callard 1997) and that *C. ocellatus* differs from it in structural features that likely reflect differences in functional capability (Ibrahim 1977; Ghiara *et al.* 1987; Corso *et al.* 2000).

In contrast to oviparous scincids, a thin eggshell is present in early stages of gestation of both *Chalcides chalcides* and *C. ocellatus* but it occurs as discontinuous fragments over the embryonic hemisphere of the egg by mid-gestation and is absent in this region during late gestation (Ibrahim 1977; Blackburn and Callard 1997; Corso *et al.* 2000). In both species, the shell membrane is shed, and pieces of it accumulate at the abembryonic pole of the egg (Ibrahim 1977; Blackburn 1993c).

Although there have been multiple studies of reproductive biology of *Chalcides chalcides* and *C. ocellatus*, pattern of embryonic nutrition has not been quantified for either species. Neonates of *C. chalcides* have a considerably greater wet mass (900 mg) than ovulated eggs (100–130 mg) (Ghiara *et al.* 1987), which suggests that placental transfer provides a major nutrient source. The diameter of ovulated ova of *C. chalcides* (3 mm) is substantially smaller than that of *C. ocellatus* (10 mm) (Ghiara *et al.* 1987) but the precise contribution of yolk to neonate is not known for either species.

The sequence of placental development is similar for *Chalcides chalcides* and *C. ocellatus* but differences in structural specialization arise during late stages of gestation that likely confer different functional properties (ten Cate-Hoedemaker 1933; Ibrahim 1977; Ghiara *et al.* 1987; Blackburn 1993c; Blackburn and Callard 1997; Corso *et al.* 2000). Morphogenesis of the yolk cleft is similar to other lizards, both oviparous and viviparous species, and the choriovitelline membrane is confined to the embryonic hemisphere of the egg by development of the yolk cleft–isolated yolk mass complex. The choriovitelline membrane splits and the allantois grows into the resulting extraembryonic coelom. However, in contrast to oviparous scincid species, the chorioallantoic membrane is confined to the embryonic hemisphere of the egg and does not replace the bilaminar omphalopleure. During formation of the yolk cleft of *C. chalcides*, a linear array of cells develops, which persists as the inter-omphalopleuric membrane to separate the yolk cleft from the extraembryonic coelom of all later embryonic stages (ten Cate-Hoedemaker 1933; Blackburn 1993c; Blackburn and Callard 1997) (Figs. 13.13, 13.14A). The egg is thus partitioned into two hemispheres with the chorioallantoic membrane over the embryonic hemisphere and the yolk sac placenta, an omphaloplacenta, over the abembryonic hemisphere (Fig. 13.13).

The chorioallantoic placenta of both *Chalcides chalcides* and *C. ocellatus* is regionally diversified. In *C. ocellatus*, uterine and embryonic tissues are folded in the region of the embryonic pole of the egg, underlying the mesometrium (Ibrahim 1977; Corso *et al.* 2000). The uterine epithelium contains large cuboidal or columnar cells interspersed with a thin epithelium overlaying blood vessels. This morphology exposes the embryonic tissue to secretory cells with a capacity for histotrophy, in addition to sites of close apposition to the maternal vascular system with a potential for hemotrophic exchange. Allantoic blood vessels are covered by thin, squamous chorionic cells, which facilitate transfer to the embryo. The remainder of the chorioallantoic placenta is smooth with thin uterine and chorionic epithelial cells lying between maternal and embryonic vascular systems. In contrast

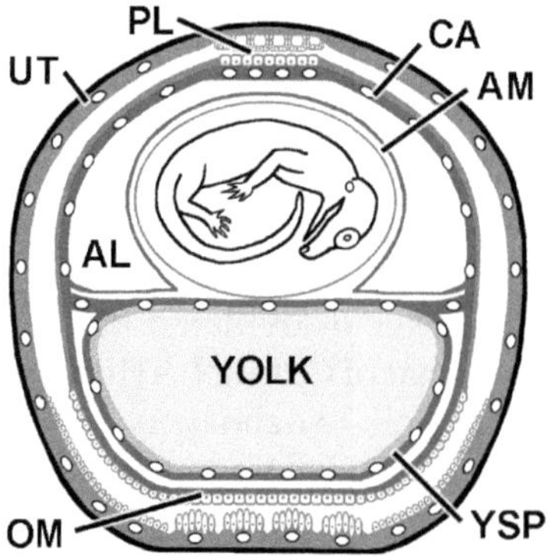

Fig. 13.13 Topology of the extraembryonic membranes of a stage 40 embryo of *Chalcides chalcides* or *Pseudemoia* sp. AL, allantois; AM, amnion; CA, chorioallantoic membrane; PL, placentome; OM, omphalopleure; UT, uterus; YSP, yolk splanchnopleure.

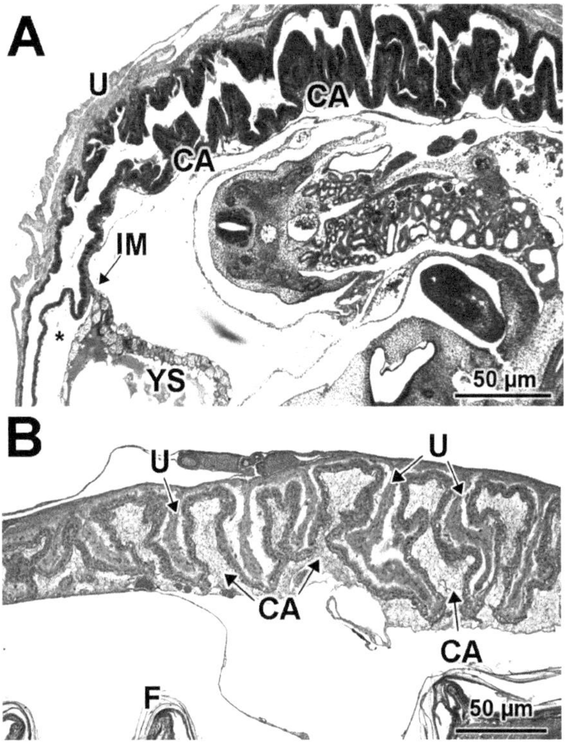

Fig. 13.14 Chorioallantoic placenta in *Chalcides chalcides*. **(A)** Limb bud stage. A placentome is developing through interdigitation of chorioallantois (CA) and lining of the uterus (U). The interomphalopleuric membrane (IM) prevents expansion of allantois into the yolk cleft (asterisk). **(B)** Placentome in a stage 40 embryo. F, fetus; YS, yolk sac cavity.

Color image of this figure appears in the color plate section at the end of the book.

to *C. ocellatus*, most of the chorioallantoic placenta of *C. chalcides* consists of a placentome of deep interdigitating folds centered under the mesometrium (ten Cate-Hoedemaker 1933; Blackburn 1993c; Blackburn and Callard 1997) (Fig. 13.14A,B). The uterine epithelium is columnar, as are the chorionic epithelial cells, and a rich vascular network supplies both embryonic and maternal tissues. A narrow zone of chorioallantoic placentation between the placentome and the omphaloplacenta, the paraplacentomal region, lacks folds. Thin uterine and chorionic epithelial cells of this region overlay extensive uterine and allantoic vascular networks, features that are presumed to function in gas exchange.

The terminal yolk sac placenta is an omphaloplacenta in both species (Blackburn 1993c; Corso *et al.* 2000). The outer epithelium of the omphalopleure is cuboidal or columnar (Ibrahim 1977; Blackburn 1993c;

Blackburn and Callard 1997; Corso *et al.* 2000). These cells form ridge-like projections in *C. ocellatus* (Ibrahim 1977). The omphalopleure of *C. chalcides* is deeply folded and the hypertrophied epithelial cells form papillae that project into a mass of debris (shed shell membrane and degenerating cells) that lies in the uterine lumen at the abembryonic pole (Blackburn 1993c; Blackburn and Callard 1997) (Fig. 13.15A). There is a difference between species in the fate of the isolated yolk mass that has implications for the proximity of uterine and embryonic vascular systems. Early in development, the isolated yolk mass of both species is a barrier between the uterine lumen and the closest embryonic blood vessels in the yolk splanchnopleure. In *C. ocellatus*, the isolated yolk mass is present until at least embryonic stage 39 (Corso *et al.* 2000), whereas in *C. chalcides* it regresses and the omphalopleure contacts the vascular yolk splanchnopleure late in development through reduction in width of the yolk cleft (ten Cate-Hoedemaker 1933; Blackburn 1993c) (Fig. 13.15B). Cells of the uterine epithelium of the omphaloplacenta of both species are cuboidal or columnar. These cells form folds in *C. ocellatus* (Ibrahim 1977; Corso *et al.* 2000) but the epithelium has a smooth surface in *C. chalcides* (Blackburn 1993c; Blackburn and Callard 1997). During late stages of *C. ocellatus*, the uterine cells enlarge further (Ibrahim 1977). An extensive uterine vascular network lies at the base of the epithelium.

Niveoscincus. The Australian scincid genus *Niveoscincus* promises to be an important model for the evolution of placentotrophy because all species are viviparous, placental structure is variable, and both interspecific and intraspecific variation in pattern of embryonic nutrition have been reported. The genus consists of eight species, six of which are endemic to the island of

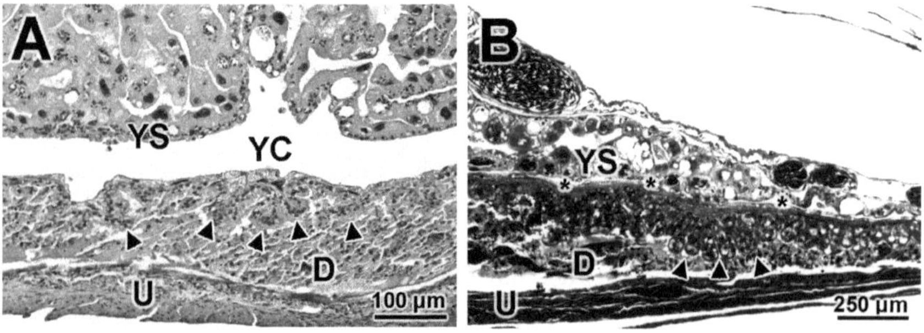

Fig. 13.15 Yolk sac placenta in *Chalcides chalcides* in late gestation. **(A)** The bilaminar omphalopleure bears papillae (arrowheads) that extend into a mass of detritus (D), formed from shell membrane and degenerating cells. Yolk cleft (YC) separates omphalopleure from the yolk sac splanchnopleure (YS). Isolated yolk mass is no longer present. **(B)** The omphalopleure contacts the yolk sac splanchnopleure; remnants of the yolk cleft are marked by asterisks. U, uterus.

Color image of this figure appears in the color plate section at the end of the book.

Tasmania (Hutchinson *et al.* 1990). Of the remaining two species, the Metallic Skink, *N. metallicus*, which has an extensive distribution in Tasmania, also occurs on adjacent mainland Australia in southeastern Victoria and on islands in the intervening Bass Strait. *Niveoscincus coventryi*, Coventry's Skink, is absent from Tasmania and occurs in the southeastern Australian states of Victoria and New South Wales. Studies of single populations of several species have revealed variation in sources of nutrients for embryos (Swain and Jones 1997, 2000; Jones *et al.* 1998; Thompson *et al.* 1999a, 2001a, 2001b; Atkins *et al.* 2006) and in placental structure (Stewart and Thompson 1994, 1998, 2004, 2009b), suggesting that a comprehensive comparative research program would be informative. As with the two species of *Chalcides*, uterine and embryonic epithelia of species of *Niveoscincus* are in direct contact during late developmental stages. A thin eggshell is secreted by the uterus but this structure is absent by mid-gestation (Stewart and Thompson 2009).

Matrotrophic indices, i.e., neonate dry mass/egg dry mass, have been estimated for single populations of *Niveoscincus coventryi* (Thompson *et al.* 2001b), the Ocellated Skink, *N. ocellatus* (Thompson *et al.* 2001a), the Southern Snow Skink, *N. microlepidotus* (Atkins *et al.* 2006) and for two populations of *N. metallicus* (Jones *et al.* 1998; Thompson *et al.* 1999a) (Table 13.6). In comparison to oviparous species, these data indicate that considerable placental transport of organic nutrients is likely for *N. ocellatus* and for a population of *N. metallicus* from southeastern Tasmania. Further, there is a positive relationship between maternal nutritional condition and offspring size in this population of *N. metallicus* (Swain and Jones 2000) and placental uptake of the amino acid leucine is correlated with embryonic stages of high growth (Swain and Jones 1997). In contrast, patterns of embryonic nutrition of *N. coventryi*, *N. microlepidotus* and the Flinders Island population of *N. metallicus* do not differ markedly from oviparous species.

Placental transport of calcium occurs in *Niveoscincus coventryi* (Thompson *et al.* 2001b), *N. ocellatus* (Thompson *et al.* 2001a), and a population of *N. metallicus* on Flinders Island (Thompson *et al.* 2000) (Table 13.6) but has not been studied in other species. The proportion of offspring calcium from placental transfer is comparable to the proportion of calcium extracted from the eggshell of oviparous species. However, provision of calcium to embryos may be responsive to environmental variables because pattern of embryonic calcium nutrition varies geographically in *N. metallicus*. Embryos from the Flinders Island population receive more than half of their calcium from placental transport, whereas embryos of the southeastern Tasmanian population obtain all calcium from yolk (Table 13.6).

There is also evidence that placental transport is selective. Although embryos of the southeastern Tasmanian population of *Niveoscincus metallicus* obtain all calcium from yolk, a considerable quantity of organic nutrients is transported across the placenta. In addition, the population of *N. ocellatus*, which is substantially placentotrophic for organic molecules, is as dependent on yolk for embryonic calcium as are oviparous species. These data indicate that the mechanisms for placental transport of organic nutrients are not coupled with those for calcium transport in these two species.

Placental structure has been investigated for six species of *Niveoscincus*; *N. metallicus*, *N. coventryi*, *N. ocellatus*, the Northern Snow Skink, *N. greeni*, *N. microlepidotus*, and the Tasmanian Tree Skink, *N. pretiosus* (Weekes 1930; Stewart and Thompson 1994, 1998, 2004, 2009b; Ramírez-Pinilla *et al.* 2011b). Development of the yolk sac and growth of the chorioallantoic membrane is similar to species of *Chalcides*. An inter-omphalopleuric membrane develops in association with the yolk cleft and the egg is apportioned into two hemispheres (Fig. 13.16). Late stage placentation includes a chorioallantoic placenta and an omphaloplacenta. The chorioallantoic placenta is regionally diversified and the omphaloplacenta contains specializations for histotrophic transport.

Chorioallantoic placentas of species of *Niveoscincus* are differentiated into areas or zones of structural specializations that are unique among squamates (Stewart and Thompson 2009b). The uterine contribution to the chorioallantoic placenta has a similar morphology in all species of *Niveoscincus*; a simple squamous epithelium faces the embryonic tissue. An extensive network of blood vessels is closely apposed to the epithelium, resulting in distinct ridges that protrude into the uterine lumen in some regions (Fig. 13.17A). There are also areas of the chorioallantoic placenta with a smooth interface (Fig. 13.17B,C). The distribution and prevalence of

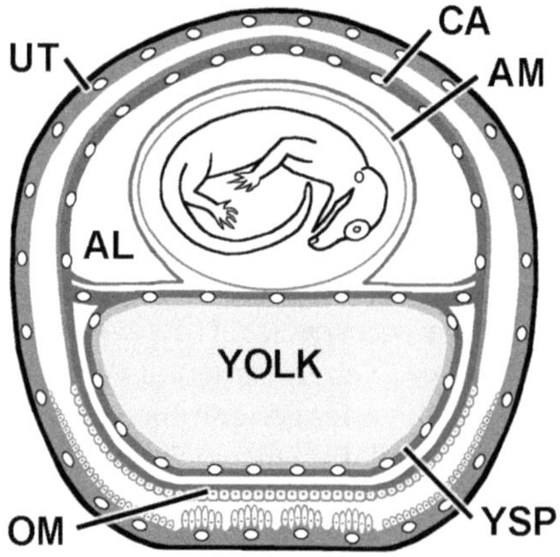

Fig. 13.16 Topology of the extraembryonic membranes of a stage 40 embryo of *Niveoscincus* sp. AL, allantois; AM, amnion; CA, chorioallantoic membrane; OM, omphalopleure; UT, uterus; YSP, yolk splanchnopleure.

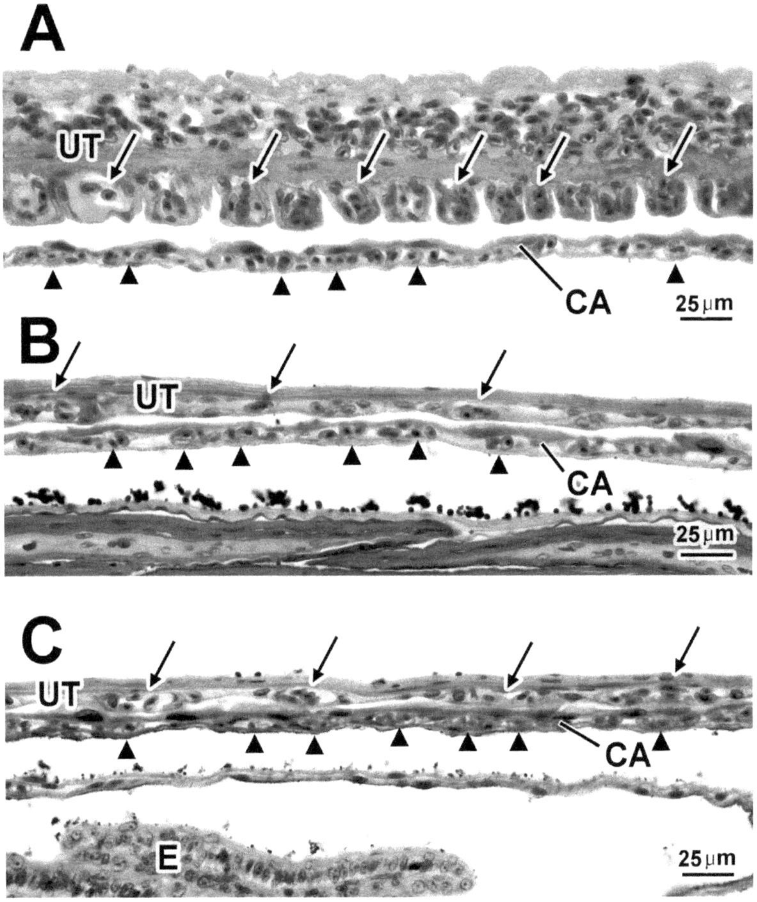

Fig. 13.17 Structural variation in the chorioallantoic placentas of the scincid genus *Niveoscincus*. **A, B**. *Niveoscincus metallicus*. Embryonic stage 39. **C**. *Niveoscincus ocellatus*. Embryonic stage 39. CA, chorioallantoic membrane; E, embryo; UT, uterus. Arrows, uterine blood vessels; Arrowheads, allantoic blood vessels.

Color image of this figure appears in the color plate section at the end of the book.

uterine ridges varies among species. In contrast to the uterus, variation in the chorioallantoic membrane among species is pronounced. Some regions of the chorioallantoic membrane of all species are similar to oviparous skinks with a thin bilayer of squamous chorionic epithelial cells underlain by a dense network of allantoic blood vessels (Fig. 13.17A,B,C). Variations in the chorioallantoic membrane, not seen in oviparous skinks, include a cuboidal outer chorionic epithelium and an outer chorionic epithelium with large irregular-shaped cells interspersed with squamous cells. The distribution and prevalence of these hypertrophied chorionic epithelial cells is a primary difference among species of *Niveoscincus*. For all species, there are regions containing patches of enlarged chorionic epithelial cells aligned contiguously. This morphology is prominent adjacent to the inter-omphalopleuric membrane but there are differences between species in the size of the chorionic epithelial cells and in the total area of their distribution. In *N. metallicus*, there is a distinctive region adjacent to the inter-omphalopleuric membrane with large chorionic epithelial cells that often have irregular shapes (Fig. 13.18A). The epithelium is folded in some regions to form deep crypts (Fig. 13.18B). The uterine epithelium is distinctly ridged in apposition to the enlarged chorionic epithelium. This chorioallantoic morphology is unique among studied squamates. The chorioallantoic membranes of *N. greeni, N. microlepidotus* and *N. ocellatus* also exhibit a specialization that has not been seen in other lineages. Large irregular-shaped chorionic epithelial cells interdigitate with the troughs between uterine ridges, while squamous cells overlay blood vessels apposed to the uterine ridges (Fig. 13.18C,D). Patches of this structural interface occur throughout the chorioallantoic placenta. The interdigitating regions of the chorioallantoic placenta, which have a structural profile characteristic of both histotrophic and hemotrophic exchange, distinguish these species from other placental squamates.

In contrast to the bilaminar omphalopleure of oviparous skinks (Section 13.3.5.2, above), the epithelium of the omphalopleure of the omphaloplacenta of the six species of *Niveoscincus* is hypertrophied and cells of the outer layer are enlarged (Fig. 13.19A,B,C). The omphalopleure of *N. greeni, N. microlepidotus* and *N. ocellatus* is highly folded (Fig. 13.19C), which distinguishes it from *N. coventryi* and *N. metallicus*. The uterine epithelium of *N. greeni, N. microlepidotus* and *N. ocellatus* is high cuboidal or columnar and ridges form over blood vessels, whereas the uterine epithelium of *N. coventryi* and *N. metallicus* is squamous or low cuboidal and is not ridged.

Pseudemoia. Pattern of embryonic nutrition and placentation have been studied for three of the six species of *Pseudemoia* (Smith 2001), *P. entrecasteauxii*, the Woodland Tussock-Skink, *P. pagenstecheri* the Grassland Tussock-Skink, and *P. spenceri*, Spencer's Skink (Weekes 1929, 1930, 1935; Stewart and Thompson 1993, 1996, 1998; Thompson and Stewart 1994; Thompson *et al.* 1999b,c; Adams *et al.* 2005; Stewart *et al.* 2009a; Stinnett *et al.* 2012). In contrast to *Chalcides* and *Niveoscincus*, all three of these species of *Pseudemoia* are highly placentotrophic (Table 13.6). This nutritional pattern likely is characteristic of all species in the genus because *P. spenceri* is the sister taxon to a clade that includes all other members of the genus (Smith 2001). Variation in egg size, clutch size and habitat preference among the three species suggest research focusing on life history patterns could provide insight into the evolution of placentotrophy.

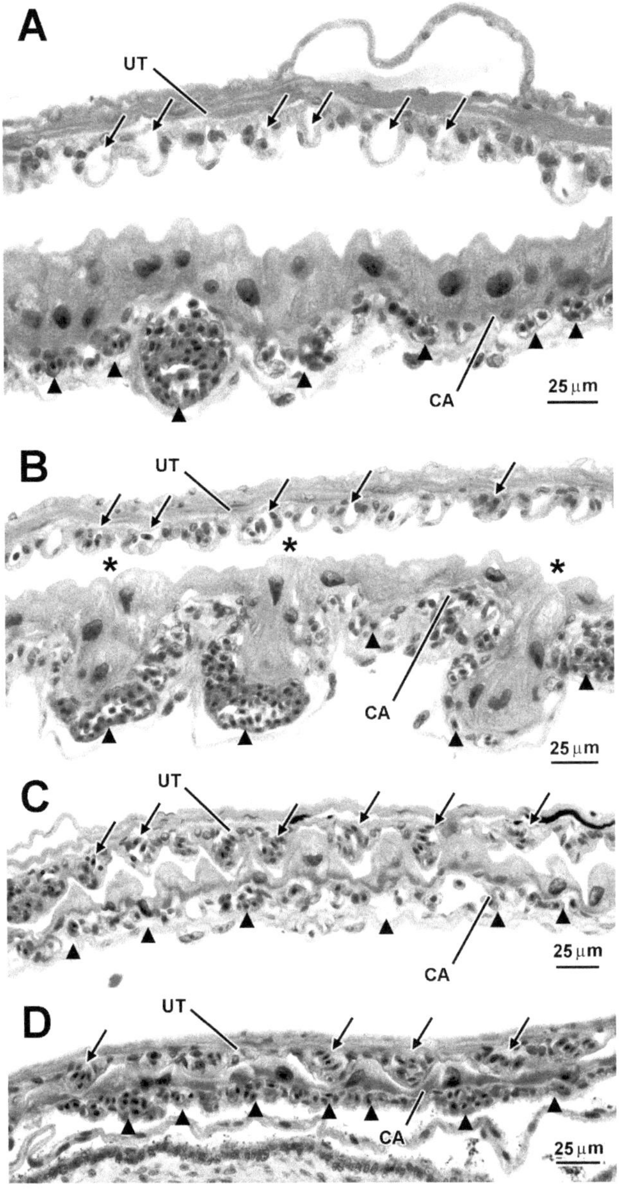

Fig. 13.18 Structural variation in the chorioallantoic placentas of the scincid genus *Niveoscincus*. **A, B**. *Niveoscincus metallicus*. Embryonic stage 36. **C**. *Niveoscincus greeni*. Embryonic stage 37. **D**. *Niveoscincus ocellatus*. Embryonic stage 37. CA, chorioallantoic membrane; UT, uterus. Arrows, uterine blood vessels; Arrowheads, allantoic blood vessels; Asterisks, openings into deep folds in the chorioallantoic membrane.

Color image of this figure appears in the color plate section at the end of the book.

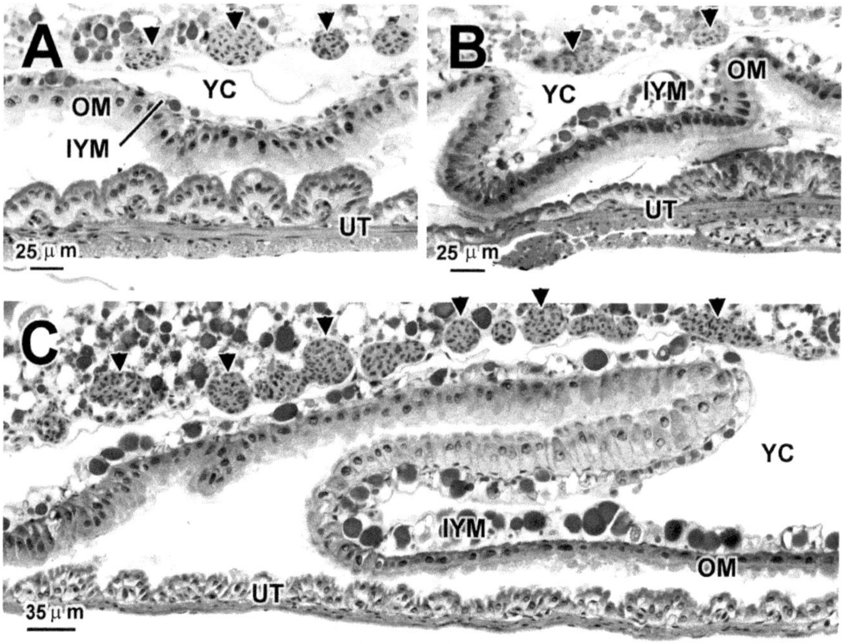

Fig. 13.19 Omphaloplacentas of species of the genus *Niveoscincus*. **(A)** *Niveoscincus greeni*. Embryonic stage 40. **(B)** *Niveoscincus microlepidotus*. Embryonic stage 37. **(C)** *Niveoscincus ocellatus*. Embryonic stage 37. IYM, isolated yolk mass; OM, omphalopleure; UT, uterus; YC, yolk cleft. Arrowheads, omphalomesenteric blood vessels in the yolk splanchnopleure.

As with species of *Chalcides* and *Niveoscincus*, a thin eggshell is secreted around the egg by the oviduct of the three species of *Pseudemoia* and this structure disappears early in gestation (Stewart and Thompson 1996, 1998).

Matrotrophic indices for the three species reveal that placental transport contributes substantially to size of neonates, including both organic and inorganic molecules (Table 13.6) (Stewart and Thompson 1993; Thompson and Stewart 1994; Thompson *et al.* 1999b,c). However, differences between species in size of neonates are strongly influenced by egg size (Fig. 13.20). Pattern of embryonic calcium nutrition is also influenced by egg size and egg calcium content. The relative contribution of yolk calcium to neonatal calcium content of *Pseudemoia spenceri*, which has the highest yolk calcium mass, does not differ from closely related oviparous skinks (Table 13.6). In contrast, eggs of *P. entrecasteauxii* and *P. pagenstecheri* have relatively small calcium masses and placental transport delivers a greater proportion of total calcium to neonates. Placental provision of calcium compensates for the low yolk calcium content because calcium concentrations of neonates of *P. pagenstecheri* are greater than those of a closely related oviparous skink (Stewart *et al.* 2009a).

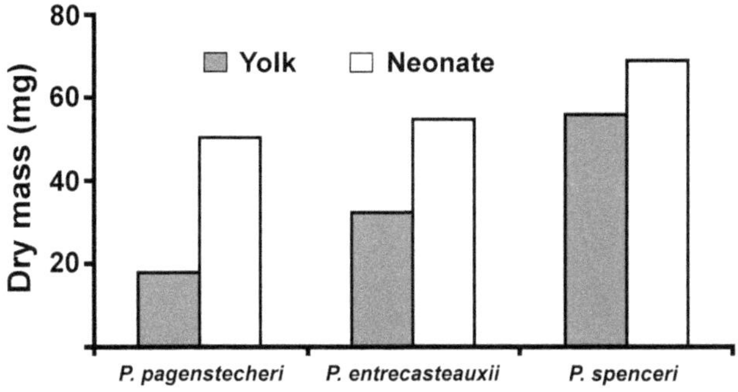

Fig.13.20 Size of eggs and neonates of three species of *Pseudemoia*. Each species is placentotrophic, yet size of eggs contributes significantly to size of neonates.

The pattern of development of the extraembryonic membranes of the three species of *Pseudemoia* is similar to that of *Chalcides* and *Niveoscincus*, although there are differences in placental morphology. Intravitelline mesoderm grows into the yolk sac cavity to form the yolk cleft and also contributes to an inter-omphalopleuric membrane. The inter-omphalopleuric membrane is a persistent vascularized septum separating the egg into two hemispheres. Development of the yolk cleft–isolated yolk mass complex is similar to species of *Chalcides* and *Niveoscincus* and the terminal stage of placentation includes both a chorioallantoic placenta and an omphaloplacenta.

The chorioallantoic placentas of species of *Pseudemoia* are regionally diversified (Fig. 13.13). A placentome occurs as an ellipsoid-shaped zone along the mesometrial axis of the embryonic hemisphere of the egg in each of the species. There are prominent folds in the uterine epithelium and an extensive vascular bed (Fig. 13.21A,B). Uterine epithelial cells are cuboidal and the chorionic epithelium consists of large cuboidal or columnar cells with a high density of allantoic blood vessels in support. The chorionic epithelium over the remainder of the chorioallantoic membrane is squamous, as is the apposing uterine epithelium (Fig. 13.22A,B).

The omphalopleure of the omphaloplacenta is folded and the epithelium is a bilayer with large columnar cells in the outer layer and an inner layer of thin squamous cells (Fig. 13.23). The isolated yolk mass is initially thin and regresses early in development, which positions the squamous epithelium lining the yolk cleft adjacent to the inner aspect of the omphalopleure (Fig. 13.23). This membrane, composed of ectoderm and mesoderm, forms the outer wall of the yolk cleft and abuts the vascular yolk splanchnopleure, which forms the inner wall of the yolk cleft. The uterus of the omphaloplacenta is highly vascular with a cuboidal epithelium, which frequently forms folds or ridges over blood vessels.

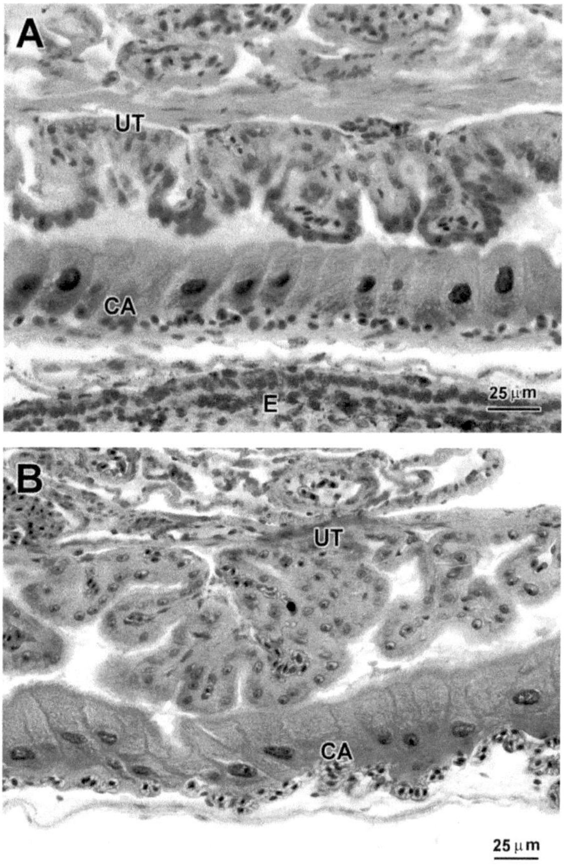

Fig. 13.21 Chorioallantoic placentomes of species of the scincid genus *Pseudemoia*. **(A)** *Pseudemoia spenceri*. Embryonic stage 38. **(B)** *Pseudemoia pagenstecheri*. Embryonic stage 40. CA, chorioallantoic membrane; E, embryo; UT, uterus.

Color image of this figure appears in the color plate section at the end of the book.

Ultrastructure of the uterine epithelium of *Pseudemoia entrecasteauxii* reveals regional variation in functional profiles (Adams *et al.* 2005). Uterine epithelial cells are thin with attenuated cytoplasmic extensions over uterine capillaries over most of the surface area of the chorioallantoic placenta, indicating the capacity for efficient gas exchange and/or hemotrophic transport. In contrast, the hypertrophied uterine epithelium of the chorioallantoic placentome has structural attributes characteristic of histotrophic transport. Hypertrophied uterine epithelial cells of the omphaloplacenta also are histotrophic but these cells secrete vesicles into the uterine lumen, whereas cells of the placentome do not. The uterine pattern

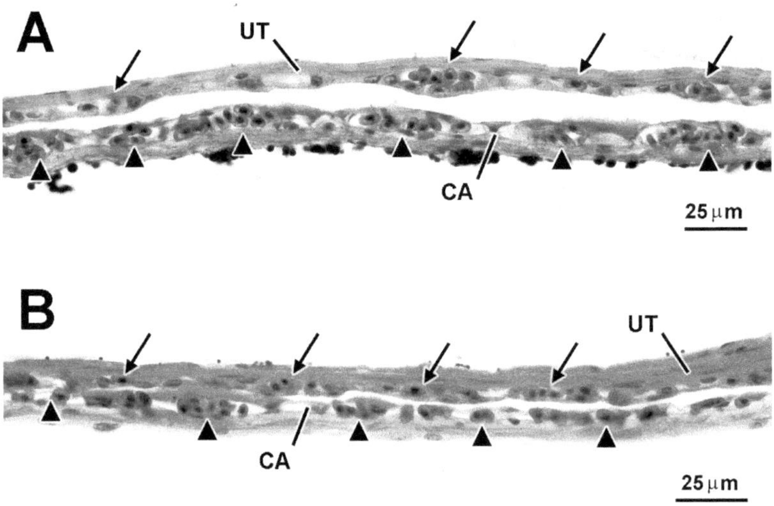

Fig. 13.22 Chorioallantoic placentas of species of the scincid lizard genus *Pseudemoia*. **(A)** *Pseudemoia spenceri*. Embryonic stage 40. **(B)** *Pseudemoia pagenstecheri*. Embryonic stage 40. CA, chorioallantoic membrane; UT, uterus. Arrows, uterine blood vessels; Arrowheads, allantoic blood vessels.

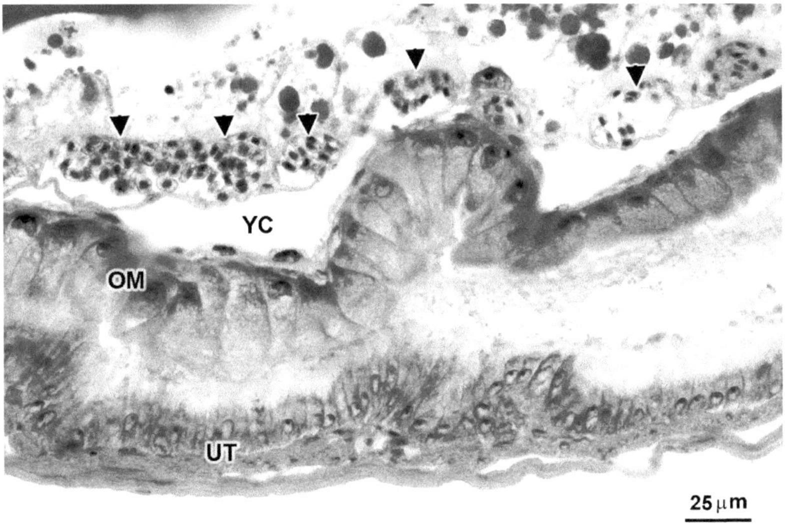

Fig. 13.23 Omphaloplacenta of *Pseudemoia entrecasteauxii*. Embryonic stage 32. OM, omphalopleure; UT, uterus; YC, yolk cleft. Arrowheads, omphalomesenteric blood vessels in the yolk splanchnopleure.

of expression of a gene coding for the lipid transport enzyme, lipoprotein lipase, indicates that the omphaloplacenta is the major site of lipid transport (Griffith *et al.* 2013). Further support for a histotrophic function for the placentome and omphaloplacenta is indicated by the capacity for embryonic epithelial cells of *P. entrecasteauxii* to take up relatively large molecules by endocytosis (Stewart *et al.* 2006). These two placental regions also function in the transport of calcium. Calbindin-D_{28K} is localized in the chorioallantoic placentome and epithelium of the omphalopleure of *P. pagenstecheri*, indicating that these tissues are specialized for calcium uptake (Fig. 13.24A,B) (Stinnett *et al.* 2012). The uterine epithelium apposed

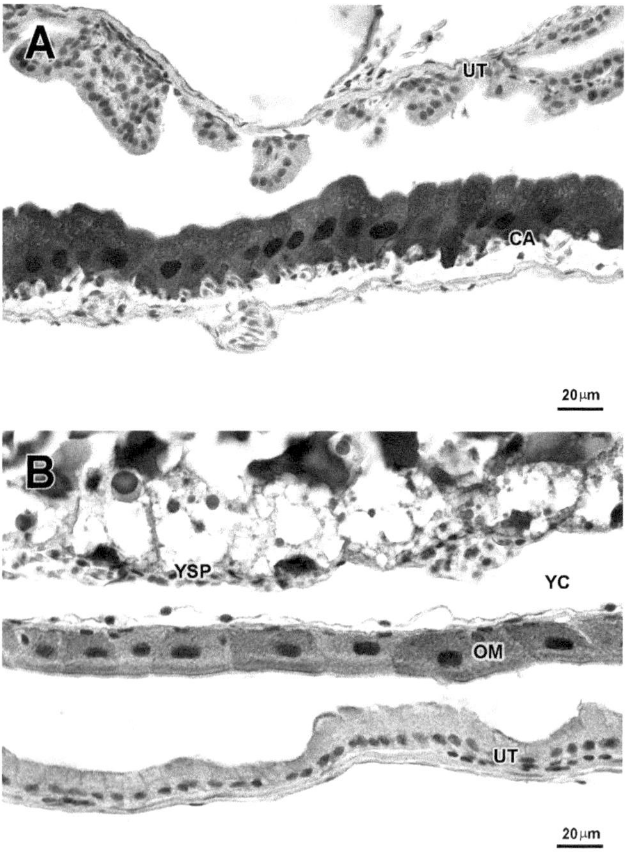

Fig. 13.24 Immunohistological localization (brown precipitate) of calbindin D_{28K} in the **(A)** chorioallantoic placentome and **(B)** omphaloplacenta of *Pseudemoia pagenstecheri*. Embryonic stage 34. CA, chorioallantoic membrane; OM, omphalopleure; UT, uterus; YC, yolk cleft; YSP, yolk splanchnopleure.

Color image of this figure appears in the color plate section at the end of the book.

to the chorioallantoic membrane and omphalopleure of *P. entrecasteauxii* and *P. spenceri* expresses plasma membrane calcium ATPase, a marker for calcium transport (Herbert *et al.* 2006, 2010).

Histochemical evidence indicates that uterine functional profiles vary during gestation and thus may be responsive to embryonic development. The expression of proteins associated with epithelial cell tight junctional complexes increases late in gestation in *P. entrecasteauxii* and *P. spenceri* at a time when alkaline phosphatase activity in uterine epithelial cells is also high (Biazik *et al.* 2007, 2008, 2010a). These results suggest an increased regulation of transport through a transcellular route during the interval when embryonic gains in mass are greatest.

Summary. Species of *Chalcides*, *Niveoscincus* and *Pseudemoia* share four characteristics that distinguish them from oviparous skinks: (1) absence of an eggshell during late embryonic stages, (2) presence of an inter-omphalopleuric membrane, (3) a persistent bilaminar omphalopleure with hypertrophied epithelium, and (4) a chorioallantoic membrane with regional differentiation. These structural modifications enhance the potential for physiological exchange, alter the topology of the extraembryonic membranes and increase the capacity for nutrient uptake.

The eggshell, a thin acellular structure secreted by the oviduct soon after ovulation, differs from oviparous species in thickness and in the absence of a calcareous outer layer. Disruption of the eggshell early in gestation allows intimate contact between maternal and embryonic epithelia.

The structure of the inter-omphalopleuric membrane and its relationship to the yolk sac and outer perimeter of the egg indicates that it functions as a conduit for blood vessels and likely serves as a vascular communication between omphalomesenteric and allantoic circulatory systems (ten Cate-Hoedemaker 1933; Blackburn 1993c; Stewart and Thompson 1994, 2004; Blackburn and Callard 1997). One of the consequences of this developmental pattern is that the inter-omphalopleuric membrane forms a barrier between the extraembryonic coelom of the embryonic hemisphere of the egg and the yolk cleft. Thus, the allantois cannot expand beyond the inter-omphalopleuric membrane. The yolk cleft, which is transitory in oviparous species, persists throughout gestation in these viviparous skinks. The yolk cleft of oviparous species has a germinal epithelium that populates the yolk cleft, and likely the yolk sac cavity, with cells. Elaborate modifications of the yolk cleft of oviparous species allow the germinal epithelium to continue production when the yolk cleft is disrupted by expansion of the allantois around the abembryonic pole of the egg. The three viviparous lineages lack specializations of the yolk cleft indicating different functional attributes of the yolk sac tissue.

The extended life span of the yolk cleft in viviparous species preserves the bilaminar omphalopleure, which is transformed into a specialized transport epithelium. Placentotrophic species of *Chalcides*, *Niveoscincus* and *Pseudemoia* have greatly enlarged epithelial cells in the omphalopleure. In addition, the bilaminar omphalopleure is folded extensively in all species of *Pseudemoia* and in three species of *Niveoscincus*. The primary mechanism for nutrient transport of the omphaloplacenta is histotrophy because the uterine epithelium also consists of large cells with cytological features characteristic of transporting cells.

Regional differentiation of the chorioallantoic membrane is universal among the viviparous species but structural specializations vary. The chorionic epithelium contains both thin squamous cells and large cuboidal or columnar cells in all species except *Chalcides ocellatus*. The pattern of distribution of cell types varies among species. In *Chalcides chalcides* and all three studied species of *Pseudemoia*, enlarged chorionic epithelial cells form an elliptical pattern in the chorioallantoic membrane over the embryonic pole of the egg. The chorioallantoic membrane is folded and has an extensive network of allantoic blood vessels. The uterine epithelium interdigitates with folds of the chorioallantoic membrane and consists of enlarged cells with a dense network of supportive blood vessels. This pattern of chorioallantoic placentation is termed a placentome. The remainder of the chorioallantoic placenta has a smooth interface between squamous epithelial cells. The chorionic epithelium of *Niveoscincus* is also a composite of thin, squamous cells and enlarged cells. However, chorionic epithelial cell hyperplasia occurs either in a zone adjacent to the inter-omphalopleuric membrane or distributed as a mosaic over the surface of the chorioallantoic membrane. The uterine epithelium of species of *Niveoscincus* is squamous and forms ridges overlying blood vessels in some regions. The ridges are most commonly apposed to the regions of enlarged chorionic cells. In comparison to oviparous species, the chorioallantoic membranes of all three genera of viviparous species are structurally diverse with multiple cell types that likely support an equally diverse set of functions.

13.3.5.3 Placentotrophic species with extremely small eggs

New World *Mabuya*. *Mabuya* is a Neotropical scincid genus of >25 species that are distributed throughout South and Central America, southern Mexico, and various Caribbean islands. Although the taxon traditionally has included many species of Africa and Asia, upon its partition the generic name was restricted to species of the Americas (Mausfeld *et al.* 2002). A recent monograph has proposed that the New World *Mabuya* be reclassified into their own family as 61 species in 16 genera (Hedges and Conn 2012), a controversial proposal that has elicited resistance (Vences *et al.* 2013). Here, we follow the taxonomy of Mausfeld *et al.* (2002) as amended by Bauer (2003).

The unusual nature of reproduction in this genus came to light from studies of the Brazilian species *Mabuya heathi* (Vitt and Blackburn 1983; Blackburn *et al.* 1984). Females of this species ovulate minuscule 1 mm ova yet give birth to offspring of ~31 mm SVL, with placental transport accounting for more than 99% of the fetal dry mass. Gestation lasts 8–12 months, and is timed such that offspring are born at the outset of the wet season. Through the precocial maturation of gonads, females first become pregnant at only 3–4 months of age while still at juvenile body size. During early gestation, females grow rapidly, reaching a size large enough for their fetuses just before the latter undergo a dramatic increase in size. Studies of other New World *Mabuya* indicate that this unusual suite of reproductive specializations is widely shared (and probably universal) throughout the genus (Vitt and Blackburn 1991; Blackburn and Vitt 1992; Ramírez-Pinilla *et al.* 2002; Vrcibradic and Rocha 2011).

Reproductive characteristics related to placentation are best known for three species, *Mabuya bistriata* and *M. heathi* from Brazil (Blackburn *et al.* 1984; Blackburn and Vitt 1992, 2002; Vitt and Blackburn 1991), and populations of an undescribed species, *Mabuya* sp. from the Andean region of Columbia (Ramírez-Pinilla 2006; Ramírez-Pinilla *et al.* 2002, 2006, 2011a; Jerez and Ramírez-Pinilla 2001, 2003; Vieira *et al.* 2007). These species ovulate small eggs (1–2 mm in diameter) (Blackburn *et al.* 1984; Vitt and Blackburn 1991; Ramírez-Pinilla *et al.* 2002). The recently ovulated egg is enclosed in a thin eggshell that is disrupted early in development (Blackburn *et al.* 1984; Jerez and Ramírez-Pinilla 2003) and placental nutrient transport provides nearly all of the nourishment for embryonic development (Table 13.6). The ontogeny of nutrient uptake by embryos has been studied in detail for *Mabuya* sp. (Ramírez-Pinilla 2006; Ramírez-Pinilla *et al.* 2011a). Although recently ovulated eggs lack yolk platelets (Gomez *et al.* 2004; Vieira *et al.* 2010; also see Hernández-Franyutti *et al.* 2005), they do contain small quantities of organic (lipid, protein) and inorganic molecules. Nutrients are transported to embryos throughout development with the greatest gains in embryonic mass occurring during the latest embryonic stages.

Development of the extraembryonic membranes of *Mabuya* sp. from the Andean region of Columbia (Jerez and Ramírez-Pinilla 2003) differs from other viviparous squamates in pattern of growth of extraembryonic mesoderm and timing of formation of the vascular yolk sac splanchnopleure relative to that of the chorioallantoic membrane. A bilaminar omphalopleure, the primary yolk sac, surrounds the entire yolk sac cavity and is converted to a vascular choriovitelline membrane in the embryonic hemisphere of the egg as it does in all reptiles. Development of the yolk sac of the abembryonic hemisphere of the egg is similar to non-squamate reptiles but differs from other squamates, because mesoderm continues to grow within the bilaminar omphalopleure and does not invade the yolk as intravitelline mesoderm. Thus, the choriovitelline membrane surrounds the entire yolk sac cavity and a yolk cleft does not form. This developmental pattern is most likely derived from ancestors with a yolk cleft (see Section 13.3.2) and the similarity with turtles and archosaurs is independently derived, i.e., an homoplasy.

Expansion of the extraembryonic coelom within the choriovitelline membrane and the associated extension of the chorion around the entire egg also are unusual for squamates. In most squamates, oviparous and viviparous, the extraembryonic coelom initially appears adjacent to the embryo, as it does in *Mabuya* sp. but extends around the perimeter of the egg only as the allantois grows outward from the embryo. As the chorion forms, it is quickly fused to the allantois to form the chorioallantoic membrane and large expanses of independent chorion are never present. In *Mabuya* sp., the extraembryonic coelom expands independently of the growth of the allantois and the chorion surrounds the entire egg. Whether these developmental differences are related to placental specializations or to the small size of the eggs of this species is not clear.

One of the prominent features of placental development in this species is that specializations for nutrient transport are apparent early in development (Jerez and Ramírez-Pinilla 2003). Structure of the bilaminar omphalopleure is similar to other squamates; enlarged endodermal cells front the yolk sac cavity and squamous ectodermal cells face the uterine lumen. In contrast to other viviparous lizards, the uterine epithelium surrounding most of the surface of the bilaminar omphalopleure is cuboidal, there are active glands in the uterine mucosa and the uterus is modified for nutrient transport near the embryonic pole of the egg. The latter structure, underlying the mesometrium, is a region of folded, columnar uterine epithelium with extensive vascular support. As the chorion develops above the embryo, the chorionic contribution to the chorioplacenta in this region is a "dorsal absorptive plaque" consisting of columnar chorionic epithelial cells with two structural profiles. During this phase, the chorioplacenta occupies a limited area above the embryo, a non-vascular trilaminar yolk sac placenta is present between the chorioplacenta and the equatorial region of the egg, and the bilaminar yolk placenta persists over the abembryonic region of the egg. Additional specializations for placentation include "absorptive plaques" containing columnar epithelial cells in the non-vascular trilaminar yolk sac placenta.

Vascularization of the trilaminar omphalopleure and the subsequent disruption of this membrane, the choriovitelline membrane, to form yolk sac splanchnopleure (endoderm, mesoderm) and chorion (mesoderm, ectoderm) occur quickly in this species. By the late neurula embryonic stage, a highly vascular yolk sac splanchnopleure completely surrounds the yolk sac cavity and the entire placental interface is chorioplacenta (Jerez and Ramírez-Pinilla 2003). It is not unusual for the choriovitelline placenta to be transitory in viviparous lizards but in most species it is supplanted by the vascular chorioallantoic placenta and the chorionic epithelium retains vascular support during the transition. The prominence of the chorioplacenta in *Mabuya* sp. is curious because of the absence of close vascular support for the embryonic tissue. However, the epithelium of the chorioplacenta is apparently specialized for uptake of nutrients because the dorsal absorptive plaque is more prominent than in earlier stages and additional absorptive plaques stud the surface of the chorion over the remainder of the placenta. Nutrients moving into the egg across the chorioplacenta would enter the extraembryonic coelom and could then be taken up by blood vessels of the yolk sac splanchnopleure.

When the allantois expands from the embryonic hindgut, the outer allantoic membrane initially contacts and fuses with the chorionic epithelium of the dorsal absorptive plaque providing close vascular support for this tissue (Jerez and Ramírez-Pinilla 2003). Likewise, as the allantois expands to fill the extraembryonic coelom, the outer allantoic membrane incorporates the additional chorionic absorptive plaques into specialized regions of chorioallantoic placentation. The chorioallantoic placenta ultimately surrounds the entire egg contents.

During late embryonic stages, the chorioallantoic placentas of *Mabuya* sp. (Jerez and Ramírez-Pinilla 2001; Ramírez-Pinilla *et al.* 2006) and *Mabuya heathi* (Blackburn and Vitt 2002) are regionally diversified and contain specialized sites for histotrophic nutrient exchange. A prominent placentome is associated with the mesometrium (Fig. 13.25A) and chorionic areolae (Fig. 13.25B) are scattered over the surface of the chorioallantoic membrane in the embryonic hemisphere of the egg. Two additional structurally distinct regions occur in the abembryonic hemisphere of the egg of Mabuya sp., absorptive plaques and respiratory segments (Jerez and Ramírez-Pinilla 2001; Ramírez-Pinilla *et al.* 2006).

The placentome is a region of deep invaginations in the chorioallantoic membrane that interdigitate with folds in the uterine epithelium (Fig. 13.25A). The chorionic epithelium of the placentome consists of two cell types, giant cells and tall, thin interstitial cells (Fig. 13.25C) (Jerez and Ramírez-Pinilla 2001; Blackburn and Vitt 2002; Ramírez-Pinilla *et al.* 2006). Each cell type is binucleate and has microvilli on the apical surface. The uterine epithelium of the placentome is syncytial (Jerez and Ramírez-Pinilla 2001; Blackburn and Vitt 2002; Ramírez-Pinilla *et al.* 2006) with apical microvilli. Microvilli of chorionic and uterine cells interdigitate (Ramírez-

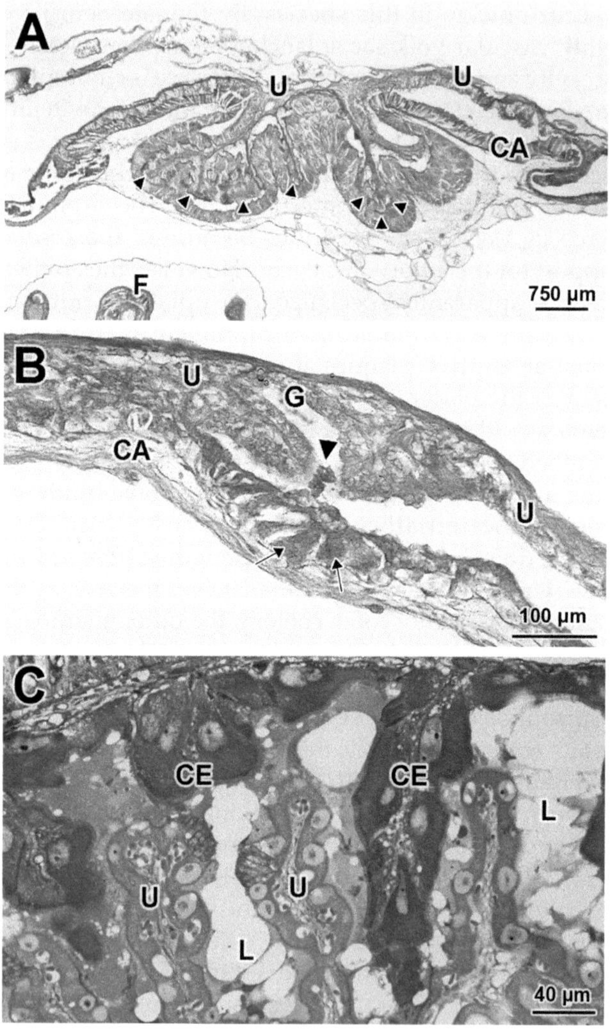

Fig. 13.25 Placentation in the South American skink *Mabuya heathi*, in late gestation. **(A)** Placentome, formed from interdigitation of tissues of the uterus (U) and chorioallantois (CA). Arrowheads mark the uterine- fetal interface. F, fetus. **(B)** Chorionic areola, showing a uterine gland (G) apposed to absorptive cells of the chorionic epithelium (arrows). Arrowhead, secretory material. **(C)** Placentome cytology. Epithelium of the uterus (U) is syncytial and overlies a rich vasculature. Chorionic epithelium (CE) consists of enlarged binucleated cells interspersed with thin interstitial cells; both bear microvilli; L, uterine lumen.

Color image of this figure appears in the color plate section at the end of the book.

Pinilla *et al.* 2006). The uterine folds include both the epithelium and the lamina propria; secretory glands are present at the base of the folds and blood vessels extend into the folds.

The placentome of Andean populations of *Mabuya* sp. also contains cells that originate in the chorion and disperse in clusters into the uterine epithelial syncytium and associate with maternal capillaries (Ramírez-Pinilla *et al.* 2006; Vieira *et al.* 2007). These cells are structurally and functionally distinct from the chorionic epithelial cells implicated in nutrient transport. Cytoplasmic organelles present in the invasive cells suggest that they are specialized for nutrient metabolism and possibly secretion (Vieira *et al.* 2007). Although the function of these cells is unknown, they illustrate that the relationship between maternal and fetal tissues in the complex placentation of this species is highly intimate, a placental feature often considered to be restricted to eutherian mammals.

In *Mabuya* sp., there is a zone extending circumferentially from the placentome, which contains some features characteristic of the placentome but lacks folds. This region, called the paraplacentome, has binucleated giant and interstitial chorionic epithelial cells and low columnar uterine epithelial cells (Jerez and Ramírez-Pinilla 2001; Ramírez-Pinilla *et al.* 2006). In contrast to the placentome, both uterine and chorionic epithelia lack microvilli (Ramírez-Pinilla *et al.* 2006). There are abundant uterine glands in this region. Chorionic epithelial cells are rich in stored lipid droplets suggesting that, among other functions, this region is an important site of lipid transport (Ramírez-Pinilla *et al.* 2006).

Chorionic areolae are invaginations lined by columnar cells with apical microvilli (Jerez and Ramírez-Pinilla 2001; Blackburn and Vitt 2002; Ramírez-Pinilla *et al.* 2006; Blackburn and Flemming 2009). The areolae lie opposite multicellular uterine glands lined by columnar epithelial cells. In addition to secretions from uterine glands, apocrine secretions of luminal epithelial cells of *Mabuya* sp. are released into the uterine lumen (Ramírez-Pinilla *et al.* 2006).

Absorptive plaques are present in the chorioallantoic membrane of the abembryonic hemisphere of the egg of Andean populations of *Mabuya* sp. (Jerez and Ramírez-Pinilla 2001; Ramírez-Pinilla *et al.* 2006). The chorionic epithelium of these structures is similar to that of the paraplacentomal region and uterine epithelial cells are columnar. Uterine and chorionic epithelial cells are adherent (Ramírez-Pinilla *et al.* 2006).

Regions of thin squamous fetal and maternal epithelia are interspersed between the plaques in the abembryonic hemisphere of the egg (Jerez and Ramírez-Pinilla 2001; Blackburn and Vitt 2002; Ramírez-Pinilla *et al.* 2006). In these areas, thin uterine and embryonic epithelia position fetal and maternal vascular systems in close proximity (Ramírez-Pinilla *et al.* 2006).

Species of *Mabuya* are highly placentotrophic with the greatest embryonic increases in organic and inorganic mass during late developmental stages when regional differentiation of the chorioallantoic placenta is distinct (Blackburn and Vitt 1992; Ramírez-Pinilla *et al.* 2011a). Ultrastructural evidence suggests that the placentome, paraplacentome, chorionic areolae and absorptive plaques are specialized sites for histotrophic nutrient transfer and that among other functions, all transport lipids (Ramírez-Pinilla *et al.* 2006). Immunolocalization of antibodies to proteins implicated in transport of water, glucose and calcium supports the inference from structural studies that the placentome, paraplacentome and absorptive plaques have significant roles in nutrient uptake by embryos (Wooding *et al.* 2010).

Eumecia. The Western Serpentiform Skink *Eumecia anchietae* is a large, semi-aquatic lizard that inhabits swampy and floodplain areas of central and western Africa. Its distribution extends from the upper Zambezi region into western Tanzania and southern Kenya, and its very rare congener *E. johnstoni* is known from Malawi (Flemming and Branch 2001). Undeveloped eggs of *E. anchietae* measure no more than 1 mm in diameter. Near term fetuses measure ~160 mm in total length and ~204 mg in dry mass, figures that reflect an extreme degree of placentotrophy.

Details of development and placentation are provided by Flemming and Branch (2001; for a summary see Flemming and Blackburn 2003). The yolk of undeveloped and early eggs contains acidophilic material and yellow granules but apparently no platelets. Although a thin shell membrane is deposited around the egg, it disappears early in development. Early development of the yolk sac has not been described and whether an isolated yolk mass forms is not certain. If an isolated yolk mass does develop, it is no longer present by the early limb bud stage (Dufaure and Hubert ["D&H"] stages 25–28), at which point a choriovitelline membrane covers the abembryonic surface of the yolk sac cavity.

In early pregnancy, the incubation chamber expands substantially, such that fetal membranes are not forced into direct contact with the uterine lining. During the early limb bud stages (stages 25–28), the uterine epithelium releases massive amounts of secretory material that accumulates in the space between the uterine lining and the fetal membranes (Fig. 13.26A). This material emanates from a highly-specialized columnar epithelium as well as small, rounded cells that degenerate and spill their contents into the uterine lumen. The fetal membranes are correspondingly specialized for absorption. A chorioallantoic membrane covers the embryonic hemisphere of the egg and extends into the abembryonic hemisphere where it contacts a choriovitelline membrane. The choriovitelline membrane forms invaginations that are filled with the uterine secretions, and lined externally by hypertrophied absorptive cells that bear microvilli (Fig. 13.26B). The chorioallantoic membrane also is lined by enlarged, absorptive cells with microvilli (Fig. 13.26C). Thus, nutrition in early pregnancy involves histotrophic secretion by the uterus and absorption by both yolk sac and chorioallantoic membranes.

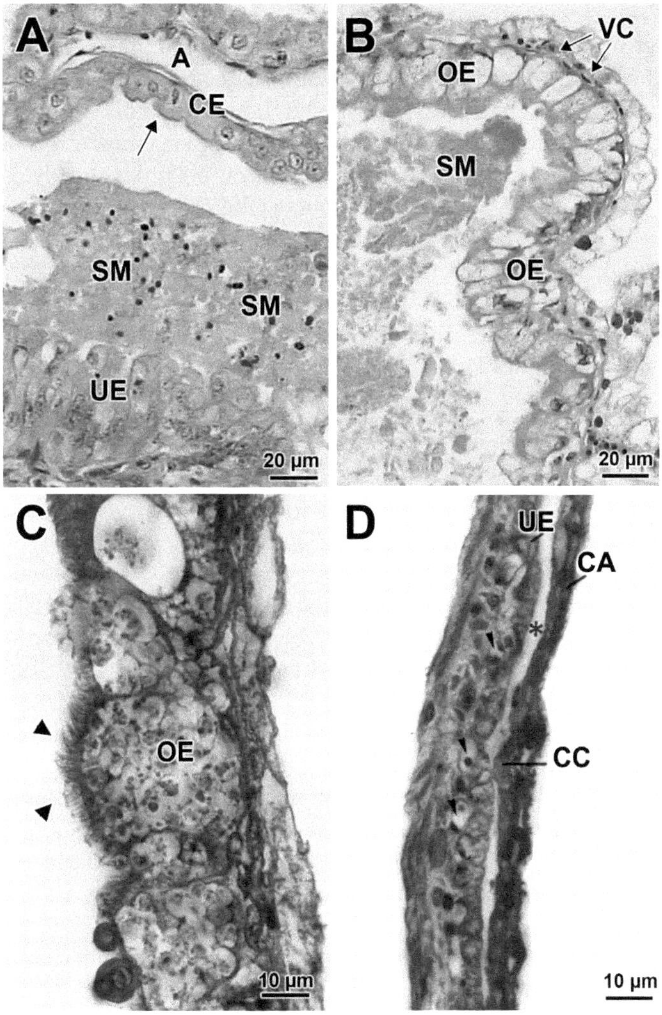

Fig. 13.26 Placentation in *Eumecia anchietae*. **(A)** Chorioallantoic placenta, stage 28. The uterine epithelium (UE) produces abundant secretory material (SM) that is absorbed by the chorionic epithelium (CE); note the chorionic microvilli (arrows). A = allantois. **(B)** Yolk sac, stage 28. The omphalopleure shows invaginations filled with uterine secretory material. OE, omphalopleure epithelium; VC = vitelline capillaries. **(C)** Cells of the omphalopleure epithelium showing their prominent microvilli. Secretory material is absorbed by these cells. **(D)** Chorioallantoic placenta, stage 33. The space (asterisk) between the chorionic cells (CC) of the chorionic epithelium (CE) and uterine epithelium is artifactual. Vascularity of the uterine tissue is indicated by the arrowheads. CA, chorioallantoic membrane. Original figures are provided through the courtesy of Alexander F. Flemming.

Later in development (stage 33/34), the uterine epithelium loses its secretory properties, regressing to a monolayer of cuboidal to squamous cells that overlie a highly vascular stroma. The chorioallantois expands in extent and becomes more vascularized, whereas the yolk sac loses contact with the uterine lining. Thus, the chorioallantoic placenta entirely supplants the choriovitelline placenta. The chorioallantois becomes closely associated with the uterine lining, and epithelial cells of the chorion and uterus decrease in height, reducing the distance between maternal and fetal capillaries (Fig. 13.26D). This arrangement may signify a switch from histotrophic to hemotrophic (blood-blood) nutrient transfer. Later stages of development have not been examined, and whether further specializations for nutrient transfer develop is not known.

Trachylepis. *Trachyleps ivensii* is a semi-aquatic, lygosomine skink from Central Africa (Branch and Haagner 1993; Haagner *et al.* 2000). It belongs to a genus of >70 species that was established for African and Madagascan species formerly referred to *Mabuya* (Bauer 2003). *Trachylepis* contains both oviparous and viviparous species (FitzSimons 1943; Simbotwe 1980; Branch 1998; Flemming 1994; Goldberg 2007), and viviparity has evolved multiple times in the genus (Blackburn 1982, 1999a; Shine 1985). Previously studied viviparous species appear to be relatively lecithotrophic, based on size of the vitellogenic oocytes and ovulated eggs (Robertson *et al.* 1965; Farag 1983; Flemming 1994).

Placental specializations in *Trachyleps ivensii* are among the most extreme to be discovered in viviparous reptiles. They include features long assumed to be confined to eutherian mammals, notably invasive implantation and a placenta in which maternal blood vessels and fetal tissues lie in direct contact (Blackburn and Flemming 2012). Development and placentation in this species is documented in two recent papers (Blackburn and Flemming 2010, 2012).

Female *Trachylepis ivensii* ovulate yolks of about 1 mm in diameter (Fig. 13.27A). The early eggs contain minuscule granules but lack yolk platelets and droplets. Cleavage is meroblastic and yields a small embryonic disk at one pole of the egg. A thin eggshell membrane (of 2.5 μm in thickness) is deposited around the newly fertilized egg; however, it has entirely disappeared (having been shed) by the early neurula stage. The yolk is enveloped precociously by the vascular choriovitelline membrane (formed of extraembryonic ectoderm, mesoderm, and endoderm). Thus, no isolated yolk mass complex develops. The exocoelom forms in mesoderm adjacent to the embryo, and as it expands, it splits the choriovitelline membrane into an outer chorion and an inner yolk sac splanchnopleure. As a result of this process, the conceptus is nearly surrounded by chorion by the late neurula stage. In addition, blood vessels of the yolk sac splanchnopleure, with

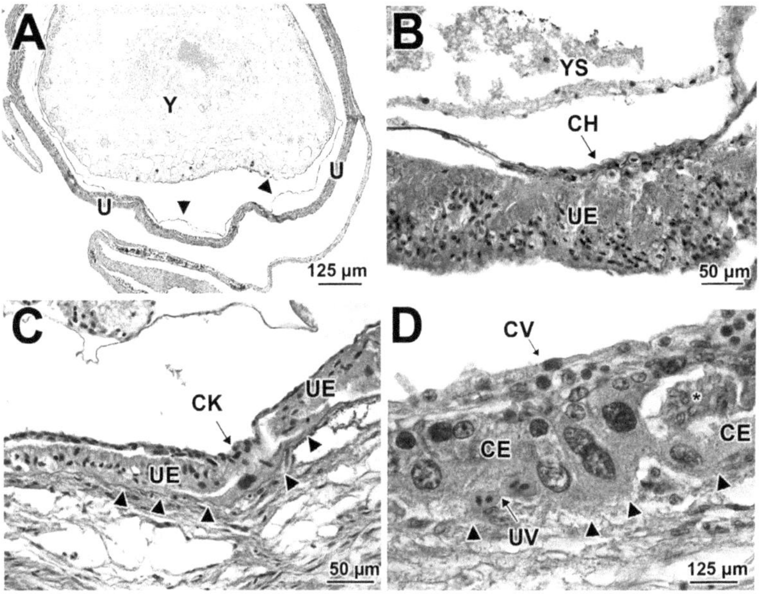

Fig.13.27 Placentation in *Trachylepis ivensii.* **(A)** Early pregnancy, showing a cleavage stage egg in the uterus (U). The shell membrane is indicated by arrowheads. Y = yolk. **(B)** Chorionic placenta in the neurula stage. The chorion (CH) lies apposed to uterine epithelium (UE); the latter contains abundant secretory material, evident as acidophilic and basophilic granules. YS = yolk sac. **(C)** Implantation in the late neurula stage. At the site of a chorionic knob (CK), fetal tissue has penetrated the uterine lining, giving rise to issue that is undercutting the uterine epithelium (UE) (arrowheads). **(D)** Placental interface. Enlarged cells of the chorionic ectoderm (CE) directly contact the uterine vessels (UV) and uterine stroma (arrowheads). The asterisk marks a piece of stripped uterine epithelium. CV, choriovitelline membrane. Figures are from research by D. Blackburn with Alexander F. Flemming.

Color image of this figure appears in the color plate section at the end of the book.

vascular connections to the embryo, surround the entire yolk sac cavity. In the pharyngula stage, the yolk sac splanchnopleure expands, obliterating the exocoelom and thereby coming in contact with the chorion. Through this process, the yolk sac vessels vascularize the chorion. Thus, the conceptus becomes largely surrounded by vascularized fetal membranes. The process is analogous to the way that the allantois of typical amniotes expands to vascularize the chorion. However, in *T. ivensii* at this stage, the allantois has just begun to protrude from the embryonic hind gut; thus placentation is entirely choriovitelline and chorionic in nature.

During very early pregnancy, cells of the uterine epithelium grow from cuboidal to columnar, and release secretory material that is taken up by the fetal membranes. The secretions are released into the uterine lumen in the form of acidophilic material as well as basophilic granules with a carbohydrate moiety. In addition, the uterine lamina propria swells with fluid, bringing the uterine epithelium into contact with the fetal membranes. Concomitantly, the incubation chambers themselves expand dramatically in size.

In the neurula stage, contact of the chorion and choriovitelline membranes with the uterine lining form placental arrangements (Fig. 13.27B). Throughout the neurula and pharyngula stages, ectoderm of these fetal membranes progressively establishes an intimate relationship with the apposed uterine tissues, in a complex process of implantation. The external chorionic cells initially attach to the uterine epithelium by sending cellular extensions between adjacent epithelial cells that eventually penetrate to the level of the uterine basement membrane. The penetrating tissue then develops into syncytial (multinucleated) "chorionic knobs", which constitute sites of tissue proliferation (Fig. 13.27C). From these knobs, the proliferating chorionic tissue extends beneath the uterine epithelium, stripping it away from the basement membrane. Through this implantation process, fetal ectoderm has replaced the uterine epithelium throughout the incubation chamber by the late pharyngula stage (Fig. 13.27D). Degeneration of the uterine epithelium then brings the chorionic ectoderm back into contact with the inner layers of the chorion and choriovitelline membrane. A major consequence of the implantation process is that capillaries of the uterine stroma lie in direct contact with fetal extraembryonic tissue (Fig. 13.27D). Under the system of classification of mammal placentas (Mossman 1987), this tissue arrangement can be characterized as "endotheliochorial", a term that refers to the contact between maternal endothelial cells and fetal chorion. While *Mabuya* sp. also shows contact between fetal cells and uterine blood vessels (Vieira *et al.* 2007), the arrangement is different in the two groups. Development of the placental membranes in later embryonic stages of *T. ivensii* has not been described.

Summary. *Eumecia anchietae* and *Trachylepis ivensii* share reproductive and developmental characteristics with South American species of *Mabuya* that are notable because they have not been described in other species. The small eggs (1–2 mm diameter) lack yolk platelets, yet are surrounded by a thin eggshell, which is disrupted early in intrauterine development. Yolk sac development differs from other squamates, because a choriovitelline membrane grows around the entire yolk sac cavity. Thus, in contrast to other squamates, neither a yolk cleft, nor an isolated yolk mass is formed.

Although formation of a choriovitelline membrane occurs commonly in both oviparous and viviparous squamates, the development of a hypertrophied epithelium, an apparent specialization for histotrophy, as in *Eumecia anchietae* is unique. Following development of the choriovitelline membrane, the pattern of yolk sac formation in *E. anchietae* is similar to other reptiles. Growth of the allantois around the perimeter of the egg is coincident with disruption of the choriovitelline membrane and vascular support for the chorionic epithelium shifts from omphalomesenteric blood vessels of the yolk sac to allantoic blood vessels of the chorioallantoic membrane. The chorioallantoic membrane expands to completely encircle the egg.

In contrast, disruption of the choriovitelline membrane in *Trachylepis ivensii* and species of *Mabuya* is unique for reptiles. The choriovitelline membrane is disrupted independently of contact with the allantois and vascular support is transferred from the perimeter of the egg to the newly formed yolk sac splanchnopleure. The chorion, which lacks direct vascular support, covers the outer perimeter of the egg. There is no developmental stage in other investigated reptiles in which the chorion surrounds the entire egg.

Following its establishment around the egg, the chorion of *Trachylepis ivensii* and species of *Mabuya* each receives vascular support but through different developmental pathways. In *Mabuya* sp., the allantois expands into the exocoelom between the yolk sac splanchnopleure and chorion and the outer allantoic membrane fuses with the chorion, forming the chorioallantoic membrane. Vascular support for the chorion of *T. ivensi* is curious and unique. The yolk sac splanchnopleure expands to contact the chorion, which is vascularized by omphalomesenteric blood vessels in the yolk sac. This establishes a structure with a similar composition to the original choriovitelline membrane (ectoderm, vascularized mesoderm, endoderm) but with a developmental pattern unique to amniotes.

An additional feature of placental development of *Trachylepis ivensii* and populations of *Mabuya* sp. from the Andean region of Columbia is the presence of cells originating from the chorion that invade the uterine epithelium. The tissue of origin and pattern of growth of these cells differs between the two lineages but in each lineage they arise from chorionic epithelium. They originate from the chorioallantoic placentome in *Mabuya* sp. and from the chorion and choriovitelline membrane of *T. ivensii*. The cells invade the uterine epithelium and establish contact with uterine blood vessels, an endotheliochorial relationship (Ramírez-Pinilla *et al.* 2006; Vieira *et al.* 2007; Blackburn and Flemming 2012).

Nutritional provision to all of these species is almost entirely through placental transport and during early stages of embryonic development the uterus of each is secretory and releases nutritive material into the uterine lumen. This nutritional pattern does not occur in other viviparous squamates, in which nearly all placental transport of nutrients occurs during late developmental stages. This unusual pattern is undoubtedly related to the small size of the eggs. During early stages of embryonic development, embryos do require a nutrient source but the increase in embryonic mass is slight. In contrast, the extraembryonic membranes are undergoing substantial growth and likely are dependent on placental nutrient transport.

Unfortunately, specimens of *Eumecia anchietae* and *Trachylepis ivensii* are difficult to obtain and late embryonic stages have not been described. During late stages of development of species of *Mabuya*, the chorioallantoic placenta is the sole interface between mothers and embryos and it has an impressive suite of specializations for placental exchange unique to squamates.

13.3.5.4 Lecithotrophic viviparous species

The category of lecithotrophic viviparous lizards includes species with a low matrotrophic index as well as species with large eggs but no further information on pattern of embryonic nutrition. Details of placental development and structure vary because sampling was not uniform among studies. However, a broad phylogenetic distribution is represented and these species contribute to understanding lizard placental diversity. Comparisons among these species are a foundation for hypotheses on the evolution of placental structure but the absence of data for oviparous outgroups for some lineages leaves transformation details in doubt. Our discussion will emphasize structural characteristics: 1) eggshell, 2) yolk sac placentation, 3) growth and topology of the allantois, and 4) chorioallantoic placentation. We will follow with information on functional attributes.

A thin eggshell is secreted by the uterus in all of the species. The eggshell is present to the latest stage sampled in the Northern Alligator Lizard, *Elgaria coerulea* (Stewart 1985) but is disrupted in the embryonic hemisphere early in development and accumulates in the abembryonic region in the Desert Night Lizard, *Xantusia vigilis* (Yaron 1977), *Eulamprus* sp. (Murphy *et al.* 2012), the Cleft Lizard, *Sceloporus mucronatus* (Villagran *et al.* 2005), Yarrow's Spiny Lizard, *S. jarovii* (Blackburn *et al.* 2010; Anderson *et al.* 2011) and *Woodworthia maculatus* (Boyd 1942; Girling *et al.* 1997).

A yolk cleft–isolated yolk mass complex develops in a similar pattern in all of the species as described in Section 13.3.2. The choriovitelline placenta, which is presumably a transitory stage in yolk sac placentation for each of these species, develops in the embryonic hemisphere of the egg. The distribution of this membrane is limited by the yolk cleft–isolated yolk mass complex. The omphalopleure of the isolated yolk mass, which lacks direct vascular support, contributes the embryonic tissue to the omphaloplacenta; the primary yolk sac placenta of the abembryonic hemisphere of the egg in each of these species. Subsequent development of placentation in the abembryonic hemisphere is determined by growth of the allantois.

In all of the species, the allantois grows into the extraembryonic coelom resulting from splitting of the sheet of mesoderm within the choriovitelline membrane. In this manner, the chorioallantoic membrane replaces the choriovitelline membrane in the embryonic hemisphere of the egg. Growth of the allantois in the abembryonic hemisphere of the egg varies among species and thus topology of the extraembryonic membranes in late embryonic stages differs. Four developmental patterns occur. The allantois remains at the margin of the yolk cleft and the omphaloplacenta persists throughout gestation in *Elgaria coerulea* (Stewart 1985). In the gekkonid lizard *Woodworthia maculatus*, the allantois advances under the yolk sac as the isolated yolk mass regresses (Boyd 1942). The outer allantoic membrane disrupts the omphalopleure and fuses with the outer epithelial layer to form a chorioallantoic membrane. The allantois also extends under the yolk sac as the isolated yolk mass regresses in *Eulamprus* sp. but does so by growing into the yolk cleft and fusing with intravitelline mesoderm (Weekes 1927b; Murphy *et al.* 2012). Yolk endoderm is still present in the remnant of the isolated yolk mass as the allantois advances and a transitory omphalallantoic membrane is formed in at least some species (Weekes 1927a). The isolated yolk mass and endoderm are depleted and the final structure is a chorioallantoic membrane. In *Sceloporus mucronatus* (Villagran *et al.* 2005) and *S. jarovii* (Blackburn *et al.* 2010; Anderson *et al.* 2011), the allantois grows into the yolk cleft but the omphalopleure is present throughout gestation and a chorioallantoic placenta does not form across the abembryonic hemisphere of the egg. Allantoic blood vessels provide the nearest vascular support to the omphalopleure. In *S. mucronatus*, the allantois enters the yolk cleft when the isolated yolk mass is relatively thick, prior to stage 37. Late embryonic stages of *Xantusia vigilis* have not been described but earlier embryos have a chorioallantoic placenta and an omphaloplacenta.

Patterns of embryonic nutrition have been assessed for, the Eastern Water Skink, *Eulamprus quoyii* (Thompson 1977a,b, 1981, 1982), the Highland Water Skink, *Eulamprus tympanum* (Thompson *et al.* 2001d) and *Elgaria coerulea* (Stewart and Castillo 1984). Matrotrophic indices for the three species are similar (Table 13.6) and indicate that yolk is the primary source of organic nutrients. However, the placenta is a major source of inorganic nutrients for each species and a substantial amount of water is supplied to embryos across the placenta. The placenta is the primary source of calcium for *E. quoyii* and *E. tympanum*.

Functional attributes of specific placental regions have not been studied but details of placental structure provide a solid platform for hypotheses. The chorioallantoic placentas of all of these species are similar; thin uterine and chorionic epithelia supported by an abundance of blood vessels. For all except *Elgaria coerulea*, the shell membrane is absent in late developmental stages and uterine and chorionic epithelial cells are in direct apposition. With the exception of *Eulamprus quoyii*, the chorioallantoic placentas have a uniform morphology throughout. The chorionic epithelium of *E. quoyii* is similar to the other species but the uterine epithelium is regionally diversified. Uterine epithelial cells are columnar in a zone associated with the mesometrium and the epithelium is folded in late developmental stages.

Ultrastructure of the uterine epithelium of *Eulamprus tympanum* (Adams *et al.* 2007) and *Sceloporus jarovii* (Blackburn *et al.* 2010) reveals thinning of cytoplasm over capillaries and apical microvilli, both features that would enhance diffusion. Chorionic epithelial cells of *S. jarovii* are also thin where they overlie allantoic blood vessels (Blackburn *et al.* 2010). The chorioallantoic placenta covers most of the perimeter of the egg and the combination of close apposition of the maternal and fetal capillaries and large surface area indicate a high potential for diffusion from one vascular system to the other. Thus, respiratory gases and small, diffusible molecules are candidates for movement across the chorioallantoic placentas. In addition, cytological features of uterine and chorionic epithelial cells of *S. jarovii* indicate the potential for maternal nutrient secretion and fetal uptake (Blackburn *et al.* 2010; Anderson *et al.* 2011).

The morphology of the omphaloplacenta suggests a different functional profile. The uterine epithelium of the omphaloplacenta contains enlarged cells and the outer epithelium of the omphalopleure is cuboidal or columnar in *Woodworthia maculatus, Elgaria coerulea, Sceloporus mucronatus, S. jarovii, Xantusia vigilis*, and some species of *Eulamprus*. The shell membrane or its remnants are present between the two epithelia and the uterine lumen often contains additional material in the region of the abembryonic pole of the egg. The maternal and embryonic epithelia differ in vascular support. Uterine blood vessels lie at the base of the epithelium in the lamina propria but the nearest vascular supply to the omphalopleure is either yolk sac splanchnopleure on the inside of the yolk cleft or the outer allantoic membrane within the yolk cleft. The uterine epithelium is folded in *S. mucronatus*. The enlarged epithelial cells indicate that secretion and uptake of molecules occurs by histotrophy and that substances must traverse multiple cell layers to reach the embryonic circulatory system. The functional capability of the yolk sac placenta may be reduced late in gestation. The size of epithelial cells in both the uterus and omphalopleure are reduced in late developmental stages in *W. maculatus*.

The most thorough analysis of morphology of the yolk sac placenta among lecithotrophic lizards is a study of *Sceloporus jarovii* (Blackburn *et al.* 2010; Anderson *et al.* 2011). The yolk sac placenta is differentiated into two regions, an area where uterine and embryonic epithelia are closely apposed and an area where they are separated by material within the uterine lumen. Cytological characteristics of the uterine epithelium and omphalopleure of the area of apposition reveal specializations for placental transport. Cells of both epithelia are enlarged, cuboidal or columnar, and uterine epithelial cells contain two types of inclusions. The surface area of embryonic epithelial cells is increased substantially by the presence of channels surrounding the apices of the cells. These channels are continuous with the uterine lumen. In contrast, structure of uterine epithelial cells in the area with debris in the uterine lumen indicates the capacity for phagocytosis of this material, which includes cellular material from the degeneration of the omphalopleure.

Summary. The maternal–fetal interface of each of these viviparous lecithotrophic species is complex in that there are regional differences, which are most distinctive between chorioallantoic placentas and yolk sac placentas. Structural differences between the chorioallantoic placenta and yolk sac placenta within species are greater than differences between species within a given placental type. This pattern of variation indicates regional differences in placental function within individuals and functional similarity of placental type, chorioallantoic or yolk sac, across lineages. Whether these similarities represent conservation of ancestral traits or parallel specializations for placental exchange cannot be assessed because we lack developmental data for appropriate oviparous species. In addition, there is little information on functional aspects of oviparous and viviparous lizard development. One recognizable derived characteristic that these species share is reduction in thickness of the eggshell and, for all but one species, disruption of the eggshell relatively early in development. Oviparous species in each of these lineages have thick eggshells that are present throughout incubation.

13.4 PATTERNS OF PLACENTAL EVOLUTION

Hazel Claire Weekes (1935) summarized ten years of research on placental morphology of Australian skinks in the form of the first model for the evolution of squamate placentation. Weekes' material included species within at least six genera, including *Eulamprus*, *Niveoscincus* and *Pseudemoia* and she was aware of the work of Giacomini (1891, 1906) on *Chalcides chalcides* and *C. ocellatus*. Two primary observations provide the foundation for Weekes' model: 1) species of *Pseudemoia* and *C. chalcides* have eggs with relatively little yolk content and also share a distinctive structure in the chorioallantoic placenta (Giacomini 1891; Weekes 1935), elsewhere called a placentome (ten Cate-Hoedemaker 1933; Luckett 1977), and 2) eggs of other species have a large yolk mass relative to offspring size and lack this specialization of the chorioallantoic placenta. Weekes (1935) speculated that the placentome was a specialization for nutritional transport and that chorioallantoic placentas lacking a placentome, denoted as simple chorioallantoic placentas, functioned in respiration. Further, the morphology of simple chorioallantoic placentas varies, designated as either type I or type II (Weekes 1935). The unique features described for the chorioallantoic placentas of species of *Niveoscincus* (Weekes 1930) were assigned to type II (Weekes 1935), yet were considered to be functionally equivalent to type I, which occurred in all other species. The evolutionary scenario proposed to explain this structural variation was that the type III chorioallantoic placenta, i.e., with a placentome, evolved from type I via the intermediate condition exemplified by type II (Weekes 1935). The squamate yolk sac placenta, which was only described for a few species with type I chorioallantoic placentation, was given no nutritive function in this scenario but was considered to function in transport of water (Weekes 1935).

Weekes' (1927a,b, 1929, 1930, 1935) contribution to an understanding of reptilian placentation is monumental. She championed a comparative approach to the study of evolutionary patterns and generated hypotheses that remain relevant and are now subject to test within phylogenetic frameworks. Her work has had a dominant influence on more recent research of reptilian placentation both as a theoretical foundation and because she identified species that have continued to be useful models.

Whereas Weekes (1935) focused on the evolution of specific placental structures, reflection on the relationship between reproductive mode and pattern of embryonic nutrition in the evolution of placentation has also proven fruitful (Blackburn 1985b, 1992, 1995; Stewart 2013). An important component of this approach is the recognition that reproductive mode and pattern of embryonic nutrition are independent, yet interactive, variables (Blackburn 1992, 1995). Conceptually, selective factors favoring transitions between reproductive modes may be different from those favoring a particular pattern of embryonic nutrition. As noted above, most oviparous squamates retain eggs in the uterus for a considerable period of embryonic development (Shine 1983a; DeMarco 1993; Blackburn 1995; Andrews and Mathies 2000). This reproductive pattern is termed oviparity because embryos undergo further development outside the female before hatching. These embryos are exposed to two different environments. During the initial phase of development, they are surrounded by a wealth of potential nourishment, the availability of which is dependent on the existence of embryonic mechanisms to access it. Blackburn (1985b, 1992, 1995) proposed that the uterine environment presents a suitable condition for opportunistic uptake of nutrients by embryos and that the initial stage in evolution of placentation involves incipient placentotrophy. Longer intervals of uterine egg retention provide greater opportunity for embryos to capture maternal nutrients. The components of this model are: 1) Oviparous embryos have the machinery to acquire materials from their immediate environment, 2) the lengthy period of intrauterine egg retention, as is characteristic of many squamates, results in incidental transfer of nutrients to embryos, i.e., incipient placentotrophy, and 3) incidental placental transport of nutrients provides a template for selection to enhance placental transport through specializations for placentotrophy. A key feature of this model is the prediction that placentotrophy evolves coincident with viviparity. Further, it does not exclude the possibility that placentotrophy can evolve in oviparous species with extended intrauterine egg retention.

Testing hypotheses for evolution of placentation requires a comparative methodology; placental tissues are unlikely to be fossilized. An ideal model would be a lineage with a continuum of reproductive modes and patterns of embryonic nutrition. This combination of traits will be rare, yet we can identify three candidate genera, *Chalcides*, *Niveoscincus*, and *Trachylepis* that may exhibit these characteristics. Patterns of embryonic nutrition within these viviparous genera include both lecithotrophy and placentotrophy and placental structure differs among species. One suggestion for future research would be a comprehensive program aimed at developmental morphology, physiology and genomics of each of these lineages.

We can also seek patterns from comparisons across lineages and the large number of independent origins of viviparity among squamates provides numerous examples for such comparisons. Throughout this review, we have organized species accounts into four categories based on similarity in pattern of embryonic nutrition: 1) intraspecific geographic variation in reproductive mode, 2) lecithotrophic viviparous species, 3) placentotrophic lineages with interspecific variation and, 4) placentotrophic species with extremely small eggs. Each of theses categories represents a specialized reproductive pattern and collectively they illustrate why lizards are appropriate models to study the evolution of viviparity and placentation. Patterns of embryonic nutrition can be clearly defined and each has evolved on multiple occasions; they are homoplastic traits. Comparisons among lineages within each reproductive category will reveal patterns of phenotypic evolution and identify specific traits for further study of the underlying developmental mechanisms.

13.4.1 Intraspecific Geographic Variation in Mode of Reproduction

Reproductively bimodal species not only bring to light characteristics associated with reproductive mode but also are important experimental models to study the evolutionary transition. Similarities between reproductive modes identify traits that are functionally viable independent of the embryonic environment. Hypotheses concerning functional capability of similar traits and the relevance of differences can be tested experimentally. Unfortunately, developmental comparisons have not been done for all of the known reproductively bimodal lizards. Where we do have information, there are no apparent differences in pattern of embryonic growth or development of the extraembryonic membranes, at least at the histological level (Stewart *et al.* 2004). This suggests that at least minimal embryonic metabolic needs are met both in nest and uterine environments by structures inherent in oviparous forms.

There are, however, structural and functional specializations of the uterus of viviparous females. Eggshell glands responsible for secretion of the eggshell are reduced in viviparous females. The selective pressure for this specialization presumably is a response to differences in nest and uterine environments (Packard *et al.* 1977). Thicker, calcareous eggshells provide protection, yet allow gaseous exchange, in the nest. Thinner, non-calcareous eggshells are favored for gaseous exchange in the uterus. The primary nutritional cost to embryos in the evolution of viviparity is loss of the eggshell. Conversely, this is a nutritional gain to females, because the large quantities of calcium mobilized during egg shelling are no longer required. In the absence of a compensatory source of calcium, viviparity would be unlikely to evolve in either *S. equalis* or *Z. vivipara* because of the substantial dependence on eggshell calcium for embryonic development. The loss of eggshell calcium is ameliorated by a second modification of uterine function in viviparous females. Calcium secretion is down-regulated during the early phase of intrauterine egg retention in viviparous females but up-regulated late in development during the phase of highest embryonic growth. Viviparity is accompanied by calcium placentotrophy. More pointedly, evolution of viviparity is contingent on placental calcium transport. The dependence of embryos on calcium secreted by the uterus also would preclude a transition from viviparity to oviparity in the absence of a calcareous eggshell.

The unusual reproductive pattern of some populations of *S. equalis* reveals another feature of the relationship between reproductive mode and pattern of calcium nutrition (Linville *et al.* 2010). The uterus of females with prolonged egg retention secretes calcium that becomes incorporated into the eggshell early in intrauterine gestation. They produce calcareous eggshells. The uterus of these females also secretes calcium later in embryonic development and this calcium is taken up by the embryo prior to oviposition of the egg. Following oviposition, the embryo extracts additional calcium from the eggshell. Thus, embryonic uptake of placental calcium precedes egg laying and subsequent calcium extraction from the eggshell. Specifically, calcium placentotrophy occurs in an oviparous lizard, albeit one that retains eggs in the uterus for most of embryonic development (Linville *et al.* 2010; Stewart 2013). This reproductive condition also indicates how variation in the timing of uterine calcium secretion could function in transitions between viviparity and oviparity to maintain adequate embryonic calcium nutrition.

Comparison of these two lineages reveals that phenotypic similarity in the evolution of reproductive mode involves both retention of plesiomorphic traits and independent evolution of derived traits. Viviparity is accompanied by reduction in eggshell thickness, an independently derived similarity, which results from two different developmental patterns, reduction in either size or number of uterine eggshell glands (Heulin *et al.* 2005; Stewart *et al.* 2010). The chorioallantoic membrane retains two major functions, transport of gases and active uptake of calcium. Whereas embryonic acquisition of calcium is a plesiomorphic trait, modification of the timing of uterine secretion of calcium is homoplastic. As a result, placental transport of calcium is coincident with the transition to viviparity or occurs during prolonged intrauterine egg retention, i.e., prior to the transition to viviparity.

However, calcium nutrition of viviparous offspring of both *Saiphos equalis* and *Zootoca vivipara* is deficient compared to oviparous offspring (Stewart *et al.* 2009b; Linville *et al.* 2010). In addition, total mass of offspring of viviparous females is lower than for oviparous females of *Z. vivipara*. These data indicate a nutritional cost to viviparity. Modification of the timing of uterine calcium secretion in viviparous females of *Saiphos equalis* and *Zootoca vivipara* does not fully compensate for loss of eggshell calcium. Either uterine secretion is inadequate or the embryonic mechanism of calcium recovery, which is specialized for extraction from the eggshell, does not function as effectively in mobilizing calcium directly from uterine secretion. If higher embryonic calcium concentrations promote neonatal fitness, selection could act on variation in placental delivery mechanisms to enhance embryonic calcium nutrition. Thus, placental specializations for calcium transport could be promoted early in the evolution of viviparity.

13.4.2 Lecithotrophic Viviparous Species

If the evolution of viviparity promotes specializations for intrauterine gestation, comparisons between predominantly lecithotrophic viviparous species and closely related oviparous species should be instructive. Comparisons of viviparous species from several lineages do reveal interesting patterns of variation (Section 13.3.5.4) but we know little about oviparous counterparts. One consistent similarity that we can confidently associate with viviparity is secretion of an eggshell that is much thinner than that of oviparous species. Eggshell thickness is reduced in all lecithotrophic viviparous species and the eggshell is absent over the embryonic hemisphere of the egg during late developmental stages in most of these species. The absence of an acellular structure within the placenta results in greater intimacy for physiological exchange than exists for reproductively bimodal species. How this structural alignment influences the functional relationship between mothers and embryos has not been studied.

The most thoroughly studied lecithotrophic viviparous lizards are species within the scincid genera *Eulamprus* and *Niveoscincus*. Placental ontogeny is known for multiple species of each of these genera and, as indicated by matrotrophic indices, with the exception of the highly placentotrophic *Niveoscincus ocellatus*, both genera are predominantly lecithotrophic. Yolk is a primary source of calcium, as it is for organic molecules, yet placental transport provides a substantial quantity of calcium to developing embryos. Comparisons with oviparous species in the same subfamily indicate that placental transport compensates for loss of calcium from the eggshell but neither placental sites of calcium transport, nor calcium transport mechanisms, have been studied. There is indirect evidence that placental uptake of calcium is influenced by yolk calcium content in *Niveoscincus metallicus*. The pattern of embryonic calcium nutrition varies between populations. Yolk supplies all embryonic calcium in a population with a high yolk calcium concentration, whereas more than half of calcium in neonates of a second population is provided by the placenta (Thompson *et al.* 1999a, 2000) (Table 13.6). The mechanism of placental calcium uptake may be facultative in this species. This aspect of placental function in lecithotrophic species may be an important property of placental variation and evolution (Stewart 1989). Geographic or temporal variation in pattern of embryonic calcium nutrition has not been investigated in other species of lizards but has been reported for a viviparous snake (Sangha *et al.* 1996; Fregoso *et al.* 2010).

One of the lessons from these two genera, *Eulamprus*, *Niveoscincus*, is that specializations within placental categories (Section 13.3.3) can occur in predominantly lecithotrophic species. Species of *Eulamprus* have both yolk sac placentas and chorioallantoic placentas and the chorioallantoic placenta of *E. quoyii* contains a zone of specialization (Murphy *et al.* 2011). All species of *Niveoscincus* have elaborate yolk sac placentas and regional specializations of the chorioallantoic placenta (Stewart and Thompson 2009b). Estimates of placental nutrient provision based on matrotrophic indices indicate that *Niveoscincus ocellatus* is highly placentotrophic (Thompson *et al.* 2001a), whereas its congener *N. microlepidotus* is not (Atkins *et al.* 2006). Curiously, these species have similar complex chorioallantoic and yolk sac placentas. Additionally, placental transport of organic molecules differs between populations of *Niveoscincus metallicus* (Swain and Jones 1997, 2000; Jones *et al.* 1998; Thompson *et al.* 1999a). The differences between populations in patterns of embryonic nutrition, calcium and organics, and between species with similar placental morphology suggest that the complex placentas of these species function facultatively. Thus, placental nutrient provision would only be detected under specific conditions. This hypothesis is consistent with the existence of elaborate placental structures in predominantly lecithotrophic species.

Specialized placental structures, in the form of yolk sac placentas, also occur in the lecithotrophic species *Woodworthia maculatus* (Boyd 1942), *Xantusia vigilis* (Yaron 1977, 1985), *Elgaria coerulea* (Stewart 1985), *Sceloporus mucronatus* (Villagran *et al.* 2005) and *Sceloporus jarovii* (Blackburn *et al.* 2010). Unfortunately, details of the pattern of embryonic nutrition are unknown for these species and data are unavailable for suitable oviparous species for comparison. The prevalence of specialized yolk sac placentation in lecithotrophic viviparous species is one of the most intriguing aspects of the evolution of squamate placentation. Development of these structures is superimposed on the yolk cleft–isolated yolk mass complex, a primary characteristic of which is the absence of direct vascular support for the embryonic epithelium, the omphalopleure. The abembryonic region of squamate eggs would seem to be an unsuitable site for efficient transport of materials into the embryonic circulatory system. Unfortunately, functional characteristics of the squamate yolk sac have not been studied. However, the broad phylogenetic distribution of the yolk cleft–isolated yolk mass system among oviparous squamates and the retention of these structures in lecithotrophic viviparous species suggests a critical role in squamate yolk sac function. The evolution of squamate viviparity and placentation will not be fully understood in the absence of an explanation for the function of the yolk sac in oviparous and in lecithotrophic viviparous species.

Comparisons between lecithotrophic viviparous species and oviparous species indicate that reduction in eggshell thickness is homoplastic, as is disruption of the thin eggshell early in embryonic development. The pattern of eggshell loss, i.e., in the embryonic hemisphere of the egg, is similar among lineages. The chorioallantoic membrane retains plesiomorphic traits in some lineages (Guillette and Jones 1985; Stewart 1985) but exhibits a derived phenotype in others (Stewart and Thompson 2009a; Murphy *et al.* 2011). In each of these lineages, the close apposition of embryonic and maternal blood vessels of the chorioallantoic placenta, allows diffusion of gases, and perhaps incidental transport of other molecules. Comparisons between oviparous skinks and lecithotrophic viviparous species of *Eulamprus* and *Niveoscincus* reveal similar independently derived characteristics of the omphalopleure and apposing uterine epithelium. Hyperplasia of epithelial cells of the omphalopleure of the isolated yolk mass and of the apposing uterine epithelium transforms the yolk sac placenta into a site of histotrophic nutrient transport in both lineages.

The increased intimacy over part of the placental interface, coupled with specializations of the yolk sac placenta and chorioallantoic placenta in some lineages, suggests that communication and molecular transport have been established between mothers and embryos. Eggs of predominantly lecithotrophic species are not merely oviparous eggs that have been retained in the uterus.

13.4.3 Placentotrophic Lineages with Interspecific Variation

Viviparous species with obligatory placentotrophy have been the focus of a variety of studies aimed at understanding the breadth of specializations for embryonic nutrition (Section 13.3.5.2). Placentotrophic viviparous species secrete a thin egg shell which is disrupted early in development; this is an independently derived characteristic shared with many lecithotrophic viviparous species. The regionally diversified chorioallantoic placentas are also homoplasious but with differences in structural similarity. The placentomes of *Chalcides* and *Pseudemoia* are structurally similar, yet distinguishable at the microscopic level. The independent evolution of tissues specialized for histotrophic transport, in association with the mesometrial uterine region, are likely a response to similar selection pressures operating on a shared genetic–developmental system. In contrast to species of *Chalcides* and *Pseudemoia*, evolution of histotrophic transport in the chorioallantoic placenta of *Niveoscincus ocellatus* does not take the form of a placentome. Chorioallantoic placentation of *N. ocellatus* arises through a different developmental pathway resulting in variation in placental morphology between lineages. Chorioallantoic placentation of *N. ocellatus* is convergent on that of the other two lineages, yet is structurally distinct with a different pattern of development. Whether the two forms of regional structural diversification have similar functional properties is unknown.

Structure of the yolk sac and yolk sac placentas of *N. ocellatus* and of species of *Chalcides* and *Pseudemoia* are independently derived yet arise through similar developmental patterns. The independent evolution of an inter-omphalopleuric membrane in each of these lineages suggests an important function in intrauterine gestation expressed through similar modifications of the pattern of embryonic vascular flow. Similarity in development and morphology of yolk sac placentation among these three lineages suggests that these tissues are functionally equivalent.

13.4.4 Placentotrophic Species with Extremely Small Eggs

In the continuum of fetal nutritional patterns, *Eumecia anchietae*, *Trachylepis ivensii* and species of the South American genus *Mabuya* provide the most extreme examples of placentotrophy described for squamate reptiles. Many of the reproductive specializations of these species are convergent on those of eutherian mammals. They ovulate eggs that are extremely small (1–2 mm diameter) compared to other reptiles, yet retain the reptilian pattern of meroblastic cleavage. Secretion of a thin eggshell and its subsequent loss early in intrauterine gestation is a homoplastic trait they share with other placentotrophic species and most lecithotrophic viviparous species.

Because of the small mass of yolk, embryos are dependent on placental nutrition at an early developmental stage for differentiation of embryonic tissues and development of the extraembryonic membranes. The sources of nutrients early in embryonic development are glandular and histotrophic secretions from the uterus, which are transported into embryonic tissues by chorionic epithelial cells in *Mabuya* sp. and *Trachylepis ivensii*. Specialized regions of the chorionic epithelium, absorptive plaques, participate in nutrient transport in *Mabuya* sp. populations from the Andean region of Columbia. Chorioplacentas have been described as transitory structures early in development in a diversity of viviparous squamates but these two lineages are the only ones with chorioplacentas that occupy the entire perimeter of the egg and that have specializations for nutrient transport. For most viviparous squamates, early stages of embryonic development are fueled by nutrients from yolk.

Yolk sac development in these three lineages differs from all other squamates in several additional features. A choriovitelline membrane grows around the entire yolk sac cavity and neither a yolk cleft nor an isolated yolk mass develops. Based on structural specializations for histotrophic nutrient transport the choriovitelline membrane of *Eumecia anchietae* contributes significantly to placentotrophy; a unique feature of squamate placentation. The choriovitelline membrane of *Trachylepis ivensii* also is unique. Chorionic epithelial cells of this tissue invade the uterine epithelium and contact uterine blood vessels. The displaced uterine epithelial cells degenerate and chorionic cells re-establish contact with the yolk sac to form a secondary vascular trilaminar omphalopleure in direct contact with endothelial cells of the uterine vascular system. Although major differences in structure and function of the choriovitelline membranes of *E. anchietae* and of *T. ivensii* identify them each as uniquely derived characteristics, prominence of choriovitelline placentation is a shared pattern unknown in other reptiles.

The absence of data for later developmental stages of *E. anchietae* and *T. ivensii* precludes a detailed comparison but placental structures that have been described reveal divergent traits among the three lineages. Development of the chorion is homoplastic in *T. ivensii* and *Mabuya* sp. but specializations of the chorioplacenta, the absorptive plaques, of species of *Mabuya* are novel. The absence of a yolk cleft and growth of a choriovitelline membrane around the perimeter of the yolk sac cavity in each of the three lineages are similar independently derived traits that likely co-vary with the small size of the eggs. However, the choriovitelline membranes of the three lineages are structurally distinct. The choriovitelline membrane of *E. anchietae* has elaborate specializations for placental transport that are absent in *T. ivensii* and *Mabuya* sp. and development of the choriovitelline membrane of *T. ivensii* is unique among amniotes. Although some of the derived characteristics of development are similar in these species, structural specializations for yolk sac placentation are divergent among the three lineages.

Details of chorioallantoic placentation for late stages of gestation are not known for *E. anchietae* or *T. ivensii*. However, by mid-gestation, the chorioallantoic membrane surrounds the perimeter of the egg in *E. anchietae* and both the embryonic and uterine tissues are well vascularized and have thin epithelia. There are prominent epithelial specializations for histotrophic transport in the chorioallantoic placenta of *Mabuya* sp. by this stage. Thus, development of the chorioallantoic placentas of these two lineages is divergent. The regionally diversified chorioallantoic placentation of species of *Mabuya* is similar to that of other viviparous skinks, i.e., *Chalcides chalcides*, *Niveoscincus* sp., *Pseudemoia* sp., in that all include areas with structural profiles that indicate the potential for gas exchange and hemotrophy, in conjunction with specializations for histotrophy. However, chorioallantoic placentation in *Mabuya* sp. is unique in the extensive interdigitation of fetal and maternal tissues and histology of the placentome and in the novel specializations for placental transport, the chorionic areolae and absorptive plaques.

Given that the three lineages in which extensive reduction in egg yolk mass has evolved (*Eumecia anchietae*, *Trachylepis ivensii* and species of the South American genus *Mabuya*) are all members of the same family of lizards, Scincidae, the diversity of patterns of development and specializations for placentation is striking.

13.4.5 Model for the Evolution of Lizard Placentotrophy

Squamate patterns of embryonic nutrition have often been represented as lying along a continuum with lecithotrophy and placentotrophy as extremes (Blackburn 2005, 2006). The four reproductive patterns (above) each are identifiable by traits shared by several lineages and may represent relatively stable complexes along that continuum. Although more than one mechanism may be shaping each pattern, the existence of common characteristics suggests that the evolution of viviparity among different lineages of lizards may follow similar pathways. Based on this assumption, these reproductive patterns provide a useful template for a hypothesis-based model for the evolution of placentotrophy.

Hypothesis 1. Selection for prolonged intrauterine egg retention promotes reduction in protein and calcium secretion by uterine shell glands, through a reduction in gland size or quantity. It has long been recognized that one of the primary functions of a placenta in the transition from oviparity to viviparity must be to fulfill respiratory demands of developing embryos (Weekes 1935; Packard *et al.* 1977; Guillette 1982b; Blackburn 1992). One mechanism to enhance respiratory exchange during intrauterine gestation is reduction in thickness of the eggshell, an evolutionary response characteristic of all viviparous squamates. Selection for a novel reproductive mode, prolonged intrauterine egg retention, establishes a novel embryonic environment which favors reduction in protein and calcium secretion by uterine shell glands. Two mechanisms for reduction in eggshell thickness have been exploited by viviparous lizards, reduction in gland size (Heulin *et al.* 2005) and reduction in number of glands (Stewart *et al.* 2010).

Hypothesis 2. Prolonged intrauterine egg retention stimulates uterine calcium secretion late in gestation. Whereas reduction in eggshell thickness benefits respiration of embryos, the resulting absence of a calcareous eggshell creates a nutritional deficit. Oviparous lizard embryos extract calcium from the eggshell during the latest stages of development, which contributes substantially to embryonic nutrition. Uterine secretion of calcium occurs soon after ovulation and calcium is stored in the eggshell until embryos develop the capacity to mobilize it. Although viviparous females do secrete a thin eggshell, the structure contains little calcium (Stewart *et al.* 2009b; Linville *et al.* 2010). Thus, uterine calcium secretion during prolonged intrauterine gestation would be favored to diminish the nutritional deficit (Stewart 2013). The pattern of calcium mobilization by embryos of species with geographic variation in reproductive mode reveals a mechanism to maintain embryonic calcium nutrition during the transition between oviparity and viviparity; uterine secretions provide calcium to embryos of *Saiphos equalis* and *Zootoca vivipara* during intrauterine gestation (Stewart *et al.* 2009b; Linville *et al.* 2010). This property of the uterus indicates that uterine delivery of calcium, i.e., calcium placentotrophy, can be coincident with evolution of viviparity, or as illustrated by *S. equalis*, precede the evolution of viviparity. Embryos of *S. equalis* with prolonged intrauterine egg retention mobilize calcium from three sources: yolk, uterine secretions late in development and eggshell deposits following oviposition.

Hypothesis 3. Loss of the eggshell in the embryonic hemisphere of the egg early in development enhances respiratory exchange across the chorioallantoic placenta and the increased intimacy between mothers and embryos allows incidental transfer of nutrients to embryos. In contrast to calcium, all organic nutrients to oviparous embryos are supplied in yolk and, as long as the nutritional content of yolk remains stable, the evolution of viviparity does not entail a deficit in molecules supplied solely in yolk. Placental transport of these nutrients evolves as a supplement to yolk nourishment as incidental transport across the thin eggshell of lecithotrophic species, or directly to embryos in lecithotrophic species in which the eggshell is lost early in development (Blackburn 2005, 2006).

Hypothesis 4. Placental structural specializations that enhance embryonic calcium transport and incidental nutrient uptake are favored. These structures respond to developmental cues and supply nutrients facultatively. Incipient placentotrophy, incidental transport across placental tissues, establishes a functional relationship that favors enhancements for nutrient transport (Blackburn 1992). Replacement of eggshell calcium and supplementation of yolk nutrients promote structural specializations for placental transport that are responsive to developmental cues and supply nutrients facultatively. For example, facultative placentotrophy is suggested for *Niveoscincus metallicus* (Jones *et al.* 1998; Thompson *et al.* 1999a, 2000), a species with placental specializations for histotrophic transport that are not present in oviparous skinks.

Hypothesis 5. Selection favors greater placental nutrient provision to embryos of species with histotrophic specializations for facultative placentotrophy. Novel structures may arise because specializations for facultative placentotrophy may differ between lineages. Placentotrophic lizards exhibit a diversity of placental structural specializations, some of which may reflect unique attributes of ancestors. However, functional differences and relaxation of developmental constraints may help shape placental structure and contribute to placental diversity. Whereas placentas may initially arise through minor modification of individual elements with a shared ancestry responding to basic physiological needs of embryos, transport of organic nutrients is a novel function, which may arise through novel pathways and produce novel complex structures. Thus, phenotypic evolution often would be divergent in highly placentotrophic lineages. Evolution of novel placental features may be rapid and unpredictable because of antagonistic co-evolution between embryos and mothers, as predicted by the maternal–offspring conflict hypothesis (Crespi and Semeniuk 2004).

Hypothesis 6. Specializations for placentotrophy release females from dependence on production of large yolked eggs allowing a seasonal shift in nutrient provisioning. Selection favors smaller eggs. A consequence of the evolution of specializations for placental nutrient transport is reduction in mass of ovulated yolk and obligative placentotrophy. Yolk reduction engenders alteration of pathways for yolk sac development and contributes to divergent phenotypic evolution as specializations for placentation arise in novel interfaces between mothers and embryos.

13.5 ACKNOWLEDGMENTS

Grant support for research in our respective labs has been provided by the National Science Foundation, Howard Hughes Medical Institute, Australian Research Council, East Tennessee State University, Trinity College, University of Tulsa, and the Thomas S. Johnson Research Professorship funds. Many students have contributed to research on which this review has drawn, including Kristie Anderson, Richard Castillo, Jessica Chin, David Crotzer, Doug Florian, Santiago Fregoso, Courtney Garland, Greg Gavelis, Siobhan Knight, Brent Linville, Rachel Lorenz, Ashley Mathieson, Shauna McKinney, Elizabeth Price, Soni Sangha, Craig Shadrix, Michael Smola, and Haley Stinnett. We thank Alex Flemming for providing photomicrographs of *Eumecia* placenta.

13.6 LITERATURE CITED

Adams, S. M., Biazik, J. M., Thompson, M. B. and Murphy, C. R. 2005. Cyto-epitheliochorial placenta of the viviparous lizard *Pseudemoia entrecasteauxii*: a new placental morphotype. Journal of Morphology 264: 264–276.

Adams, S. M., Liu, S., Jones, S. M., Thompson, M. B. and Murphy, C. R. 2007. Uterine epithelial changes during placentation in the viviparous skink *Eulamprus tympanum*. Journal of Morphology 268: 385–400.

Agassiz, L. 1857. *Contributions to the natural history of the United States of America.* Vol. II, Part III. Embryology of the turtle. pp. 451–643. Little, Brown, & Co., Boston.

Albergotti, L. C. and Guillette, L. J., Jr. 2011. Viviparity in reptiles: evolution and endocrine physiology. pp. 247–275. In D. O. Norris and K. H. Lopez (eds.), *Hormones and reproduction of vertebrates. Volume 3, Reptiles*. Academic Press, San Diego.

Albergotti, L. C., Hamlin, H. J., McCoy, M. W. and Guillette, L. J., Jr. 2009. Endocrine activity of extraembryonic membranes extends beyond placental amniotes. Plos One 4(5): e5452.

Anderson, K. E., Blackburn, D. G. and Dunlap, K. D. 2011. Scanning EM of the placental interface in the mountain spiny lizard, *Sceloporus jarrovi*. Journal of Morphology 272: 465–484.

Andrews, R. M. 1997. Evolution of viviparity: variation between two sceloporine lizards in the ability to extend egg retention. Journal of Zoology 243: 579–595.

Andrews, R. M. 2000. Evolution of viviparity in squamate reptiles (*Sceloporus* spp.): a variant of the cold-climate model. Journal of Zoology 250: 243–253.

Andrews, R. M. 2002. Low oxygen: a constraint on the evolution of viviparity in reptiles. Physiological and Biochemical Zoology 75: 145–154.

Andrews, R. M. and Mathies, T. 2000. Natural history of reptilian development: constraints on the evolution of viviparity. Bioscience 50: 227–238.

Andrews, R. M. and Rose, B. R. 1994. Evolution of viviparity: constraints on egg retention. Physiological Zoology 67: 1006–1024.

Andrews, R. M., Mendez de la Cruz, F. R. and Villagran Santa Cruz, M. 1997. Body temperatures of female *Sceloporus grammicus*: thermal stress or impaired mobility? Copeia 1997: 108–115.

Andrews, R. M., Mendez de la Cruz, F. R., Villagran Santa Cruz, M. and Rodriguez Romero, F. 1999. Field and selected body temperatures of *Sceloporus aeneus* and *Sceloporus bicanthalis*. Journal of Herpetology 33: 93–100.

Angelini, F. and Ghiara, G. 1984. Reproductive modes and strategies in vertebrate evolution. Bollettino di Zoologia 51: 121–203.

Arrayago, M. J., Bea, A. and Heulin, B. 1996. Hybridization experiment between oviparous and viviparous strains of *Lacerta vivipara*: a new insight into the evolution of viviparity in reptiles. Herpetologica 52: 333–342.

Atkins, N., Swain, R. and Jones, S. M. 2006. Does date of birth or a capacity for facultative placentotrophy influence offspring quality in a viviparous skink, *Niveoscincus microlepidotus*? Australian Journal of Zoology 54: 369–374.

Ballinger, R. E. 1973. Comparative demography of two viviparous iguanid lizards (*Sceloporus jarrovi* and *Sceloporus poinsetti*). Ecology 54: 269–283.

Ballinger, R. E., Lemos-Espinal, J. A. and Smith, G. R. 2000. Reproduction in females of three species of crevice-dwelling lizards (genus *Xenosaurus*) from Mexico. Studies on Neotropical Fauna and Environment 35: 179–183.

Bauer, A. M. 2003. On the identity of *Lacerta punctate* Linnaeus 1758, the type species of the genus *Euprepis* Wagler, 1830, and the generic assignment of Afro-Malagasy skinks. African Journal of Herpetology 52: 1–7.

Bauwens, D. and Thoen, C. 1981. Escape tactics and vulnerability to predation associated with reproduction in the lizard *Lacerta vivipara.* Journal of Animal Ecology 50: 733–743.

Bell, T. 1839. *A history of British reptiles*, 2nd ed., Van Voorst, London, pp. 142.

Beuchat, C. A. 1986. Reproductive influences on thermoregulatory behavior of a live-bearing lizard. Copeia 1986: 971–979.

Beuchat, C. A. and Ellner, S. 1987. A quantitative test of life history theory: thermoregulation by a viviparous lizard. Ecological Monographs 57: 45–60.

Beuchat, C. A. and Vleck, D. 1990. Metabolic consequences of viviparity in a lizard, *Sceloporus jarrovi*. Physiological Zoology 63: 555–570.

Bhatia, M. L. and Dayal, J. 1933. On the arterial system of the lizard *Hemidactylus flaviviridis* Rüppel (the wall lizard). Anatomischer Anzeiger 76: 417–437.

Biazik, J. M., Thompson, M. B. and Murphy, C. R. 2007. The tight junctional protein occludin is found in the uterine epithelium of squamate reptiles. Journal of Comparative Physiology B: Biochemical, Systemic, and Environmental Physiology 177: 935–943.

Biazik, J. M., Thompson, M. B. and Murphy, C. R. 2008. Claudin-5 is restricted to the tight junction region of uterine epithelial cells in the uterus of pregnant/gravid squamate reptiles. Anatomical Record 291: 547–556.

Biazik, J. M., Thompson, M. B. and Murphy, C. R. 2009. Lysosomal and alkaline phosphatase activity indicate macromolecule transport across the uterine epithelium in two viviparous skinks with complex placenta. Journal of Experimental Zoology Part B: Molecular and Developmental Evolution 312: 817–826.

Biazik, J. M., Thompson, M. B. and Murphy, C. R. 2010a. Desmosomes in the uterine epithelium of noninvasive skink placentae. Anatomical Record: Advances in Integrative Anatomy and Evolutionary Biology 293: 502–512.

Biazik, J. M., Thompson, M. B. and Murphy, C. R. 2010b. Paracellular and transcellular transport across the squamate uterine epithelium. Herpetological Conservation and Biology 5: 257–262.

Biazik, J. M., Parker, S. L., Murphy, C. R. and Thompson, M. B. 2012. Uterine epithelial morphology and progesterone receptors in a Mifepristone-treated viviparous lizard *Pseudemoia entrecasteauxii* (Squamata: Scincidae) during gestation. Journal of Experimental Zoology Part B: Molecular and Developmental Evolution 318: 148–158.

Birchard, G. F., Black, C. P., Schuett, G. W. and Black, V. 1984a. Influence of pregnancy on oxygen consumption, heart rate, and hematology in the garter snake: implications for the 'cost of reproduction' in livebearing reptiles. Comparative Biochemistry and Physiology 77: 519–523.

Birchard, G. F., Black, C. P., Schuett, G. W. and Black, V. 1984b. Foetal-maternal blood respiratory properties of an ovoviviparous snake the cottonmouth, *Agkistrodon piscivorus*. Journal of Experimental Biology 108: 247–255.

Birchard, G. F., Walsh, T., Rosscoe, R. and Reiber, C. L. 1995. Oxygen uptake by Komodo dragon *(Varanus komodoensis)* eggs: the energetics of prolonged development in a reptile. Physiological Zoology 68: 622–633.
Blackburn, D. G. 1982. Evolutionary origins of viviparity in the Reptilia. I. Sauria. Amphibia-Reptilia 3: 185–205.
Blackburn, D. G. 1985a. Evolutionary origins of viviparity in the Reptilia. II. Serpentes, Amphisbaenia, and Ichthyosauria. Amphibia-Reptilia 5: 259–291.
Blackburn, D. G. 1985b. *The evolution of viviparity and matrotrophy in vertebrates, with special reference to reptiles.* Ph.D. Diss., Cornell University, Ithaca, New York, pp. 665.
Blackburn, D. G. 1992. Convergent evolution of viviparity, matrotrophy, and specializations for fetal nutrition in reptiles and other vertebrates. American Zoologist 32: 313–321.
Blackburn, D. G. 1993a. Standardized criteria for the recognition of reproductive modes in squamate reptiles. Herpetologica 49: 118–132.
Blackburn, D. G. 1993b. Chorioallantoic placentation in squamate reptiles: structure, function, development, and evolution. Journal of Experimental Zoology 266: 414–430.
Blackburn, D. G. 1993c. Histology of the late-stage placentae of the matrotrophic skink *Chalcides chalcides* (Lacertilia: Scincidae). Journal of Morphology 216: 179–195.
Blackburn, D. G. 1994a. Standardized criteria for the recognition of developmental nutritional patterns in squamate reptiles. Copeia 1994: 925–935.
Blackburn, D. G. 1994b. Discrepant usage of the term "ovoviviparity" in the herpetological literature. Herpetological Journal 4: 65–72.
Blackburn, D. G. 1995. Saltationist and punctuated equilibrium models for the evolution of viviparity and placentation. Journal of Theoretical Biology 174: 199–216.
Blackburn, D. G. 1998a. Resorption of oviductal eggs and embryos in squamate reptiles. Herpetological Journal 8: 65–71.
Blackburn, D. G. 1998b. Structure, function, and evolution of the oviducts of squamate reptiles, with special reference to viviparity and placentation. Journal of Experimental Zoology 282: 560–617.
Blackburn, D. G. 1998c. Reconstructing the evolution of viviparity and placentation. Journal of Theoretical Biology 192: 183–190.
Blackburn, D. G. 1999a. Viviparity and oviparity: evolution and reproductive strategies. pp. 994–1003. In T. E. Knobil and J. D. Neill (eds.), *Encyclopedia of reproduction, Vol. 4.* Academic Press, London, U.K.
Blackburn, D. G. 1999b. Are viviparity and egg-guarding evolutionarily labile in squamates? Herpetologica 55: 556–573.
Blackburn, D. G. 2000a. Viviparity: past research, future directions, and appropriate models. Comparative Biochemistry and Physiology—Part A: Molecular and Integrative Physiology 127: 391–409.
Blackburn, D. G. 2000b. Classification of the reproductive patterns of amniotes. Herpetological Monographs 14: 371–377.
Blackburn, D. G. 2005. Amniote perspectives on the evolution of viviparity. pp. 319–342. In M. C. Uribe and H. J. Grier (eds.), Viviparous fishes. New Life Publications, Homestead, Florida.
Blackburn, D. G. 2006. Squamate reptiles as model organisms for the evolution of viviparity. Herpetological Monographs 20: 131–146.
Blackburn, D. G. 2014. Evolution of vertebrate viviparity and specializations for fetal nutrition: a quantitative and qualitative analysis. Journal of Morphology. DOI: 10.1002/jmor.20272.
Blackburn, D. G. and Callard, I. P. 1997. Morphogenesis of the placental membranes in the viviparous, placentotrophic lizard *Chalcides chalcides* (Squamata: Scincidae). Journal of Morphology 232: 35–55.
Blackburn, D. G. and Flemming, A. F. 2009. Morphology, development, and evolution of fetal membranes and placentation in squamate reptiles. Journal of Experimental Zoology B. Molecular and Developmental Evolution 312B: 579–589.

Blackburn, D. G. and Flemming, A. F. 2010. Reproductive specializations in a viviparous African skink: implications for evolution and biological conservation. Herpetological Conservation and Biology 5: 263–270.

Blackburn, D. G. and Flemming, A. F. 2012. Invasive ovo-implantation and intimate placental associations in a placentotrophic African lizard. Journal of Morphology 273: 137–159.

Blackburn, D. G., Kleis-San Francisco, S. and Callard, I. P. 1998. Histology of abortive egg sites in the uterus of a viviparous, placentotrophic lizard, the skink *Chalcides chalcides*. Journal of Morphology 235: 97–108.

Blackburn, D. G. and Lorenz, R. 2003a. Placentation in garter snakes. Part II. Transmission EM of the chorioallantoic placenta of *Thamnophis radix* and *T. sirtalis*. Journal of Morphology 256: 171–186.

Blackburn, D. G. and Lorenz, R. 2003b. Placentation in garter snakes. Part III. Transmission EM of the omphalallantoic placenta of *Thamnophis radix* and *T. sirtalis*. Journal of Morphology 256: 187–204.

Blackburn, D. G. and Stewart, J. R. 2011. Viviparity and placentation in snakes. pp. 119–181. In D. Sever and R. Aldridge (eds.), *Reproductive biology and phylogeny of snakes*. CRC Press, Boca Raton, Florida.

Blackburn, D. G. and Vitt, L. J. 1992. Reproduction in viviparous South American lizards of the genus *Mabuya*. pp. 150–164. In W. Hamlett (ed.), *Reproductive biology of South American vertebrates: aquatic and terrestrial*. Springer-Verlag, New York.

Blackburn, D. G. and Vitt, L. J. 2002. Specializations of the chorioallantoic placenta in the Brazilian scincid lizard, *Mabuya heathi*: a new placental morphotype for reptiles. Journal of Morphology 254: 121–131.

Blackburn, D. G., Vitt, L. J. and Beuchat, C. A. 1984. Eutherian-like reproductive specializations in a viviparous reptile. Proceedings of the National Academy of Sciences (Washington) 81: 4860–4863.

Blackburn, D. G., Evans, H. E. and Vitt, L. J. 1985. The evolution of fetal nutritional adaptations. pp. 437–439. In H. -R. Duncker and G. Fleischer (eds.), *Vertebrate morphology*. Fortschritte der Zoologie 30. Gustav Fischer, Stuttgart.

Blackburn, D. G., Stewart, J. R., Baxter, D. C. and Hoffman, L. H. 2002. Placentation in garter snakes. Scanning EM of the placental membranes of *Thamnophis ordinoides* and *T. sirtalis*. Journal of Morphology 252: 263–275.

Blackburn, D. G., Weaber, K. K., Stewart, J. R. and Thompson, M. B. 2003a. Do pregnant lizards resorb or abort inviable eggs and embryos? Morphological evidence from an Australian skink, *Pseudemoia pagenstecheri*. Journal of Morphology 256: 219–234.

Blackburn, D. G., Johnson, A. R. and Petzold, J. L. 2003b. Histology of the extraembryonic membranes of an oviparous snake: towards a reconstruction of basal squamate patterns. Journal of Experimental Zoology 290A: 48–58.

Blackburn, D. G., Anderson, K. E., Johnson, A. R., Knight, S. R. and Gavelis, G. S. 2009. Histology and ultrastructure of the placental membranes of the viviparous brown snake, *Storeria dekayi* (Colubridae: Natricinae). Journal of Morphology 270: 1137–1154.

Blackburn, D. G., Gavelis, G. S., Anderson, K. E., Johnson, A. R. and Dunlap, K. D. 2010. Placental specializations in the mountain spiny lizard *Sceloporus jarrovi*. Journal of Morphology 271: 1153–1175.

Bleu, J., Massot, M., Haussy, C. and Meylan, S. 2012a. Experimental litter size reduction reveals costs of gestation and delayed effects on offspring in a viviparous lizard. Proceedings of the Royal Society B: Biological Sciences 279: 489–498.

Bleu, J., Heulin, B., Haussy, C., Meylan, S. and Massot, M. 2012b. Experimental evidence of early costs of reproduction in conspecific viviparous and oviparous lizards. Journal of Evolutionary Biology 25: 1264–1274.

Botte, V. 1973. Morphology and histochemistry of the oviduct in the lizard *Lacerta sicula:* the annual cycle. Bollettino di Zoologia 40: 305–314.

Boyd, M. M. M. 1940. The structure of the ovary and formation of the corpus luteum in *Hoplodactylus maculatus*. Quarterly Journal of Microscopic Science 82: 337–376.

Boyd, M. M. M. 1942. The oviduct, foetal membranes, and placentation in *Hoplodactylus maculatus* Gray. Proceedings of the Zoological Society of London, Series A 112: 65–104.
Bragdon, D. E. 1951. The non-essentiality of the corpus luteum for the maintenance of gestation in certain live-bearing snakes. Journal of Experimental Biology 118: 419–435.
Braña, F. and Bea, A. 1987. Bimodalite de la réproduction chez *Lacerta vivipara* (Reptilia, Lacertidae). Bulletin de la Société Herpétologique de France 44: 1–5.
Braña, F., Bea, A. and Arrayago, M. J. 1991. Egg retention in lacertid lizards: relationships with reproductive ecology and the evolution of viviparity. Herpetologica 47: 218–226.
Branch, B. 1998. *Field guide to snakes and other reptiles of southern Africa.* 3rd edition. Ralph Curtis Books, Sanibel Island, Florida, pp. 399.
Branch, W. R. and Haagner, G. V. 1993. The skink *Mabuya ivensii:* new records from Zambia and Zaire, and the status of the subspecies *septemlineata* Laurent 1964 and the genus *Lubuya* Horton 1972. Amphibia-Reptilia 14: 105–115.
Brandley, M. C., Ota, H., Hikida, T., Nieto-Montes De Oca, A., Fería-Ortíz, M., Guo, X. and Wang, Y. 2012. The phylogenetic systematics of blue-tailed skinks *(Plestiodon)* and the family Scincidae. Zoological Journal of the Linnean Society 165: 163–189.
Brandley, M. C., Young, R. L., Warren, D. L., Thompson, M. B. and Wagner, G. P. 2012. Uterine gene expression in the live-bearing lizard, *Chalcides ocellatus*, reveals convergence of squamate reptile and mammalian pregnancy mechanisms. Genome Biology & Evolution 4: 394–411.
Brattstrom, B. H. 1951. The number of young in *Xantusia*. Herpetologica 7: 143–144.
Brown, W. S. 1991. Female reproductive ecology in a northern population of the timber rattlesnake, *Crotalus horridus.* Herpetologica 47: 101–115.
Brown Wessels, H. L. 1989. Bimodal reproductive strategy in *Mabuya capensis* (Gray) (Squamata: Scincidae). Journal of the Herpetological Association of Africa 36: 46–50.
Bull, J. J. 1980. Sex determination in reptiles. Quarterly Review of Biology 55: 3–21.
Bull, J. J. and Shine, R. 1979. Iteroparous animals that skip opportunities for reproduction. American Naturalist 114: 296–316.
Bustard, H. R. 1966. Observations on the life history and behavior of *Chameleo bitaeniatus* Fischer. Herpetologica 22: 13–23.
Calderón-Espinosa, M. L., Andrews, R. M. and Méndez-de la Cruz, F. R. 2006. Evolution of egg retention in the *Sceloporus spinosus* group: exploring the role of physiological, environmental, and phylogenetic factors. Herpetological Monographs 20: 147–158.
Caldwell, M. W. and Lee, M. S. Y. 2001. Live birth in Cretaceous marine lizards (mosasauroids). Proceedings of the Royal Society of London B 268: 2397–2401.
Caputo, V., Guarino, F. M. and Angelini, F. 2000. Body elongation and placentome evolution in the scincid lizard genus *Chalcides* (Squamata, Scincidae). Italian Journal of Zoology 67: 385–391.
Carranza, S., Arnold, E. N., Geniez, P., Roca, J. and Mateo, J. A. 2008. Radiation, multiple dispersal and parallelism in the skinks, *Chalcides* and *Sphenops* (Squamata: Scincidae), with comments on *Scincus* and *Scincopus* and the age of the Sahara Desert. Molecular Phylogenetics and Evolution 46: 1071–1094.
Chapple, D. G., Ritchie, P. A. and Daugherty, C. H. 2009. Origin, diversification, and systematics of the New Zealand skink fauna (Reptilia: Scincidae). Molecular Phylogenetics and Evolution 52: 470–487.
Clausen, H. J. 1940. Studies on the effects of ovariotomy and hypophysectomy on gestation in snakes. Endocrinology 27: 700–704.
Cordero-López, N. and Morales, M. H. 1995. Lack of proteins of oviductal origin in the egg of a tropical anoline lizard. Physiological Zoology 68: 512–523.
Corso, G., Delitala, G. M. and Carcupino, M. 2000. Uterine morphology during the annual cycle in *Chalcides ocellatus tiligugu* (Gmelin) (Squamata: Scincidae). Journal of Morphology 243: 153–165.
Cree, A. 1994. Low annual reproductive output in female reptiles from New Zealand. New Zealand Journal of Zoology 21: 351–372.

Cree, A. C. and Guillette, L. J., Jr. 1991. Effect of b-adrenergic stimulation on uterine contraction in response to arginine vasotocin and prostaglandin F2a in the gecko *Hoplodactylus maculatus*. Biology of Reproduction 44: 499–510.
Cree, A. and Guillette, L. J., Jr. 1995. Biennial reproduction with a fourteen-month pregnancy in the gecko *Hoplodactylus maculatus* from southern New Zealand. Journal of Herpetology 29: 163–173.
Crespi, B. and Semeniuk, C. 2004. Parent-offspring conflict in the evolution of vertebrate reproductive mode. American Naturalist 163: 635–653.
Corso, G., Delitala, G. M. and Carcupino, M. 2000. Uterine morphology during the annual cycle in *Chalcides ocellatus tiligugu* (Gmelin) (Squamata: Scincidae). Journal of Morphology 243: 153–165.
Crocco, M., Ibargüengoytía, N. R. and Cussac, V. 2008. Contributions to the study of oviparity-viviparity transition: placentary structures of *Liolaemus elongatus* (Squamata: Liolaemidae). Journal of Morphology 269: 865–874.
Cruze, L., Kohno, S., McCoy, M. W. and Guillette, Jr., L. J. 2012. Towards an understanding of the evolution of the chorioallantoic placenta: steroid biosynthesis and steroid hormone signaling in the chorioallantoic membrane of an oviparous reptile 1. Biology of Reproduction 87: 1–11.
Cruze, L., Hamlin, H. J., Kohno, S., McCoy, M. W. and Guillette Jr., L. J. 2013. Evidence of steroid hormone activity in the chorioallantoic membrane of a turtle (*Pseudemys nelsoni*). General and Comparative Endocrinology 186: 50–57.
Cuvier, G. 1829. *Le règne animal distribué d'après son organisation: Les reptiles et les poisons.* Déterville, Paris.
Das, S. 1960. A comparative functional study of the urinogenital system in *Uromastix hardwickii* Gray (sand lizard) *Ptyas mucosus* (rat snake) and *Eryx conicus* Bonlenger [*sic*] (sand boa). Proceedings of the National Academy of Sciences, India 30: 59–78.
DeMarco, V. 1993. Estimating egg retention times in sceloporine lizards. Journal of Herpetology 27: 453–458.
DeMarco, V. and Guillette, L. J., Jr. 1992. Physiological cost of pregnancy in a viviparous lizard (*Sceloporus jarrovi*). Journal of Experimental Zoology 262: 383–390.
DeNardo, D. F., Lourdais, O. and Stahlschmidt, Z. R. 2012. Are females maternal manipulators, selfish mothers, or both? Insight from pythons. Herpetologica 68: 299–307.
Domini, G. 1928. Osservazioni sui riassorbimento degli embrioni nella *Seps chalcides*. Archivio Zoologico Italiano (Torino) 12: 191–218.
Duellman, W. E. 1965. A biogeographic account of the herpetofauna of Michoacan, Mexico. University of Kansas Publications of the Museum of Natural History 15: 629–708.
Dufaure, J. P. and Hubert, J. 1961. Table de développement du lézard vivipare: *Lacerta (Zootoca) vivipara* Jacquin. Archives d'Anatomie Microscopique et de Morphologie Expérimentale 50: 309–328.
Duméril, A. M. C. and Bibron, G. 1834–1854. *Erpétologie générale ou histoire naturelle complète des reptiles.* 10 vols. Roret, Paris.
Dunham, A. E. and Miles, D. B. 1985. Patterns of covariation in life history traits of squamate reptiles: the effects of size and phylogeny reconsidered. American Naturalist 126: 231–257.
Dunham, A. E., Miles, D. B. and Reznick, D. 1988. Life history patterns in squamate reptiles. pp. 441–522. In C. Gans and R. B. Huey (eds.), *Biology of the Reptilia, vol. 16B*. Alan B. Liss, New York.
Edwards, A., Jones, S. M. and Wapstra, E. 2002. Multiennial reproduction in females of a viviparous, temperate-zone skink, *Tiliqua nigrolutea*. Herpetologica 58: 407–414.
Fairbairn, J., Shine, R., Moritz, C. and Frommer, M. 1998. Phylogenetic relationships between oviparous and viviparous populations of an Australian lizard (*Lerista bougainvillii*, Scincidae). Molecular Phylogenetics and Evolution 10: 95–103.
Farag, A. A. 1983. Histology of the ovary of the viviparous Wiegmann's skink *Mabuya brevicollis* during its gestation period. Annals of Zoology (Agra) 20: 147–165.

Fenwick, A. M., Greene, H. W. and Parkinson, C. L. 2012. The serpent and the egg: unidirectional evolution of reproductive mode in vipers? Journal of Zoological Systematics and Evolutionary Research 50: 59–66.
Ferguson, M. W. J. 1982. The structure and composition of the eggshell and embryonic membranes of *Alligator mississippiensis*. Transactions of the Zoological Society of London 36: 99–152.
Ferguson, M. W. J. 1985. Reproductive biology and embryology of the crocodilians. pp. 329–491. In C. Gans, F. Billet and P. F. A. Maderson (eds.), *Biology of the Reptilia, Vol. 14*. John Wiley & Sons, New York.
Fergusson, B. and Bradshaw, S. D. 1992. *In vitro* uterine contractions in the viviparous lizard *Tiliqua rugosa*: effects of gestation and steroid pretreatment *in vivo*. General and Comparative Endocrinology 86: 203–210.
Ferner, K. and Mess, A. 2011. Evolution and development of fetal membranes and placentation in amniote vertebrates. Respiratory Physiology & Neurobiology 178: 39–50.
Fitch, H. S. 1935. Natural history of the alligator lizards. Transactions of the Academy of Science, St. Louis 29: 3–38.
Fitch, H. S. 1970. Reproductive cycles in lizards and snakes. University of Kansas Museum of Natural History, Miscellaneous Publications 52: 1–247.
FitzSimons, V. F. M. 1943. The lizards of South Africa. Transvaal Museum Memoirs 1: 1–528.
Flemming, A. F. 1994. Male and female reproductive cycles of the viviparous lizard, *Mabuya capensis* (Sauria: Scincidae) from South Africa. Journal of Herpetology 28: 334–341.
Flemming, A. F. and Blackburn, D. G. 2003. Evolution of placentotrophy in viviparous African and South American lizards. Journal of Experimental Zoology 299A: 33–47.
Flemming, A. F. and Branch, W. R. 2001. Extraordinary case of matrotrophy in the African skink *Eumecia anchietae*. Journal of Morphology 246: 264–287.
Fox, H. 1977. The urogenital system of reptiles. pp. 1–157. In C. Gans and T. S. Parsons (eds.), *Biology of the Reptilia, vol. 6*. Academic Press, London.
de Fraipont, M., Clobert, J. and Barbault, R. 1996. The evolution of oviparity with egg-guarding and viviparity in lizards and snakes: a phylogenetic analysis. Evolution 50: 391–400.
Fregoso, S. P., Stewart, J. R. and Ecay, T. W. 2010. Embryonic mobilization of calcium in a viviparous reptile: evidence for a novel pattern of placental calcium secretion. Comparative Physiology and Biochemistry A: Molecular & Integrative Physiology 156: 147–150.
Gadow, H. 1910. The effect of altitude upon the distribution of Mexican amphibians and reptiles. Zoologischer Jahrbuch 29: 689–714.
Gerrard, A. M. 1974. *Placental transfer of steroids at different stages of development and its possible implications in the sexual differentiation of Thamnophis radix haydenii embryos*. Ph.D. Diss., University of Colorado, Boulder, pp. 81.
Ghiara, G., Angelini, F., Zerani, M., Gobbetti, A., Cafiero, G. and Caputo, V. 1987. Evolution of viviparity in Scincidae (Reptilia, Lacertilia). Acta Embryologiae Experimentalis n.s. 8: 187–201.
Giacomini, E. 1891. Matériaux pour l'étude du developpement du *Seps chalcides*. Archives Italiennes De Biologie 16: 332–359.
Giacomini, E. 1893. Sull' ovidutto del Sauropsidi. Monitore Zoologico Italiano 4: 202–265.
Giacomini, E. 1906. Sulla maniera di gestazione e sugli annessi embrionali del *Gongylus ocellatus* Forsk. Memorie Accademia delle Scienze dell' Istituto di Bologna 3: 401–445.
Giersberg, H. 1922. Untersuchungen über Physiologie und Histologie des Eileiters der Reptilien und Vögel; nebst einem Beitrag zue Fasergenese. Zeitschrift für Wissenschaftliche Zoologie 70: 1–97.
Girling, J. E. 2002. The reptilian oviduct: a review of structure and function and directions for future research. Journal of Experimental Zoology 293: 141–170.
Girling, J. E. and Jones, S. M. 2003. *In vitro* progesterone production by maternal and embryonic tissues during gestation in the southern snow skink (*Niveoscincus microlepidotus*). General and Comparative Endocrinology 133: 100–108.

Girling, J. E., Cree, A. and Guillette Jr, L. J. 1997. Oviductal structure in a viviparous New Zealand gecko, Hoplodactylus maculatus. Journal of Morphology 234: 51–68.
Girling, J. E., Cree, A. and Guillette Jr., L. J. 1998. Oviducal structure in four species of gekkonid lizard differing in parity mode and eggshell structure. Reproduction, Fertility, and Development 10: 139–154.
Goldberg, S. R. 1971. Reproductive cycle of the ovoviviparous iguanid lizard *Sceloporus jarrovi* Cope. Herpetologica 27: 123–131.
Goldberg, S. R. 2007. Reproductive cycle of the Namibian striped skink, *Trachylepis sparsa* (Squamata: Scincidae) from southern Africa. African Zoology 42: 289–293.
Goldberg, S. R. 2009. Notes on the reproduction of the knob-scaled lizard, *Xenosaurus grandis* (Squamata: Xenosauridae), from Veracruz, Mexico. Texas Journal of Science 61: 317–322.
Goldberg, S. R. and Bezy, R. L. 1974. Reproduction in the island night lizard, *Xantusia riversiana.* Herpetologica 30: 350–360.
Gomez, D. and Ramírez-Pinilla, M. P. 2004. Ovarian histology of the placentotrophic *Mabuya mabouya* (Squamata, Scincidae). Journal of Morphology 259: 90–105.
Greene, H. W. 1970. Mode of reproduction in lizards and snakes of the Gomez Farias region, Tamaulipas, Mexico. Copeia 1970: 565–568.
Greer, A. E. 1968. Mode of reproduction in the squamate faunas of three altitudinally correlated life zones in East Africa. Herpetologica 24: 229–232.
Greer, A. E. 1970. A subfamilial classification of scincid lizards. Bulletin of the Museum of Comparative Zoology 139: 151–183.
Greer, A. E. 1976. On the adaptive significance of the loss of an oviduct in reptiles. Proceedings of the Linnaean Society of New South Wales 101: 242–249.
Griffith, O.W., Ujvari, B., Belov, K. and Thompson, M. B. 2013. Placental lipoprotein lipase (LPL) gene expression in a placentotrophic lizard, Pseudemoia entrecasteauxii. Journal of Experimental Zoology Part B: Molecular and Developmental Evolution 320: 465–470.
Grigg, G. C. and Harlow, P. 1981. A fetal-maternal shift of blood oxygen affinity in an Australian viviparous lizard, *Sphenomorphus quoyii* (Reptilia, Scincidae). Journal of Comparative Physiology 142: 495–499.
Guarino, F. M., Paulesu, L., Cardone, A., Bellini, L., Ghiara, G. and Angelini, F. 1998. Endocrine activity of the corpus luteum and placenta during pregnancy in *Chalcides chalcides* (Reptilia, Squamata). General and Comparative Endocrinology 111: 261–270.
Guillette, L. J., Jr. 1981. *Reproductive strategies and the evolution of viviparity in two allopatric populations of the Mexican lizard, Sceloporus aeneus*. Ph.D. Dissertation, University of Colorado, Boulder, Colorado, pp. 120.
Guillette, L. J., Jr. 1982a. Effects of gravidity on the metabolism of the reproductively bimodal lizard, *Sceloporus aeneus*. Journal of Experimental Zoology 223: 33–36.
Guillette, L. J., Jr. 1982b. The evolution of viviparity and placentation in the high-altitude, Mexican lizard *Sceloporus aeneus*. Herpetologica 38: 94–103.
Guillette, L. J., Jr. 1983. Notes concerning reproduction of the montane skink, *Eumeces copei.* Journal of Herpetology 17: 141–148.
Guillette, L. J., Jr. 1987. The evolution of viviparity in fish, amphibians, and reptiles: an endocrinological approach. pp. 523–562. In D. O. Norris and R. E. Jones (eds.), *Hormones and reproduction in fishes, amphibians, and reptiles.* Plenum Press, New York.
Guillette, L. J., Jr. 1989. The evolution of vertebrate viviparity: morphological modifications and endocrine control. pp. 219–233. In D. B. Wake and G. Roth (eds.), *Complex organismal functions: integration and evolution in vertebrates.* John Wiley & Sons, Chichester.
Guillette, L. J., Jr. and Casas-Andreu, G. 1980. Fall reproductive activity in the high altitude Mexican lizard, *Sceloporus grammicus microlepidotus*. Journal of Herpetology 14: 143–147.
Guillette, L. J., Jr. and Casas-Andreu, G. 1987. The reproductive biology of the high elevation Mexican lizard *Barisia imbricata.* Herpetologica 43: 29–38.
Guillette, L. J., Jr. and Jones, R. E. 1985. Ovarian, oviductal, and placental morphology of the reproductively bimodal lizard, *Sceloporus aeneus*. Journal of Morphology 184: 85–98.

Guillette Jr., L. J. and Méndez-de la Cruz, F. R. 1993. The reproductive cycle of the viviparous Mexican lizard *Sceloporus torquatus*. Journal of Herpetology 27: 168–174.
Guillette, L. J., Jr., Fox, S. L. and Palmer, B. D. 1989. Oviductal morphology and egg shelling in the oviparous lizards *Crotaphytus collaris* and *Eumeces obsoletus*. Journal of Morphology 201: 145–159.
Guillette, L. J., Jr., Dubois, D. H. and Cree, A. 1991. Prostaglandins, oviducal function, and parturient behavior in nonmammalian vertebrates. American Journal of Physiology 260: R854–R861.
Guillette, L. J., Jr., DeMarco, V., Palmer, B. and Masson, G. R., 1992. Effects of arachidonic acid, prostaglandin F2a, prostaglandin E2, and arginine vasotocin on induction of birth *in vivo* and *in vitro* in a viviparous lizard (*Sceloporus jarrovi*). General and Comparative Endocrinology 85: 477–485.
Haacke, W. 1885. Uber eine neue Art uterinlaer Brutpflege bei Reptilien. Zoologischer Anzeiger 8: 435–439.
Haagner, G. V., Branch, W. R. and Haagner, A. J. F. 2000. Notes on a collection of reptiles from Zambia and adjacent areas of the Democratic Republic of the Congo. Annals of the Eastern Cape Museums 1: 1–25.
Hall, B. K. 2003. Descent with modification: the unity underlying homology and homoplasy as seen through an analysis of development and evolution. Biological Reviews 78: 409–433.
Hamilton, H. L. 1952. *Lillie's development of the chick,* 3rd ed. Henry Holt & Co., New York, pp. 624.
Harrison, L. and Weekes, H. C. 1925. On the occurrence of placentation in the scincid lizard, *Lygosoma entrecasteauxi*. Proceedings of the Linnean Society of New South Wales 50: 472–486.
Hedges, S. B. and Conn, C. E. 2012. A new skink fauna from Caribbean islands (Squamata, Mabuyidae, Mabuyinae). Zootaxa 3288: 1–244.
Heimlich, E. M. and Heimlich, M. G. 1950. Uterine changes and placentation in the yucca night lizard. Journal of Entomology and Zoology 42: 5–12.
Herbert, J. F., Lindsay, L. A., Murphy, C. R. and Thompson, M. B. 2006. Calcium transport across the uterine epithelium of pregnant lizards. Herpetological Monographs 20: 205–211.
Herbert, J. F., Murphy, C. R. and Thompson, M. B. 2010. Calcium ATPase localization in the uterus of two species of *Pseudemoia* (Lacertilia: Scincidae) with complex placentae. Herpetological Conservation and Biology 5: 290–296.
Hernández-Franyutti, A., Uribe Aranzábal, M. C. and Guillette, L. J. 2005. Oogenesis in the viviparous matrotrophic lizard *Mabuya brachypoda*. Journal of Morphology 265: 152–164.
Heulin, B. 1990. Étude comparative de la membrane coquillère chez les souches ovipare et vivipare du lézard *Lacerta vivipara*. Canadian Journal of Zoology 68: 1015–1019.
Heulin, B. and Guillaume, C. 1989. Extension géographique des populations ovipares de *Lacerta vivipara*. Revue d'Écologie 44: 283–289.
Heulin, B., Arrayago M.- J. and Bea, A. 1989. Expérience d'hybridation entre les souches ovipare et vivipare du lézard *Lacerta vivipara*. Comptes rendus de l'Académie des sciences Paris 308: 341–346.
Heulin, B., Osenegg, K. and Lebouvier, M. 1991. Timing of embryonic development and birth dates in oviparous and viviparous strains of *Lacerta vivipara*: testing the predictions of an evolutionary hypothesis. Acta Oecologica 12: 517–528.
Heulin, B., Guillaume, C., Bea, A. and Arrayago, M. J. 1993. Interpretation biogéographique de la bimodalité de reproduction du lézard *Lacerta vivipara* Jacquin (Sauria, Lacertidae): un modèle pour l'étude de l'évolution de la viviparité. Compte Rendu des Séances de la Société de Biogéographie 69: 3–13.
Heulin, B., Osenegg-Leconte, K. and Michel, D. 1997. Demography of a bimodal reproductive species of lizard (*Lacerta vivipara*): survival and density characteristics of oviparous populations. Herpetologica 53: 432–444.
Heulin, B., Ghielmi, S., Vogrin, N., Surget-Groba, Y. and Guillaume, C. P. 2002. Variation in eggshell characteristics and in intrauterine egg retention between two oviparous clades of

the lizard *Lacerta vivipara*: insight into the oviparity-viviparity continuum in squamates. Journal of Morphology 252: 255–262.

Heulin, B., Stewart, J. R., Surget-Groba, Y., Bellaud, B., Jouan, F., Lancien, G. and Deunff, J. 2005. Development of the uterine shell glands during the preovulatory and early gestation periods in oviparous and viviparous *Lacerta vivipara*. Journal of Morphology 266: 80–96.

Heulin, B., Surget-Groba, Y., Sinervo, B., Miles, D. and Guiller, A. 2011. Dynamics of haplogroup frequencies and survival rates in a contact zone of two mtDNA lineages of the lizard *Lacerta vivipara*. Ecography 34: 436–447.

Hodges, W. L. 2004. Evolution of viviparity in horned lizards (*Phrynosoma*): testing the cold-climate hypothesis. Journal of Evolutionary Biology 17: 1230–1237.

Hoffman, L. H. 1970. Placentation in the garter snake, *Thamnophis sirtalis*. Journal of Morphology 131: 57–88.

Hosie, M. J., Adams, S. M., Thompson, M. B. and Murphy, C. R. 2003. Viviparous lizard, *Eulamprus tympanum*, shows changes in the uterine surface epithelium during early pregnancy that are similar to the plasma membrane transformation of mammals. Journal of Morphology 258: 346–357.

Hrabowski, H. 1926. Das Dotterorgan der Eidechsen. Zeitschrift für wissenschaftliche Zoologie 128: 305–382.

Hughes, R. L. 1993. Monotreme development with particular reference to the extraembryonic membranes. Journal of Experimental Zoology 266: 480–494.

Hutchinson, M. N., Donnellan, S. C., Baverstock, P. R., Krieg, M., Simms, S. and Burgin, S. 1990. Immunological relationships and generic revision of the Australian lizards assigned to the genus *Leiolopisma* (Scincidae: Lygosominae). Australian Journal of Zoology 38: 535–554.

Ibargüengoytía, N. R. 2004. Prolonged cycles as a common reproductive pattern in viviparous lizards from Patagonia, Argentina: reproductive cycle of *Phymaturus patagonicus*. Journal of Herpetology 38: 73–79.

Ibargüengoytía, N. R. and Casalins, L. M. 2007. Reproductive biology of the southernmost gecko *Homonota darwini*: convergent life-history patterns among southern hemisphere reptiles living in harsh environments. Journal of Herpetology 41: 72–80.

Ibargüengoytía, N. R. and Cussac, V. E. 1998. Reproductive biology of the viviparous lizard, *Liolaemus pictus* (Tropiduridae): biennial female reproductive cycle? Herpetological Journal 6: 137–143.

Ibrahim, M. M. 1977. Studies on viviparity in *Chalcides ocellatus* Forsk. Ph.D. Diss., University College of North Wales, Bangor, United Kingdom.

Ingermann, R. L., Berner, N. J. and Ragsdale, F. R. 1991. Effect of pregnancy and temperature on red cell oxygen-affinity in the viviparous snake *Thamnophis elegans*. Journal of Experimental Biology 156: 399–406.

Itonaga, K., Jones, S. M. and Wapstra, E. 2012a. Effects of maternal basking and food quantity during gestation provide evidence for the selective advantage of matrotrophy in a viviparous lizard. PLoS-One 7: 1–9.

Itonaga, K., Wapstra, E. and Jones, S. M. 2012b. A novel pattern of placental leucine transfer during mid to late gestation in a highly placentotrophic viviparous lizard. Journal of Experimental Zoology B: Molecular and Developmental Evolution 318: 308–315.

Jacobi, L. 1936. Ovoviviparie bei einheimischen Eidechsen. Vergleichende Untersuchungen an den Eiern und am Ovidukt von *Lacerta agilis, Lacerta vivipara* und *Anguis fragilis*. Zeitschrift für wissenschaftliche Zoologie 148: 401–464.

de Jacquin, J. F. 1787. Lacerta vivipara, observatio Jos. Francisci de Jacquin. Nota Acta Helvetica 1: 33–34.

Jaksić, F. M. and Schwenk, K. 1983. Natural history observations on *Liolaemus magellanicus*, the southernmost lizard in the world. Herpetologica 39: 457–461.

Jerez, A. and Ramírez-Pinilla, M. P. 2001. The allantoplacenta of *Mabuya mabouya* (Sauria, Scincidae). Journal of Morphology 249: 132–146.

Jerez, A. and Ramírez-Pinilla, M. P. 2003. Morphogenesis of extraembryonic membranes and placentation in *Mabuya mabouya* (Squamata, Scincidae). Journal of Morphology 258: 158–178.

Ji, X. and Braña, F. 1999. The influence of thermal and hydric environments on embryonic use of energy and nutrients, and hatchling traits, in the wall lizards (*Podarcis muralis*). Comparative Biochemistry and Physiology-Part A: Molecular & Integrative Physiology 124: 205–213.

Ji, X., Fu, S.- Y., Zhang, H.-S. and Sun, P. Y. 1996. Material and energy budget during incubation in a Chinese skink, *Eumeces chinensis.* Amphibia-Reptilia 17: 209–216.

Ji, X., Lin, L. H., Luo, L. G., Lu, H. L., Gao, J. F. and Han, J. 2006. Gestation temperature affects sexual phenotype, morphology, locomotor performance, and growth of neonatal brown forest skinks, *Sphenomorphus indicus.* Biological Journal of the Linnean Society 88: 453–463.

Ji, X., Lin, C. X., Lin, L. H., Qiu, Q. B. and Du, Y. 2007. Evolution of viviparity in warm-climate lizards: an experimental test of the maternal manipulation hypothesis. Journal of Evolutionary Biology 20: 1037–1045.

Jones, C. J. P., Cateni, C., Guarino, F. M. and Paulesu, L. R. 2003. Glycosylation of the materno-foetal interface in the pregnant viviparous placentotrophic lizard *Chalcides chalcides :* a lectin histochemical study. Placenta 24: 489–500.

Jones, R. E. and Baxter, D. C. 1991. Gestation, with emphasis on corpus luteum biology, placentation, and parturition. pp. 205–301. In P. K. T. Pang, M. P. Schreibman and R. Jones (eds.), *Vertebrate endocrinology: fundamentals and biomedical implications, vol. 4, part A.* Academic Press, New York.

Jones, R. E. and Guillette, L. J., Jr. 1982. Hormonal control of oviposition and parturition in lizards. Herpetologica 38: 80–93.

Jones, S. M. and Swain, R. 2010. Placental transfer of 3H-oleic acid in three species of viviparous lizards: a route for supplementation of embryonic fat bodies? Herpetological Monographs 20: 186–193.

Jones, S. M., Bennett, E. J. and Swadling, K. M. 1998. Lipids in yolks and neonates of the viviparous lizard *Niveoscincus metallicus.* Comparative Biochemistry and Physiology Part B: Biochemistry and Molecular Biology 121: 465–470.

Kasturirangan, L. R. 1951a. Placentation in the sea-snake, *Enhydrina schistosa* (Daudin). Proceedings of the Indian Academy of Sciences B 34: 1–32.

Kasturirangan, L. R. 1951b. The allantoplacenta of the sea-snake, *Hydrophis cyanocinctus* Daudin. Journal of the Zoological Society of India 3: 277–290.

King, M. A. 1977. Reproduction in the Australian gecko *Phyllodactylus marmoratus* (Gray). Herpetologica 33: 7–13.

Kupriyanova, L. A., Mayer, W. and Böhme, W. 2006. Karyotype diversity of the Eurasian lizard *Zootoca vivipara* (Jacquin 1787) from Central Europe and the evolution of viviparity. pp. 67–72. In M. Vences, J. Köhler, T. Ziegler and W. Böhme (eds.), Herpetologia Bonnensis II. *Proceedings of the 13th Congress of the Societas Europaea Herpetologica.*

Labra, A. and Bozinovic, F. 2002. Interplay between pregnancy and physiological thermoregulation in *Liolaemus* lizards. Ecoscience 9: 421–426.

Lamarck, J.- B. 1835. *Histoire naturelle des animaux sans vertèbres*, 2nd ed. J. B. Bailliere, Paris, pp. 440.

Lambert, S. M. and Wiens, J. J. 2013. Evolution of viviparity: a phylogenetic test of the cold-climate hypothesis in phrynosomatid lizards. Evolution 67: 2614–2630.

Leal, F. and Ramírez-Pinilla, M. P. 2008. Morphological variation in the allantoplacenta within the genus *Mabuya* (Squamata: Scincidae). Anatomical Record 291: 1124–1139.

Lee, M. S. Y. and Shine, R. 1998. Reptilian viviparity and Dollo's law. Evolution 52: 1441–1450.

Lemos-Espinal, J. A., Smith, G. R. and Ballinger, R. E. 2003. Ecology of *Xenosaurus grandis agrenon,* a knob-scaled lizard from Oaxaca, México. Journal of Herpetology 37: 192–196.

Lemus, D. and Badinez, O. 1967. Placentacion en *Liolaemus gravenhorsti.* Biologica 41: 55–68.

Li, H., Qu, Y. -F., Hu, R. -B. and Ji, X. 2009. Evolution of viviparity in cold-climate lizards: testing the maternal manipulation hypothesis. Evolutionary Ecology 23: 777–790.

Lin, C. -X., Zhang, L. and Ji, X. 2008. Influence of pregnancy on locomotor and feeding performances of the skink, *Mabuya multifasciata:* why do females shift thermal preferences when pregnant? Zoology 111: 188–195.

Lindtke, D., Mayer, W. and Böhme, W. 2010. Identification of a contact zone between oviparous and viviparous common lizards (*Zootoca vivipara*) in central Europe: reproductive strategies and natural hybridization. Salamandra 46: 73–82.

Linville, B. J., Stewart, J. R., Ecay, T. W., Herbert, J. F., Parker, S. L. and Thompson, M. B. 2010. Placental calcium provision in a lizard with prolonged oviductal egg retention. Journal of Comparative Physiology B 180: 221–227.

Lorioux, S., Lisse, H. and Lourdais, O. 2013. Dedicated mothers: predation risk and physical burden do not alter thermoregulatory behaviour of pregnant vipers. Animal Behaviour 86: 401–408.

Lovern, M. B. 2011. Hormones and reproductive cycles in lizards. pp. 321–353. In D. O. Norris and K. H. Lopez (eds.), *Hormones and reproduction of vertebrates. Volume 3, Reptiles.* Academic Press, San Diego.

Luckett, W. P. 1977. Ontogeny of amniote fetal membranes and their application to phylogeny. pp. 439–516. In M. K. Hecht, P. C. Goody and B. M. Hecht (eds.), *Major patterns in vertebrate evolution.* Plenum Press, New York.

Lynch, V. J. 2010. Live-birth in vipers (Viperidae) is a key innovation and adaptation to global cooling during the Cenozoic. Evolution 63: 2457–2465.

Lynch, V. J. and Wagner, G. P. 2010. Did egg-laying boas break Dollo's Law? Phylogenetic evidence for reversal to oviparity in sand boas (*Eryx:* Boidae). Evolution 64: 207–216.

Marion, K. R. and Sexton, O. J. 1971. The reproductive cycle of the lizard *Sceloporus malachiticus* in Costa Rica. Copeia 1971: 517–526.

Martínez-Torres, M., Hernández-Caballero, M. E., Luis-Díaz, J. A., Ortiz-López, G., Cárdenas-León, M. and Moreno-Fierros, L. 2010. Effects of luteectomy in early pregnancy on the maintenance of gestation and plasma progesterone concentrations in the viviparous temperate lizard *Barisia imbricata imbricata.* Reproductive Biology and Endocrinology 8: 19.

Masson, G. R. and Guillette, L. J., Jr. 1987. Changes in oviducal vascularity during the reproductive cycle of three oviparous lizards (*Eumeces obsoletus*, *Sceloporus undulatus* and *Crotaphytus collaris*). Journal of Reproduction and Fertility 80: 361–371.

Mathies, T. and Andrews, R. M. 1995. Thermal and reproductive biology of high and low elevation populations of the lizard *Sceloporus scalaris*: implications for the evolution of viviparity. Oecologia 104: 101–111.

Mathies, T. and Andrews, R. M. 1996. Extended egg retention and its influence on embryonic development and egg water balance: implications for the evolution of viviparity. Physiological Zoology 69: 1021–1035.

Mathies, T. and Andrews, R. M. 1997. Influence of pregnancy on the thermal biology of the lizard, *Sceloporus jarrovi*: why do pregnant females exhibit low body temperatures? Functional Ecology 11: 498–507.

Mathies, T. and Andrews, R. M. 2000. Does reduction of the eggshell occur concurrently with or subsequent to the evolution of viviparity in phrynosomatid lizards? Biological Journal of the Linnean Society 71: 719–736.

Matthews, L. H. 1955. The evolution of viviparity in vertebrates. Memoirs of the Society of Endocrinology 4: 129–148.

Mausfeld, P., Schmitz, A., Böhme, W., Misof, B., Vrcibradic, D. and Rocha, C. F. D. 2002. Phylogenetic affinities of *Mabuya atlantica* Schmidt, 1945, endemic to the Atlantic Ocean archipelago of Fernando de Noronha (Brazil): necessity of partitioning the genus *Mabuya* Fitzinger, 1826 (Scincidae: Lygosominae). Zoologischer Anzeiger 241: 281–293.

McCoy, C. J. 1968. Reproductive cycles and viviparity in Guatemalan *Corytophanes percarinatus* (Reptilia: Iguanidae). Herpetologica 24: 175–178.

Medina, M. and Ibargüengoytía, N. R. 2010. How do viviparous and oviparous lizards reproduce in Patagonia? A comparative study of three species of *Liolaemus.* Journal of Arid Environments 74: 1024–1032.

Mell, R. 1929. *Beiträge zur Fauna sinica. IV. Grundzüge einer Ökologie der chinesischen Reptilien und einer herpetologischen Tiergeographie Chinas.* Walter de Gruyter & Co., Liepzig, Germany, pp. 282.
Méndez-de la Cruz, F. R., Cruz, M. V. S. and Andrews, R. M. 1998. Evolution of viviparity in the lizard genus *Sceloporus*. Herpetologica 54: 521–532.
Miller, M. R. 1951. Some aspects of the life history of the yucca night lizard, *Xantusia vigilis*. Copeia 1951: 114–120.
Miller, M. R. 1954. Further observations on reproduction in the lizard *Xantusia vigilis*. Copeia 1954: 38–40.
Mingazzini, P. 1892. L'oolisi della *Seps chalcides*. Atti della Accademia Nazionale dei Lincei: Rendiconti V 1: 41–45.
Mossman, H. W. 1937. Comparative morphogenesis of the fetal membranes and accessory uterine structures. Carnegie Institute Contributions in Embryology 26: 129–246.
Mossman, H. W. 1987. *Vertebrate fetal membranes.* Rutgers University Press, New Brunswick, New Jersey, pp. 383.
Mulaik, D. D. M. 1946. A comparative study of the urinogenital systems of an oviparous and two ovoviviparous species of the lizard genus *Sceloporus*. Bulletin of the University of Utah 37: 1–24.
Murphy, B. F. and Thompson, M. B. 2011. A review of the evolution of viviparity in squamate reptiles: the past, present and future role of molecular biology and genomics. Journal of Comparative Physiology B: Biochemical, Systemic, and Environmental Physiology 181: 575–594.
Murphy, B. F., Belov, K. and Thompson, M. B. 2009. Evolution of viviparity and uterine angiogenesis: vascular endothelial growth factor (VEGF) in oviparous and viviparous skinks. Journal of Experimental Zoology B: Molecular and Developmental Evolution 314: 148–156.
Murphy, B. F., Parker, S. L., Murphy, C. R. and Thompson, M. B. 2010. Angiogenesis of the uterus and chorioallantois in the eastern water skink *Eulamprus quoyii*. Journal of Experimental Biology 213: 3340–3347.
Murphy, B. F., Parker, S. L., Murphy, C. R. and Thompson, M. B. 2011. Placentation in the eastern water skink *(Eulamprus quoyii)*: a placentome-like structure in a lecithotrophic lizard. Journal of Anatomy 218: 678–689.
Murphy, B. F., Brandley, M. C., Murphy, C. R. and Thompson, M. B. 2012. Morphology and development of the placentae in *Eulamprus quoyii* group skinks (Squamata: Scincidae). Journal of Anatomy 220: 454–471.
Neill, W. T. 1964. Viviparity in snakes: some ecological and zoogeographical considerations. American Naturalist 98: 35–55.
Odierna, G., Aprea, G., Capriglione, T. and Puky, M. 2004. Chromosomal evidence for the double origin of viviparity in the European common lizard, *Lacerta (Zootoca) vivipara*. Herpetogical Journal 14: 157–160.
Olsson, M. and Shine, R. 1998. Timing of parturition as a maternal care tactic in an alpine lizard species. Evolution 52: 1861–1864.
Olsson, M. and Shine, R. 1999. Plasticity in frequency of reproduction in an alpine lizard, *Niveoscincus microlepidotus*. Copeia 1999: 794–796.
Olsson, M., Shine, R. and Bak-Olsson, E. 2000. Locomotor impairment of gravid lizards: is the burden physical or physiological? Journal of Evolutionary Biology 13: 263–268.
Ortega, A. and Barbault, A. 1984. Reproductive cycles in the mesquite lizard *Sceloporus grammicus*. Journal of Herpetology 18: 168–175.
Ortiz, E. and Morales, M. 1974. Development and function of the female reproductive tract of the tropical lizard, *Anolis pulchellus*. Physiological Zoology 47: 207–217.
Owen, R. 1866. *On the anatomy of vertebrates. Vol. 1, Fishes and reptiles.* Longmans, Green, & Co., London, pp. 650.
Packard, G. C. 1966. The influence of ambient temperature and aridity on modes of reproduction and excretion of amniote vertebrates. American Naturalist 100: 677–682.

Packard, G. C. and Packard, M. J. 1988. The physiological ecology of reptilian eggs and embryos. pp. 523–605. In C. Gans and R. B. Huey (eds.), *Biology of the Reptilia. vol. 16.* Liss, New York.
Packard, G. C., Tracy, C. R. and Roth, J. J. 1977. The physiological ecology of reptilian eggs and embryos, and the evolution of viviparity within the Class Reptilia. Biological Reviews 52: 71–105.
Packard, M. J. 1994. Patterns of mobilization and deposition of calcium in embryos of oviparous, amniotic vertebrates. Israel Journal of Zoology 40: 481–492.
Packard, M. J. and DeMarco, V. G. 1991. Eggshell structure and formation in eggs of oviparous reptiles. pp. 53–70. In D. C. Deeming and M. W. J. Ferguson (eds.), *Egg incubation: its effects on embryonic development in birds and reptiles.* Cambridge University Press, Cambridge, U.K.
Packard, M. J., Packard, G. C., Miller, J. D., Jones, M. E. and Gutze, W. H. N. 1985. Calcium mobilization, water balance, and growth in embryos of the agamid lizard *Amphibolurus barbatus*. Journal of Experimental Zoology 235: 349–357.
Packard, M. J., Thompson, M. B., Goldie, K. N. and Vos, M. 1988. Aspects of shell formation in eggs of the tuatara, *Sphenodon punctatus*. Journal of Morphology 197: 147–157.
Packard, M. J., Phillips, J. A. and Packard, G. C. 1992. Sources of mineral for green iguanas (Iguana iguana) developing in eggs exposed to different hydric environments. Copeia 1992: 851–858.
Painter, D. L. and Moore, M. C. 2005. Steroid hormone metabolism by the chorioallantoic placenta of the mountain spiny lizard *Sceloporus jarrovi* as a possible mechanism for buffering maternal-fetal hormone exchange. Physiological and Biochemical Zoology 78: 364–372.
Palmer, B. D., Demarco, V. G. and Guillette Jr., L. J. 1993. Oviductal morphology and eggshell formation in the lizard, *Sceloporus woodi*. Journal of Morphology 217: 205–217.
Panigel, M. 1951. Rapports anatomo-histologiques établis au cours de la gestation entre l'oeuf et l'oviducte maternel chez le lézard ovovivipare *Zootoca vivipara* W. (*Lacerta vivipara* J.) Bulletin de la Société zoologique de France 76: 163–170.
Panigel, M. 1953. Rôle des corps jaunes au cours de la gestation chez le lézard vivipare, *Zootoca vivipara*. Comptes Rendus de l'Académie des Sciences (Paris) Series D 236: 849–851.
Panigel, M. 1956. Contribution a l'étude de l'ovoviviparité chez les reptiles: gestation et parturition chez le lézard vivipare *Zootoca vivipara*. Annales des sciences naturelles. Zoologie et biologie animale. (XI) 18: 569–668.
Parameswaran, K. N. 1962. The foetal membranes and placentation of *Enhydris dussumieri* (Smith). Proceedings of the Indian Academy of Science B 56: 302–327.
Parker, S. L. and Andrews, R. M. 2006. Evolution of viviparity in sceloporine lizards: *in utero* PO_2 as a developmental constraint during egg retention. Physiological and Biochemical Zoology 79: 581–592.
Parker, S. L., Murphy, C. R. and Thompson, M. B. 2010a. Uterine angiogenesis in squamate reptiles: implications for the evolution of viviparity. Herpetological Conservation and Biology 5: 330–334.
Parker, S. L., Manconi, F., Murphy, C. R. and Thompson, M. B. 2010b. Uterine and placental angiogenesis in the Australian skinks, *Ctenotus taeniolatus*, and *Saiphos equalis*. Anatomical Record 293: 829–838.
Parra-Olea, G. and Wake, D. B. 2001. Extreme morphological and ecological homoplasy in tropical salamanders. Proceedings of the National Academy of Sciences 98: 7888–7891.
Paulesu, L., Romagnoli, R., Marchetti, M., Cintorino, M., Ghiara, P., Guarino, F. M. and Ghiara, G. 1995. Cytokines in the viviparous reproduction of squamate reptiles: Interleukin-1α (IL-1α) and IL-1β in placental structures of a skink. Placenta 16: 193–205.
Paulesu, L., Bigliardi, E., Paccagnini, E., Ietta, F., Cateni, C., Guillaume, C. P. and Heulin, B. 2005. Cytokines in the oviparity/viviparity transition: evidence of the interleukin-1 system in a species with reproductive bimodality, the lizard *Lacerta vivipara*. Evolution & Development 7: 282–288.
Perkins, M. J. and Palmer, B. D. 1996. Histology and functional morphology of the oviduct of an oviparous snake, *Diadophis punctatus*. Journal of Morphology 227: 67–79.

Picariello, O., Ciarcia, G. and Angelini, F. 1989. The annual cycle of oviduct in *Tarentola m. mauritanica* L. (Reptilia, Gekkonidae). Amphibia-Reptilia 10: 371–386.
Pincheira-Donoso, D., Tregenza, T., Witt, M. J. and Hodgson, D. J. 2013. The evolution of viviparity opens opportunities for lizard radiation but drives it into a climatic cul-de-sac. Global Ecology and Biogeography 22: 857–867.
Pyron, R. A. and Burbrink, F. T. 2013. Early origin of viviparity and multiple reversions to oviparity in squamate reptiles. Ecology Letters 17: 13–21.
Pyron, R. A., Burbrink, F. T. and Wiens, J. J. 2013. A phylogeny and revised classification of Squamata, including 4161 species of lizards and snakes. BMC Evolutionary Biology 13: 93.
Qualls, C. P. 1996. Influence of the evolution of viviparity on eggshell morphology in the lizard, *Lerista bougainvilli*. Journal of Morphology 228: 119–125.
Qualls, C. P. and Andrews, R. M. 1999a. Maternal body volume constrains water uptake by lizard eggs *in utero*. Functional Ecology 13: 845–851.
Qualls, C. P. and Andrews, R. M. 1999b. Cold climates and the evolution of viviparity in reptiles: cold incubation temperatures produce poor-quality offspring in the lizard, *Sceloporus virgatus*. Biological Journal of the Linnean Society 67: 353–376.
Qualls, C. P. and Shine, R. 1995. Maternal body-volume as a constraint on reproductive output in lizards: evidence from the evolution of viviparity. Oecologia 103: 73–78.
Qualls, C. P. and Shine, R. 1996. Reconstructing ancestral reaction norms: an example using the evolution of reptilian viviparity. Functional Ecology 10: 688–697.
Qualls, C. P. and Shine, R. 1998. Costs of reproduction in conspecific oviparous and viviparous lizards, *Lerista bougainvillii*. Oikos 82: 539–551.
Qualls, C. P. and Shine, R. 2002. *Lerista bougainvillii*, a case study for the evolution of viviparity in reptiles. Journal of Evolutionary Biology 11: 63–78.
Qualls, C. P., Shine, R., Donnellan, S. and Hutchinson, M. 1995. The evolution of viviparity in the Australian scincid lizard, *Lerista bougainvilli*. Journal of Zoology (London) 237: 13–26.
Qualls, C. P., Andrews, R. M. and Mathies, T. 1997. The evolution of viviparity and placentation revisted [sic]. Journal of Theoretical Biology 185: 129–135.
Rafferty, A. R., Evans, R. G., Scheelings, T. F. and Reina, R. D. 2013. Limited oxygen availability *in utero* may constrain the evolution of live birth in reptiles. American Naturalist 181: 245–253.
Ragsdale, F. R. and Ingermann, R. L. 1993. Biochemical bases for difference in oxygen affinity of maternal and fetal red blood cells of rattlesnake. American Journal of Physiology 264: R481–R486.
Ragsdale, F. R., Imel, K. M., Nilsson, E. E. and Ingermann, R. L. 1993. Pregnancy-associated factors affecting organic phosphate levels and oxygen affinity of garter snake red cells. General and Comparative Endocrinology 91: 181–188.
Rahn, H. 1939. Structure and function of placenta and corpus luteum in viviparous snakes. Proceedings of the Society Experimental Biology and Medicine 40: 381–382.
Rahn, H. 1942. The reproductive cycle of the prairie rattler. Copeia 1942: 233–240.
Ramírez-Pinilla, M. P. 2006. Placental transfer of nutrients during gestation in an Andean population of the highly matrotrophic lizard genus *Mabuya* (Squamata: Scincidae). Herpetological Monographs 20: 194–204.
Ramírez-Pinilla, M. P., Serrano, V. H. and Galeano, J. C. 2002. Annual reproductive activity of *Mabuya mabouya* (Squamata, Scincidae). Journal of Herpetology 36: 667–677.
Ramírez-Pinilla, M. P., De Pérez, G. and Carreño-Escobar, J. F. 2006. Allantoplacental ultrastructure of an Andean population of *Mabuya* (Squamata, Scincidae). Journal of Morphology 267: 1227–1247.
Ramírez-Pinilla, M. P., Rueda, E. D. and Stashenko, E. 2011a. Transplacental nutrient transfer during gestation in the Andean lizard *Mabuya* sp. (Squamata, Scincidae). Journal of Comparative Physiology B 181: 249–268.
Ramírez-Pinilla, M. P., Parker, S. L., Murphy, C. R. and Thompson, M. B. 2011b. Uterine and chorioallantoic angiogenesis and changes in the uterine epithelium during gestation in

the viviparous lizard, *Niveoscincus coventryi* (Squamata: Scincidae). Journal of Morphology 273: 8–23.
Raynaud, A. 1962. Lés ebauches des membres de l'embryon d'orvet (*Anguis fragilis* L.). Comptes Rendus de l'Academie des Sciences (Paris) 254: 3449–3451.
Robert, K. A. and Thompson, M. B. 2001. Viviparous lizard selects sex of embryos. Nature 412: 698–699.
Robert, K. A. and Thompson, M. B. 2009. Viviparity and temperature-dependent sex determination. Sexual Development 4: 119–128.
Robertson, I. A. D., Chapman, D. M. and Chapman, R. F. 1965. Notes on the biology of the lizards *Agama cyanogaster* and *Mabuya striata* collected in the Rukwa Valley, Southwest Tanganyika. Proceedings of the Zoological Society of London 145: 305–320.
Rock, J. and Cree, A. 2003. Intraspecific variation in the effect of temperature on pregnancy in the viviparous gecko *Hoplodactylus maculatus*. Herpetologica 59: 8–22.
Rodríguez-Díaz, T. and Braña, F. 2011a. Shift in thermal preferences of female oviparous common lizards during egg retention: insights into the evolution of reptilian viviparity. Evolutionary Biology 38: 352–359.
Rodríguez-Díaz, T. and Braña, F. 2011b. Plasticity and limitations of extended egg retention in oviparous *Zootoca vivipara* (Reptilia: Lacertidae). Biological Journal of the Linnean Society 102: 75–82.
Rodríguez-Díaz, T. and Braña, F. 2012. Altitudinal variation in egg retention and rates of embryonic development in oviparous *Zootoca vivipara* fits predictions from the cold-climate model on the evolution of viviparity. Journal of Evolutionary Biology 25: 1877–1887.
Rodríguez-Díaz, T., González, F., Ji, X. and Braña, F. 2010. Effects of incubation temperature on hatchling phenotypes in an oviparous lizard with prolonged egg retention: are the two main hypotheses on the evolution of viviparity compatible? Zoology 113: 33–38.
Roitberg, E. S., Kuranova, V. N., Bulakhova, N. A., Orlova, V. F., Eplanova, G. V., Zinenko, O. I., Shamgunova, R. R., Hofmann, S. and Yakovlev, V. A. 2013. Variation of reproductive traits and female body size in the most widely-ranging terrestrial reptile: testing the effects of reproductive mode, lineage, and climate. Evolutionary Biology 40: 420–438.
Rollinat, R. 1904. Observations sur la tendance vers l'ovoviviparité chez quelques sauriens et ophidiens de la France centrale. Mémoires de la Société zoologique de France 17: 30–41.
Romanoff, A. L. 1960. *The avian embryo: structural and functional development.* MacMillan, New York.
Saint Girons, H. 1957. Le cycle sexuel chez *Vipera aspis* (L.) dans l'ouest de la France. Bulletin biologique de la France et de la Belgique 91: 284–350.
Sangha, S., Smola, M. A., McKinney, S. L., Crotzer, D. R., Shadrix, C. A. and Stewart, J. R. 1996. The effect of surgical removal of oviductal eggs on placental function and size of neonates in the viviparous snake *Virginia striatula*. Herpetologica 52: 32–36.
Schuett, G. W., Repp, R. A., Amarello, M. and Smith, C. F. 2013. Unlike most vipers, female rattlesnakes *(Crotalus atrox)* continue to hunt and feed throughout pregnancy. Journal of Zoology 289: 101–110.
Schulte, J. A. and Moreno-Roark, F. 2010. Live birth in iguanian lizards predates the Pliocene-Pleistocene. Biological Letters 6: 216–218.
Schulte, J. A., Macey, J. R., Espinoza, R. E. and Larson, A. 2000. Phylogenetic relationships in the iguanid lizard genus *Liolaemus*: multiple origins of viviparous reproduction and evidence for recurring Andean vicariance and dispersal. Biological Journal of the Linnaean Society 2000: 69–102.
Schwarzkopf, L. 1996. Decreased food intake in reproducing lizards: a fecundity-dependent cost of reproduction. Australian Journal of Ecology 21: 355–362.
Schwarzkopf, L. and Andrews, R. M. 2012a. Are moms manipulative or just selfish? Evaluating the "maternal manipulation hypothesis" and implications for life-history studies of reptiles. Herpetologica 68: 147–159.

Schwarzkopf, L. and Andrews, R. M. 2012b. "Selfish mothers" use "maternal manipulation" to maximize lifetime reproductive success. Herpetologica 68: 308–311.
Schwarzkopf, L. and Shine, R. 1991. Thermal biology of reproduction in viviparous skinks, *Eulamprus tympanum*: why do gravid females bask more? Oecologia 88: 562–569.
Semon, R. 1894. Die Embryonalhüllen der Monotremen und Marsupialier. Denkschriften der Medicinisch-Naturwissenschaftlichen Gesellschaft zu Jena 5: 19–58.
Sergeev, A. M. 1940. Researches in the viviparity of reptiles. Moscow Society of Naturalists 13: 1–34.
Shadrix, C. A., Crotzer, D. R., McKinney, S. L. and Stewart, J. R. 1994. Embryonic growth and calcium mobilization in oviposited eggs of the scincid lizard, *Eumeces fasciatus*. Copeia 1994: 493–498.
Shine, R. 1980. "Costs" of reproduction in reptiles. Oecologia 46: 92–100.
Shine, R. 1983a. Reptilian reproductive modes: the oviparity-viviparity continuum. Herpetologica 39: 1–8.
Shine, R. 1983b. Reptilian viviparity in cold climates: testing the assumptions of an evolutionary hypothesis. Oecologia 57: 397–405.
Shine, R. 1984. Physiological and ecological questions on the evolution of reptilian viviparity. pp. 147–154. In R. Seymour (ed.), *Respiration and metabolism of embryonic vertebrates*. Junk Press, The Hague, Netherlands.
Shine, R. 1985. The evolution of viviparity in reptiles: an ecological analysis. pp. 605–694. In C. Gans and F. Billet (eds.), *Biology of the Reptilia, vol. 15*. John Wiley & Sons, New York.
Shine, R. 1995. A new hypothesis for the evolution of viviparity in reptiles. American Naturalist 145: 809–823.
Shine, R. 2002. Reconstructing an adaptationist scenario: what selective forces favor the evolution of viviparity in montane reptiles? American Naturalist 160: 582–593.
Shine, R. 2003. Effects of pregnancy on locomotor performance: an experimental study on lizards. Oecologia 136: 450–456.
Shine, R. 2004. Does viviparity evolve in cold climate reptiles because pregnant females maintain stable (not high) body temperatures? Evolution 58: 1809–1818.
Shine, R. 2006. Is increased maternal basking an adaptation or a pre-adaptation to viviparity in lizards? Journal of Experimental Zoology 305A: 524–535.
Shine, R. 2012. Manipulative mothers and selective forces: the interplay between reproduction and thermoregulation in reptiles. Herpetologica 68: 289–298.
Shine, R. 2014. Evolution of an evolutionary hypothesis: a history of changing ideas about the adaptive significance of viviparity in reptiles. Journal of Herpetology 48: 147–161.
Shine, R. and Berry, J. F. 1978. Climatic correlates of live bearing in squamate reptiles. Oecologia (Berlin) 33: 261–268.
Shine, R. and Bull, J. J. 1979. The evolution of live-bearing in lizards and snakes. American Naturalist 113: 905–923.
Shine, R. and Harlow, P. 1993. Maternal thermoregulation influences offspring viability in a viviparous lizard. Oecologia 96: 122–127.
Shine, R. and Harlow, P. S. 1996. Maternal manipulation of offspring phenotypes via nest site selection in an oviparous lizard. Ecology 77: 1808–1817.
Shine, R. and Lee, M. S. Y. 1999. A reanalysis of the evolution of viviparity and egg-guarding in squamate reptiles. Herpetologica 55: 538–549.
Shine, R. and Olsson, M. 2003. When to be born? Prolonged pregnancy or incubation enhances locomotor performance in neonatal lizards (Scincidae). Journal of Evolutionary Biology 16: 823–832.
Shine, R., Elphick, M. J. and Barrott, E. G. 2003. Sunny side up: lethally high, not low, nest temperatures may prevent oviparous reptiles from reproducing at high elevations. Biological Journal of the Linnean Society 78: 325–334.
Siegel, D. S., Miralles, A., Chabarria, R. E. and Aldridge, R. D. 2011. Female reproductive anatomy: cloaca, oviduct, and sperm storage. pp. 347–409. In R. D. Aldridge and D. M. Sever (eds.), *Reproductive biology and phylogeny of snakes*. Science Publishers, Enfield, New Hampshire.

Simbotwe, M. P. 1980. Reproductive biology of the skinks *Mabuya striata* and *Mabuya quinquetaeniata* in Zambia. Herpetologica 36: 99–104.
Sites Jr., J. W., Reeder, T. W. and Wiens, J. J. 2011. Phylogenetic insights on evolutionary novelties in lizards and snakes: sex, birth, bodies, niches, and venom. Annual Review of Ecology, Evolution, and Systematics 42: 227–244.
Skinner, A., Hugall, A. F. and Hutchinson, M. N. 2011. Lygosomine phylogeny and the origins of Australian scincid lizards. Journal of Biogeography 38: 1044–1058.
Smith, H. M. 1939. The Mexican and Central American lizards of the genus *Sceloporus*. Zoological Series of the Field Museum of Natural History 26: 1–397.
Smith, S. A. 2001. *A molecular phylogenetic study of the Eugongylus group of skinks*. Ph.D. Dissertation, University of Adelaide, pp. 492.
Smith, S. A. and Shine, R. 1997. Intraspecific variation in reproductive mode within the scincid lizard *Saiphos equalis*. Australian Journal of Zoology 45: 435–445.
Smith, S. A., Austin, C. C. and Shine, R. 2001. A phylogenetic analysis of variation in reproductive mode within an Australian lizard (*Saiphos equalis*, Scincidae). Biological Journal of the Linnean Society 74: 131–139.
Stanley, E. L., Bauer, A. M., Jackman, T. R., Branch, W. R. and Mouton, P. L. F. N. 2011. Between a rock and a hard polytomy: rapid radiation in the rupicolous girdled lizards (Squamata: Cordylidae). Molecular Phylogenetics and Evolution 58: 53–70.
Stebbins, R. C. 1954. *Amphibians and reptiles of western North America*. McGraw Hill, New York, pp. 528.
Stewart, J. R. 1984. Thermal biology of the live bearing lizard *Gerrhonotus coeruleus*. Herpetologica 40: 349–355.
Stewart, J. R. 1985. Placentation in the lizard *Gerrhonotus coeruleus* with a comparison to the extraembryonic membranes of the oviparous *Gerrhonotus multicarinatus* (Sauria, Anguidae). Journal of Morphology 185: 101–114.
Stewart, J. R. 1989. Facultative placentotrophy and the evolution of squamate placentation: quality of eggs and neonates in *Virginia striatula*. American Naturalist 133: 111–137.
Stewart, J. R. 1990. Development of the extraembryonic membranes and histology of the placentae in *Virginia striatula* (Squamata: Serpentes). Journal of Morphology 205: 1–11.
Stewart, J. R. 1992. Placental structure and nutritional provision to embryos in predominantly lecithotrophic viviparous reptiles. American Zoologist 32: 303–312.
Stewart, J. R. 1993. Yolk sac placentation in reptiles: structural innovation in a fundamental vertebrate nutritional system. Journal Experimental Zoology 266: 431–449.
Stewart, J. R. 1997. Morphology and evolution of the egg of oviparous amniotes. pp. 291–326. In S. S. Sumida and K. L. M. Martin (eds.), *Amniote Origins*. Academic Press, San Diego.
Stewart, J. R. 2013. Fetal nutrition in lecithotrophic squamate reptiles: toward a comprehensive model for evolution of viviparity and placentation. Journal of Morphology 274: 824–843.
Stewart, J. R. and Blackburn, D. G. 1988. Reptilian placentation: structural diversity and terminology. Copeia 1988: 838–851.
Stewart, J. R. and Brasch, K. R. 2003. Ultrastructure of the placentae of the natricine snake, *Virginia striatula* (Reptilia: Squamata). Journal Morphology 255: 177–201.
Stewart, J. R. and Castillo, R. E. 1984. Nutritional provision of the yolk of two species of viviparous reptiles. Physiological Zoology 57: 377–383.
Stewart, J. R. and Ecay, T. W. 2010. Patterns of maternal provision and embryonic mobilization of calcium in oviparous and viviparous squamate reptiles. Herpetological Conservation and Biology 5: 341–359.
Stewart, J. R. and Florian, J. D., Jr. 2000. Ontogeny of the extraembryonic membranes of the oviparous lizard, *Eumeces fasciatus* (Squamata: Scincidae). Journal of Morphology 244: 81–107.
Stewart, J. R. and Thompson, M. B. 1993. A novel pattern of embryonic nutrition in a viviparous reptile. Journal of Experimental Biology 174: 97–108.
Stewart, J. R. and Thompson, M. B. 1994. Placental structure of the Australian lizard, *Niveoscincus metallicus* (Squamata: Scincidae). Journal of Morphology 220: 223–236.

Stewart, J. R. and Thompson, M. B. 1996. Evolution of reptilian placentation: development of extraembryonic membranes of the Australian scincid lizards *Bassiana duperreyi* (oviparous) and *Pseudemoia entrecasteauxii* (viviparous). Journal of Morphology 227: 349–370.
Stewart, J. R. and Thompson, M. B. 1998. Placental ontogeny of the Australian scincid lizards *Niveoscincus coventryi* and *Pseudemoia spenceri*. Journal of Experimental Zoology 282: 535–559.
Stewart, J. R. and Thompson, M. B. 2000. Evolution of placentation among squamate reptiles: recent research and future directions. Comparative Biochemistry and Physiology A: Molecular and Integrative Physiology 127: 411–431.
Stewart, J. R. and Thompson, M. B. 2003. Evolutionary transformations of the fetal membranes of viviparous reptiles: a case study in two lineages. Journal of Experimental Zoology 299A: 13–32.
Stewart, J. R. and Thompson, M. B. 2004. Placental ontogeny of the Tasmanian scincid lizard, *Niveoscincus ocellatus* (Reptilia: Squamata). Journal of Morphology 259: 214–237.
Stewart, J. R. and Thompson, M. B. 2009a. Parallel evolution of placentation in Australian scincid lizards. Journal of Experimental Zoology. Part B. Molecular and Developmental Evolution 312B: 590–602.
Stewart, J. R. and Thompson, M. B. 2009b. Placental ontogeny in Tasmanian snow skinks (genus *Niveoscincus*) (Lacertilia: Scincidae). Journal of Morphology 270: 485–516.
Stewart, J. R., Blackburn, D. G., Baxter, D. C. and Hoffman, L. H. 1990. Nutritional provision to the embryos in *Thamnophis ordinoides* (Squamata: Colubridae), a predominantly lecithotrophic placental reptile. Physiological Zoology 63: 722–734.
Stewart, J. R., Heulin, B. and Surget-Groba, Y. 2004. Extraembryonic membrane development in a reproductively bimodal lizard, *Lacerta (Zootoca) vivipara*. Zoology 107: 289–314.
Stewart, J. R., Thompson, M. B., Attaway, M. B., Herbert, J. F. and Murphy, C. R. 2006. Uptake of dextran-FITC by epithelial cells of the chorioallantoic placentome and the omphalopleure of the placentotrophic lizard, *Pseudemoia entrecasteauxii*. Journal of Experimental Zoology Part A: Comparative Experimental Biology 305: 883–889.
Stewart, J. R., Ecay, T. W., Garland, C. P., Fregoso, S. P., Price, E. K., Herbert, J. F. and Thompson, M. B. 2009a. Maternal provision and embryonic uptake of calcium in an oviparous and a placentotrophic viviparous Australian lizard (Lacertilia: Scincidae). Comparative Biochemistry and Physiology-Part A: Molecular & Integrative Physiology 153: 202–208.
Stewart, J. R., Ecay, T. W. and Heulin, B. 2009b. Calcium provision to oviparous and viviparous embryos of the reproductively bimodal lizard *Lacerta (Zootoca) vivipara*. Journal of Experimental Biology 212: 2520–2524.
Stewart, J. R., Mathieson, A. N., Ecay, T. W., Herbert, J. F., Parker, S. L. and Thompson, M. B. 2010. Uterine and eggshell structure and histochemistry in a lizard with prolonged uterine egg retention (Lacertilia, Scincidae, *Saiphos*). Journal of Morphology 271: 1342–1351.
Stewart, J. R., Ecay, T. W., Heulin, B., Fregoso, S. P. and Linville, B. J. 2011. Developmental expression of calcium transport proteins in extraembryonic membranes of oviparous and viviparous *Zootoca vivipara* (Lacertilia, Lacertidae). Journal of Experimental Biology 214: 2999–3004.
Stewart, J. R., Russell, K. J. and Thompson, M. B. 2012. Development of yolk sac and chorioallantoic membranes in the Lord Howe Island skink, *Oligosoma lichenigerum*. Journal of Morphology 273: 1163–1184.
Stinnett, H. K., Stewart, J. R., Ecay, T. W., Pyles, R. A., Herbert, J. F. and Thompson, M. B. 2012. Placental development and expression of calcium transporting proteins in the extraembryonic membranes of a placentotrophic lizard. Journal of Morphology 273: 347–359.
Sun, Y. Y., Du, Y., Yang, J., Fu, T. B., Lin, C. X. and Ji, X. 2012. Is the evolution of viviparity accompanied by a relative increase in maternal abdomen size in lizards? Evolutionary Biology 39: 388–399.
Surget-Groba, Y., Heulin, B., Guillaume, C. P., Thorpe, R. S., Kupriyanova, L., Vogrin, N., Maslak, R., Mazzotti, S., Venczel, M., Ghira, I., Odierna, G., Leontyeva, O., Monney, J. C.

and Smith, N. 2001. Intraspecific phylogeography of *Lacerta vivipara* and the evolution of viviparity. Molecular Phylogenetics and Evolution 18: 449–459.
Surget-Groba, Y., Heulin, B., Guillaume, C. -P., Puky, M., Semenov, D., Orlova, V., Kupriyanova, L., Ghira, I. and Smajda, B. 2006. Multiple origins of viviparity, or reversal from viviparity to oviparity? The European common lizard (*Zootoca vivipara*, Lacertidae) and the evolution of parity. Biological Journal of the Linnean Society 87: 1–11.
Swain, R. and Jones, S. M. 1997. Maternal-fetal transfer of 3H-labelled leucine in the viviparous lizard *Niveoscincus metallicus* (Scincidae: Lygosominae). Journal of Experimental Zoology 77: 139–145.
Swain, R. and Jones, S. M. 2000. Facultative placentotrophy: half-way house or strategic solution? Comparative Biochemistry and Physiology-Part A: Molecular & Integrative Physiology 127: 441–451.
Taylor, E. H. 1963. The lizards of Thailand. University of Kansas Science Bulletin 44: 687–1077.
ten Cate-Hoedemaker, N. J. 1933. Beiträge zur Kenntnis der Plazentation bei Haien und Reptilien: Der Bau der reifen Plazenta von *Mustelus laevis* Risso und *Seps chalcides* Merr (*Chalcides tridactylus* Laur.). Zeitschrift für Zellforschung und Mikroskopische Anatomie 18: 299–345.
Thompson, J. 1977a. The transfer of amino acids across the placenta of a viviparous lizard, *Sphenomorphus quoyi*. Theriogenology 8: 158.
Thompson, J. 1977b. Embryo-maternal relationships in a viviparous skink *Sphenomorphus quoyi* (Lacertilia: Scincidae). pp. 279–280. In J. H. Calaby and C. H. Tyndale-Biscoe (eds.), *Reproduction and evolution*. Australian Academy of Science, Canberra, New South Wales.
Thompson, J. 1981. A study of the sources of nutrients for embryonic development in a viviparous lizard *Sphenomorphus quoyii*. Comparative Biochemistry and Physiology A, Comparative Physiology 70: 509–518.
Thompson, J. 1982. Uptake of inorganic ions from the maternal circulation during development of the embryo of a viviparous lizard, *Sphenomorphus quoyii*. Comparative Biochemistry and Physiology A, Comparative Physiology 71: 107–112.
Thompson, M. B. and Blackburn, D. G. 2006. Evolution of viviparity in reptiles: introduction to the symposium. Herpetological Monographs 20: 129–130.
Thompson, M. B. and Speake, B. K. 2006. A review of the evolution of viviparity in lizards: structure, function, and physiology of the placenta. Journal of Comparative Physiology B 176: 179–189.
Thompson, M. B. and Stewart, J. R. 1994. Egg and clutch size of the viviparous Australian skink, *Pseudemoia pagenstecheri* and the identity of species with type III allantoplacentae. Journal of Herpetology 28: 519–521.
Thompson, M. B. and Stewart, J. R. 1997. Embryonic metabolism and growth in lizards of the genus *Eumeces*. Comparative Biochemistry and Physiology A 118: 647–654.
Thompson, M. B., Speake, B. K., Stewart, J. R., Russell, K. J., McCartney, R. J. and Surai, P. F. 1999a. Placental nutrition in the viviparous lizard *Niveoscincus metallicus*: the influence of placental type. Journal of Experimental Biology 202: 2985–2992.
Thompson, M. B., Stewart, J. R., Speake, B. K., Russell, K. J., McCartney, R. J. and Surai, P. F. 1999b. Placental nutrition in a viviparous lizard (*Pseudemoia pagenstecheri*) with a complex placenta. Journal of Zoology 248: 295–305.
Thompson, M. B., Stewart, J. R., Speake, B. K., Russell, K. J. and McCartney, R. J. 1999c. Placental transfer of nutrients during gestation in the viviparous lizard, *Pseudemoia spenceri*. Journal of Comparative Physiology B: Biochemical, Systemic, and Environmental Physiology 169: 319–328.
Thompson, M. B., Stewart, J. R. and Speake, B. K. 2000. Comparison of nutrient transport across the placenta of lizards with different placental complexities. Comparative Biochemistry and Physiology A: Molecular and Integrative Physiology 127: 469–479.
Thompson, M. B., Speake, B. K., Stewart, J. R., Russell, K. J. and McCartney, R. J. 2001a. Placental nutrition in the Tasmanian skink, *Niveoscincus ocellatus*. Journal of Comparative Physiology B: Biochemical, Systemic, and Environmental Physiology 171: 155–160.

Thompson, M. B., Stewart, J. R., Speake, B. K., Russell, K. J. and McCartney, R. J. 2001b. Utilisation of nutrients by embryos of the enigmatic Australian viviparous skink *Niveoscincus coventryi*. Journal of Experimental Zoology 290: 291–298.
Thompson, M. B., Speake, B. K., Russell, K. J. and McCartney, R. J. 2001c. Utilisation of lipids, protein, ions and energy during embryonic development of Australian oviparous skinks in the genus *Lampropholis*. Comparative Biochemistry and Physiology-Part A: Molecular & Integrative Physiology 129: 313–326.
Thompson, M. B., Speake, B. K., Russell, K. J. and McCartney, R. J. 2001d. Nutrient uptake by embryos of the Australian viviparous lizard *Eulamprus tympanum*. Physiological and Biochemical Zoology 74: 560–567.
Thompson, M. B., Stewart, J. R., Speake, B. K., Hosie, M. J. and Murphy, C. R. 2002. Evolution of viviparity: what can Australian lizards tell us? Comparative Biochemistry and Physiology Part B: Biochemistry and Molecular Biology 131: 631–643.
Thompson, M. B., Lindsay, L. A., Herbert, J. F. and Murphy, C. R. 2007. Calcium ATPase expression in the oviducts of the skink, *Lampropholis guichenoti*. Comparative Biochemistry and Physiology-Part A: Molecular & Integrative Physiology 147: 1090–1094.
Thompson, M. B., Blackburn, D. G. and Parker, S. L. 2010. Reproduction in reptiles, from genes to ecology: a retrospective and prospective vision. Herpetological Conservation and Biology 5: 252–256.
Tilbury, C. R. and Tolley, K. A. 2009. A re-appraisal of the systematics of the African genus *Chamaeleo* (Reptilia: Chamaeleonidae). Zootaxa 2079: 57–68.
Tinkle, D. W. 1967. The life and demography of the side-blotched lizard, *Uta stansburiana*. Miscellaneous Publications of the Museum of Zoology, University of Michigan 132: 1–182.
Tinkle, D. W. 1969. The concept of reproductive effort and its relation to the evolution of life histories of lizards. American Naturalist 103: 501–516.
Tinkle, D. W. and Gibbons, J. W. 1977. The distribution and evolution of viviparity in reptiles. Miscellaneous Publications of the Museum of Zoology, University of Michigan 154: 1–55.
Tinkle, D. W. and Hadley, N. F. 1975. Lizard reproductive effort: caloric estimates and comments on its evolution. Ecology 56: 427–434.
Tinkle, D. W., Wilbur, H. M. and Tilley, S. G. 1970. Evolutionary strategies in lizard reproduction. Evolution 24: 55–74.
Uribe-Aranzábal, M. C., Hernández-Franyutti, A. and Guillette, L. J. 2006. Interembryonic regions of the uterus of the viviparous lizard *Mabuya brachypoda* (Squamata: Scincidae). Journal of Morphology 267: 404–414.
Van Damme, R., Bauwens, D. and Verheyen, R. F. 1989. Effect of relative clutch mass on sprint speed in the lizard *Lacerta vivipara*. Journal of Herpetology 23: 459–461.
Van Dyke, J. U. and Beaupre, S. J. 2012. Stable isotope tracer reveals that viviparous snakes transport amino acids to offspring during gestation. Journal of Experimental Biology 215: 760–765.
Van Dyke, J. U., Beaupre, S. J. and Kreider, D. L. 2012. Snakes allocate amino acids acquired during vitellogenesis to offspring: are capital and income breeding consequences of variable foraging success? Biological Journal of the Linnean Society 106: 390–404.
Van Dyke, J. U., Brandley, M. C. and Thompson, M. B. 2014. The evolution of viviparity: molecular and genomic data from squamate reptiles advance understanding of live birth in amniotes. Reproduction 147(1): R15–R26.
Vences, M., Guayasamin, J. M., Miralles, A. and De La Riva, I. 2013. To name or not to name: criteria to promote economy of change in Linnaean classification schemes. Zootaxa 3636: 201–244.
Vial, J. L. and Stewart, J. R. 1985. The reproductive cycle of *Barisia monticola*: a unique variation among viviparous lizards. Herpetologica 41: 51–57.
Vieira, S., de Perez, G. and Ramírez-Pinilla, M. P. 2007. Invasive cells in the placentome of Andean populations of *Mabuya:* an endotheliochorial contribution to the placenta? Anatomical Record 290: 1508–1518.

Vieira, S., de Pérez, G. R. and Ramírez-Pinilla, M. P. 2010. Ultrastructure of the ovarian follicles in the placentotrophic Andean lizard of the genus *Mabuya* (Squamata: Scincidae). Journal of Morphology 271: 738–749.

Villagrán, M., Méndez, F. R. and Stewart, J. R. 2005. Placentation in the Mexican lizard *Sceloporus mucronatus* (Squamata: Phrynosomatidae). Journal of Morphology 264: 286–297.

Vitt, L. J. 1981. Lizard reproduction: habitat specificity and constraints on relative clutch mass. American Naturalist 117: 506–514.

Vitt, L. J. and Blackburn, D. G. 1983. Reproduction in the lizard *Mabuya heathi* (Scincidae): a commentary on viviparity in New World *Mabuya*. Canadian Journal of Zoology 61: 2798–2806.

Vitt, L. J. and Blackburn, D. G. 1991. Ecology and life history of the viviparous lizard *Mabuya bistriata* (Scincidae) in the Brazilian Amazon. Copeia 1991: 916–927.

Vitt, L. J. and Caldwell, J. P. 2013. *Herpetology: an introductory biology of amphibians and reptiles.* 4th edition. Academic Press, London, pp. 757.

Vitt, L. J. and Congdon, J. D. 1978. Body shape, reproductive effort, and relative clutch mass in lizards: resolution of a paradox. American Naturalist 112: 595–608.

Vitt, L. J. and Price, H. J. 1982. Ecological and evolutionary determinants of relative clutch mass in lizards. Herpetologica 1982: 237–255.

Voituron, Y., Heulin, B. and Surget-Groba, Y. 2004. Comparison of the cold hardiness capacities of the oviparous and viviparous forms of *Lacerta vivipara.* Journal of Experimental Zoology Part A: Comparative Experimental Biology 301: 367–373.

Vrcibradic, D. and Rocha, C. F. D. 2011. An overview of female reproductive traits in South American *Mabuya* (Squamata, Scincidae), with emphasis on brood size and its correlates. Journal of Natural History 45: 813–825.

Wake, D. B. 1991. Homoplasy: the result of natural selection, or evidence of design limitations? American Naturalist 138: 543–567.

Wake, D. B. 2003. Homology and homoplasy. pp. 191–201. In B. K. Hall and W. M. Olson (eds.), *Keywords and concepts in evolutionary developmental biology.* Harvard University Press, Cambridge, Massachusetts.

Wake, D. B., Wake, M. H. and Specht, C. D. 2011. Homoplasy: from detecting pattern to determining process and mechanism of evolution. Science 331: 1032–1035.

Wake, M. H. 1985. Oviduct structure and function in non-mammalian vertebrates. In H. -R. Duncker and G. Fleischer (eds.), *Functional morphology of vertebrates.* Gustav Fischer Verlag, Stuttgart. Fortschritte der Zoologie 30: 427–435.

Wake, M. H. 2002. Viviparity and oviparity. pp. 1141–1143. In M. Pagel (ed.), *Encyclopedia of evolution, vol. 2.* Oxford Univ. Press, New York.

Wang, Y. and Evans, S. E. 2011. A gravid lizard from the Cretaceous of China and the early history of squamate viviparity. Naturwissenschaften 98: 735–743.

Wang, Z., Lu, H. L., Ma, L. and Ji, X. 2014. Viviparity in high-altitude *Phrynocephalus* lizards is adaptive because embryos cannot fully develop without maternal thermoregulation. Oecologia. 174: 639–649.

Wapstra, E. 2000. Maternal basking opportunity affects juvenile phenotype in a viviparous lizard. Functional Ecology 14: 345–352.

Wapstra, E., Olsson, M., Shine, R., Edwards, A., Swain, R. and Joss, J. M. 2004. Maternal basking behaviour determines offspring sex in a viviparous reptile. Proceedings of the Royal Society of London. Series B: Biological Sciences 271(Suppl. 4): S230–S232.

Wapstra, E., Uller, T., While, G. M., Olsson, M. and Shine, R. 2010. Giving offspring a head start in life: field and experimental evidence for selection on maternal basking behaviour in lizards. Journal of Evolutionary Biology 23: 651–657.

Warner, D. A. and Andrews, R. M. 2003. Consequences of extended egg retention in the Eastern Fence Lizard (*Sceloporus undulatus*). Journal of Herpetology 37: 309–314.

Webb, J. K., Shine, R. and Christian, K. A. 2006. The adaptive significance of reptilian viviparity in the tropics: testing the maternal manipulation hypothesis. Evolution 60: 115–122.

Weekes, H. C. 1927a. Placentation and other phenomena in the scincid lizard *Lygosoma* (*Hinulia*) *quoyi*. Proceedings of the Linnean Society of New South Wales 52: 499–554.
Weekes, H. C. 1927b. A note on reproductive phenomena in some lizards. Proceedings of the Linnean Society of New South Wales 52: 25–32.
Weekes, H. C. 1929. On placentation in reptiles. I. Proceedings of the Linnean Society of New South Wales 54: 34–60.
Weekes, H. C. 1930. On placentation in reptiles. II. Proceedings of the Linnean Society of New South Wales 55: 550–576.
Weekes, H. C. 1933. On the distribution, habitat and reproductive habits of certain European and Australian snakes and lizards with particular regard to their adoption of viviparity. Proceedings of the Linnean Society of New South Wales 58: 270–274.
Weekes, H. C. 1934. The corpus luteum in certain oviparous and viviparous reptiles. Proceedings of the Linnean Society of New South Wales 69: 380–391.
Weekes, H. C. 1935. A review of placentation among reptiles, with particular regard to the function and evolution of the placenta. Proceedings of the Zoological Society of London 2: 625–645.
Wooding, F. B. P., Ramírez-Pinilla, M. P. and Forhead, A. S. 2010. Functional studies of the placenta of the lizard *Mabuya* (Scincidae) using immunocytochemistry. Placenta 31: 675–685.
Wourms, J. P. 1981. Viviparity: the maternal-fetal relationship in fishes. American Zoologist 21: 473–515.
Wourms, J. P., Grove, B. D. and Lombardi, J. 1988. The maternal embryonic relationship in viviparous fishes. pp. 2–134. In W. S. Hoar and D. J. Randall (eds.), *Fish physiology vol. XI: The physiology of developing fish. B. Viviparity and posthatching juveniles*, Academic Press, San Diego.
Wu, Q., Thompson, M. B. and Murphy, C. R. 2011. Changing distribution of cadherins during gestation in the uterine epithelium of lizards. Journal of Experimental Zoology Part B: Molecular and Developmental Evolution 316: 440–450.
Yan, X. F., Tang, X. L., Yue, F., Zhang, D. J., Xin, Y., Wang, C. and Chen, Q. 2011. Influence of ambient temperature on maternal thermoregulation and neonate phenotypes in a viviparous lizard, *Eremias multiocellata*, during the gestation period. Journal of Thermal Biology 36: 187–192.
Yaron, Z. 1977. Embryo-maternal interrelationships in the lizard *Xantusia vigilis*. pp. 271–277. In J. H. Calaby and C. H. Tyndale-Biscoe (eds.), *Reproduction and evolution*. Australian Academy of Science, Canberra, New South Wales.
Yaron, Z. 1985. Reptile placentation and gestation: structure, function, and endocrine control. pp. 527–603. In C. Gans and F. Billet (eds.), *Biology of the Reptilia, vol. 15*. John Wiley & Sons, New York.
Zamudio, K. R., Parra-Olea, G. and Douglas, M. E. 2000. Reproductive mode and female reproductive cycles of two endemic Mexican horned lizards (*Phrynosoma taurus* and *Phrynosoma braconnieri*). Copeia 2000: 222–229.

CHAPTER 14

The Evolution of Polyandry and Patterns of Multiple Paternity in Lizards

Erik Wapstra[1,*] and *Mats Olsson*[2,3]

14.1 INTRODUCTION

The huge advances made during the 1990s in the use of molecular techniques to assign paternity within clutches/litters has led to a corresponding acceptance of the complexity of mating systems and patterns of paternity both within and between taxonomic groups. Perhaps most importantly this has led to a recognition of a disconnect between observed social systems (e.g., monogamous pair bonds) and mating systems (e.g., high rates of extra-pair paternity). It has also become clear that mating by females with multiple males leading to multiple paternity is taxonomically widespread (Birkhead and Møller 1998 and references therein; Jennions and Petrie 2000; Griffith *et al.* 2002; Simmons 2005; Slatyer *et al.* 2012; Pizzari and Wedell 2013) including in lizards (Olsson and Madsen 1998; Uller and Olsson 2008). In Fig. 14.1, we have conceptualized the complex dynamics that lead to multiple paternity. At its simplest level, multiple paternity can only result if females have a mating system in which they mate with multiple different males (i.e., polyandry). However, the level to which multiple paternity is reflective of multiple mating will depend on post-copulatory processes

[1] School of Biological Sciences, Private Bag 55, University of Tasmania, Hobart, 7001, Australia.
[2] School of Biological Sciences, University of Sydney, Australia.
[3] Department of Biology and Environmental Sciences, University of Gothenburg, Sweden.
* Corresponding author

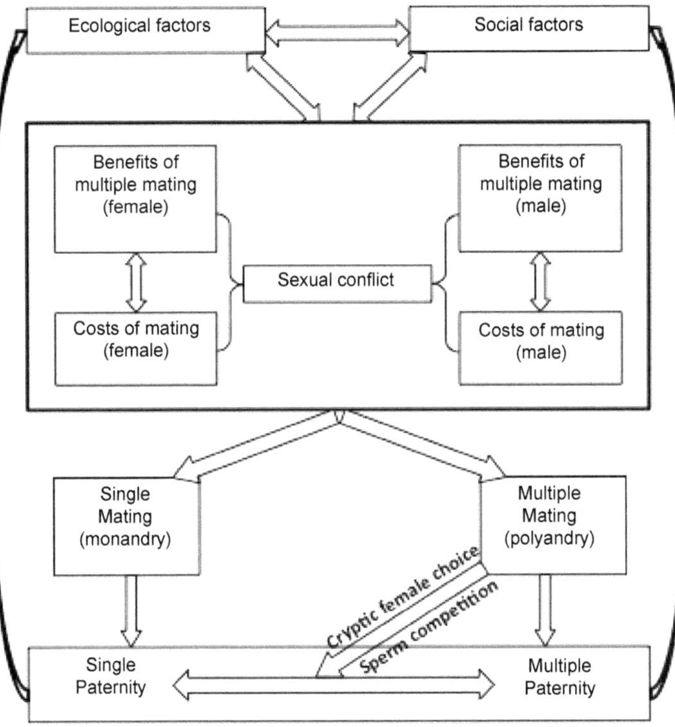

Fig. 14.1 Conceptual diagram depicting the complex evolutionary and ecological landscape leading to multiple paternity. Multiple paternity is a consequence of ecological and social factors that dictate the relative costs and benefits of multiple mating males and females. The resolution of the sexual conflict between optimal mating rates largely dictates whether mating will be monandrous or polyandrous. If mating is polyandrous, the post-copulatory phenomena of sperm competition and cryptic female choice will dictate the degree to which polyandry is reflected in multiple paternity.

including sperm competition and cryptic female choice (see Birkhead and Pizzari 2002 for review). The idea of post-copulatory female choice (i.e., cryptic female choice) is a natural extension of the vast literature on pre-copulatory female choice that occurs for both a variety of phenotypic and genetic characters (Birkhead and Pizzari 2002; see Cox and Kahrl Chapter 4, this volume). Single paternity arises if females choose, or are constrained, to mate with a single male within a reproductive cycle (i.e., monandry) and is also the only outcome for species with a clutch or litter size of one. Paternity can also be skewed towards single paternity when females are polyandrous and post-copulatory mechanisms of sperm competition and cryptic female choice bias fertilization success to a single male (see Fig. 14.1).

Polyandrous mating was traditionally explained through classic sexual selection theory (e.g., Bateman's Principle; Bateman 1948) based on the premise that a male's reproductive rate is limited by access to females willing to mate, whereas a female's reproductive success is limited by access to material resources she can convert into offspring rather than access to males/sperm (Bateman 1948; Trivers 1972; Andersson 1994; Simmons 2005; Parker and Birkhead 2013; Pizzari and Wedell 2013; see also Cox and Karhl Chapter 4, this volume). Under this view, polyandrous mating is a consequence of sexual conflict where males seek to mate multiple times to maximize their reproductive output with females passively accepting male mating with little to no benefit in order to reduce the potential cost of the conflict (Fig. 14.1; Andersson 1994). However, with the growing empirical evidence that female multiple mating is widespread this view has changed and there is growing appreciation that females do benefit from mating with several males within a reproductive cycle (i.e., polyandry). While multiple paternity across all taxonomic groups studied to date points to widespread polyandry, understanding the evolutionary and ecological conditions under which polyandry benefits the female remains a major challenge of evolutionary biology (Simmons 2005; Eizaguirre *et al.* 2007; Uller and Olsson 2008; Alonzo 2010; Slatyer *et al.* 2012; Pizzari and Wedell 2013).

The majority of work aimed at addressing the questions of why females mate with multiple males has revolved around three main hypotheses. These non mutually exclusive hypotheses suggest that polyandry evolves because it: 1) directly increases female fitness through increases in fecundity or survival; 2) indirectly increases female fitness through genetic benefits to offspring; or 3) evolves through sexual conflict over optimal mating rates where females mate multiply to minimize negative costs associated with avoiding multiple matings (convenience polyandry) (see below). To sort between the relative importance of these hypotheses, we need to appreciate that mating is the outcome of a two player game, a male and a female, and thus will ultimately be dictated by the costs and benefits to each player (Arnqvist and Kirkpatrick 2005).

When the cost and benefits differ between the sexes, sexual conflict arises (Fig. 14.1). The resolution of these costs will invariably be context-dependent and result in variation in levels of multiple paternity. Thus, understanding patterns of paternity will benefit from a perspective that takes into account direct and indirect fitness costs and benefits of mating strategies and the way they feed back with ecological and social factors (Badyaev and Qvarnstrom 2002; Cornwallis and Uller 2010; Oh 2011; Holman and Kokko 2013; Fig. 14.1). Furthermore, we need to recognize that evolutionary and ecological drivers of male and female multiple mating (polygyny and polyandry) operate at multiple scales from the individual female (shaping variation in polyandry among females), populations (shaping variation in polyandry among years within populations and between populations) and species (shaping variation in polyandry between species) (see Botero and Rubenstein 2012; Bonier *et al.* 2014).

Ultimately, providing a unified framework for understanding patterns of multiple paternity across taxa is of fundamental importance to evolutionary biologists. This is because female polyandry, and female mating behavior more generally, can have significant implications for the evolutionary trajectory of populations (Price *et al.* 2010; Cornwallis *et al.* 2010; Holman and Kokko 2013). For example, explaining the variation in patterns of multiple paternity is being increasingly recognized as important because of its role in driving within family/group relatedness and the consequences this has for understanding the emergence and diversification of social systems and behavior, including investment in parental care and cooperation (Hughes *et al.* 2008; Cornwallis *et al.* 2010; Griffin *et al.* 2013; Lukas and Clutton-Brock 2013; Pizzari and Wedell 2013).

14.2 PATTERNS OF PATERNITY IN LIZARDS

While documentation of patterns of paternity in lizards (and reptiles more generally) falls well behind many other taxa, such as birds and invertebrates, in terms of numbers of studies or diversity of species (compare for example Griffith *et al.* 2002 with Uller and Olsson 2008), multiple paternity has been shown in all lizards studied to date (see Table 14.1 and Uller and Olsson 2008) and this reflects similar data for other reptilian taxa (e.g., natricine snakes; Wusterbarth *et al.* 2010). What is also clear is that levels of multiple paternity in lizards are generally high but vary among individuals within populations, among populations within species, and among species (Table 14.1 and references therein; Uller and Olsson 2008). These patterns reflect the outcomes of multiple ecological and social selective pressures as well as the outcomes of sexual conflict between optimal mating rates (Fig. 14.1). In field correlative studies, demonstration of multiple paternity is an outcome of females mating with different males but otherwise the processes remain largely a "black box" representing a range of processes with mate encounter rate, mate choice, sperm competition, cryptic female choice, and potential embryo loss all potentially contributing to the observed patterns. To address this, research has moved towards more experimental approaches where competing hypotheses on costs and benefits of polyandry (especially from the female perspective), along with proximate and physiological mechanisms behind it, can begin to be teased apart (e.g., Fitze *et al.* 2005; LaDage *et al.* 2008; Le Galliard *et al.* 2008; Keogh *et al.* 2013; Noble *et al.* 2013).

Why is multiple paternity common in lizards (and reptiles more generally)? Are there particular features of their biology that make them likely to undertake multiple mating leading to multiple paternity? With

Table 14.1 Patterns of multiple paternity in lizards from natural populations. Only studies where all offspring were known are included (typically because all eggs from a clutch were collected or females gave birth in the laboratory). In some cases multiple paternity is reported but paternity of individual offspring is unknown. Table based on Uller and Olsson (2008), and extensive literature searches conducted to update the information to 2013. Experimental manipulations (even in large field enclosures) where patterns of paternity are expected to deviate from natural conditions (adult sex ratios, unnatural densities, restricted mate choice, costs of mating) have not been included.

Taxon	% multiple paternity	# clutch examined	Reference
Scincidae			
White's skink, *Egernia whitii*	12	50	Chapple and Keogh 2005
White's skink, *Egernia whitii*	17	90	While *et al.* 2009a
Spiny-tailed skink, *Egernia stokesii*	25	16	Gardner *et al.* 2000, 2002
Cunningham's skink, *Egernia cunninghami*	3	38	Stow and Sunnucks 2004
Sleepy lizard, *Tiliqua rugosa*	19	21	Bull *et al.* 1998
Southern water skink, *Eulamprus heatwolei*	65	17	Morrison *et al.* 2002
Blue Mountains water skink, *Eulamprus leuraensis*	27	11	Dubey *et al.* 2011
Southern snow skink, *Niveoscincus microlepidotus*	75	8	Olsson *et al.* 2005c
Grand skink, *Oligosoma grande*	47	15	Berry 2006
Mt log skink, *Pseudomoia eurecateuixii*	53	17	Stapley *et al.* 2003
Spanish rock lizard, *Iberolacerta cyreni*	48	33	Salvador *et al.* 2008
Common five lined skinks, *Plestiodon fasciatus*	65	20	Bateson *et al.* 2011
Lacertidae			
Common lizard, *Lacerta vivipara*[1]	67	46	Eizaguirre *et al.* 2007
Common lizard, *Lacerta vivipara*[2]	47	51	Eizaguirre *et al.* 2007
Common lizard, *Lacerta vivipara*[3]	55	38	Eizaguirre *et al.* 2007
Common lizard, *Lacerta vivipara*	65	26	Hofmann and Henle 2006

Common lizard, *Lacerta vivipara*[p1]	68	458	Laloi *et al.* 2004
Common lizard, *Lacerta vivipara*[p2]	50	15	Laloi *et al.* 2004
Common lizard, *Lacerta vivipara*	54	54	Richard *et al.* 2012
Common lizard, *Lacerta vivipara*	unreported	unreported	Laloi *et al.* 2009
Sand lizard, *Lacerta agilis*	80	5	Gullberg *et al.* 1997
Sand lizard, *Lacerta agilis*	unreported	unreported	Olsson *et al.* 2011a,b
Common wall lizard, *Podarcis muralis*	87	31	Oppliger *et al.* 2007
Agamidae			
Ornate dragon, *Ctenophorus ornatus*	25	20	Lebas 2001
Painted dragon, *Ctenophorus pictus*	18	51	Olsson *et al.* 2007
Phrynosomatidae			
Striped plateau lizard, *Sceloporus virgatus*	61	13	Abell 1997
Side-blotched lizard, *Uta stansburiana*	72	123	Zamudio and Sinervo 2000
Iguanidae			
Black spiny tailed iguana, *Ctenosaura pectinata*	11	10	Faria *et al.* 2010

[1] Refers to first year of study; [2]Refers to second year of study; [3]Refers to third year of study; [p1]Refers to population 1; [p2]Refers to population 2.

the exception of a restricted number of lizard taxa that show complex forms of long term social bonds (see While *et al.* Chapter 15, this volume), which in other taxa (including mammals and birds) reduces selection and opportunity for multiple mating, the majority of lizard mating systems can be typified by intense male-male competition for females via direct male-male conflict or resource-driven polygyny (see Olsson and Madsen 1998; Cox and Kahrl Chapter 4, this volume). Another feature of lizard mating systems is the near ubiquitous presence of sperm storage by females (see Olsson and Madsen 1998; Sever and Hamlett 2002). Sperm storage provides an extended temporal window for multiple mating that leads to opportunities for multiple mating (by dissociating to some extent the temporal window of receptivity with fertilization). Sperm storage following polyandrous mating leads to enhanced opportunities for sperm competition and cryptic female choice, both of which are central to hypotheses for the evolution of polyandry (see below).

As storage of sperm of multiple males by females is common, there should be concomitant selection on males to minimize their paternity loss to other males (see Olsson and Madsen 1998; Simmons 2005). These processes can occur pre- or post-copulatory and include male mate choice, sophisticated sperm expenditure including altered ejaculate size, copulatory plugs, male mate guarding, increased testis size, and selection on sperm morphology (Olsson and Madsen 1998). What is clear is that despite ongoing selection on males to maximize their share of paternity through such mechanisms as mate guarding (including social monogamy), high levels of multiple paternity within clutches/litters is still the common pattern (Bull *et al.* 1998; Olsson *et al.* 2005a; While *et al.* 2009a,b; Ancona *et al.* 2010).

Table 14.1 shows the limited number of studies (up to and including 2013) that report multiple paternity under field conditions. Further experimental work examining the fitness benefits of polyandry are discussed in further sections addressing support for alternative hypotheses for the evolution of polyandry. Despite the growing number of studies detailing patterns of paternity in lizards in both field and experimental studies, there are still considerable gaps in the literature for the majority of species. Indeed, current research into multiple paternity in lizards exhibits a strong taxonomic bias (Table 14.1). This presents significant challenges when trying to make broad phylogenetic inferences on variation in patterns of paternity across lizard lineages. For example, more than 75% of the studies are restricted to three families of lizards (skinks, lacertids and agamids) and for the vast majority of families we have next to no data on patterns of paternity, or the ecological or evolutionary factors that may promote it. Furthermore, nearly a quarter of the studies to date have centered on the *Egernia*-group of skinks in Australia (Table 14.1; Uller and Olsson 2008). The *Egernia*-group has been utilized heavily as model organisms for studies related to the evolution of sociality, mating systems and parental care; traits which are expected to co-vary with benefits to multiple mating (e.g., Bull *et al.* 1998; Chapple 2003; Chapple and Keogh 2005; While *et al.* 2009a,b, 2011, 2014; see While *et al.* Chapter 15, this volume). For example, patterns of paternity and parental care are predicted to co-evolve because the former dictates the cost–benefit ratio of the latter by affecting genetic relatedness between adults and offspring (Griffin *et al.* 2013). It is unfortunately premature to undertake a sufficiently rigorous meta-analysis on the limited number of empirical studies (or experimental studies) that document patterns of paternity in lizards to separate out competing hypotheses for the evolution of polyandry and paternity in lizards. What is clear is that lizard social and mating systems and reproductive cycles are diverse (more so than has been traditionally accepted; Doody *et al.* 2013) and thus we would expect a variety of patterns of paternity to emerge with potentially a variety of selective forces shaping those patterns.

As outlined in other chapters in this book, lizards have already made an impact on our understanding of key processes including sexual selection, mating systems, reproductive allocation, sex allocation and parental care. Below we show that lizards have also provided good model systems to further our understanding of the benefits of polyandry and multiple paternity. Given the limited taxonomic coverage and only recent move to experimental approaches, valuable models for understanding the evolution of polyandry and multiple paternity are still left ready for exploitation. Below we briefly review alternative hypotheses for multiple paternity and indicate where lizards have either contributed to our understanding or have the potential to do so.

14.3 ASSESSING EVIDENCE FOR THE HYPOTHESES FOR MULTIPLE PATERNITY IN LIZARDS

Patterns of polyandry reflect the resolution of the sexual conflict between male mating rates (driven by costs and benefits to males) and female mating rates (driven by cost and benefits to females). In principle, the reasons males mate multiply is non-contentious; the more females they mate with the higher their reproductive output. While this may be balanced to some extent by high costs of mating and may result in careful mate choice and sophisticated sperm expenditure, in general males benefit strongly from multiple mating. Sexual selection in reptiles has been examined previously (Olsson and Madsen 1998) and has been recently re-examined in detail for lizards by Cox and Kahrl (Chapter 4, this volume). Given their concentration on sexual selection from the male perspective including both pre- and post-copulatory phenomena, where appropriate in this chapter we refer readers to their chapter. To prevent overlap between chapters, we concentrate on hypotheses explaining polyandry and multiple paternity from the female perspective.

Given the potential costs associated with multiple mating (e.g., energetic cost, loss of feeding opportunity, disease transmission, injury and increased risk of predation (Olsson and Madsen 1998; Watson *et al.* 1998; Eizaguirre *et al.* 2007; Madsen 2011; Slatyer *et al.* 2012), and that females immediate reproductive output is not enhanced as directly as males by multiple mating, explanations for female multiple mating are typically divided into three main adaptive hypotheses (Jennions and Petrie 2000; Eberhard and Cordero 2003; Simmons 2005; Uller and Olsson 2008; Slatyer *et al.* 2012). We address these hypotheses below firstly by describing the underlying theory and empirical support from other taxa, and then by submitting the hypotheses to scrutiny using field and experimental studies from lizards.

14.3.1 Polyandry Through Direct Benefits

Females may benefit directly from multiple mating because they receive direct material benefits from males (Fedorka and Mousseau 2002). These can include nuptial gifts, nutrients in ejaculates, or increased male provisioning to either the female or her offspring (see Andersson 1994; Birkhead and Møller 1998; Gwynne 2008). In this scenario, multiple mating results in increased transfer or access to male resources that increase either a female's fecundity or survival (increasing reproductive potential into future years). Such direct benefits are especially common in birds and insects (see Vahed 1998; Arnqvist and Nilsson 2000). A key to understanding the situations where this scenario leads to polyandry (i.e., multiple mating with different males rather the same male) is to separate out the effects of multiple mating *per se* from the benefits of mating with multiple males (LaDage *et al.* 2008). The potential for direct fitness gains to females from mating with different males must be balanced against potential fitness costs. In birds, for example, mating with different males could benefit the female if it leads to enhanced offspring care because there are more males to provide the care but there may be costs associated with the cuckolded male potentially reducing provisioning (Kokko 1999; Griffin *et al.* 2013 and references therein).

In lizards, we can largely rule out hypotheses centered on females receiving direct material benefits from nuptial gifts, ejaculate nutrients and extended paternal assistance: males do not provide nuptial gifts and there is no evidence in lizards that energy contained in ejaculates is either available to females, or sufficient in terms of energy, to increase their reproductive efforts (Olsson *et al.* 2004b; Uller and Olsson 2008; Eizaguirre *et al.* 2007). Furthermore, in reptiles male parental care of offspring is rare (but see While *et al.* Chapter 15, this volume) and where rudimentary care of offspring through protection from conspecifics or predators does occur (mostly in *Egernia* skinks; O'Connor and Shine 2004; Sinn *et al.* 2008), it appears to select against multiple mating by females (While *et al.* 2009b). Therefore, direct benefits as a potential factor influencing the evolution of polyandry is unlikely in lizards.

An extension of the hypothesis that polyandry is driven by direct benefits to females is that multiple mating may confer a direct fitness gain if it maximizes the provision of fertile sperm or reduces the risk of infertile sperm (e.g., assurance of fertilization, Sheldon 1994; Wedell *et al.* 2002; García-González and Simmons 2005). While levels of infertility in lizards have not been examined extensively, in sand lizards (*Lacerta agilis*), levels of infertility in males (which would select for female multiple mating) is very low in the natural population and is unlikely to play a role in the evolution of polyandry (Olsson and Shine 1997). However, males may be infertile (immature sperm) early in the season when they emerge from hibernation and this may drive temporal patterns of mating as well as patterns of sex-specific emergence in sand lizards (Olsson and Madsen 1996) and Southern snow skinks (*Niveoscincus microlepidotus*) (Olsson *et al.* 1999). In other taxa, there is evidence that females can benefit from multiple mating through increased fecundity or fertility; in common lizards (*Lacerta vivipara*), Eastern water skinks (*Eulamprus quoyii*) and in leopard geckos (*Eublepharis macularius*) multiple mating leads to higher reproductive output and this has been implicated in these species in the evolution of polyandry (Uller and Olsson 2005; LaDage *et al.* 2008; Noble *et al.* 2013). However, as an explanation for the widespread occurrence of polyandry in lizards it remains unsupported because of poor taxonomic coverage exploring this phenomenon in general (Sheldon 1994; Slatyer *et al.* 2012) and in lizards (Noble *et al.* 2013). In lizards, the scenarios in which females would be most susceptible to inadequate sperm transfer are predicted to include those with short mating seasons, those where encounter rates are low or unpredictable, when copulations result in inadequate sperm transfer, and/or males vary in sperm quality (Noble *et al.* 2013): these data are either not available for many lizards or have not been used to test this hypothesis more broadly.

14.3.2 Polyandry Through Indirect Genetic Benefits

The second set of hypotheses for the evolution of polyandry center on the indirect genetic benefits derived from elevated mean offspring fitness (and therefore parental fitness) potentially because the presence of sperm of multiple males opens up the opportunity for sperm competition and cryptic female choice (Jennions and Petrie 2000; Zeh and Zeh 2003; Slatyer *et al.* 2012). This can lead to increased female fitness via increased genetic diversity of offspring (genetic bet-hedging), increased offspring quality through fertilization by higher quality males (e.g., sexy sperm hypothesis or "trade-up" hypothesis), avoidance of genetic incompatibility through egg/sperm incompatibility and/or cryptic female choice, increased levels of genetic compatibility/complementarity, or inbreeding avoidance (Zeh and Zeh 1996; Jennions and Petrie 2000; Neff and Pitcher 2005, 2008; Uller and Olsson 2008; Puurtinen *et al.* 2009; Slatyer *et al.* 2012; Noble *et al.* 2013). Many of these hypotheses are logical extensions of the hypotheses for female pre-copulatory mate choice (over one partner) and may be especially important where initial female choice is constrained by social or ecological factors. Importantly, these hypotheses on indirect genetic benefits predict that offspring from polyandrous females should have increased fitness, on average, compared with singly mated females. Some of these effects could arise simply through sperm competition; however, multiple mating provides the opportunity for females to use cryptic mate choice to bias paternity towards males that elevate offspring fitness.

Indirect genetic benefits of polyandry remains controversial (e.g., Yasui 1998) with both empirical studies and recent meta-analyses providing mixed support (e.g., Simmons 2005; Fisher *et al.* 2006; Hettyey *et al.* 2010; Slatyer *et al.* 2012). For example, recent meta-analyses on birds provided weak support for indirect benefits as an explanation for multiple mating (Arnqvist and Kirkpatrick 2005; Akçay and Roughgarden 2007), while a meta-analyses on insects suggested that polyandry was associated with increased hatching success (Simmons 2005). The equivocal support might be because of taxonomic differences in life history and reproductive strategies. For example, in birds and mammals, sophisticated (bi-)parental care may make it comparatively more difficult to pick-up subtle differences in offspring fitness arising through indirect genetic effects than in insects where maternal provisioning usually ends at hatching or birth (but see Simmons 2005). Therefore, tests of polyandry-derived genetic effects on offspring fitness may be compromised (except in artificially controlled situations) by patterns of parental investment (i.e., maternal and paternal effects) where higher levels of investment, for example, may increase offspring fitness masking or exaggerating the genetic effects. Parental investment can vary according to genetic or phenotypic quality of partner, relatedness, and levels of paternity assuredness in the litter/clutch (e.g., Senar *et al.* 2002; Horvathova *et al.* 2012).

In some taxa, these adjustments can occur at multiple stages. In birds, for example, altered allocation patterns can be achieved by adjusting investment (i.e., egg size) within and between clutches and/or adjusting post-hatching feeding rate and food quality within and between clutches. Similarly in mammals, investment can be altered at several developmental stages including pre-birth, during gestation and certainly via maternal provisioning post birth (Hewison and Gaillard 1999). Clearly, the multitude of levels of investment may allow parental adaptive control over offspring fitness, but it also severely compromises the potential to make *a priori* predictions regarding the direction and magnitude of indirect genetic effects at a given level of investment. Female allocation patterns are more straightforward in lizards (but see Uller and While Chapter 12, this volume) and complications arising from male parental effort are largely non-existent (see below). We suggest that lizards could be valuable models for disentangling the importance of indirect genetic effects of male and female multiple mating which is of utmost importance for further development of this field.

Broadly, there is support for the importance of indirect genetic effects in lizards (and more broadly in reptiles). Reptiles (particularly adders, *Vipera berus*, sand lizards, *Lacerta agilis* and side-blotched lizards, *Uta stansburiana*) certainly played a key role in formulating our thinking that multiple mating can have a positive effect on female fitness (Madsen *et al.* 1992, 1999, 2004; Olsson *et al.* 1994a,b, 1996, 2005b; Zamudio and Sinervo 2000; Olsson and Madsen 2001; Calsbeek and Sinervo 2002, 2004; Sinervo *et al.* 2006) through enhanced offspring viability or survival. However, it has since been debated whether these indirect genetic effects were detected (or even occurred) more readily because the populations used in these studies have low genetic diversity (see Madsen 2008; Olsson and Uller 2009) or potentially in the case of side-blotched lizards because polymorphisms create strong sire effects. Does this mean that indirect genetic benefits as a result of polyandrous mating are rare in lizards? Recent work would suggest not, with evidence from other lizard species indicating that multiple mating by females increases offspring fitness.

Recent studies of the European common lizard (*Lacerta vivipara*) established that while polyandry and monandry coexist among females (Laloi *et al.* 2004), polyandry confers fitness benefits (Fitze *et al.* 2005). Specifically, Eizaguirre *et al.* (2007) demonstrated that polyandrous females produce larger clutches than monandrous females (controlling for number of matings) and that embryo mortality during late stages of development was also reduced in polyandrous clutches. With their field-based study, they were not able to separate out the competing hypotheses of genetic compatibility, intrinsic male quality or inbreeding avoidance as the causal mechanism. In field studies, other potential fitness effects of multiple mating have been demonstrated with dispersal being greater in polyandrous litters than in monandrous litters (Laloi *et al.* 2009), which may affect a range of demographic processes including female-offspring competition (see Chapple and Keogh 2005 for links between polyandry versus monandry and sex biased dispersal in the White's skink, *Egernia whitii*). Recent work on the common lizard has also provided evidence consistent with the hypothesis that multiple mating provides an opportunity for females to genetically "trade-up" (Laloi *et al.* 2011; see also Fitze *et al.* 2010). They found that females preferentially accepted males of higher heterozygosity (presumably resulting in higher quality offspring) for second matings. Similarly, While *et al.* (2014) showed that female multiple mating in White's skink may serve as an inbreeding avoidance mechanism. Specifically, there were high levels of relatedness between male and female pairs as a result of strong genetic structure within the population. Females alleviate the constraints of social mate choice imposed by population viscosity by mating outside their pair bond with less related males. As a result offspring from extra-pair males exhibited a significantly increased genetic diversity compared to within-pair offspring.

Recent models predict such female choosiness and "trade-up" decisions (Bleu *et al.* 2012) but also that choosiness should be context-dependent: females gain greater benefits from such choosiness when encounter rates or the cost of mating is high. Context-dependent mate choice (especially context-dependence of choice of second or third mates) remains largely untested in lizards. Until the mid 90s, widespread evidence for pre-copulatory mate choice was considered weak in reptiles (Olsson and Madsen 1995; Tokarz 1995), however, since then, increasing evidence for choice using olfactory cues, including potentially at the Major Histocompatibility Complex (MHC), has been demonstrated (e.g., Lopez *et al.* 2003; Olsson *et al.* 2003) suggesting that mechanisms for sophisticated female (and male) decisions exist. For example, Martin and Lopez (2013) demonstrated (in an experimental context) that female Carpetan rock lizards (*Iberolacerta cyreni*) preferred scents of large territorial males over smaller males but they also preferred areas with scents of multiple males. They suggested this latter result may increase the probability of obtaining multiple copulations with different males, which may favor sperm competition and cryptic female choice. For a more extensive discussion of the role of pheromones in lizard communication, including its important role in mate choice and intersexual competition (see Martin and Lopez Chapter 3, this volume).

Cryptic female choice is the logical extension of female pre-copulatory choice and has been argued as one of the most likely benefits of female multiple matings in lizards (see Olsson *et al.* 1994a, 1996; Olsson and Madsen 2001), especially when evidence of female pre-copulatory choice in lizards was lacking. While there is compelling evidence for cryptic female choice, including sophisticated matching of sperm from different males to offspring sex (Calsbeek and Sinervo 2004; Corl *et al.* 2012; Calsbeek and Bonneaud 2008; Olsson *et al.* 1994a,b, 1996, 2004b, 2005a), it is difficult to reconcile the ubiquitous nature of multiple paternity in lizards (at very high levels) with this idea (Uller and Olsson 2008). If, in general, cryptic female choice was strong (and effective), paternity skews should be high (biased towards the female's favored genetic partner) yet this is not the generally observed pattern with levels of multiple paternity generally high (Uller and Olsson 2008; Table 14.1). Furthermore, without information on what the paternity skew is expected to be from the female's fitness perspective (e.g., based on theoretical or empirical predictions of fitness effects for different sires), it is difficult to easily demonstrate cryptic female choice in lizards because paternity skews can arise from a host of other factors that affect the outcomes of sperm competition.

In order to separate out evidence for cryptic female choice from outcomes of sperm competition, there are some logistical constraints to overcome that are challenging for lizards. Firstly, we need information on how many times a female has copulated; secondly we need information on which males she has copulated; thirdly we need to have information on how many sperm each male transferred during copulation; and, finally we need information on variation in sperm morphology and behavior between males (see for example extensive review by Snook 2005). The first is often easy in lizards as copulation can leave distinct copulation marks from the male biting the female (e.g., Fitze *et al.* 2005; While and Wapstra 2009). The second challenge in assessing which males a female has mated with presents greater challenges because this information must not be inferred from molecular techniques but must be directly obtained from observations of copulations. In the majority of lizards, this information is difficult to obtain in the field or semi-natural enclosures because copulations are generally cryptic and occur relatively quickly (Olsson and Madsen 1998; While and Wapstra 2009). In the laboratory, these processes can be more readily observed and laboratory studies have the added advantage that male-female relatedness, female pre-copulatory choice and mating order can be controlled (but with a loss of realism). Knowing the amount of sperm transferred by males during copulation presents the greatest logistical challenges in lizards because of internal fertilization (compare the ease with which paternity skews can be assessed with external fertilizers such as frogs for example; Sherman *et al.* 2008, 2009). In reptiles, there are limited data to suggest that copulation duration is related to sperm transfer (Olsson and Madsen 1998; Olsson 2001; Shine *et al.* 2000; Olsson *et al.* 2004a), but this does not take into account inter-male differences in sperm count nor the sophisticated nature of sperm allocation (even between ejaculates) that can occur (e.g., Olsson 2001; Olsson *et al.* 2004a; Uller *et al.* 2013; see Snook 2005 for an extensive review).

14.3.3 Evolution of Polyandry Through Sexual Conflict

This hypothesis suggests that because there is sexual conflict over mating rates, a female is faced with a trade-off between resisting male mating and incurring the costs associated with harassment and even forced copulations or minimizing these costs by accepting the mating ("convenience polyandry"; Slatyer *et al.* 2012). In this situation, polyandry can arise without any fitness gain to the female in terms of direct or indirect fitness gains, however, fitness "loss" is minimized. This explanation requires that the costs to males mating multiply are significantly lower than those of females and that the reduction in fitness from multiple mating to females (through, for example, increased disease risk, injury, predation, and energy use) are less than the costs associated with avoidance of multiple mating. Understanding the adaptive scenario where this may arise relies on detailed knowledge of the costs of multiple mating to the female (e.g., energy, predation risk, disease transmission, poor genetic compatibility) against the costs of male harassment and/or potential forced copulation (see Arnqvist and Rowe 2005 and references therein). Embedded within the logic of this hypothesis is that any factor that alters the optimal mating rate for males (including for example density effects or contact rates), will, in the absence of altered costs or benefits to females, affect the mating rate between males and females (i.e., the degree of polyandry).

In a review of multiple mating in reptiles, Uller and Olsson (2008) suggested that despite the emerging evidence for indirect genetic effects in reptiles, compelling evidence for indirect genetic effects driving polyandry in reptiles was weak. They suggested that the most parsimonious explanation for the frequent and high levels of polyandry was that it represented a combined effect of mate-encounter frequency and conflict over mating rates between males and females driven by large male benefits and relatively small costs to female mating. In part, their view is based on the evolutionary origin of polyandry via male-driven selection and the idea that female benefits arise later (see Olsson and Uller 2009 for a more detailed argument). The key to understanding the role of convenience polyandry is an acceptance that additional mating invariably will carry a cost but that cost, in relation to the costs of harassment and the potential gains in indirect fitness, is low (see also Fitze *et al.* 2005; Uller and Olsson 2008; Le Galliard *et al.* 2008; Madsen 2011). Potential costs to females with increased contact with males could include disease transmission. In sleepy lizards (*Tiliqua rugosa*) and in Tuatara (*Sphenodon punctatus*), there is evidence that contact between individuals, including potentially during mating, is related to parasitic and bacterial infection (Godfrey *et al.* 2009, 2010; Leu *et al.* 2010; Fenner *et al.* 2011; Bull *et al.* 2012). In addition, multiple mating may confer increased risks of predation to females, as has been demonstrated in adders (*Vipera berus*; Madsen 2011), but data in lizards are currently lacking.

Experimental approaches on common lizards (*Lacerta vivipara*) using large semi-natural enclosures have been instrumental in advancing our understanding of the role of sexual conflict and the costs of mating and female benefits. By manipulating adult sex ratios, a series of studies showed that while overall polyandry was associated with larger clutches (as also detailed above), male biased populations led to higher rates of copulation through sexual harassment and that females from these populations had lowered reproductive success and survival (Fitze *et al.* 2005; Le Galliard *et al.* 2005). In follow-up work, Le Galliard *et al.* (2008) examined whether the direct costs of increased male harassment on female reproductive success and survival were mitigated by indirect benefits through offspring growth, offspring survival or mating success; they were not, leading them to conclude that their observations of conflict over mating rates arise because fitness gains are greater in males than the net costs to females.

Our own work in natural populations of sand lizards (*Lacerta agilis*) also emphasizes the importance of proximate factors in determining male and female encounter rates and concomitant rates of multiple paternity. In a decade-long study in their natural environment in Sweden we showed that increased temperature during the spring mating season increased mating rates and the numbers of sires per clutch. In our case, this had positive effects on indirect measures of fitness including a reduction in malformations within clutches and increased offspring survival (Olsson *et al.* 2011a,b). Thus, in warm years when activity and encounters between males and females are high, there was evidence of positive indirect fitness benefits to females and in contrast to the work in outdoor enclosures, direct costs to females were not evident probably because harassment costs could be avoided by avoidance and fleeing. Recently, Keogh *et al.* (2013) also emphasized the importance of male encounter rates with females as a predictor of reproductive success in the Australian Southern water skink (*Eulamprus heatwolei*) concluding that female multiple mating was best explained by the combined effects of mate encounter frequency and high benefits to males but low costs to females.

14.4 TOWARDS A HOLISTIC UNDERSTANDING OF PATTERNS OF PATERNITY IN LIZARDS

Polyandry leading to multiple paternity is widespread in lizards but its explanation remains controversial (Uller and Olsson 2008; Madsen 2008; Olsson and Uller 2009; Madsen 2011; Noble *et al.* 2013). This controversy is driven, in part, by the ongoing debate more broadly on what selective forces lead to polyandry and why patterns may differ among females, populations, species, and episodes of selection (e.g., years/mating seasons) (e.g., Simmons 2005; Slatyer *et al.* 2012; Parker and Birkhead 2013). We would argue that the central question of whether females mate multiply to gain direct fitness gains, indirect genetic fitness gains or to reduce the costs of male harassment is unsatisfactory—in many taxa all of these explanations are relevant and involved in explaining observed patterns. As Parker and Birkhead (2013) elegantly argue the "why polyandry" for a given population (or species) can only be addressed by analysis of the distribution of all possible state-dependent encounters between pairs of males and females in a population (see also Alonzo and Sinervo 2001). Each of these encounters then has a probability of resulting in a mating depending on the sum of costs and benefits to the male and female involved and the resolution of the conflict over these. The sum of the probabilities of encounters and the mating outcomes will result observed mean level of polyandry in the population (Parker and Birkhead 2013). The level to which polyandry then results in multiple paternity will then be dependent on the post-copulatory phenomena of sperm competition and cryptic female choice.

This verbal argument is necessarily individual, population, and species-specific from the outset and its simplicity hides a multitude of ecological and evolutionary factors (Fig. 14.1). For example, in lizards, a host of behavioral (e.g., movement rates, social and mating systems), demographic (e.g., density, frequency of reproduction, operational sex ratios) and ecological factors (e.g., habitat complexity, weather dependent activity patterns) affect encounter rates between males and females and these feedback on each other in complex ways. Demonstrating the relative costs and benefits of mating to each sex is equally complex, taxon-specific, and will covary with mating system and level of inbreeding. There is little doubt that both males and females incur mating costs that are more than offset by the benefits of multiple mating (hence the ubiquitous patterns of multiple paternity) but we are in our infancy in understanding what affects the sum of these for either of the sexes (which determines the asymmetry in the optimal mating frequency), let alone the combined sum for these which resolves the sexual conflict. We agree with Pizzari and Wedell (2013) that studies considering the evolutionary ecology of polyandry (and the outcome of multiple paternity) with a wider, interdisciplinary context have much to offer in advancing our understanding. Specifically we require a tighter connect between proximate studies of costs and benefits with studies measuring fitness consequences under realistic conditions.

14.5 FUTURE DIRECTIONS

Below we provide some key future directions for studying patterns of paternity in lizards that will broaden our taxonomic understanding and move to address the complexity of factors that influence multiple mating and multiple paternity in lizards.

14.5.1 Broader Taxonomic Coverage

As has been advocated by the recent surge in reviews and meta-analyses within and across taxa, a broad taxonomic coverage is required to provide insights that are hidden from a specific taxonomic perspective, particularly the contexts in which polyandry may arise through selection on males versus females. Specifically, in lizards we lack data on a large number of families and within the more well studied families (e.g., lacertids and skinks), research is biased towards a limited number of species (e.g., *Lacerta vivipara* and *Lacerta agilis* in the lacertids and the *Egernia*-group in skinks).

14.5.2 More Long-Term Field Studies and Assessment of Offspring Fitness in the Wild

Ultimately, addressing the adaptive hypotheses for polyandry/multiple paternity should ideally occur under natural (preferably) field conditions because field patterns will reflect the ecological and social factors that influence encounter rates, and the sum of the costs and benefits to both sexes. If multiple paternity is in part driven by offspring fitness (which we predict them to be), field studies can also provide realistic tests of fitness under realistic conditions. This has proven a major obstacle in many insect, bird, and mammal systems because it is often difficult to assess long-term fitness of offspring (e.g., Cockburn *et al.* 2002; Komdeur and Pen 2002). Many lizards are relatively conspicuous animals with restricted dispersal and territoriality (and geographically small home ranges) therefore it is relatively easy to assess offspring survival and male and female reproductive success using a combination of capture-mark-recapture, field observations, and molecular determinants of paternity (e.g., Zamudio and Sinervo 2000; Pen *et al.* 2010; Olsson *et al.* 2011a,b; Uller *et al.* 2011; While *et al.* 2014).

One of the major challenges to understanding patterns of multiple mating in lizards is the lack of long-term studies which report variation in patterns of paternity between years and between populations (but for recent examples see Eizaguirre *et al.* 2007; While *et al.* 2009a, 2011; Olsson *et al.* 2011a,b). Long term field studies have the potential to reveal ongoing (and fluctuating) selection on mating patterns as well as potential external factors that may affect encounter rates. Currently, too many of the field studies (again noting the relative paucity of work on lizards in general) represent snapshots in time.

14.5.3 Greater Emphasis on Linking Experimental Approaches with Field Approaches

Field studies (with a greater taxonomic coverage) will provide us with the patterns of polyandry and multiple paternity that are required to establish taxon-relevant hypotheses best tested with experimental manipulations. Because of the ease with which lizards can be monitored in the field, they have the added advantage that they are suitable for experimental field work which will be crucial as we advance from correlative field approaches. Lizards offer a suite of potentially powerful techniques. Firstly, in many species it is relatively easy to control encounter rates mating in the laboratory or in semi-natural field enclosures (see above for excellent examples with *Lacerta vivipara*). By directly manipulating female mating rates and combining with field tests with, for example, offspring survival, we can expect to understand the contexts in which indirect genetic effects are important. As advocated by most researchers in this field, carefully controlled breeding designs will also allow the crucial tests of the benefits of multiple mating from mating with several males (see extensive discussion in Slatyer *et al.* 2012). Researchers using lizard systems are yet to realize this potential as demonstrated by the fact that only one experimental study (LaDage *et al.* 2008) met the criteria for inclusion in the meta-analyses by Slatyer *et al.* (2012). In their recent meta-analysis across a broader range of taxa examining the genetic benefits of polyandry, Slatyer *et al.* (2012) found polyandry had a broadly beneficial effect on offspring fitness traits. The strength of this meta-analysis was that it was confined to studies using an experimental approach specifically designed to quantify the potential genetic effects after controlling for the number of matings of monandrous and polyandrous females, i.e., the rigorous experimental design of Tregenza and Weddell (1998) as exemplified, for example, by Fisher *et al.* (2006). With carefully designed breeding experiments that control for genetic relatedness between partners it will also be possible to advance our understanding of the importance of "good genes" versus "compatible genes", including whether their effects are context-dependent on levels of inbreeding-outbreeding.

Simmons (2005) emphasized that experimental protocols or taxa where the effects of maternal effects can be controlled (which can confound our analyses of genetic effects) are important. One key maternal effect that is often hard to control is maternal allocation to offspring (as discussed earlier). For example, larger females may attract more partners but also produce larger offspring. In lizards, separating out effects of offspring size from offspring quality led to the development of 'allometric engineering' techniques, allowing the experimenter to efficiently manipulate offspring size in both oviparous and viviparous species (Sinervo *et al.* 1992; Olsson *et al.* 2002). Thus far, this technique has not been used to examine size-dependence of indirect genetic effects but its potential is clear, especially when combined with release of offspring into the wild.

14.5.4 Greater Understanding of Costs of Multiple Mating for Females (and Males)

Understanding of costs of multiple mating for females (and males) is still poor for lizards which is one of the reasons for the current debate on the role that such costs play in explaining polyandry. This is despite the important role that lizards (and snakes) have played in understanding costs of reproduction in general and the role that such costs play in explaining life history variation. We suggest increased attention examining costs associated with multiple mating (especially to females), especially those associated with the risk of injury, predation and/or disease will provide important insights and help resolve the contexts in which polyandry is favored (e.g., especially if combined with concurrent examination of the indirect genetic benefits). As discussed earlier, it is crucial that we carefully (probably through experimental control initially) separate out costs associated with multiple mating with the same male versus mating with multiple males.

14.6 ACKNOWLEDGMENTS

We thank Justin Rheubert, Dustin Siegel, Stanley Trauth, and Barrie Jamieson for inviting us to contribute to this publication. We would like to acknowledge the ongoing support of the Australian Research Council to our work. Jo McEvoy's assistance with literature searches and data extraction for the Table and Geoff While contributions to discussions and drafts of this chapter are gratefully acknowledged.

14.7 LITERATURE CITED

Abell, A. J. 1997. Estimating paternity with spatial behaviour and DNA fingerprinting in the striped plateau lizard, *Sceloporus virgatus* (Phrynosomatidae). Behavioral Ecology and Sociobiology 41: 217–226.

Akçkay, E. and Roughgarden, J. 2007. Extra-pair paternity in birds: review of the genetic benefits. Evolutionary Ecological Research 9: 855–868.

Alonzo, S. H. 2010. Social and coevolutionary feedbacks between mating and parental investment. Trends in Ecology and Evolution 25: 99–108.

Alonzo, S. H. and Sinervo, B. 2001. Mate choice games, context-dependent good genes, and genetic cycles in the side-blotched lizard, *Uta stansburiana*. Behavioral Ecology and Sociobiology 49: 176–186.

Ancona, S., Drummond, H. and Zaldivar-Rae, J. 2010. Male whiptail lizards adjust energetically costly mate guarding to male-male competition and female reproductive value. Animal Behaviour 79: 75–82.

Andersson, M. 1994. *Sexual Selection*. Princeton University Press: New Jersey, pp. 519.
Arnqvist, G. and Kirkpatrick, M. 2005. The evolution of infidelity in socially monogamous passerines: the strength of direct and indirect selection on extrapair copulation behavior in females. The American Naturalist 165: S26–S37.
Arnqvist, G. and Nilsson, T. 2000. The evolution of polyandry: multiple mating and female fitness in insects. Animal Behaviour 60: 145–164.
Arnqvist, G. and Rowe, L. 2005. *Sexual conflict*. Princeton University Press, New Jersey, pp. 330.
Badyaev, A. V. and Qvarnstrom, A. 2002. Putting sexual traits into the context of an organism: a life-history perspective in studies of sexual selection. Auk 119: 301–310.
Bateman, A. J. 1948. Intra-sexual selection in *Drosophila*. Heredity 2: 349–368.
Bateson, Z. W., Krenz, J. D. and Sorensen, R. E. 2011. Multiple paternity in the common five-lines skink (*Plestiodon fasciatus*). Journal of Herpetology 45: 504–510.
Berry, O. F. 2006. Inbreeding and promiscuity in the endangered grand skink. Conservation Genetics 7: 427–437.
Birkhead, T. R. and Møller, A. P. 1998. Sperm competition, sexual selection and different routes to fitness. pp. 757–781. In T. R. Birkhead and A. P. Møller (eds.), *Sperm Competition and Sexual Selection*. Academic Press: Cambridge.
Birkhead, T. R. and Pizzari, T. 2002. Postcopulatory sexual selection. Nature Reviews Genetics 3: 262–273.
Bleu, J., Bessa-Gomes, C. and Laloi, D. 2012. Evolution of female choosiness and mating frequency: effects of mating cost, density and sex ratio. Animal Behaviour 83: 131–136.
Bonier, F., Eikenaar, C., Martin, P. R. and Moore, I. T. 2014. Extrapair paternity rates vary with latitude and elevation in Emberizid sparrows. The American Naturalist 183: 54–61.
Botero, C. A. and Rubenstein, D. R. 2012. Fluctuating environments, sexual selection and the evolution of flexible mate choice in birds. Plos One 7: e32311.
Bull, C. M., Cooper, S. J. B. and Baghurst, B. C. 1998. Social monogamy and extra-pair fertilization in an Australian lizard, *Tiliqua rugosa*. Behavioral Ecology and Sociobiology 44: 63–72.
Bull, C. M., Godfrey, S. S. and Gordon, D. M. 2012. Social networks and the spread of Salmonella in a sleepy lizard population. Molecular Ecology 21: 4386–4392.
Calsbeek, R. and Bonneaud, C. 2008. Postcopulatory fertilization bias as a form of cryptic sexual selection. Evolution 62: 1137–1148.
Calsbeek, R. and Sinervo, B. 2002. Uncoupling direct and indirect components of female choice in the wild. Proceedings of the National Academy of Sciences of the United States of America 99: 14897–14902.
Calsbeek, R. and Sinervo, B. 2004. Within clutch variation in offspring sex determined by differences in sire body size: cryptic mate choice in the wild. Journal of Evolutionary Biology 17: 464–470.
Chapple, D. G. 2003. Ecology, life-history, and behavior in the Australian scincid genus *Egernia*, with comments on the evolution of complex sociality in lizards. Herpetological Monographs 17: 145–180.
Chapple, D. G. and Keogh, J. S. 2005. Complex mating system and dispersal patterns in a social lizard, *Egernia whitii*. Molecular Ecology 14: 1215–1227.
Cockburn, A., Legge, S. and Double, M. C. 2002. Sex ratios in birds and mammals: can the hypotheses be disentangled? pp. 266–286. In I. C. W. Hardy (ed.), *Sex ratios: concepts and research methods*. University Press: Cambridge.
Corl, A., Lancaster, L. T. and Sinervo, B. 2012. Rapid formation of reproductive isolation between two populations of side-blotched lizards, *Uta stansburiana*. Copeia 2012: 593–602.
Cornwallis, C. K. and Uller, T. 2010. Towards an evolutionary ecology of sexual traits. Trends in Ecology and Evolution 25: 145–152.
Cornwallis, C. K., West, S. A. Davis, K. E. and Griffin, A. S. 2010. Promiscuity and the evolutionary transition to complex societies. Nature 466: 969–U91.
Doody, J. S., Burghardt, G. M. and Dinets, V. 2013. Breaking the social-non-social dichotomy: a role for reptiles in vertebrate social behavior research? Ethology 119: 95–103.

Dubey, S., Chevalley, M. and Shine, R. 2011. Sexual dimorphism and sexual selection in a montane scincid lizard (*Eulamprus leuraensis*). Austral Ecology 36: 68–75.
Eberhard, W. G. and Cordero, C. 2003. Sexual conflict and female choice. Trends in Ecology and Evolution 18: 438–439.
Eizaguirre, C., Laloi, D., Massot, M., Richard, M., Federici, P. and Clobert, J. 2007. Condition dependence of reproductive strategy and the benefits of polyandry in a viviparous lizard. Proceedings of the Royal Society of London B 274: 425–430.
Faria, C. M. A., Zarza, E., Reynoso, V. H. and Emerson, B. C. 2010. Predominance of single paternity in the black spiny-tailed iguana: conservation genetic concerns for female-biased hunting. Conservation Genetics 11: 1645–1652.
Fedorka, K. M. and Mousseau, T. A. 2002. Material and genetic benefits of female multiple mating and polyandry. Animal Behaviour 64: 361–367.
Fenner, A. L., Godfrey, S. S. and Bull, M. C. 2011. Using social networks to deduce whether residents or dispersers spread parasites in a lizard population. Journal of Animal Ecology 80: 835–843.
Fisher, D. O., Double, M. C., Blomberg, S. P., Jennions, M. D. and Cockburn, A. 2006. Post-mating sexual selection increases lifetime fitness of polyandrous females in the wild. Nature 444: 89–92.
Fitze, P. S., Le Galliard, J. F., Federici, P., Richard, M. and Clobert, J. 2005. Conflict over multiple-partner mating between males and females of the polygynandrous common lizards. Evolution 59: 2451–2459.
Fitze, P. S., Cote, J. and Clobert, J. 2010. Mating order-dependent female mate choice in the polygynandrous common lizard *Lacerta vivipara*. Oecologia 162: 331–341.
Garcia-Gonzalez, F. and Simmons, L. W. 2005. Sperm viability matters in insect sperm competition. Current Biology 15: 271–275.
Gardner, M., Bull, C. M., Cooper, S. J. B. and Duffield, G. A. 2000. Microsattelite mutations in litters of the Australian lizard *Egernia stokesii*. Journal of Evolutionary Biology 13: 551–560.
Gardner, M. G., Bull, C. M., Cooper, S. J. B. and Duffield, G. A. 2002. High levels of genetic monogamy in the social Australian lizard *Egernia stokesii*. Molecular Ecology 11: 1787–1794.
Godfrey, S. S., Bull, C. M., James, R. and Murray, K. 2009. Network structure and parasite transmission in a group living lizard, *Egernia stokesii*. Behavioral Ecology and Sociobiology 63: 1045–1056.
Godfrey, S. S., Moore, J. A., Nelson, N. J. and Bull, C. M. 2010. Social network structure and parasite infection patterns in a territorial reptile, the tuatara (*Sphenodon punctatus*). International Journal for Parasitology 40: 1575–1585.
Griffin, A. S., Alonzo, S. H. and Cornwallis, C. K. 2013. Why do cuckolded males provide paternal care? Plos Biology 11: e1001520.
Griffith, S., Owens, I. and Thuman, K. 2002. Extrapair paternity in birds: a review of interspecific variation and adaptive function. Molecular Ecology 11: 2195–2212.
Gullberg, A., Olsson, M. and Tegelström, H. 1997. Male mating success, reproductive success and multiple paternity in a natural population of sand lizards: behavioural and molecular genetics data. Molecular Ecology 6: 105–112.
Gwynne, D. T. 2008. Sexual conflict over nuptial gifts in insects. Annual Review of Entomology 53: 83–101.
Hettyey, A., Hegyi, G., Puurtinen, M., Hoi, H., Torok, J. and Penn, D. J. 2010. Mate choice for genetic benefits: time to put the pieces together. Ethology 116: 1–9.
Hewison, A. J. M. and Gaillard, J. M. 1999. Successful sons or advantaged daughters? The Trivers-Willard model and sex-biased maternal investment in ungulates. Trends in Ecology and Evolution 14: 229–234.
Hofmann, S. and Henle, K. 2006. Male reproductive success and intrasexual selection in the common lizard determined by DNA microsatellites. Journal of Herpetology 40: 1–6.
Holman, L. and Kokko, H. 2013. The consequences of polyandry for population viability, extinction risk and conservation. Philosophical Transaction of the Royal Society of London B 368: 20120053.

Horvathova, T., Nakagawa, S. and Uller, T. 2012. Strategic female reproductive investment in response to male attractiveness in birds. Proceedings of the Royal Society of London B 279: 163–170.
Hughes, W. O. H., Oldroyd, B. P., Beekman, M. and Ratnieks, F. L. W. 2008. Ancestral monogamy shows kin selection is key to the evolution of eusociality. Science 320: 1213–1216.
Jennions, M. D. and Petrie, M. 2000. Why do females mate multiply? A review of the genetic benefits. Biological Reviews 75: 21–64.
Keogh, J. S., Umbers, K. D. L., Wilson, E., Stapley, J. and Whiting, M. J. 2013. Influence of alternate reproductive tactics and pre- and postcopulatory sexual selection on paternity and offspring performance in a lizard. Behavioral Ecology and Sociobiology 67: 629–638.
Kokko, H. 1999. Cuckoldry and the stability of biparental care. Ecology Letters 2: 247–255.
Komduer, J. and Pen, I. 2002. Adaptive sex allocation in birds: the complexities of linking theory and practice. Philosophical Transaction of the Royal Society of London B 357: 373–380.
LaDage, L. D., Gutzke, W. H. N., Simmons, R. A., II and Ferkin, M. H. 2008. Multiple mating increases fecundity, fertility and relative clutch mass in the female leopard gecko (*Eublepharis macularius*). Ethology 114: 512–520.
Laloi, D., Eizaguirre, C., Federici, P. and Massot, M. 2011. Female choice for heterozygous mates changes along successive matings in a lizard. Behavioural Processes 88: 149–154.
Laloi, D., Richard, M., Federici, P., Clobert, J., Teillac-Deschamps, P. and Massot, M. 2009. Relationship between female mating strategy, litter success and offspring dispersal. Ecology Letters 12: 823–829.
Laloi, D., Richard, M., Lecomte, J., Massot, M. and Clobert, J. 2004. Multiple paternity in clutches of common lizard *Lacerta vivipara*: data from microsatellite markers. Molecular Ecology 13: 719–723.
Lebas, N. R. 2001. Microsatellite determination of male reproductive success in a natural population of the territorial ornate dragon lizard, *Ctenophorus ornatus*. Molecular Ecology 10: 193–203.
Le Galliard, J. F., Fitze, P. S., Ferriere, R. and Clobert, J. 2005. Sex ratio bias, male aggression, and population collapse in lizards. Proceedings of the National Academy of Sciences of the United States of America 102: 18231–18236.
Le Galliard, J. F., Cote, J. and Fitze, P. S. 2008. Lifetime and intergenerational fitness consequences of harmful male interactions for female lizards. Ecology 89: 56–64.
Leu, S. T., Bashford, J., Kappeler, P. M. and Bull, C. M. 2010. Association networks reveal social organization in the sleepy lizard. Animal Behaviour 79: 217–225.
Lopez, P., Aragon, P. and Martin, J. 2003. Responses of female lizards, *Lacerta monticola*, to males' chemical cues reflect their mating preference for older males. Behavioral Ecology and Sociobiology 55: 73–79.
Lukas, D. and Clutton-Brock, T. H. 2013. The evolution of social monogamy in mammals. Science 341: 526–530.
Madsen, T. 2008. Female nonavian reptiles benefit from multiple matings. Molecular Ecology 17: 3753–3753.
Madsen, T. 2011. Cost of multiple matings in female adders (*Vipera berus*). Evolution 65: 1823–1825.
Madsen, T., Shine, R., Loman, J. and Håkansson, T. 1992. Why do female adders copulate so frequently? Nature 355: 440–441.
Madsen, T., Shine, R., Olsson, M. and Wittzell, H. 1999. Conservation biology: restoration of an inbred adder population. Nature 402: 34–35.
Madsen, T., Ujvari, B. and Olsson, M. 2004. Novel genes continue to enhance population growth in adders (*Vipera berus*). Biological Conservation 120: 145–147.
Martin, J. and Lopez, P. 2013. Responses of female rock lizards to multiple scent marks of males: effects of male age, male density and scent over-marking. Behavioural Processes 94: 109–114.

Morrison, S., Keogh, S. and Scott, A. 2002. Molecular determination of paternity in a natural population of the multiply mating polygynous lizard *Eulamprus heatwolei*. Molecular Ecology 11: 535–545.

Neff, B. D. and Pitcher, T. E. 2005. Genetic quality and sexual selection: an integrated framework for good genes and compatible genes. Molecular Ecology 14: 19–38.

Neff, B. D. and Pitcher, T. E. 2008. Mate choice for non-additive genetic benefits: a resolution to the lek paradox. Journal of Theoretical Biology 254: 147–155.

Noble, D. W. A., Keogh, J. S. and Whiting, M. J. 2013. Multiple mating in a lizard increases fecundity but provides no evidence for genetic benefits. Behavioral Ecology 24: 1128–1137.

O'Connor, D. and Shine, R. 2004. Parental care protects against infanticide in the lizard *Egernia saxatilis* (Scincidae). Animal Behaviour 68: 1361–1369.

Oh, K. P. 2011. Inclusive fitness of 'kissing cousins': new evidence of a role for kin selection in the evolution of extra-pair mating in birds. Molecular Ecology 20: 2657–2659.

Olsson, M. 2001. 'Voyeurism' prolongs copulation in the dragon lizard *Ctenophorus fordi*. Behavioral Ecology and Sociobiology 50: 378–381.

Olsson, M. and Madsen, T. 1995. Female choice on male quantitative traits in lizards—why is it so rare? Behavioral Ecology and Sociobiology 36: 179–184.

Olsson, M. and Madsen, T. 1996. Costs of mating with infertile males selects for late emergence in female sand lizards (*Lacerta agilis* L.). Copeia 2: 462–464.

Olsson, M. and Madsen, T. 1998. Sexual selection and sperm competition in reptiles. pp. 503–578. In T. R. Birkhead and A. P. Møller (eds.), *Sperm Competition and Sexual Selection*. Academic Press, Cambridge.

Olsson, M. and Madsen, T. 2001. Promiscuity in sand lizards (*Lacerta agilis*) and adder snakes (*Vipera berus*): causes and consequences. Journal of Heredity 92: 190–197.

Olsson, M. and Shine, R. 1997. Advantages of multiple matings to females: a test of the infertility hypothesis using lizards. Evolution 51: 1684–1688.

Olsson, M. and Uller, T. 2009. On parsimonious paternity and scientific rigor: a reply to Madsen. Molecular Ecology 18: 25–27.

Olsson, M., Gullberg, A., Tegelstrom, H., Madsen, T. and Shine, R. 1994a. Can female adders multiply. Nature 369: 528–528.

Olsson, M., Madsen, T., Shine, R., Gullberg, A. and Tegelstrom, H. 1994b. Rewards of promiscuity—a reply. Nature 372: 230–230.

Olsson, M., Shine, R., Madsen, T., Gullberg, A. and Tegelstrom, H. 1996. Sperm selection by females. Nature 383: 585–585.

Olsson, M., Birkhead, T. and Shine, R. 1999. Can relaxed time constraints on sperm production eliminate protandry in an ectotherm? Biological Journal of the Linnean Society 66: 159–170.

Olsson, M., Wapstra, E. and Olofsson, C. 2002. Offspring size-number strategies: experimental manipulations of offspring size in a viviparous lizard (*Lacerta vivipara*). Functional Ecology 16: 135–141.

Olsson, M., Madsen, T., Nordby, J., Wapstra, E., Ujvari, B. and Wittsell, H. 2003. Major histocompatibility complex and mate choice in sand lizards. Proceedings of the Royal Society of London B 270: S254–S256.

Olsson, M., Madsen, T., Ujvari, B. and Wapstra, E. 2004a. Fecundity and MHC affects ejaculation tactics and paternity bias in sand lizards. Evolution 58: 906–909.

Olsson, M., Ujvari, B., Madsen, T., Uller, T. and Wapstra, E. 2004b. Haldane rules: costs of outbreeding at production of daughters in sand lizards. Ecology Letters 7: 924–928.

Olsson, M., Madsen, T., Uller, T., Wapstra, E. and Ujvari, B. 2005a. The role of Haldane's rule in sex allocation. Evolution 59: 221–225.

Olsson, M., Madsen, T., Wapstra, E., Silverin, B., Ujvari, B. and Wittsell, H. 2005b. MHC, health, colour, and reproductive success in sand lizards. Behavioural Ecology and Sociobiology 58: 289–294.

Olsson, M., Ujvari, B., Wapstra, E., Madsen, T., Shine, R. and Bensch, S. 2005c. Does mate guarding prevent rival mating in snow skinks? A test using AFLP. Herpetologica 61: 389–394.

Olsson, M., Healey, M., Wapstra, E., Schwartz, T., Lebas, N. and Uller, T. 2007. Mating system variation and morph fluctuations in a polymorphic lizard. Molecular Ecology 16: 5307–5315.
Olsson, M., Schwartz, T., Wapstra, E., Uller, T., Ujvari, B., Madsen, T. and Shine, R. 2011a. Climate change, multiple paternity and offspring survival in lizards. Evolution 65: 3323–3326.
Olsson, M., Wapstra, E., Schwartz, T., Madsen, T., Ujvari, B. and Uller, T. 2011b. In hot pursuit: fluctuating mating system and sexual selection in sand lizards. Evolution 65: 574–583.
Oppliger, A., Degen, L., Bouteiller-Reuter, C. and John-Alder, H. B. 2007. Promiscuity and high level of multiple paternity in common wall lizards, *Podarcis muralis*: data from microsatellite markers. Amphibia-Reptilia 28: 301–303.
Parker, G. A. and Birkhead, T. R. 2013. Polyandry: the history of a revolution. Philosophical Transactions of the Royal Society of London B 368: 20120335.
Pizzari, T. and Wedell, N. 2013. Introduction: the polyandry revolution. Philosophical Transactions of the Royal Society of London B 368: 20120041.
Pen, I., Uller, T., Feldmeyer, B., Harts, A., While, G. M. and Wapstra, E. 2010. Climate-driven population divergence in sex-determining systems. Nature 468: 436–U262.
Price, T. A. R., Hurst, G. D. D. and Wedell, N. 2010. Polyandry prevents extinction. Current Biology 20: 471–475.
Puurtinen, M., Ketola, T. and Kotiaho, J. S. 2009. The good-genes and compatible-genes benefits of mate choice. The American Naturalist 174: 741–752.
Richard, M., Massot, M., Clobert, J. and Meylan, S. 2012. Litter quality and inflammatory reponse are dependent on mating strategy in a reptile. Oecologia 170: 39–46.
Salvador, J., Diaz, J. A., Veija, J. P., Bloor, P. and Brown, R. P. 2008. Correlates of reproductive success in male lizards of the alpine species *Iberolacerta cyreni*. Behavioral Ecology 19: 169–176.
Senar, J. C., Figuerola, J. and Pascual, J. 2002. Brighter yellow blue tits make better parents. Proceedings of the Royal Society of London B 269: 257–261.
Sever, D. M. and Hamlett, W. C. 2002. Female sperm storage in reptiles. Journal of Experimental Zoology 292: 187–199.
Sheldon, B. C. 1994. Male phenotype, fertility, and the pursuit of extra-pair copulations by female birds. Proceedings of the Royal Society of London B 257: 25–30.
Sherman, C. D. H., Wapstra, E., Uller, T. and Olsson, M. 2008. Males with high genetic similarity to females sire more offspring in sperm competition in Peron's tree frog *Litoria peronii*. Proceedings of the Royal Society of London B 275: 971–978.
Sherman, C. D. H., Wapstra, E. and Olsson, M. 2009. Consistent male-male paternity differences across female genotypes. Biology Letters 5: 232–234.
Shine, R., Olsson, M. and Mason, R. T. 2000. Chastity belts in gartersnakes: the functional significance of mating plugs. Biological Journal of the Linnean Society 70: 377–390.
Simmons, L. W. 2005. The evolution of polyandry: sperm competition, sperm selection, and offspring viability. Annual Review of Ecology Evolution and Systematics 36: 125–146.
Sinervo, B., Doughty, P., Huey, R. B. and Zamudio, K. 1992. Allometric engineering—a causal analysis of natural selection on offspring size. Science 258: 1927–1930.
Sinervo, B., Calsbeek, R., Comendant, T., Both, C., Adamopoulou, C. and Clobert, J. 2006. Genetic and maternal determinants of effective dispersal: the effect of sire genotype and size at birth in side-blotched lizards. The American Naturalist 168: 88–89.
Sinn, D. L., While, G. M. and Wapstra, E. 2008. Maternal care in a social lizard: links between female aggression and offspring fitness. Animal Behaviour 76: 1249–1257.
Slatyer, R. A., Mautz, B. S., Backwell, P. R. Y. and Jennions, M. D. 2012. Estimating genetic benefits of polyandry from experimental studies: a meta-analysis. Biological Reviews 87: 1–33.
Snook, R. R. 2005. Sperm in competition: not playing by the numbers. Trends in Ecology and Evolution 20: 46–53.

Stapley, J., Hayes, C. M. and Keogh, J. S. 2003. Population genetic differentiation and multiple paternity determined by novel microsatellite markers from the Mountain Log Skink (*Pseudemoia entrecasteauxii*). Molecular Ecology Notes 3: 291–293.
Stow, A. J. and Sunnucks, P. 2004. High mate and site fidelity in Cunningham's skinks (*Egernia cunninghami*) in natural and fragmented habitat. Molecular Ecology 13: 419–430.
Tokarz, R. R. 1995. Mate choice in lizards—a review. Herpetological Monographs 9: 17–40.
Tregenza, T. and Wedell, N. 1998. Benefits of multiple mates in the cricket *Gryllus bimaculatus*. Ethology 52: 1726–1730.
Trivers, R. L. 1972. Parental investment and sexual selection. pp. 136–179. In B. Campbell (ed.), *Sexual selection and the descent of man, 1871–1971*. Aldine: Chicago.
Uller, T. and Olsson, M. 2005. Multiple copulations in natural populations of lizards: evidence for the fertility assurance hypothesis. Behaviour 142: 45–56.
Uller, T. and Olsson, M. 2008. Multiple paternity in reptiles: patterns and processes. Molecular Ecology 17: 2566–2580.
Uller, T., While, G. M., Cadby, C. D., Harts, A., O'Connor, K., Pen, I. and Wapstra, E. 2011. Altitudinal divergence in maternal thermoregulatory behaviour may be driven by differences in selection on offspring survival in a viviparous lizard. Evolution 65: 2313–2324.
Uller, T., Schwartz, T., Koglin, T. and Olsson, M. 2013. Sperm storage and sperm competition across ovarian cycles in the dragon lizard, *Ctenophorus fordi*. Journal of Experimental Zoology Part A-Ecological Genetics and Physiology 319: 404–408.
Vahed, K. 1998. The function of nuptial feeding in insects: review of empirical studies. Biological Reviews of the Cambridge Philosophical Society 73: 43–78.
Watson, P. J., Arnqvist, G. and Stallman, R. R. 1998. Sexual conflict and the energetic costs of mating and mate choice in water striders. The American Naturalist 151: 46–58.
While, G. M. and Wapstra, E. 2009. Snow skinks (*Niveoscincus ocellatus*) do not shift their sex allocation patterns in response to mating history. Behaviour 146: 1405–1422.
While, G. M., Uller, T. and Wapstra, E. 2009a. Within-population variation in social strategies characterize the social and mating system of an Australian lizard, *Egernia whitii*. Austral Ecology 34: 938–949.
While, G. M., Uller, T. and Wapstra, E. 2009b. Family conflict and the evolution of sociality in reptiles. Behavioral Ecology 20: 245–250.
While, G. M., Uller, T. and Wapstra, E. 2011. Variation in social organization influences the opportunity for sexual selection in a social lizard. Molecular Ecology 20: 844–852.
While, G. M., Uller, T., Bordogna, G. and Wapstra, E. 2014. Promiscuity resolves constraints on social mate choice imposed by population viscosity. Molecular Ecology 23: 721–732.
Wusterbarth, T. L., King, R. B., Duvall, M. R., Grayburn, W. S. and Burghardt, G. M. 2010. Phylogenetically widespread multiple paternity in new world natricine snakes. Herpetological Conservation and Biology 5: 86–93.
Wedell, N., Gage, M. J. G. and Parker, G. A. 2002. Sperm competition, male prudence and sperm-limited females. Trends in Ecology and Evolution 17: 313–320.
Yasui, Y. 1998. The 'genetic benefits' of female multiple mating reconsidered. Trends in Ecology and Evolution 13: 246–250.
Zamudio, K. R. and Sinervo, E. 2000. Polygyny, mate-guarding, and posthumous fertilization as alternative male mating strategies. Proceedings of the National Academy of Sciences of the United States of America 97: 14427–14432.
Zeh, J. A. and Zeh, D. W. 1996. The evolution of polyandry I: intragenomic conflict and genetic incompatibility. Proceedings of the Royal Society of London B 263: 1711–1717.
Zeh, J. A. and Zeh, D. W. 2003. Toward a new sexual selection paradigm: polyandry, conflict and incompatibility (Invited article). Ethology 109: 929–950.

CHAPTER 15

The Evolutionary Ecology of Parental Care in Lizards

Geoffrey M. While,[1,2,*] *Ben Halliwell*[1] and *Tobias Uller*[2,3]

15.1 INTRODUCTION

Parental care is any investment by a parent that increases the survival or reproductive success of its offspring, typically at a cost to the parent's growth, survival or future reproduction (Clutton-Brock 1991; Royle *et al.* 2012). This broad definition of care covers a tremendous diversity of behavior exhibited by organisms across the animal kingdom. Parental care can occur prior to fertilization, during embryo development, and can continue well after nutritional dependence. It can be provided by either the mother or father (uniparental care) or by both parents (biparental care).

Needless to say, the diversity in form, function and taxonomic distribution of parental care raises a suite of interesting questions regarding the factors responsible for its emergence, maintenance and evolutionary diversification. The classical explanation (sensu Clutton-Brock 1991) maintains that each parental care trait is the result of an optimal balance of the trade-off between the cost of expressing the trait and the long-term inclusive fitness advantage conferred by it. Ultimately, the outcome of these cost-benefit tradeoffs will be contingent on the ecology, life-history and evolutionary history of a given species (Stearns 1976; Clutton-Brock 1991; Klug and Bonsall 2010). The evolution of care behavior is of profound interest to evolutionary biologists because it represents a nexus point for the

[1] School of Zoology, University of Tasmania, Tasmania, Australia.
[2] Edward Grey Institute, Department of Zoology, University of Oxford, Oxford, UK.
[3] Department of Biology, Lund University, Lund, Sweden.
* Corresponding author

adaptive radiation of kin relationships and sociality. Thus, studying parental care provides general insights into the social dynamics of family groups including sexual conflict, parent-offspring conflict, sibling rivalry, kin competition, and kin-mediated cooperation (e.g., Trivers 1974; Wilson 1980; Mock and Parker 1997; Hatchwell 1999; Field and Brace 2004; Cockburn 2006; Kilner and Hinde 2008).

15.2 PARENTAL CARE IN NON-AVIAN REPTILES

The majority of research on parental care has focused on mammals and birds (Clutton-Brock 1991; Koenig and Dickinson 2004; Balshine 2012). These include familiar examples of care such as lactation (Kunz and Hood 2000), parental food provisioning of young (Wright and Cuthill 1990; Brotherton *et al.* 2001), and prolonged juvenile retention and social initiation and learning (Clay and de Waal 2013). In contrast, non-avian reptiles have largely been ignored when it comes to parental care. The one exception to this rule are crocodiles, where care has been found to be ubiquitous and diverse, including behaviors such as nest excavation, transporting hatchlings and eggs to water from the nest, post-partum feeding and protection of offspring, and even vocal communication between offspring and parents (Vergne *et al.* 2009; Whitaker 2011; Campos *et al.* 2012; Charruau and Hénaut 2012).

In contrast, with the exception of maternal investment in terms of egg size (Uller and While Chapter 12, this volume) and transitions between oviparity and viviparity (Shine 1985), research into squamate reptile (lizards and snakes) parental care is still in its infancy, with the majority of research being restricted to one or two key groups (e.g., *Egernia*; Chapple 2003; *Plestiodon*; Vitt and Cooper 1989; *Pythonidae*; Stahlschmidt and DeNardo 2011). However, recent research has identified complex forms of parental care in a growing number of species, suggesting that parental care is more common in squamates than previously thought (Somma 2003a; Shine 2013). Indeed, some form of parental behavior has now been documented in at least 30 of the 63 lizard families (Somma 2003a). Parental care in squamates can serve a variety of functions including protection of eggs from predators and conspecifics, provisioning of gametes and embryos, and the maintenance of optimal embryonic development (Shine 1988; Shine and Harlow 1996; Somma 2003a).

The functional and taxonomic diversity of parental care in squamates suggests that this group has the potential to play a valuable role in helping us to understand the early steps in the evolution of complex parent-offspring relationships. Importantly, lizards in particular exhibit a number of characteristics that make them valuable model systems for this task. First, while lizards exhibit functional diversity in parental care, care itself is relatively simple. This means that care can be easily quantified with limited need to control for trade-offs between different forms of care or at different stages in ontogeny (e.g., parental provisioning before and after hatching or birth). Second, the relatively sedentary behavior of many lizards coupled with the ease with which they can be monitored in the wild means that the costs (condition, survival, future reproductive success) and benefits (offspring survival, growth and reproductive success) of parental care can be assessed using both experimental techniques and long-term field data (see Fitze *et al.* 2005; Pen *et al.* 2010; While *et al.* 2011 for similar approaches). Third, the simplicity of parental care coupled with an ancestral state of no care means that we can address questions regarding factors which have influenced the initial origins, and not just the maintenance, of parental care. Advanced parental behavior, such as parental provisioning and care after nutritional independence, which define complex animal societies have their origins in relatively simple traits (e.g., attendance and guarding of offspring), which were subsequently modified into the elaborate and complex forms of parental care observed in many animal taxa (Clutton-Brock 1991). Therefore, focusing on the conditions that allowed increased investment in offspring from a state of no care is crucial if we are to understand the evolution of parental care, since, once it has evolved, the forces that maintain care may differ from those that promoted its origin.

Our aim in this chapter is to highlight the diversity of care types observed in lizards and provide insights into their functional significance and taxonomic distribution. In doing so, we do not attempt to provide an exhaustive list of the modes of lizard parental care (for this we direct the reader towards the comprehensive review of the literature in Somma 2003a). Rather we identify the most common forms and emphasize the complexity of care behaviors exhibited across the taxon. We finish by exploring the potential evolutionary origins of parental care in lizards and discuss current hypotheses concerning the diversification and maintenance of care before offering directions for future study.

15.2.1 Functional Diversity of Parental Care in Lizards

To understand and appreciate the adaptive significance and diversity of care in the lizards it is useful to categorize parental traits into specific forms. For the purposes of this chapter we differentiate between pre-partum care, defined as all parental investment in offspring from the onset of vitellogenesis up to and including the oviposition of eggs/birth of offspring, and post-partum care, defined as all parental investment after oviposition/birth until disassociation between parents and young. Using this distinction, we present the major forms of care exhibited by lizards over the course of ontogeny. We describe the behavioral and physiological investment by the parent in each form of care, provide relevant examples, and discuss their adaptive significance.

15.2.2 Provisioning of Gametes

Provisioning gametes with nutrient stores that exceed the minimum requirements for fertilization represents parental care in its most basic form. The vast majority of lizards deposit nutrient rich yolk into the oocyte before ovulation. Lizard yolk contains proteins (vitellogenins) synthesized in the liver, as well as various lipids remobilized from maternal fat stores (Ho *et al.* 1982). These include polar lipids, sterols, and large amounts of triacylglycerols, the primary lipid source made available to embryos via the yolk (Jones 1998; Lance and Morafka 2001). The provisioning of yolk varies substantially both within and between species (Ar *et al.* 2004; Belinsky *et al.* 2004) and between individuals of a species (Troyer 1983; Rhen and Lang 1999; Roosenburg and Dennis 2005), which partly reflects selective pressures associated with ecological, life-history, and phylogenetic characteristics (see Uller and While Chapter 12, this volume for further discussion)

During incubation, yolk stores are metabolized and depleted by the embryos to aid in growth and development. Some studies indicate that yolk proteins and lipids are depleted at equal rates during embryogenesis (Troyer 1983), while others demonstrate that lipids are the main source of embryonic energy (Stewart 1989). In addition to nutritional inputs, mothers also provision embryos with hormones, minerals, anti-oxidants, and immunological factors that facilitate development and prime offspring for life after birth (Speake *et al.* 2003; Thompson and Speake 2003).

Many species complete embryological development without metabolizing the entire yolk reserve (Pezaro *et al.* 2013) and a portion often remains unused when offspring are ready to emerge. This 'residual yolk' is drawn into the hatchling's body (internalized) and provides a postnatal ration to supplement the metabolic demands of early life (in addition to stored lipids in adipose tissue; Speake *et al.* 2003; Pandav *et al.* 2006). Because the majority of reptiles provide no form of post-partum care, this initial yolk provisioning is a crucial contribution to offspring fitness and offers an alternative strategy to contribute to post-partum nutrition in lieu of post-partum feeding behavior. Indeed, maternal investment in vitellogenesis is often substantial; up to 75% of the yolk is reserved for internalization in some species (Lee *et al.* 2007) and can serve as the exclusive source of nutrition for the first six months of postnatal life (Lance and Morafka 2001). Experimental manipulation of yolk during embryogenesis suggests that residual yolk may be as (or more) important for offspring fitness than the size at hatching. For example, surgical removal of yolk from periparturient eggs of the changeable lizard, *Calotes versicolor*, did not influence the amount of residual yolk at hatching but significantly decreased the size of hatchlings (Radder *et al.* 2004). This conservatism in residual yolk volume indicates the potential importance of preserving a portion of parental investment for postnatal nutrition. Despite this, there is considerable variation both among and within species in the proportion of the yolk reserved and even among individuals within a clutch (Radder *et al.* 2007). This variation is thought to be influenced by differences in life history, ecology, and the energetic needs of post-emergence offspring. For example, internalized yolk may be particularly important for species in which offspring face a considerable energetic challenge before they have the opportunity to feed, or before gut microflora are properly established (as in Chelonians; Troyer 1983; Tucker *et al.* 1998; Booth and Astill 2001; Lance and Morafka 2001; Ischer *et al.* 2009; Booth and Evans 2011).

Alternatively, variation in residual yolk deposition may be a non-adaptive consequence of incubation conditions. Specifically, both thermal and hydric conditions have been shown to influence the amount of residual yolk available (Packard *et al.* 1988; Packard 1991; Ji and Du 2001; Belinsky *et al.* 2004; Warner *et al.* 2012). Furthermore, surgical removal of internalized yolk from newly hatched jacky dragons, *Amphibolurus muricatus*, had no effect on growth, survival, or locomotor performance of neonates (Radder *et al.* 2007).

15.2.3 Oviposition Site Selection and Nest Building

Perhaps the most primitive form of behavioral parental care is the choice or preparation of a suitable site in which to receive eggs or young. The investment and complexity displayed in this form of care is highly variable, ranging from depositing eggs on the soil surface to judicious site selection and complex nest construction. It is also often associated with sexual dimorphism in morphology (e.g., Warner *et al.* 2006). Parental control of the ovipositing site helps buffer eggs against exposure to environmental hazards including drought, desiccation, temperature extremes, hypoxia, predation, and parasitism. Such site selection confers distinct advantages to offspring fitness. Specifically, environmental conditions experienced during incubation not only impact embryo survivorship itself (Shine and Elphick 2001) but can also influence hatchling size (Elphick and Shine 1998; Qualls and Andrews 1999; Shine and Elphick 2001; Reedy *et al.* 2012), performance (Shine and Harlow 1996), growth (Qualls and Andrews 1999; Robbins and

Warner 2010), behavior (Elphick and Shine 1998), and sex (Harlow and Shine 1999; Harlow 2000; Harlow and Taylor 2000; Harlow 2004).

Temperature and moisture are generally considered to be the most important factors in nest site selection in lizards. Indeed, variability in nest temperature and, to a lesser degree, moisture have been found to affect various fitness-related phenotypic traits of offspring in a wide range of lizard species (Deeming 2004). Appropriate site selection may even enable females to exploit offspring developmental plasticity to optimize the phenotype of offspring to their birthing environment (Shine and Wall 2005). For example, manipulating mean temperatures and temperature variance during incubation considerably alters rates of embryogenesis in *Bassiana duperreyi* and also influences fitness related traits such as body shape, activity levels, thermoregulatory behavior, and running speeds of neonates (Shine and Harlow 1996). However, it should be noted that many such maternal effects may simply reflect weak selection on, and hence limited canalization of, offspring phenotype (Uller *et al.* 2011). The importance of nest temperature for successful completion of development means that females sometimes move far away from their home range for oviposition. For example, eastern fence lizards (*Sceloporus undulatus*) move from their home range in the forest to open areas with low canopy cover and high soil temperatures to oviposit (Angilletta *et al.* 2009; see also Shine *et al.* 2002).

Water availability can also be important for successful incubation and sometimes affects offspring phenotype, suggesting that females may choose nest site locations based on hydric conditions. For example, when given a choice of nesting sites differing in moisture content, brown anoles (*Anolis sagrei*) choose nest sites that result in the highest hatching success, largest offspring, and highest offspring survival (Socci *et al.* 2005; Reedy *et al.* 2012). Additionally, female eastern fence lizards, *Sceloporus undulatus*, choose nest sites with both thermal and moisture conditions most conducive to hatching success (Warner and Andrews 2002).

The choice of oviposition site can also be based on social factors. Specifically, a large number of lizard species (255 species (7% of all lizard species) across 15 families) are known to preferentially oviposit in areas that contain the eggs of other individuals (Graves and Duvall 1995; Doody *et al.* 2009). Although some studies have suggested that such communal nesting results from constraints on the availability of suitable nesting sites (Vitt 1993; Graves and Duvall 1995) several studies have shown that egg clustering confers adaptive advantages to embryonic development (reviewed in Doody *et al.* 2009). For example, communal nesting can infer thermal benefits to offspring via increased metabolic output from neighboring eggs (as in snakes and frogs; Blouin-Demers *et al.* 2004; Håkansson and Loman 2004). As above, this has the potential to speed up embryonic development and alter offspring phenotype. Alternatively, communal nesting can influence the hydric environment of the eggs by increasing (or decreasing) water uptake. For example, *Lacerta schreiberi* and *Bassiana duperreyi* eggs incubated within a cluster of other eggs take up less water than do solitarily incubated eggs (Marco *et al.* 2004; Radder and Shine 2007). In *B. duperreyi* this has significant implications for offspring development via increased size and performance of hatchlings (Radder and Shine 2007; see also Brown and Shine 2009). Finally, communal nesting can decrease risk of predation to individual nestlings during incubation or upon emergence from the nest (i.e., the selfish herd; Hamilton 1971).

15.2.4 Care of Fertilised Eggs

For the majority of oviparous lizard species, parental care ends at nest site selection as females abandon their eggs shortly after laying (Doody *et al.* 2013). However, an increasing number of species are being shown to exhibit a prolonged association with their eggs beyond that required to complete oviposition. This parental care behavior is known collectively as egg attendance and encompasses all behaviors related to the defense and manipulation or brooding of eggs by the parent. Egg attendance may increase offspring fitness in a number of ways. Specifically, it can guard eggs against environmental hazards such as desiccation, flooding, and hypoxia as well as pathogens, parasites, predators, and oophagic conspecifics (Groves 1982; Hasegawa 1985; Shine 1988; Somma and Fawcett 1989; Lang 1990). The degree of egg attendance varies between species and in many cases is maintained throughout the incubation period (Shine 1988).

In its simplest form, egg attendance consists of the protection of eggs from predators. Egg defense has been observed in at least 60 different lizard species from 12 different families (e.g., Iguanidae, Varanidae, Scincidae; Somma 2003a). The majority of egg defense is often directed towards oophagic conspecifics. For example, most *iguanids* and *Plestiodon* skinks defend their nests from conspecifics during nest construction and following oviposition (Wiewandt 1982; Somma 2003a). A number of varanids lay their eggs in termite mounds and actively guard these nest sites for several weeks following oviposition (e.g., *Varanus rosenbergi*; Rismiller *et al.* 2010). However, egg defense can also function as protection against heterospecific predators. Perhaps the best-documented case of this is the long-tailed skink, *Eutropis longicaudata*. In some island populations *E. longicaudata* occur sympatrically with an oophagous snake, *Oligodon formosanus*. Egg loss from attended nests is due to predation by *O. formosanus* (Huang 2006) and mothers actively defend against nest intrusion by biting the head or neck of the intruding snake and pushing them out of the nest (Huang 2006). Such nest defense may be more common than previously expected. However, it should be noted that there is a lack of experimental tests that unequivocally show that lizards defend their eggs and not simply their territory (or nest sites). Indeed, even in *Eutropis* females defend nest sites irrespective of whether or not they have eggs in them (Huang and Pike 2011). Studies that show that female behavior is directed towards protection of eggs or offspring would be an important step towards understanding the adaptive nature of parental behavior in lizards.

In some species the attendance of eggs goes beyond protection and towards the facultative control of the environmental conditions to which the eggs are exposed. While this is well documented in some snakes (e.g., egg brooding in pythons, Stahlschmidt and DeNardo 2011) it is relatively rare in lizards. However, there is some evidence that egg brooding does occur. This has been most well studied in the genus *Plestiodon* (previously part of *Eumeces*). A number of species in this genus exhibit long term association with eggs and complex brooding behavior in which females actively manipulate eggs throughout incubation. The function of this behavior is yet to be fully elucidated but it is thought to revolve around regulating the thermal and hydric conditions experienced by the developing embryos. For example, brooding behavior can regulate the moisture levels of the nest (Fitch 1954), the humidity of the nest (Vitt and Cooper 1986), or the extent of water exchange (Fitch 1954; Campbell and Simmons 1961). This regulation of moisture levels can have significant effects on offspring fitness. For example, brooding behavior in female prairie skinks (*Eumeces septentrionalis*) under extreme hydric conditions has been shown to reduce egg mortality (Somma and Fawcett 1989; see also Hasegawa 1985).

In contrast to, for example, bird incubation, the costs of these behavioral forms of pre-partum care appear to be relatively low in lizards and snakes. The energetic investment in eggs typically exceeds substantially the costs associated with behavioral forms of egg attendance and brooding. For example, egg production in water pythons, *Liasis fuscus*, is associated with a loss of 60% body mass while the estimated loss due to brooding is less than 5% (Stahlschmidt *et al.* 2012). In *Eutropis longicaudata*, all mothers guard egg clutches for >7 days yet neither body mass, survival, fecundity, timing of a second clutch, or clutch frequencies differ between females that invest in long-term egg guarding (>28 days) and those that only provided short-term egg guarding (<16 days) (Huang 2007). Thus, many care behaviors are energetically undemanding and may lead to a situation in which increased expression of the behavior does not necessarily translate to increased costs to parents. However, while energetic costs may be low, there could be significant costs associated with increased predation risk. In *E. longicaudata* egg attendance is highly plastic varying both within and between populations. For example, the extent of egg defense by mothers is influenced by the presence of a predator (Huang and Wang 2009; Huang and Pike 2013), the type of predator (egg vs. lizard predator (Huang 2006; Huang 2008)), and the frequency of predation attempts (Huang and Wang 2009). When the costs (and risk) of defending a clutch from predation are perceived to outweigh the reproductive benefits represented by that clutch, for example when predation levels are high or the threat is directed to the mother, mothers will cut their losses and abandon their eggs. Although stopping short of demonstrating actual costs to the mothers, these patterns are consistent with predictions that an optimal defense strategy will consider the risk to future reproduction of the parent weighed against the investment represented by current offspring and the expected benefit to offspring from the defense behavior (e.g., Ghalambor and Martin 2001).

15.2.5 Viviparity

Egg attendance has been taken to extremes in many lizards, resulting in egg retention *in utero*. This is accompanied by a reduction in shell thickness, increased gas and nutrient exchange and, if the offspring are retained until completion of development, viviparity (Shine 1983; Shine 1985; Blackburn 1999). Viviparity enables several behavioral forms of care similar to nest selection and egg attendance in oviparous species (Packard *et al.* 1977; De Fraipont *et al.* 1996). Therefore, viviparity can be viewed as a form of parental investment that serves a number of different care functions and that can have significant implications for offspring fitness.

First, viviparity gives females greater control of the embryonic environment, particularly with respect to climate and predation. There is substantial evidence that cool climate, and hence suboptimal soil temperature for embryonic development, is the key factor that has promoted the retention of embryos *in utero* (e.g., 'cool climate hypothesis'; Shine 1985; Shine 2004; Pyron and Burbrink 2013). Once eggs are retained, females can influence the developmental trajectory of fitness-related phenotypes in an analogous way to nest site selection in oviparous species. These effects can be a passive consequence of the micro-habitat the female occupies during gestation. For example, a large number of studies have shown that keeping viviparous females under different thermal conditions during gestation results in a suite of effects on offspring, including the timing of birth (i.e., gestation length), offspring size, growth, and survival (Wapstra 2000; Caley and Schwarzkopf 2004; While and Wapstra 2009; Wapstra *et al.* 2010). However, females can also actively manipulate the environment that embryos develop in through facultative micro-habitat choice and behavioral thermoregulation. While there is relatively limited data on the latter, recent research has shown that facultative behavioral thermoregulation can buffer offspring from sub-optimal thermal conditions during gestation and thereby prevent negative effects on offspring phenotype (Uller *et al.* 2011; also see Schwarzkopf and Shine 1991; Shine and Harlow 1993).

Second, viviparity enables embryonic provisioning of offspring via the placenta and yolk stores. In oviparous species, all nutrient inputs from the mother are deposited into the oocyte before ovulation. In contrast, viviparity allows mothers to supplement embryonic nutrition via placentotrophy; the transfer of maternal nutrients across a placenta (Stewart and Castillo 1984). The extent of placentation in lizards is highly variable, ranging from predominantly lecithotrophic (dependent on yolk nutrition) in some species (Stewart and Castillo 1984; Stewart 1989; Stewart *et al.* 1990) to predominantly placentotrophic in others (Thompson *et al.* 2000; Blackburn and Vitt 2002; Thompson and Speake 2003). Indeed, lizards exhibit some of the most extreme examples of placentotrophy outside mammals, with placental hypertrophy and mammal-like embryonic development and pregnancy (Thompson and Speake 2006). For example, the Brazilian skink, *Mabuya heathi*, has a true epitheliochorial placenta and placentotrophic nutrients account for >99% of increases in fetus dry mass during gestation (Blackburn *et al.* 1984; Blackburn and Vitt 2002). The ability of females to supplement embryonic provisioning throughout gestation has a number of important implications for both offspring and parents. First and foremost it frees females from the demanding resource requirements during vitellogeneiss as they are able to spread the nutritional investment into offspring over time (Trexler and DeAngelis 2003). Placentation also allows for the transport of inorganic ions (Blackburn 1993; Blackburn and Lorenz 2003) and histotrophic transfer (Stewart and Brasch 2003). For example, the skink *Mabuya heathi*, has a highly modified chorioallantois which is responsible for transport of organic nutrients (Blackburn *et al.* 1984; Blackburn and Vitt 2002). Southern grass skinks, *Pseudemoia entrecasteauxii*, transfer dietary β-carotenes ingested during gestation to developing embryos, which increase the innate immune response of neonates post-partum (Itonaga *et al.* 2011).

Although viviparity gives females greater control over offspring quality compared to oviparity (Shine 1995) it also confers several considerable costs to the mother carrying the eggs. The most significant cost is the physical burden of carrying offspring throughout gestation. This can have several negative impacts on female performance during pregnancy including reduced sprint speed (Sinervo *et al.* 1991; Olsson *et al.* 2000; Shine 2003), endurance (Miles *et al.* 2000; Zani *et al.* 2008), and acceleration (Scales and Butler 2007). It has been argued that this will have significant effects on female fitness through increased predation risk (Shine 1980) and reduced foraging ability/intake (Shine 1980; Brodie 1989; Lin *et al.* 2008). Viviparity also incurs several significant physiological costs to the mother (Webb 2004) as well as potentially to the offspring (e.g., suboptimal hormone exposure: Uller *et al.* 2004; Warner *et al.* 2009; exposure to toxic substances ingested by the mother: Wiklund and Sundelin 2001). For example, pregnancy can retard physiological homeostasis, leading to an array of additional performance deficits (Olsson *et al.* 2000). It also increases metabolic demand of the mother, due both to embryo metabolism itself as well as increased cardiac output (Birchard *et al.* 1984), oxygen consumption (DeMarco 1993), and renal function (to compensate for embryonic excretion; Clark and Sisken 1956). Finally, viviparity can significantly impact respiratory function due to compression of the lungs by developing embryos. For example, in the sleepy lizard, *Tiliqua rugosa*, respiratory function was reduced by 30%, resulting in a 3 fold increase in the energetic costs of breathing in gravid compared to non-gravid females (Munns 2013).

15.2.6 Care of Neonates

Post-partum parental care is not as widespread among lizards as pre-partum care. However, it has now been documented in several taxa and appears to be widespread in some taxonomic groups. At least 60 oviparous and viviparous lizard species from 13 families (e.g., agamids, cordylids, geckos, skinks, varanids, xantusiids, xenosaurids) care for their neonates after hatching/birth (Somma 2003a). While the functional costs and benefits of post-partum care in lizards remain largely unexplored, typical explanations for its occurrence are centered around assisting offspring to emerge from eggs or egg membranes and offspring protection as opposed to food provisioning (but see Evans 1959; Masters and Shine 2003 for anecdotal evidence of offspring provisioning or assistance during feeding). This is perhaps because offspring mortality depends more on predation and inter- and intra-specific competition than failing to supply the metabolic demands of life. Indeed, lizard neonates are born or hatch precocial, being able to locate and consume prey and find suitable shelter immediately after hatching or birth (Vitt and Cooper 1986) alleviating the need for parental provisioning to sustain growth and development.

The simplest form of post-partum care is the assistance of offspring during hatching or parturition. Perhaps the best example of this comes from the varanids, where females of a number of species have been shown to return to their nest cavities upon hatching to release their hatchlings (Cogger 1992; Carter 1999). Early studies indicated evidence of similar behavior in oviparous skinks, specifically *Plestiodon*, where mothers of *P. obsoletus* and *P. septentrionalis* have been observed assisting in the hatching process (Evans 1959; Somma 1987). Assistance in emergence has also been observed in viviparous species where females may assist offspring to emerge from their vitellogenic membranes. For example, *Egernia stokesii* mothers assist offspring to emerge from their embryonic sac and then consume the yolk sac and birth membranes (Lanham and Bull 2000).

The majority of incidences of post-partum care of offspring in lizards involve longer-term parent/offspring association post birth/hatching. This parent offspring association exhibits diversity in form and duration as well as the extent to which it is provided by the female only or by both parents (Box 15.1). For example, parent-offspring association after birth or hatching can reflect passive tolerance of offspring within a parent's territory or active defense of offspring from conspecifics and predators (Masters and Shine 2003; O'Connor and Shine 2004; While *et al.* 2009a). The duration of care can be as short as a few days or weeks following birth or continue for several years. The most advanced and diverse forms of post-partum parental care in lizards come from the Australian lizard group *Egernia*. In many species of *Egernia*, males and females form long-term monogamous pair bonds and juveniles delay dispersal and remain in their parents' core range following birth, sometimes for several years (Chapple 2003). This parental investment

BOX 15.1 Female Only vs. Bi-parental Care.

The majority of work on parental care in lizards has emphasized the expression of parental behavior in females. Male care in contrast has been assumed to be absent, with little evidence that males exhibit any form of parental behavior (Shine 1988). This is particularly true for pre-partum care, although anecdotal evidence does exist of males attending nests and guarding eggs (Rismiller *et al.* 2010). In contrast, post-partum parental care provides a foundation for which males can play a greater role. Indeed, while exclusive male care has never been recorded in a reptile (Shine 1988), recent research has identified many examples of bi-parental care. For example, males of the genus *Egernia* provide paternal care to offspring by tolerating them within their home-range, which may increase offspring growth or survival. Interestingly, evidence that this form of care is specifically targeted own offspring come from *Egernia whitii*, where extra-pair offspring are excluded from paternal home ranges (While *et al.* 2009a). Similar patterns of male investment in care are observed across the *Egernia* group and also in a number of other phylogenetically independent lizard lineages (e.g., *Xantusia*; Davis *et al.* 2010, *Phrynocephalus*; Qi *et al.* 2012). As with birds, in the majority of cases bi-parental care in lizards is associated with long-term male female pair bonds (Davis *et al.* 2010). These patterns suggest that evolutionary transitions to bi-parental care in lizards have occurred from a primitive state of female-only care and in association with long term male-female association, similar to suggestions for bi-parental care in general (Reynolds *et al.* 2002). Once such associations have formed, selection may favor increasingly advanced forms of paternal care if there is joint selection for low levels of female promiscuity or constraints on female ability to undertake extra-pair matings (While *et al.* 2009a).

in *Egernia* shows considerable diversity between species, ranging from species that do not exhibit any parental tolerance of offspring (e.g., *Egernia inornata*; Daniel 1998, *Tiliqua rugosa*; Bull and Baghurst 1998) to those that tolerate a single offspring or cohort of offspring (e.g., *Egernia whitii*; Chapple and Keogh 2006; While *et al.* 2009a, *Egernia saxatilis*; O'Connor and Shine 2003), and to species that tolerate multiple cohorts of offspring resulting in the formation of multi-generational family groups (e.g., *Egernia stokesii*; Gardner et al. 2001, *Egernia cunninghamii*; Stow *et al.* 2001).

Extended parent offspring association can have a number of functional benefits to the offspring. First and foremost it can protect offspring from predation and aggressive conspecifics. Indeed, post partum parent offspring association is often accompanied by aggressive defense of young in both geckos and skinks (Somma 1987; Somma 2003a). For example, aggressive protection of hatchlings is displayed by both parents in *Gekko gecko* (van de Hulst 2001) and by female crocodile skinks, *Tribolonotus gracilis* (Hartdegen *et al.* 2001). In the *Egernia* group, the primary benefit of delayed juvenile dispersal is protection from infanticidal conspecifics. Aggression towards conspecifics, including juveniles, is common in *Egernia* and infanticide is a key cause of offspring mortality (Lanham and Bull 2000; O'Connor and Shine 2004). Parents of most species vigorously and aggressively defend their home range from conspecifics (Chapple 2003; O'Connor and Shine 2004), providing a significant benefit to the offspring who reside within. Experimental evidence from *Egernia saxatilis* has shown that the presence of a parent reduces (almost eliminates) the amount of aggression displayed towards its offspring by unrelated adults (O'Connor and Shine 2004). Work on both *Egernia saxatilis* and *Egernia whitii* have shown that female aggression is heightened during periods of post-partum parent offspring association, when offspring are presumably most at risk (O'Connor and Shine 2004; Sinn *et al.* 2008). Specifically, *Egernia whitii* females exhibit a two-fold increase in aggression following birth (Sinn *et al.* 2008). As a consequence, the extent of aggression a female displays towards a conspecific is the key predictor of offspring survival in this species (Sinn *et al.* 2008).

While reduction in conspecific aggression is the primary benefit of parental tolerance of offspring, anecdotal evidence suggests that this defense may extend to defense of offspring against predatory attack. For example, Masters and Shine (2003) reported incidence of adult Kings skinks (*Egernia kingii*) aggressively attacking a predatory tiger snake (*Notechis scutatus*) in the presence of offspring. Empirical confirmation that parent-offspring association specifically functions in protection from predators is needed in order to confirm this suggestion.

Parental tolerance of offspring may also help young maintain access to high quality resources defended by their parents that they would otherwise be unable to defend. In *Egernia saxatilis*, offspring from dominant families spent more time exploring and basking compared to those from subordinate families in staged laboratory encounters (O'Connor and Shine 2004). Therefore, this indirect parental care confers valuable benefits to offspring in the form of increased availability of key resources, such as basking sites, and promotes safe feeding opportunities (i.e., increased exploration). However, it should be noted that some research has found no differences in habitat characteristics (sun exposure, vegetation cover) of home ranges occupied by family groups and those occupied by solitary orphaned juveniles (Langkilde *et al.* 2007).

There are several important caveats regarding the diversity and function of post-partum care in lizards. First, while molecular assignment of kinship has been conducted in a number of species (particularly *Egernia*, but also *Gnypetoscincus* and *Xantusia;* Gardner *et al.* 2001; Stow *et al.* 2001; O'Connor and Shine 2003; Chapple and Keogh 2005; Chapple and Keogh 2006; Sumner 2006; While *et al.* 2009a) the majority of species where offspring are associated with adults parentage are yet to be confirmed with molecular methods. This shortcoming needs to be addressed if we are to fully appreciate the extent to which parent offspring association represents a behavioral trait directed towards kin, which is crucial for understanding its evolutionary significance. Linked to this a second important caveat is that, as with egg attendance, the actual function of parental behavior is unknown in the majority of cases. Indeed, the extent to which parents actively defend offspring as opposed to defend the home range is unclear (even for most *Egernia* species). Nevertheless, if the presence of a parent decreases the offspring risk of predation or increases its access to resources this can be viewed as a form of parental care. Indeed, such indirect parental care is likely to have been the initial first step to more advanced care behavior in this and other taxa (Clutton-Brock 1991).

15.3 EVOLUTIONARY TRANSITIONS BETWEEN DIFFERENT CARE TYPES

Because of the paucity of studies examining parental care in lizards we are currently restricted with respect to the extent to which we can explore evolutionary transitions between different care types and the key factors that underpin these transitions (with the exception of viviparity; see Shine 1985; Shine 2004; Pyron and Burbrink 2013). However, it is clear from the taxonomic and functional diversity of parental care in lizards that the evolution of parental care has most likely involved multiple independent transitions. This provides a good opportunity to begin to think about the potential factors which may have facilitated these transitions. Theory predicts that parental care will evolve from an ancestral state of non-care when the benefits of care in terms of offspring fitness outweigh the costs of providing care (Clutton-Brock 1991). Therefore, differences in care between species typically arise from interspecific differences in ecology, life history, and/or phylogeny that mediate the value of benefit and cost functions involved in the trade-off.

15.3.1 Life History Traits

Life history theory plays a central role in evolutionary models predicting the emergence and diversification of parental care (Kokko and Jennions 2008; Klug and Bonsall 2010). Classic theory suggests that parental care will be associated with relatively long life spans and low fecundity (e.g., *K*-selected species) due to the high expected fitness returns on parental investment when each offspring represents substantial fraction of lifetime reproductive success (Stearns 1976; Wilson 1980; but see Winemiller and Rose 1992 for an alternative hypothesis). Indeed, prolonged parent offspring association in birds and mammals has been suggested to be associated with increased longevity and delayed maturity, both of which increase the benefits of prolonged parental care to offspring while simultaneously decreasing the cost to parents (Covas and Griesser 2007). Interestingly, many lizard species that display the most complex forms of post-partum parental care are also relatively long-lived and exhibit delayed maturation (e.g., *Egernia*, Chapple 2003; *Xantusia*, Miller 1951; Lee 1975; *Cordylus*, Bowler 1977), suggesting that the *K*-selected strategies may also facilitate the evolution of increased investment in offspring post hatching/birth in lizards. Furthermore, the extent of egg care covaries with egg size in lizards, with those species that produce few, large eggs providing comparatively more care than those producing many, small eggs (Shine 1988). While this relationship also makes sense in light of K- and r-selection theory, casual links must be made cautiously. For example, in lizards, patterns of the relationship between egg size and number of eggs in a clutch are confounded by seasonal and inter-annual variation as well as plastic clutch frequency and resource availability (Uller and While Chapter 12, this volume). Therefore, relationships between egg size and parental care may not be quite this straight forward. Finally, egg guarding has been suggested to be associated with larger body sizes and the advantage of larger body sizes for deterring predators (Shine 1988; Huang and Wang 2009). Indeed, egg guarding oviparous species are significantly larger than non-guarding cogeners, and also larger than viviparous cogeners (De Fraipont *et al.* 1996).

15.3.2 Ecological Factors

Parental care is also predicted to evolve when environmental conditions are harsh, leading to high competition for key resources and a relative increase in the benefits of care to offspring fitness (Klug *et al.* 2012). Indeed, it has been suggested that relatively simple forms of parental care (e.g., egg attendance) usually evolve in response to a specific ecological threat (Smiseth *et al.* 2012). In lizards, this has been well studied with respect to the evolution of female nest site choice, viviparity and facultative thermoregulatory behavior (see discussion above) all of which have evolved in order to cope with harsh or suboptimal ecological conditions (e.g., the thermal or hydric environment; see references above). However, environmental conditions may also influence more advanced forms of care. Predation has been repeatedly suggested to play an important role in the evolution of vertebrate parental care (Wilson 1980; Tallamy and Denno 1981). Predation risk, specifically to eggs and juveniles, may select for parental care if the parents suffer substantially less from their exposure than the offspring. In *Eutropis longicaudata,* plasticity in egg attendance between populations is the result of differences in egg predation risk (Huang and Pike 2013). Furthermore, where females suffer increased risk from exposure to the predator (e.g., where the predator is a lizard predator) females abandon their eggs thus not compromising future reproduction (Huang 2006; Huang 2008).

Parent-offspring association following hatching/birth is also likely to be the result of ecological factors. For example, variation in both predation risk and resource availability will mediate the benefits of delayed dispersal for offspring as well as the costs of prolonged care for the parents (Covas and Griesser 2007). This has been suggested as the major driving force behind the variation in parent-offspring association in *Egernia* (Duffield and Bull 2002; While *et al.* 2009a). Because of a reliance on permanent shelter and crevice sites, *Egernia* populations are typically highly saturated with intense competition over a limited number of patchily distributed retreat sites, high levels of conspecific aggression, and high juvenile mortality (Chapple 2003; O'Connor and Shine 2004; Langkilde *et al.* 2005). Given that large scale dispersal is generally low within the genus (Duffield and Bull 2002; Langkilde *et al.* 2005; While *et al.* 2009b), a lack of available crevice sites within an outcrop or an increased risk of encountering infanticidal conspecifics may promote offspring to delay dispersal and share crevices (Duffield and Bull 2002; While *et al.* 2009a). Kin selection could then influence tolerance of offspring within the natal home range, as opposed to the crevices of other individuals, resulting in the formation of closely related family groups. Furthermore, increased pair stability (and selection against female promiscuity) across years will increase the average relatedness of offspring within and among cohorts, which reduces competition and could lead to the evolution of larger social colonies as found in some species of *Egernia* (While *et al.* 2009a). These arguments are analogous to those put forward to explain the evolution of advanced forms of social behavior (e.g., cooperative breeding, eusociality) in birds, mammals and social insects (Hughes *et al.* 2008; Cornwallis *et al.* 2010; Lukas and Clutton-Brock 2013). Thus, differences between and within species in ecological characteristics could explain inter- and intra-specific differences in parental tolerance of offspring. However, few studies, correlative or experimental, have tested this hypothesis (but see Gardner *et al.* 2007).

15.3.3 Pre-existing Traits/Parent-Offspring Association

The extent to which ecological and life history factors influence the evolution of parental care will also depend on the presence of pre-existing traits and social interactions that can be shaped by natural selection to enhance offspring survival (Tallamy 1984; Klug *et al.* 2012). First, care is more likely to evolve when parents recognize or regularly encounter their offspring (Lion and van Baalen 2007). This is obviously the case in viviparous animals who provision their young before birth. Interestingly, the occurrence of post hatching care is much more common in viviparous lizards (occurring in ~6% of species for which data is available) than it is in oviparous species (occurring in <1% of species for which data is available) (Halliwell and While, unpublished data). Unsurprisingly, 17 out of the 19 lizard species that have been shown to exhibit complex kin based sociality, which is largely the result of delayed juvenile dispersal and prolonged parent offspring association, are viviparous (Davis *et al.* 2010). Likewise, in many species, eggs are laid (or offspring born) in breeding territories, creating a close physical association between parents and offspring. Thus, the evolution of attendance and guarding of offspring against predators is likely to have evolved from defensive or aggressive behaviors found in ancestral non-caring species (Huang and Pike 2011). As lizards often display strict territoriality, and the localization of eggs/offspring within the parental home range is likely to reduce the costs of providing care, territoriality may be a particularly important trait in the evolution of lizard parental care. Such simple ancestral forms of care may set the stage for the rapid unidirectional evolution of more elaborate forms of care (Clutton-Brock 1991; Gardner and Smiseth 2011).

15.4 FUTURE DIRECTIONS

Research into lizard (and reptilian) parental care has produced intriguing insights to date but is still in its infancy and provides exciting areas for future study. Advances in molecular techniques have made genetic analyses such as paternity assignment, kinship and molecular phylogenies more accessible and have vastly increased the hypothesis testing power of empirical studies. Below we outline several key areas which we believe would be fruitful avenues for future research and for which lizards have the potential to act as particularly good model organisms.

15.4.1 Documentation of the Occurrence of Parental Care Strategies

One of the key factors restricting our ability to fully incorporate reptiles into a broad synthesis of the evolution of parental care is the paucity of studies documenting patterns of parental care (or lack thereof). For example, current evidence suggests that post-partum parental care exists in less than 1% of species. However, reliable documentation of this form of care behavior is available for only ~30% of lizard and snake species in total. Furthermore, data obtained by empirical or experimental methods only exist for ~11% of these species (Somma 2003a). The only form of parental behavior that has been relatively well studied is mode of parity, which is still only known for ~44% of squamate species (Pyron and Burbrink 2013). Thus, we are likely to have grossly underestimated the true extent of parental care in reptiles therefore restricting our ability to make broad inferences regarding the factors which have contributed to its evolution.

To rectify this, it is clear that future work needs to document the occurrence of parental care more thoroughly. Lizards provide some intriguing challenges in this regard. First, parental care in lizards is often more cryptic than it is in other taxa. The primary reason for this is the central role that chemical communication, as opposed to visual or auditory cues, plays in the interactions between individuals (Doody *et al.* 2013; Shine 2013; see also Martin and Lopez Chapter 3, this volume). Even for lizard species with the most overt parental care behavior (such as *Egernia*), parent-offspring associations are relatively cryptic and often only identified using molecular techniques and detailed field studies (Stow *et al.* 2001; Chapple and Keogh 2005; While *et al.* 2009a; While *et al.* 2009b). Lizard oviposition and birth sites are also often more difficult to observe than in birds and mammals, making observations of parent-offspring associations during incubation or immediately following birth largely anecdotal. However, the growing appreciation that lizards can play a fundamental role in understanding the early evolution of parental care and the advent of more sophisticated molecular and field techniques (e.g., pit tagging, data loggers), should encourage herpetologists, and behavioral and evolutionary ecologists in general, to pay greater attention to documenting the diversity of parental care exhibited by reptiles.

15.4.2 Variation in Parental Care Strategies Between Individuals

Key to understanding the factors responsible for the diversification of parental care between species is to identify the factors that influence this costs-benefit trade-off at the individual level. However, even in those lizard taxa for which we have a good understanding of the prevalence of different care types we currently know very little about the costs or benefits of different modes of care. Future work could rectify this utilizing a number of different approaches. Specifically, long-term observational studies could be combined with experimental approaches to explicitly target hypotheses regarding the ecological conditions under which we would expect parental care to evolve. In this context, lizards offer a number of pragmatic advantages over other vertebrate systems. First, age-, size- and sex-specific social and reproductive strategies can be easily determined (e.g., Fitze *et al.* 2005; Pen *et al.* 2010; While *et al.* 2011), meaning the costs (condition, survival, future reproductive output) and benefits (offspring survival) of care can be assessed under natural conditions (e.g., While *et al.* 2009a). Second, the simple structure of parental care in lizards has the potential to provide a greater connect between empirical and theoretical research as most models of the evolution of parental care treat parental care as a single trait rather than a combination of several functionally integrated traits (as is observed in many mammals and birds; Clutton-Brock 1991; Royle *et al.* 2012). Finally, the ease with which environmental traits can be manipulated using field enclosures makes it possible to do so using rigorous controlled experiments. Importantly, this has the potential to provide direct tests of the causal relationships between ecology and the costs and benefits of parental care.

15.4.3 Comparisons Between Populations and Species

Studying the costs and benefits of care within species is useful for inferring the causal relationships that maintain care states. However, to infer broader evolutionary patterns, comparative analyses of care traits across species are required. This can be achieved in several ways using lizard model systems.

Comparisons between species within lineages will reveal how closely related species can take widely divergent parental care paths. Importantly, there are a number of lizard groups which have the potential to be particularly important within this context. For example, the *Plestiodon* group exhibit many different care types (viviparity, oviparity with egg attendance, oviparity without egg attendance, post-partum care; Somma 2003a) and a well resolved phylogeny (Brandley *et al.* 2012), making it an excellent candidate for examining the ecological and life history characteristics that promote parental care. Similarly, the *Egernia* group exhibits a highly diverse range of parental care behaviors from the absence of parental care to investment in multiple cohorts of young by both parents.

Identification of broad phylogenetic patterns across lizards (and squamates more generally) could also permit testing of ecological and life history factors that favor the initial emergence of parental care. While some studies have endeavored to achieve this for particular care modes (e.g., Gans 1996; Shine and Lee 1999) these are currently limited. Indeed, it has been suggested that the considerable diversity of care behaviors exhibited by lizards and the potential multiple evolutionary origins and trajectories will make such analyses problematic (Somma 2003b). However, we believe advanced comparative methods combined with increased data on the functional and taxonomic diversity of care within lizards means that such a goal is achievable in the near future. Indeed, the distribution of different modes of care across lizards highlights a number of interesting patterns that require more formal testing (e.g., the role of viviparity in the evolution of post-partum care; Davis *et al.* 2010). Integrating such phylogenetic analyses with the aforementioned experimental data from within populations will allow us to link the mechanisms of change with the process of phylogenetic diversification.

Undertaking such a broad comparative approach both within and between lineages will not only be important to understanding the evolution of parental care but also has the potential to offer valuable insights into more complex forms of social behavior. Darwin (1871) suggested that parental care is likely the foundation of more complex social behavior and modern research places a strong emphasis on the evolutionary connections between the origins of parental care and other social behaviors (e.g., Field and Brace 2004). For example, the formation of stable social groups associated with cooperative breeding in birds and mammals probably evolved in response to prolonged parental care and juvenile retention/delayed dispersal (Hatchwell and Komdeur 2000; Covas and Griesser 2007). Similarly in lizards, the indirect fitness conferred to parents by providing parental care may promote (or contribute to) the evolution of inter-individual tolerance, social aggregation and social complexity (Duffield and Bull 2002). Evidence for this comes from the documentation of kin based social aggregations in a number of phylogenetically independent lizard lineages. Fundamental to these social aggregations is delayed juvenile dispersal, and prolonged offspring-parent associations (While *et al.* 2009a). Therefore, understanding the factors that influence the evolution of delayed juvenile dispersal and prolonged parental tolerance of offspring across species will be crucial for elucidating the factors responsible for the emergence of complex kin based social systems more generally.

15.5 SUMMARY

With a few notable exceptions, more complex forms of parental care in lizards have generally been assumed to be absent. Here we argue that this notion is outdated and that parental care in lizards is both functionally and taxonomically diverse. For example, lizards (and squamates in general; Stahlschmidt *et al.* 2011) exhibit all but the most advanced forms of care exhibited by other ectothermic taxa including amphibians (e.g., Corben *et al.* 1974) and fish (e.g., Gross and Sargent 1985), taxa which are often championed as model systems for exploring the evolutionary origin and diversification of parental care (Amundsen 2003; Mank *et al.* 2005; Summers *et al.* 2006; Brown *et al.* 2010). Furthermore, the taxonomic distribution of parental care in squamates often exceeds these groups. For example, 27% of squamate reptile species exhibit some form of parental care in addition to egg investment, 43% at the family level, compared to 30% of fish families, 6–15% of anuran and 20% of salamander species (Summers *et al.* 2006; Wells 2007; Balshine 2012).

This functional and taxonomic diversity in association with the various pragmatic benefits of studying parental care in lizards (Doody *et al.* 2013) and an increasingly well resolved phylogeny (Pyron *et al.* 2013) provide some real benefits to using lizards to address fundamental questions regarding the evolution of parental care. Specifically, lizards offer an opportunity to examine relatively simple forms of parental care, which generate insights into its evolutionary origin and provide a robust comparison with the conditions that support maintenance of complex care in endothermic vertebrates. Indeed, conditions that give rise to the origin of care are not necessarily similar to the conditions that maintain care. The shared features of lizards and the diversity between species offer outstanding opportunities to address the evolutionary and ecological causes and consequences of parental care and we anticipate that lizards will feature more prominently in the parental care literature in the future.

15.6 ACKNOWLEDGMENTS

We would like to thank Justin Rheubert, Dustin Siegel, Stanley Trauth and Barrie Jamieson for inviting us to contribute to this publication. GW is supported by a Marie Curie research fellowship. TU is supported by the Royal Society of London and the Knut and Alice Wallenberg Foundation.

15.7 LITERATURE CITED

Amundsen, T. 2003. Fishes as models in studies of sexual selection and parental care. Journal of Fish Biology 63: 17–52.

Angilletta, M. J. M., Sears, M. W. M. and Pringle, R. M. R. 2009. Spatial dynamics of nesting behavior: lizards shift microhabitats to construct nests with beneficial thermal properties. Ecology 90: 2933–2939.

Ar, A., Belinsky, A., Dmi'el, R. and Ackerman, R. A. 2004. Energy provision and utilization. pp. 143–185. In D. C. Deeming (ed.), *Reptilian Incubation: Environment, Evolution and Behavior*. Nottingham University Press, Nottingham, U.K.

Balshine, S. 2012. Patterns of parental care in vertebrates. pp. 62–80. In N. J. Royle, P. T. Smiseth and M. Kölliker (eds.), *The Evolution of Parental Care*. Oxford University Press, Oxford, U.K.

Belinsky, A., Ackerman, R. A., Dmi'el, R. and Ar, A. 2004. Water in reptilian eggs and hatchlings. pp. 125–141. In D. C. Deeming (ed.), *Reptilian Incubation: Environment, Evolution and Behavior*. Nottingham University Press, Nottingham, U.K.

Birchard, G. F., Black, C. P., Schuett, G. W. and Black, V. 1984. Influence of pregnancy on oxygen consumption, heart rate and hematology in the garter snake: implications for the 'cost of reproduction' in live bearing reptiles. Comparative Biochemistry and Physiology Part A 77: 519–523.

Blackburn, D. G. 1993. Chorioallantoic placentation in squamate reptiles: structure, function, development, and evolution. Journal of Experimental Zoology 266: 414–430.

Blackburn, D. G. 1999. Are viviparity and egg-guarding evolutionarily labile in squamates? Herpetologica 55: 556–573.

Blackburn, D. G. and Vitt, L. J. 2002. Specializations of the chorioallantoic placenta in the Brazilian scincid lizard, *Mabuya heathi*: A new placental morphotype for reptiles. Journal of Morphology 254: 121–131.

Blackburn, D. G. and Lorenz, R. L. 2003. Placentation in garter snakes II. Transmission EM of the chorioallantoic placenta of *Thamnophis radix* and *T. sirtalis*. Journal of Morphology 256: 171–186.

Blackburn, D. G., Vitt, L. J. and Beuchat, C. A. 1984. Eutherian-like reproductive specializations in a viviparous reptile. Proceedings of the National Academy of Sciences, USA 81: 4860–4863.

Blouin-Demers, G., Weatherhead, P. J. and Row, J. R. 2004. Phenotypic consequences of nest-site selection in black rat snakes (*Elaphe obsoleta*). Canadian Journal of Zoology 82: 449–456.

Booth, D. T. and Astill, K. 2001. Incubation temperature, energy expenditure and hatchling size in the green turtle (*Chelonia mydas*), a species with temperature-sensitive sex determination. Australian Journal of Zoology 49: 389–396.

Booth, D. T. and Evans, A. 2011. Warm water and cool nests are best. How global warming might influence hatchling green turtle swimming performance. PLoS ONE 6: e23162.

Bowler, J. K. 1977. *Longevity of Reptiles and Amphibians in North American Collections as of 1 November, 1975*. Society for the Study of Amphibians and Reptiles, Milwuakee, USA, pp. 32.

Brandley, M. C., Ota, H., Hikida, T., de Oca, A. N. -M., Feria-Ortiz, M., Guo, X. and Wang, H. -Y. 2012. The phylogenetic systematics of blue-tailed skinks (*Plestiodon*) and the family Scincidae. Zoological Journal of the Linnaean Society 165: 163–189.
Brodie, E. D. III. 1989. Behavioral modification as a means of reducing the cost of reproduction. American Naturalist 134: 225–238.
Brotherton, P., Clutton-Brock, T. H., O'Riain, M. J., Gaynor, D., Sharpe, L., Kansky, R. and McIlrath, G. M. 2001. Offspring food allocation by parents and helpers in a cooperative mammal. Behavioral Ecology 12: 590–599.
Brown, G. P. and Shine, R. 2009. Beyond size-number trade-offs: clutch size as a maternal effect. Philosophical Transactions of the Royal Society B: Biological Sciences 364: 1097–1106.
Brown, J. L., Morales, V. and Summers, K. 2010. A key ecological trait drove the evolution of biparental care and monogamy in an amphibian. American Naturalist 175: 436–446.
Bull, C. M. and Baghurst, B. C. 1998. Home range overlap of mothers and their offspring in the sleepy lizard, *Tiliqua rugosa*. Behavioral Ecology and Sociobiology 42: 357–362.
Caley, M. J. and Schwarzkopf, L. 2004. Complex growth rate evolution in a latitudinally widespread species. Evolution 58: 862–869.
Campbell, H. W. and Simmons, R. S. 1961. Notes on the eggs and young of *Eumeces callicephalus Bocourt*. Herpetologica 17: 212–213.
Campos, Z., Sanaiotti, T., Muniz, F., Farias, I. and Magnusson, W. E. 2012. Parental care in the dwarf caiman, *Paleosuchus palpebrosus* Cuvier 1807 (Reptilia: Crocodilia: Alligatoridae). Journal of Natural History 46: 2979–2984.
Carter, D. B. 1999. Nesting and evidence of parental care by the lace monitor *Varanus varius*. Mertensiella 11: 137–147.
Chapple, D. G. 2003. Ecology, life-history, and behavior in the Australian Scincid genus *Egernia*, with comments on the evolution of complex sociality in lizards. Herpetological Monographs 17: 145–180.
Chapple, D. G. and Keogh, S. J. 2005. Complex mating system and dispersal patterns in a social lizard, *Egernia whitii*. Molecular Ecology 14: 1215–1227.
Chapple, D. G. and Keogh, S. J. 2006. Group structure and stability in social aggregations of White's skink, *Egernia whitii*. Ethology 112: 247–257.
Charruau, P. and Hénaut, Y. 2012. Nest attendance and hatchling care in wild American crocodiles (*Crocodylus acutus*) in Quintana Roo, Mexico. Animal Biology 62: 29–51.
Clark, H. and Sisken, B. F. 1956. Nitrogenous excretion by embryos of the viviparous snake *Thamnophis s. sirtalis* (L.). Journal of Experimental Biology 33: 384–393.
Clay, Z. and de Waal, F. B. M. 2013. Development of socio-emotional competence in bonobos. Proceedings of the National Academy of Sciences, USA 110: 18121–18126.
Clutton-Brock, T. H. 1991. *The Evolution of Parental Care*. Princeton University Press, Princeton, U.K., pp. 368.
Cockburn, A. 2006. Prevalence of different modes of parental care in birds. Proceedings of the Royal Society B: Biological Sciences 273: 1375–1383.
Cogger, H. G. 1992. *Reptiles and Amphibians of Australia*. Comstock Publishing Associates, Ithaca, NY, pp. 584.
Corben, C. J., Ingram, G. J. and Tyler, M. J. 1974. Gastric brooding: unique form of parental care in an Australian frog. Science 186: 946–947.
Cornwallis, C. K., West, S. A., Davis, K. E. and Griffin, A. S. 2010. Promiscuity and the evolutionary transition to complex societies. Nature 466: 969–972.
Covas, R. and Griesser, M. 2007. Life history and the evolution of family living in birds. Proceedings of the Royal Society B: Biological Sciences 274: 1349–1357.
Daniel, M. C. 1998. Aspects of the ecology of Rosen's Desert Skink, *Egernia inornata*, in the Middleback Ranges, Eyre Peninsula. Honours Thesis, University of Adelaide, Roseworthy, Australia.
Darwin, C. 1871. *The Descent of Man, and Selection in Relation to Sex*. John Murray, London, U.K., pp. 424.

Davis, A. R., Corl, A., Surget-Groba, Y. and Sinervo, B. 2010. Convergent evolution of kin-based sociality in a lizard. Proceedings of the Royal Society B: Biological Sciences 278: 1507–1514.
De Fraipont, M., Clobert, J. and Barbault, R. 1996. The evolution of oviparity with egg guarding and viviparity in lizards and snakes: a phylogenetic analysis. Evolution 50: 391–400.
Deeming, D. C. 2004. *Reptilian Incubation: Environment, Evolution and Behavior*. Nottingham University Press, Nottingham, U.K., pp. xiii + 349.
DeMarco, V. 1993. Metabolic rates of female viviparous lizards (*Sceloporus jarrovi*) throughout the reproductive cycle: do pregnant lizards adhere to standard allometry? Physiological Zoology 66: 166–180.
Doody, J. S., Burghardt, G. M. and Dinets, V. 2013. Breaking the social–non-social dichotomy: a role for reptiles in vertebrate social behavior research? Ethology 119: 95–103.
Doody, J. S. J., Freedberg, S. S. and Keogh, S. J. 2009. Communal egg-laying in reptiles and amphibians: evolutionary patterns and hypotheses. The Quarterly Review of Biology 84: 229–252.
Duffield, G. A. and Bull, C. M. 2002. Stable social aggregations in an Australian lizard, *Egernia stokesii*. Naturwissenschaften 89: 424–427.
Elphick, M. J. and Shine, R. 1998. Longterm effects of incubation temperatures on the morphology and locomotor performance of hatchling lizards (*Bassiana duperreyi*, Scincidae). Biological Journal of the Linnean Society 63: 429–447.
Evans, L. T. 1959. A motion picture study of maternal behavior of the lizard, *Eumeces obsoletus* Baird and Girard. Copeia 1959: 103–110.
Field, J. and Brace, S. 2004. Pre-social benefits of extended parental care. Nature 428: 650–652.
Fitch, H. S. 1954. Life History and Ecology of the Five-lined Skink, *Eumeces fasciatus*. University of Kansas Publication of the Museum of Natural History 8: 1–156.
Fitze, P. S., Le Galliard, J. F., Federici, P., Richard, M. and Clobert, J. 2005. Conflict over multiple-partner mating between males and females of the polygynandrous common lizards. Evolution 59: 2451–2459.
Gans, C. 1996. An overview of parental care among the Reptilia. Advances in the Study of Behavior 25: 145–157.
Gardner, A. and Smiseth, P. T. 2011. Evolution of parental care driven by mutual reinforcement of parental food provisioning and sibling competition. Proceedings of the Royal Society B: Biological Sciences 278: 196–203.
Gardner, M. G., Bull, C. M., Cooper, S. J. B. and Duffield, G. A. 2001. Genetic evidence for a family structure in stable social aggregations of the Australian lizard *Egernia stokesii*. Molecular Ecology 10: 175–183.
Gardner, M. G., Bull, C. M., Fenner, A., Murray, K. and Donnellan, S. C. 2007. Consistent social structure within aggregations of the Australian lizard, *Egernia stokesii* across seven disconnected rocky outcrops. Journal of Ethology 25: 263–270.
Ghalambor, C. K. and Martin, T. E. 2001. Fecundity-survival trade-offs and parental risk-taking in birds. Science 292: 494–497.
Graves, B. M. and Duvall, D. 1995. Aggregation of squamate reptiles associated with gestation, oviposition, and parturition. Herpetological Monographs 9: 102–119.
Gross, M. R. and Sargent, R. C. 1985. The evolution of male and female parental care in fishes. American Zoologist 25: 807–822.
Groves, J. D. 1982. Egg-eating behavior of brooding five-lined skinks, *Eumeces fasciatus*. Copeia 1982: 969–971.
Hamilton, W. D. 1971. Geometry for the selfish herd. Journal of Theoretical Biology 31: 295–311.
Harlow, P. S. 2000. Incubation temperature determines hatchling sex in Australian rock dragons (Agamidae: Genus *Ctenophorus*). Copeia 2000: 958–964.
Harlow, P. S. 2004. Temperature-dependent sex determination in lizards. pp. 42–52. In N. Valenzuela and V. A. Lance (eds.), *Temperature-dependent Sex Determination in Vertebrates*. Smithson Books, Washington DC, USA.
Harlow, P. S. and Shine, R. 1999. Temperature-dependent sex determination in the frillneck lizard, *Chlamydosaurus kingii* (Agamidae). Herpetologica 55: 205–212.

Harlow, P. S. and Taylor, J. E. 2000. Reproductive ecology of the jacky dragon (*Amphibolurus muricatus*): an agamid lizard with temperature-dependent sex determination. Austral Ecology 25: 640–652.

Hartdegen, R. W., Russell, M. J., Young, B. and Reams, R. D. 2001. Vocalization of the crocodile skink, *Tribolonotus gracilis* (De Rooy 1909), and evidence of parental care. Contemporary Herpetology 2: 1094–2246.

Hasegawa, M. 1985. Effect of brooding on egg mortality in the lizard *Eumeces okadae* on Miyake-jima, Izu Islands, Japan. Copeia 1985: 497–500.

Hatchwell, B. J. 1999. Investment strategies of breeders in avian cooperative breeding systems. American Naturalist 154: 205–219.

Hatchwell, B. J. and Komdeur, J. 2000. Ecological constraints, life history traits and the evolution of cooperative breeding. Animal Behavior 59: 1079–1086.

Håkansson, P. and Loman, J. 2004. Communal spawning in the common frog Rana temporaria– egg temperature and predation consequences. Ethology 110: 665–680.

Ho, S. -M., Kleis, S., McPherson, R., Heisermann, G. J. and Callard, I. P. 1982. Regulation of vitellogenesis in reptiles. Herpetologica 38: 40–50.

Huang, W. -S. 2006. Parental care in the long-tailed skink, *Mabuya longicaudata*, on a tropical Asian island. Animal Behavior 72: 791–795.

Huang, W. -S. 2007. Costs of egg caring in the skink, *Mabuya longicaudata*. Ecological Research 22: 659–664.

Huang, W. S. 2008. Predation risk of whole-clutch filial cannibalism in a tropical skink with maternal care. Behavioral Ecology 19: 1069–1074.

Huang, W. -S. and Wang, H. -Y. 2009. Predation risks and anti-predation parental care behavior: an experimental study in a tropical skink. Ethology 115: 273–279.

Huang, W. S. and Pike, D. A. 2011. Does maternal care evolve through egg recognition or directed territoriality? Journal of Evolutionary Biology 24: 1984–1991.

Huang, W. -S. and Pike, D. A. 2013. Testing cost-benefit models of parental care evolution using lizard populations differing in the expression of maternal care. PLoS ONE 8: e54065.

Hughes, W. O. H., Oldroyd, B. P., Beekman, M. and Ratnieks, F. L. W. 2008. Ancestral monogamy shows kin selection is key to the evolution of eusociality. Science 320: 1213–1216.

Ischer, T., Ireland, K. and Booth, D. T. 2009. Locomotion performance of green turtle hatchlings from the Heron Island Rookery, Great Barrier Reef. Marine Biology 156: 1399–1409.

Itonaga, K., Jones, S. M. and Wapstra, E. 2011. Effects of variation in maternal carotenoid intake during gestation on offspring innate immune response in a matrotrophic viviparous reptile. Functional Ecology 25: 1318–1326.

Ji, X. and Du, W. -G. 2001. The effects of thermal and hydric environments on hatching success, embryonic use of energy and hatchling traits in a colubrid snake, *Elaphe carinata*. Comparative Biochemistry and Physiology, Part A 129: 461–471.

Jones, S. M. 1998. Lipids in yolks and neonates of the viviparous lizard *Niveoscincus metallicus*. Comparative biochemistry and physiology. Biochemistry and Molecular Biology 121: 465–470.

Kilner, R. M. and Hinde, C. A. 2008. Information warfare and parent-offspring conflict. Advances in the Study of Behavior 38: 283–336.

Klug, H. and Bonsall, M. B. 2010. Life history and the evolution of parental care. Evolution 64: 823–835.

Klug, H., Alonzo, S. H. and Bonsall, M. B. 2012. Theoretical foundations of parental. pp. 21–39. In N. J. Royle, P. T. Smiseth and M. Kölliker (eds.), *The Evolution of Parental Care*. Oxford University Press, Oxford, U.K.

Koenig, W. D. and Dickinson, J. L. 2004. *Ecology and Evolution of Cooperative Breeding in Birds*. Cambridge University Press, Cambridge, U.K., pp. 293.

Kokko, H. and Jennions, M. D. 2008. Parental investment, sexual selection and sex ratios. Journal of Evolutionary Biology 21: 919–948.

Kunz, T. H. and Hood, W. R. 2000. Lactation, milk and related nutritional effort. pp. 415–468. In T. H. Kunz and W. Hood (eds.), *Parental Care and Postnatal Growth in the Chiroptera*. Academic Press London, London, U.K.
Lance, V. A. and Morafka, D. J. 2001. Post natal lecithotroph: a new age class in the ontogeny of reptiles. Herpetological Monographs 15: 124–134.
Lang, J. W. 1990. Behavioral thermoregulation of eggs by prairie skinks. American Zoologist 30: A108.
Langkilde, T., Lance, V. A. and Shine, R. 2005. Ecological consequences of agonistic interactions in lizards. Ecology 86: 1650–1659.
Langkilde, T., O'Connor, D. and Shine, R. 2007. Benefits of parental care: Do juvenile lizards obtain better-quality habitat by remaining with their parents? Austral Ecology 32: 950–954.
Lanham, E. J. and Bull, C. M. 2000. Maternal care and infanticide in the Australian skink, *Egernia stokesii*. Herpetological Review 31: 151–152.
Lee, J. C. 1975. The autecology of *Xantusia bensbmui* (Sauria: Xantusiidae). Transactions of the San Diego Society of Natural History 17: 259–277.
Lee, T. N., Plummer, M. V. and Mills, N. E. 2007. Use of posthatching yolk and external forage to maximize early growth in *Apalone Mutica* hatchlings. Journal of Herpetology 41: 492–500.
Lin, C. -X., Zhang, L. and Ji, X. 2008. Influence of pregnancy on locomotor and feeding performances of the skink, *Mabuya multifasciata*: Why do females shift thermal preferences when pregnant? Zoology 111: 188–195.
Lion, S. and van Baalen, M. 2007. From infanticide to parental care: why spatial structure can help adults be good parents. American Naturalist 170: E26–E46.
Lukas, D. and Clutton-Brock, T. H. 2013. The evolution of social monogamy in mammals. Science 341: 526–530.
Mank, J. E., Promislow, D. E. L. and Avise, J. C. 2005. Phylogenetic perspectives in the evolution of parental care in ray-finned fishes. Evolution 59: 1570–1578.
Marco, A., Diaz-Paniagua, C. and Hidalgo-Vila, J. 2004. Influence of egg aggregation and soil moisture on incubation of flexible-shelled lacertid lizard eggs. Canadian Journal of Zoology 82: 60–65.
Masters, C. and Shine, R. 2003. Sociality in lizards: family structure in free-living King's Skinks *Egernia kingii* from southwestern Australia. Australian Zoologist 32: 377–380.
Miles, D. B., Sinervo, B. and Frankino, W. A. 2000. Reproductive burden, locomotor performance, and the cost of reproduction in free ranging lizards. Evolution 54: 1386–1395.
Miller, M. R. 1951. Some aspects of the life history of the yucca night lizard, *Xantusia vigilis*. Copeia 1951: 114–120.
Mock, D. W. and Parker, G. A. 1997. *The Evolution of Sibling Rivalry*. Oxford University Press, Oxford, U.K., pp. 464.
Munns, S. L. 2013. Gestation increases the energetic cost of breathing in the lizard *Tiliqua rugosa*. Journal of Experimental Biology 216: 171–180.
O'Connor, D. and Shine, R. 2003. Lizards in 'nuclear families': a novel reptilian social system in *Egernia saxatilis* (Scincidae). Molecular Ecology 12: 743–752.
O'Connor, D. E. and Shine, R. 2004. Parental care protects against infanticide in the lizard *Egernia saxatilis* (Scincidae). Animal Behavior 68: 1361–1369.
Olsson, M., Shine, R. and Bak-Olsson, E. 2000. Locomotor impairment of gravid lizards: is the burden physical or physiological? Journal of Evolutionary Biology 13: 263–268.
Packard, G. C., Packard, M. J., Miller, K. and Boardman, T. J. 1988. Effects of temperature and moisture during incubation on carcass composition of hatchling snapping turtles (*Chelydra serpentina*). Journal of Comparative Physiology B 158: 117–125.
Packard, G. C. 1991. Physiological and ecological importance of water to embryos of oviparous reptiles. pp. 213–228. In D. C. Deeming and M. W. J. Ferguson (eds.), *Egg Incubation: Its Effects on Embryonic Development in Birds and Reptiles*.
Packard, G. C., Tracy, C. R. and Roth, J. J. 1977. The physiological ecology of reptilian eggs and embryos, and the evolution of viviparity within the class reptilia. Biological Reviews 52: 71–105.

Pandav, B. N., Shanbhag, B. A. and Saidapur, S. K. 2006. Functional significance of posthatching residual yolk in the lizard, *Calotes versicolor*. Journal of Herpetology 40: 385–387.

Pen, I., Uller, T., Feldmeyer, B., Harts, A., While, G. M. and Wapstra, E. 2010. Climate-driven population divergence in sex-determining systems. Nature 468: 436–439.

Pezaro, N., Doody, J. S., Green, B. and Thompson, M. B. 2013. Hatching and residual yolk internalization in lizards: evolution, function and fate of the amnion. Evolution & Development.

Pyron, R. A. and Burbrink, F. T. 2013. Early origin of viviparity and multiple reversions to oviparity in squamate reptiles. Ecology Letters 17: 13–21.

Pyron, R., Burbrink, F. T. and Wiens, J. J. 2013. A phylogeny and revised classification of Squamata, including 4161 species of lizards and snakes. BMC Evolutionary Biology 13: 93.

Qi, Y., Noble, D. W. A., Fu, J. and Whiting, M. J. 2012. Spatial and social organization in a burrow-dwelling lizard (*Phrynocephalus vlangalii*) from China. PLoS ONE 7: e41130.

Qualls, C. P. and Andrews, R. M. 1999. Cold climates and the evolution of viviparity in reptiles: cold incubation temperatures produce poor-quality offspring in the lizard, *Sceloporus virgatus*. Biological Journal of the Linnean Society 67: 353–376.

Radder, R. S. and Shine, R. 2007. Why do female lizards lay their eggs in communal nests? Journal of Animal Ecology 76: 881–887.

Radder, R. S., Shanbhag, B. A. and Saidapur, S. K. 2004. Yolk partitioning in embryos of the lizard, *Calotes versicolor*: Maximize body size or save energy for later use? Journal of Experimental Zoology 301A: 783–785.

Radder, R. S., Warner, D. A., Cuervo, J. J. and Shine, R. 2007. The functional significance of residual yolk in hatchling lizards *Amphibolurus muricatus* (Agamidae). Functional Ecology 21: 302–309.

Reedy, A. M., Zaragoza, D. and Warner, D. A. 2012. Maternally chosen nest sites positively affect multiple components of offspring fitness in a lizard. Behavioral Ecology 24: 39–46.

Reynolds, J. D., Goodwin, N. B. and Freckleton, R. P. 2002. Evolutionary transitions in parental care and live bearing in vertebrates. Philosophical Transactions of the Royal Society B: Biological Sciences 357: 269–281.

Rhen, T. and Lang, J. W. 1999. Incubation temperature and sex affect mass and energy reserves of hatchling snapping turtles, *Chelydra serpentina*. Oikos: 311–319.

Rismiller, P. D., McKelvey, M. W. and Green, B. 2010. Breeding Phenology and Behavior of Rosenberg's Goanna (*Varanus rosenbergi*) on Kangaroo Island, South Australia. Journal of Herpetology 44: 399–408.

Robbins, T. R. and Warner, D. A. 2010. Fluctuations in the incubation moisture environment affect growth but not survival of hatchling lizards. Biological Journal of the Linnean Society 100: 89–102.

Roosenburg, W. M., Dennis, T. and Beaupre, S. J. 2005. Egg component comparisons within and among clutches of the diamondback terrapin, *Malaclemys terrapin*. Copeia 2005: 417–423.

Royle, N. J., Smiseth, P. T. and Kölliker, M. 2012. *The Evolution of Parental Care*. Oxford University Press, Oxford, U.K., pp. 356.

Scales, J. and Butler, M. 2007. Are powerful females powerful enough? Acceleration in gravid green iguanas (*Iguana iguana*). Integrative and Comparative Biology 47: 285–294.

Schwarzkopf, L. and Shine, R. 1991. Thermal biology of reproduction in viviparous skinks, *Eulamprus tympanum*: why do gravid females bask more? Oecologia 88: 562–569.

Shine, R. 1980. 'Costs' of reproduction in reptiles. Oecologia 46: 92–100.

Shine, R. 1983. Reptilian reproductive modes: the oviparity-viviparity continuum. Herpetologica 39: 1–8.

Shine, R. 1985. The evolution of viviparity in reptiles: an ecological analysis. Biology of the Reptilia 15: 605–694.

Shine, R. 1988. Parental care in reptiles. pp. 275–330. In C. Gans and R. B. Huey (eds.), *Biology of the Reptilia, Volume 16*. Alan R. Liss Incorporated, New York, USA.

Shine, R. 1995. A new hypothesis for the evolution of viviparity in reptiles. American Naturalist 145: 809–823.

Shine, R. 2003. Locomotor speeds of gravid lizards: placing 'costs of reproduction' within an ecological context. Functional Ecology 17: 526–533.
Shine, R. 2004. Does viviparity evolve in cold climate reptiles because pregnant females maintain stable (not high) body temperatures? Evolution 58: 1809–1818.
Shine, R. 2013. The reptiles. Current Biology 23: 227–231.
Shine, R. and Elphick, M. J. 2001. The effect of short-term weather fluctuations on temperatures of lizard nests, and on the phenotypic traits of hatchling lizards. Biological Journal of the Linnean Society 72: 555–565.
Shine, R. and Harlow, P. 1993. Maternal thermoregulation influences offspring viability in a viviparous lizard. Oecologia 96: 122–127.
Shine, R. and Harlow, P. S. 1996. Maternal manipulation of offspring phenotypes via nest-site selection in an oviparous lizard. Ecology 77: 1808–1817.
Shine, R. and Lee, M. S. 1999. A reanalysis of the evolution of viviparity and egg-guarding in squamate reptiles. Herpetologica 55: 538–549.
Shine, R. and Wall, M. 2005. Ecological divergence between the sexes in reptiles. pp. 221–253. In K. E. Ruckstuhl and P. Neuhaus (eds.), *Sexual Segregation in Vertebrates: Ecology of the Two Sexes*. Cambridge University Press, Cambridge, U.K.
Shine, R., Barrott, E. G. and Elphick, M. J. 2002. Some like it hot: effects of forest clearing on nest temperatures of montane reptiles. Ecology 83: 2808–2815.
Sinervo, B., Hedges, R. and Adolph, S. C. 1991. Decreased sprint speed as a cost of reproduction in the lizard *Sceloporus occidentalis*: variation among populations. Journal of Experimental Biology 155: 323–336.
Sinn, D. L., While, G. M. and Wapstra, E. 2008. Maternal care in a social lizard: links between female aggression and offspring fitness. Animal Behavior 76: 1249–1257.
Smiseth, P. T., Kölliker, M. and Royle, N. J. 2012. What is parental care? pp. 1–17. In P. T. Smiseth, M. Kölliker and N. J. Royle (eds.), *The Evolution of Parental Care*. Oxford University Press, Oxford, U.K.
Socci, A. M., Schlaepfer, M. A. and Gavin, T. A. 2005. The importance of soil moisture and leaf cover in a female lizard's (*Norops polylepis*) evaluation of potential oviposition sites. Herpetologica 61: 233–240.
Somma, L. and Fawcett, J. D. 1989. Brooding behavior of the prairie skink, *Eumeces septentrionalis*, and its relationship to the hydric environment of the nest. Zoological Journal of the Linnean Society 95: 254–256.
Somma, L. A. 1987. Maternal care of neonates in the prairie skink, *Eumeces septentrionalis*. Western North American Naturalist 47: 536–537.
Somma, L. A. 2003a. *Parental Behavior in Lepidosaurian and Testudinian Reptiles: A Literature Survey*. Krieger Publishing Company, Malabar, USA, pp. x + 174.
Somma, L. A. 2003b. Reptilian Parental Behavior. The Linnean 19: 42–46.
Speake, B. K., Thompson, M. B., Thacker, F. E. and Bedford, G. S. 2003. Distribution of lipids from the yolk to the tissues during development of the water python (*Liasis fuscus*). Journal of Comparative Physiology B 173: 541–547.
Stahlschmidt, Z. R. and DeNardo, D. F. 2011. Parental care in snakes. pp. 673–702. In R. D. Aldridge and D. M. Sever (eds.), *Reproductive Biology and Phylogeny of Snakes*. Science Publishers, New Hamshire, USA.
Stahlschmidt, Z. R., Brashears, J. and DeNardo, D. F. 2011. The role of temperature and humidity in python nest site selection. Animal Behavior 81: 1077–1081.
Stahlschmidt, Z. R., Shine, R. and DeNardo, D. F. 2012. The consequences of alternative parental care tactics in free-ranging pythons in tropical Australia. Functional Ecology 26: 812–821.
Stearns, S. C. 1976. Life-history tactics: a review of the ideas. Quarterly Review of Biology 51: 3–47.
Stewart, J. R. 1989. Facultative placentotrophy and the evolution of squamate placentation: quality of eggs and neonates in *Virginia striatula*. American Naturalist 133: 111–137.
Stewart, J. R. and Brasch, K. R. 2003. Ultrastructure of the placentae of the natricine snake, *Virginia striatula* (Reptilia: Squamata). Journal of Morphology 255: 177–201.

Stewart, J. R. and Castillo, R. E. 1984. Nutritional provision of the yolk of two species of viviparous reptiles. Physiological Zoology 57: 377–383.

Stewart, J. R., Blackburn, D. G., Baxter, D. C. and Hoffman, L. H. 1990. Nutritional provision to embryos in a predominantly lecithotrophic placental reptile, *Thamnophis ordinoides* (Squamata: Serpentes). Physiological Zoology 63: 722–734.

Stow, A. J., Sunnucks, P., Briscoe, D. A. and Gardner, M. G. 2001. The impact of habitat fragmentation on dispersal of Cunningham's skink (*Egernia cunninghami*): evidence from allelic and genotypic analyses of microsatellites. Molecular Ecology 10: 867–878.

Summers, K., McKeon, C. S. and Heying, H. 2006. The evolution of parental care and egg size: a comparative analysis in frogs. Proceedings of the Royal Society B: Biological Sciences 273: 687–692.

Sumner, J. 2006. Higher relatedness within groups due to variable subadult dispersal in a rainforest skink, *Gnypetoscincus queenslandiae*. Austral Ecology 31: 441–448.

Tallamy, D. W. 1984. Insect parental care. BioScience 34: 20–24.

Tallamy, D. W. and Denno, R. F. 1981. Maternal care in *Gargaphia solani* (Hemiptera: Tingidae). Animal Behavior 29: 771–778.

Thompson, M. B. and Speake, B. K. 2003. Energy and nutrient utilisation by embryonic reptiles. Comparative Biochemistry and Physiology, Part A 133: 529–538.

Thompson, M. B. and Speake, B. K. 2006. A review of the evolution of viviparity in lizards: structure, function and physiology of the placenta. Journal of Comparative Physiology B 176: 179–189.

Thompson, M. B., Stewart, J. R. and Speake, B. K. 2000. Comparison of nutrient transport across the placenta of lizards differing in placental complexity. Comparative Biochemistry and Physiology, Part A 127: 469–479.

Trexler, J. C. and DeAngelis, D. L. 2003. Resource allocation in offspring provisioning: an evaluation of the conditions favoring the evolution of matrotrophy. American Naturalist 162: 574–585.

Trivers, R. L. 1974. Parent-offspring conflict. American Zoologist 14: 249–264.

Troyer, K. 1983. Posthatching yolk energy in a lizard: utilization pattern and interclutch variation. Oecologia 58: 340–344.

Tucker, J. K., Filoramo, N. I., Paukstis, G. L. and Janzen, F. J. 1998. Residual yolk in captive and wild-caught hatchlings of the red-eared slider turtle (*Trachemys scripta elegans*). Copeia 1998: 488–492.

Uller, T., Massot, M., Richard, M., Lecomte, J. and Clobert, J. 2004. Long-lasting fitness consequences of prenatal sex ratio in a viviparous lizard. Evolution 58: 2511–2516.

Uller, T., While, G. M., Cadby, C. D., Harts, A., O'Connor, K., Pen, I. and Wapstra, E. 2011. Altitudinal divergence in maternal thermoregulatory behavior may be driven by differences in selection on offspring survival in a viviparous lizard. Evolution 65: 2313–2324.

van de Hulst, J. 2001. De Tokkeh (*Gekko gecko*), een gekko vol verrassingen. Lacerta 593: 108–110.

Vergne, A. L., Pritz, M. B. and Mathevon, N. 2009. Acoustic communication in crocodilians: from behavior to brain. Biological Reviews 84: 391–411.

Vitt, L. J. 1993. Ecology of isolated open-formation *Tropidurus* (Reptilia: Tropiduridae) in Amazonian lowland rain forest. Canadian Journal of Zoology 71: 2370–2390.

Vitt, L. J and Cooper, W. E., Jr. 1986. Skink reproduction and sexual dimorphism: *Eumeces fasciatus* in the southeastern United States, with notes on *Eumeces inexpectatus*. Journal of Herpetology 20: 65–76.

Vitt, L. J. and Cooper, W. E., Jr. 1989. Maternal care in skinks (*Eumeces*). Journal of Herpetology 23: 29–34.

Wapstra, E. 2000. Maternal basking opportunity affects juvenile phenotype in a viviparous lizard. Functional Ecology 14: 345–352.

Wapstra, E., Uller, T., While, G. M., Olsson, M. and Shine, R. 2010. Giving offspring a head start in life: field and experimental evidence for selection on maternal basking behavior in lizards. Journal of Evolutionary Biology 23: 651–657.

Warner, D. A. and Andrews, R. M. 2002. Nest-site selection in relation to temperature and moisture by the lizard *Sceloporus undulatus*. Herpetologica 58: 399–407.

Warner, D. A., Moody, M. A., Telemeco, R. S. and Kolbe, J. J. 2012. Egg environments have large effects on embryonic development, but have minimal consequences for hatchling phenotypes in an invasive lizard. Biological Journal Linnaean Society 105: 25–41.

Warner, D. A., Radder, R. S. and Shine, R. 2009. Corticosterone Exposure during Embryonic Development Affects Offspring Growth and Sex Ratios in Opposing Directions in Two Lizard Species with Environmental Sex Determination. Physiological and Biochemical Zoology 82: 363–371.

Warner, D. A., Tucker, J. K., Filoramo, N. I. and Towey, J. B. 2006. Claw function of hatchling and adult red-eared slider turtles (*Trachemys scripta elegans*). Chelonian Conservation and Biology 5: 317–320.

Webb, J. K. 2004. Pregnancy decreases swimming performance of female northern death adders (*Acanthophis praelongus*). Copeia 2004: 357–363.

Wells, K. D. 2010. *The Ecology and Behavior of Amphibians*. University of Chicago Press, Chicago, USA, pp. 1400.

While, G. M. and Wapstra, E. 2009. Effects of basking opportunity on birthing asynchrony in a viviparous lizard. Animal Behavior 77: 1465–1470.

While, G. M., Uller, T. and Wapstra, E. 2009a. Family conflict and the evolution of sociality in reptiles. Behavioral Ecology 20: 245–250.

While, G. M., Uller, T. and Wapstra, E. 2009b. Within-population variation in social strategies characterize the social and mating system of an Australian lizard, *Egernia whitii*. Austral Ecology 34: 938–949.

While, G. M., Uller, T. and Wapstra, E. 2011. Variation in social organization influences the opportunity for sexual selection in a social lizard. Molecular Ecology 20: 844–852.

Whitaker, N. 2011. Extended parental care in the Siamese Crocodile (*Crocodylus siamensis*). Russian Journal of Herpetology 14: 203–206.

Wiewandt, T. A. 1982. Evolution of nesting patterns in iguanine lizards. pp. 119–141. In G. M. Burghardt and A. S. Rand (eds.), *Iguanas of the World: their Behavior, Ecology and conservation.* Noyes Park Ridge, New Jersey.

Wiklund, A. -K. E. and Sundelin, B. 2001. Impaired reproduction in the amphipods *Monoporeia affinis* and *Pontoporeia femorata* as a result of moderate hypoxia and increased temperature. Marine Ecology Progress Series 222: 131–141.

Wilson, E. 1980. *Sociobiology: The New Synthesis*. Belknap Press, Cambridge, pp. 378.

Winemiller, K. O. and Rose, K. A. 1992. Patterns of life-history diversification in North American fishes: implications for population regulation. Canadian Journal of Fisheries and Aquatic Sciences 49: 2196–2218.

Wright, J. and Cuthill, I. 1990. Biparental care: short-term manipulation of partner contribution and brood size in the starling, *Sturnus vulgaris*. Behavioral Ecology 1: 116–124.

Zani, P. A., Neuhaus, R. A., Jones, T. D. and Milgrom, J. E. 2008. Effects of Reproductive Burden on Endurance Performance in Side-Blotched Lizards (*Uta stansburiana*). Journal of Herpetology 42: 76–81.

CHAPTER 16

Reproductive Anatomy and Cycles of Tuatara (*Sphenodon punctatus*), an Intriguing Non-squamate Lepidosaur

Alison Cree

16.1 INTRODUCTION

Like the lizards that they resemble, tuatara (*Sphenodon punctatus*) are members of Lepidosauria, a major group of reptiles accounting for over 9000 species (http://www.reptile-database.org, accessed 1 January 2014). Unlike lizards and their sub-clade the snakes, tuatara are not squamates. Instead, tuatara are the only living representatives of Rhynchocephalia, the sister group of Squamata (Gauthier *et al.* 1988; Jones *et al.* 2009; Evans and Jones 2010). As such, they hold considerable evolutionary importance.

Tuatara survive in geographic isolation within the archipelago of New Zealand in the South Pacific. From earliest descriptions of external appearance, a superficial resemblance of tuatara to agamid lizards has been noted (Gray 1842; Dawbin 1982). Both sexes of tuatara are of medium size (males to about 1 kg in mass and 300 mm in snout-vent length, females to about 500 g in mass and 250 mm in snout-vent length, though some populations have smaller maximum sizes). Dorsal and nuchal crests of soft white spines (to which the Māori name tuatara refers) are present in both sexes, although more prominent in males. In addition to their larger body

Department of Zoology, University of Otago, Dunedin 9054, New Zealand.

mass and more obvious crest, males also possess a relatively longer and wider head and a more prominent dark stripe on the sides of the throat than in females (Fig. 16.1).

Following the arrival of humans in New Zealand about 730 years ago (Wilmshurst *et al.* 2008), tuatara have been dramatically reduced in distribution to a few dozen offshore islands. In total these constitute probably less than 0.5% of the area formerly inhabited (Cree and Butler 1993). The relict distribution of tuatara (on islands that are virtually or completely uninhabited by humans, and difficult to access), combined with the species' strict legal protection since 1895 and strong cultural significance have constrained the nature and extent of research. Tuatara are not currently recognized as a threatened species (their IUCN status is Lower Risk/least

Fig. 16.1 Sexual dimorphism in live adult tuatara on Stephens Island. **(A)** female, 179 mm snout-vent length (SVL). **(B)** Male, 185 mm SVL. These small adults are just above the size of sexual maturity, which occurs at about 170–180 mm SVL. Females have a less obvious nuchal and dorsal crest with shorter spines, a paler throat, a less distinct stripe on the side of the throat and a relatively smaller head than in males, differences that become more dramatic as males grow in size. Photos: Alison Cree.

concern; IUCN 2013; their national status is At Risk/relict; Hitchmough *et al.* 2013). However, they have high cultural significance within the indigenous Māori population (New Zealand Waitangi Tribunal 2011) and also general significance as a national icon for many New Zealanders (Cree in press).

This chapter puts the evolutionary significance of tuatara in brief context. It then summarizes current knowledge regarding reproductive anatomy and associated cycles of hormonal and behavioral activity. The emphasis is on the most intensively studied population—that on Stephens Island in Cook Strait (40° 40′ S), known also by the Māori name Takapourewa. For more detail on reproduction and life history, including geographic variation among the 32 natural populations on islands, as well as reproduction in captivity and in populations that have recently been established on other islands and in fenced mainland ecosanctuaries, see Cree (in press).

16.2 TUATARA: LAST OF THE RHYNCHOCEPHALIANS

Although it has been popular to view tuatara as a 'living fossil' that retains features of basal lepidosaurs, the species is a modern reptile with many derived features relative to its long-extinct Mesozoic ancestors (Jones and Cree 2012). A wealth of palaeontological research over recent decades has revealed that Mesozoic rhynchocephalians were a diverse group (Carroll and Wild 1994; Evans 2003; Evans and Jones 2010; Jones *et al.* 2013). The lineage first appeared in the fossil record in the Middle Triassic, about 238–240 million years ago, and the separation of rhynchocephalian and squamate lineages is estimated to only slightly precede that date (Jones *et al.* 2013). Across the Mesozoic Era as a whole, at least 37 genera of rhynchocephalians have been described. Compared with tuatara, Mesozoic taxa include species that were smaller and species that were larger, as well as species that were more slender and those that were more robust; furthermore, some Mesozoic rhynchocephalians were herbivorous and some were aquatic, unlike the terrestrial and carnivorous tuatara (Carroll and Wild 1994; Reynoso and Clark 1998; Reynoso 2000; Apesteguía and Novas 2003; Evans and Jones 2010; Rauhut *et al.* 2012).

Mesozoic rhynchocephalians were clearly a widespread and initially successful group, occupying the landmasses of what are now Europe, Asia, Africa, North America and South America (Jones *et al.* 2009). Reasons for the eventual demise of rhynchocephalians are unclear; possible factors include competition with squamates and/or small mammals, predation and environmental change (Apesteguía and Rougier 2007; Evans and Jones 2010; Apesteguía and Jones 2012). The last-known Mesozoic taxa are from the Late Cretaceous of South America (Apesteguía and Novas 2003; Apesteguía and Jones 2012). The fossil record for rhynchocephalians is then silent until the Early Miocene about 19–16 million years ago, when partial jawbones of an unnamed sphenodontine from southern New Zealand were deposited (Jones *et al.* 2009). The oldest bones referable to *Sphenodon*, the tuatara genus, date to between 35,000 and 100,000 years ago (Worthy and Grant-Mackie 2003).

Much has been recorded about the reproductive anatomy, cycles and behavior of tuatara since studies began in the late 1800s. However, nothing is known about equivalent aspects in the species' Mesozoic relatives. Most fossil rhynchocephalians have been described only from partial skeletons (typically skulls) with no information on soft tissues. Whether all Mesozoic rhynchocephalians were egg-laying species (as is the modern tuatara) is, for example, unknown. Viviparity has evolved repeatedly within the squamate lineage (Stewart and Blackburn Chapter 13, this volume), possibly with an early first occurrence and transitions in parity mode from viviparity to oviparity and vice versa on multiple occasions (Pyron and Burbrink 2013). The possibility of similar diversity within the rhynchocephalians in this and other aspects of reproduction cannot be discounted. As tuatara are the only rhynchocephalians for which information on reproductive biology is available, the rest of this chapter is concerned solely with the extant species. The chapter follows the reproductive process from egg laying and embryonic development to the anatomy and hormonal cycles of adult females and males. The possible phylogenetic and functional significance of differences from squamates is considered in the final section, along with suggestions for future research.

16.3 CLUTCH SIZE, SEX DETERMINATION AND SEXUAL DIFFERENTIATION

Tuatara lay clutches of about 1–18 ovoid eggs with a parchment-like eggshell (Fig. 16.2). The eggs are deposited in specially dug nests in soil (Schauinsland 1898, 1899; Dendy 1899; Thilenius 1899; Thompson *et al.* 1996). On Stephens Island, clutch size averages about 9.4 eggs at nesting, but varies with snout-vent length of the female (Newman *et al.* 1994; Cree in press). Several populations on smaller islands have smaller body sizes and correspondingly smaller mean clutch sizes (e.g., 5.9 eggs on Ruamahuaiti Island; Tyrrell *et al.* 2000).

The structure of the eggshell in tuatara has the usual main layers for squamate (and all reptilian) eggs: an inner layer (the shell membrane) composed of protein fibres and an outer, calcareous layer. However, the shell is unusual in that columns of calcium carbonate, within which fibres

Fig. 16.2 A female tuatara on Stephens Island lays an ovoid, parchment-shelled egg in response to an injection of arginine vasotocin in November. Photo: Alison Cree.

are embedded, penetrate deeply into the fibrous layer (Packard *et al.* 1988; Cree *et al.* 1996). A similar structure has been noted in at least one agamid lizard (Packard *et al.* 1991). Although this structure is of uncertain functional significance in tuatara, it clearly allows the eggs to swell; indeed, it is possible that water uptake from the surrounding soil is essential for successful hatching (Cree *et al.* 1996; Thompson 1990; Cree in press).

In most oviparous squamates, embryos are at least at the limb bud stage of development at laying (Stewart and Blackburn Chapter 13, this volume). By comparison, the embryos inside newly laid eggs of tuatara are poorly developed, with no visible embryo (at gastrula stage; Schauinsland 1898, 1899; Thompson 1990), a situation more typical of turtles (Andrews 2004). This very limited development in tuatara is particularly remarkable when the long period of egg processing *in utero* (6–8 months; see below) is considered. However, a few squamate species have an equivalent stage of development at laying (certain chameleons; Andrews and Karsten 2010).

A distinctive characteristic of embryonic development in squamates is the formation of a "yolk cleft-isolated yolk mass complex", a pattern of yolk sac development that differs from that in all other main groups of amniotes (Stewart and Blackburn Chapter 13, this volume). Although aspects of embryonic development in tuatara were studied in the late 1800s (by Dendy, in English, and by Schauinsland and Thilenius, in German), and some summary information in English is available (Moffat 1985), it is not clear how the pattern of yolk sac development in tuatara compares with that in squamates. Another feature of development that may be of phylogenetic interest is the relative time of appearance of the allantois and limb buds (Andrews 2004); although few reptile species have been examined as yet, the sequence in tuatara (allantois before limb buds; Dendy 1899) is consistent with the pattern in lizards rather than turtles or crocodilians (Andrews 2004).

As with many egg-laying reptiles (Warner 2011), tuatara have a form of sex determination in which sex is established not at fertilization but later during embryonic development, in response to temperatures in the nest. The specific form of temperature-dependent sex determination (TSD) in tuatara appears to be the rare FM pattern, in which females arise from low incubation temperatures and males from high (Cree *et al.* 1995; Mitchell *et al.* 2006; Cree in press); however, sexes arising from the full range of successful egg incubation temperatures have yet to be described.

Gonadal sex in tuatara becomes apparent during about the middle third of embryonic development, a pattern typical of other reptiles, including lizards (Ramirez-Pinilla *et al.* Chapter 8, this volume). Sexual differentiation in tuatara has not been described in detail but some information is available from sex-determination studies in which sex was identified histologically (Cree *et al.* 1995; Nelson *et al.* 2004; Besson *et al.* 2012; unpubl. obs.). The only staging scheme available for embryos of tuatara is the 17-point series of letter codes (C to S) established by Dendy (1899). In this scheme, gonadal sex becomes evident at about early stage Q (when the five digits appear on the limbs; roughly equivalent to stage 34–35 of the 40-point staging scheme for the lizard *Zootoca vivipara* (formerly *Lacerta vivipara*) developed by Dufaure and Hubert; see Moffat 1985 for equivalences). As in squamates (Raynaud and Pieau 1985; Austin 1988; Neaves *et al.* 2006), the early differentiation of ovaries in tuatara is marked by the accumulation of germ cells (presumptive oogonia) in the more peripheral, cortical region of the gonad (Fig. 16.3A). Conversely, the early differentiation of testes is marked by the accumulation of germ cells (presumptive spermatogonia) in the more central, medullary region of the gonad, with the testes exhibiting weakly defined seminiferous tubules by stage R (Fig. 16.3B). In juveniles that are at least a few months old and that have grown by 1–2 cm in snout-vent length after hatching, somatic cells have surrounded the largest oocytes to form primordial follicles and the seminiferous tubules of the enlarging testes are more distinct (see Fig. 6.1 of Nelson *et al.* 2004).

As in squamates, Müllerian ducts (presumptive oviducts) are initially present in embryos of both sexes of tuatara (Fig. 16.3). The Müllerian ducts develop on the dorsolateral edge of the mesonephric kidneys alongside the Wolffian ducts. By about a few months after hatching, the Müllerian ducts of female tuatara have enlarged into ribbon-like structures with a well-differentiated epithelium and central lumen, whereas those of males have partly regressed (see Fig. 6.1 of Nelson *et al.* 2004). The mesonephric kidneys remain present in both sexes at this stage. Although the mesonephric kidneys will later regress in females and become incorporated into the

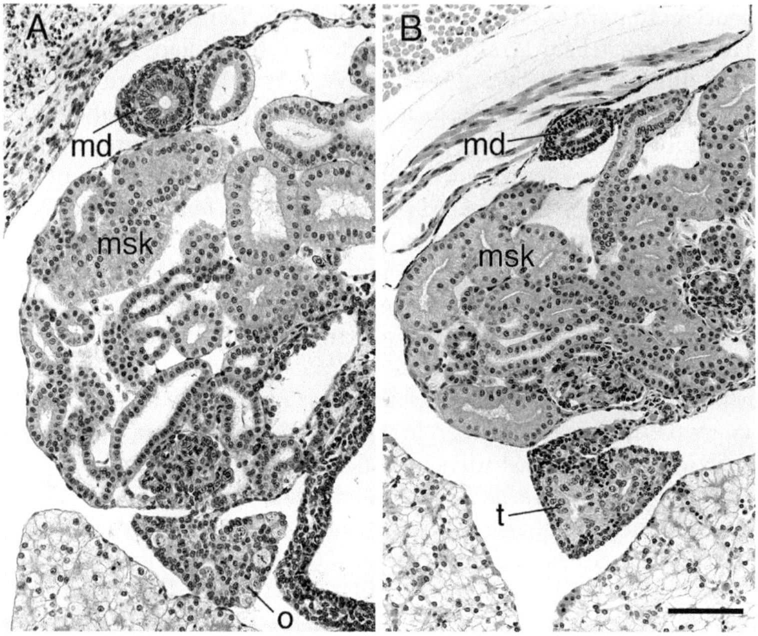

Fig. 16.3 Developing gonads in embryos of tuatara from Stephens Island. **(A)** Ovary and associated structures from female embryo at Stage Q. **(B)** Testis and associated structures from male embryo at stage R. Germ cells are accumulating in the gonadal cortex in A, and in developing seminiferous tubules of the gonadal medulla in B. Müllerian ducts are present in both sexes. Abbreviations: md = Müllerian duct; msk = mesonephric kidney; o = ovary; t = testis. Scale bar =100 μm in both.

genital ducts of males, traces of the Müllerian ducts may remain in adult males as vestigial structures (Osawa 1897).

Overall, the general pattern of differentiation of the gonads and Müllerian ducts in tuatara appears similar to that in lizards, in which gonadal differentiation typically becomes evident in embryos between about stages 32–35 of development and in which Müllerian ducts are initially present in both sexes. Variation exists among lizards in whether, by the time of birth or hatching, the Müllerian ducts of males have completely regressed and the primordial follicles of females have developed (Raynaud and Pieau 1985; Austin 1988; Neaves *et al.* 2006); in these respects tuatara resemble the more slowly differentiating species of lizards.

16.4 MORPHOLOGY OF THE ADULT OVARIES

The paired ovaries of mature female tuatara are located in the abdomen lateral to the adrenal glands (Fig. 16.4). Like those of lizards (Ramirez-Pinilla

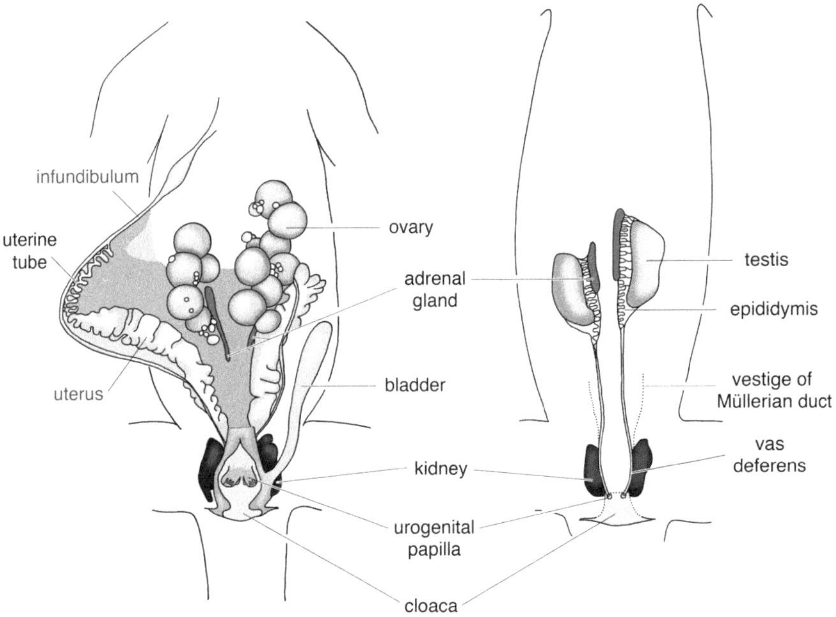

Fig. 16.4 Diagrams of the reproductive systems of adult tuatara. In the female in mid-late vitellogenesis **(left)**, the right oviduct (as viewed dorsally) and the bladder have been reflected to the sides. The cloacal walls are shown, cut open ventrally to reveal the urogenital papillae inside. The anterior end of the left oviduct and both vaginas are concealed. In the male **(right)**, the cloacal walls and urinary bladder are not shown. Vestiges of Müllerian ducts are present in only some males. Based on dissections and drawings by Claudine Tyrrell, with supporting information from Günther (1867) and Osawa (1897, 1898). Modified from Cree (in press).

et al. Chapter 8, this volume), the ovaries are hollow sacs containing lymph spaces. A framework formed from strands of connective tissue penetrates the interior. The most immature stages (oogonia and naked oocytes) are found within the ovarian epithelium (Gabe and Saint Girons 1964), predominantly on the dorsal surface but apparently also on the ventral surface in the interstices between vitellogenic follicles (Osawa 1898; Fig. 16.5A). Developing follicles move into the interior lymph spaces but remain attached to the ovary wall by a strand of connective tissue (Osawa 1898).

As in other reptiles in which yolk provides the major source of nutrition for developing embryos, the ovaries of tuatara vary greatly in size depending on the stage of yolk-protein production and incorporation (vitellogenesis). In one female from Stephens Island at the end of vitellogenesis, two weeks after mating and just prior to ovulation, the combined mass of the two ovaries was equivalent to 10.6% of body mass, and the pre-ovulatory follicles had diameters up to 19 mm (Fig. 16.5A; Cree in press). Atretic follicles (those in

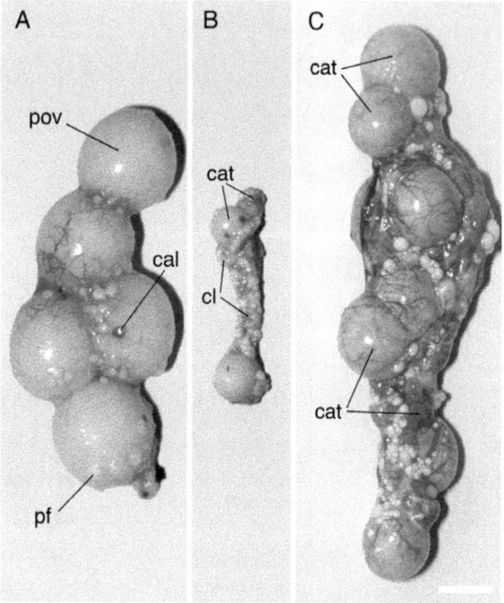

Fig. 16.5 Ventral views of ovaries from three adult tuatara of Stephens Island origin. **(A)** Right ovary from wild female in pre-ovulatory condition (184 mm SVL). This female was seen mating on 4 March 1988, collected immediately afterwards, and dissected on 18 March 1988. The ovary contained five enlarged, pre-ovulatory follicles 18–19 mm in diameter (pale circles about 5 mm in diameter, the presumed stigmata or sites of impending follicular rupture, were present on these follicles). Immature white or creamy-white follicles up to 4 mm diameter (pre-vitellogenic follicles) and small flat brown scars up to 3 mm in length (presumed corpora albicantia) are also visible. **(B)** Left ovary from wild female in gravid condition collected on 17 July 1988 and dissected on 4 August 1988 (218 mm SVL). The ovary contained seven corpora lutea up to 5 mm diameter with visible ovulation apertures, several yolked follicles 5–12 mm in diameter (presumed corpora atretica), numerous pre-vitellogenic follicles and several presumed corpora albicantia. For histology of the corpora lutea of this female, see female M52 of Guillette and Cree (1997). Uterine eggs in this female were radio-opaque but incompletely calcified on the surface (see Cree *et al.* 1996). **(C)** Abnormal right ovary of overweight female that died after several decades in captivity, including at least several years without access to a male (225 mm SVL). The ovary contained many enlarged follicles up to 17 mm diameter with signs of atresia (uneven or brown coloration, sometimes with internal fluid), as well as numerous pre-vitellogenic follicles. Abnormal fatty deposits were present on the liver and peritoneum. Abbreviations: cal = corpus albicans; cat = corpus atreticum; cl = corpus luteum; pf = pre-vitellogenic follicle; pov = pre-ovulatory follicle. Scale bar = 1 cm in all. Photos: A and B: Alison Cree; C: Marcus Simons.

Color image of this figure appears in the color plate section at the end of the book.

the process of being resorbed after aborting development) are sometimes seen in wild females, at both pre-vitellogenic (Gabe and Saint Girons 1964) and vitellogenic stages (pers. obs.; Fig. 16.5B). Atresia is far more common and conspicuous in the follicles of females maintained for years in captivity, in which case it is often associated with obesity, failure to ovulate and unusual fatty deposits within the abdomen (Fig. 16.5C).

In one major respect the ovarian follicles of female tuatara are different from those of apparently all squamates. During the pre-vitellogenic stage, the granulosa of squamate follicles becomes multi-layered, containing small, intermediate and pyriform (pear-shaped) cells. Pyriform cells are considered as 'nurse cells', assisting oocyte growth by transporting various cellular components such as mitochondria, ribosomes and macromolecules into the oocyte via intercellular bridges (Ramirez-Pinilla *et al.* Chapter 8, this volume). Although detailed studies have not been made of tuatara, the available information strongly suggests that pyriform cells are lacking and that the granulosa remains more-or-less single-layered (Osawa 1898; Gabe and Saint Girons 1964; Cree in press; Fig. 16.6). A (possibly transient) change in cell height and shape was noted in one follicle (diameter slightly larger than 1.0 mm) by Osawa (1898): the granulosa was taller and the cells more cylindrical. In remaining essentially single-layered and in lacking intermediate and pyriform cells, the granulosa of tuatara resembles that of turtles and crocodilians more than that of squamates. The functional implications of the absence of pyriform cells in certain reptiles, including tuatara, warrant closer attention.

At ovulation, the enlarged oocytes of tuatara rupture their follicles and pass into the oviducts for fertilization and shell deposition. The follicular tissues that remain in the ovary develop into corpora lutea during the post-ovulatory period. Each corpus luteum is a pinkish-white disc about 5–7 mm in diameter (Guillette and Cree 1997). The corpora lutea are unusual

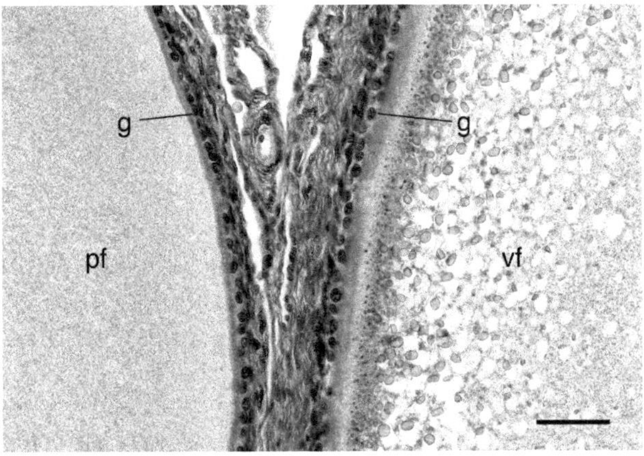

Fig. 16.6 Section through two ovarian follicles from a female tuatara. Note the single-layered granulosa composed of cuboidal cells in both previtellogenic and early vitellogenic follicles. Abbreviations: g = granulosa; pf = pre-vitellogenic follicle; vf = vitellogenic follicle. Scale bar = 50 μm. Photo: Alison Cree.

compared with those of turtles and squamates in having an ovulation aperture that remains open and a central cavity that is never completely filled by the luteal cell mass (Guillette and Cree 1997; see Ramirez-Pinilla *et al.* Chapter 8, this volume for lizards). These features are, however, shared with crocodilians, and perhaps connected with the theca externa being relatively thick-walled (Guillette and Cree 1997). The corpora lutea of tuatara remain similar in size and histological appearance during the 6–8 months that eggs remain in the oviducts, and have steroidogenic features suggesting that they are the source of progesterone that remains detectable in the plasma during gravidity. Once eggs are laid, the corpora lutea regress rapidly (within 1–2 months). However, they probably never completely disappear. Dozens of small, orange-brown-black scars (1–3 mm in diameter) are present in the ovaries of large females and probably represent corpora albicantia, the scars of corpora lutea (Guillette and Cree 1997; Cree in press). Corpora albicantia are also thought to persist for life in some squamates (Fox 1977; Ramirez-Pinilla *et al.* Chapter 8, this volume).

16.5 MORPHOLOGY OF THE ADULT OVIDUCTS

Paired oviducts are found within the abdomen of female tuatara. Each oviduct begins just posterior to the axillae (armpits) and adjacent to the lungs, runs past the ovary and extends to the cloaca. The oviducts can conveniently be divided longitudinally into four regions, here labeled infundibulum, uterine tube, uterus and vagina (Osawa 1898; Gabe and Saint Girons 1964; Cree in press; Fig. 16.4). A cleft-shaped ostium (or ostium abdominale) leads into the delicate, thin-walled and translucent infundibulum, which is followed posteriorly by a thicker-walled, more pleated uterine tube. Preliminary descriptions suggest that, as in squamates, albumen-secreting glands are lacking from these regions in tuatara (Osawa 1898; Gabe and Saint Girons 1964), which is consistent with the apparent absence or near-absence of albumen ("egg white") in the eggs (Cree in press). In this respect lepidosaurs differ from turtles and crocodilians, in which glands of the uterine tube are the presumed source of albumen present in the eggs (Girling 2002).

Ovulated oocytes of tuatara move quickly through the infundibulum and uterine tube and spend a lengthy 6–8 months in the uterus, where egg shelling occurs. The uterus accounts for at least half the length of the oviduct and, as expected for an egg-laying reptile, is relatively thick-walled and convoluted, especially during vitellogenesis. It contains numerous tubular glands, mostly single but sometimes branching, that open to the lumen (Osawa 1898; Gabe and Saint Girons 1964); these are likely to produce proteins incorporated into the shell membrane, as in other egg-laying reptiles (Siegel *et al.* Chapter 6, this volume). In female tuatara in mid-late vitellogenesis, the glands were extremely numerous and encompassed almost the entire thickness of the uterine wall, their cells being filled with secretory granules that had strong histochemical affinities for red stains (Osawa 1898; Gabe and Saint Girons 1964). Similarly, the uterus of oviparous lizards has dense aggregations of simple tubular glands, in the epithelium of which are secretory granules that stain intensely with eosin (Siegel *et al.* Chapter 6, this volume).

Although the uterine epithelium in tuatara has received little detailed attention, it contains both ciliated and non-ciliated cells and evidence of secretory activity, as in other egg-laying reptiles including lizards (Palmer *et al.* 1993; Girling *et al.* 1998). In a female tuatara in mid-late vitellogenesis, cells of the luminal epithelium contained a secretory product staining positive for mucopolysaccharides of average acidity (Gabe and Saint Girons 1964). In a female in mid-gravidity, in which the uterine eggs were incompletely calcified on the surface (Cree *et al.* 1996), scanning electron micrography revealed ciliated and non-ciliated microvillous cells, as well as secretions and/or cellular protrusions of an unidentified nature that extended into the lumen (Fig. 16.7). In other oviparous reptiles the uterine epithelium is the presumed source of calcium in the eggshell (Palmer *et al.* 1993), and there is no reason to doubt this for tuatara.

Fig. 16.7 Scanning electron micrographs of uterine epithelium from female M52 in mid-gravidity in August (for ovary of this female, see Fig. 16.5B). **(A)** Lower-power view of uterine epithelium. Scale bar = 5 µm. **(B)** Enlargement of area shown within the box in A. Note presence of ciliated cells, non-ciliated microvillous cells as well as secretions or cellular protrusions extending into the lumen. Photo: Alison Cree.

Posterior to the uterus lies a very short, muscular and relatively straight-walled vagina about 4 mm in length. In some squamates, sperm storage tubules are present in the vagina or the infundibulum (Siegel *et al.* Chapter 6, this volume); no such structures have been detected in the oviduct of tuatara, in which ovulation happens within a few weeks of mating (see Section 16.6). The vagina enters the dorsal wall of the urodaeum (the middle chamber of the cloaca) alongside the ureter at a urogenital papilla. The vaginal opening on the papilla is slightly anterior to that of the ureter (Osawa 1898). The urogenital papilla of females can be noticeably larger than that of males, with fingerlike projections extending into the cloaca (C. L. Tyrrell pers. comm.; compare also Fig. 1 of Plate 23 in Osawa (1898) with Fig. 6 of Plate 8 in Osawa (1897)). Female tuatara apparently differ from female squamates in possessing a urogenital papilla; in most female squamates, although the oviduct may open at a genital tubercle (= oviducal papilla), this structure is well separated from the opening of the ureters (Gabe and Saint Girons 1965; Sánchez-Martínez *et al.* 2013; Siegel *et al.* Chapter 6, this volume). Given that male tuatara (unlike male squamates) lack intromittent organs (see Section 16.8 below), it is tempting to speculate that the urogenital papillae of female tuatara serve some functional role in assisting sperm to enter the oviducts; alternatively or additionally, the papillae may have some function at a different stage of the reproductive cycle, such as in helping to orientate the eggs at oviposition. Endoscopic examination of urogenital papillae at different stages of reproduction in living tuatara, as well as comparisons with the oviducal papillae of lizards, could provide evidence to refine these hypotheses.

The cloaca of tuatara opens to the exterior at a slit-like vent orientated in a transverse direction (as in squamates) rather than in an antero-posterior direction (as in turtles and crocodilians; Gauthier *et al.* 1988). As in many lizards, a urinary bladder opens on the ventral face of the coprodaeum. Tubular glands are present in the urodaeum of both sexes, a situation also seen in many squamates (Sánchez-Martínez *et al.* 2013; Siegel *et al.* Chapter 6, this volume). Alongside the ventro-lateral walls of the proctodaeum of tuatara are found conspicuous glands of a sebaceous nature (Günther 1867; Osawa 1897, 1898; Gabe and Saint Girons 1964; Saint Girons and Newman 1987). These sebaceous glands, which secrete a thick, creamy-white material through about eight small openings into the cloaca (see illustration in Cree in press), are present in both sexes and appear to differ from any cloacal or paracloacal glands seen in squamates (Gabe and Saint Girons 1964, 1965; Siegel *et al.* Chapter 6, this volume). The secretion varies in composition among individual tuatara and possibly enables different individuals to identify each other (Flachsbarth *et al.* 2009).

16.6 CYCLES OF REPRODUCTIVE ACTIVITY IN FEMALES

Unusually for lepidosaurs, the ovarian cycle of tuatara is spread over several years. On Stephens Island, nesting occurs no more frequently than once every two years, and on average about once every 4–5 years (Cree *et al.* 1991a, 1992; Cree 1994; Newman *et al.* 1994; Refsnider *et al.* 2010). Although less-than-annual reproduction is unusual for oviparous lizards, biennial reproduction is known in some species including the marine iguana (*Amblyrhynchus cristatus*), in which it is associated with swings in availability of marine algal food (Vitousek *et al.* 2010), and in the gecko *Homonota darwini*, in which it is associated with a cool Patagonian environment (Ibargüengoytía and Casalins 2007).

In New Zealand, less-than-annual reproduction is also seen in females of several small-bodied New Zealand geckos in the family Diplodactylidae, although, unlike tuatara, these species are viviparous (Cree 1994; Cree and Guillette 1995). In contrast with marine iguanas neither tuatara nor New Zealand geckos experience obvious variation between years in food availability. On the other hand, tuatara and these New Zealand geckos inhabit a cool environment and all are extremely long-lived: tuatara to at least 80–90 years on Stephens Island (Nicola Nelson, Victoria University of Wellington, pers. comm.; Cree in press) and some geckos to at least several decades (Cree 1994; Cree and Guillette 1995; Lettink and Whitaker 2006). Thus, these characteristics of infrequent reproduction and great longevity are not unique in New Zealand to tuatara, and may be common responses of semi-nocturnal lepidosaurs in the southern hemisphere to temperate or cool-temperate climates (Cree and Daugherty 1990; Cree 1994 and in press; Ibargüengoytía and Casalins 2007).

Cycles in plasma concentrations of the steroid hormones estradiol, testosterone and progesterone have been examined in female tuatara in relation to reproductive activity. As in amniotes generally, vitellogenesis in tuatara is stimulated by estradiol (Cree 1991b). Plasma concentrations of endogenous estradiol rise gradually over the several years of vitellogenesis (Brown *et al.* 1991; Cree *et al.* 1992). During the final year of the cycle, plasma concentrations of estradiol and testosterone peak in the austral late-summer to early autumn (February–March) at the time that mating occurs (Brown *et al.* 1991; Cree *et al.* 1992). Rising concentrations of these hormones potentially contribute to hypertrophy of the oviducts and to sexual receptivity, as in other reptiles (Whittier and Tokarz 1992; Girling 2002).

Plasma concentrations of progesterone (a hormone typically associated with ovulation and early gravidity in reptiles) peak soon after those of estradiol and testosterone (see also Cree *et al.* 1991b). Females are gravid from early-mid autumn (March–April) until mid-late spring or early summer (October–December) when nesting occurs. Throughout the gravid period, plasma concentrations of estradiol and testosterone are low; progesterone also drops soon after ovulation but remains detectable during gravidity, and probably helps maintain quiescence of the uterine muscles (Cree *et al.* 1992; Guillette and Cree 1997). Overall, cycles in plasma concentrations of these three steroid hormones show relationships with ovarian and oviducal activity that are consistent with those for squamates and other reptiles (Gist 2011; Jones 2011; Lovern 2011).

In tuatara, natural variation in plasma concentrations of several additional hormones has been examined in relation to reproductive activity. Elevated concentrations of arginine vasotocin and prostaglandin F are seen at oviposition, consistent with evidence from tuatara and other reptiles that these hormones stimulate contractions of uterine muscle that lead to egg expulsion (Guillette *et al.* 1990, 1991; Fig. 16.2). In addition, variation in plasma concentrations of the adrenal glucocorticoid corticosterone in relation to nesting has been noted. This is intriguing, in that corticosterone is typically considered a "stress hormone" in reptiles given its obvious involvement in the physiological response to capture and confinement (Tokarz and Summers 2011; see Tyrrell and Cree 1998 for tuatara). In tuatara, baseline concentrations of corticosterone are high during late gravidity and the early stages of nesting but have fallen dramatically in females that have just laid and are still guarding their nests (Tyrrell and Cree 1998; Cree and Tyrrell 2001). A similar pattern has been reported in the marine iguana, *Amblyrhynchus cristatus* (Rubenstein and Wikelski 2005). In the case of tuatara, these observations have led to the suggestion that corticosterone plays a role in the metabolic demands of nesting, including migration to nest sites and nest construction (Cree and Tyrrell 2001; Cree in press).

16.7 MORPHOLOGY OF THE TESTES

The paired, creamy-white testes of male tuatara are located in the mid-abdomen close to the adrenal glands (Osawa 1897; Fig. 16.4). The testes vary somewhat in shape and size, both within and among individuals, and have been variously described as elongate-ovate (Günther 1867), elliptical and sharp-edged (Osawa 1897) and as ovoid or elongate discs (Cree in press). Among 11 preserved males that I examined, the testes averaged about 18 mm in the anterior-posterior direction, about 9 mm in width, and were dorso-ventrally flattened (depth up to about 5 mm; Cree in press). Sampling of healthy individuals at known times of the year is limited as yet, but none of the variation in shape and size is obviously related to season (Cree *et al.* 1992).

The internal structure of the testes is broadly similar to that considered typical for reptiles and other amniotes (Gribbins 2011; Gribbins and

Rheubert Chapter 11, this volume). Seminiferous tubules contain the developing sperm cells in the expected stages (spermatogonia, primary and secondary spermatocytes, spermatids and spermatozoa: Hogben 1921; Saint Girons and Newman 1987; Cree *et al.* 1992; Fig. 16.8), according to the seasonal cycle described below. Sertoli cells lie within the tubules, though with their nuclei often placed in the second or third row of the seminiferous epithelium rather than against the basal membrane (Saint Girons and Newman 1987; Healy and Jamieson 1994). Interstitial or Leydig cells, though small and apparently rare, are present in the interstitial tissue (Gabe and Saint Girons 1964; Saint Girons and Newman 1987).

Spermatozoa from the testes of tuatara are about 140 μm in length and have a filiform (threadlike) shape. Although they have the expected components in terms of gross morphology (an acrosome, a nucleus and a flagellum), ultrastructural studies have revealed some unexpected features compared with those of squamate reptiles (Healy and Jamieson 1992, 1994; Jamieson and Healy 1992; Jamieson Chapter 17, this volume). For example, the spermatozoa of tuatara contain two endonuclear canals within the nucleus, structures that are also present (in numbers from 1–3) in the spermatozoa of crocodilians and turtles but that have apparently been lost from the spermatozoa of squamates. Mitochondria within the spermatozoa of tuatara also resemble those of turtles and crocodilians, rather than squamates, in developing concentric cristae (Healy and Jamieson 1992; Jamieson and Healy 1992). Conversely, the spermatozoa of squamates have some features not seen in tuatara, including a paracrystalline structure of the

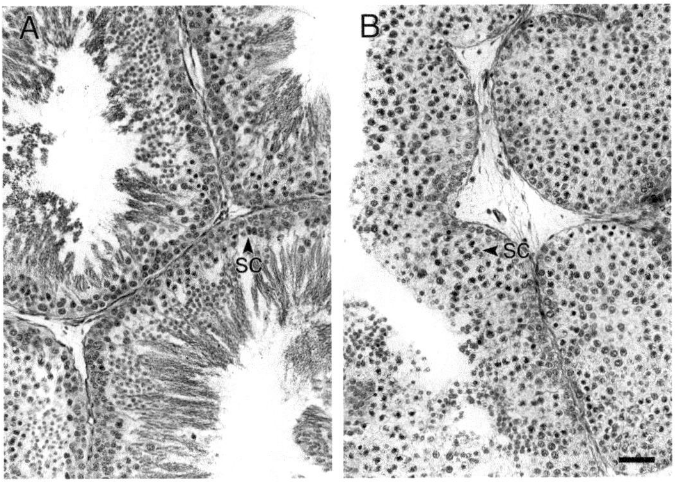

Fig. 16.8 Cross-sections of the testes of male tuatara from Stephens Island. **(A)** December (summer), with seminiferous epithelium in full spermiogenesis. **(B)** July (winter), with spermatocytogenesis still apparent. Abbreviation: sc = Sertoli cell. Scale bar = 50 μm in both.

subacrosomal cone, and deep penetration of the fibrous sheath anteriorly into the midpiece (Jamieson Chapter 17, this volume). The functional significance of these differences between the sperm of tuatara and squamates is not entirely clear.

16.8 MORPHOLOGY OF THE SPERM DUCTS AND ASSOCIATED STRUCTURES

The reproductive ducts of male tuatara have been described in several older studies (Gabe and Saint Girons 1964; Saint Girons and Newman 1987; Cree *et al.* 1992). However, terminology for the ducts in lepidosaurs has recently changed to match that for other amniotes (Sever 2010; Gist 2011; Rheubert *et al.* 2013; Rheubert *et al.* Chapter 9, this volume), leading to the changes followed here. In many of its general features, the duct system described for male tuatara is similar to those of squamates (Rheubert *et al.* Chapter 9, this volume). Testicular spermatozoa collect within tubules of the rete testis, exiting the testis at anterior and posterior positions (Rheubert *et al.* 2013). The rete testis tubules then divide into a network of fine tubules (ductuli efferentes, or efferent tubules) to reach the epididymis. Within the epididymis, the highly coiled and ciliated efferent tubules eventually lead into a central epididymal duct, which straightens and narrows posteriorly to become the vas deferens, a relatively short structure lying alongside the kidney (Saint Girons and Newman 1987; Rheubert *et al.* 2013). The vas deferens and ureter open separately in the urodaeum of the cloaca at a urogenital papilla (Günther 1867; Osawa 1897). There are preliminary indications that the urogenital papilla of males is (or can be) smaller than that of females (Osawa 1897 cf. Osawa 1898; C. L. Tyrrell pers. comm.), sufficiently so that it may not be detected (Rheubert *et al.* 2013). Laterally placed urogenital papillae (as in tuatara) or a single, medially placed urogenital papilla are also seen in male squamates (Rheubert *et al.* Chapter 9, this volume).

Within the epididymis of tuatara, the epithelium of the central duct is pseudo-stratified and secretory (Gabe and Saint Girons 1964; Saint Girons and Newman 1987; Rheubert *et al.* 2013; Fig. 16.9). Secretions within this region probably contribute to changes in the ultrastructure of sperm from the epididymis compared with those taken directly from the testes. For example, spermatozoa within the epididymides have a sheath of flocculent material around the principal piece (Healy and Jamieson 1994). The epididymis and vas deferens provide sites for at least short-term sperm storage in males (Saint Girons and Newman 1987; Cree *et al.* 1992; Rheubert *et al.* 2013).

Remarkably, as was first recognized by Günther (1867), male tuatara have no intromittent organ. In this respect, tuatara are unique not just

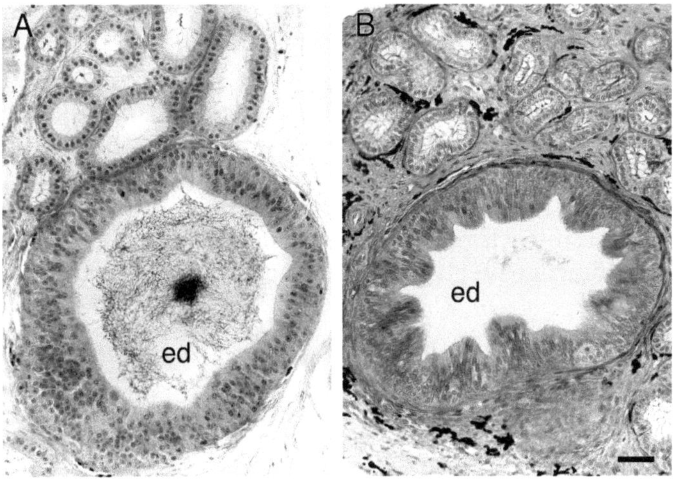

Fig. 16.9 Cross-sections of the epididymides of male tuatara from Stephens Island (the same males as in Fig. 16.8). **(A)** December (summer), with large mass of sperm in the lumen of the central epididymal duct. **(B)** July (winter), with trace of sperm in the epididymal duct. Note the lack of complete regression of the luminal epithelium in winter. Abbreviation: ed = epididymal duct. Scale bar = 50 μm in both.

among lepidosaurs but among all living reptiles. There are neither paired, laterally placed hemipenes (as in squamates) nor a central, ventrally placed penis (as in turtles and crocodilians), though the possibility that folds or out-pouchings of the cloaca in tuatara might aid in copulation has been raised (Gadow 1887; Arnold 1984).

Several additional features have been noted about the reproductive tract of male tuatara. The relative extent to which the efferent tubules continue posteriorly (lying alongside the anterior third of the kidney, located deep in the pelvis) is longer than known for any squamate (Saint Girons and Newman 1987; Rheubert *et al.* 2013). The spermatozoa have an unusual alignment within the epididymal duct: they are aggregated with their heads directed into the centre of spherical bundles, and it has been suggested that this arrangement might somehow benefit sperm transfer in the absence of an intromittent organ (Rheubert *et al.* 2013).

Male tuatara are also exceptional among lepidosaurs in lacking a renal sex segment (Gabe and Saint Girons 1964; Saint Girons and Newman 1987; Rheubert *et al.* 2013). This androgen-sensitive, accessory sex organ is present in the kidneys of male squamates but not in those of other reptiles (Kumar *et al.* 2011; Rheubert *et al.* Chapter 9, this volume). Secretions from the renal sex segment of squamates may activate sperm and help form a copulatory plug (Fox 1977; Gist 2011; Rheubert *et al.* Chapter 9, this volume).

On the other hand, male tuatara (along with females) are distinctive among lepidosaurs in having paired sebaceous glands in the wall of the proctodaeum (Günther 1867; Gadow 1887; Osawa 1897; Gabe and Saint Girons 1964, 1965; Saint Girons and Newman 1987). It is possible that these glands are consistently larger in males relative to females (Gabe and Saint Girons 1964, 1965), but studies in which gland size is properly scaled to body size are needed. The creamy-white secretions from males smell strongly to humans (especially in late summer, in the months that tuatara mate) and a pheromonal function seems probable (Flachsbarth *et al.* 2009).

16.9 CYCLES OF REPRODUCTIVE ACTIVITY IN MALES

Like other temperate-zone reptiles (Gribbins 2011; Lovern 2011; Gribbins and Rheubert Chapter 11, this volume), male tuatara reproduce on a seasonal basis. As noted above, mating occurs during mid-summer until early autumn (January–March on Stephens Island). Although this mating season may seem unusual compared with the cycles of many temperate-zone lizards, in which reproduction in spring is widely perceived as the norm (Lovern 2011; Gribbins and Rheubert Chapter 11, this volume), mating in late summer or autumn is also seen in several New Zealand geckos and skinks (Robinson 1985; Rock *et al.* 2000; Cree *et al.* 2003; Molinia *et al.* 2010), as well as in skinks from similar cool climates in Tasmania, Australia (Jones *et al.* 1997).

In tuatara, spermiogenesis immediately precedes mating, when plasma testosterone concentrations are maximal (Cree *et al.* 1992). Tuatara can therefore be said to have a pre-nuptial testicular cycle (Saint Girons and Newman 1987; see Lovern 2011 for explanation), which is typical of lizards (Gribbins and Rheubert Chapter 11, this volume). Unusually for reptiles, however, mating is not followed by a period of complete testicular regression. Instead, spermatocytogenesis continues on a limited basis over winter, with spermiogenesis resuming again by mid-spring (Saint Girons and Newman 1987; Cree *et al.* 1992). This continuous spermatocytogenic activity in tuatara (though probably very slow over winter) helps explain the lack of distinct seasonal cycles noted above in testicular mass or size. The epithelium of the epididymal duct, which is hypertrophied in summer, is reduced but also not completely inactive over winter (Saint Girons and Newman 1987; Cree *et al.* 1992). Although continuous spermatocytogenesis is unusual for lepidosaurs, a similar phenomenon is seen in several species of New Zealand geckos in the genus *Woodworthia* (until recently referred to as *Hoplodactylus maculatus*) (Robinson 1985; Saint Girons and Newman 1987). As with autumn mating, this similarity suggests that the pattern is an adaptation to the local climate, with winters not cold enough to completely suppress spermatocytogenesis and summers not hot enough to enable early completion of spermiogenesis and mating (Saint Girons and Newman 1987; Cree *et al.* 1992).

Testosterone, the main circulating androgen of tuatara (Bradshaw *et al.* 1988), probably originates in the plasma from the interstitial cells, as in other reptiles (Kumar *et al.* 2011). Although very low (≤0.7 ng/ml) over winter, plasma testosterone rises in male tuatara in spring to reach maximum concentrations of about 11–16 ng/ml during mid-summer until early autumn, before falling rapidly by late autumn (Cree *et al.* 1992; see also Bradshaw *et al.* 1988). This seasonal pattern is consistent with experimental studies on other reptiles showing that androgens generally, or testosterone specifically, stimulate hypertrophy of the sperm ducts and sexual behavior (Gist 2011; Kumar *et al.* 2011).

16.10 CONCLUSIONS

As the only surviving species of the ancient and once-diverse lineage Rhynchocephalia, the tuatara *Sphenodon punctatus* holds a special place in the evolution of reptiles. However, interpreting the evolutionary significance of particular traits in phylogenetic relicts such as tuatara, including those associated with reproduction, remains challenging. Although it has been tempting for biologists of the past to *assume* that unusual features of tuatara must be ancestral to those of squamates, in the absence of compelling evidence this temptation must be resisted.

One reason why caution is required is that the species *S. punctatus* cannot and does not represent the full biological diversity of rhynchocephalians. The tuatara genus is in fact a relatively derived member of the lineage (Evans and Jones 2010; Jones *et al.* 2013). Tuatara are the product of about a quarter of a billion years of phylogenetic independence from squamates, including evolution through changing climates, land masses and species relationships (Cree in press). Thus, some features in *S. punctatus* will inevitably be different from those of its Mesozoic ancestors and other relatives from the Rhynchocephalia, as well as from the last common ancestor of rhynchocephalians and squamates.

Second, it is apparent that some features of tuatara that might once have been thought to be distinctive (and perhaps ancestral for lepidosaurs) are also present in some squamates from the same or similar climates, including some New Zealand lizards. Reproductive features in this category include late summer mating, continuous spermatocytogenesis in males, and less-than-annual reproduction in females. The existence of such similarities means that independent evolution in response to a common climate cannot be ruled out. In future studies of reproductive activity in tuatara, representative lizards from New Zealand (especially geckos with similar activity patterns and thermal preferences) should be routinely included to help identify what may be common but independently evolved responses.

The interpretation of tuatara and squamates as each other's sister groups within the living lepidosaurs is strongly supported by many morphological and molecular analyses (Gauthier *et al.* 1988; Evans and Jones 2010; Jones *et al.* 2013; Pyron *et al.* 2013). It may therefore initially seem surprising that tuatara present such an intriguing composite of reproductive features, including some typical of squamates and others more typical of turtles and crocodilians. For example, the transverse vent, the apparent lack of an albumen-producing region of the oviduct (or significant albumen in the egg), and the parchment-like nature of the eggshell are shared with some or all squamates, whereas the single-layered granulosa of the ovarian follicles, the incompletely filled ovulation aperture of the corpora lutea, the absence of a renal sex segment in males and the presence of endonuclear canals in the sperm are features in common with turtles and/or crocodilians. It is also worth noting that tuatara are oviparous and lay eggs in which embryos are poorly developed at laying, like all turtles and (to a lesser extent) crocodilians. In contrast, most oviparous squamates lay eggs in which embryos are at stage 30 or more of development, with a further 17–19% of squamates being viviparous (Pyron and Burbrink 2013; Stewart and Blackburn Chapter 13, this volume). Body size of tuatara is also moderately large for a lepidosaur, especially in comparison with lizards (almost 80% of lizards are estimated as <20 g; Pough 1980). Perhaps some of the features that differ between tuatara plus turtles plus crocodilians on the one hand and squamates on the other are functionally related to each other and to body size, including the length of the female's oviducts and the size of the oocytes.

One aspect of reproductive morphology in tuatara that has long been recognised as distinctive among reptiles is the absence of an intromittent organ (or organs) in males. Despite this recognition, functional studies of sperm transfer in tuatara are lacking. Such studies would be of considerable interest and could help establish, for example, whether the cloaca and urogenital papillae of either or both sexes have evolved in unique ways to ensure effective insemination in the absence of intromittent organs.

In conclusion, one of the outcomes from studies of tuatara is to more sharply throw into light the impressive number of reproductive novelties that apparently distinguish all squamates among the lepidosaurs. In males, these squamate traits include a laterally placed pair of intromittent organs, a renal sex segment and features of sperm structure such as a paracrystalline subacrosomal cone and the loss of endonuclear canals. In females, they include the evolution of a multilayered granulosa with pyriform cells during the pre-vitellogenic growth of ovarian follicles. There may also be a new pattern of yolk sac development in embryos. Although the frequency of evolution of these structures within squamates is not known, a parsimonious interpretation could be that each feature evolved only once, early in the evolution of squamates.

The order of evolution of these distinctive features within squamates has apparently received little if any attention. Could there be functional relationships among these features such that one feature facilitated the evolution of another, or the evolution of viviparity? For example, are the origins of paired intromittent organs, a renal sex segment and changes in sperm structure connected? Was small body size a precondition to the evolution of any of these features? The impressive success of the squamate lineage has understandably been viewed in light of changes to skull and jaw structure and chemosensory detection related to feeding (Vitt and Pianka 2003). Although the diversity in squamates of reproductive parity modes, sex-determination systems and origins of parthenogenesis has also been recognized (e.g., Sites *et al.* 2011; Warner 2011; Stewart and Blackburn Chapter 13, this volume), perhaps it is timely to consider the broader role that a suite of reproductive innovations, presumably not shared with the last common ancestor with rhynchocephalians, might have contributed to this success.

16.10 ACKNOWLEDGMENTS

I thank my past collaborators for their involvement in work reviewed here, Claudine Tyrrell for dissections and drawings, Matthew Downes and the late Gerald Stokes for histology, Sophie Penniket for research assistance, Marcus Simons for a photograph and Ken Miller for graphical assistance. Translations provided by Bernard Goetz, Emmanuelle Martinez, Donald Newman, Ilka Söhle, Bob Zuur and Doris Zuur were especially appreciated.

16.11 LITERATURE CITED

Andrews, R. M. 2004. Patterns of embryonic development. pp. 75–102. In D. C. Deeming (ed.), *Reptilian incubation: environment, evolution and behaviour,* Nottingham University Press: Nottingham.

Andrews, R. M. and Karsten, K. B. 2010. Evolutionary innovations of squamate reproductive and developmental biology in the family Chamaeleonidae. Biological Journal of the Linnean Society 100: 656–668.

Apesteguía, S. and Jones, M. E. H. 2012. A Late Cretaceous "tuatara" (Lepidosauria: Sphenodontinae) from South America. Cretaceous Research 34: 154–160.

Apesteguía, S. and Novas, F. E. 2003. Large Cretaceous sphenodontian from Patagonia provides insight into lepidosaur evolution in Gondwana. Nature 425: 609–612.

Apesteguía, S. and Rougier, G. W. 2007. A Late Campanian sphenodontid maxilla from northern Patagonia. American Museum Novitates No. 3581: 1–11.

Arnold, E. N. 1984. Variation in the cloacal and hemipenial muscles of lizards and its bearing on their relationships. Symposium of the Zoological Society, London 52: 47–85.
Austin, H. B. 1988. Differentiation and development of the reproductive system in the iguanid lizard, *Sceloporus undulatus*. General and Comparative Endocrinology 72: 351–363.
Besson, A. A., Nelson, N. J., Nottingham, C. M. and Cree, A. 2012. Is cool egg incubation temperature a limiting factor for the translocation of tuatara to southern New Zealand? New Zealand Journal of Ecology 36: 90–99.
Bradshaw, S. D., Owen, F. J. and Saint Girons, H. 1988. Seasonal changes in plasma sex steroid levels in the male tuatara, *Sphenodon punctatus*, from Stephens Island, New Zealand. General and Comparative Endocrinology 70: 460–465.
Brown, M. A., Cree, A., Chambers, G. K., Newton, J. D. and Cockrem, J. F. 1991. Variation in plasma constituents during the natural vitellogenic cycle of tuatara, *Sphenodon punctatus*. Comparative Biochemistry and Physiology 100B: 705–710.
Carroll, R. L. and Wild, R. 1994. Marine members of the Sphenodontia. pp. 70–83. In N. C. Fraser and H. -D. Sues (eds.), *In the shadow of the dinosaurs*. Cambridge University Press: Cambridge.
Cree, A. 1994. Low annual reproductive output in female reptiles from New Zealand. New Zealand Journal of Zoology 21: 351–372.
Cree, A. In press. *Tuatara: biology and conservation of a venerable survivor*. Canterbury University Press: Christchurch, New Zealand.
Cree, A. and Butler, D. 1993. Tuatara recovery plan (*Sphenodon* spp.). *Threatened Species Recovery Plan No. 9*. Department of Conservation: Wellington.
Cree, A. and Daugherty, C. 1990. Tuatara sheds its fossil image. New Scientist No. 1739: 22–26.
Cree, A. and Guillette, L. J. Jr. 1995. Biennial reproduction with a fourteen-month pregnancy in the gecko *Hoplodactylus maculatus* from southern New Zealand. Journal of Herpetology 29: 163–173.
Cree, A. and Tyrrell, C. L. 2001. Patterns of corticosterone secretion in tuatara (*Sphenodon*): comparisons with other reptiles, and applications in conservation management. pp. 433–441. In H. J. T. Goos, R. K. Rastogi, H. Vaudry and R. Pierantoni (eds.), *Perspective in Comparative Endocrinology: Unity and Diversity (14th International Congress of Comparative Endocrinology, Sorrento, Italy, May 26–30, 2001)*. Monduzzi Editore: Bologna.
Cree, A., Cockrem, J. F., Brown, M. A., Watson, P. R., Guillette, L. J. Jr., Newman, D. G. and Chambers, G. K. 1991a. Laparoscopy, radiography, and blood analyses as techniques for identifying the reproductive condition of female tuatara. Herpetologica 47: 238–249.
Cree, A., Guillette, L. J. Jr., Brown, M. A., Chambers, G. K., Cockrem, J. F. and Newton, J. D. 1991b. Slow estradiol-induced vitellogenesis in the tuatara, *Sphenodon punctatus*. Physiological Zoology 64: 1234–1251.
Cree, A., Cockrem, J. F. and Guillette, L. J. Jr. 1992. Reproductive cycles of male and female tuatara (*Sphenodon punctatus*) on Stephens Island, New Zealand. Journal of Zoology, London 226: 199–217.
Cree, A., Guillette, L. J. Jr. and Reader, K. 1996. Eggshell formation during prolonged gravidity of the tuatara *Sphenodon punctatus*. Journal of Morphology 230: 129–144.
Cree, A., Tyrrell, C. L., Preest, M. R., Thorburn, D. and Guillette, L. J. Jr. 2003. Protecting embryos from stress: corticosterone effects and the corticosterone response to capture and confinement during pregnancy in a live-bearing lizard (*Hoplodactylus maculatus*). General and Comparative Endocrinology 134: 316–329.
Dawbin, W. H. 1982. The tuatara *Sphenodon punctatus* (Reptilia: Rhynchocephalia): a review. pp. 149–181. In D. G. Newman (ed.), *New Zealand Herpetology*. New Zealand Wildlife Service Occasional Publication No. 2: Wellington.
Dendy, A. 1899. Outlines of the development of the tuatara, *Sphenodon* (*Hatteria*) *punctatus*. Quarterly Journal of Microscopical Science 42: 1–87 + plates 1–10.
Evans, S. E. 2003. At the feet of the dinosaurs: the early history and radiation of lizards. Biological Reviews 78: 513–551.

Evans, S. E. and Jones, M. E. H. 2010. The origin, early history and diversification of lepidosauromorph reptiles. pp. 27–44. In S. Bandyopadhyay (ed.), *New aspects of Mesozoic diversity*. Springer-Verlag: Berlin Heidelberg.

Flachsbarth, B., Fritzsche, M., Weldon, P. J. and Schulz, S. 2009. Composition of the cloacal gland secretion of tuatara, *Sphenodon punctatus*. Chemistry and Biodiversity 6: 1–37.

Fox, H. 1977. The urogenital system of reptiles. pp. 1–157 + addendum pp. 463–464. In C. Gans and T. S. Parsons (eds.), *Biology of the Reptilia. Vol. 6, Morphology E*. Academic Press: London.

Gabe, M. and Saint Girons, H. 1964. *Contribution a l'histologie de Sphenodon punctatus Gray*. Éditions du Centre National de la Recherche Scientifique: Paris.

Gabe, M. and Saint Girons, H. 1965. Contribution a la morphologie comparée du cloaque et des glandes épidermoïdes de la région cloacale chez les lépidosauriens. Mémoires du Muséum National d'Histoire Naturelle, Paris Série A Zoologie 33: 149–292 + 15 plates.

Gadow, H. 1887. II. Remarks on the cloaca and on the copulatory organs of the Amniota. Philosophical Transactions of the Royal Society of London Series B 178: 5–37 + plates 2–5.

Gauthier, J., Estes, R. and de Queiroz, K. 1988. A phylogenetic analysis of Lepidosauromorpha. pp. 15–98. In R. Estes and G. Pregill (eds.), *Phylogenetic relationships of the lizard families*. Stanford University Press: Stanford.

Girling, J. E. 2002. The reptilian oviduct: a review of structure and function and directions for future research. Journal of Experimental Zoology 293: 141–170.

Girling, J. E., Cree, A. and Guillette, Jr., L. J. 1998. Oviducal structure in four species of gekkonid lizard differing in parity mode and eggshell structure. Reproduction, Fertility and Development 10: 139–154.

Gist, D. H. 2011. Hormones and the sex ducts and sex accessory structures of reptiles. pp. 117–139. In D. O. Norris and K. H. Lopez (eds.), *Hormones and reproduction of vertebrates. Vol. 3. Reptiles*. Academic Press, London.

Gray, J. E. 1842. Descriptions of two hitherto unrecorded species of reptiles from New Zealand; presented to the British Museum by Dr. Dieffenbach. Zoological Miscellany 4: 72.

Gribbins, K. M. 2011. Reptilian spermatogenesis: a histological and ultrastructural perspective. Spermatogenesis 1: 250–269.

Guillette, L. J. Jr. and Cree, A. 1997. Morphological changes in the corpus luteum of tuatara (*Sphenodon punctatus*) during gravidity. Journal of Morphology 232: 79–91.

Guillette, L. J. Jr., Cree, A. and Gross, T. S. 1990. Endocrinology of oviposition in the tuatara (*Sphenodon punctatus*): I. Plasma steroids and prostaglandins during natural nesting. Biology of Reproduction 43: 285–289.

Guillette, L. J. Jr., Propper, C. R., Cree, A. and Dores, R. M. 1991. Endocrinology of oviposition in the tuatara (*Sphenodon punctatus*): II. Plasma arginine vasotocin concentrations during natural nesting. Comparative Biochemistry and Physiology 100A: 819–822.

Günther, A. 1867. Contribution to the anatomy of *Hatteria* (*Rhynchocephalus*, Owen). Philosophical Transactions of the Royal Society of London 157: 595–629 + plates XXVI–XXVIII.

Healy, J. M. and Jamieson, B. G. M. 1992. Ultrastructure of the spermatozoon of the tuatara (*Sphenodon punctatus*) and its relevance to the relationships of the Sphenodontida. Philosophical Transactions of the Royal Society of London B 335: 193–205.

Healy, J. M. and Jamieson, B. G. M. 1994. The ultrastructure of spermatogenesis and epididymal spermatozoa of the tuatara *Sphenodon punctatus* (Sphenodontida, Amniota). Philosophical Transactions of the Royal Society of London B 344: 187–199.

Hitchmough, R., Anderson, P., Barr, B., Monks, J., Lettink, M., Reardon, J., Tocher, M. and Whitaker, T. 2013. Conservation status of New Zealand reptiles, 2012. In *New Zealand Threat Classification Series 2*. Department of Conservation, Wellington.

Hogben, L. T. 1921. A preliminary account of the spermatogenesis of *Sphenodon*. Journal of the Royal Microscopical Society 1921: 341–352.

Ibargüengoytía, N. R. and Casalins, L. M. 2007. Reproductive biology of the southernmost gecko *Homonota darwini*: convergent life-history patterns among southern hemisphere reptiles living in harsh environments. Journal of Herpetology 41: 72–80.

IUCN. 2013. *Red list of threatened species*. Version 2013.2. www.iucnredlist.org [accessed 18 February 2014].

Jamieson, B. G. M. and Healy, J. M. 1992. The phylogenetic position of the tuatara, *Sphenodon* (Sphenodontida, Amniota), as indicated by cladistic analysis of the ultrastructure of spermatozoa. Philosophical Transactions of the Royal Society of London B 335: 207–219.

Jones, M. E. H. and Cree, A. 2012. Tuatara. Current Biology 22: R986–987.

Jones, M. E. H., Tennyson, A. J. D., Worthy, J. P., Evans, S. E. and Worthy, T. H. 2009. A sphenodontine (Rhynchocephalia) from the Miocene of New Zealand and palaeobiogeography of the tuatara (*Sphenodon*). Proceedings of the Royal Society B 276: 1385–1390.

Jones, M. E. H., Anderson, C. L., Hipsley, C. A., Müller, J., Evans, S. E. and Schoch, R. R. 2013. Integration of molecules and new fossils supports a Triassic origin for Lepidosauria (lizards, snakes, and tuatara). BMC Evolutionary Biology 13(208): 1–21.

Kumar, S., Roy, B. and Rai, U. 2011. Hormonal regulation of testicular functions in reptiles. pp. 63–88. In D. O. Norris and K. H. Lopez (eds.), *Hormones and reproduction of vertebrates. Vol. 3. Reptiles.* Academic Press, London.

Lettink, M. and Whitaker, T. 2006. *Hoplodactylus maculatus* (common gecko): longevity. Herpetological Review 37: 223–224.

Lovern, M. B. 2011. Hormones and reproductive cycles in lizards. pp. 321–353. In D. O. Norris and K. H. Lopez (eds.), *Hormones and reproduction of vertebrates. Vol. 3. Reptiles.* Academic Press, London.

Mitchell, N. J., Nelson, N. J., Cree, A., Pledger, S., Keall, S. N. and Daugherty, C. H. 2006. Support for a rare pattern of temperature-dependent sex determination in archaic reptiles: evidence from two species of tuatara (*Sphenodon*). Frontiers in Zoology 3: 9 (doi: 10.1186/1742-9994-3-9).

Moffat, L. A. 1985. Embryonic development and aspects of reproductive biology in the tuatara, *Sphenodon punctatus*. pp. 493–521. In C. Gans, F. Billet and P. Maderson (eds.), *Biology of the Reptilia, Vol. 14, Development A.* John Wiley and Sons: New York.

Molinia, F. C., Bell, T., Norbury, G., Cree, A. and Gleeson, D. M. 2010. Assisted breeding of skinks or how to teach a lizard old tricks! Herpetological Conservation and Biology 5: 311–319.

Neaves, L., Wapstra, E., Birch, D., Girling, J. E. and Joss, J. M. P. 2006. Embryonic gonadal and sexual organ development in a small viviparous skink, *Niveoscincus ocellatus*. Journal of Experimental Zoology 305A: 74–82.

Nelson, N. J., Cree, A., Thompson, M. B., Keall, S. N. and Daugherty, C. H. 2004. Temperature-dependent sex determination in tuatara. pp. 53–58. In N. Valenzuela and V. Lance (eds.), *Temperature-dependent sex determination in vertebrates*. Smithsonian: Washington.

New Zealand Waitangi Tribunal. 2010. *Ko Aotearoa tēnei: a report into claims concerning New Zealand law and policy affecting Māori culture and identity.* Wai 262 Waitangi Tribunal Report. Legislation Direct: Wellington.

Newman, D. G., Watson, P. R. and McFadden, I. 1994. Egg production by tuatara on Lady Alice and Stephens Island, New Zealand. New Zealand Journal of Zoology 21: 387–398.

Osawa, G. 1897. Beiträge zur Lehre von den Eingeweiden der *Hatteria punctata*. Archiv für Mikroskopische Anatomie und Entwicklungsgeschichte 49: 113–226 + Tafeln VIII–XIV.

Osawa, G. 1898. Nachtrag zur Lehre von den Eingeweiden der *Hatteria punctata*. Die weiblichen Geschlechtsorgane. Archiv fur Mikroskopische Anatomie und Entwicklungsmechanik 51: 764–794 + Tafeln XXIII–XXV.

Packard, M. J., Hirsch, K. F., Packard, G. C., Miller, J. D. and Jones, M. E. 1991. Structure of shells from eggs of the Australian lizard *Amphibolurus barbatus*. Canadian Journal of Zoology 69: 303–310.

Packard, M. J., Thompson, M. B., Goldie, K. N. and Vos, M. 1988. Aspects of shell formation in eggs of the tuatara, *Sphenodon punctatus*. Journal of Morphology 197: 147–157.

Palmer, B. D., DeMarco, V. G. and Guillette, L. J. Jr. 1993. Oviductal morphology and eggshell formation in the lizard, *Sceloporus woodi*. Journal of Morphology 217: 205–217.

Pough, F. H. 1980. The advantages of ectothermy for tetrapods. The American Naturalist 115: 92–112.

Pyron, R. A. and Burbrink, F. T. 2013. Early origin of viviparity and multiple reversions to oviparity in squamate reptiles. Ecology Letters (doi 10.1111/ele.12168).

Pyron, R. A., Burbrink, F. T. and Wiens, J. J. 2013. A phylogeny and revised classification of Squamata, including 4161 species of lizards and snakes. BMC Evolutionary Biology 13(93): 1–53.

Rauhut, O. W. M., Heyng, A. M., López-Arbarello, A. and Hecker, A. 2012. A new rhynchocephalian from the Late Jurassic of Germany with a dentition that is unique among tetrapods. PLoS ONE 7(10): e46839, 1–9.

Raynaud, A. and Pieau, C. 1985. Embryonic development of the genital system. pp. 149–300. In C. Gans and B. Billett (eds.), *Biology of the Reptilia, Vol. 15, Development B*. John Wiley and Sons: New York.

Refsnider, J. M., Daugherty, C. H., Keall, S. N. and Nelson, N. J. 2010. Nest-site choice and fidelity in tuatara on Stephens Island, New Zealand. Journal of Zoology, London 280: 396–402.

Reynoso, V. H. 2000. An unusual aquatic sphenodontian (Reptilia: Diapsida) from the Tlayua Formation (Albian), Central Mexico. Journal of Paleontology 74: 133–148.

Reynoso, V. H. and Clark, J. M. 1998. A dwarf sphenodontian from the Jurassic La Boca Formation of Tamaulipas, México. Journal of Vertebrate Paleontology 18: 333–339.

Rheubert, J. L., Cree, A., Downes, M. and Sever, D. M. 2013. Reproductive morphology of the male tuatara, *Sphenodon punctatus*. Acta Zoologica (Stockholm) 94: 454–461.

Robinson, M. 1985. *Sexual cycles of New Zealand lizards, with particular reference to the gecko Hoplodactylus maculatus (Boulenger)*. MSc thesis, Victoria University of Wellington.

Rock, J., Andrews, R. M. and Cree, A. 2000. Effects of reproductive condition, season and site on selected temperatures of a viviparous gecko. Physiological and Biochemical Zoology 73: 344–355.

Rubenstein, D. R. and Wikelski, M. 2005. Steroid hormones and aggression in female Galápagos marine iguanas. Hormones and Behavior 48: 329–341.

Saint Girons, H. and Newman, D. G. 1987. The reproductive cycle of the male tuatara, *Sphenodon punctatus*, on Stephens Island, New Zealand. New Zealand Journal of Zoology 14: 231–237.

Sánchez-Martínez, P. M., Ramírez-Pinilla, M. P. and Miranda-Esquivel, D. R. 2013. Comparative histology of the vaginal cloacal region in Squamata and its phylogenetic implications. Acta Zoologica (Stockholm) 88: 289–307.

Schauinsland, H. 1898. Beiträge zur Biologie der *Hatteria* Sitzungsberichte der Akademie der Wissenschaften zu Berlin 44: 701–704.

Schauinsland, H. 1899. Beiträge zur Biologie und Entwickelung der *Hatteria* nebst Bemerkungen über die Entwickelung der Sauropsiden. Anatomischer Anzeiger 15: 309–334 + Tafeln II–III.

Sever, D. M. 2010. Ultrastructure of the reproductive system of the black swamp snake (*Seminatrix pygaea*). VI. Anterior testicular ducts and their nomenclature. Journal of Morphology 271: 104–115.

Thilenius, G. 1899. Vorläufiger Bericht über die Eiablage und erste Entwickelung der *Hatteria punctata*. Sitzungsberichte der Königlich Preussischen Akademie der Wissenschaften zu Berlin 1: 247–256.

Thompson, M. B. 1990. Incubation of eggs of tuatara, *Sphenodon punctatus*. Journal of Zoology, London 222: 303–318.

Thompson, M. B., Packard, G. C., Packard, M. J. and Rose, B. 1996. Analysis of the nest environment of tuatara *Sphenodon punctatus*. Journal of Zoology, London 238: 239–251.

Tokarz, R. R. and Summers, C. H. 2011. Stress and reproduction in reptiles. pp. 169–213. In D. O. Norris and K. H. Lopez (eds.), *Hormones and reproduction of vertebrates. Vol. 3. Reptiles*. Academic Press: London.

Tyrrell, C. L. and Cree, A. 1998. Relationships between corticosterone concentration and season, time of day and confinement in a wild reptile (tuatara, *Sphenodon punctatus*). General and Comparative Endocrinology 110: 97–108.
Tyrrell, C. L., Cree, A. and Towns, D. R. 2000. Variation in reproduction and condition of northern tuatara (*Sphenodon punctatus punctatus*) in the presence and absence of kiore. Science for Conservation Vol. 153. Department of Conservation: Wellington.
Vitousek, M. N., Mitchell, M. A., Romero, L. M., Awerman, J. and Wikelski, M. 2010. To breed or not to breed: physiological correlates of reproductive status in a facultatively biennial iguanid. Hormones and Behavior 57: 140–146.
Vitt, L. J., Pianka, E. R., Cooper, W. E. and Schwenk, K. 2003. History and the global ecology of squamates. The American Naturalist 162: 44–60.
Warner, D. A. 2011. Sex determination in reptiles. pp. 1–38. In D. O. Norris and K. H. Lopez (eds.), *Hormones and reproduction of vertebrates. Vol. 3. Reptiles*. Academic Press: London.
Wilmshurst, J. M., Anderson, A. J., Higham, T. F. G. and Worthy, T. H. 2008. Dating the late prehistoric dispersal of Polynesians to New Zealand using the commensal Pacific rat. Proceedings of the National Academy of Sciences 105: 7676–7680.
Worthy, T. H. and Grant-Mackie, J. A. 2003. Late-Pleistocene avifaunas from Cape Wanbrow, Otago, South Island, New Zealand. Journal of the Royal Society of New Zealand 33: 427–485.

CHAPTER 17

The Ultrastructure of Spermatogenesis and Spermatozoa of the Tuatara *Sphenodon punctatus* (Sphenodontida, Amniota)

Barrie G. M. Jamieson

17.1 INTRODUCTION

The reptilian order Sphenodontida (=Rhynchocephalia) is represented by only one living species of the genus *Sphenodon* (*S. punctatus*, including its junior synonym *S. guntheri*), popularly known as the New Zealand Tuatara; a second recent species is extinct (Daugherty *et al.* 1991). The nominate subspecies *Sphenodon punctatus punctatus* is the subject of this account. Anatomically, *Sphenodon* shares numerous features with other amniote groups as well as possessing its own unique characteristics. Many recent workers have regarded the Sphenodontida as the sister-group of the Squamata (snakes, lizards) (Fraser 1985; Evans 1984, 1988; Gauthier, Estes and de Quieroz 1988; Gauthier, Kluge *et al.* 1988; Rest *et al.* 2003); or as advanced squamates (Whiteside 1986). In contrast, cladistic analysis of sperm morphology (Jamieson and Healy 1992) suggested that sphenodontids were basal within the Amniota and were not the sister-group of squamates, only

Department of Zoology and Entomology, School of Biological Sciences, University of Queensland, Brisbane, Queensland 4072, Australia.

the turtles having fewer spermatozoal apomorphies. Relationship with squamates will be re-examined in this chapter and molecular evidence for this will be briefly discussed.

The fine structure of spermatogenesis and spermiogenesis is known for representatives of major groups of the 'Reptilia' including the Sphenodontida (Healy and Jamieson 1992, 1994; Jamieson and Healy 1992). Hogben (1921) studied changes in chromosome morphology in spermatogonia and spermatocytes of *Sphenodon punctatus* but did not include observations on spermiogenic stages or spermatozoa. The basic histology and seasonal activity of the testis of *S. punctatus* has been investigated (Saint Girons and Newman 1987), including the effects of varying sex hormone levels on gametogenesis (Cree *et al.* 1992) and a major work on the male reproductive system is that of Rheubert *et al.* (In press; see also Cree Chapter 16, this volume). Healy and Jamieson (1992) detailed the ultrastructure of testicular spermatozoa and also presented some relevant observations on developing cells in order to determine the origins and therefore the homology of certain key features subsequently used in cladistic analysis (Jamieson and Healy 1992). The ultrastructure of spermatogenesis and of mature epididymal spermatozoa of specimens of *S. punctatus* were investigated by Healy and Jamieson (1994) who also discussed comparative spermatogenesis and sperm morphology within the Amniota. As no new investigations of the ultrastructure of spermatogenesis in this endangered species have occurred since these studies the descriptions in this chapter will be based on those accounts, which may be consulted for much additional detail.

17.2 SPERMATOGENESIS

17.2.1 Spermatocytes

In spermatocytes, the nucleus is ovate to spheroidal, approximately 9.0–11.0 μm in diameter and, depending on the division stage reached, may exhibit irregular patches of dense chromatin or chromosomal cores (sometimes associated with synaptinemal complexes). Small, round to oblong mitochondria and endoplasmic reticular cisternae are scattered throughout the cytoplasm. A pair of very elongate (8.0 μm long), parallel, rod-shaped centrioles is present within the spermatocyte cytoplasm (Fig. 17.1A-C). Each centriole shows triplet microtubular substructure, which may be partly obscured by the enveloping, granular matrix (Fig. 17.1B,C) and each is associated, near one end, with a spheroidal deposit of dense material and with a short centriole at right angles. Central microtubules are absent from the centrioles at this stage.

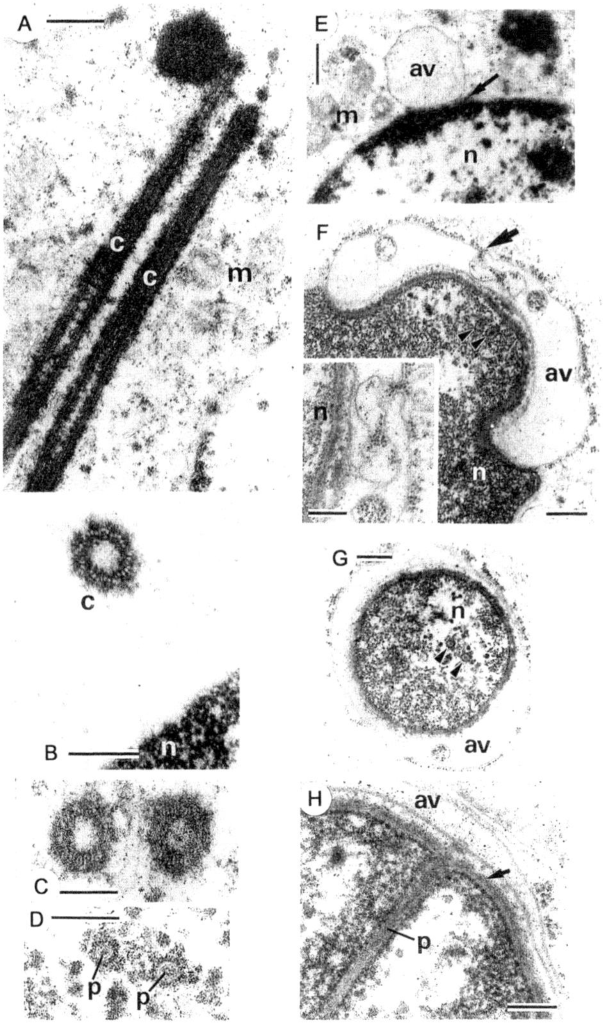

Fig. 17.1 *Sphenodon punctatus,* spermiogenesis. **(A)** Dense material associated with a pair of very elongate centrioles in the secondary spermatocyte. **(B)** Triplet substructure of spermatid centriole. **(C)** A pair of rod-shaped centrioles in spermatocyte, shown in transverse section. **(D)** Two endonuclear canals in spermatid in transverse section. **(E)** Early spermatid with acrosome vesicle newly attached to nucleus and (arrow) curved layer of subacrosomal material. **(F-H)** Middle stage spermatid. **(F)** Acrosome vesicle; arrow heads indicating endonuclear canals, arrow indicating budding off of vesicles of dense granules. Inset: detail of budding vesicle. **(G)** Transverse section through acrosome and nuclear rostrum; arrowheads indicate endonuclear canals. **(H)** Apical region of endonuclear canal, containing perforatorium, and showing confluence with subacrosomal material. Abbreviations: av, acrosome vesicle; c, centriole; m, mitochondria; n, nucleus; p, perforatorium. Scale bars: A, E, F, G, 0.5 µm; B, C, D, H, 0.25 µm. Adapted from Healy, J. M. and Jamieson, B. G. M. 1994. Philosophical Transactions of the Royal Society of London B Biological Sciences 344: 187–199. Figs. 1 f, g, h and 2.a, j, f, h, i.

In the testes, Sertoli cells are recognizable by presence of a polygonal, oblong nucleus (14.0 x 10.0 μm) and an extensive cytoplasm featuring numerous spherical, dense vesicles. Typically the Sertoli cells are positioned among the spermatocytes but not always in close proximity to the basal epithelium of the seminiferous tubules. It was not determined whether desmosome-like junctions between these cells and spermatogenic cells, described by Gribbins and Rheubert (2011) for snakes, occur.

17.2.2 Spermatid Stages

Four stages of spermatid development can be arbitrarily distinguished by a number of features, discussed below, of which the state of the nucleus is diagnostic: (i) the early stage (nucleus round; nuclear contents granular with a thin, condensed periphery); (ii) the middle stage (nucleus pyriform with two endonuclear canals; nuclear contents fibrogranular with thick periphery); (iii) the advanced stage (nucleus elongate and rod shaped; nuclear contents coarsely granular); (iv) the late stage (nucleus very elongate and associated with a longitudinally arranged microtubular sheath; nuclear contents very condensed).

17.2.2.1 Early spermatids

The nucleus of early spermatids is spheroidal, with dense chromatin distributed as irregular patches and a thin peripheral layer. Only one elongate centriole (and associated short centriole and dense deposit) is present in the cytoplasm of early spermatids, suggesting that elongate centriolar pairs observed in spermatocytes (see above) are partitioned between the two spermatids during the final meiotic division of' the spermatocyte. The triplet microtubular substructure of the centrioles is well defined (Fig. 17.1B). The single, elongate distal centriole of each spermatid clearly must be a persistence of an elongate centriole of the spermatocyte. The central pair of microtubules (also termed central singlets) within the distal centriole of the spermatid is presumably acquired during formation of the axoneme for which the centriole forms the basal body.

Mitochondria, endoplasmic reticular cisternae and numerous ribosomal granules are present throughout the cytoplasm, the mitochondria exhibiting typical (i.e., linear) cristae (see Fig. 3c in Healy and Jamieson 1994). Cytoplasmic bridges, reinforced by dense collars, connect spermatids. The Golgi complex appears to be poorly developed throughout spermiogenesis (probably as an artifact of formalin fixation), but the acrosomal vesicle is clearly visible even at the early spermatid stage (Fig. 17.1E). Acrosomal development in *Sphenodon* resembles most closely the pattern reported in crocodiles (Saita *et al.* 1987), turtles (Sprando and Russell 1988) and non-passerine birds (see Jamieson 2007; see also a comprehensive review of reptilian spermatogenesis by Gribbins 2011). On first contacting the nucleus (Fig. 17.1E) the acrosomal vesicle is spheroidal and its contents sparse, finely granular and only slightly electron-dense. A thin, curved, subacrosomal layer lies between the curved nuclear surface and the base of the acrosomal vesicle. In squamates, in contrast with *Sphenodon*, during spermiogenesis the acrosomal vesicle exhibits an obvious basal granule (within the vesicle). Moreover, endonuclear (perforatorial) canals, such as those observed in *Sphenodon*, turtles, crocodiles and non-passerine birds, have not been found in squamates (Healy and Jamieson 1992).

17.2.2.2 Middle stage spermatids

In middle stage spermatids, the nucleus is initially ovoid and its contents very dense peripherally and granular internally. The anterior and posterior poles of the spermatid nucleus become identifiable by the attached acrosomal complex (anteriorly) and newly attached centriolar apparatus (posteriorly). As development proceeds, the nucleus becomes pyriform and the acrosomal complex drapes over the nuclear apex (Fig. 17.1F). Small, highly electron-dense granules cover the outside of the acrosomal vesicle, particularly on the anterior surface (Fig. 17.1H). Vesicles containing such granules are usually seen within the acrosomal vesicle and appear to arise by budding off internally from the acrosomal vesicle membrane (Fig. 17.1F and inset). Contents of the acrosomal vesicle are fibro-granular and sparse (Fig. 17.1 F,G,H). The subacrosomal material, like the acrosomal vesicle, gradually adopts a conical shape and eventually transforms into the subacrosomal cone of late spermatids and mature spermatozoa. Commencing anteriorly and progressing posteriorly, the inner granular contents of the nucleus are gradually converted to a fibro-granular consistency. During this phase two narrow, eccentric invaginations into the nucleus, the incipient endonuclear canals, develop (Fig. 17.1D,G,H) immediately beneath the attached acrosomal complex and they each contain a dense perforatorial deposit. These canals extend posteriorly past the acrosomal region, deeper into the nucleus in middle stage spermatids.

A loose network, composed of moderately electron-dense material and a less dense background matrix, gradually develops around the centriolar region. The elongate, rod-shaped centriole, now the distal centriole or basal body, gives rise to the flagellar (9+2 substructure) axoneme and becomes penetrated by the pair of singlet microtubules from the axoneme. Throughout the middle spermatid phase, mitochondria remain small with unmodified cristae and begin to collect posteriorly around the distal centriole prior to midpiece development.

17.2.2.3 Advanced spermatids

Advanced (elongating) spermatids (Fig. 17.2A, C, 3F) can be recognized by their narrow (1.0 μm) cylindrical nucleus, the contents of which have been converted to coarse granules (Fig. 17.2A) that eventually fuse into a loose network.

Development of endonuclear canals. Development of the endonuclear canals in *Sphenodon* probably commences at the early spermatid stage as in the Caiman, *Caiman crocodilus* (Saita *et al.* 1987), and illustrated for turtles and certain non-passerine birds (Sprando and Russell 1988). Gribbins *et al.* (2010), however, ascribe formation of the canals to the elongating spermatid in *Alligator mississipiensis*. Similar canals occur in sperm of sarcopterygian fish (coelacanth, Mattei *et al.* 1988; and lungfish (*Neoceratodus*), Jespersen 1971; Jamieson 2009) and primitive frogs (for example, *Ascaphus*, Jamieson *et al.* 1993). This confirms the endonuclear canals as symplesiomorphies of tetrapods. Almost certainly these canals and their granular contents perform some perforatorial function (for further discussion see Healy and Jamieson 1992), although it has also been suggested that they may be involved with nuclear metabolism (Yasuzumi *et al.* 1971). The homology of endonuclear filaments occurring in lampreys and sturgeon (see Jamieson 1991, 2009) with those of sarcopterygians is questionable. It has been questioned that those in the sturgeon, although containing actin, are perforatorial (Cherr and Clark 1984) but Wei *et al.* (2007) demonstrated an "actin filament" projecting from the tip of the acrosome in the Chinese Sturgeon, *Acipenser sinensis*, and Psenicka *et al.* (2011) showed that those spermatozoa of the Sterlet, *Acipenser ruthenus*, which were able to activate the acrosomal reaction were more successful in fertilization (see also Jamieson 2009: 222).

Perinuclear microtubules. Microtubules are arranged in a low angle helical sheath around the nucleus. The distal and proximal centrioles have become closely attached to and partly inserted within the nuclear base. The changing arrangement of perinuclear microtubules during spermiogenesis in *Sphenodon*, that is, from a low angle helical orientation in advanced spermatids to longitudinal in late spermatids, has been observed in all other groups of amniotes (turtles: Yasuzumi and Yasuda 1968; crocodiles: Saita *et al.* 1987; alligator: Gribbins *et al.* 2010; squamates: Clark 1967; Cruz-Hofling and Cruz-Landim 1978; Al-Hajj *et al.* 1987; Saita *et al.* 1988a,b; Dehlawi *et al.* 1992; non-passerine birds: Phillips and Asa 1989; monotreme mammals: Carrick and Hughes 1982). Although most authors postulate that the low-

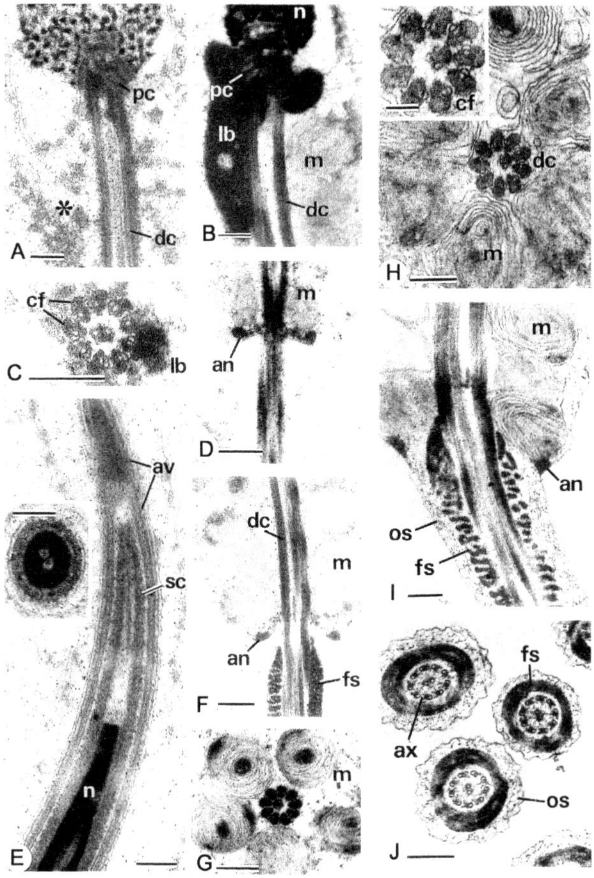

Fig. 17.2 *Sphenodon punctatus* spermiogenesis. **(A)** Advanced spermatid, detail of attached centrioles and basal nuclear fossa. Asterisk = pericentriolar network. **(B)** Late spermatid in region of nuclear fossa, showing lateral body and attached centrioles. Note concentric cristae of mature mitchondria. **(C)** Detail of distal centriole in transverse section, showing coarse fibers, developing lateral body and open C tubules of triplets in advanced spermatid. **(D)** Junction of midpiece and principal piece of late spermatid (annulus present but fibrous sheath not yet formed). **(E)** Acrosomal complex of almost mature testicular spermatozoon. Inset, transverse section showing acrosome, nuclear apex and endonuclear canals. **(F)** Junction of midpiece and principal piece of testicular spermatozoon; note presence of fibrous sheath. **(G)** Transverse section of midpiece of testicular spermatozoon; note concentric cristae surrounding dense intramitochondrial body. **(H)** Mature spermatozoon; Transverse section of midpiece; inset, detail of distal centriole. **(I)** Junction of midpiece and principal piece, showing annulus, fibrous sheath and outer sheath. Unlike squamates, the fibrous sheath does not extend throughout the midpiece though there is slight intrusion. **(J)** Transverse through principal piece. Abbreviations: an, annulus; ax, axoneme; cf, coarse fibers; dc, distal centriole; fs, fibrous sheath; lb, lateral body; m, mitochondrion; n, nucleus; os, outer sheath; pc, proximal centriole. Scale bars all 0.25 μm, excepting D and F, 0.5 μm and inset in H which is 0.1 μm. Adapted from Healy, J. M. and Jamieson, B. G. M. 1994. Philosophical Transactions of the Royal Society of London B Biological Sciences 344: 187–199. Figs. 3b, e, d, h, f, I, j, 4g, h, j.

angle helically arranged microtubules simply reorientate longitudinally late in spermiogenesis, Saita *et al.* (1987) presented evidence to suggest that two distinct generations of microtubules are being produced, at least in *Caiman crocodilus*; low-angle helical and longitudinal microtubules of differing diameter coexist in the cytoplasm of many cells.

The C microtubules of the proximal and distal centriole are open (i.e., incomplete). The distal centriole also exhibits a central pair of singlet microtubules (continuous with those of the axoneme), nine coarse fibers and a lateral body. The lateral body is deduced to be derived from condensation of the loose network of pericentriolar material first observed in mid-spermatids and shows traces of periodic striation (see Healy and Jamieson 1992) and sometimes, electron-lucent lacunae. It seems likely that the lateral body of *Sphenodon* is homologous with the 'connecting collar' surrounding the base of the nucleus in turtles (Hess *et al.* 1991). In advanced spermatids, mitochondrial cristae are arranged parallel to each other, while within the matrix an eccentrically positioned intramitochondrial granule is usually observed.

17.2.2.4 Late spermatids

On reaching the late spermatid stage (Fig. 17.2B,D), the acrosome has become attenuated and conical although still associated with traces of cytoplasm. A sheath of longitudinally aligned microtubules has surrounded the nucleus. The nucleus exhibits highly condensed contents as well as two rod-filled endonuclear canals, and, like the acrosomal complex, has reached its final form. The lateral body is well developed (Fig. 17.2B). Mitochondria are spherical and are arranged around the elongate distal centriole to form the midpiece. A poorly defined sheath of microtubules, probably continuous with that surrounding the condensing nucleus, is associated with midpiece development.

The transformation of unmodified spermatid mitochondria, with linear cristae, into the specialized mitochondria of late spermatids and mature sperm, with concentric cristae surrounding a dense intramitochondrial body, has elsewhere been observed only in turtles (Yasuzumi and Yasuda 1968), crocodilians (Saita *et al.* 1987) and, in mammals, in the Woolly Opossum *Caluromys philander* (Fawcett 1970; Phillips 1970), Virginia Opossum, *Didelphis virginiana* (Temple-Smith *et al.* 1980) and the Rufous Hare Wallaby *Lagorchestes hirsutus* (Jamieson 1999; Johnston *et al.* 2003); it should, however, be noted that these marsupial mitochondria are arranged around the axoneme in contrast with *Sphenodon*, turtles and crocodiles in which mitochondria are rounded, and are stacked longitudinally around the distal centriole. The significance of the substructural changes to a concentric condition in mitochondrial cristae is still unclear, although Yasuzumi and Yasuda (1968) related these changes to a possible need for increased mitochondrial efficiency (and therefore cristal surface area) during sperm storage. Similar changes were reported by Hess *et al.* (1991) for turtle spermatozoa. Although the mitochondria of squamates also undergo substantial structural changes during spermiogenesis, these involve an increase in mitochondrial length, helical coiling of mitochondria around the axoneme and the production of dense intra-mitochondrial rings or bodies (Jamieson and Healy 1992; Jamieson and Scheltinga 1993). Although a well-developed annulus is present in late spermatids of *Sphenodon* at the posterior limit of the mitochondria, the fibrous sheath of' the principal piece has only just begun to form.

17.3 THE TESTICULAR SPERMATOZOON

The fine structure of testicular spermatozoa from *Sphenodon punctatus* (Figs. 17.2E,F,G; 17.3A-D,G-P) was described by Healy and Jamieson (1992) and that of epididymal spermatozoa (Fig. 17.2H-J) by Healy and Jamieson (1994).

Light microscopy (Fig. 1A in Jamieson and Healy 1992) reveals that testicular spermatozoa of *Sphenodon punctatus* are filiform, approximately 140–144 μm long, with a head region (acrosome plus nucleus) which forms a loose helix 58–60 μm long, a midpiece 7–8 μm long, a further 74–78 μm for the principal piece and 2–4 μm for the endpiece. By transmission electron microscopy the acrosomal complex is 4 μm long, the elongate, helical nucleus 54–56 μm long.

17.3.1 Acrosomal Complex

The acrosomal complex of the testicular sperm (Figs. 17.2E; 17.3A-E) consists of three elements: (i) a conical and membrane-bound acrosomal vesicle (length approximately 4 μm), (ii) a hollow subacrosomal cone and (iii) perforatorial (subacrosomal) material occupying the paired endonuclear canals. The acrosomal vesicle and subacrosomal cone are curved and ensheath the tapered, anterior region of the nucleus (rostrum) for a distance of 1.6 μm (Fig. 17.3A). An electron-lucent layer separates the acrosomal vesicle from the subacrosomal cone (Figs. 17.3A,B,D). The base of the acrosome rests on a shoulder-like widening of the nucleus (Fig. 17.3E).

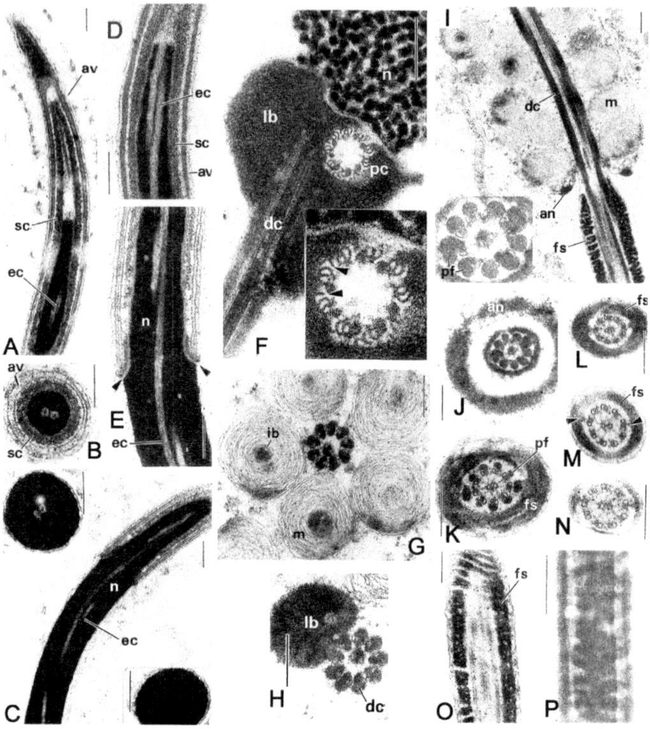

Fig. 17.3 *Sphenodon punctatus.* **(A-D, G-P)**. Testicular spermatozoon; **F**. Advanced spermatid. **(A)** Longitudinal section of the acrosomal complex ensheathing the nuclear apex; note the subacrosomal cone and endonuclear canals. **(B)** Acrosomal complex and nucleus in transverse section, showing endonuclear canals containing putative perforatorial material. **(C)** Penetration of the endonuclear canals posterior to the acrosomal complex. Upper inset: Transverse section of endonuclear canals posterior to acrosome. Lower inset: nucleus posterior to endonuclear canals. **(D)** Anterior extremity of nucleus. **(E)** Basal region of acrosomal complex with arrow heads indicating nuclear shoulders. Note twisting of endonuclear canals. **(F)** Proximal centriole and elongate distal centriole embedded in the matrix of the lateral body at the base of the condensing nucleus of an advanced spermatid. Inset, detail of proximal centriole showing open C tubules (arrow heads). **(G)** Transverse section through posterior region of midpiece, showing concentric cristae and dense intramitochondrial bodies. **(H)** Lateral body and distal centriole of late spermatid. **(I)** Longitudinal section through annulus at the junction of the midpiece, showing mitochondria and principal piece with fibrous sheath. Inset, Peripheral fibers associated with axonemal doublets. **(J)** Transverse section through annulus. **(K)** Anterior region of principal piece, showing fibrous sheath surrounding axoneme and peripheral fibers. **(L)** Anterior region of principal piece posterior to K, showing loss of seven peripheral fibers. **(M)** Posterior region of principal piece; arrowheads, peripheral fibers near doublets 3 and 8. **(N)** Distal region of principal piece. **(O)** Portion of principal piece in oblique longitudinal section. **(P)** Negative stained principal piece, showing anastomosing fibers of fibrous sheath. Abbreviations: an, annulus; av, acrosome vesicle; dc, distal centriole; ec, endonuclear canal; fs, fibrous sheath; lb, lateral body; m, mitochondrion; n, nucleus; pc, proximal centriole; pf, peripheral fibers; sc, subacrosomal cone. Scale bars all 0.25 µm. Adapted from Healy, J. M. and Jamieson, B. G. M. 1992. Philosophical Transactions of the Royal Society of London B Biological Sciences 335: 193–205. Figs. 1b, c, f; 2b; 3e, b; 4a-g, h.

17.3.2 The Nucleus

The nucleus is very elongate, 54–56 μm long, coiled in a loose helix, and tapers to a point anteriorly (Fig. 17.3A). A small, cuboidal fossa containing dense material and a portion of the proximal centriole is present at the base of the nucleus of late spermatids and presumably also occurs in mature sperm. The anterior, tapered portion of the nucleus is ensheathed by the posterior portion of' the acrosomal complex (Fig. 17.3A,C,D,E). Two narrow, endonuclear canals (diameter 40–45 nm) are present anteriorly, each filled with a perforatorial deposit and opening within the subacrosomal cone (Fig. 17.3A, D). Longitudinal sections indicate that the canals, though closely parallel, are helically twisted around each other. Although the length of the canals was not determined, they were followed in longitudinal sections to at least 2.5 μm beyond the base of the acrosomal complex (Fig. 17.3C) and possibly extend much deeper. Nuclear contents of mature spermatozoa are highly electron-dense, are invested by nuclear and plasma membranes and exhibit occasional electron-lucent lacunae distinct from the endonuclear canals. Nuclear condensation proceeds in late spermatids by aggregation of coarse (44 nm) granules and gradual, although not total, elimination of lacunae.

17.3.3 Neck Region, Centrioles

The morphology of the nucleus-midpiece junction or `neck' region was established only for the late spermatid (Fig. 17.2B) and advanced spermatid (Fig. 17.3F) as favorable sections were not obtained for fully mature sperm. Longitudinal sections of late spermatids show the presence of a nuclear fossa; a proximal centriole consisting of nine triplets of microtubules of which the C microtubule often appears open, as in the advanced spermatid (Fig. 17.3F); and an extensive deposit of pericentriolar material, here termed the lateral body, which extends into the nuclear fossa and posteriorly lies in contact with the elongate distal centriole (Figs. 17.2B, 17.3F,H). The lateral body is collar-shaped anteriorly and cylindrical distally (within the midpiece) with occasional electron-lucent lacunae (late spermatid, Fig. 17.2B). In the immediate vicinity of the proximal centriole, faint periodic striations (repeat distance 30–40 nm) are seen within the lateral body.

The distal centriole of the mature spermatozoon is unusual relative to squamate sperm in its great length and in containing central singlets. Proximal and distal centrioles, despite their marked difference in length, both show an open C microtubule within the triplets.

17.3.4 Midpiece

Following the nucleus and neck is the midpiece, occupying only 7–8 μm of the length of the sperm but nevertheless long in comparison with `primitive', externally fertilizing sperm. The midpiece consists of spherical mitochondria (each with concentric cristae, Fig. 17.3G) which surround an elongate, distal centriole and the posterior portion of the lateral body (Fig. 17.3H). Eight or nine spherical mitochondria, 0.7–0.9 μm in diameter, are present in cross section anteriorly, decreasing to five or six posteriorly (Fig. 17.3G). Each mitochondrion has 11 to 14 concentric cristae around a dense central body (Fig. 17.3G). Mitochondria of early and middle stage spermatids (and spermatocytes), by contrast, contain only regular (linear) cristae, indicating that transformation from regular to concentric cristae takes place relatively late in spermiogenesis. The elongate distal centriole runs the entire length of the midpiece. It consists of nine triplets of microtubules, often showing an open C tubule; nine peripheral fibers which partly ensheath the triplets and project into the centriolar lumen; and the pair of central microtubules which are embedded in and partly obscured by dense material (Fig. 17.3G, H). The distal and proximal centrioles are embedded in the lateral body near the nuclear fossa (advanced spermatid, Fig. 17.3F). Posteriorly the distal centriole is continuous with the axoneme of the principal piece (Fig. 17.3I), its triplets giving way to the doublets of the axoneme.

The presence of a central pair of microtubules within the lumen of the distal centriole in turtles and crocodiles has led some authors (for example, Furieri 1970; Saita *et al.* 1987) to interpret this elongate structure as the anterior portion of the axoneme. The triplet substructure of this centriole has since been convincingly demonstrated in turtles and *Sphenodon* (see Hess *et al.* 1991; Healy and Jamieson 1992, 1994; Jamieson 1999). In addition, Phillips and Asa (1991) and Gribbins *et al.* (2010) demonstrate a rod-shaped centriole in spermatids and sperm of the American alligator (*Alligator mississipiensis*). Furthermore, presence of the two centrioles as separate entities has been demonstrated in the spermatocyte (Fig. 17.1A). Aside from crocodilians, turtles and *Sphenodon*, penetration of the elongate distal centriole by paired singlet microtubules late in spermiogenesis has also been recorded in palaeognath birds (Phillips and Asa 1989; Baccetti *et al.* 1991; see review by Jamieson 2007). Presumably these axonemal singlets within the centriolar lumen strengthen the centriole-axoneme connection. The presence of a rod-shaped distal centriole running the full length of the sperm midpiece must be considered an amniote symplesiomorphy and not, as suggested by Baccetti *et al.* (1991), an autapomorphy of palaeognath (ratite) birds. In the spermatid and spermatozoon of these amniotes only one of the two rod-shaped centrioles seen in the spermatocyte persists.

No ultrastructural data exist on centrioles in turtle spermatocytes but Risley (1936) demonstrated, by light microscopy, centrioles as long rods in germ cells of five species of turtles. In prophase of spermatogonial mitosis in the Musk Turtle, *Sternotherus odoratus*, there are initially two rods each of which duplicates itself so that two bivalent V-shaped centrioles are produced. This process is repeated in the primary spermatocyte. Each secondary spermatocyte receives a single bivalent centriole and this separates into two rods which lie parallel to each other and tangential to the nuclear surface; the arrangement described by Healy and Jamieson (1994) for *Sphenodon*. Further, Risley showed that only one rod enters each spermatid in *S. odoratus* and in the Diamond-Back Terrapin, *Malaclemys terrapin*, again as observed for *Sphenodon*, although there, by TEM, shown to be accompanied by a short centriole.

17.3.5 Annulus

A dense ring, constituting a well-developed annulus, is present at the distal extremity of the midpiece. This structure is electron-dense, approximately 1.0 µm in diameter and is closely applied to the inside surface of the plasma membrane (Fig. 17.3I). Electron-dense material is also observed at the point of reflection of the plasma membrane giving an inner periaxonemal ring in transverse sections (Fig. 17.3J). Transverse sections also show that the transition from distal centriole to axoneme occurs within the annular region and that peripheral fibers associated with doublets 3 and 8 are appreciably thicker than the remaining seven fibers (Fig. 17.3K,L). The thicker fibers at doublets 3 and 8 continue posteriorly in contact with the inner surface of the fibrous sheath of the principal piece.

17.3.6 Principal Piece

The principal piece occurs posterior to the midpiece (3I). This is the longest region (74–78 µm) of the spermatozoon and consists of the 9+2 axoneme surrounded by an electron-dense fibrous sheath and the plasma membrane (Fig. 17.3I,K-M). The A tubule of each doublet is filled with electron-dense material, and bears two dynein arms which project towards the B tubule of the succeeding doublet. Initially the axoneme emerges from the midpiece accompanied by all nine peripheral fibers (Fig. 17.3K) but these rapidly decrease in diameter and, with the exception of fibers 3 and 8, terminate within the anterior region of the principal piece (Fig. 17.3M). Diameter of the principal piece decreases from 0.73 µm anteriorly (Fig. 17.3K) to 0.38 µm approaching the endpiece (Fig. 17.3N). The anastomosing, and probably helical, arrangement of the fibrous sheath can best be observed in oblique longitudinal sections (Fig. 17.3O) and in negative-stained preparations (Fig. 17.3P).

The fibrous sheath of the principal piece is a synapomorphy of the Amniota (Healy and Jamieson 1992). Absence of this sheath in the sperm of all passerine and some non-passerine birds (Asa and Phillips 1987; Jamieson 2007) appears secondary and therefore apomorphic relative to the plesiomorphic retention in palaeognath birds. According to Fawcett and Phillips (1970) the sheath appears to arise through the accretion of materials around the axonemal complex late in spermiogenesis.

17.3.7 Endpiece

The endpiece (length 2–4 μm) of the spermatozoon was observed only by light microscopy.

17.4 EPIDIDYMAL SPERM

Spermatozoa in the epididymis (Fig. 17.2H-J), which should be considered more mature than those of the testes, closely resemble the testicular spermatozoa but show some significant differences. First, in epididymal sperm, the lateral body completely ensheathes the distal centriole anteriorly, whereas the lateral body is of lesser extent in testicular sperm, and becomes a half-sheath posteriorly. Secondly, the anterior extremity of the fibrous sheath projects just inside the annulus (Fig. 17.2I) whereas the annulus and fibrous sheath are well separated in testicular sperm. Thirdly, the anterior region of the principal piece not only possesses a fibrous sheath but also has an outer layer of granulo-fibrous material (Fig. 17.2J); such material is absent in testicular sperm. A similar layer has been observed in sperm of the Crested Tinamou, *Edromia elegans* (Asa *et al.* 1986; Asa and Phillips 1987; Phillips and Asa 1989) and the Woolly Opossum (Phillips 1970) but is absent in sperm of most palaeognath birds (Baccetti *et al.* 1991; see review by Jamieson 2007), most mammals (Fawcett 1970), turtles (Furieri 1970; Hess *et al.* 1991; Healy and Jamieson 1992), squamates (Furieri 1970; references in Jamieson 1999) and crocodilians (Saita *et al.* 1987; Jamieson *et al.* 2003; Gribbins 2010). Asa and Phillips (1987) showed this outer layer within the principal piece of the Crested Tinamou was structurally similar to glycogen. The corresponding (homologous?) layer in *Sphenodon*, however, shows no morphological resemblance to glycogen granules.

A significant proportion of the epididymal sperm observed were still associated with a cytoplasmic droplet, indicating that the epididymis acts not only as a sperm storage organ but also as a site where maturation is completed.

Rheubert *et al.* (2012) report that large amounts of spermatozoa are present within the epididymis even when the testes are spermatogenically inactive. This suggested that male Tuatara are capable of sperm storage, at least for a short period of time, as appears to be true of lizards, whereas in snakes sperm are stored in the ductus deferens. In contrast to squamates, the spermatozoa of *Sphenodon punctatus* are arranged in spherical bundles with their heads pointing toward the center of each bundle.

17.5 COMPARATIVE AMNIOTE SPERMATOZOAL ULTRASTRUCTURE AND PHYLOGENETIC DISCUSSION

Sphenodon has traditionally been regarded as a little changed survivor of Permo-Triassic "stem reptiles" of the Thecodontia (von Wettstein 1931; Crook 1975) or Eosuchia (Carroll 1969) since erection of the order Rhynchocephalia by Günther in 1867 to receive it. While endorsing its primitive status, several investigators have placed *Sphenodon*, on the basis of somatic morphology, with the squamates in the Lepidosauria (Rieppel 1978; Schwenk 1986; Peterson 1984; Laurin 1991; Schaerlaeken *et al.* 2008; Jones and Lappin 2009) or Lepidosauromorpha (Gauthier, Estes and de Quieroz 1988). Alternatively, a derived status for *Sphenodon* relative to lizards, within the Lepidosauria, has been argued on the basis of the head skeleton (Whiteside 1986).

The effectiveness of phylogenetic analysis of the structure of spermatozoa (spermiocladistics, Jamieson 1987) in resolving hitherto intractable or taxonomic and phylogenetic problems or in arbitrating between conflicting views of relationship has been demonstrated by many workers, some of which are listed by Jamieson and Healy (1992). In an attempt to elucidate the relationships of the Tuatara, Healy and Jamieson (1991) described the ultrastructure of the spermatozoon of *Sphenodon punctatus punctatus*. They concluded that the spermatozoa of the Tuatara strongly resemble those of turtles and crocodiles and unequivocally indicated that the living Tuatara (and by association other sphenodontids) is truly primitive and is a basal amniote with no special relationship to the Squamata investigated to date (lizards, skinks, geckos and snakes). This conclusion was based on an intuitive consideration of comparative sperm ultrastructure in *Sphenodon* and other amniotes but also on a computerized cladistic analysis which sought to find the most parsimonious phylogenetic tree from a set of 16 spermatozoal characters.

A consideration of these characters, which are labeled in Fig. 17.4, provides a comparison of the ultrastructure of the spermatozoa of *Sphenodon* with other amniotes.

Attention has previously been drawn (Healy and Jamieson 1992; Jamieson and Healy 1992) to the remarkable similarities between

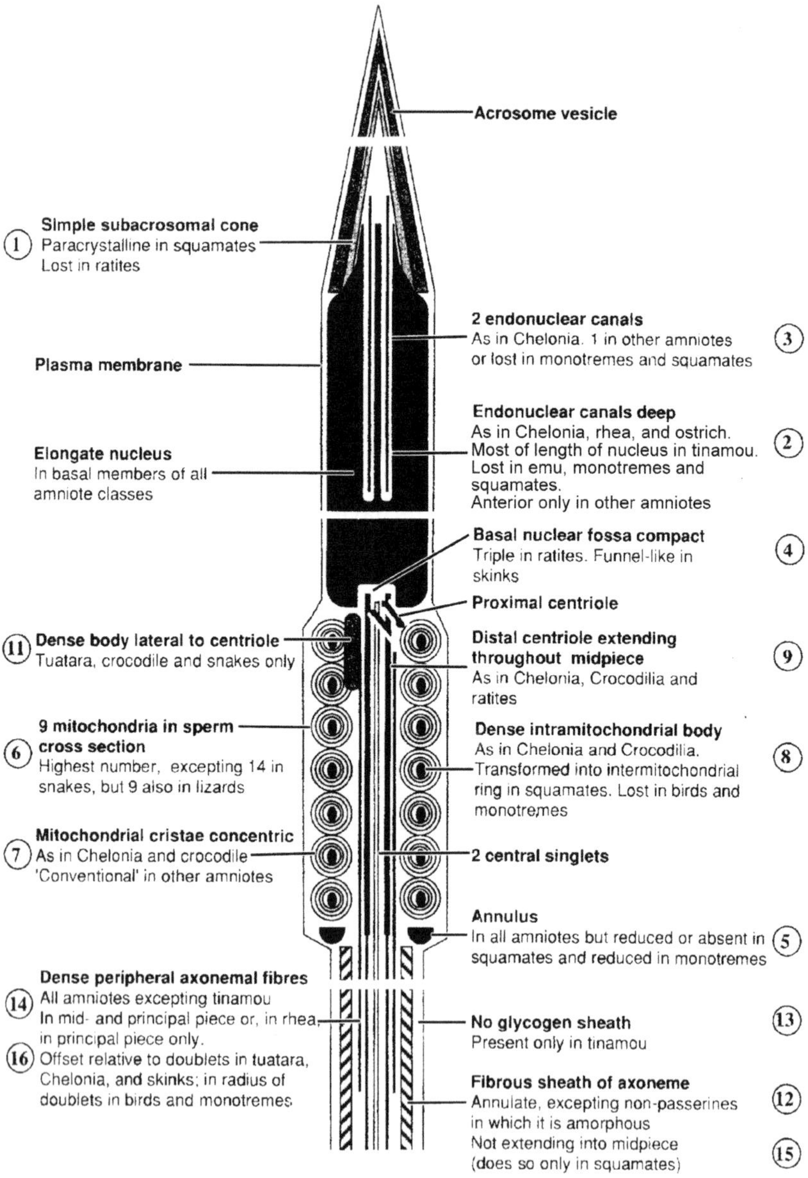

Fig. 17.4 Diagram of the spermatozoon of *Sphenodon punctatus* based on transmission electron micrographs. Major characters are labeled and numbered and their condition in other amniotes is annotated.
Modified from Jamieson, B. G. M. and Healy, J. M. 1992. Philosophical Transactions of the Royal Society of London B Biological Sciences 335: 207–219. Fig. 1.

spermatozoa of *Sphenodon* and those of' turtles (Furieri 1970; Hess *et al.* 1991; Healy and Jamieson 1992), the crocodiles (Saita *et al.* 1987; Jamieson *et al.* 1997) and to palaeognath birds (Jamieson 2007). All of these groups share the following features: (i) an acrosomal complex associated with one or more narrow endonuclear canals (absence of an endonuclear canal in the Emu (*Dromaius novaehollandiae*) (Baccetti *et al.* 1991) is almost certainly due to secondary loss because all other examined palaeognath birds exhibit a canal (See, however, du Plessis and Soley 2014, who argue from spermatozoal ultrastructure for an independent origin of Australasian ratites); (ii) a midpiece containing an elongate distal centriole; (iii) non-intrusion of the fibrous sheath into the midpiece. Among reptiles, only *Sphenodon*, turtles and crocodiles possess rounded sperm mitochondria characterized by concentric cristae and a prominent intramitochondrial body (see Furieri 1970; Yasuzumi and Yasuda 1968; Saita *et al.* 1987; Hess *et al.* 1991; Healy and Jamieson 1992; Jamieson and Healy 1992; Jamieson 1999). Further, of these three groups, only *Sphenodon* and turtles possess multiple (usually paired), elongate, helically twisted nuclear canals. The fact that morphological differences between sperm of *Sphenodon* and turtles are relatively minor (turtle sperm lacking the outer fibrous layer in the principal piece and occasionally showing three rather than two endonuclear canals) led Jamieson and Healy (1994) to question the widely held view that sphenodontids are closely allied to the Squamata. Figure 17.5 compares the spermatozoa of the chelonian Krefft's Short-necked Turtle (*Emydura kreffti*), the Tuatara (*Sphenodon punctatus*), the Australian Freshwater Crocodile (*Crocodylus johnstoni*), the Ostrich (*Struthio camelus*) and a generalized snake spermatozoon representing the Squamata. These may be compared with the ultrastructure of the hypothetical plesiomorphic amniote spermatozoon (Fig. 17.6) which was deduced by comparison of sperm throughout the Amniota, in the Lissamphibia, particularly the primitive Tailed Frog, *Ascaphustruei*, and in the Dipnoi among other taxa (Jamieson 1999). Features of the basal amniote spermatozoon recognized in the latter work have been strongly endorsed by Gribbins and Rheubert (2011) and will be discussed here before considering pertinent modifications it has undergone in evolution.

Tetrapod features of amniote spermatozoa. Features of the hypothetical amniote sperm deduced to have been retained from their tetrapod ancestry, and still seen in Chelonia and Sphenodontida and to varying extents in other amniotes, are as follows. The spermatozoon (Fig. 17.6) is elongated and filiform, with a hollow anterior conical acrosome vesicle overlying a simple subacrosomal cone (both comprising the acrosomal complex). The base of the acrosome vesicle invests the tapered anterior tip (rostrum) of the nucleus and rests on pronounced nuclear shoulders. The subacrosomal

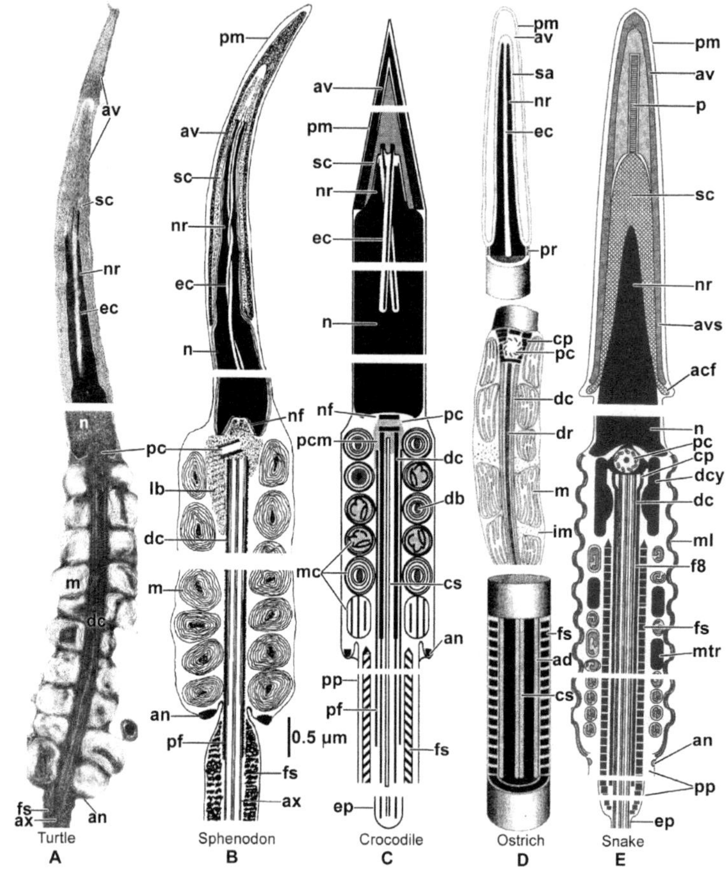

Fig. 17.5 Comparison of longitudinal section of the spermatozoa of some reptiles and the ostrich. **(A)** a turtle, *Emydura krefftii*. **(B)** Tuatara, *Sphenodon punctatus*. **(C)** Crocodile, *Crocodylus johnstoni*. **(D)** Ostrich, *Struthio camelus*. **(E)** Squamates, generalized diagram for the snake families Colubridae, Elapidae and Boidae. Adapted from the following: A and B. Healy, J. M. and Jamieson, B. G. M. 1992. Philosophical Transactions of the Royal Society of London B Biological Sciences 335: 193–205, Fig. 5b,f. B, Same, Fig. 6B. C. Jamieson, B. G. M., Scheltinga, D. M. and Tucker, A. D. 1997. Journal of Submicroscopic Cytology and Pathology 29: 265–274, Fig. 1. D. Soley, J. 1999. Reproduction. pp. 129–158. In D. C. Deeming (ed.), *The Ostrich: Biology, Production and Health*. CAB International, Fig. 6.5. E. Jamieson, B. G. M. and Koehler, L. 1994. Canadian Journal of Zoology 72: 1648–1652, Fig. 1. Abbreviations: ad, axonemal doublets; an, annulus; acf, flange of acrosome cone; av, acrosome vesicle; avs, sleeve of acrosome vesicle; ax, axoneme; cp, connecting piece; cs, central singlet microtubules; db, dense body in mitochondrion; dc, distal centriole; dcy, dense collar (neck cylinder); dr, dense rod in distal centriole; ec, endonuclear canal(s); ep, endpiece; f8, enlarged peripheral fiber at 8; fs, fibrous sheath; lb, lateral body; m, mitochondrion; mc, mitochondrial cristae; ml, multilaminar membrane; mtr, mitochondrial transformation; n, nucleus; nf, nuclear fossa; nr, nuclear rostrum; p, perforatorium; pc, proximal centriole; pcm, pericentriolar material; pf, nine peripheral fibers; pm, plasma membrane; pp, principal piece; pr, posterior ring; sa, subacrosomal space; sc, subacrosomal cone.

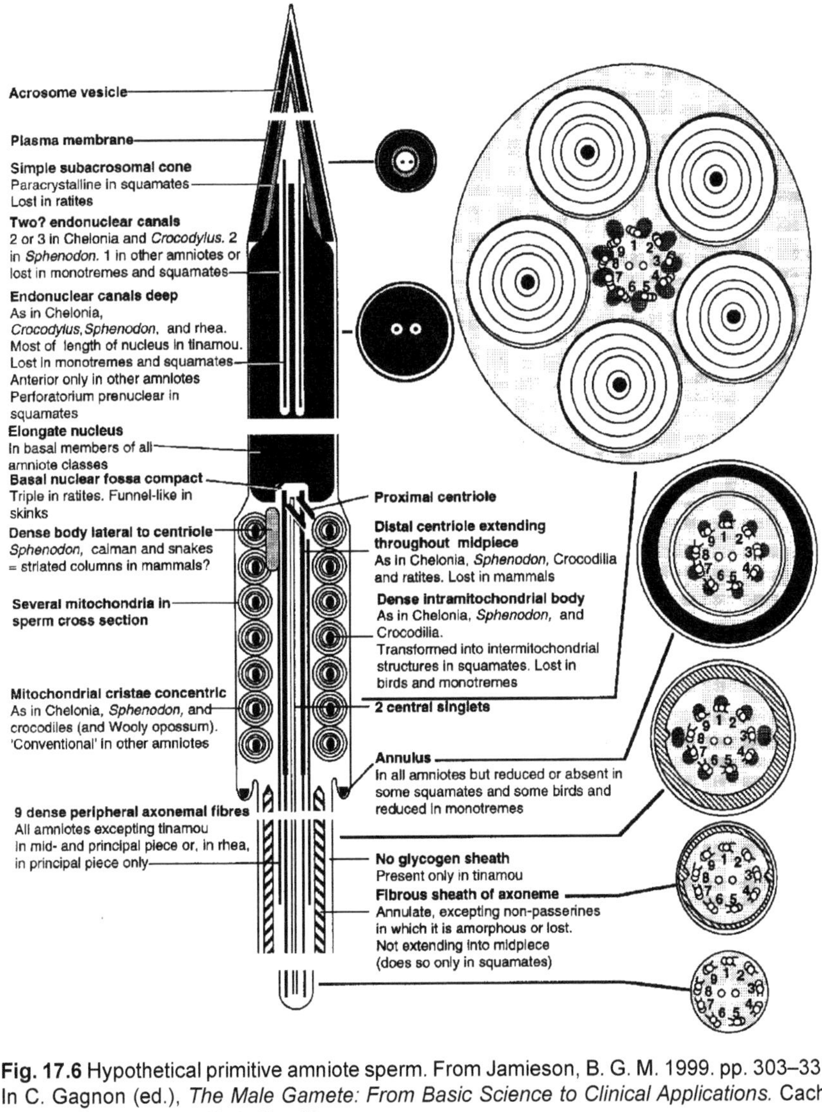

Fig. 17.6 Hypothetical primitive amniote sperm. From Jamieson, B. G. M. 1999. pp. 303–331. In C. Gagnon (ed.), *The Male Gamete: From Basic Science to Clinical Applications*. Cache River Press, Vienna (USA). Fig. 10.

space within the acrosomal complex contains two or three axial rods (putative perforatoria) or, less likely, only one rod. These penetrate the nucleus deeply, almost to its base, in endonuclear canals. Plesiomorphically, the nucleus is elongated and cylindrical in amniotes from Chelonia through *Sphenodon*, crocodiles, squamates, birds, monotremes and, in therian mammals, the Pangolin, *Manis pentadactyla*, alone (Leung and Cummins

1988), as in lissamphibians. At the base of the nucleus there is a compact fossa (implantation fossa). Associated with this are two triplet centrioles. The distal centriole forms the basal body of the flagellar axoneme. Whether the presence of an annulus is plesiomorphic or apomorphic is debatable. The terminal portion of the 9+2 axoneme forms a short endpiece distinguished from the principal piece by the absence of the fibrous sheath.

Derived features of amniote spermatozoa. Features of the hypothetical amniote sperm (Fig. 17.6) that are derived relative to other tetrapods and are seen in turtles (Fig. 17.5A), the Tuatara (Fig. 17.5B), crocodiles (Fig. 17.5C), and ratites (Fig. 17.5D) are as follows. The distal centriole is extremely elongated and extends the entire length of the long midpiece (the latter defined by its mitochondria), an apparent basal synapomorphy of amniotes. In turtles, Tuatara (Healy and Jamieson 1992, 1994; Jamieson 1995a), *Caiman crocodilus* (Saita *et al.* 1987) and *Crocodylus johnstoni* (Jamieson 1995a; Jamieson *et al.* 1997), the mitochondria have concentric cristae, or these surround tubular cristae in *Alligator missipiensis* (Gribbins *et al.* 2010). Concentric cristae are reported elsewhere in amniotes in the sperm of some marsupials, as noted above. The marsupial condition may be homoplasic but it was considered possible (Jamieson 1999) that it is evidence of an ancient synapsid and, therefore, mammalian link. The cristae also tend to a circular arrangement in monotremes (Bedford and Rifkin 1979; Carrick and Hughes 1982). The mitochondrial cristae in these three "reptilian" taxa usually surround a large central dense body. In all other amniotes studied, the cristae have a "conventional" appearance, being linear or curved, as in Lissamphibia, but never concentric, and do not surround a dense body. In spermatids of *Sphenodon* (Healy and Jamieson 1992; Jamieson and Healy 1992), the cristae have the linear appearance usual for metazoan sperm and the concentric arrangement is a late development. "Reversion" of concentric cristae to the linear condition seen in other amniotes would need only suppression of this final transformation which occurs during spermiogenesis (Jamieson and Healy 1992). The spermatozoon of *Crocodylus johnstoni* and *Alligator mississipiensis* is apomorphic relative to Chelonia and *Sphenodon* in reduction of concentric mitochondrial cristae but is less similar to that of ratites than is that of *Caiman crocodilus*, differing from ratites in the longer, multiplied perforatoria. The compact dense sheath around the central singlets of the distal centriole is a possible crocodilian autapomorphy (Jamieson *et al.* 1997) but is more likely a synapomorphy with ratites as exemplified by the centriolar rod in Ostrich described by Soley (1999).

A dense ring, the annulus, at the posterior end of the midpiece is a feature of many metazoan sperm. It is clearly plesiomorphic for amniotes, occurring in all classes (Jamieson and Healy 1992) but its absence in Dipnoi questionably indicates apomorphic re-acquisition in tetrapods. It is well developed in Chelonia, *Sphenodon* (Healy and Jamieson 1992, 1994, 1995a), *Caiman crocodilus* (Saita *et al.* 1987), *Alligator mississipiensis* (Phillips and Asa 1993; Gribbins *et al.* 2010) and *Crocodylus johnstoni* (Jamieson *et al.* 1991).

A dense fibrous sheath must, clearly, have developed as an annulated structure in the earliest amniotes, as it is present in all amniote classes. With the exception of squamates, it commences immediately behind the midpiece, as in *Sphenodon* (testicular sperm), Chelonia, Crocodylia and ratites.

The endonuclear canals and putative perforatoria extend deeply into the nucleus in the turtle *Emydura*. In *Sphenodon* they have been traced for about 2.5 μm and extend well behind the nuclear rostrum. In *Crocodylus johnstoni*, where there are one to three canals, they also extend behind the rostrum, having been traced for at least 5 μm, about half the length of the nucleus (Jamieson *et al.* 1997). However, in *Caiman crocodilus* (Saita *et al.* 1987) and *Alligator mississipiensis* (Gribbins *et al.* 2010) there is only one endonuclear canal which is said to be limited to the nuclear rostrum. The ratites Ostrich (Soley and Roberts 1994; Soley 1989, 1993, 1994, 1996, 1999) and Rhea (Phillips 1989) also have a single canal but this is lost in the Emu, *Dromaius novaehollandiae* (Baccetti *et al.* 1991). For a detailed review of palaeognath birds (Struthioniformes and Tinamiformes) see Jamieson (2007).

Spermatozoa of the Squamata. It is evident from Fig. 17.5E that the squamate spermatozoon differs greatly from the basal amniote ultrastructure exemplified by *Sphenodon, Emydura, Crocodylus* and Ostrich. A major distinction is that the perforatorial rod in the Squamata is wholly prenuclear (commonly with a basal plate) and endonuclear canals are absent (Jamieson 1995b; Jamieson *et al.* 1996; Oliver *et al.* 1996). The anterior restriction and reduction to a single perforatorium are clearly apomorphic relative to the endonuclear condition in Chelonia, Sphenodontida and Crocodylia. Presence of a well-developed epinuclear electron-lucent region is a squamate autapomorphy (Jamieson 1995a,b; Jamieson *et al.* 1996; Oliver *et al.* 1996). The intermitochondrial rings or dense bodies of squamate sperm are regarded as derivations of the intramitochondrial dense bodies (Carcupino *et al.* 1989; Healy and Jamieson 1992; Jamieson and Healy 1992) seen in basal amniotes (chelonians and sphenodontids). Origin of intermitochondrial material from mitochondria has been confirmed ontogenetically in the sperm of some squamates (Oliver *et al.* 1996). In addition, extramitochondrial dense bodies are almost limited to squamates though are seen, poorly developed, in the doves *Geopelia striata* (Jamieson 1995a) and *Ocyphaps lophotes* (Jamieson 1999) in which, although appearing homoplasic, they may well indicate persistence of a genetic basis laid down in early amniotes. In squamates alone, in the Amniota, the fibrous sheath extends anteriorly into the midpiece, a striking squamate autapomorphy (Healy and Jamieson 1992;

Jamieson and Healy 1992; Jamieson and Scheltinga 1993). In the Squamata, the subacrosomal cone has a paracrystalline substructure (Butler and Gabri 1984; Carcupino *et al.* 1989; Furieri 1970), which constitutes a basal synapomorphy of the Squamata (Jamieson and Healy 1992). Shortening of the distal centriole from the elongated basal amniote character is also a basal apomorphy of squamates (Jamieson 1995a, 1999).

The many internal apomorphic changes within the Squamata are discussed in some detail by Jamieson (1995b; see also Gribbins and Rheubert 2011) and consideration of these, and the various sub-groups, is beyond the scope of this chapter. However, as snake sperm have been chosen to exemplify the Squamata (Fig. 17.5E) it may be noted that they are characterized, apomorphically, by multilaminar membranes in place of the normal plasma membrane of the midpiece and axoneme (Jamieson and Koehler 1994; Oliver *et al.* 1996). Snake sperm are unique in the Squamata in the immense elongation of the midpiece (Jamieson 1995a,b) and they show reduction of the epinuclear electron-lucent region and reduction or loss of the perforatorial base plate usual in squamates and greater development of extracellular tubules than is known in any other squamate (Jamieson 1995a,b, 1999; Gribbins and Rheubert 2011).

***Sphenodon* and the origin of squamates.** The cladistic analysis of spermatozoal ultrastructure in *Sphenodon*, and some other amniotes (Jamieson and Healy 1992) (Fig. 17.7), unequivocally confirmed the exceedingly primitive status of *Sphenodon*. Squamates formed the sister-group of a bird + monotreme clade while the three sister-groups successively below the bird + monotreme + squamate assemblage were the caiman, the tuatara and the outgroup (turtles). The monotreme + bird couplet, supported the concept of the Haemothermia but could only be regarded heuristically. It, nevertheless, indicated the very distinctive form of monotreme sperm, with their elongate nuclei, compared with those of other mammals excepting Pangolin in which the nucleus is also extremely elongated (Leung and Cummins 1988). All spermiocladistic analyses made, and a separate consideration of apomorphies, suggested that *Sphenodon* is spermatologically the most primitive amniote, excepting the Chelonia. *Sphenodon* is derived (apomorphic) for the amniotes (Fig. 17.7) in only two of the 16 spermatozoal characters considered in the analysis, namely presence of a dense body lateral to the distal centriole and putative increase in the number of mitochondria in transverse section to nine. A close, sister-group relationship of *Sphenodon* with squamates was not endorsed.

In reconsidering the analysis it must be stressed that the turtle-tuatara-caiman assemblage was paraphyletic and that although it strongly indicated the basal position of these taxa it did not imply monophyly of the assemblage. If a sister-group relationship of the Crocodylia with Aves were

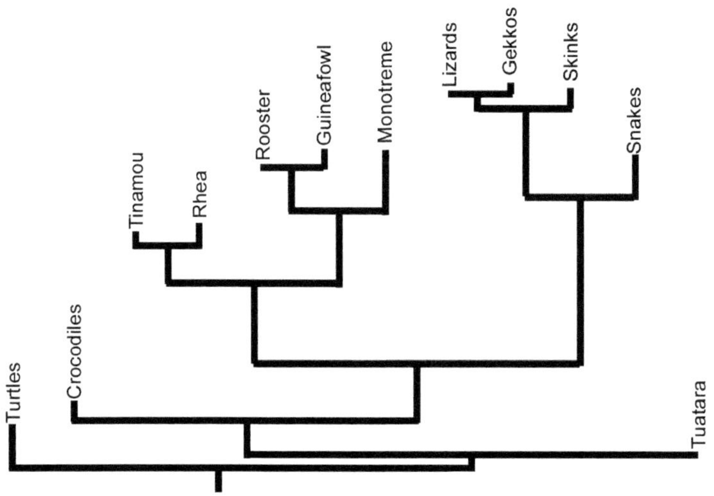

Fig. 17.7 Most parsimonious tree, with no phylogenetic constraints, for *Sphenodon* and other amniotes, based on spermatozoal ultrastructure alone. From Jamieson, B. G. M. and Healy, J. M. 1992. Philosophical Transactions of the Royal Society of London B Biological Sciences 335: 207–219. Fig. 10.

accepted (computed as less parimonious), as is embraced in the concept of the Archosauria, *Sphenodon* would appear basal to the Squamata (see discussion of Molecular Evidence, below). In a further (Macclade) analysis for the present chapter, if the tree in Fig. 17.7 is modified to give monotremes the most basal position, the resultant tree is considerably less parsimonious and the same length is obtained if turtle is moved to below the crocodile through squamate + aves clade. On the other hand, placing tuatara below turtle in Fig. 17.7 results in no change in parsimony relative to the latter.

The changes necessary to derive the squamate spermatozoon from the basal amniote spermatozoon of an ancestor shared with *Sphenodon* would be very profound, as evidenced by the above brief account for squamates, but the features of squamate sperm cannot have originated *de novo*. A possible clue to an origin of the squamate sperm in common with *Sphenodon* might be the observation by Healy and Jamieson (1994; see Fig. 17.2I) that in epididymal *Sphenodon* sperm the "anterior extremity of the fibrous sheath projects just inside the annulus versus annulus and fibrous sheath well separated in testicular sperm". Deep penetration of the sheath into the midpiece is a major autapomorphy of squamates.

Molecular evidence. Molecular analyses have provided conflicting results as to the relationships of reptiles and of Sphenodontida within these. For a detailed discussion, see Wiens and Lambert (Chapter 2, this volume). Lyson *et al.* (2011) stated that an analysis of microRNAs unambiguously supported a turtle + lepidosaur group. Phylogenetic analysis of six nuclear protein-coding loci (where, we may note, alignment is less problematical than for non-protein-coding sequences) suggested clustering of the tuatara with archosaurs (crocodilians and birds) or with turtles, rather than with squamates (Hedges and Poling 1999). In contrast, from other molecular analyses, strong evidence was presented that turtles evolved from a common ancestor of birds and crocodilians (Rest *et al.* 2003; Iwabe *et al.* 2005; Hugall *et al.* 2007; Shen *et al.* 2011; Crawford 2012). In those analyses which included *Sphenodon*, it was found to be the sister of squamates (Rest *et al.* 2003; Hugall *et al.* 2007; Crawford *et al.* 2012) (Fig. 17.8). The study of Crawford *et al.* (2012) was at the genomic level. Hugall *et al.* (2007) considered *Sphenodon* to be a remarkable phylogenetic relic, being the sole survivor of a lineage more than a quarter of a billion years old.

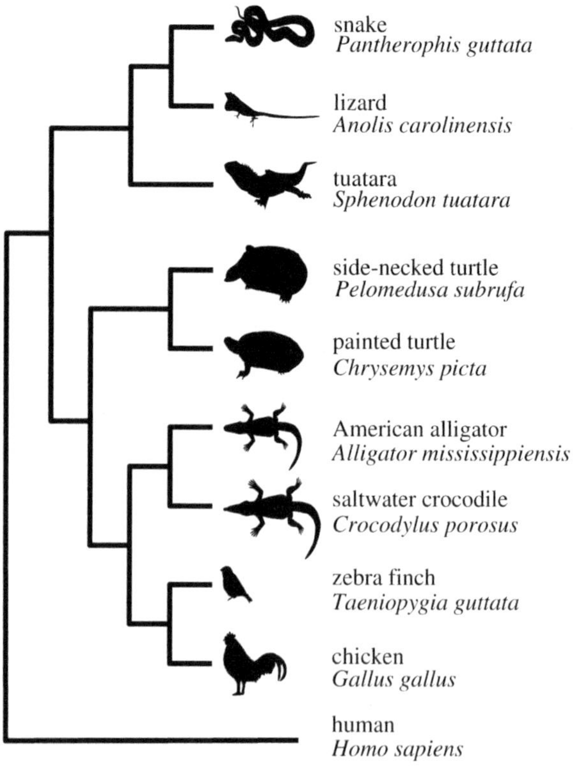

Fig. 17.8 Reptilian phylogeny estimated from 1145 ultra-conserved loci using Bayesian analysis of concatenated date and species-tree methods, yielding identical topologies. After Crawford *et al.* 2012. Biology Letters 8: 783–786. Fig. 2.

More recently, the epoch making paper of Pyron *et al.* (2013) examined the molecular phylogeny of no less than 4161 species of snakes and lizards. *Sphenodon* was placed outside the Squamata, as its sister group. No turtles, crocodiles or birds were included in the analysis and therefore an even more basal position of *Sphenodon* was not tested.

17.6 CHAPTER SUMMARY

Using transmission electron microscopy (TEM) the events of spermatogenesis are reviewed for Tuatara, *Sphenodon punctatus punctatus* (Gray), a representative of the reptilian order Sphenodontida. Secondary spermatocytes contain two greatly elongate (8.0 µm), rod-shaped centrioles which lie parallel to one another and are each associated with a small deposit of dense material and a short centriole. Spermatids contain only one rod-shaped centriole (associated with a short centriole) which gives rise to the flagellar axoneme thereby becoming the distal centriole. Four stages of spermatid development can be distinguished: (i) the early stage (nucleus round; nuclear contents granular with a thin, condensed periphery; mitochondria scattered; acrosomal vesicle spheroidal, slightly depressed onto nuclear surface); (ii) the middle stage (nucleus pyriform with two endonuclear canals are formed; nuclear contents fibro-granular with thick periphery: mitochondria chiefly posterior; acrosomal vesicle flattened; centriolar complex attached to nucleus); (iii) the advanced stage (nucleus elongate and rod shaped; nuclear contents coarsely granular; mitochondria (containing linear cristae) clustered around the distal centriole; acrosomal vesicle conical; centriolar complex attached to posterior fossa of nucleus); (iv) the late stage (nucleus very elongate and associated with a longitudinally arranged microtubular sheath; nuclear contents very condensed midpiece fully formed and featuring mitochondria with concentric cristae and a dense intramitochondrial body; centrioles associated with dense, lateral body). Testicular sperm have a conical acrosome complex consisting of an acrosome vesicle (length 4 µm) and subacrosomal cone, an elongate (length 54–56 µm) helical nucleus, a midpiece (length 8 µm, featuring spheroidal mitochondria containing concentric cristae and a dense body), an annulus, an elongate principal piece (length 74–78 µm, featuring a dense, fibrous sheath) and a short endpiece (length 2–4 µm). Epididymal sperm differ from those in the testis by having a more developed lateral body in the midpiece and a sheath of flocculent material surrounding the fibrous sheath in the principal piece. The relatively large number of epididymal sperm still associated with a cytoplasmic droplet suggests that sperm spend a significant period maturing within the epididymis. The features of spermatogenesis and mature sperm suggest that the Sphenodontida are primitive amniotes; only chelonians have fewer spermatozoal apomorphies while the crocodilians are little more apomorphic. Spermatologically, origin of squamate sperm from the basal amniote type seen in *Sphenodon*, though previously rejected, is possibly indicated by partial extension of the fibrous sheath into the midpiece as a precursor to the deep penetration of the sheath, which is a major autapomorphy of squamates.

17.7 ACKNOWLEDGMENTS

I continue to thank Dr. Alison Cree (Department of Zoology, University of Otago, New Zealand) for her great generosity in supplying formalin-fixed samples (portions of testis and epididymis) from *Sphenodon* for my previous studies with John Healy which are the subject of this review. I am grateful to Justin Rheubert for running the additional, Macclade, analysis referred to above. David Sever, Kevin Gribbins and Justin Rheubert are thanked for their careful scrutiny of the manuscript and their constructive suggestions which have improved the quality of this chapter.

17.8 LITERATURE CITED

Al-Hajj, H., Janakat, S. and Mahmoud, F. 1987. Electron microscopic study of sperm head differentiation in thc lizard *Agama stellio*. Canadian Journal of Zoology 65: 2959–2968.

Asa, C., Phillips, D. M. and Stover, J. 1986. Ultrastructure of spermatozoa of the Crested Tinamou. Journal of Ultrastructure and Molecular Structure Research 94: 170–175.

Asa, C. S. and Phillips, D. M. 1987. Ultrastructure of avian spermatozoa: a short review. pp. 365–373. In H. Mohri (ed.), *New horizons in sperm cell research*. Japan Scientific Society Press, Gordon and Breach Scientific Publications, Tokyo/New York.

Baccetti, B., Burrini, A. G. and Falchetti, E. 1991. Spermatozoa and relationships in paleognath birds. Biology of the Cell (Paris) 71: 209–216.

Carrrick, F. N. and Hughes, R. L. 1982. Aspects of the structure and development of monotreme spermatozoa and their relevance to evolution of mammalian sperm morphology. Cell and Tissue Research 222: 127–141.

Cherr, G. N. and Clark, W. H., Jr. 1984. An acrosome reaction in sperm from the white sturgeon *Acipensertransmontanus*. Journal of Experimental Zoology 232: 129–139.

Clark, A. W. 1967. Some aspects of spermiogenesis in a lizard. American Journal of Anatomy 121: 369–399.

Crawford, N. G., Faircloth, B. C., Mccormack, J. E., Brumfield, R. T., Winker, K. and Glen, T. C. 2012. More than 1000 ultraconserved elements provide evidence that turtles are the sister group of archosaurs. Biology Letters 8: 783–786.

Cree, A., Cockrem, J. F. and Guillette, L. J., Jr. 1992. Reproductive cycles of male and female tuatara (*Sphenodon punctatus*) on Stephens Island, New Zealand. Journal of Zoology 226: 199–217.

Cruz-Hofling, da M. A. and Cruz-Landim, da C. 1978. The fine structure of nuclei during spermiogenesis in the lizard *Tropidurus torquatus* (Lacertilia). Cytologia 43: 61–68.

Dehlawi, G.Y., Ismail, M. F., Hamdi, S.A. and Jamjoon, M. B. 1992. Ultrastructure of spermiogenesis of Saudian rcptiles 6. 'The sperm head differentiation in *Agama adramitana*. Archives of Andrology 28: 223–234.
du Plessis, L. and Soley, J. T. 2014. A re-evaluation of sperm ultrastructure in the emu, *Dromaius novaehollandiae*. Theriogenology 81(8): 1073–84. doi: 10.1016/j.theriogenology.2014.01.034. Epub 2014 Jan 31.
Evans, S. E. 1984. The classification of the Lepidosauria. Zooloogical Journal of the Linnean Society 82: 87–100.
Evans, S. E. 1988. The early history and relationships of the Diapsida. pp. 221–260. In M. J. Benton (ed.), *The phylogeny and classificaton of the tetrapods, Vol. 1.* Clarendon Press, Oxford.
Fraser, N. C. 1985. New Triassic Sphenodontids from Southwest England UK and a Review of Their Classification. Palaeontology (Oxford) 29: 165–186.
Gauthier, J. A., Estes, R. and De Queiroz, K. 1988. A phylogenetic analysis of Lepidosauromorpha. pp. 15–98. In R. Estes and G. Pregill (ed.), *Phylogenetic relationships of the lizard. families.* Clarendon Press, Oxford.
Gauthier, J. A., Kluge, A. J. and Rowe, T. 1988. The early evolution of the Amniota. pp. 103–155. In M. J. Benton (ed.), *Phylogenetic relationships of the lizard. families.* Clarendon Press, Oxford.
Gribbins, K. M. 2011. Reptilian spermatogensis. A histological and ultrastructural perspective. Spermatogenesis 1, 3: 1–20.
Gribbins, K. M. and Rheubert, J. L. 2011. The ophidian testis, spermatogenesis, and mature spermatozoa. pp. 183–264. In R. D. Aldridge and D. M. Sever (eds.), *Reproductive Biology and Phylogeny of Snakes, vol. 9.* Science Publishers, Enfield, New Hampshire and CRC Press, Boca Raton, Florida, and Science Publishers, Enfield, New Hampshire.
Gribbins, K. M., Siegel, D. S., Anzalone, M. L., Jackson, D. P., Venable, K. J., Rheubert, J. L. and Elsey, R. M. 2010. Ultrastructure of spermiogenesis in the American Alligator, *Alligator mississippiensis* (Reptilia, Crocodylia, Alligatoridae). Journal of Morphology 271: 1260–1271.
Healy, J. M. and Jamieson, B. G. M. 1992. Ultrastructure of the Spermatozoon of the Tuatara *Sphenodon punctatus* and its relevance to the relationships of the Sphenodontida. Philosophical Transactions of the Royal Society of London B Biological Sciences 335: 193–205.
Healy, J. M. and Jamieson, B. G. M. 1994. The ultrastructure of spermatogenesis and epididymal spermatozoa of the tuatara *Sphenodon punctatus* (Sphenodontida, Amniota). Philosophical Transactions of the Royal Society of London B Biological Sciences 344: 187–199.
Hedges, S. B. and Poling, L. L. 1999. A molecular phylogeny of reptiles. Science 283: 998–1001.
Hess, R. A., Thurston, R. J. and Gist, D. H. 1991. The Anatomical Record 229(4): 473–81.
Hogben, L. 1921. A preliminary account of the spermatogenesis of *Sphenodon*. Transactions of the Microscopical Society of London 341–352.
Hugall, A. F., Foster, R. and Lee, M. S. Y. 2007. Calibration choice, rate smoothing, and the pattern of tetrapod diversification according to the long nuclear gene RAG-1. Systematic Biology 56: 543–563.
Iwabe, N., Hara, Y., Kumazawa, Y., Shibamoto, K., Saito, Y., Miyata, T. and Katoh, K. 2005. Sister group relationship of turtles to the bird-crocodilian clade revealed by nuclear DNA-coded proteins. Molecular Biology and Evolution 22: 810–813.
Jamieson, B. G. M. 1991. *Fish evolution and systematics; evidence from spermatozoa.* Cambridge University Press, pp. 319.
Jamieson, B. G. M. 1999. Spermatozoal phylogeny of the Vertebrata. pp. 303–331. In C. Gagnon (ed.), *The Male Gamete: From Basic Science to Clinical Applications.* Cache River Press, Vienna (USA).
Jamieson, B. G. M. 2007. Avian spermatozoa: structure and phylogeny. In B. G. M. Jamieson (ed.), *Reproductive Biology and Phylogeny of Birds, vol. 6A.* Science Publishers, Enfield, New Hampshire, U.S.A. Jersey, Plymouth, U.K.
Jamieson, B. G. M. 2009. *Reproductive biology and phylogeny of Fishes (Agnathans and bony fishes). Volume 8A.* Science Publishers, Enfield, New Hampshire, USA, pp. 788.

Jamieson, B. G. M. and Healy, J. M. 1992. The phylogenetic position of the tuatara, *Sphenodon* (Sphenodontida, Amniota) as indicated by cladistic analysis of the ultrastructure of spermatozoa. Philosophical Transactions of the Royal Society of London B Biological Sciences 335: 207–219.

Jamieson, B. G. M. and Koehler, L. 1994. The ultrastructure of the spermatozoon of the northern water snake, *Nerodia sipedon* (Colubridae, Serpentes), with phylogenetic considerations. Canadian Journal of Zoology 72: 1648–1652.

Jamieson, B. G. M., Lee, S. Y. and Long, K. 1993. Ultrastructure of the spermatozoon of the internally fertilizing frog *Ascaphus truei* (Ascaphidae: Anura: Amphibia) with phylogenetic considerations. Herpetologica 49: 52–65.

Jamieson, B. G. M. and Scheltinga, D. M. 1993. The ultrastructure of spermatozoa of *Nangura spinosa* (Scincidae, Reptilia). Memoirs of the Queensland Museum 34: 169–179.

Jamieson, B. G. M., Scheltinga, D. M. and Tucker, A. D. 1997. The ultrastructure of spermatozoa of the Australian freshwater crocodile, *Crocodylus johnstoni* Krefft 1873 (Crocodylidae, Reptilia). Journal of Submicroscopic Cytology and Pathology 29: 265–274.

Jespersen, A. 1971. Fine structure of the spermatozoon of the Australian Lungfish *Neoceratodus forsteri*. Journal of Ultrastructure Research 37: 178–185.

Johnston, S. D., Daddow, L., Carrick, F. N., Jamieson, B. G. M. and Smith, L. 2003. Cauda epididymidal spermatozoa of the rufous hare wallaby, *Lagorchestes hirsutus* (Metatheria, Mammalia) imaged by electron and confocal microscopy. Acta Zoologica 84: 139–143.

Jones, M. E. H. and Lappin, A. K. 2009. Bite-force performance of the last rhynchocephalian (Lepidosauria: Sphenodon). Journal of the Royal Society of New Zealand 39: 71–83.

Leung, L. K. P. and Cummins, J. M. 1988. Morphology of immature spermatozoa of the Chinese Pangolin (*Manuspentadactyla*: Pholidota). Proceedings of the Australian Society of Reproductive Biology (Newcastle, Australia, 1988): 94.

Lyson, T. R., Sperling, E. A., Gauthier, J. A., Heimberg, A. M. and Peterson, K. J. 2011. MicroRNAs support a Testudines-Lepidosaur clade. Integrative and Comparative Biology 51: E222.

Mattei, X., Siau, Y. and Seret, B. 1988. Ultrastructural Study of Coelacanth spermatozoon *Latimeria chalumnae*. Journal of Ultrastructure and Molecular Structure Research 101: 243–251.

Phillips, D. M. and Asa, C. S. 1989. Development of spermatozoa in the Rhea. Anatomical Record 223: 276–282.

Pyron, R. A., Burbrink, F. T. and Wiens, J. J. 2013. A phylogeny and revised classification of Squamata, including 4161 species of lizards and snakes. BMC Evolutionary Biology 13: 93. doi:10.1186/1471-2148-13-93.

Psenicka, M., Kaspar, V., Alavi, S. M. H., Rodina, M., Gela, D., Li, P., Borishpolets, S., Cosson, J., Linhart, O. and Ciereszko, A. 2011. Potential role of the acrosome of sturgeon spermatozoa in the fertilization process. Journal of Applied Ichthyology 27: 678–682.

Rest, J. S., Ast, J. C., Austin, C. C., Waddell, P. J., Tibbetts, E. A., Hay, J. M. and Mindell, D. P. 2003. Molecular systematics of primary reptilian lineages and the tuatara mitochondrial genome. Molecular Phylogenetics and Evolution 29: 289–297.

Rheubert, J. L., Cree, A., Downes, M. and Sever, D. M. In press. Reproductive morphology of the male Tuatara, *Sphenodon punctatus*. Acta Zoologica (Stockholm).

Risley, P. 1936. Centrioles in germ cells of turtles including observation on the "manchette" in spermatogenesis. Zeitschrift fwissenschaftliche Zoologie A 148(1): 133–158.

Saint Girons, H. and Newman, D. G. 1987. The reproductive cycle of thc male tuatara, *Sphenodon punctatus*, on Stephens Island, Ncw Zealand. New Zealand Journal of Zoology 14: 231–237.

Saita, A., Comazzi, M. and Perrotta, E. 1987. Electron microscope study of spermiogenesis in *Caiman crocodilus* L. Bollettino di Zoologia 54: 307–318.

Schaerlaeken, V., Herrel, A., Aerts, P. and Ross, C. F. 2008. The functional significance of the lower temporal bar in *Sphenodon punctatus*. Journal of Experimental Biology 211: 3908–3914.

Shen, X. -X., Liang, D., Wen, J. -Z. and Zhang, P. 2011. Multiple genome alignments facilitate development of NPCL markers: a case study of tetrapod phylogeny focusing on the position of turtles. Molecular Biology and Evolution 28: 3237–3252.

Soley, J. T. 1989. Transmission electron microscopy of ostrich (*Struthio camelus*) sperm. Electron Microscopy Society of Southern Africa Proceedings 19: 145–146.

Soley, J. T. 1993. Ultrastructure of ostrich (*Struthio camelus*) spermatozoa: I. Transmission electron microscopy. Onderstepoort Journal of Veterinary Research 60: 119–130.

Soley, J. T. 1994. Ostrich sperm ultrastructure—evidence of a close link between the ratites and tinamous. pp. 5.27–25.28. In N. M. Bradley and J. M. Cummins (eds.), Proceedings of the VIIth International Symposium on Spermatology, Cairns, Australia. Murdoch University, Western Australia.

Soley, J. T. 1996. Differentiation of the acrosomal complex in ostrich (*Struthio camelus*) spermatids. Journal of Morphology 227: 101–111.

Soley, J. 1999. Reproduction. pp. 129–158. In D. C. Deeming (ed.), *The Ostrich: Biology, Production and Health*. CAB International.

Soley, J. T. and Roberts, D. C. 1994. Ultrastructure of ostrich (*Struthio camelus*) spermatozoa: II. Scanning electron microscopy. Onderstepoort Journal of Veterinary Research 61: 239–246.

Sprando, R. I. and Russell, L. D. 1988. Spermiogenesis in the Red-Ear Turtle *Pseudemys scripta* and the Domestic Fowl *Gallus domesticus* a study of cytoplasmic events Including cell volume changes and cytoplasmic elimination. Journal of Morphology 198: 95–118.

Wei, Q., Li, P., Psenicka, M., Alavi, S. M. H., Shen, L., Liu, J., Peknicova, J. and Linhart, O. 2007. Ultrastructure and morphology of spermatozoa in Chinese sturgeon (*Acipenser sinensis* Gray 1835) using scanning and transmission electron microscopy. Theriogenology 67: 1269–1278.

Whiteside, D. I. 1986. The Head Skeleton of the Rhaetian sphenodontid *Diphydontosaurus avonis* new genus new species and the modernizing of a living fossil. Philosophical Transactions of the Royal Society of London B Biological Sciences 312: 379–430.

Yasuzumi, G. and Yasuda, M. 1968. Spermatogenesis in animals as revealed by electron microscopy. XVIII Fin structure of developing spermatids of the Japanese fresh-water turtle with potassium permanganate. Zeitschrift fuer Zellforsschung und Mikrosopische Anatomie 85: 18–33.

Yasuzumi, G., Yasuda, M. and Shirai, T. 1971. Spermatogenesis in animals as revealed by electron microscopy Part 25 Fine structure of spermatids and nutritive cells during spermiogenesis in the Japanese fresh Water turtle. Monitore Zoologico Italiano 5: 117–132.

Index

B

F

I

J

P

R

S

U

V

About the Editors

VOLUME EDITORS

Justin L. Rheubert received his Bachelor's degree at Wittenberg University in 2008 under the direct supervision of Dr. Kevin M. Gribbins. It is during this time that Justin became infatuated with reproductive biology and amphibians and reptiles. Justin continued his education at Southeastern Louisiana University under Dr. David M. Sever. Dr. Sever allowed Justin to explore his interests in reproductive biology and teaching. Upon receiving his masters degree in 2011 Justin moved to St. Louis Missouri where he performed research with Dr. Robert D. Aldridge. In his first 6 years within scientific research Justin has become extremely accomplished coauthoring over 35 research articles, giving over 30 presentations at local, national, and international meetings, and developing an extensive collaborative network with researchers in multiple states within the U.S., Mexico, and New Zealand. Justin is also an associate editor for the Journal of North American Herpetology, a member of the Society for the Study of Amphibians and Reptiles and a member of the American Society of Ichthyologists and Herpetologists. Justin is currently a faculty member at The University of Findlay teaching Anatomy and Physiology and focusing his research on morphological evolution in vertebrate reproductive systems and using his research as an avenue to interest young scientists in the world of biology.

Dr. Dustin S. Siegel is currently an Assistant Professor of Biology in the Department of Biology at Southeast Missouri State University in Cape Girardeau, Missouri. He teaches animal physiology, histology, and a variety of lower level courses. He received a Ph.D. degree from Saint Louis University in 2011 under the direction of Dr. Robert Aldridge. Dustin became interested in a career in academia with a herpetological focus after a vertebrate zoology course with Dr. Lowell Orr at Kent State University, where he completed his B.S. in Zoology in 2005. Dr. David Sever further developed Dustin's interest in "herp" reproductive biology during Dustin's M.S. degree work at Southeastern Louisiana University, which culminated in 2007. Dustin is currently an associate editor for the journals *Copeia* and *Herpetological Conservation and Biology*, a member of the Board of Governors for the American Society of Ichthyologists and Herpetologists, and has not missed a Joint Meeting of Ichthyologists and Herpetologists since 2005. Dustin has coauthored nearly 40 publications on a variety of amphibians and reptiles, and he is currently enamored with the evolution of primary and secondary sexual function of the kidneys in amphibians.

Dr. Stanley E. Trauth is Senior Faculty Member and Professor of Zoology in the Department of Biological Sciences at Arkansas State University in Jonesboro, Arkansas, where he teaches comparative vertebrate anatomy, animal histology, electron microscopy, natural history of vertebrates, and herpetology. He received a Ph.D. degree from Auburn University in 1980 under the guidance of Dr. Robert H. Mount. He is also the curator of the ASU herpetological collection that numbers over 33,000 catalogued specimens, most of which were collected during Trauth's 30-year tenure at ASU. He is past-president of *The Herpetologists' League* (2012–2013) and is co-founder and Special Publications Editor for *Herpetological Conservation and Biology*. He is the senior author of the book, *The Amphibians and Reptiles of Arkansas*, which is the state's only comprehensive guide for these animals. He has authored or coauthored nearly 300 publications over a broad range of topics. As a histo-herpetologist, his primary research interests include squamate reproductive anatomy, sperm morphology in reptiles and amphibians, and Rathke's glands in turtles.

SERIES EDITOR

Dr. Barrie G. M. Jamieson is Emeritus Professor of Zoology in the School of Biological Sciences, University of Queensland. He holds a Ph.D. from the University of Bristol, England, a D. Sc from the University of Queensland, and is a former Visiting Fellow of, and member of the Association of Corpus Christi College, Cambridge. Among other institutions in which he has worked are Makerere College, The British Museum (Natural History), Sydney University, Bedford College, the Smithsonian Institution, Wenner-Gren Institute Stockholm, working with Bjorn Afzelius, and the Muséum National d'Histoire Naturelle Paris, working with Danièle Guinot and Jean-Lou Justine. In 1990 he was awarded the Clarke Medal for Research in Natural Sciences, early recipients of which were Thomas Henry Huxley and Richard Owen. His chief field of research is spermatozoal ultrastructure and its relevance to phylogeny but he is also an authority on taxonomy of earthworms and has published on bioluminescence, trematode taxonomy and life cycles, and DNA-based phylogenetics. He has named some 170 invertebrate species, has published more than 200 scientific papers and is the author, coauthor or editor of twenty one books.

Color Plate Section

Chapter 1

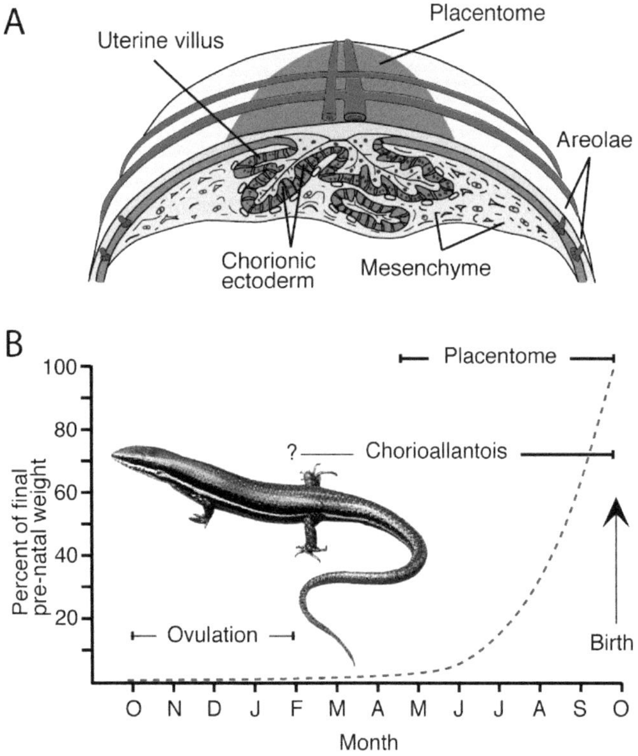

Fig. 1.3 (A) Stylized representation of the structure of the chorioallantoic placenta in *Brasiliscincus* (*Mabuya*) *heathi*. The placenta, which lies above the embryo, consists of hypertrophied uterine (maternal) and chorionic (fetal) tissue forming the placentome, the joint structure for nutrient transfer to the embryo, waste transfer to the female, and gaseous exchange. Transfer and exchange occur in interdigitating chorionic areolae. **(B)** Seasonal growth of implanted embryos (red dashed line). The embryo increases more than 74,000% of its freshly ovulated mass as the result of nutrient uptake from the female. Adapted from Blackburn and Vitt 1992, pp. 150–164 in W. C. Hamlett (ed.). *Reproductive biology of South American vertebrates*. Springer-Verlag: New York, Fig. 11.4. and Vitt and Caldwell 2014, *Herpetology, An Introductory Biology of Amphibians and Reptiles, Fourth Edition*. Academic Press, San Diego, Figs. 5.13 and 5.14.

Chapter 5

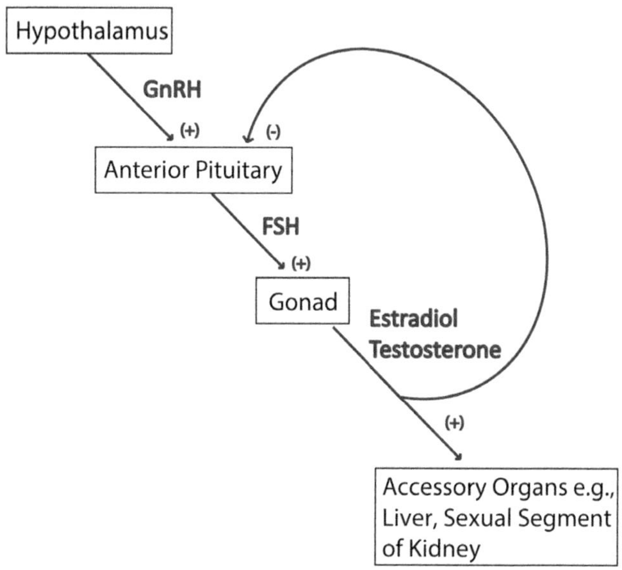

Fig. 5.1 Generalized structure of the vertebrate hypothalamic-pituitary-gonadal axis (HPGA). The HPGA forms the backbone of a heuristic model of the signaling mechanisms that communicate reproductive cues to the brain and result in reproductive decisions. Organs are represented by black boxes, while hormones are blue. Addition signs (+) indicate that a hormone stimulates upregulation of the receiving organ, while subtraction symbols (–) indicate that it stimulates downregulation of the receiving organ. It remains unclear how the HPGA behaves differently in associated or dissociated breeders.

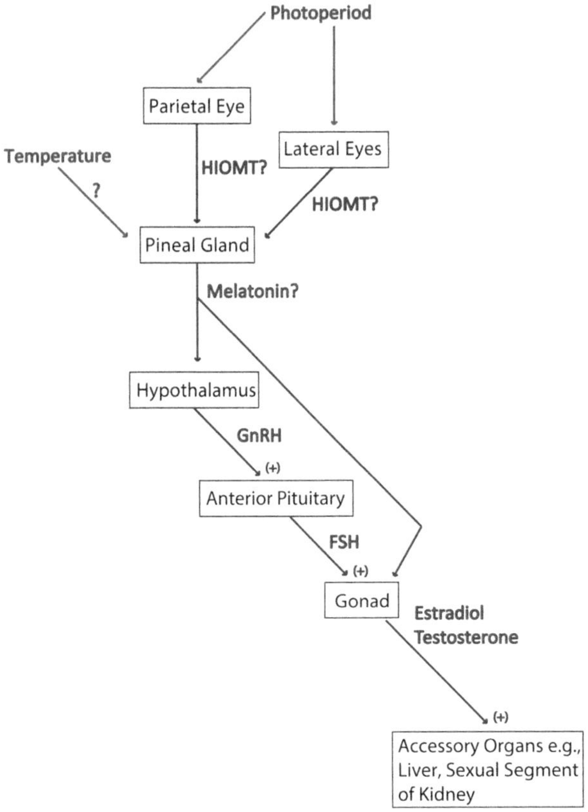

Fig. 5.2 Hypothetical mechanisms squamates might use to detect environmental conduciveness for reproduction are added to the generalized HPGA. Question marks indicate that detection or signaling mechanisms are unknown (temperature), are hypothetical (HIOMT), or are not fully understood (melatonin). Temperature is likely to be more important than photoperiod in stimulating reproduction, but the mechanisms for detecting and communicating temperature to the pineal gland are unknown. HIOMT represents hydroxyindole-O-methyltransferase, but serotonin or norepinephrine could also act as signaling factors between both the parietal and lateral eyes, and the pineal gland. The pineal also directly innervates the pretectal and tegmental areas of the brain, which could provide a mechanism for neuronal signaling to the HPGA. The estradiol/testosterone negative feedback loop to the anterior pituitary is removed for clarity. Organs are represented by black boxes, hormones are blue, and cues are green.

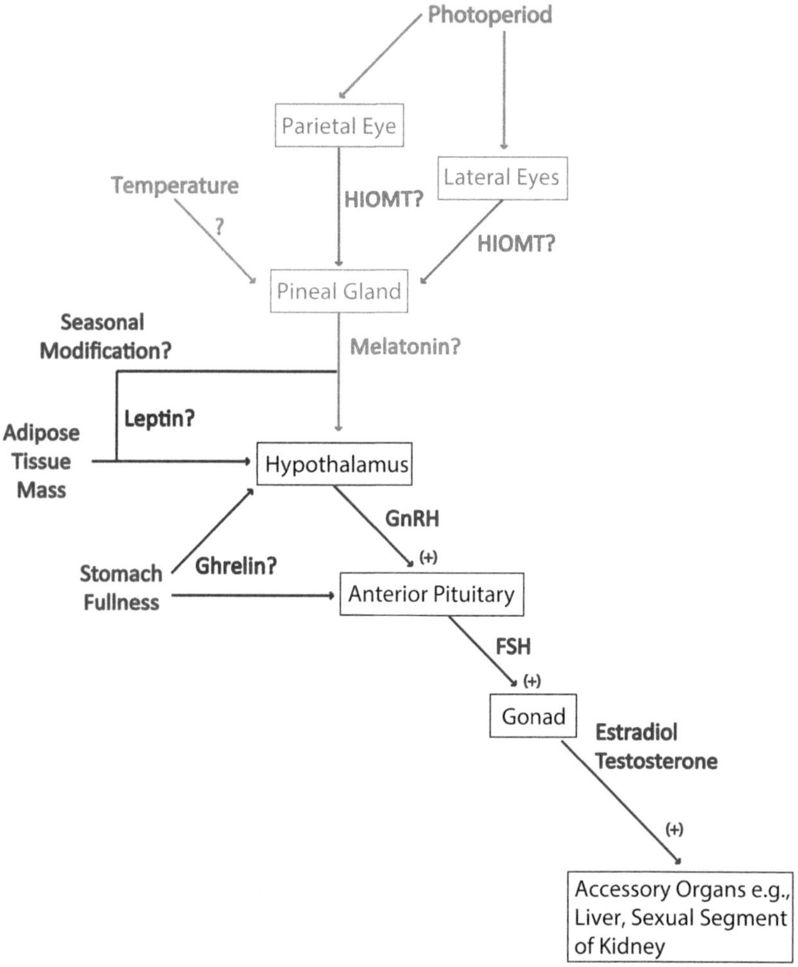

Fig. 5.3 Mechanisms squamates might use to detect resource availability are added to the generalized HPGA. Question marks indicate that leptin and ghrelin signaling mechanisms are hypothetical indicators of resource availability to the HPGA, and that melatonin may seasonally modify the action of leptin. Organs are represented by black boxes, hormones are blue, and cues are green. Mechanisms of detecting environmental conduciveness, introduced in Fig. 2, are obscured to enhance clarity yet emphasize that multiple detection systems might interact to inform the decision to reproduce. The pathway linking melatonin directly to the gonad has been removed for clarity.

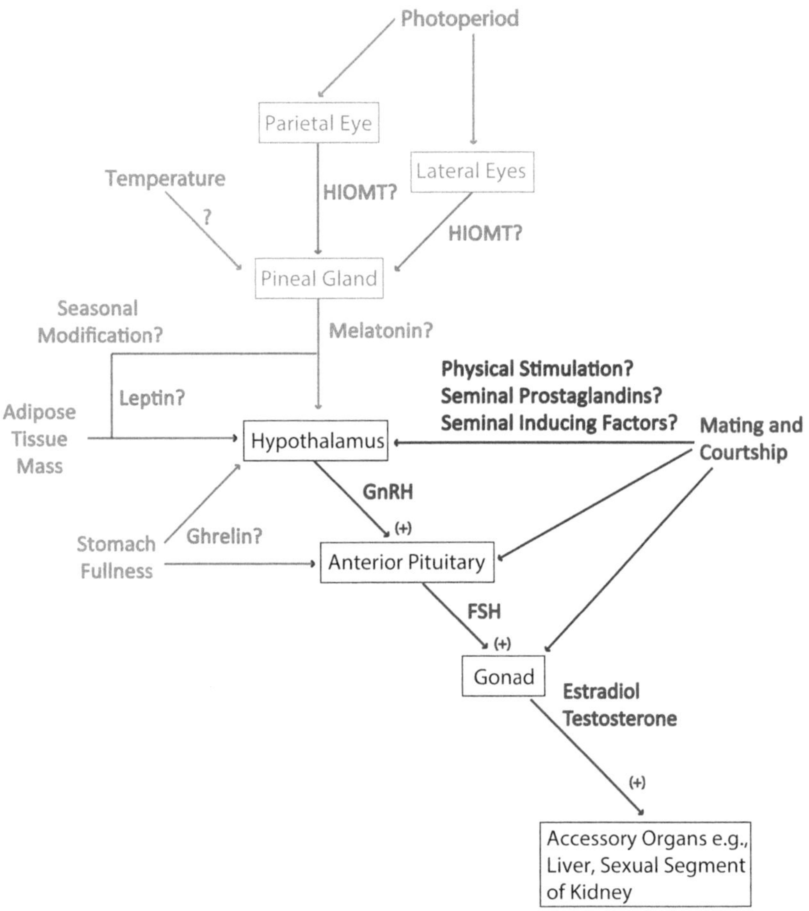

Fig. 5.4 Mechanisms female squamates might use to determine whether fertilization is likely are added to the generalized HPGA. Question marks indicate that all signaling mechanisms that might communicate mating and/or courtship to the HPGA are hypothetical. It is also unclear whether any of these hypothetical mechanisms act directly on the hypothalamus, anterior pituitary, or gonads. Organs are represented by black boxes, hormones and neural signaling are blue, and cues are green. Previously introduced cue detection mechanisms are obscured to enhance clarity yet emphasize that multiple detection systems might interact to inform the decision to reproduce.

Chapter 6

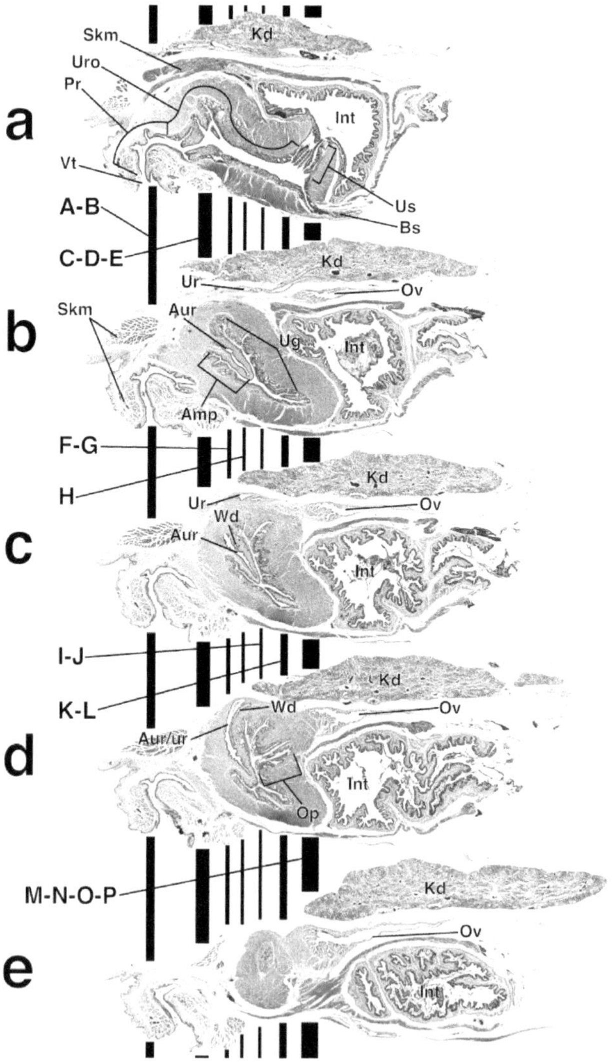

Fig. 6.1 Representative sagittal sections of the cloaca of female *Sceloporus jarrovi*. **a-e.** Sections from the mid-sagittal plane of the cloaca **(a)** to the left periphery **(e)** of the cloaca (hematoxylin and eosin; scale bar [top right horizontal line] = 1,000 μm). Sections are orientated so that their left extreme represents the caudal extreme of the cloacal apparatus and the right extreme represents the cranial extreme of the cloacal apparatus. See text for description of figure. Amp, ampullary papilla; Aur, ampulla ureter; Aur/ur, ampulla ureter/ureter communication; Bs, bladder stalk; Int, intestine; Kd, kidney; Op, oviducal papilla; Ov, oviduct; Pr, proctodaeum; Skm, skeletal muscle; Ug, urodaeal gland; Uro, urodaeum; Us, urodaeal sphincter; Vt, vent; Wd, Wolffian duct.

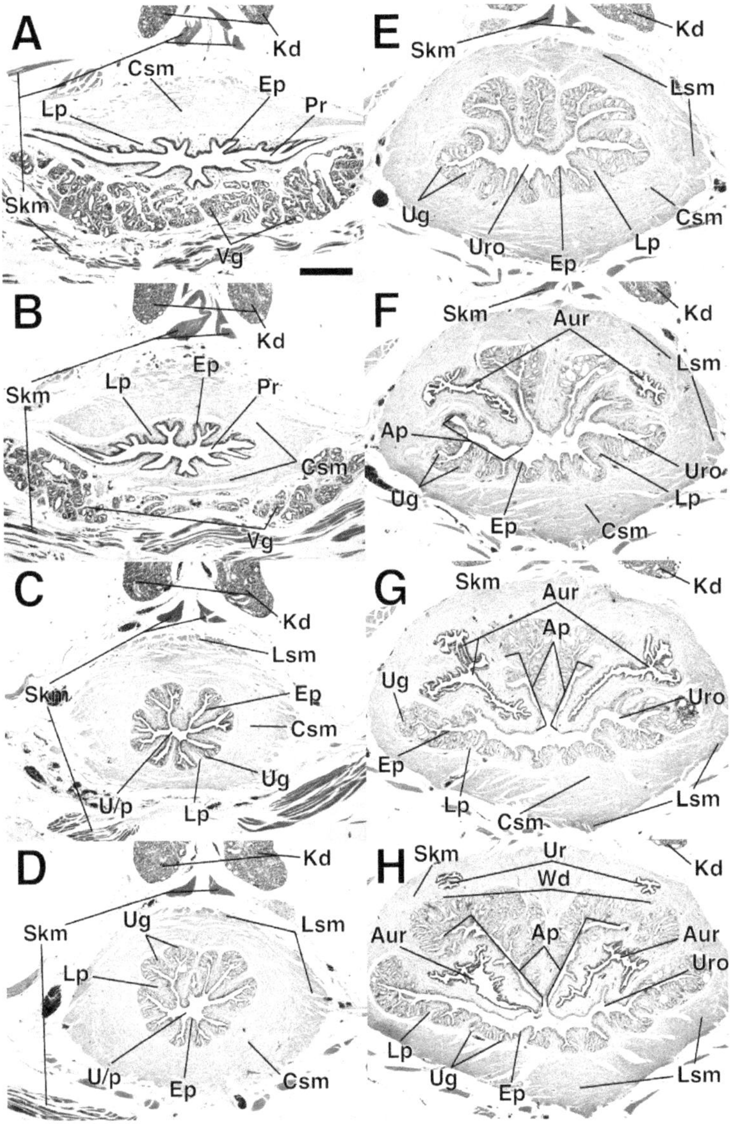

Fig. 6.2 Representative transverse sections from the cloaca of female *Sceloporus jarrovi*. Sections are organized from most caudal **(A)** to most cranial **(H)**. Individual micrograph lettering represents the capital lettering on Fig. 9.1, as demonstrative of the approximate location that each representative transverse section originated (hematoxylin and eosin; scale bar = 500 µm). See text for description of figure. Ap, ampullary papilla; Aur, ampulla ureter; Csm, circular smooth muscle; Ep, epithelium; Kd, kidney; Lp, lamina propria; Lsm, longitudinal smooth muscle; Pr, proctodaeum; Skm, skeletal muscle; Ug, urodaeal gland; U/p, urodaeal/proctodaeal transition; Ur, ureter; Uro, urodaeum; Vg, ventral gland; Wd, Wolffian duct.

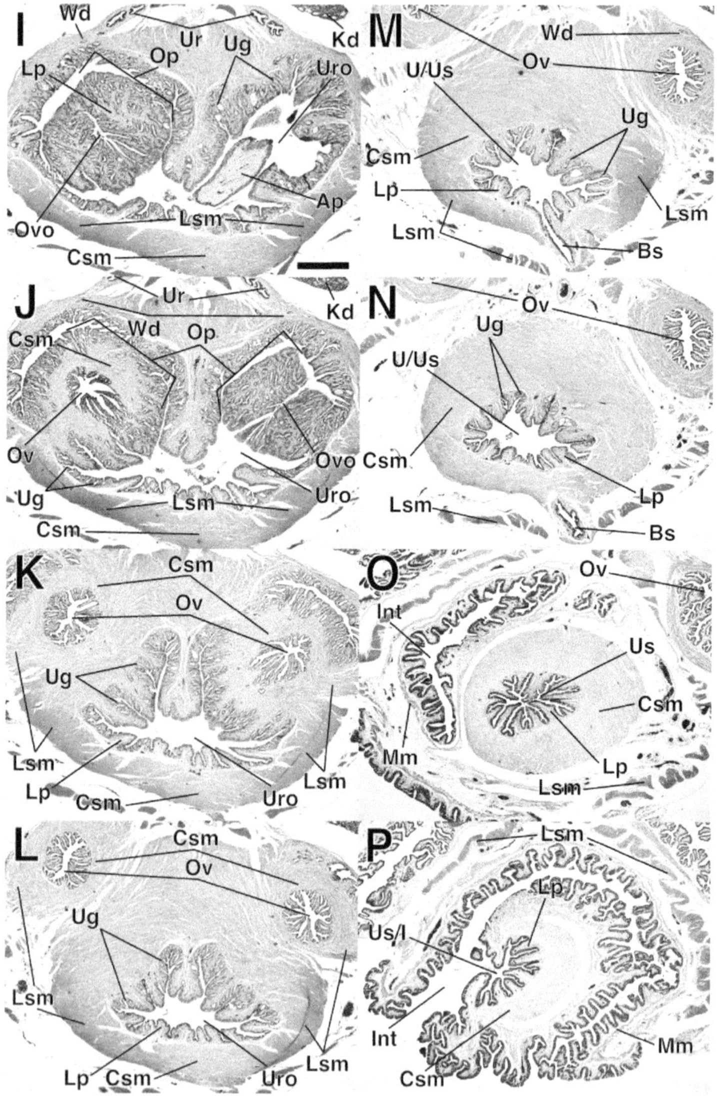

Fig. 6.3 Representative transverse sections from the cloaca of female *Sceloporus jarrovi*. Sections are organized from most caudal **(I)** to most cranial **(P)**. Individual micrograph lettering represents the capital lettering on Fig. 9.1, as demonstrative of the approximate location that each representative transverse section originated (hematoxylin and eosin; scale bar = 500 µm). See text for description of figure. Ap, ampullary papilla; Bs, bladder stalk; Csm, circular smooth muscle; Int, intestine; Lsm, longitudinal smooth muscle; Lp, lamina propria; Mm, muscularis mucosae; Op, oviducal papilla; Ov, oviduct; Ovo, opening of oviduct; Ug, urodaeal gland; Ur, ureter; Uro, urodaeum; U/Us, urodaeal/urodael sphincter transition; Us/I, urodaeal sphincter opening to intestine; Wd, Wolffian duct.

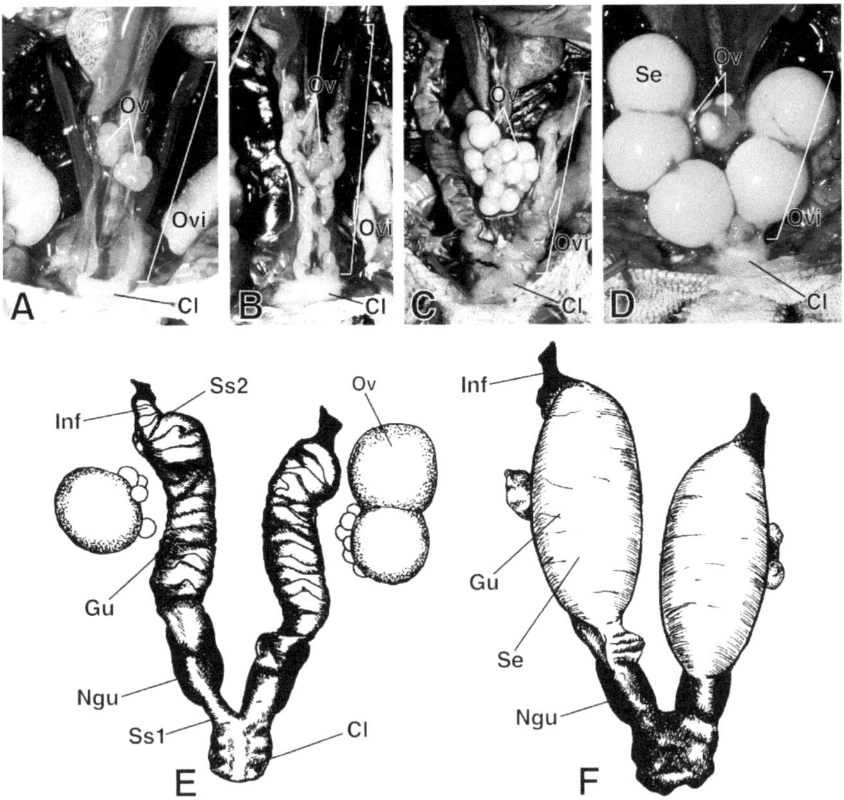

Fig. 6.4 Gross morphology of the oviduct from immaturity (**A**; *Sceloporus virgatus*) to adulthood **(B-F)**. Adult micrographs represent non-reproductive (**B**; *S. virgatus*), vitellogenic (**C** [*S. consobrinus*], **E** [*Scincella lateralis*]), and post-ovulation (**D** [*S. jarrovi*], **F** [*S. lateralis*]) oviducal morphology. Note that Ss2 (cranial sperm storage location) is for diagrammatic purposes only, as *S. lateralis* only stores sperm in the Ss1 (caudal sperm storage location) region. Cl, cloaca; Gu, glandular uterus; Inf, infundibulum; Ngu, non-glandular uterus; Ov, ovary; Ovi, oviduct; Se, eggs shelling in the uterus; Ss1, site of caudal sperm storage location; Ss2, site of cranial sperm storage location.

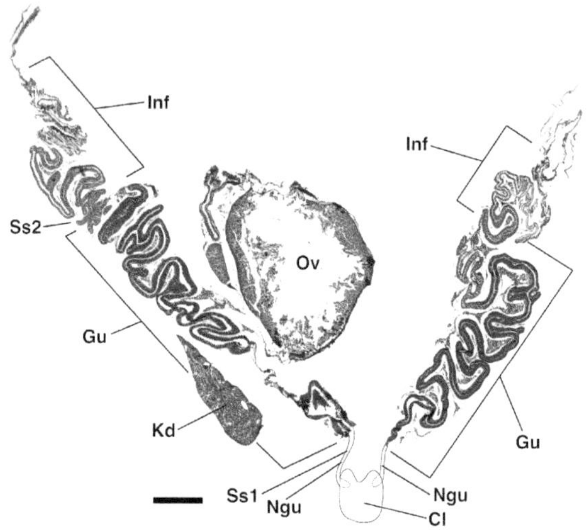

Fig. 6.5 Frontal sections from the oviduct of female *Hemidactylus turcicus* demonstrating the histological regions of the squamate oviduct (hematoxylin and eosin; scale bar = 1350 μm). Note that Ss1 (caudal sperm storage location) is for diagrammatic purposed only, as *H. turcicus* only stores sperm in the Ss2 (cranial sperm storage location) region. Cl, cloaca; Gu, glandular uterus; Inf, infundibulum; Kd, kidney; Ngu, non-glandular uterus; Ov, ovary; Ss1, site of caudal sperm storage location; Ss2, site of cranial sperm storage location.

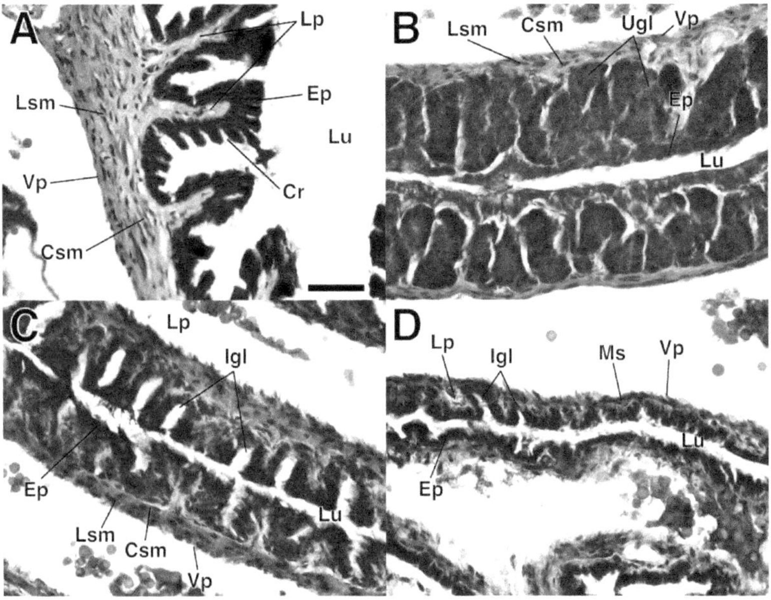

Fig. 6.6 High magnification histological sections from the oviduct (caudal to cranial) of *Hemidactylus turcicus* demonstrating four histologically unique regions (hematoxylin and eosin; scale bar = 50 μm). **(A)** The non-glandular uterus demonstrating a simple columnar epithelium with regular crypts lining the lumen, and a rather thick muscularis compared to more cranial regions of the oviduct. **(B)** The glandular uterus demonstrating numerous uterine glands in the lamina propria, a simple epithelium lining the lumen, and a thin muscularis compared to that of the non-glandular uterus. **(C)** The posterior infundibulum demonstrating infundibular glands, a simple epithelium lining the lumen, and a muscularis of similar thickness to that of the glandular uterus. **(D)** The anterior infundibulum demonstrating a decrease in infundibular glands, a simple epithelium lining the lumen, and a thin muscularis compared to more caudla regions of the oviduct. Cr, crypt; Csm, circular smooth muscle; Ep, epithelium; Igl, infundibular gland; Lp, lamina propria; Lsm, longitudinal smooth muscle; Lu, lumen; Ms, muscularis; Ugl, uterine gland; Vp, visceral pleuroperitoneum.

A

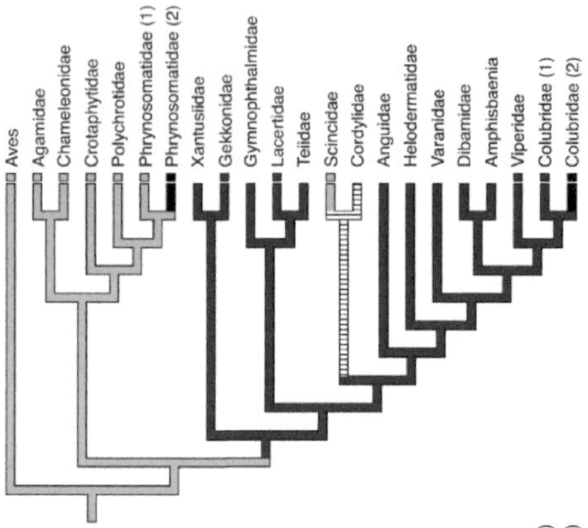

B

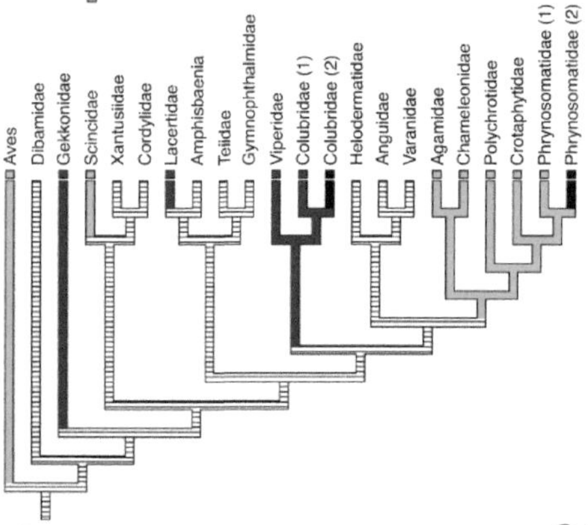

C

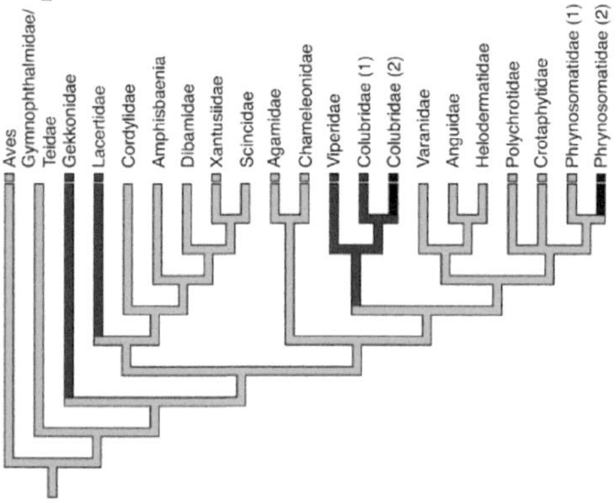

Fig. 6.7 Location of SSTs in squamate reptiles mapped on **(A)** a morphological phylogeny from Lee (2005), **(B)** a molecular phylogeny from Vidal and Hedges (2005), and **(C)** a molecular phylogeny from Eckstut *et al.* (2009b). Green indicates Ssts in the non-glandular uterus, blue indicates infundibular Ssts, and black indicates both non-glandular uterus and infundibular Ssts. Cross hatching represents equivocal ancestral state reconstructions. Modified from Eckstut *et al.* (2009a, Amphibia-Reptilia, 30: 45–56, Fig. 4).

Chapter 7

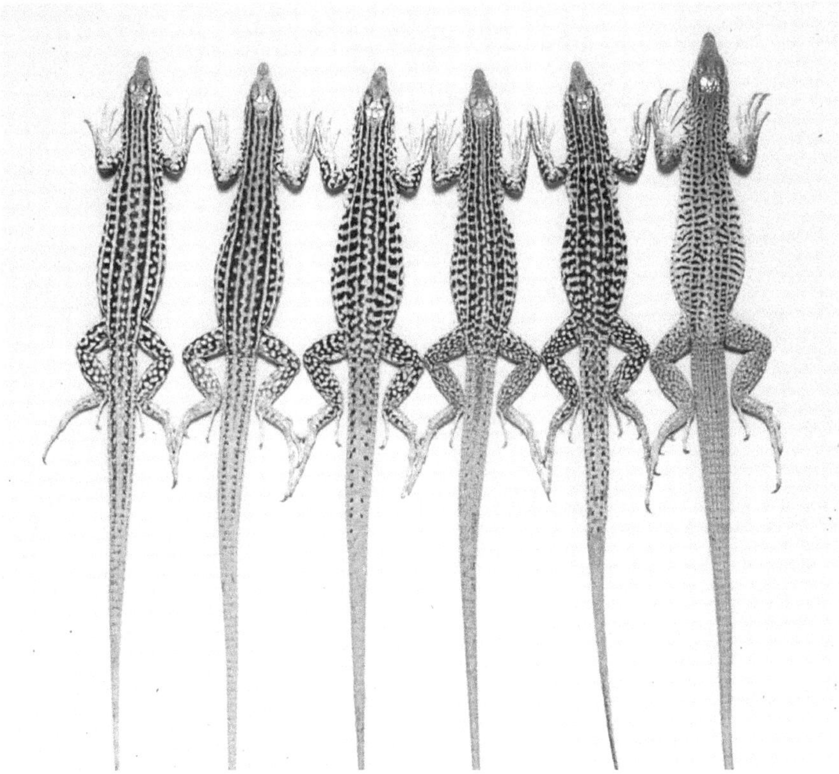

Fig. 7.1 *Cnemidophorus tesselatus* color pattern classes **A, B, C, D, E**, & **F** (Zweifel 1965). Left to Right: Color pattern class **A** captured 7/4/1968 in Pueblo County, Colorado; **B** captured 7/2/1968 in Otero County, Colorado; **C** captured 7/10/1968 in San Miguel County, New Mexico; **D** captured 7/3/1968 in Otero County, Colorado; **E** captured 8/22/1968 in Debaca County, New Mexico; **F** captured 8/1/1968 in Hidalgo County, New Mexico. Figs. 1–3 photographed in sunlight using Ektachrome B sheet film 10/1/1968 in Boston, Massachusetts after chilling lizards in refrigerator one hour at 4° Fahrenheit.

Fig. 7.2 Bisexual parental ancestral species of diploid *Cnemidophorus tesselatus* as proposed (Wright and Lowe 1967b). Left to Right: *C. septemvitattus*, parental ancestral species, captured 8/8/1968 in Brewster County, Texas; *C. tesselatus*, color pattern class E, captured 8/22/1968 in Debaca County, New Mexico; *C. tigris*, parental ancestral species, captured 8/4/1968 in Luna County, New Mexico.

Fig. 7.3 Parental ancestral species of triiploid *Cnemidophorus tesselatus* as proposed (Wright and Lowe 1967b). Left to Right: *C. sexlineatus*, parental ancestral species, captured 7/1/1968 in Las Animas County, Colorado; *C. tesselatus*, color pattern class A, captured 7/4/1968 in Pueblo County, Colorado; *C. tesselatus*, color pattern class E, parental ancestral species, captured 8/22/1968 in Debaca County, New Mexico.

Chapter 8

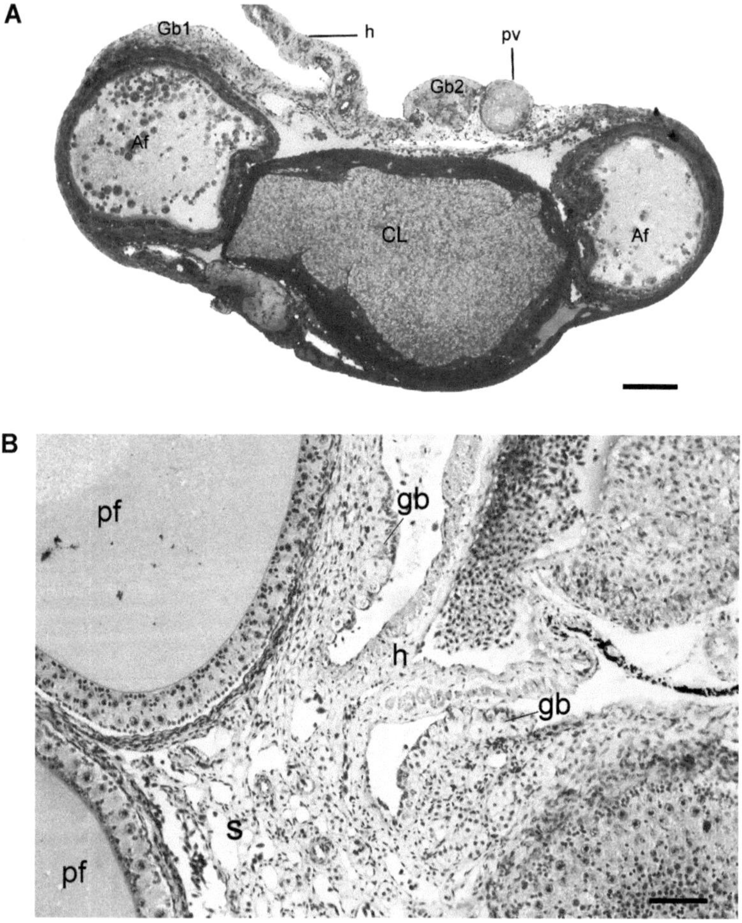

Fig. 8.1 (A) General view of the *Mabuya* sp. ovary during pregnancy. h, ovarian hilum; pv, an early previtellogenic follicle; Gb1 and Gb2, two germinal beds; CL, an active corpus luteum and two atretic follicles (Af). Scale bar = 80 µm. **(B)** General view of the *Liolaemus scapularis* ovary. Two germinal beds (gb) near to the ovarian hilum (h) and two previtellogenic follicles (pf) are observed. s, ovarian stroma. Scale bar = 50 µm.

A

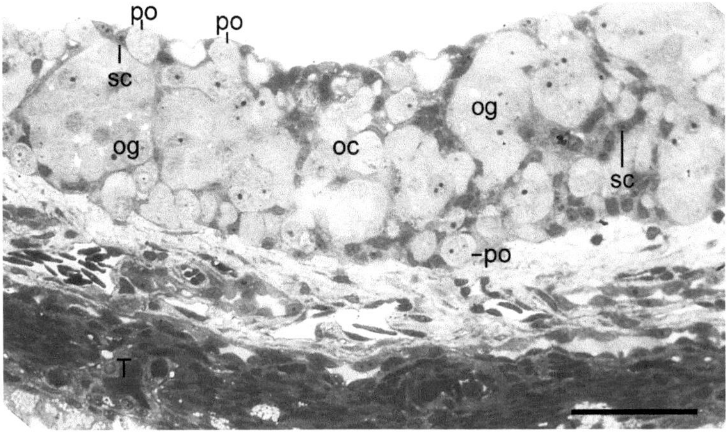

B

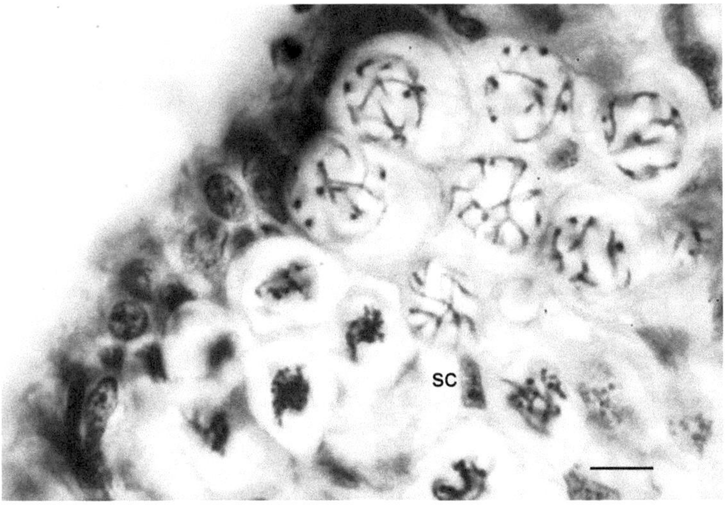

Fig. 8.2 (A) Germinal bed of a pregnant female *Mabuya* sp. exhibiting the prevalence of clusters of secondary oogonia (og) that are discernible by their larger size and pale cytoplasm. oc, primary oocytes; po, primary oogonia; sc, somatic cells (darker cytoplasm); T, theca of an adjacent corpus luteum. Scale bar = 50 μm. **(B)** Germinal bed of a reproductively active female of *Liolaemus scapularis* with prevalence of oocytes that are discernible by their nuclei in different stages of prophase I (zygotene and pachytene). Somatic cells (sc) have a darker cytoplasm and small nuclei. Scale bar = 8 μm.

A

B

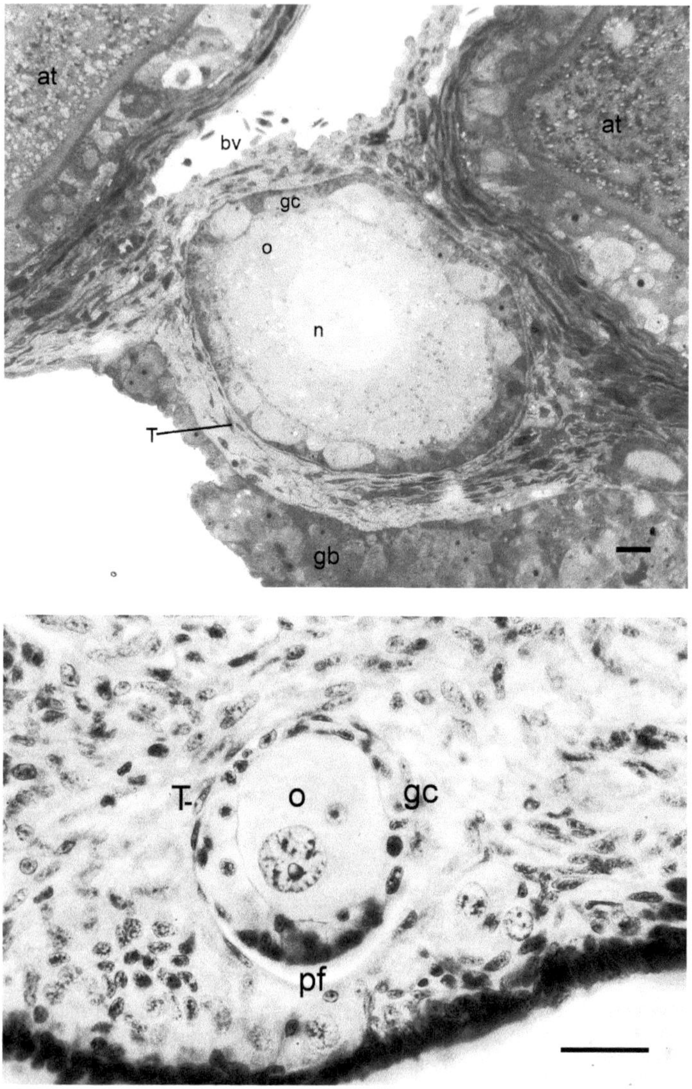

Fig. 8.3 (A) Young previtellogenic follicule of *Mabuya* sp. The theca (T) is thin. The follicular epithelium is monolayered and composed of cuboidal epithelial cells (gc), some of which are larger and possess a clear cytoplasm. bv, blood vessel; gb, germinal bed; n, nucleus, o, oocyte; at, adjacent atretic follicles Scale bar = 20 µm. **(B)** Primary follicule within the germinal bed of *Liolaemus scapularis*. The theca (T) is very thin. The follicular epithelium is monolayered and composed of cuboidal epithelial cells (gc), some of which are larger and possess a clear cytoplasm. o, diplotene oocyte. Scale bar = 30 µm.

A

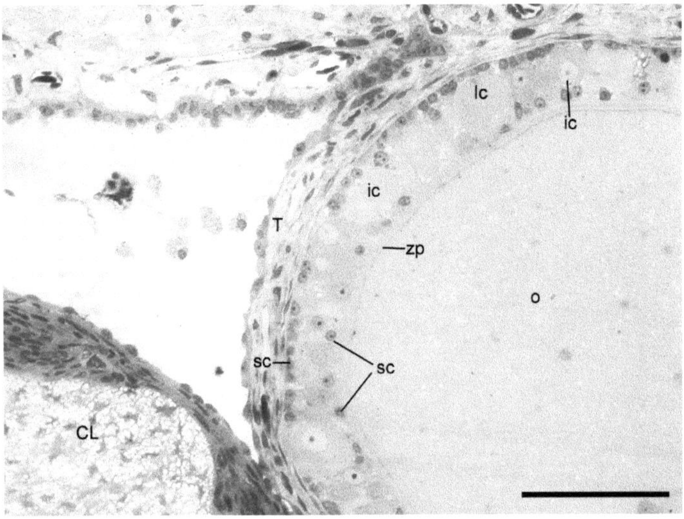

B

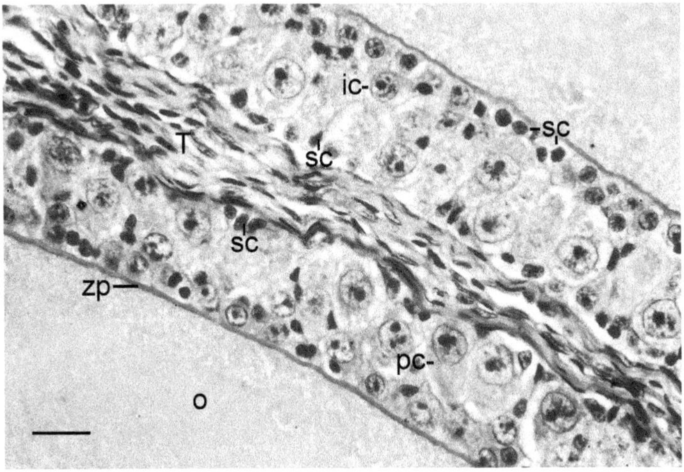

Fig. 8.4 (A) Partial view of a corpus luteum (CL) and a previtellogenic follicle (right) of *Mabuya* sp. The previtellogenic follicle is enveloped by the theca (T). Three types of characteristic follicular epithelial cells are observed at this stage: small (sc), intermediate (ic) and large (lc). In the apical region of the follicular cells the zona pellucida (zp) is conspicuous. O, oocyte; CL, corpus luteum in early luteolysis. Scale bar = 50 μm. **(B)** Partial view of the follicular wall of two adjacent previtellogenic follicles of *Liolaemus scapularis*. Three types of characteristic follicular epithelial cells are observed at this stage: small (sc), intermediate (ic) and pyriform (pc). In the apical region of the follicular cells the zona pellucida (zp) is conspicuous. O, oocyte; T, thecae. Scale bar = 12 μm.

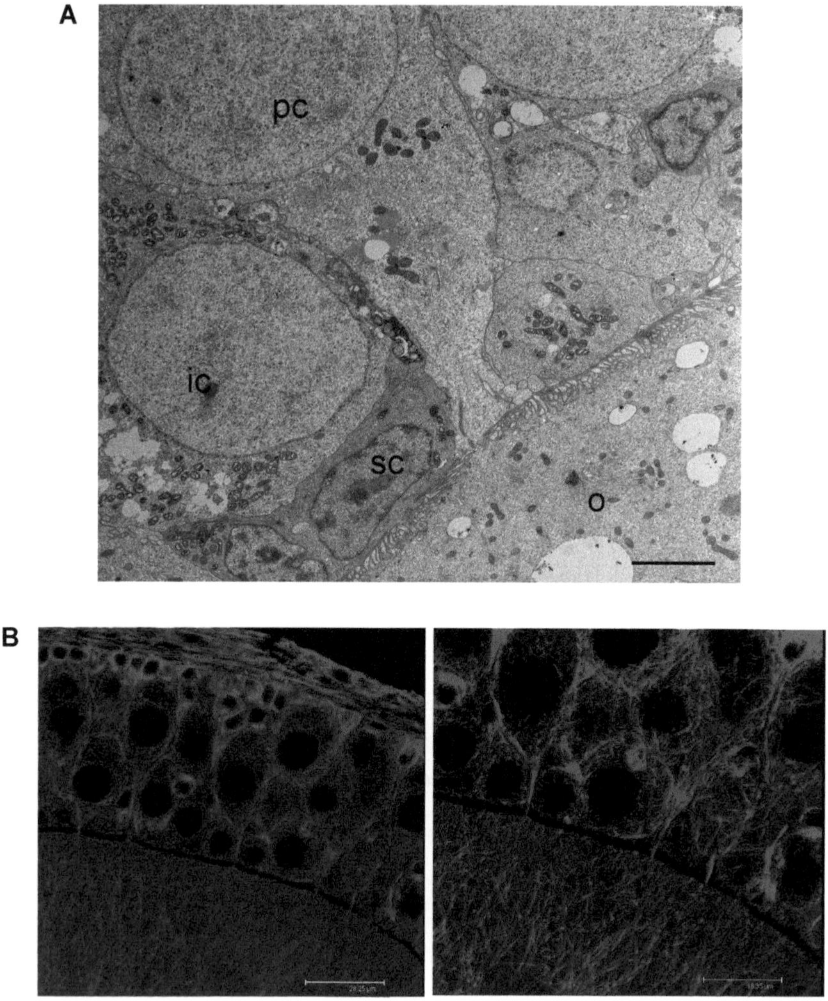

Fig. 8.6 (A) Transmission electron micrograph of a pyriform cell with the apex pointed to the oocyte surface in a follicle of about 350 μm in diameter of *Podarcis siculus*. Observe the pyriform cell (pc), an intermediate (ic) and a small apical cell (sc), and the ooplasm of the oocyte (o). Scale bar = 2 μm. Image provided by Prof. Carlo Taddei. **(B)** Confocal images of the follicular epithelium and oocyte of *Podarcis siculus* immunostained for α-Tubulin with DM1A monoclonal antibody. The image at the right is an enlargement of the left image showing the filaments crossing the intercellular bridges to the oocyte cytoplasm. Scale bar left image = 28.3 μm; right image = 18.4 μm. Images provided by Prof. Carlo Taddei.

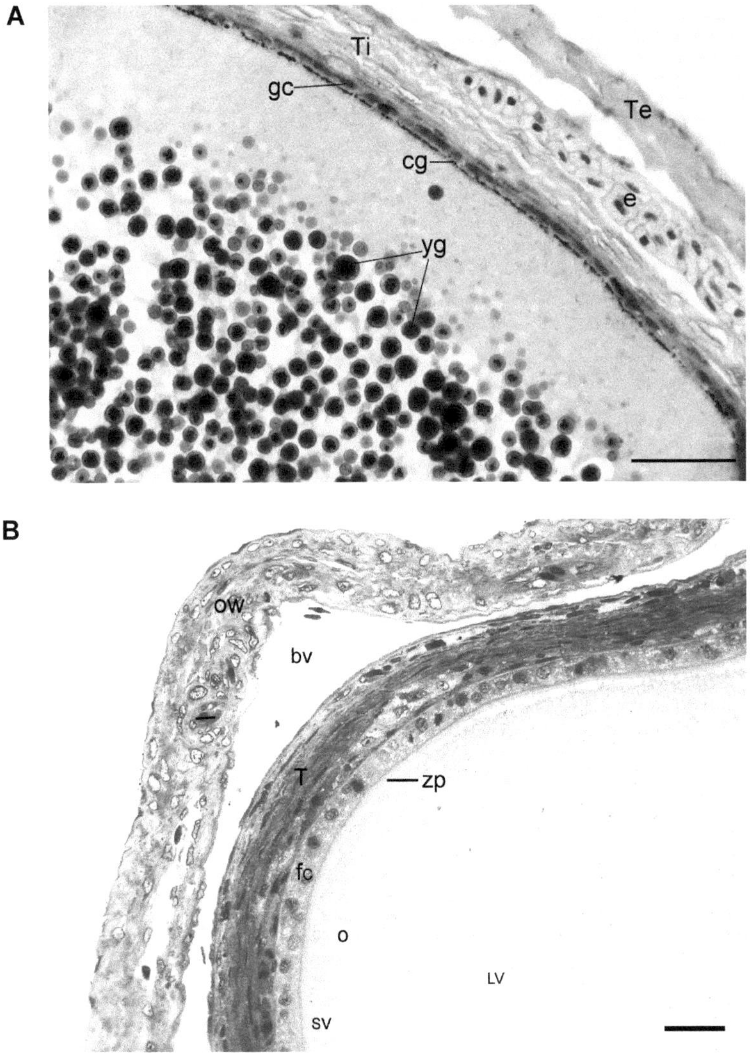

Fig. 8.8 (A) Vitellogenic follicle of the oviparous gymnophtalmid *Ptychoglossus bicolor*. The granulosa is monolayered with squamous epithelial cells (gc). Dark cortical granules (cg) are observed near oolemma. Abundant yolk granules (yg) are observed in the ooplasm. e, erythrocytes; Ti, theca interna; Te, theca externa. Scale bar = 50 μm. **(B)** Early vitellogenic follicle of *Mabuya* sp. The granulosa cells (fc) constitute a cubic monolayered epithelium surrounded by a thick fibrous theca (T). A thin zona radiata is present. The clear vesicles (SV) are smaller near the zona radiata and larger at the center (LV) of the ooplasm (o). No yolk platelets are observed. bv, blood vessel; ow, ovarian wall; zp, zona pellucida. Scale bar = 80 μm.

Chapter 9

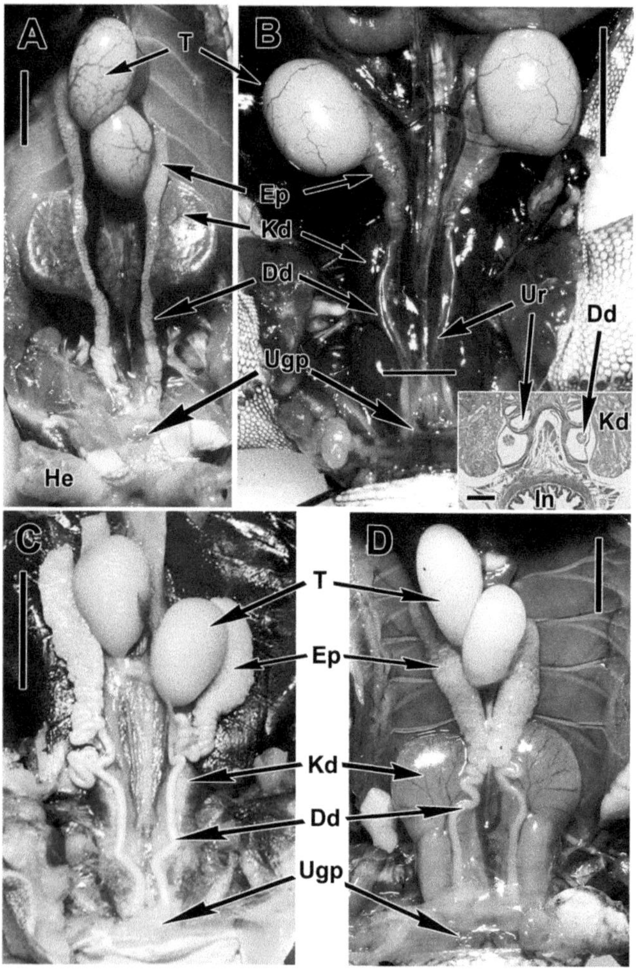

Fig. 9.1 Gross morphology of the urogenital systems in reproductively active representatives of four lizard families. Scale bar = 5 mm for **A**-**D**. **(A)** *Aspidoscelis sexlineata* (family Teiidae). **(B)** *Crotaphytus collaris* (family Crotaphytidae). Inset: histosection at transverse plane level indicated by bar; inset scale bar = 500 µm. **(C)** *Holbrookia propinqua* (family Phrynosomatidae). **(D)** *Plestiodon fasciatus* (family Scincidae). Ep, ductus epididymis; Dd, ductus deferens; He, hemipenis; In, intestine; Kd, kidney; T, testis; Ugp, urogenital papilla; Ur, ureter.

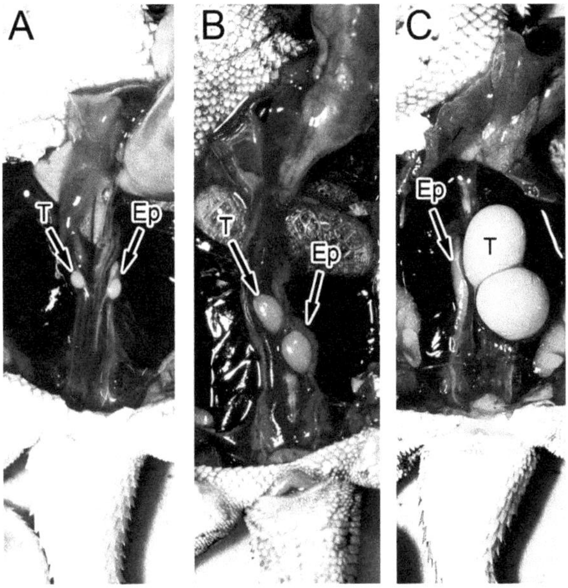

Fig. 9.2 Gross morphology of the reproductive system in **(A)** an immature individual of *Sceloporus undulatus*, **(B)** mature but reproductively inactive individual of *Podarcis muralis*, and **(C)** a mature and reproductively active individual of *Sceloporus virgatus*.

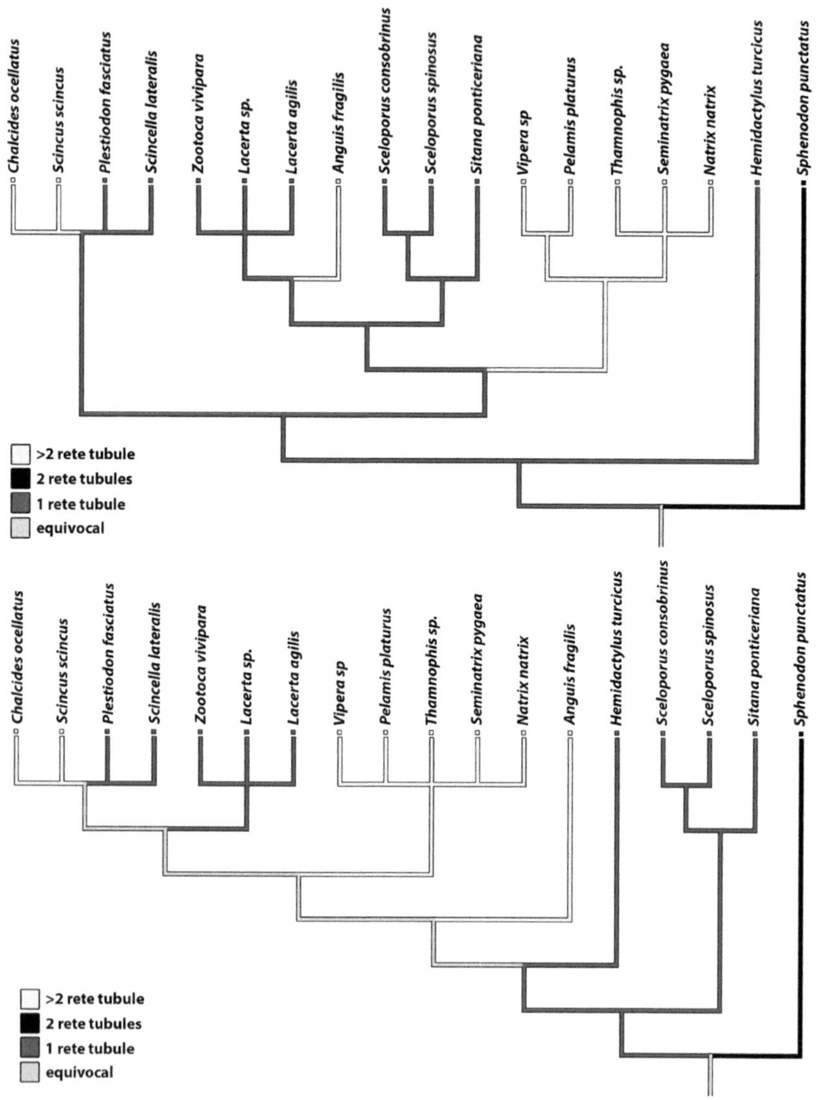

Fig. 9.6 Character optimization of the number of rete tubules present on the **(A)** morphological phylogenetic hypothesis provided by Conrad (2008) and **(B)** the molecular phylogenetic hypothesis provided by Pyron *et al.* (2013).

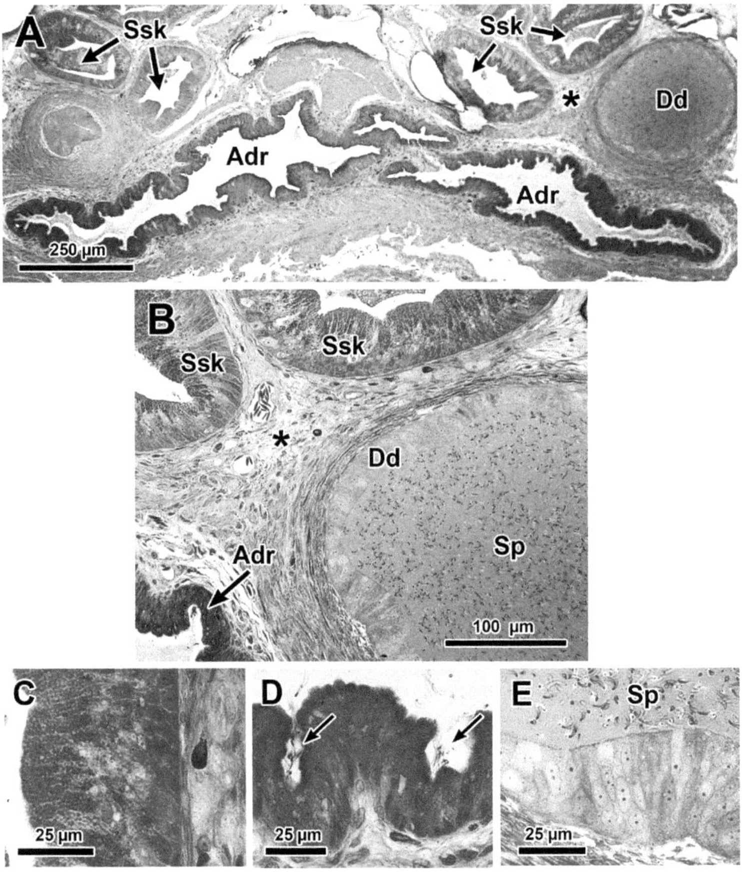

Fig. 9.17 Light micrographs (plastic sections) of the distal urogenital anatomy in *Plestiodon fasciatus* revealing the sexual segment of the kidney (Ssk), the posterior ductus deferens (Dd) and the anterior dorsal recess of the urodaeum (Adr). **(A)** Section through cloacal region similar to the plane shown in Fig. 15**D**. Asterisk denotes position shown in **B**. **(B)** Magnification of region around asterisk of **A**. Sp, sperm. **(C)** Tall columnar epithelium of the Ssk shown in **B** (upper area). **(D)** Pseudostratified columnar epithelium of the Adr; arrows point to sperm within epithelial folds. **(E)** Pseudostratified columnar epithelium of the Dd shown in **B** (lower middle area).

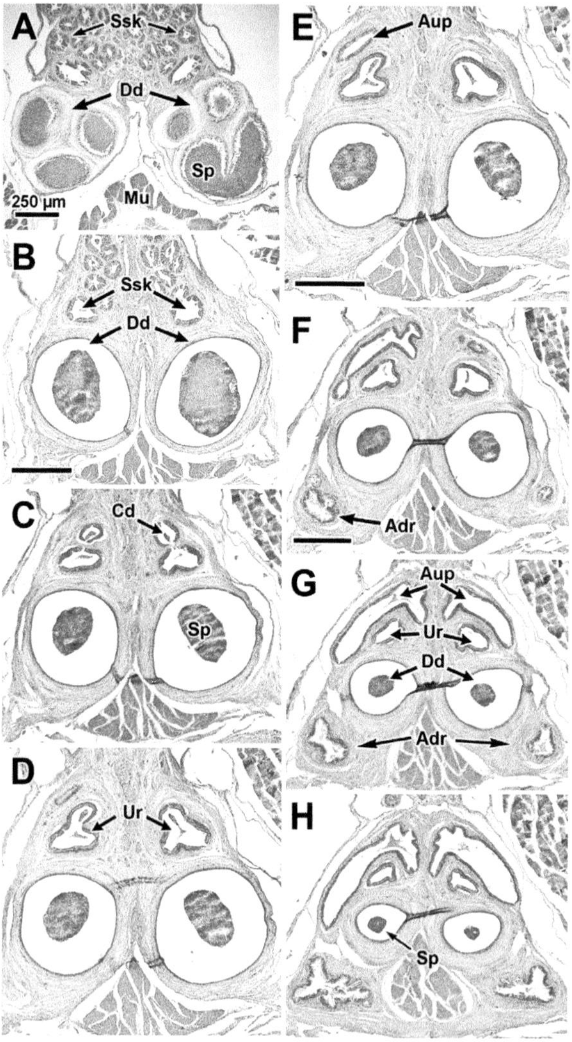

Fig. 9.19 Light micrographs (paraffin sections) of cranial-to-caudal serial histosections of the distal urogenital anatomy of *Aspidoscelis gularis* stained with hematoxylin and eosin. **A**. Section through the posterior region of the kidney revealing numerous tubules and associated sexual segments of the kidney (Ssk) lying dorsal to the ductus deferens (Dd) packed with sperm (Sp). Mu, medial striated muscle. **B**. Kidney tubules appear to merge, adjoining a large ventral collecting duct. The ductus deferens has now become greatly enlarged compared to its condition in **A**. **C**. The posteriormost collecting ducts (Cd) on both sides merge into single ducts. **D**. The ureter (Ur) now drains kidney products. **E**. The ampulla urogenital papilla (Aup) appears dorsolateral to the ureter. **F**. The anterior dorsal recess of the urodaeum (Adr) appears ventrolateral to the ductus deferens. **G-H**. The Aup greatly expands to occupy the dorsolateral region of the distal urogenital anatomy complex.

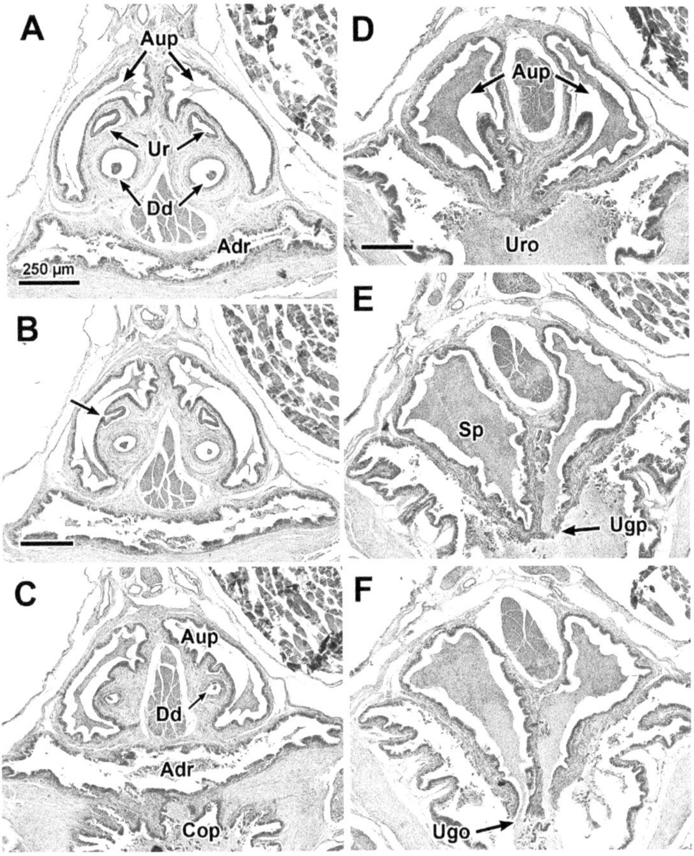

Fig. 9.20 *Aspidoscelis gularis*, continuation of Fig. 9.19. **(A)** Section through the posterior region of the distal urogenital complex; the paired ampulla urogenital papillae (Aup) dominate the dorsolateral region, whereas the paired anterior dorsal recesses (Adr) have greatly expanded within the ventral region. **(B)** A ureter (arrow) is shown merging with an Aup; the diameter of the ductus deferens has become greatly reduced. Both Adr have merged into a single cavity of the anterior urodaeum. **(C)** The Aup have expanded to occupy most of the dorsolateral, lateral, and ventrolateral region around the much-reduced ductus deferens; the Adr of the urodaeum nears its confluence with the coprodaeum (Cop). **(D)** Section at the level of the urodaeum proper (Uro), just posterior to the merging of the ductuli deferentia with the greatly expanded Aup. Sperm (Sp) and other urogenital products fill the Aup. **(E)** The Aup appears elongated in the region of the urogenital papilla (Ugp). **(F)** Release of sperm and urogenital products from the paired orifices (Ugo) of the urogenital papilla.

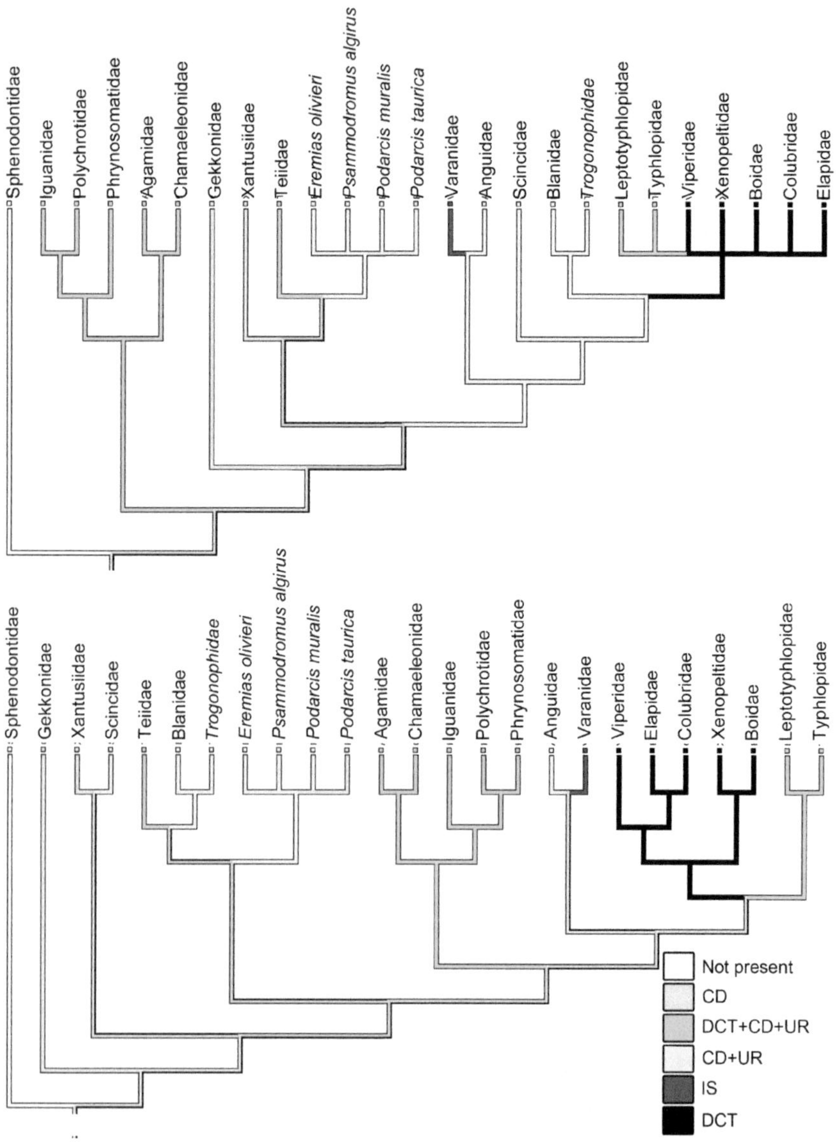

Fig. 9.22 Character mapping of region of kidney occupied by the SSK onto a morphological and molecular phylogenetic hypothesis for squamates. **(Top)** Morphological topology proposed by Conrad (2008). Bulletin of the American Museum of Natural History 2008: 1–182. **(Bottom)** Molecular topology proposed by Pyron *et al.* (2013). BMC Evolutionary Biology 13: 93.

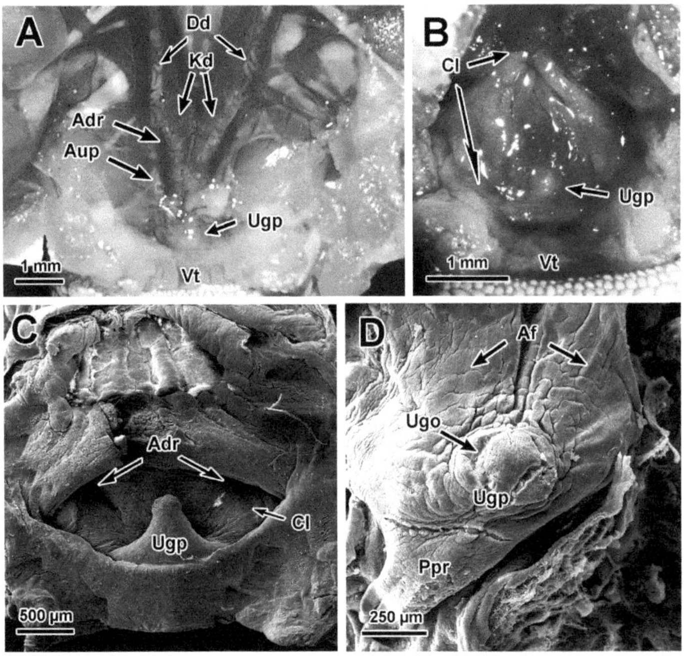

Fig. 9.24 Ventral views of the cloaca and distal urogenital anatomy of *Aspidoscelis sexlineata*. **A**. Macroscopic view of urogenital structures revealing the ductus deferens (Dd) lying ventrolateral to paired posterior extensions of the kidney (Kd), the right ampulla urogenital papilla (Aup) shown lateral to the right anterior dorsal recess (Adr) of the urodaeum (filled with blood on both sides), the urogenital papilla (Ugp), and the vent (Vt). **B**. Macroscopic view of the cloaca (Cl) showing the urogenital papilla (Ugp) situated posteriorly in the cloacal cavity just anterior to the vent (Vt). **C**. Scanning electron micrograph of the cloaca (similar to view as shown in **B**) revealing the paired anterior dorsal recesses (Adr) of the urodaeum and an unpaired urogenital papilla (Ugp). **D**. Scanning electron micrograph of an unpaired urogenital papilla (Ugp) along with its right urogenital orifice (Ugo) as well as the paired anterior ampullary folds (Af) and the single posterior papillary ridge (Ppr).

Chapter 10

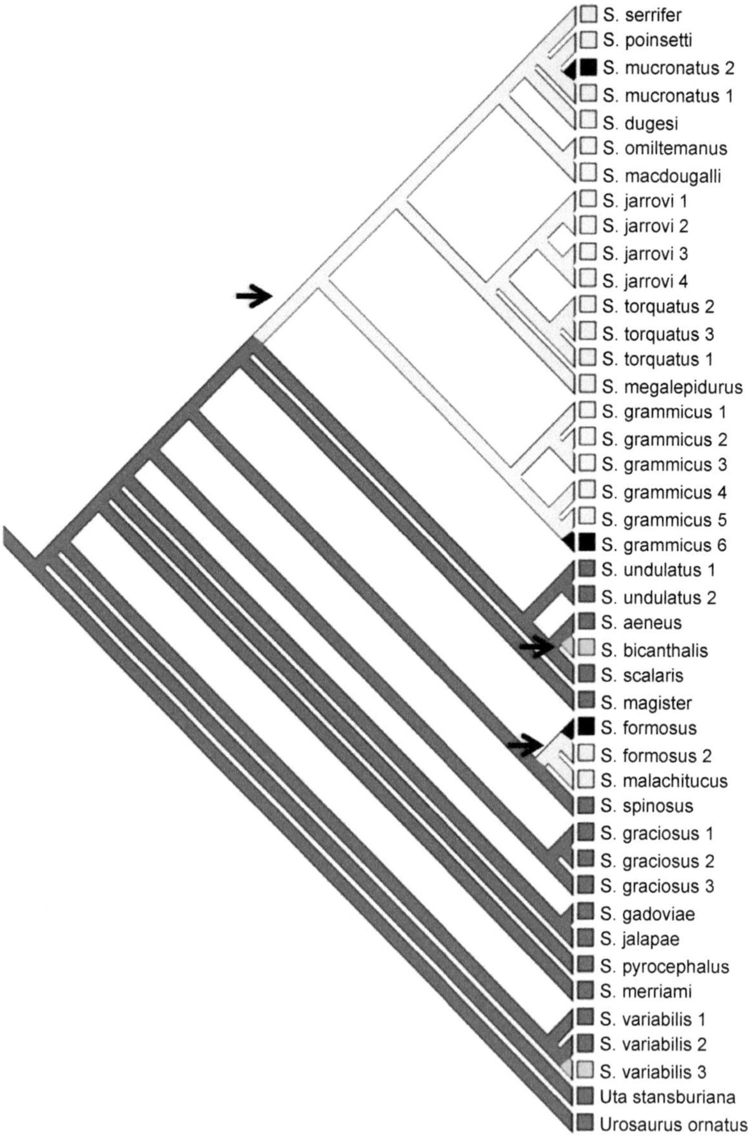

Fig. 10.6 Evolution of reproductive cycles in the species of the genus *Sceloporus*. **Blue**—Associated cycles with spring-summer reproductive activity. **Green**—Continuous reproduction. **Yellow**—Associated cycles with fall reproductive activity. **Black**—Dissociated cycles with fall reproductive activity of females and spring-summer of males. **Arrows** indicate events of evolution of viviparity. Reconstruction of ancestral states using parsimony. Phylogenetic information was based on Wiens *et al.* 2010.

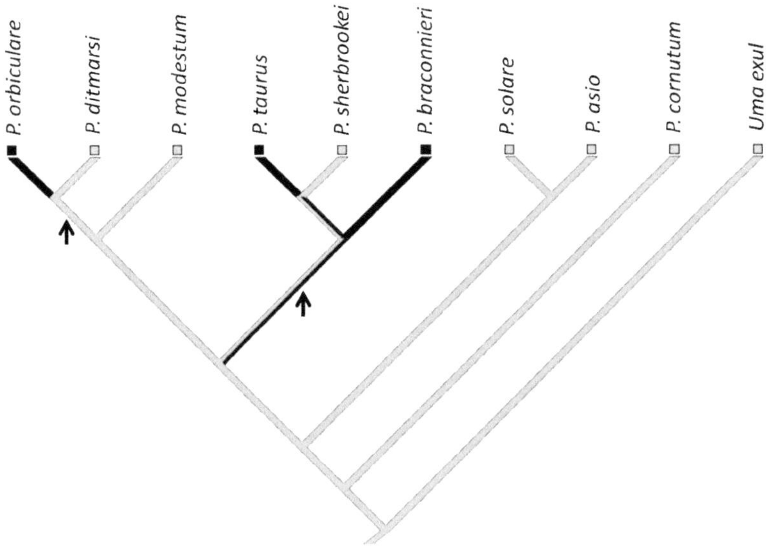

Fig. 10.10 Evolution of reproductive cycles in the species of the genus *Phrynosoma*. **Green**—Associated cycles with spring-summer reproductive activity. **Black**—Associated cycles with fall reproductive activity. **Arrows** indicate events of evolution of viviparity. Phylogenetic information was based on Nieto Montes de Oca *et al.* 2014.

Chapter 11

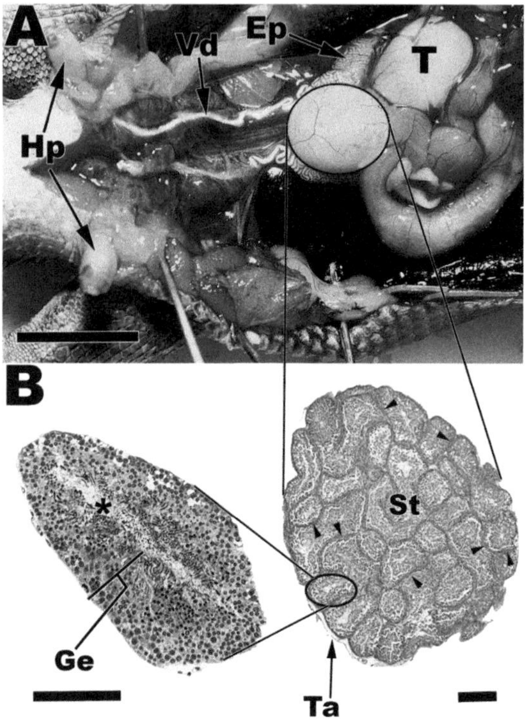

Fig. 11.1 (A) Gross dissection of the urogenital tract of *Sceloporus bicanthalis*. Scale = 10 mm. Note the location of the testes (T) and excurrent ducts (Ep, Vd) of the reproductive system. Ep, epididymis; Vd, vas deferens; Hp, hemipenis. **(B)** Cross sections of the entire testis (*Sceloporus consobrinus*), right, and of an individual seminiferous tubule (left) within *Barisia imbricata*. Bar (testis) = 10 mm, Bar (tubule) = 75 μm. A thin tunica albuginea (Ta) encases the highly convoluted seminiferous tubules (ST). Septa (black arrows) originating from the tunica albuginea and penetrate the testis providing the framework of the interstitial space interiorly. Individual seminiferous tubules have a thick germinal epithelium (GE) and a lumen (*) where spermatozoa are released once spermatogenesis is complete. Photo A: Oswaldo Hernández-Gallegos, Photo B: (CS Testis) Justin Rheubert (CS Tubule), Kevin Gribbins.

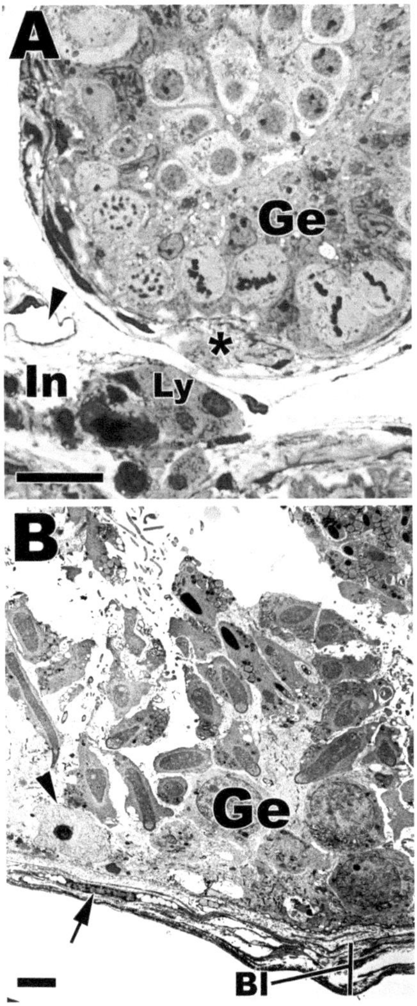

Fig. 11.10 (A) *Podarcis muralis*. Light micrograph showing the boundary layer (*) and interstitial tissue (In) outside of the germinal epithelium (Ge) of a seminiferous tubule within the August testis. Bar = 20 μm. Leydig cells (Ly), blood vessel (black arrowhead). **(B)** Low power electron micrograph depicting the boundary layer (Bl) below the germinal epithelium (Ge) in a late April testis. Bar = 5 μm. Myofibroblasts (black arrow), Sertoli cell nucleus (black arrowhead).

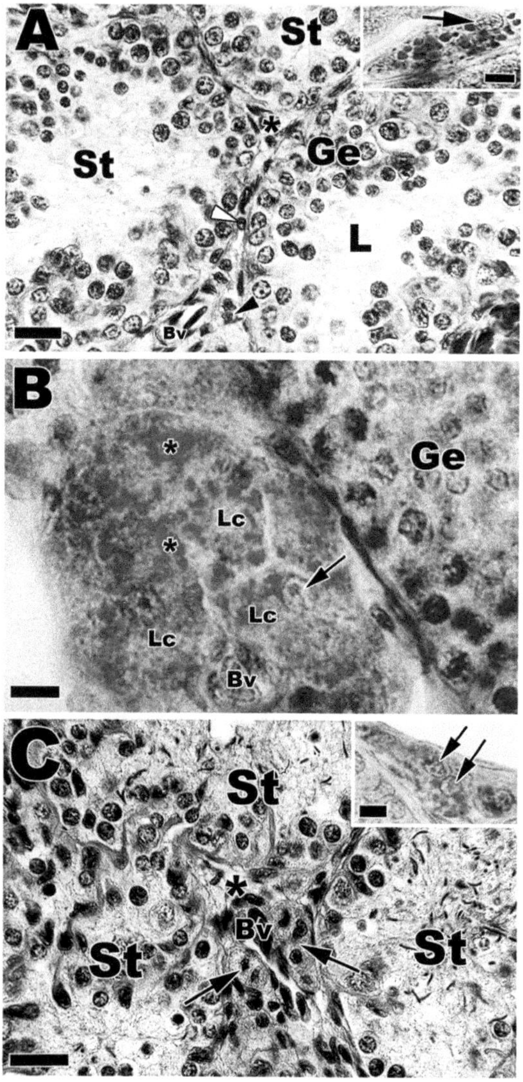

Fig. 11.16 *Sceloporus mucronatus*. The interstitium (*) between seminiferous tubules (St) within the February **(A)** and September **(B)** testes showing Leydig cells (Lc) that are located both within the boundary layer (white arrowhead) and the interstitial space (black arrowhead). The inset in A and C show sudanophilic lipid inclusions that are moderate in February versus hypertrophied lipid droplets (* in B) within Leydig cells (Lc) of September lizards. Blood vessels (Bv), germinal epithelium (Ge), lumen (L), Leydig cell nucleus (black arrow). **(C)** Leydig cells (black arrows) located between November seminiferous tubules (St). Inset reveals smaller sudanophilic lipid droplets around Leydig cell nuclei (black arrows) than during recrudescence in February. Blood vessel (Bv), Interstitium (*). Bar in A/C = 25 μm; Insets A/C = 10 μm and Bar in B = 10 μm.

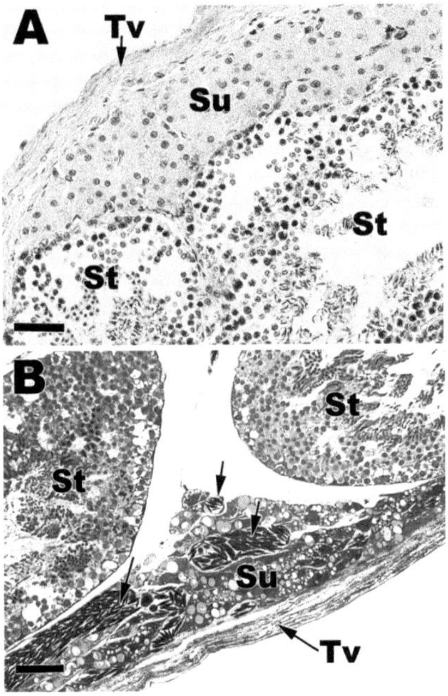

Fig. 11.17 Light micrographs of the subtunic layer (Su) of Leydig cells in *Aspidoscelis costata* **(A)** and *Phrynosoma cornutum* **(B)**. These are very large layers of Leydig cells just under the tunica vaginialis (Tv) that have ample vascularization (black arrows). Bar = 50 μm. Seminiferous tubules (St).

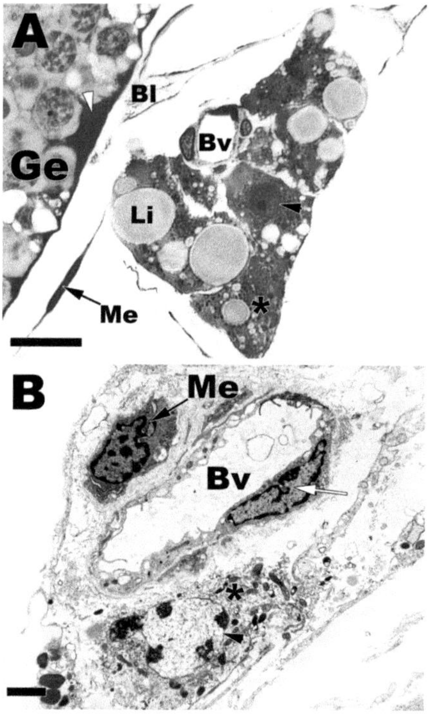

Fig. 11.18 *Podarcis muralis*. Leydig cells (*) near blood vessels (Bv) in April **(A)** and August **(B)**. Note the numerous lipid droplets (Li) found in the April Leydig cells, which are lacking in August (B, *). Bar in A = 25 μm and in B = 5 μm. Germinal epithelium (Ge), Sertoli cell nucleus (white arrowhead), boundary layer (Bl), myofibroblasts (Me), Leydig cell nucleus (black arrowheads), endothelial cell (white arrow).

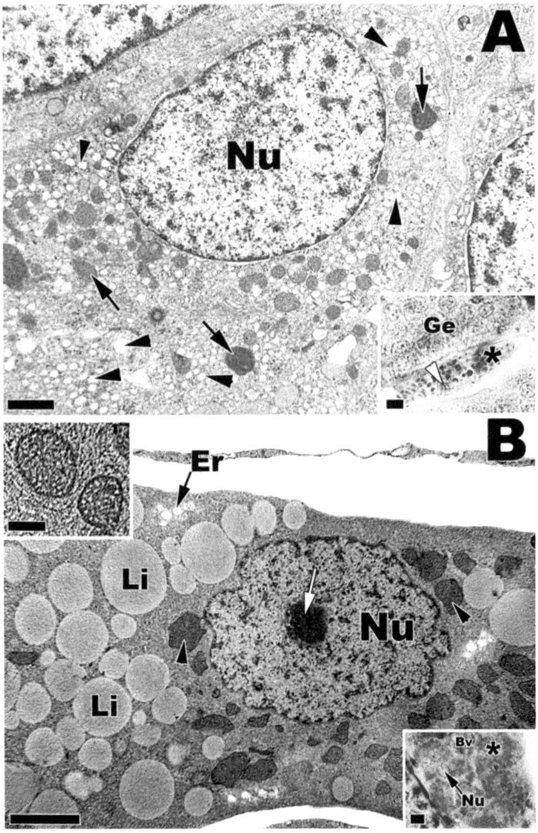

Fig. 11.19 (A) Ultrastructure of a Leydig cell in the August testis of *Podarcis muralis*. Lipid inclusions (black arrowheads) are small, numerous, variable in size, and evenly distributed. Bar = 2 μm. Mitochondria with small tubular cristae (black arrows) Inset: Shows a similar trend in early recrudescence in February in *Sceloporus mucronatus* with small sudanophilic lipid droplets (*) and heterochromatic nucleus (white arrowhead). Bar = 10 μm. Germinal epithelium (Ge). **(B)** Ultrastructure of a Leydig cell near the climax of spermatogenesis (July) in *Phrynosoma cornutum*. Bar = 1 μm Lipid droplets (Li) are large as are the mitochondria (black arrowheads), which have dilated tubular cristae (upper left corner inset, Bar = 0.2 μm). Dialated smooth ER (Er), Nuceolus (white arrow). Lower right corner inset: November Leydig cells in *Sceloporus mucronatus* have large sudanophilic lipid droplets (*) with large nuclei (Nu). Bar = 2 μm. Blood vessel (Bv).

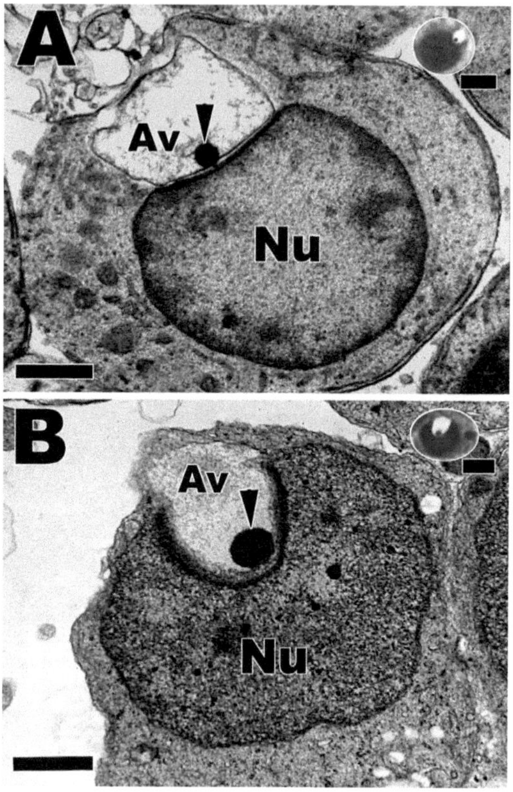

Fig. 11.31 Comparison of acrosome vesicle (AV) indentation depth into the apical nucleus (NU) of round spermatids in *Sceloporus variabilis* **(A)** and *S. bicanthalis* **(B)**. Bar = 2 µm for TEM and 5 µm for light micrographs (insets). Note how much deeper *S. bicanthalis* indentation, which is typical of the lizards studied to date, is compared to *S. variabilis*. Acrosome granule (black arrowhead).

Table 11.1 Comparison of spermatid ultrastructural characters for the ten species of lizards within the literature that have extensive ontogenetic details of spermatid development. Red colored letters represent unique characters of a species compared with the other lizards within the table. Sources: *Barisia imbricata*, Gribbins *et al.* 2013b; *Lacerta vivipara*, Courtens and depeiges 1985; *Chalcides ocellatus*, Carcupino *et al.* 1989; *Tropidurus torquatus*, Vieira *et al.* 2001; *Sceloporus bicanthalis*, Rheubert *et al.* 2012; *Sceloporus variabilis*, Gribbins *et al.* 2013a; *Scincella lateralis*, Gribbins *et al.* 2007; *Anolis lineatopus*, Rheubert *et al.* 2010; *Iguana iguana*, Ferreira and Dolder 2002; *Hemidactylus turcicus*, Rheubert *et al.* 2011.

Acrosome Complex

	ALR	PE	BP	ELZ	ER/MI COM	NUCR	EMY	LB	ER	SGA	NR	AI	PV
Barisia Imbricata	1	B	1	B	0	1	1	1	0	1	0	D	0
Zootoca vivipara	1	N	NR	N	0	1	S	1	0	Poss	1	D	0
Chalcides ocellatus	1	NR	NR	N	0	1	0	1	0	0	0	D	0
Tropidurus torquatus	1	N	NR	N	0	1	0	1	0	1	0	D	0
Sceloporus bicanthalis	1	N	1	N	0	1	S	1	0	1	0	D	0
Sceloporus variabilis	1	N	1	N	1	1	0	1	0	1	0	S	0
Scincella lateralis	1	S	1	0	0	1	0	1	0	1	0	D	0
Iguana Iguana	1	N	1	N	0	1	0	NR	1	NR	0	D	1
Anolis Lineatopus	1	N	1	N	0	1	0	1	0	1	0	D	0
Hemidactylus turcicus	1	N	1	N	0	1	S	1	0	1	0	D	0

ALR	Acrosomal Lucent Ridge	1: Present; 0: Absent
PE	Peforatorium	N: Normal; B: Bulbous; S: Short
BP	Basal Plate	1: Present; 0: Absent
ELZ	Epinuclear Lucent Zone	N: Normal; B: Bulbous; 0: Asent
ER/MI COM	ER/Mitochondria Complex	1: Present; 0: Absent
NUCR	Nuclear Rostrum	1: Present; 0: Absent
EMY	Extra Myelin Acrosomal Material	1: Present; 0: Absent; S: Some
LB	Lamellar Body (ER)	1: Present; 0: Absent
ER	ER Aid in Acrosome Dev.	1: Present; 0: Absent
SGA	Subacrosomal Granule	1: Present; 0: Absent
NR	Nuclear Ribbon	1: Present; 0: Absent
AI	Acrosomal Nuclear Indentation	D: Deep; S: Shallow
PV	Proacrosomal Vesicles	1: Present; 0: Absent
		****NR: Not Reported**
		**** Poss: Possibly Found (Not reported but can observe in micrographs)**

Nucleus

	CH	MA	LIP	LA	NP	NF	CHS
Barisia Imbricata	F	L/C	1	0	NS	C	0
Zootoca vivipara	G/F	L/C	1	Poss	1/NS	N	1
Chalcides ocellatus	F	L/C	NR	0	NS	N	0
Tropidurus torquatus	G/F	L/C	Poss	NR	1/NS	N	NR
Sceloporus bicanthalis	G/F	L/C	1	1	1/NS	N	1
Sceloporus variabilis	F	l/C	0	0	1/NS	N	0
Scincella lateralis	G/F	L/c	0	1	1/Poss NS	N	0
Iguana Iguana	G/F	L/C	NR	Poss	Poss Both	N	1
Anolis Lineatopus	G/F	0	1	0	IR/NS	N	0
Hemidactylus turcicus	G/F	L/C	0	0	1/NS	N	0

CH	Chromatin Condensation	F: Filamentous; G/F: Granular then Filamentous
MA	Manchette Microtubule Arrangement	L/C: Longitudinal/Circular; l/c: small longitudinal/circular; 0: Absent
LIP	Lipids surrounding late elongates	1: Present; 0: Absent
LA	Nuclear Lacuna	1: Present; 0: Absent
NP	Nuclear Membrane Pouches	1: Present; NS: Nuclear Shoulders present ; IR: Irregular pouches; 0: Absent
NF	Nuclear Implatation Fossa Shape	C: Clover Shaped; N: Cylindrical
CHS	Significant Chromatin Spiraling	1: Present; 0: Absent

Flagellum

	PC/CP	P3/8	MIL	MIR	DB	AN	MFS
Barisia Imbricata	1	M/e	4 to 5	5 to 7	1(4/3-5L)	1	3
Zootoca vivipara	Poss	NR	3	NR	1 (NR/2L)	1	NR
Chalcides ocellatus	1	NR	4	NR	1 (3/4L)	1	Poss 2
Tropidurus torquatus	1	NR	NR	NR	1 (NR)	1	NR
Sceloporus bicanthalis	1	M/p	4 to 6	4 to 6	1 (5-7/2-3L)	1	2
Sceloporus variabilis	1	M/p	4 to 6	4 to 6	1(3-5/3-4L)	1	2
Scincella lateralis	1H	M/p	4-6L	12	rare	1	Poss 1 (longNeck)
Iguana Iguana	1	M	4	NR	1 (NR)	1	poss 2
Anolis Lineatopus	1	M	4 to 5	NR	1(2/3L)	1	2
Hemidactylus turcicus	1	M	12	10 to 12	1 (8-10T/L)	1	3

AN	Annulus	1: Present; 0: Absent
P3/8	Peripheral Fibers 3 and 8	M/p: Enlarged MidP/slightly enlarged Principal P; M: Enlarged MP; e: endP
MIL	# Mitochondria in Longitudinal Plane	L: oriented elongated only
DB	Dense Bodies Present	1: Present; 0: Absent; #/#L: # in Transverse/# in Longitudinal
PC	Pericentriolar Material Present/Conn. Piece	1: Present; 0: Absent; 1H: Heavy Accumulation
MIR	# of Mitochondria in individual rings	
MFS	Tier of Mito. Where Fiberous Sheath Forms	Long Neck: Longer connecting piece than other lizards

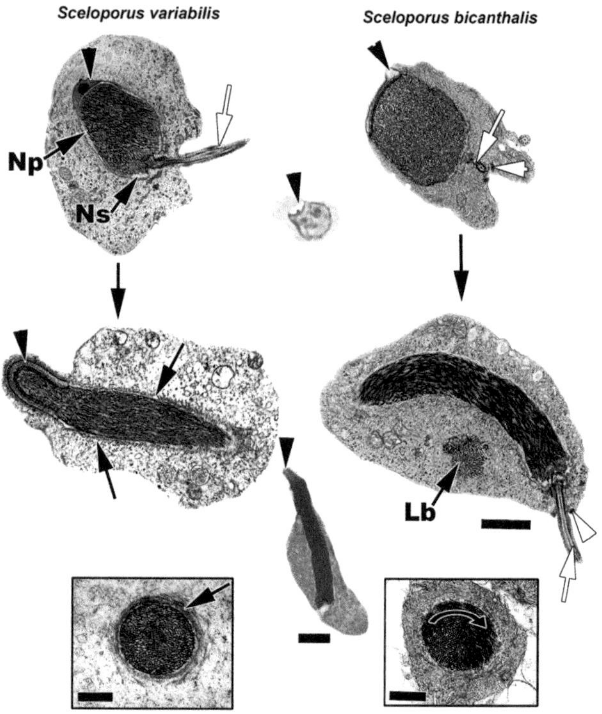

Fig. 11.32 Early and mid-elongation in *Sceloporus variabilis* and *S. bicanthalis*. Bar for TEM of sagittal elongates = 1 µm, bar for transverse elongates 0.5 µm, and bar for light micrographs 2 µm. Condensation involves spiraling of chromatin and granular condensation and then filamentous chromatin in *S. bicanthalis*. Both the spiraling and granular chromatin are absent in *S. variabilis*. Acrosome (black arrowheads), developing flagellum (white arrows), developing annulus (white arrowheads), nuclear pouches (Np), nuclear shoulders (Ns), manchette (black arrows), lamellar body (Lb), spiraling chromatin (curved black arrow).

Chapter 13

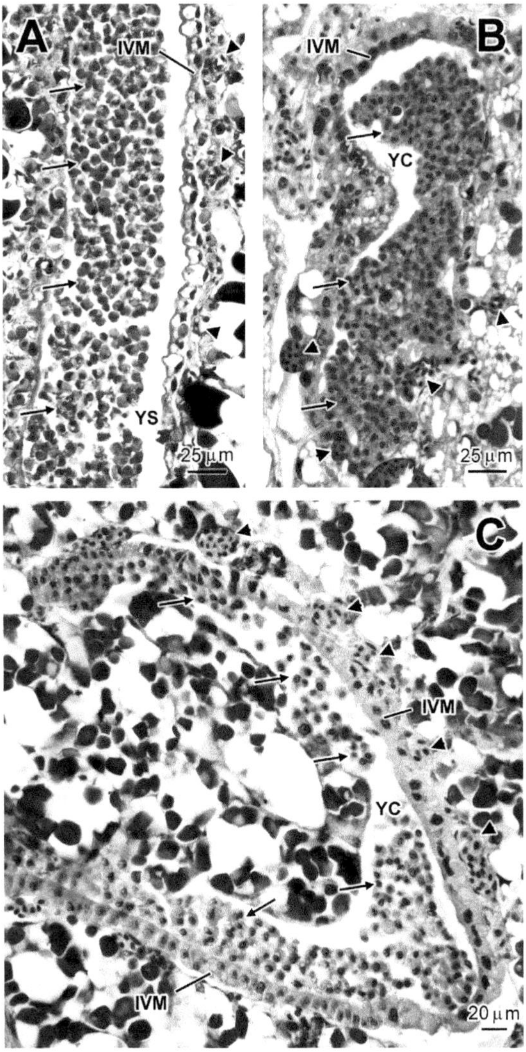

Fig. 13.4 Yolk cleft and associated structures in three species of oviparous lizards. **(A)** *Plestiodon fasciatus*, embryonic stage 38. **(B)** *Zootoca vivipara*, embryonic stage 37. **(C)** *Oligosoma lichenigerum*, embryonic stage 39. IVM, intravitelline mesoderm; YC, yolk cleft; YS, yolk sinus, derivative of the yolk cleft. Arrows, cells in yolk cleft/yolk sinus. Arrowheads, omphalomesenteric blood vessels.

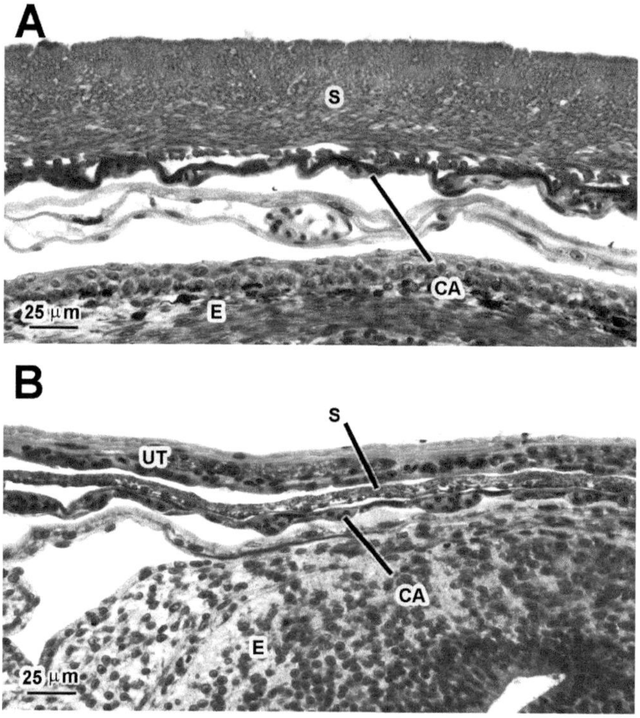

Fig. 13.10 Chorioallantoic membrane of **(A)** oviparous (embryonic stage 37) and **(B)** viviparous (embryonic stage 34) forms of *Zootoca vivipara*. CA, chorioallantoic membrane; E, embryo; S, eggshell; UT, uterus.

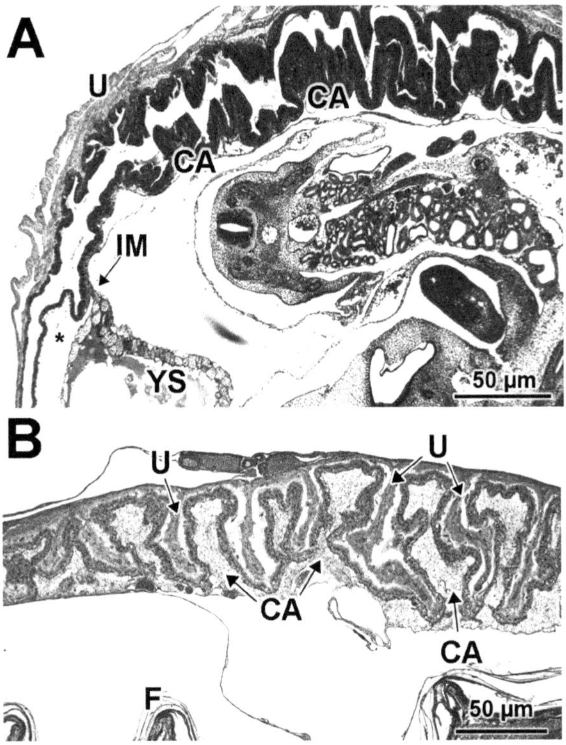

Fig. 13.14 Chorioallantoic placenta in *Chalcides chalcides.* **(A)** Limb bud stage. A placentome is developing through interdigitation of chorioallantois (CA) and lining of the uterus (U). The interomphalopleuric membrane (IM) prevents expansion of allantois into the yolk cleft (asterisk). **(B)** Placentome in a stage 40 embryo. F, fetus; YS, yolk sac cavity.

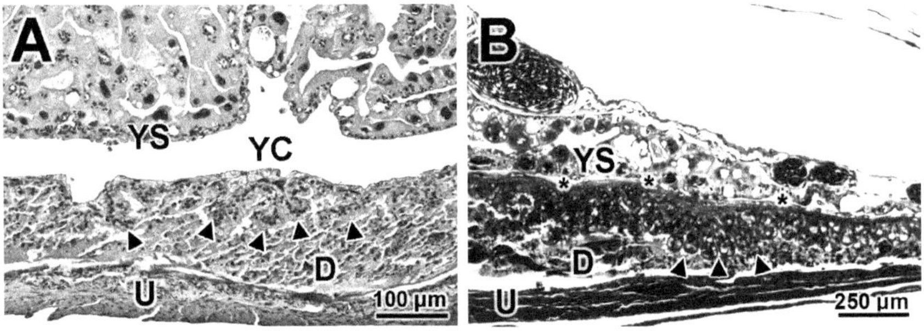

Fig. 13.15 Yolk sac placenta in *Chalcides chalcides* in late gestation. **(A)** The bilaminar omphalopleure bears papillae (arrowheads) that extend into a mass of detritus (D), formed from shell membrane and degenerating cells. Yolk cleft (YC) separates omphalopleure from the yolk sac splanchnopleure (YS). Isolated yolk mass is no longer present. **(B)** The omphalopleure contacts the yolk sac splanchnopleure; remnants of the yolk cleft are marked by asterisks. U, uterus.

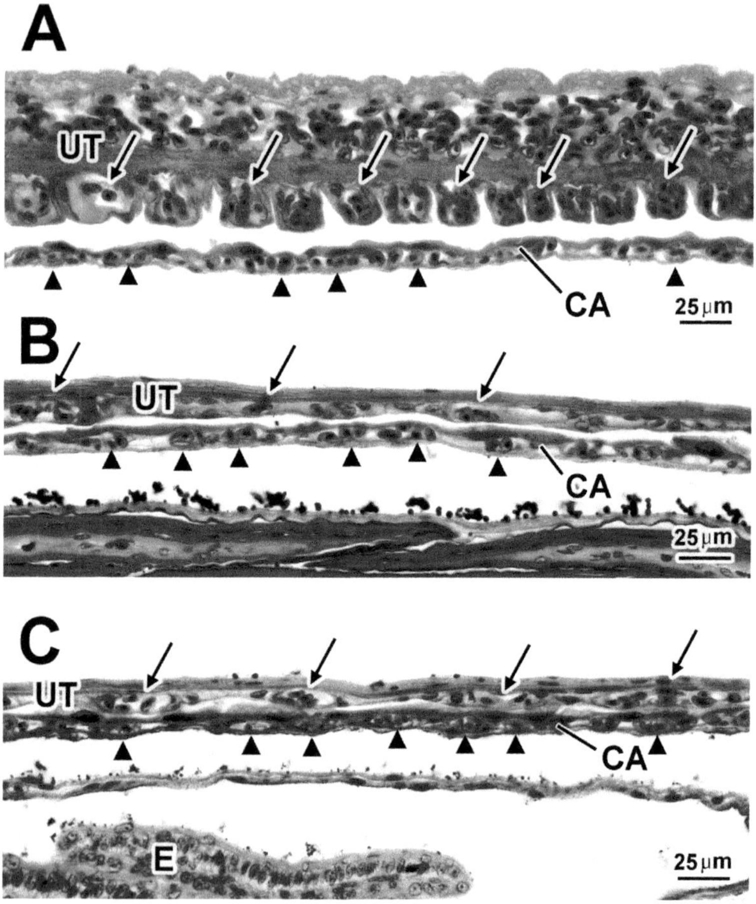

Fig. 13.17 Structural variation in the chorioallantoic placentas of the scincid genus *Niveoscincus*. **A, B**. *Niveoscincus metallicus*. Embryonic stage 39. **C**. *Niveoscincus ocellatus*. Embryonic stage 39. CA, chorioallantoic membrane; E, embryo; UT, uterus. Arrows, uterine blood vessels; Arrowheads, allantoic blood vessels.

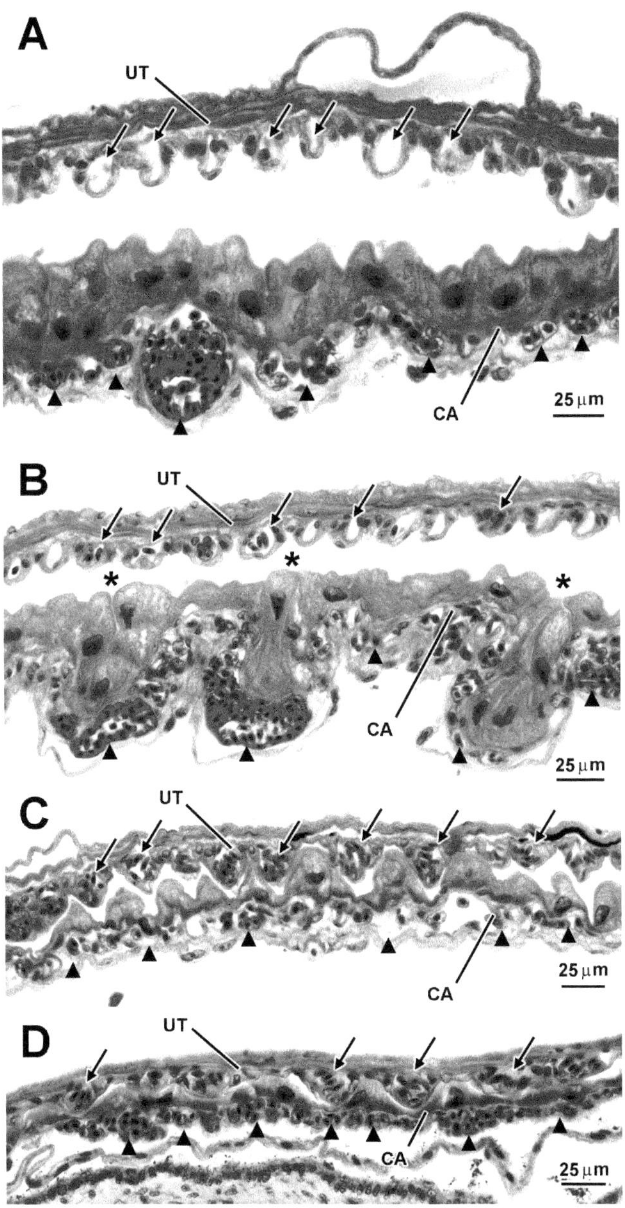

Fig. 13.18 Structural variation in the chorioallantoic placentas of the scincid genus *Niveoscincus*. **A, B**. *Niveoscincus metallicus*. Embryonic stage 36. **C**. *Niveoscincus greeni*. Embryonic stage 37. **D**. *Niveoscincus ocellatus*. Embryonic stage 37. CA, chorioallantoic membrane; UT, uterus. Arrows, uterine blood vessels; Arrowheads, allantoic blood vessels; Asterisks, openings into deep folds in the chorioallantoic membrane.

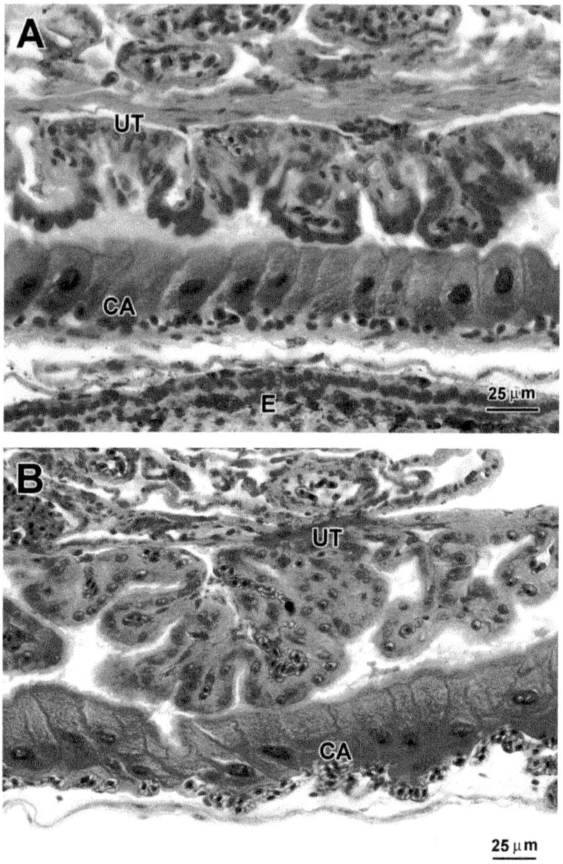

Fig. 13.21 Chorioallantoic placentomes of species of the scincid genus *Pseudemoia*. **(A)** *Pseudemoia spenceri*. Embryonic stage 38. **(B)** *Pseudemoia pagenstecheri*. Embryonic stage 40. CA, chorioallantoic membrane; E, embryo; UT, uterus.

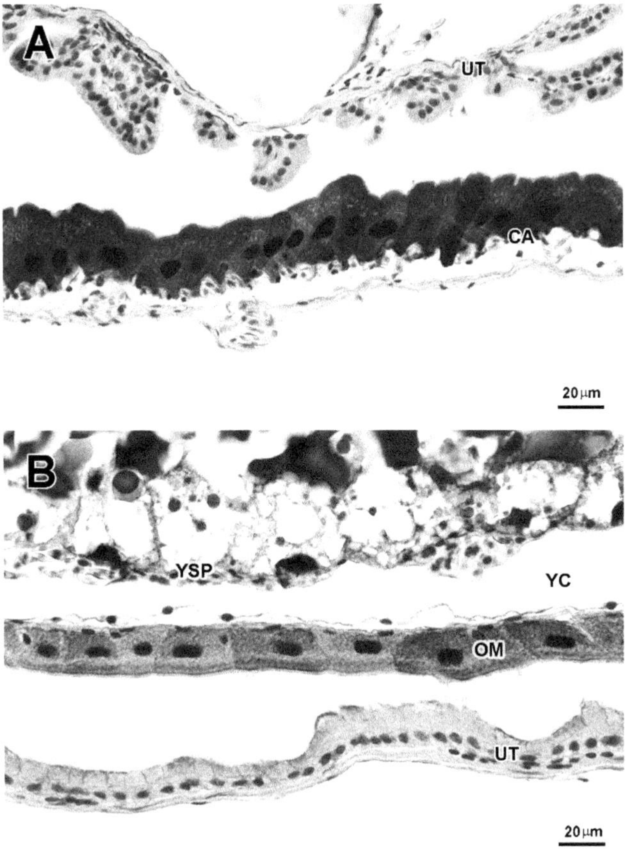

Fig. 13.24 Immunohistological localization (brown precipitate) of calbindin D_{28K} in the **(A)** chorioallantoic placentome and **(B)** omphaloplacenta of *Pseudemoia pagenstecheri*. Embryonic stage 34. CA, chorioallantoic membrane; OM, omphalopleure; UT, uterus; YC, yolk cleft; YSP, yolk splanchnopleure.

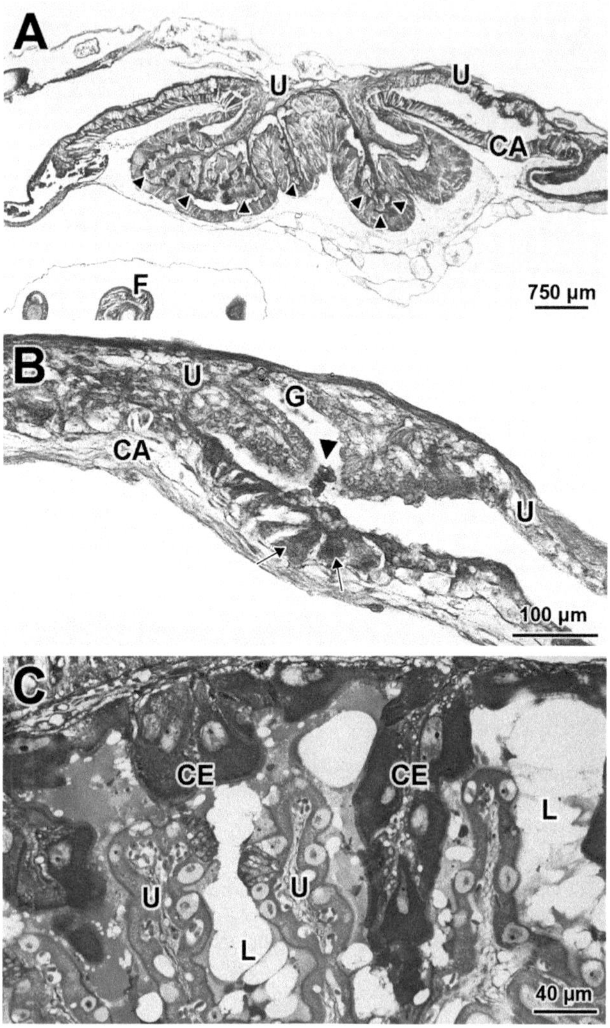

Fig. 13.25 Placentation in the South American skink *Mabuya heathi*, in late gestation. **(A)** Placentome, formed from interdigitation of tissues of the uterus (U) and chorioallantois (CA). Arrowheads mark the uterine- fetal interface. F, fetus. **(B)** Chorionic areola, showing a uterine gland (G) apposed to absorptive cells of the chorionic epithelium (arrows). Arrowhead, secretory material. **(C)** Placentome cytology. Epithelium of the uterus (U) is syncytial and overlies a rich vasculature. Chorionic epithelium (CE) consists of enlarged binucleated cells interspersed with thin interstitial cells; both bear microvilli; L, uterine lumen.

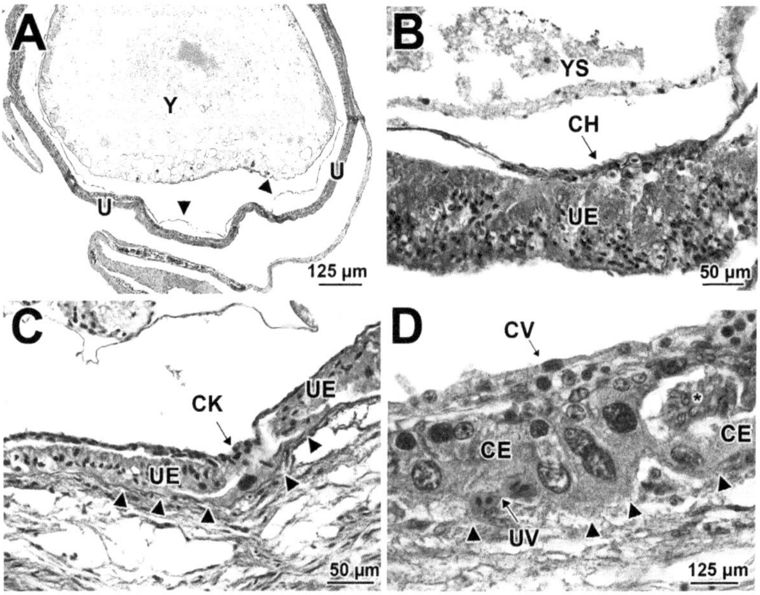

Fig.13.27 Placentation in *Trachylepis ivensii.* **(A)** Early pregnancy, showing a cleavage stage egg in the uterus (U). The shell membrane is indicated by arrowheads. Y = yolk. **(B)** Chorionic placenta in the neurula stage. The chorion (CH) lies apposed to uterine epithelium (UE); the latter contains abundant secretory material, evident as acidophilic and basophilic granules. YS = yolk sac. **(C)** Implantation in the late neurula stage. At the site of a chorionic knob (CK), fetal tissue has penetrated the uterine lining, giving rise to issue that is undercutting the uterine epithelium (UE) (arrowheads). **(D)** Placental interface. Enlarged cells of the chorionic ectoderm (CE) directly contact the uterine vessels (UV) and uterine stroma (arrowheads). The asterisk marks a piece of stripped uterine epithelium. CV, choriovitelline membrane. Figures are from research by D. Blackburn with Alexander F. Flemming.

Chapter 16

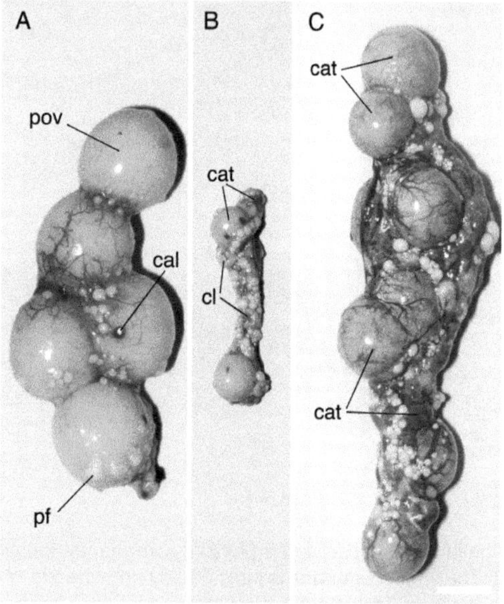

Fig. 16.5 Ventral views of ovaries from three adult tuatara of Stephens Island origin. **(A)** Right ovary from wild female in pre-ovulatory condition (184 mm SVL). This female was seen mating on 4 March 1988, collected immediately afterwards, and dissected on 18 March 1988. The ovary contained five enlarged, pre-ovulatory follicles 18–19 mm in diameter (pale circles about 5 mm in diameter, the presumed stigmata or sites of impending follicular rupture, were present on these follicles). Immature white or creamy-white follicles up to 4 mm diameter (pre-vitellogenic follicles) and small flat brown scars up to 3 mm in length (presumed corpora albicantia) are also visible. **(B)** Left ovary from wild female in gravid condition collected on 17 July 1988 and dissected on 4 August 1988 (218 mm SVL). The ovary contained seven corpora lutea up to 5 mm diameter with visible ovulation apertures, several yolked follicles 5–12 mm in diameter (presumed corpora atretica), numerous pre-vitellogenic follicles and several presumed corpora albicantia. For histology of the corpora lutea of this female, see female M52 of Guillette and Cree (1997). Uterine eggs in this female were radio-opaque but incompletely calcified on the surface (see Cree *et al*. 1996). **(C)** Abnormal right ovary of overweight female that died after several decades in captivity, including at least several years without access to a male (225 mm SVL). The ovary contained many enlarged follicles up to 17 mm diameter with signs of atresia (uneven or brown coloration, sometimes with internal fluid), as well as numerous pre-vitellogenic follicles. Abnormal fatty deposits were present on the liver and peritoneum. Abbreviations: cal = corpus albicans; cat = corpus atreticum; cl = corpus luteum; pf = pre-vitellogenic follicle; pov = pre-ovulatory follicle. Scale bar = 1 cm in all. Photos: A and B: Alison Cree; C: Marcus Simons.

For Product Safety Concerns and Information please contact our EU representative GPSR@taylorandfrancis.com Taylor & Francis Verlag GmbH, Kaufingerstraße 24, 80331 München, Germany

* 9 7 8 0 3 6 7 7 3 8 5 9 4 *

T - #0189 - 160425 - C760 - 234/156/33 - PB - 9780367738594 - Gloss Lamination